Lecture Notes in Networks and Systems

Volume 421

The series "Lecture Notes in Networks and Systems" publishes the latest developments in Networks and Systems—quickly, informally and with high quality. Original research reported in proceedings and post-proceedings represents the core of LNNS.

Volumes published in LNNS embrace all aspects and subfields of, as well as new challenges in, Networks and Systems.

The series contains proceedings and edited volumes in systems and networks, spanning the areas of Cyber-Physical Systems, Autonomous Systems, Sensor Networks, Control Systems, Energy Systems, Automotive Systems, Biological Systems, Vehicular Networking and Connected Vehicles, Aerospace Systems, Automation, Manufacturing, Smart Grids, Nonlinear Systems, Power Systems, Robotics, Social Systems, Economic Systems and other. Of particular value to both the contributors and the readership are the short publication timeframe and the world-wide distribution and exposure which enable both a wide and rapid dissemination of research output.

The series covers the theory, applications, and perspectives on the state of the art and future developments relevant to systems and networks, decision making, control, complex processes and related areas, as embedded in the fields of interdisciplinary and applied sciences, engineering, computer science, physics, economics, social, and life sciences, as well as the paradigms and methodologies behind them.

Indexed by SCOPUS, INSPEC, WTI Frankfurt eG, zbMATH, SCImago.

All books published in the series are submitted for consideration in Web of Science.

For proposals from Asia please contact Aninda Bose (aninda.bose@springer.com).

More information about this series at https://link.springer.com/bookseries/15179

Pradeep Kumar Singh · Sławomir T. Wierzchoń ·
Sudeep Tanwar · Joel J. P. C. Rodrigues ·
Maria Ganzha
Editors

Proceedings of Third International Conference on Computing, Communications, and Cyber-Security

IC4S 2021

Springer

Editors
Pradeep Kumar Singh
Department of Computer Science
KIET Group of Institutions
Ghaziabad, Delhi, India

Sudeep Tanwar
Department of Computer Science
and Engineering
Nirma University
Ahmedabad, India

Maria Ganzha
Faculty of Mathematics and Informatics
Warsaw University of Technology
Warsaw, Poland

Sławomir T. Wierzchoń
Institute of Computer Science
Polish Academy of Sciences
Warsaw, Poland

Joel J. P. C. Rodrigues
Federal University of Piauí
Teresina, Brazil

Instituto de Telecomunicações
Covilha, Portugal

ISSN 2367-3370 ISSN 2367-3389 (electronic)
Lecture Notes in Networks and Systems
ISBN 978-981-19-1141-5 ISBN 978-981-19-1142-2 (eBook)
https://doi.org/10.1007/978-981-19-1142-2

This Springer imprint is published by the registered company Springer Nature Singapore Pte Ltd.
The registered company address is: 152 Beach Road, #21-01/04 Gateway East, Singapore 189721, Singapore

Preface

The 3rd International Conference on Computing, Communications, and Cyber-Security (IC4S-2021) held on October 30–31, 2021, at Krishna Engineering College (KEC), Ghaziabad, India. The conference covered the majority of the research papers from five technical tracks; it includes (i) Communication and Networks Technologies, (ii) Advanced Computing Technologies, (iii) Data Analytics and Intelligent Learning, (iv) Latest Electrical and Electronics Trend, and (v) Security and Privacy Issues. The main idea of the conference is to provide a common platform for the scientists, researchers, policy makers to discuss the novel ideas based on architectures, algorithms, surveys, policies, design, communication challenges, open issues, and future research aspects.

The conference was hosted by the Department of Electronics and Communication Engineering of Krishna Engineering College (KEC), Ghaziabad, India. The inaugural speech along with the welcome address was given by the director and joint director of KEC, Ghaziabad, followed by the address of general chair of the conference. The first keynote talk was delivered by Dr. Vandana Bassco, Department of Electrical and Electronics Engineering, University of Mauritius, Mauritius. The vote of thanks during the inaugural address was given by Dr. A. N. Mishra, Dean (SA) and HoD (ECE), KEC, Ghaziabad, and Local Organizing Chair of IC4S-2021. Two more keynotes were also delivered by Dr. Noor Zaman Jhanjhi, Director Center for Smart Society, School of Computer Science and Engineering, Faculty of Innovation and Technology, Taylor's University, Malaysia, and by Dr. Anand Paul, Kyungpook National University, South Korea. The conference was organized with the academic support of the Knowledge University, Erbil, Iraq; Southern Federal University, Russia; and WSG University in Bydgoszcz, Poland, along with IAC, India. Many experts from these institutions helped during the conference during call for papers, review, in preparation of program schedule, during technical sessions, and for other technical support activities.

We are highly thankful to our valuable authors for their contribution and presentations. The organizing team is thankful to the Technical Program Committee for their immense support during the review process. We express our sincere thanks to the organizing team for hosting the two-day event and conducting two-day sessions very diligently. The IC4S-2021 team is thankful to all session chairs, who chaired the various technical sessions and provided wonderful suggestions to the authors. The session chairs have shared their technical expertise and enlightened the delegates of the conference during the paper presentation sessions. We express our sincere gratitude to our publication partner, Springer, LNNS Series, for believing in us.

Ghaziabad, India — Pradeep Kumar Singh
Warsaw, Poland — Sławomir T. Wierzchoń
Ahmedabad, India — Sudeep Tanwar
Teresina, Brazil — Joel J. P. C. Rodrigues
Warsaw, Poland — Maria Ganzha
October 2021

Contents

Editors and Contributors

About the Editors

Dr. Pradeep Kumar Singh is currently working as a Professor and Head, Department of Computer Science, KIET Group of Institutions, Delhi-NCR Campus, Ghaziabad, India. He is Associate Editor of the IJISMD [IJISMD is indexed by Scopus and Web of Science], IJAEC, IGI Global USA, SPY, Wiley IJISC from Romania. He is recently appointed as Section Editor, Discover IoT, Springer Journal. He has published nearly 120 research papers. He has received three sponsored research projects grant worth Rs. 25 lakhs. He has edited a total 16 books from Springer and Elsevier and also edited several special issues for SCI and SCIE Journals from Elsevier and IGI Global. He has Google Scholar citations 1551, H-index 20, and i-10 index 50.

Prof. Sławomir T. Wierzchoń received M.Sc. and Ph.D. degrees in Computer Science from Technical University of Warsaw, Poland. He holds Habilitation (D.Sc.) in Uncertainty Management from Polish Academy of Sciences. In 2003, he received the title of Professor from the President of Poland. Currently, he is Full Professor at the Institute of Computer Science of Polish Academy of Sciences. His research interests include computational intelligence, uncertainty management, information retrieval, machine learning, and data mining. He is Author/Co-author of over 100 peer-reviewed papers in international journals and international conferences. He published, as Author/Co-author, 11 monographs from the field of machine learning. In the period 2000–2013, he co-organized 13 international conferences on intelligent information systems. Co-authored proceedings from these conferences was published by Springer. He co-edited two volumes of proceedings of the international conference on computer information systems and industrial management, and he has served as Guest Co-editor of three special issues of Information and Control journal. Currently, he is Member of the editorial board for some international journals, as well as Member of many program committees for international conferences. He cooperated with medical centers in the area of statistical data analysis and knowledge discovery in databases.

Dr. Sudeep Tanwar is working as a full professor at the Nirma University, India. He is also a Visiting Professor with Jan Wyzykowski University, Poland, and the University of Pitesti, Romania. He received B.Tech in 2002 from Kurukshetra University, India, M.Tech (Honor's) in 2009 from Guru Gobind Singh Indraprastha University, Delhi, India and Ph.D. in 2016 with specialization in Wireless Sensor Network. He has authored 04 books and edited 20 books, more than 270 technical articles, including top cited journals and conferences, such as IEEE TNSE, IEEE TVT, IEEE TII, IEEE TGCN, IEEE TCSC, IEEE IoTJ, IEEE NETWORKS, ICC, IWCMC, GLOBECOM, CITS, and INFOCOM. He initiated the research field of blockchain technology adoption in various verticals, in 2017. His H-index is 52. His research interests include blockchain technology, wireless sensor networks, fog computing, smart grid, and the IoT. He is a member of the Technical Committee on Tactile Internet of IEEE Communication Society. He has been awarded the Best Research Paper Awards from IEEE IWCMC-2021, IEEE ICCCA-2021, IEEE GLOBECOM 2018, IEEE ICC 2019, and Springer ICRIC-2019. He has won Dr KW Wong Annual Best Paper Prize for 2021 sponsored by Elsevier (publishers of JISA). He has served many international conferences as a member of the Organizing Committee, such as the Publication Chair for FTNCT-2020, ICCIC 2020, and WiMob2019, and a General Chair for IC4S 2019, 2020, ICCSDF 2020, FTNCT 2021. He is also serving the editorial boards of COMCOM-Elsevier, IJCS-Wiley, Cyber Security and Applications- Elsevier, Frontiers of blockchain, and SPY, Wiley. He is also leading the ST Research Laboratory, where group members are working on the latest cutting-edge technologies.

Dr. Joel J. P. C. Rodrigues [S'01, M'06, SM'06, F'20] is Professor at the Federal University of Piauí, Brazil; Senior Researcher at the Instituto de Telecomunicações, Portugal; and Collaborator of the Post-Graduation Program on Teleinformatics Engineering at the Federal University of Ceará (UFC), Brazil. He is Leader of the Next Generation Networks and Applications (NetGNA) research group (CNPq), IEEE Distinguished Lecturer [2018–2021], Member Representative of the IEEE Communications Society on the IEEE Biometrics Council [2011–], and President of the scientific council at ParkUrbis—Covilhã Science and Technology Park [2015–]. He was Director for Conference Development—IEEE ComSoc Board of Governors [2018–2019], Technical Activities Committee Chair of the IEEE ComSoc Latin America Region Board [2018–2019], Past-Chair of the IEEE ComSoc Technical Committee on eHealth, Past-Chair of the IEEE ComSoc Technical Committee on Communications Software, Steering Committee Member of the IEEE Life Sciences Technical Community and Publications Co-Chair [2014–2017]. He is Editor-in-chief of the International Journal on E-Health and Medical Communications and Editorial Board Member of several high-reputed journals. He has been General Chair and TPC Chair of many international conferences, including IEEE ICC, IEEE GLOBECOM, IEEE HEALTHCOM, and IEEE LatinCom. He has authored or coauthored over 850 papers in refereed international journals and conferences, 3 books, 2 patents, and 1 ITU-T recommendation. He had been awarded several Outstanding Leadership and Outstanding Service Awards by IEEE Communications Society and several best papers awards. He is Member of the Internet Society, Senior Member ACM, and Fellow of IEEE.

Dr. Maria Ganzha is Associate Professor in the Faculty of Mathematics and Information Science. She has MS and Ph.D. degrees in Mathematics from the Moscow State University, Russia, and a Doctor of Science degree (in Computer Science) from the Polish Academy of Sciences. She has published more than 200 research papers, is on editorial boards of 6 journals and a book series, and was invited to program committees of more than 150 conferences. She is also Principal Investigator, of the SRIPAS team, in the INTER-IoT project. Here, her team is responsible for use of semantic technologies in the context of interoperability of IoT platforms. She has 1594 Google Scholar citations, h-index 19, and i-10 index 58 in her account. Her area of interest includes computational intelligence, distributed systems, agent-based computing, and semantic data processing.

Contributors

Mohammad Abdussami DSPM IIIT Naya Raipur, Raipur, India

Nidhi Acharya Department of Computer Engineering, Faculty of Technology and Engineering (FTE), Devang Patel Institute of Advance Technology and Research (DEPSTAR), Charotar University of Science and Technology (CHARUSAT), Changa, India

Bharat Bhushan Agarwal Department of Computer Science and Engineering, IFTM University, Moradabad, India

Smita Agrawal Computer Science and Engineering Department, Institute of Technology, Nirma University, Ahmedabad, India

Ravin Ahuja Delhi Skill and Entrepreneurship University, New Delhi, India

Akshita Department of Computer Science and Information Technology, Central University of Jammu, Bagla Suchani, J&K, India

Satish Kumar Alaria Department of Electronics & Communication, AIET, Jaipur, India

Maryam AlAyat Department of Computer Science, College of Computer Science and Information Technology, Imam Abdulrahman Bin Faisal University, Dammam, Saudi Arabia

Jumana AlGhamdi Department of Computer Science, College of Computer Science and Information Technology, Imam Abdulrahman Bin Faisal University, Dammam, Saudi Arabia

Shahad Mohammed AlOtaibi Department of Computer Science, College of Computer Science and Information Technology, Imam Abdulrahman Bin Faisal University, Dammam, Saudi Arabia

Malak AlQahtani Department of Computer Science, College of Computer Science and Information Technology, Imam Abdulrahman Bin Faisal University, Dammam, Saudi Arabia

Mona Altassan Department of Computer Science, College of Computer Science and Information Technology, Imam Abdulrahman Bin Faisal University, Dammam, Saudi Arabia

Maha AlZahrani Department of Computer Science, College of Computer Science and Information Technology, Imam Abdulrahman Bin Faisal University, Dammam, Saudi Arabia

Ruhul Amin DSPM IIIT Naya Raipur, Raipur, India

B. Amrita Department of CSE, G. Narayanamma Institute of Technology and Science, Hyderabad, India

Aditi Arora ABES Engineering College, Ghaziabad, Uttar Pradesh, India

Anansha Asthana Indian Institute of Technology, Jodhpur, Rajasthan, India

Rani Astya Department of Computer Science and Engineering, Sharda University, Greater Noida, India

Atta-ur-Rahman Department of Computer Science, College of Computer Science and Information Technology, Imam Abdulrahman Bin Faisal University, Dammam, Saudi Arabia

Joseph Bamidele Awotunde Department of Computer Science, University of Ilorin, Ilorin, Nigeria

Femi Emmanuel Ayo Department of Computer Science, McPherson University, Seriki-Sotayo, Abeokuta, Nigeria

Purnendu Bagchi Amity University, Kolkata, India

Aditya Bakshi Department of Computer Science and Engineering, Shri Mata Vaishno Devi University, Katra, Jammu & Kashmir, India

Ranjan Kumar Behera Birla Institute of Technology, Mesra, India

R. M. Belokar Department of Production and Industrial Engineering, Punjab Engineering College (PEC), Chandigarh, India

Sumit Bharadwaj Amity University, Noida, India

Akhil Bhatia Indian Institute of Technology, Jodhpur, Rajasthan, India

Soha Bhatia Department of Computer and Communication Engineering, Manipal University Jaipur, Jaipur, Rajasthan, India

Pronaya Bhattacharya Department of Computer Science and Engineering, Institute of Technology, Nirma University, Ahmedabad, Gujarat, India

Madhuri Bhavsar Institute of Technology, Nirma University, Ahmedabad, India

Priyanka Bhutani University School of Information and Communication Technology, Guru Gobind Singh Indraprastha University, Dwarka, Delhi, India

Ishan Budhiraja Bennett University, Greater Noida, Uttar Pradesh, India

Swati Chaturvedi School of Computer Science, University of Petroleum and Energy Studies, Dehradun, India

Sanjeev Chauhan Department of Production and Industrial Engineering, Punjab Engineering College (PEC), Chandigarh, India

Ameya Chawla Guru Tegh Bahadur Institute of Technology, Guru Gobind Singh Indraprastha University, New Delhi, India

Vipul Chudasama Institute of Technology, Nirma University, Ahmedabad, India

Rakesh Dani Graphic Era Deemed to be University, Dehradun, India

Mohit Dayal Dr Akhilesh Das Gupta Institute of Technology and Management, New Delhi, India

V. M. Deshmukh Kavayitri Bahinabai Chaudhari North Maharashtra University, Jalgaon, Maharashtra, India

Namrata Dhanda Department of Computer Science and Engineering, ASET, Amity University, Lucknow, Uttar Pradesh, India;
Amity School of Engineering and Technology, AUUP, Lucknow, India

Mangala R. Dhotre SSBT's College of Engineering and Technology, Bambhori, Jalgaon, India

Jaya Dubey ABES Engineering College, Ghaziabad, Uttar Pradesh, India

Himanshu Duseja SWINGER: Security, Wireless, IoT Network Group of Engineering and Research, University School of Information, Communication and Technology (USICT), Guru Gobind Singh Indraprastha University, Dwarka, Delhi, India

Geetha Ganesan Advanced Computing Research Society, Chennai, India

Ritika Gauba Zenith Ph.D. Training and Consultancy, Jaipur, India

Shubh Gaur School of Computer Science, University of Petroleum and Energy Studies, Dehradun, India

Abhavya Gautam Guild Insurance Group, Brandon, Canada

Pratiksha Gautam Amity University Gwalior, Gwalior, Madhya Pradesh, India

Parth Goel Department of Computer Science and Engineering, Faculty of Technology and Engineering (FTE), Devang Patel Institute of Advance Technology and Research (DEPSTAR), Charotar University of Science and Technology (CHARUSAT), Changa, India

Mala Goplani HVPS Ramniranjan Jhunjhunwala College Arts, Science, and Commerce, Mumbai, India

Anjana Gosain University School of Information and Communication Technology, Guru Gobind Singh Indraprastha University, Dwarka, Delhi, India

Mrinal Goswami School of Computer Science, University of Petroleum and Energy Studies, Dehradun, India

Swathi Gowroju Department of CSE, Sreyas Institute of Engineering and Technology, Hyderabad, India

Mayank Kumar Goyal Department of Computer Science & Engineering, School of Engineering & Technology, Sharda University, Greater Noida, India

Gopal Gupta ABES Engineering College, Ghaziabad, Uttar Pradesh, India

Kanchan Lata Gupta Institute of Engineering and Technology, Lucknow, India

Meenu Gupta Chandigarh University, Mohali, Punjab, India

Punit Gupta Department of Computer and Communication Engineering, Manipal University Jaipur, Jaipur, Rajasthan, India

Shiavnsh Gupta School of Computer Science, University of Petroleum and Energy Studies, Dehradun, India

Shivam Gupta ABES Engineering College, Ghaziabad, Uttar Pradesh, India

Subir Gupta Department of Computer Science and Engineering, Dr. B. C. Roy Engineering College, Durgapur, West Bengal, India

Sunanda Gupta Department of Computer Science and Engineering, Shri Mata Vaishno Devi University, Katra, Jammu & Kashmir, India

Suneet Kumar Gupta Benett University, Gautam Buddha Nagar, India

S. Harshith Department of CSE, Sreyas Institute of Engineering and Technology, Hyderabad, India

Anurag Jain Guru Gobind Singh Indraprastha University, New Delhi, New Delhi, India

Dhyanendra Jain Dr. Akhilesh Das Gupta Institute of Technology and Management, Guru Gobind Singh Indraprastha University, New Delhi, India

Khushboo Jain CSE Department, Institute of Technology, Nirma University, Ahmedabad, India

Farmanullah Jan Department of Computer Science, College of Computer Science and Information Technology, Imam Abdulrahman Bin Faisal University, Dammam, Saudi Arabia

Umesh Jangid CSE Department, Institute of Technology, Nirma University, Ahmedabad, India

Kunal Jani Santa Clara University, Santa Clara, CA, USA

Disha Jayswal Department of Computer Science and Engineering, Faculty of Technology and Engineering (FTE), Devang Patel Institute of Advance Technology and Research (DEPSTAR), Charotar University of Science and Technology (CHARUSAT), Changa, India

Abhishek Kumar Jishu DAIICT, Gandhinagar, India

Rahul Johari SWINGER: Security, Wireless, IoT Network Group of Engineering and Research, University School of Information, Communication and Technology (USICT), Guru Gobind Singh Indraprastha University, Dwarka, Delhi, India

Aarchit Joshi Shoolini University, Bajhol, HP, India

Gaurav Kansal ABES Engineering College, Ghaziabad, Uttar Pradesh, India

Meera Kansara Gujarat Vidyapith, Ahmedabad, India

Princy Kansara CSE Department, Institute of Technology, Nirma University, Ahmedabad, India

Kushal Kanwar Shoolini University, Bajhol, HP, India

Nimish Kappal Shoolini University, Bhajol, Solan, Himachal Prasdesh, India

Jasbir Kaur Chandigarh University, Mohali, Punjab, India

Mohammad Aftab Alam Khan Department of Computer Engineering, College of Computer Science and Information Technology, Imam Abdulrahman Bin Faisal University, Dammam, Saudi Arabia

Rijwan Khan ABES Institute of Technology, Ghaziabad, India

Rizwan A. Khan Department of Civil Engineering, Z.H College of Engineering and Technology, Aligarh Muslim University, Aligarh, India

Uzair Khan Department of Civil Engineering, ABES Engineering College, Ghaziabad, Uttar Pradesh, India

Pooja Khanna Amity University, Lucknow Campus, India;
MVPG College, Lucknow, India

Manish Khare DAIICT, Gandhinagar, India

Manju Khari Jawaharlal Nehru University, New Delhi, India

Bharti Khemani A. P. Shah Institute of Technology, Thane, Mumbai, India

Parul Khurana School of Computer Applications, Lovely Professional University, Phagwara, Punjab, India

Neha Kishore Chitkara University School of Engineering and Technology, Chitkara University, Solan, Himachal Pradesh, India

Ankit Kumar DAIICT, Gandhinagar, India

Ashok Kumar SWINGER: Security, Wireless, IoT Network Group of Engineering and Research, University School of Information, Communication and Technology (USICT), Guru Gobind Singh Indraprastha University, Dwarka, Delhi, India

Bhawnesh Kumar Graphic Era Deemed to be University, Dehradun, India

Gulshan Kumar School of Computer Science and Engineering, Lovely Professional University, Phagwara, Punjab, India

Manoj Kumar Amity University Gwalior, Gwalior, Madhya Pradesh, India

Nagesh Kumar School of Engineering and Technology, Chitkara University, Chitkara University, Kallujhanda, Himachal Pradesh, India

Naveen Kumar Graphic Era Deemed to be University, Dehradun, India

Neerendra Kumar Department of Computer Science and Information Technology, Central University of Jammu, Samba, Jammu and Kashmir, India

Piyush Kumar Discipline of Electrical, Electronic and Computer Engineering, School of Engineering, Howard College Campus, University of KwaZulu-Natal, Durban, South Africa;
Department of Computer Science and Engineering, ASET, Amity University, Noida, India

Rajesh Kumar North Eastern Regional Institute of Science and Technology, Nirjuli, Arunachal Pradesh, India

Sachin Kumar Amity University, Lucknow Campus, India;
MVPG College, Lucknow, India

Santosh Kumar Department of Computer Science and Engineering, ABES Engineering College, Ghaziabad, Uttar Pradesh, India

Samiksha Kumari Department of Computer Science and Engineering, Sharda University, Greater Noida, India

Rudresh V. Kurhe Symbiosis Institute of Technology, Symbiosis International (Deemed University), Lavale, Pune, Maharashtra, India

A. V. Shreyas Madhav School of Computer Science and Engineering, Vellore Institute of Technology, Chennai, Tamil Nadu, India

Aparna N. Mahajan Maharaja Agrasen University, Solan, Himachal Pradesh, India

Praveen Kumar Malik Lovely Professional University, Jalandhar, Punjab, India

Anubha Maurya Department of Computer Science and Engineering, National Institute of Technology Patna, Patna, Bihar, India

Rajesh Kumar Maurya ABES Engineering College, Ghaziabad, Uttar Pradesh, India

Swati Maurya Guru Gobind Singh Indraprastha University, New Delhi, New Delhi, India

Akshay Mewada Institute of Technology, Nirma University, Ahmedabad, India

Gouri Sankar Mishra Department of Computer Science and Engineering, Sharda University, Greater Noida, India

Sanjay Misra Department of Computer Science and Communication, Østfold University College, Halden, Norway

Jay Mittal School of Computer Science, University of Petroleum and Energy Studies, Dehradun, India

Kushagara Mittal Shoolini University, Bhajol, Solan, Himachal Prasdesh, India

Santanu Modak Department of Computer Science, The University of Burdwan, Burdwan, India

Abhoy Chand Mondal Department of Computer Science, The University of Burdwan, Burdwan, India

Biswajit Mondal Department of Computer Science and Engineering, Dr. B. C. Roy Engineering College, Durgapur, West Bengal, India

Manish Kumar Mukhija Department of Computer Science & Engineering, AIET, Jaipur, India

Sumiya Mushtaq Department of Computer Science and Information Technology, Central University of Jammu, Samba, Jammu and Kashmir, India

Vishal Nagar Department of Computer Science and Engineering, Pranveer Singh Institute of Technology, Kanpur, Uttar Pradesh, India

Husen Saifibhai Nalawala Computer Science and Engineering Department, Nirma University, Ahmedabad, India

Parma Nand Department of Computer Science and Engineering, Sharda University, Greater Noida, India

Sushil Narang School of Engineering and Technology, Chitkara University, Chitkara University, Kallujhanda, Himachal Pradesh, India

Sushil Kumar Narang Chitkara University School of Engineering and Technology, Chitkara University, Solan, Himachal Pradesh, India

Rikin Nayak V. T. Patel Department of Electronics and Communication Engineering, CHARUSAT Space Research and Technology Center, Charotar University of Science and Technology, (CHARUSAT), Changa, India

Abhijeet Negi Shoolini University, Bhajol, Solan, Himachal Prasdesh, India

Harendra Singh Negi Graphic Era Deemed to be University, Dehradun, India

Lalit Negi IT Department, Netaji Subhas University of Technology, Delhi, India

Lokesh Negi CSE Department, Netaji Subhas University of Technology, Delhi, India

Neha Department of Computer Science and Engineering, National Institute of Technology Patna, Patna, Bihar, India

Subhrendu Guha Neogi Amity University Gwalior, Gwalior, India

Radhika Nigam Department of Information Technology Engineering, Manipal University Jaipur, Jaipur, Rajasthan, India

Jonathan Oluranti Center of ICT/ICE, CUCRID, Covenant University, Ota, Nigeria

Awe M. Oluwatoyin Center of ICT/ICE, CUCRID, Covenant University, Ota, Nigeria

Parita Oza Computer Science and Engineering Department, Institute of Technology, Nirma University, Ahmedabad, India

Parashu Ram Pal SAGE University, Bhopal, Madhya Pradesh, India

Brijeshkumar Y. Panchal Department of Computer Science and Engineering, Faculty of Technology and Engineering (FTE), Devang Patel Institute of Advance Technology and Research (DEPSTAR), Charotar University of Science and Technology (CHARUSAT), Changa, India

Sharnil Pandiya Symbiosis Institute of Technology, Symbiosis International (Deemed University), Lavale, Pune, Maharashtra, India

Ajay Parikh Gujarat Vidyapith, Ahmedabad, India

Bansari Patel Department of Computer Science and Engineering, Faculty of Technology and Engineering (FTE), Devang Patel Institute of Advance Technology and Research (DEPSTAR), Charotar University of Science and Technology (CHARUSAT), Changa, India

Kushagra Pathak DAIICT, Gandhinagar, India

Pankaj Pathak Symbiosis Institute of Digital and Telecom Management Symbiosis International (Deemed University), Pune, India

Pragya Amity University, Lucknow Campus, India;
MVPG College, Lucknow, India

Vivek Kumar Prasad Institute of Technology, Nirma University, Ahmedabad, India

Mayur Rahul Department of Computer Application, CSJM University, Kanpur, India

Bharti Rana Department of Computer Science and Information Technology, Central University of Jammu, Samba, Jammu and Kashmir, India

Shikha Rastogi GD Goenka University, Sohna, Gurugram, Haryana, India;
Bharati Vidyapeeth's College of Engineering, New Delhi, India

Aarav Singh Rathor Department of Computer Science and Engineering, Institute of Technology, Nirma University, Ahmedabad, Gujarat, India

Ritika Rathore Amity Business School, Amity University, Noida, India

Yashwant Singh Rawal Amity University Rajasthan, Jaipur, India

Rinku Chitkara University School of Computer Applications, Chitkara University, Solan, Himachal Pradesh, India

Rishabh Galgotias College of Engineering and Technology, Greater Noida, India

Jewel Sabhani HVPS Ramniranjan Jhunjhunwala College Arts, Science, and Commerce, Mumbai, India

Anju Saha University School of Information and Communication Technology, Guru Gobind Singh Indraprastha University, Dwarka, Delhi, India

Gadiparthy Harika Sai School of Computer Science and Engineering, Vellore Institute of Technology, Chennai, Tamil Nadu, India

Dinesh Kumar Saini Department of Computer and Communication Engineering, Manipal University Jaipur, Jaipur, India;
DIT University, Dehradun, India

Rakesh Kumar Saini DIT University, Dehradun, India

Pallavi Sangra Department of Computer Science and Information Technology, Central University of Jammu, Samba, Jammu and Kashmir, India

N. Santhosh Ramchander Department of CSE, Sreyas Institute of Engineering and Technology, Hyderabad, India

Amit Saraswat Shoolini University, Bhajol, Solan, Himachal Prasdesh, India

Vijay Bhaskar Semwal NIT Bhopal, Bhopal, Madhya Pradesh, India;
MANIT Bhopal, Bhopal, India

Asheesh Shah Mewar University, Chittorgarh, Rajasthan, India

Jaymin Shah Department of Computer Science and Engineering, Institute of Technology, Nirma University, Ahmedabad, Gujarat, India

Visaj Nirav Shah DAIICT, Gandhinagar, India

Anupam Kumar Sharma Dr. Akhilesh Das Gupta Institute of Technology and Management, Guru Gobind Singh Indraprastha University, New Delhi, India

Ashutosh Sharma Institute of Computer Technology and Information Security, Southern Federal University, Taganrog, Russia

Bhisham Sharma School of Engineering and Technology, Chitkara University, Chitkara University, Kallujhanda, Himachal Pradesh, India

Brij Bhushan Sharma Shoolini University, Bhajol, Solan, Himachal Prasdesh, India

Dheeraj Sharma School of Electronics Engineering, Vellore Institute of Technology, Chennai, India

Gulshan Sharma Durban University of Technology, Durban, South Africa

Kiran Sharma School of Engineering and Technology, BML Munjal University, Gurugram, Haryana, India

Santosh Sharma PSIT College of Higher Education, Kanpur, India

Zakir Ahmad Sheikh Department of Computer Science and Information Technology, Central University of Jammu, Bagla Suchani, J&K, India

Anushka Shrivastava Department of Computer and Communication Engineering, Manipal University Jaipur, Jaipur, Rajasthan, India

Arjun Singh Department of Computer and Communication Engineering, Manipal University Jaipur, Jaipur, India

Arunendra Singh Pranveer Singh Institute of Technology, Kanpur, Uttar Pradesh, India

Binod Kumar Singh Structural Engineering, School of Planning and Architecture, Delhi, India

Jaspreet Singh GD Goenka University, Sohna, Gurugram, Haryana, India

Karan Kumar Singh Department of Computer Science and Engineering, Sharda University, Greater Noida, India

Pooja Singh Department of Computer Science & Engineering, AIET, Jaipur, India

Pramod Singh Meerut Institute of Engineering and Technology, Meerut, Uttar Pradesh, India

Prashant Singh Dr. Akhilesh Das Gupta Institute of Technology and Management, Guru Gobind Singh Indraprastha University, New Delhi, India

Ravendra Singh Department of Computer Science and Engineering, IFTM University, Moradabad, India

Shivani Singh ABES Engineering College, Ghaziabad, Uttar Pradesh, India

Yashwant Singh Department of Computer Science and Information Technology, Central University of Jammu, Bagla Suchani, Samba, Jammu and Kashmir, India

Rushina Singhi Amity Business School, Amity University, Noida, India

Harvinder Soni Taxila Business School, Jaipur, India

Akhilesh Kumar Srivastava ABES Engineering College, Ghaziabad, India

Anirban Sur Symbiosis Institute of Technology, Symbiosis International (Deemed University), Lavale, Pune, Maharashtra, India

Mohammad Haider Syed College of Computing and Informatics, Saudi Electronic University, Riyadh, Saudi Arabia

Rohit Tanwar School of Computer Science, University of Petroleum and Energy Studies, Dehradun, India

Sudeep Tanwar Department of Computer Science and Engineering, Institute of Technology, Nirma University, Ahmedabad, Gujarat, India

Prashant V. Thakre Kavayitri Bahinabai Chaudhari North Maharashtra University, Jalgaon, Maharashtra, India

Richa Tiwari Krishna Engineering College, Ghaziabad, UP, India

Khushboo Tripathi Department of Computer Science and Engineering, Amity University Haryana, Gurgaon, India

Rajan Prasad Tripathi Amity University, Tashkent, Uzbekistan

Udai Bhan Trivedi Pranveer Singh Institute of Technology, Kanpur, India

Alexey Tselykh Institute of Computer Technology and Information Security, Southern Federal University, Taganrog, Russia

Amit Kumar Tyagi School of Computer Science and Engineering, Centre for Advanced Data Science, Vellore Institute of Technology, Chennai, Tamil Nadu, India

Pankaj Vaidya Shoolini University, Bajhol, HP, India

Prashant Vats Dr. Akhilesh Das Gupta Institute of Technology and Management, Guru Gobind Singh Indraprastha University, New Delhi, India

Ankush Verma Amity Institute of Information Technology, AUUP, Lucknow, India

Jai Prakash Verma Department of Computer Science and Engineering, Institute of Technology, Nirma University, Ahmedabad, Gujarat, India

Rajat Verma Department of Computer Science and Engineering, Amity University Uttar Pradesh, Lucknow, India;

Department of Civil Engineering, ABES Engineering College, Ghaziabad, Uttar Pradesh, India

Deo Prakash Vidyarthi School of Computer and System Sciences, Parallel and Distributed System Lab, JNU, New Delhi, India

Vineet Vishnoi Meerut Institute of Engineering and Technology, Meerut, Uttar Pradesh, India

Satyanarayana Vollala DSPM IIIT Naya Raipur, Raipur, India

John Wejin Center of ICT/ICE, CUCRID, Covenant University, Ota, Nigeria

Akanksha Yadav Department of CSE, ASET, Amity University, Lucknow, Uttar Pradesh, India

Ram Kumar Yadav Amity University Gwalior, Gwalior, India

Vibhash Yadav Rajkiya Engineering College, Banda, India

Vikash Yadav Department of Technical Education, Kanpur, Uttar Pradesh, India

Communication and Network Technologies

Enhancement of Energy Efficiency in Wireless Sensor Network with Mobile Sink: A Survey

Akhilesh Kumar Srivastava, Suneet Kumar Gupta, and Rijwan Khan

Abstract The energy consumed by any activity taking place in WSN should be controlled such that limited energy in terms of battery backup remains focus throughout. In the case of dying nodes, battery discharge may cause the network to get disconnected. WSN design issues, e.g., location of sensor nodes, scheduling activities, routes of data flow, mobile sink route, should be dealt with keeping energy limitation in mind. The sensor nodes sense the data from the area of concern and communicate the same to the sink for processing. Sensor nodes deployed in various application areas have limited memory, computational power, and battery backup. There is no defined topology of such network and frequently changing environment, very less amount of battery, and limited storage capability of the nodes. It is essential that each node in the network has knowledge about the routing path to the sink which is energy efficient. Since random placement of the nodes restrains coders from presuming routing table data at the sensor nodes, numerous methods have been suggested to create a dynamic path up to sink. Numerous researches are performed for WSN using the mobile sink. Most of the research activities focused on energy conservation in the background while proposing approaches for clustering, data flow paths, trajectory design, etc. In the WSN with a mobile sink, the trajectory of the sink node plays a vital role. Designing of trajectory is an NP-hard problem. With the use of nature-inspired techniques, e.g., particle swarm optimization (PSO), genetic algorithm (GA), etc., can be used for generating a nearly optimal paths for the mobile sink. In this current article, the authors make attempt to present the summary of various strategies for energy-efficient data collection methodology and energy-efficient path planning of mobile sink in wireless sensor networks.

A. K. Srivastava (✉)
ABES Engineering College, Ghaziabad, India
e-mail: joinakhilesh@yahoo.com

S. K. Gupta
Benett University, Gautam Buddha Nagar, India

R. Khan
ABES Institute of Technology, Ghaziabad, India

P. K. Singh et al. (eds.), *Proceedings of Third International Conference on Computing, Communications, and Cyber-Security*, Lecture Notes in Networks and Systems 421,
https://doi.org/10.1007/978-981-19-1142-2_1

Keywords WSN · Rendezvous points · Static sink · Mobile sink · MILP · Genetic algorithm · Protocol · Cluster · Holes · Routing

1 Introduction

Wireless sensor networks (WSNs) have a dense and large quantity of sensing nodes. These nodes are placed randomly over an area of significance. These nodes sense the data from the area of concern and communicate the same to sink for processing. These sensing nodes are organized to observe the indoor and outdoor surroundings, industry and procedure mechanization under water activity monitoring, healthcare system, etc. They have the application in tracing cattle/other creatures, vehicles, etc. [1]. Sensor nodes deployed in various application areas have limited memory, computational power, and battery backup [2]. There is no defined topology of such network and frequently changing environment, very less amount of battery, and limited storage capability of the nodes. It is essential that each node in the network has the knowledge about the routing path to the sink which is energy efficient. Since random placement of the nodes restrains coders from presuming routing table data at the sensor nodes, numerous methods have been suggested to create dynamic path up to sink.

If the topology changes slowly, a proactive routing approach can be effective where topology detection is done on the periodically using broadcast of a beacon signal from the sink to the complete network [3].

Along with creating routing paths which are energy efficient, two more procedures are used in practice for realizing energy efficiency: mobility of sink [4, 5] and duty cycling of the nodes [6].

2 Challenges with WSN

- **Fault-tolerant Communication**: Since deployment of sensor nodes in any field is random, there is a fair chance of faulty sensor nodes or nodes which die down on or before observation. This may cause communication link to get broken [7].
- **Low Latency**: The events do take place rapidly in the WSN. The designed WSN needs to record and report events quickly.
- **Scalability**: The system under observation is supposed to be scalable meaning that additional nodes can be deployed in order to increase the observation area.
- **Transmission Media**: Faulty nodes can cause the broken links for communication.
- **Coverage Problems**: The quality of service in WSN is solely dependent on coverage of sensor nodes. In case of less coverage sensing, the quality of sensor nodes gets affected.

- **Sensor Holes**: Also referred to as routing hole in which the nodes are either not preset or unable to participate in routing.

It is assumed that a WSN comprises similar stationary sensor nodes. The sink is either static or mobile, and it can be located at distinct positions in the WSN. Those nodes which are closer to the sink in the case of static sink dissipate their battery level faster than the nodes which are farther. This happens because of the frequent load of data relay on nodes closer to the sink. To overcome this issue, mobile sinks were introduced, where the sink travels along a defined path in the field. It has been noticed that in majority of the situations sink mobility aids in creating a balanced load of routing and energy depletion of the nodes [8, 9].

It is however sure that mobility of sink improves balancing of load in the nodes, and it is a very important question if this improves the energy efficiency of such networks. To address this query, it is required to create a methodology for efficiency of energy.

One of the approaches for making comparison between various sink mobility schemes is to have a match in the total energy consumption of WSN nodes for a defined complete work (load) done by a WSN. This paper focuses at finding energy dissipation on average per node E_{bar}, $E_{\mathrm{bar}\,=}\sum \mathrm{N_i} = 1e_iN$, N represents the total no. of WSN nodes, and e_i is the dissipation energy of ith node while in the observation period.

Energy used by various nodes in the static sink is different for different nodes. Nodes closer to the static sink have to repeatedly do the work of relaying information to the sink because of which these nodes deplete their energy faster as compared to the other nodes. Because of this, maximum energy depletion is investigated for each node as $E_{\max\,=}\,\mathrm{Max_i} = 1, 2, \ldots, N\ e_i$.

For the purpose of load balancing, placement of static sink is usually done at the central point of WSN. If plenty of nodes in the vicinity of a static sink die because they have depleted their energy backups, the sink may get detached from remaining nodes in the network. Hence, $E_{\max}$ is one the possible parameters that indicates the lifespan of WSN [10, 11].

The energy consumption in uneven manner may cause the problem known as energy hole. This might split the N/w data transmission to the sink node will be blocked [3]. In underwater wireless sensor networks, designated gateways (DGs) collect data of sensor nodes in real time. But, underwater wireless sensor networks too suffer from energy hole phenomenon [4].

In the previous researches, researchers have focused on either the lifespan of a WSN or the avg energy loss per node (be it $E_{\max}$ or E_{bar}) [4, 7, 12]. In this paper, the sink is assumed to be mobile in nature. This paper finds various schemes of energy optimization in WSN with mobile sink.

3 Data Collection Approaches

Various methods for data collection have been proposed by researchers. Table 1 summarizes the approaches in nutshell.

Khan et al. [13] evaluated various protocols in both of these aspects highlighting that in which of the situation different information is yielded. In majority of the cases, duty cycling effects of the nodes are taken into consideration for analysis and comparison (Fig. 1).

In [13], effect of duty cycling of stationery sensor nodes and movement path of mobile sink with energy consumption was investigated. Energy efficacy of WSN model with static and mobile sink was compared with respect to E_{bar} and $E_{\max}$. Sink mobility is not the only criterion which improves energy parameters Ebar and $E_{\max}$. Mobile sink can drastically improve the energy parameters by reducing the data relay load on the nodes and congestion control.

Table 1 Data collection approaches in WSN

Approaches for data collection in WSN			
Discovery	Data transfer	Routing	Motion control
Mobility independent a. Scheduled rendezvous b. On-demand c. Asynchronous	Joint discovery and data transfer	Flat	Trajectory (either static or dynamic)
Knowledge based		Proxy based	Speed
			Hybrid

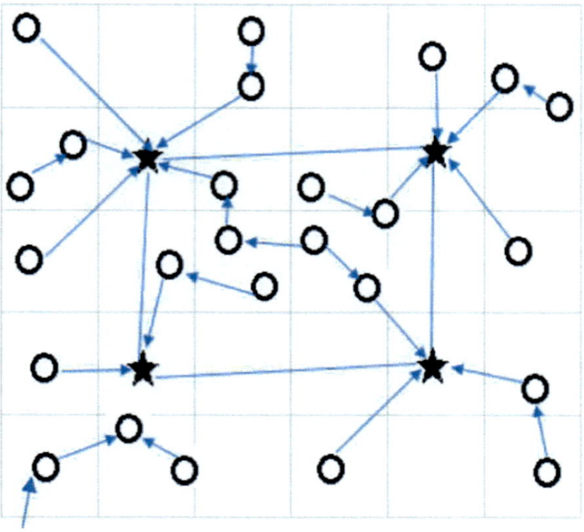

Fig. 1 Data gathering in WSN

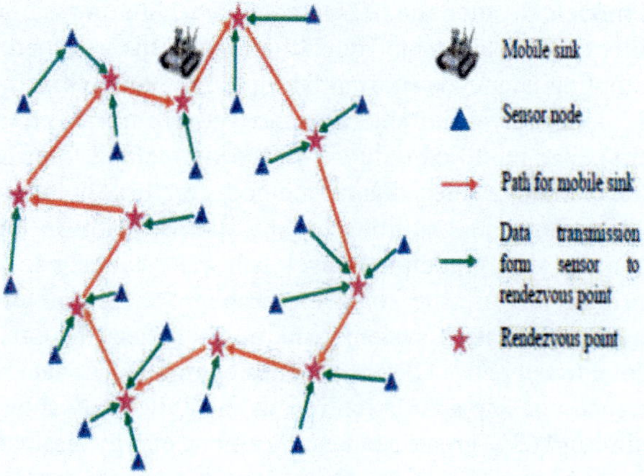

Fig. 2 Scenario of WSN with various node types

To deal with the energy hole problem and optimization of lifetime in static telluric Wireless Sensor Networks and underwater Wireless Sensor Networks (Fig. 2)

Zhu et al. [14], Shu et al. [15], Bhattacharjee and Bandyopadhyay [16] have proposed the work toward optimization of lifetime. Node positions are assumed to be fixed in these papers. It is unavoidable to limit the uneven energy dissipation and problem of energy hole. Sink nodes' mobility can resolve this issue. When the sink travels toward the positions where nodes are concentrated highly or other positions of importance, energy dissipation of nodes can be balanced and residual energy inclines toward 0.

Kumar et al. [17] proposed range-constrained clustering (RCC) technique. In RCC, the nodes in the target area are distributed into many clusters. TSP method is utilized to find the movement path of sink optimally. It travels to all centers of clusters.

Gatzianas and Georgiadis [18], Luo and Hubaux [19], Yun et al. [20], Basagni et al. [21], Zhao and Yang [22] papers explored the maximization of lifetime of WSNs. These papers considered network lifetime optimization models with one or many mobile sinks. These also tried to obtain optimal pattern.

Gatzianas and Georgiadis [18], Luo and Hubaux [19], Yun et al. [20] assumed the movement of sink node be discrete. The sink node move involves many anchors and rest time. The authors have created a model for N/w lifetime optimization assuming some energy dissipation constraints, flow balance constraint, and communication power constraint.

Basagni et al. [21] explored a linear program (LP). Its solution gives an assured upper bound on possible lifetime considering multiple sinks. The centralized and distributed heuristic were introduced to find a solution of the LP which finds the N/W lifetime which is very close to the optimum.

Zhao and Yang [22] proposed energy consumption constraint and flow balance constraint and researched into optimization of lifetime beneath two diverse situations. These were fixed rest time and varying rest time. Some methods focused on instituting and models to solve the lifespan of WSN with an assumption of known movement paths of sinks, but majority of the algorithms assumed only one sink node. Data assembling latency is substantial. In [23] Lifetime maximization of WSN with sink mobility the travel path selection and Optimization of lifetime were taken into consideration. The modified reduced clustering method, k-means clustering method, and nearest neighbor interpolation method were used to find the travel paths and obtain the near to optimal solution for the shortest path. In this, N/w lifetime optimization model with predefined travel path was established. Sub-gradient and geometric algorithm was used to solve the problem of lifetime optimization and obtain the data communication system. Sink nodes collect the data by traversing the defined optimal travel path. All sensor nodes communicate data based on the data transmission method. The simulation results in [23] clearly show that MLMS can improve lifetime of WSN, create a balancing among energy dissipation of nodes, and ease data collection time. MLMS has observed improvement over Ratio_w, TPGF, GRND, and RCC but high time complexity.

Mobile sink usually traverses each node and collects the desired data [24, 25] (known as single-hop communication) or goes to only a few positions, and nodes communicate collected data to the mobile sink [11, 26–30]. Data gathering task is faster in the multi-hop communication. Another area of concern arrives in Multi Hop Communication is the increase in energy dissipation majorly for forwarding of data.

A solution proposed to address this is to transfer the data to some intermediate sensor nodes which store the data for communicating to mobile sink as and when the mobile sink comes in their range or when the request arrives to them to send the data [31–39]. Majority of these approaches create a balance between delay in data collection and overhead in energy dissipation. Konstantopoulos et al. [31] have addressed the issue of energy holes because of intermediate data relaying nodes or cluster heads. Chen et al. [40] presented a geographic converge cast-based approach basically targeting the reconstruction of path during sink mobility.

Mamalis [41] proposed formation of many virtual circles and lines on which cluster heads are placed properly. Mobile sink approach reduces the energy usage of nodes at the cost of data collection time. In general, the mobile sink tour time upper limit is set as prerequisite for the timely collection of data. Use of multiple sink can also speed up the data collection work [21, 39, 42].

Almi'ani et al. [43] and Ekici et al. [44] proposed the hybrid approach, wherein the combination of multi-hop communication with the use of a mobile sink traverses to limited positions (called caching points—CPs). This builds direct or indirect clustering which is hierarchical in nature.

Almi'ani et al. [43] proposed the minimization of number of hops that forwards the data from sensor nodes to their nearest caching points. This method proposed a k-means node-clustering method wherein the grouping of the network in the almost equal size clusters (in terms of Sensor nodes) followed by designing a Mobile Sink trip to take one Cluster Point from every cluster, Iterating the same to cover optimum no of clusters with the limitation of maximum length of Mobile sink trip.

Mamalis [41] showed experimental setup that claimed to perform better than [44], and it was producing the optimum results. Almi'ani et al. [45, 46] presented an optimization of path for collection of data while working with the multiple sinks.

[41] proposed solution majorly based on "residual energy" of the sink nodes rather than distance and number of hops in [43]. Mamalis [41] also show the stable energy and efficient conduct presented by hierarchical clustering structures to improve lifetime of the WSN. It used node-clustering algorithm and the multi-hop clustering algorithm of [47] as its base (main criterion for formation of cluster here is the residual energy of every sensor node). It detects clusters which are balanced in terms of energy and guarantees ideal performance in terms of avg energy dissipation and lifetime of network. Mamalis [41] modified this method to satisfy the requirement of distance-restricted mobile sink trip. It also developed a data collection protocol which was based on TSP approximation path that fulfills the distance constraint. The energy holes are created around the cluster heads falling in the TSP path; Mamalis [41] used a method mix up of re-clustering phase and with alternating among various original positions of mobile sink.

Papadimitriou and Georgiadis [48] formulated the problem of maximization of network lifetime into a min–max problem in a circle considering uniform distribution of sensor nodes. Gandham et al. [49] proposed multiple mobile sinks with predefined route to gather sensing data for a particular region. It proposed an integer linear program model to find the position of the K mobile sinks in one round.

Wang et al. [50] proposed optimization of sink movement along with the rest time. It proposed the linear programming solution to the problem with an assumption of workload of a node being evenly distributed among the horizontal and vertical links.

Luo et al. [4] proposed 2-stage scheduling: (1) The mobile sink traverses the potential locations one by one and stays there at each for a small time. (2) The sink collects the buffered data of all nodes and builds the stay time profile at the potential point.

Basagni et al. [11] worked on two constraints: first, the max length at every movement of mobile sink and min stay time at each stay point. Movement length of mobile sink from one stay point to the other is bounded to ensure loss of data gathering. The paper then presented a simple and distributed heuristic considering problem in the mixed ILP.

Sugihara and Gupta [51, 52] proposed the problem to be considered as TSP, and solution suggested the updation of tour timing at each edge. This aims at collecting the maximum data by one-hop data gathering mechanism. This also aimed at reducing the energy dissipation in relay.

Xing et al. [35] suggested an approach of data gathering which is rendezvous point based, organized mobility of sink, caching of data, and limiting the tour length of the mobile sink. An approximation algorithm was designed here for minimizing the sum of energy dissipation of all participating sensor nodes. It was assumed here that prior to the transmission of data it is combined in one packet.

Guney et al. [53] proposed design of sink trajectory as an optimization problem. It aims at identifying the location of sink optimally and communication path among sinks and sensor nodes. The authors formulated this as an ILP and created many heuristics for the same.

Liang et al. [54, 55] unified the tour length of the mobile sink in the problem of maximization of network lifetime and suggested heuristics. Liang and Luo [56] have presented their work of considering multiple mobile sinks instead on 1 in their previous work. Gatzianas and Georgiadis [18] proposed the formulation of designing optimal route for a mobile sink as a LP problem and proposed a distributed solution employing Lagrangian duality principle and the sub-gradient scheme. Convergence rate of the algorithm was the basic factor in finding the run time of this distributed scheme.

Yun and Xia [57] proposed the scheme in which the sensor node does not have to transmit the data immediately after sensing it. It buffers the data until the sink reaches to the favorable location with respect to the given sensor node. It eases out the load at the sensor node, and network lifetime can be increased by this. They designed this problem as a mixed ILP given the restricted delay limit. They also proposed flow-based framework.

Xu et al. [58] aimed at discovery of a route for mobile sink to maximize the network lifetime. It has the constraints like (i) stay places of mobile sinks for data collection are fixed. The sink is allowed to stay at potential locations. This information is considered to be available as a priori. (ii) Delay on data communication is in some tolerant range. Storage space of each sensor node is limited. To avoid the loss of data because of storage overflow, there should be tolerant data communication delay and it should be fixed up.

With aim of establishing relation between lifetime of network and tolerable delay in delivery of data, a controllable parameter h "the bound on no. of hops from node to sink" was used. The selection of h played a vital role in realizing the trade-off between lifetime of network and data delivery delay; i.e., keeping h small, number of stay points and mobile sink route will be long. And delay in data delivery will be more.

Xu et al. [58] researched on achieving the trade-off between lifetime of WSN and data delivery permissible delay while mobile sink being employed for data collection. It mainly focused on designing of optimal tour path for mobile sink and formulated a protocol for transferring the sensed data to mobile sink. The research article proposed a method which minimizes the number of hops. Since this problem is NP-hard, the researchers have proposed a new framework that optimizes the trajectory.

4 Classification of WSN

WSNs can be classified depending on type of sensor used in the network. Sensor type is dependent on features like unit cost, sensing range, and communication range. Homogeneous WSNs comprise similar type of sensors, whereas heterogeneous WSNs comprise many types of sensors. Coverage requirements of the area are also a criterion for categorization. It can be equal around the sensor area, or fraction of it may be of high criticality than the other ones and better sensing observation. Sensors may either be active or sleeping. A sleeping sensor is free from sensing, transferring, and receiving data during the sleep phase. Energy dissipation during this time is minimal. Active sensor does sensing, transferring, and receiving of data and spends substantial energy during this work.

5 Designing of Wireless Sensor Network

Designing of wireless sensor network comprises major decisions like:

1. Number of sensors for deployment to satisfy the cost and coverage requirements. This should also take care of fault tolerance in case of failing sensor nodes during the observation period. Energy dissipation of sensor nodes should also be taken into consideration while designing as transmission requires substantial energy and it increases with the distance of transmission.
2. Activity scheduling of sensor nodes is the second parameter that should be thought of while designing the WSN. Keeping some of the sensor nodes active at some point of time while keeping others at rest improves the complete lifetime of WSN. While designing, we should ensure that active sensors are enough in numbers to ensure that WSN is fully functional.
3. Another major issue while designing of sensor network involves the "sink neighborhood problem" [11], "energy hole problem" [59, 60], or "the crowded center effect" [61]. A sensor node in the vicinity of the sink dissipates their battery faster as they work as accumulators for other node's sensed data. This problem can be resolved by moving sink node which enables may sensor nodes to work as accumulators for a set of sensor nodes, and responsibility of accumulation does not lie only on limited nodes.
4. The last design issue is to find the path of data flow from sensor node to sink node. A path from sensor to sink can be found easily with the prior computation, but considering the limitation of the WSN of possible failure of some sensor nodes creates the requirement of find sensor to sink node data flow path at run time with some computation.

Many papers have been written by various researches for the designing of WSN with maximization of lifetime, but most of them have taken only a subset of designing criteria written above. Because of the above, any proposed design WSN can be termed as suboptimal.

Keskin et al. [62] have given a mathematical model considering mixed integer linear programming (MILP) model is the only research which has focused on all of the above-mentioned designing criteria.

Hamida and Chelius [37] proposed the analysis of the existing data propagation protocols in mobile sinks. It focuses on the categorical decisions about movement of sinks. Majority of the papers utilized the mathematical model that deals with the optimization of WSN performance criterion, e.g., lifetime of WSN, energy dissipation in totality, complete cost incurred for known data, and total cost for given data circulation protocols.

6 Data Flow Through Optimal Path

Data is believed to move from nodes to the sinks via the smallest route in several papers, e.g., [63], in which authors assumed a mobile sink moving inside the network to gather the collected data from the sensing nodes in a single-hop fashion. This paper has sought the minimal distance route which traverses each sensing node's transmission range. The problem of finding minimal distance route is visualized as a variant of TSP.

The solution as presented in [63] is further extended for finding shortest path considering multiple mobile sink in [64]. Objective function in this paper is set to minimize the longest sink trip. At the same time, there is a constraint that each of the sensing node falls within the vicinity of sink traveling. It is also tried to fall in the vicinity of exactly one sink only such that data is sent in single-hop approach to the sink. The problem is modeled as MILP. These two papers [63] and [64] follow the optimization models, but they have little impact on the lifetime and load balancing metric. These researches propose the optimization methodology but fail to address the network lifetime, delay in travel of data, and balancing act across the nodes.

Wang et al. [50] follow different approaches from [63] and [27] where the authors propose a LP model for improving the lifespan of the network. The mobile sink is supposed to rest at some rendezvous points. The lifetime of the network is supposed to be the total of rest time at rendezvous points. The optimal solution of the linear programming model suggests the rendezvous points. In this paper, the sink route is not defined as the order of the visiting rendezvous points is undefined.

Basagni et al. [11] presented the sink route as well as the extension of work in [50]. The result of the [11] is the discovery of rendezvous points and mobile sinks schedule of vising these points. Basagni et al. [21] have extended this work further to incorporate multiple mobile sinks. The authors considered a sink configuration by randomization of the sinks along predefined set of rendezvous points. Further lifetime of the network is considered to be the accumulation of rest time of sinks.

As the objective, lifetime of the network is maximized and however the energy dissipation of the sensors should be initial level of power. In [54] sink is treated as the energy restricted device which moves in the region mechanically using some fuel. Keskin et al. [65] considered sink movement time as a part of network lifetime. The authors also took into account the time taken for data collection. The authors proposed the efficient heuristics as the solution of proposed model (MILP). Srivastava and Gupta [66, 67] proposed the genetic algorithm-based approaches for path planning of mobile sink. In the proposed work, to design the fitness function, three parameters have been used.

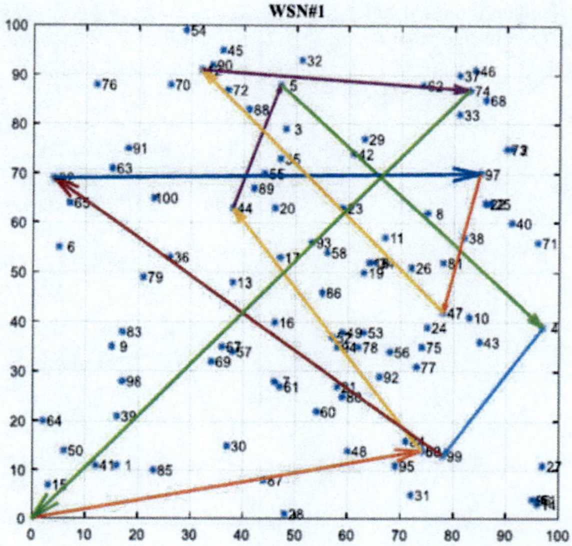

Fig. 3 Path of mobile sink using random tour

- Length of the tour of the MS
- Load of rendezvous points (RP); i.e., the RP receives the data from how many sensor nodes
- Number of nodes which forwards the data to RP using multi-hop.

The paper proposed fitness function represented as:

$$\text{Minimize } F = w_1 x T \cos t + w_2 x \{n - NW_1 H\} + w_3 x D_i f$$

where

- w_1, w_2, w_3 represent the weights and $w_1 + w_2 + w_3 = 1$.
- T_{cost} represents the cost of the tour.
- NW1H represents the numbers of nodes with 1 hop from RPs.
- $D_i f$ represents the difference between maximum and minimum values of LoadRP(i).

Figures 3 and 4 depict the comparison of random path and GA-based path of mobile sink.

7 Optimal Data Flow

In spite of assuming routing of data path apriori, data flow path for every sensing node to sink can be determined optimally. Gandham et al. [49] divided the entire time into the equal periods. In each of the durations, data routes and sinks are stationery.

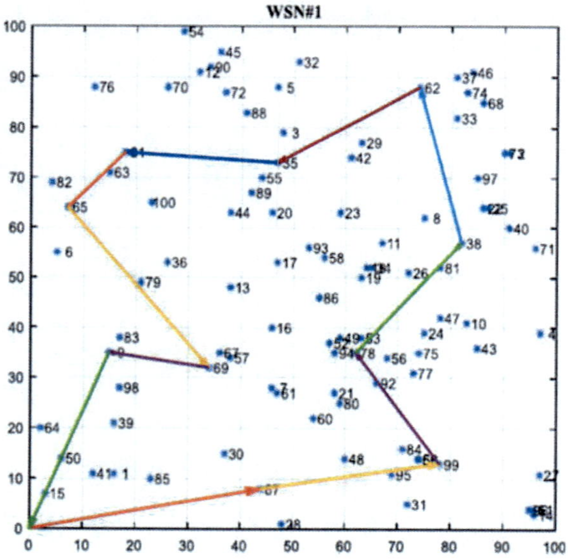

Fig. 4 Path of mobile sink using genetic algorithm

The authors proposed two different models for minimization of energy dissipation of each sensor node and cumulative energy dissipation of sensing nodes [68]. Offered 2 more heuristics along with the one in [49]. The first one assumed sinks' location at the place where neighboring nodes have maximum backup left. The other one assumed sinks' location at the place such that the difference among the minimum and maximum backups of the sensor nodes is minimized. Alsalih et al. [69] proposed similar approach to that of [68]; however, author proposed the periodic optimization of minimal backup. The paper is studied as single-period model and needs to be run separately for each period. This paper provides approximate solution rather than optimal one. Luo and Hubaux [70] consider a network field circular in nature and proposed the minimization of maximum load of sensing nodes using the MILP model. It is also assumed that the movement path of the sink is also circular in nature. Gandham et al. [49], Azad and Chockalingam [68], Alsalih et al. [69], Luo and Hubaux [70] restrict data flow in an optimal fashion, but they did not produce any approach for maximization of network lifetime directly.

Papadimitriou and Georgiadis [48] proposed a nonlinear programming model to maximize lifetime of WSN. It is defined as the cumulative rest time of moving sink at rendezvous points. Gatzianas and Georgiadis [18] revisited the model in [48] and developed a distributed algorithm. Distribution technique was proposed in [64].

Yun and Xia [57] used the [76] and [18] as base. Authors proposed the buffering of data by sensing nodes till the favorable time with respect to network lifetime. Above models are found appropriate for applications which are delay tolerant.

Yun et al. [20] proposed the algorithm for the solution of queue-oriented delay-tolerant model as given in [57]. Behdani et al. [71] proposed another algorithm for the same model as in [22] which was computationally efficient. Luo and Hubaux [19] introduced MILP model with multiple mobile sink for maximization of the lifetime of WSN. It was defined as the period lengths sum. Period here is considered as time with sink configuration given. With change in the sink location, the period changes.

Above-mentioned approaches did not touch all four major aspects of WSN. Above approaches assumed a prior defined sensor position. They did not consider activity schedules. Around 50% of the papers employed shortest path data propagation as a predefined data propagation method.

Only [62] has attempted to find the optimal WSN design with all four WSN design issues. G. 4.0. [72] presented the effect of the incorporation with comparison of lifespan of WSN gained by a combined model with the articles mentioned in earlier papers. The MILP model turns out to be unsolvable for genuinely real networks as size is huge.

8 Conclusion

Numerous researches are performed for WSN using mobile sink. Most of the research activities focused energy conservation in background while proposing approaches for clustering, data flow paths, trajectory design, etc. In the WSN with mobile sink, trajectory of sink node plays a vital role. Designing of trajectory is a NP-hard problem. With the use of nature-inspired techniques, e.g., particle swarm optimization (PSO), genetic algorithm (GA), etc., can be used for generating nearly optimal path for mobile sink.

References

1. Raghavendra, C. S., Sivalingam, K. M., & Znati, T. (2004). *Wireless sensor networks*. Kluwer Academic Publishers.
2. Akyildiz, I. F., Su, W., Sankarasubramaniam, Y., & Cayirci, E. (2002). A survey on sensor networks. *IEEE Communications Magazine., 40*, 102–114.
3. Nezhad, A. A., Makrakis, D., & Miri, A. (2007). Anonymous topology discovery for multihop wireless sensor networks. In *Proceedings of 3rd ACM Workshop on QoS and Security for Wireless and Mobile Networks, Q2SWinet '07*, Chania, Crete Island, Greece (pp. 78–85).
4. Luo, J., Panchard, J., Piorkowski, M., Grossglauser, M., & Hubaux, J.-P. (2006). Mobiroute: Routing towards a mobile sink for improving lifetime in sensor networks. In *Proceedings of IEEE International Conference on Distributed Computing in Sensor Networks (DCOSS)*, pp. 480–497 (2006)
5. Ye, F., Luo, H., Cheng, J., Lu, S., & Zhang, L. (2002). A two-tier data dissemination model for large-scale wireless sensor networks. In *Proceedings of 8th Annual international Conference on Mobile Computing and Networking, MobiCom '02*, Atlanta, Georgia, USA, September 23–28, 2002 (pp. 148–159).
6. Wang, L., & Xiao, Y. (2006). A survey of energy-efficient scheduling mechanisms in sensor networks. *Mobile Network Applications, 11*, 723–740.
7. Sankar, A., & Liu, Z. (2004). Maximum lifetime routing in wireless ad-hoc networks. In *Proceedings of 23rd IEEE INFOCOM* (pp. 1089–1097).

8. Giannakos, A., Karagiorgos, G., & Stavrakakis, I. (2009). A message-optimal sink mobility model for wireless sensor networks. In *Proceeding of 8th International Conference on, Networks* (pp. 287–291).
9. Luo, J., & Hubaux, J. P. (2005). Joint mobility and routing for lifetime elongation in wireless sensor networks. In *Proceedings of 24th IEEE INFOCOM*, Miami, USA (pp. 1735–1746).
10. Wu, Y., Zhang, L., Wu, Y., & Niu, Z. (2006). Interest dissemination with directional antennas for wireless sensor networks with mobile sinks. In *Proceedings of the 4th international Conference on Embedded Networked Sensor Systems, SenSys '06*, Boulder, Colorado, USA (pp. 99–111).
11. Basagni, S., Carosi, A., Melachrinoudis, E., Petrioli, C., & Wang, Z. M. (2008). Controlled sink mobility for prolonging wireless sensor networks lifetime. *Journal of Wireless Networks., 14*, 831–858.
12. Kinalis, A., & Nikoletseas, S. (2007). Scalable data collection protocols for wireless sensor networks with multiple mobile sinks. In *Proceedings of the 40th Annual Simulation Symposium Annual Simulation Symposium* (pp. 60–72). IEEE Computer Society.
13. Khan, M. I., Gansterer, W. N., & Haring, G., Static vs. mobile sink: The influence of basic parameters on energy efficiency in wireless sensor networks.
14. Zhu, Y.-H., Shen, D.-D., Wu, W.-D., Shen, Z.-W., & Tang, Y.-P. (2009). Dynamic routing algorithms optimizing lifetime of wireless sensor networks. *Acta Electronica Sinica, 37*(5), 1041–1045.
15. Shu, L., Zhang, Y., Zhou, Z., Hauswirth, M., Yu, Z., & Hynes, G. (2008). Transmitting and gathering streaming data in wireless multimedia sensor networks within expected network lifetime. *Mobile Networks and Applications, 13*(3–4), 306–322.
16. Bhattacharjee, S., & Bandyopadhyay, S. (2013). Lifetime maximizing dynamic energy efficient routing protocol for multi hop wireless networks. *Simulation Modelling Practice and Theory, 32*, 15–29.
17. Kumar, A. K., Sivalingam, K. M., & Kumar, A. (2013). On reducing delay in mobile data collection based wireless sensor networks. *Wireless Networks, 19*(3), 285–299.
18. Gatzianas, M., & Georgiadis, L. (2008). A distributed algorithm for maximum lifetime routing in sensor networks with mobile sink. *IEEE Transactions on Wireless Communications, 7*(3), 984–994.
19. Luo, J., & Hubaux, J. P. (2010). Joint sink mobility and routing to maximize the lifetime of wireless sensor networks: The case of constrained mobility. *IEEE/ACM Transactions on Networking, 18*(3), 871–884.
20. Yun, Y. S., Xia, Y., Behdani, B., & Smith, J. C. (2013). Distributed algorithm for lifetime maximization in a delay-tolerant wireless sensor network with a mobile sink. *IEEE Transactions on Mobile Computing, 12*(10), 1920–1930.
21. Basagni, S., Carosi, A., Petrioli, C., & Phillips, C. A. (2011). Coordinated and controlled mobility of multiple sinks for maximizing the lifetime of wireless sensor networks. *Wireless Networks, 17*(3), 759–778.
22. Zhao, M., & Yang, Y. (2012). Optimization-based distributed algorithms for mobile data gathering in wireless sensor networks. *IEEE Transactions on Mobile Computing, 11*(10), 1464–1477.
23. Maximizing Lifetime of Wireless Sensor Networks with Mobile Sink Nodes
24. Shah, R., Roy, S., Jain, S., & Brunette, W. (2003). Data MULEs: modeling and analysis of a three-tier architecture for sparse sensor networks. *Ad Hoc Networks, 1*(2–3), 215–233.
25. Sugihara, R., & Gupta, R. (2010). Optimal speed control of mobile node for data collection in sensor networks. *IEEE Transactions on Mobile Computing (TMC), 9*(1), 127–139.
26. Ammari, H., & Das, S. (2008). Promoting heterogeneity, mobility, and energy—aware Voronoi diagram in wireless sensor networks. *IEEE TPDS, 19*(7), 995–1008.
27. Valle, C.A., da Cunha, A.S., Mateus, G.R., & Aioffi, W.M. (2009). Optimization and simulation in wireless sensor networks with multiple mobile sinks, Unpublished manuscript.
28. Demirbas, M., Soysal, O., & Tosun, A. (2007). Data Salmon: A greedy mobile Basestation protocol for efficient data collection in WSNs. In *Proceedings of DCOSS'07 Conference* (pp. 267–280).

29. Vincze, Z., Vass, D., Vida, R., Vidacs, A., & Telcs, A. (2007). Adaptive sink mobility in event-driven densely deployed wireless sensor networks. *Ad Hoc & Sensor Wireless Networks (AHSWN), 3*(2–3), 255–284.
30. Friedmann, L., & Boukhatem, L. (2007). Efficient multi-sink relocation in wireless sensor network. In *Proceedings of the 3rd International Conference on Networking and Services* (p. 90).
31. Konstantopoulos, C., Mamalis, B., Pantziou, G., & Thanasias, V. (2012). Watershed-based clustering for energy efficient data gathering in wireless sensor networks with mobile collector. In *Proceedings of the Euro-Par Conference*, LNCS 7484 (pp.754–766).
32. Konstantopoulos, C., Pantziou, G., Gavalas, D., Mpitziopoulos, A., & Mamalis, B. (2012). A Rendezvous-based approach for energy-efficient sensory data collection from mobile sinks. *IEEE TPDS, 23*(5), 809–817.
33. Tirta, Y., Li, Z., Lu, Y. H., & Bagchi, S. (2004). Efficient collection of sensor data in remote fields using mobile collectors. In *Proceedings of IEEE ICCCN Conference* (pp. 515–520).
34. Ma, M., & Yang, Y. (2007). SenCar: An energy-efficient data gathering mechanism for large-scale multihop sensor networks. *IEEE TPDS, 18*(10), 1476–1488.
35. Xing, G., Wang, T., Jia, W., & Li, M. (2008). Rendezvous design algorithms for wireless sensor networks with a mobile base station. In *Proceedings of ACM MobiHoc Conference* (pp. 231–239).
36. Rao, J., & Biswas, S. (2010). Network-assisted sink navigation for distributed data gathering: Stability and delay-energy trade-offs. *Computer Communications, 33*, 160–175.
37. Hamida, E., & Chelius, G. (2008). Strategies for data dissemination to mobile sinks in wireless sensor networks. *Wireless Communications, 15*(6), 31–37.
38. Rahman, M.S., & Naznin, M. (2013). Shortening the tour-length of a mobile data collector in the WSN by the method of linear shortcut. In *Web technologies and applications* (pp. 674–685). Springer Berlin Heidelberg.
39. Gao, S., Zhang, H., & Das, S. K. (2011). Efficient data collection in wireless sensor networks with path-constrained mobile sinks. *IEEE Transactions on Mobile Computing, 10*(4), 592–608.
40. Chen, T. S., Tsai, H. W., Chang, Y. H., & Chen, T. C. (2013). Geographic converge cast using mobile sink in wireless sensor networks. *Computer Communications, 36*, 445–458.
41. Mamalis, B. G., Prolonging Network Lifetime in Wireless Sensor Networks with Path-Constrained Mobile Sink
42. Chatzigiannakis, I., Kinalis, A., Nikoletseas, S., & Rolim, J. (2007). Fast and energy efficient sensor data collection by multiple mobile sinks. In *Proceedings of MOBIWAC'07 Conference* (pp. 25–32).
43. Almi'ani, K., Viglas, A., & Libman, L. (2010). Energy-efficient data gathering with tour length-constrained mobile elements in wireless sensor networks. In *Proceedings of the 35th Conference on Local Computer Networks* (pp. 582–589).
44. Ekici, E., Gu, Y., & Bozdag, D. (2006). Mobility-based communication in WSNs. *IEEE Communications Magazine, 44*, 56–62.
45. Almi'ani, K., Viglas, A., & Libman, L. (2010). Mobile element path planning for time constrained data gathering in wireless sensor networks. In *Proceedings of the International Conference on Advanced Information Networking and Applications (AINA)* (pp. 843–850).
46. Almi'ani, K., Viglas, A., & Libman, L. (2014). Tour and path planning methods for efficient data gathering using mobile elements. *International Journal of Ad hoc and Ubiquitous Computing*, to appear.
47. Bao, X., Liu, L., Zhang, S., & Bao, F. (2010) An energy balanced multihop adaptive clustering protocol for wireless sensor networks. In *Proceedings of the 2nd IEEE ICSPS Conference* (vol. 3, pp. 47–51).
48. Papadimitriou, I., & Georgiadis, L. (2005). Maximum lifetime routing to mobile sink in wireless sensor networks. In *Proceedings of the 13th IEEE SoftCom* (pp. 1–5).
49. Gandham, S.R., Dawande, M., Prakask, R., & Venkatesan, S. (2003). Energy efficient schemes for wireless sensor networks with multiple mobile base stations. In *Proceedings of Globecom'03*. IEEE.

50. Wang, Z. M., Basagni, S., Melachrinoudis, E., & Petrioli, C. (2005). Exploiting sink mobility for maximizing sensor networks lifetime. In *Proceedings of HICSS*. IEEE.
51. Sugihara, R., & Gupta, R. K. (2008). Improving the data latency in sensor networks with controlled mobility. In *Proceedings of DCOSS'080*. IEEE.
52. Sugihara, R., & Gupta, R. K. (2009). Optimizing energy-latency trade-off in sensor networks with controlled mobility. In *Proceedings of INFOCOM'09*. IEEE.
53. Guney, E., Aras, N., Altinel, I. L., & Ersoy, C. (2010). Efficient integer programming formulations for optimum sink location and routing in heterogeneous wireless sensor networks. *Computer Networks*.
54. Liang, W., Luo, J., & Xu, X. (2010). Prolonging network lifetime via a controlled mobile sink in wireless sensor networks. In *Proceedings of Globecom'10*. IEEE.
55. Liang, W., Luo, J., & Xu, X. (2011) Network lifetime maximization for time sensitive data gathering in wireless sensor networks with a mobile sink. *Journal of Wireless Communications & Mobile Computing*.
56. Liang, W., & Luo, J. (2011). Network lifetime maximization in sensor networks with multiple mobile sinks. In *Proceedings of LCN'11*. IEEE.
57. Yun, Y., & Xia, Y. (2010). Maximizing the lifetime of wireless sensor networks with mobile sink in delay-tolerant applications. *IEEE Transactions on Mobile Computing, 9*(9), 1308–1318.
58. Xu, Z., Liang, W., & Xu, Y., Network Lifetime Maximization in Delay-Tolerant Sensor Networks with a Mobile Sink
59. Li, J., & Mohapatra, P. (2007). Analytical modeling and mitigation techniques for the energy hole problem in sensor networks. *Pervasive and Mobile Computing, 3*(3), 233–254.
60. Wu, X., Chen, G., & Das, S. K. (2008). Avoiding energy holes in wireless sensor networks with nonuniform node distribution. *IEEE Transactions on Parallel and Distributed Systems, 19*(5), 710–720.
61. Popa, L., Rostamizadeh, A., Karp, R., Papadimitriou, C., & Stoica, I. (2007). Balancing traffic load in wireless networks with curveball routing. In *Proceedings of the 8th ACM International Symposium on Mobile Ad Hoc Networking and Computing* (Vol. 9–14, pp. 170–179).
62. Keskin, M. E., Altınel, I. K., Aras, N., & Ersoy, C. (2013). *Optimal deployment, scheduling and routing for maximizing the lifetime of a wireless sensor network with multiple mobile sinks, Technical Report FBEIE- 02/2013-02*. Boğaziçi University, Istanbul.
63. Nesamony, S., Vairamuthu, M. K., Orlowska, M., & Sadiq, S. (2006). *On optimal route computation of mobile sink in a wireless sensor network*. The University Of Queensland.
64. Madan, R., & Lall, S. (2006). Distributed algorithms for maximum lifetime routing in wireless sensor networks. *IEEE Transactions on Wireless Communications, 5*(8), 2185–2193.
65. Keskin, M. E., Altınel, I. K., Aras, N., & Ersoy, C. (2011). Lifetime maximization in wireless sensor networks using a mobile sink with nonzero traveling time. *The Computer Journal, 54*(12), 1987–1999.
66. Srivastava, A. K., & Gupta, S. K. (2020). EEPMS energy efficient path planning for mobile sink in wireless sensor networks: A genetic algorithm based approach. In *Advances in Computational Intelligence and Communication Technology-2020* (pp. 101–108).
67. Srivastava, A. K., & Gupta, S. K. (2019). EERP: Energy-efficient relay node placement for *k*-connected wireless sensor networks using genetic algorithm. In Y. C. Hu, S. Tiwari, K. Mishra, & M. Trivedi (Eds.), *Ambient communications and computer systems. advances in intelligent systems and computing* (Vol. 904). Springer. https://doi.org/10.1007/978-981-13-5934-7_1
68. Azad, A., & Chockalingam, A. (2006). Mobile base stations placement and energy aware routing in wireless sensor networks. In *Wireless Communications and Networking Conference, WCNC 2006* (Vol. 1, pp. 264–269). IEEE.
69. Alsalih, W., Akl, S., & Hassanein, H. (2007). Placement of multiple mobile base stations in wireless sensor networks. In *2007 IEEE International Symposium on Signal Processing and Information Technology* (pp. 229–233). Springer.
70. Luo, J., & Hubaux, J.-P. (2005). Joint mobility and routing for lifetime elongation in wireless sensor networks. In *INFOCOM 2005, 24th Annual Joint Conference of the IEEE Computer and Communications Societies. Proceedings IEEE* (vol. 3, pp. 1735–1746). IEEE.

71. Behdani, B., Yun, Y. S., Cole Smith, J., & Xia, Y. (2012). Decomposition algorithms for maximizing the lifetime of wireless sensor networks with mobile sinks. *Computers & Operations Research, 39*(5), 1054–1061.
72. G. 4.0. (2010). Gurobi Optimizer 4.0, High-End Libraries for Math Programming. http://www.gurobi.com/

Conversion of Intermittent Water Supply to Continuous Water Supply of Chandigarh: A Case Study

Sanjeev Chauhan and R. M. Belokar

Abstract In India, where water is supplied to residents on an intermittent basis due to limited sources, implementing a continuous water supply scheme in an Indian city seems quite an arduous affair. To manage limited water resources to pump for 24 h without adding new water sources makes the project more quandary to work further. Achieving the milestone of continuous water supply in India would be a challenge worth taking and worth benefiting from. The case study prepared here provides insights regarding the importance and impact of the proposed methods and models here, based on the data received from government agencies working in this field. The goal is to prepare the outcomes of converting intermittent water supply of a city into regular water supply for the whole of Chandigarh city and not just a part of it. With the help of data collected from site visits, experimental analysis, and pilot experiments done on small scales, a case study has been prepared of how achieving continuous water supply can be made possible for a pan city with a population of more than 1 million.

Keywords Internet of things · Water treatment · District metering area · Non-revenue water · Life cycle cost · Hydraulic modelling

1 Introduction

1.1 Area, Population, and Its Present Water Source Scenario

Area: 114 km^2.

Population: 1.16 Million (Figure for 2021 as per data received from government offices provided to U.N. World Urbanization prospects).

S. Chauhan (✉) · R. M. Belokar
Department of Production and Industrial Engineering, Punjab Engineering College (PEC), Chandigarh 160012, India
e-mail: chausanju2002@gmail.com

P. K. Singh et al. (eds.), *Proceedings of Third International Conference on Computing, Communications, and Cyber-Security*, Lecture Notes in Networks and Systems 421,
https://doi.org/10.1007/978-981-19-1142-2_2

Bhakra Main Canal (BMC), situated 26 km from Chandigarh, is the primary source of potable water for Chandigarh. Water is continuously being pumped from Kajauli to Sector 39 to fulfil the water demand of Chandigarh. The pumped water is received at Sector 39, where the water treatment plant (WTP) is situated. There are currently two WTPs of 45 MGD and 25 MGD in Sector 39 and 1 WTP of 5 MGD capacity in Sector 12.

There are three seasonal streams-choe within our project area, namely N-Choe, Patiala Ki Rao Choe, and Sukhna Choe. There is no river located inside the Chandigarh area. However, the nearest river to the city is the river Ghaggar. Three man-made surface water bodies are recorded in Sukhna Lake, Dhanas Lake, and New Lake of sector 42.

Presently used method for water treatment (Fig. 1).

Table 1 is depicting the service level benchmarks as per MoUD and the existing service level in Chandigarh (Data as collected from the government departments).

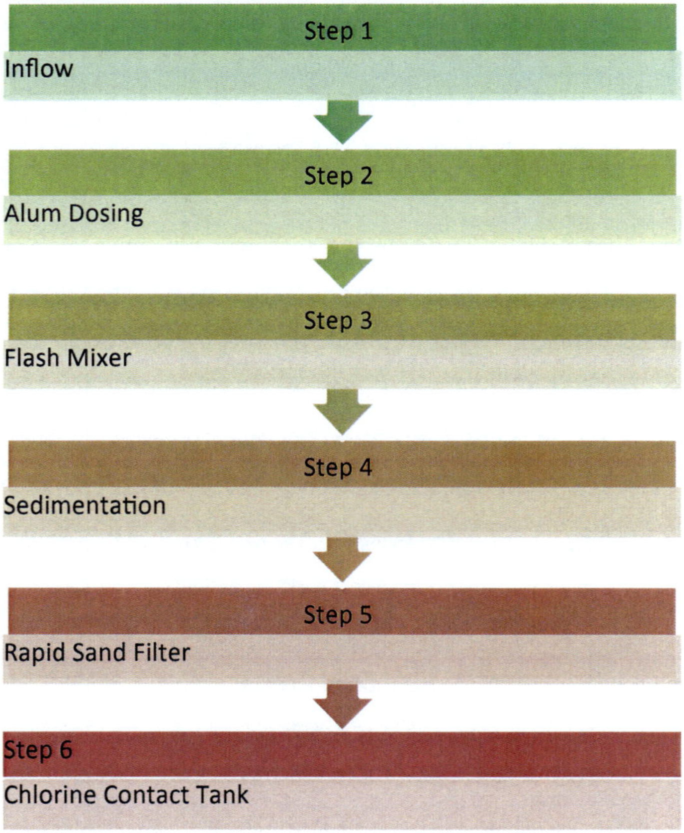

Fig. 1 Flowchart for the current method of water treatment used

Table 1 Service level benchmarks for water supply in Chandigarh

S. No.	Key performance indicators (KPI)	Service Level benchmark as per MoUD	Existing service level in Chandigarh
1	Coverage of water supply	100%	100.0%
2	Per capital supply of water	150 LPCD	227 LPCD
3	Continuity of supply	24 h	4–5 h in morning 4–5 h in evening
4	Extent of metering of water connections	100%	100%
5	Extent of non-revenue water	20%	35%
6	Quality of water	100%	Adequate and good
7	Efficiency in the collection of water-related charges	80%	92.25%

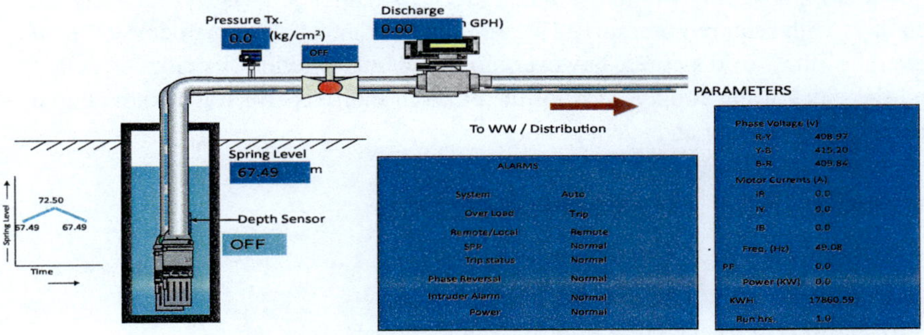

Fig. 2 Automation of water generation scheme

A figure representing the automation of the water generation scheme is shown in Fig. 2, with parameters and various alarms associated with it.

1.2 Disadvantages of Intermittent Water Supply Over Continuous Water Supply

Intermittent supply gives rise to the following deficiencies in the service and its management:

1. Severe risks to health, resulting from ingress of contaminated groundwater to the distribution system
2. High non-revenue water (NRW).
3. Lack of resources for system augmentation and remodelling.
4. Operational techniques used to control supply and demand are outdated.
5. Lack of control on the supply-side—no leak control, no data, plans, and SOPs.

6. No. of unauthorized connections causing physical and revenue leakages.
7. Low technical and financial management capacities.
8. Customer inconvenience: By imposing financial restrictions on water bills and limiting personal water usage for residents, which in many cases is below the level required for the practice of safe hygiene [3].

2 Need

Water being an essential commodity, many cities cannot even get 100 l per capita per day. With such water scarcity, it is not easy to provide 24 h water to the residents in PAN city, where most of the major cities in India are populous. Therefore, the idea was conceived with the plan that Chandigarh will implement a 24 × 7 water supply project for the whole city with a population above 1 Million. The definition of a 24 × 7 water supply would mean that every consumer gets water 24 h a day, all seven days with required pressure. This level of service has become common now in several of the world's cities; however, no Indian city provides its citizens with a round-the-clock water supply. Most of the cities in India receive water for a couple of hours during the day and evening hours.

3 Objectives

1. To study accurate metering of water and its impact.
2. To study how to avoid leakages and curb non-revenue water (NRW).
3. To assess the behavioural change of residents with continuous water supply.
4. To study optimum usage of surface water @150 LPCD (litre per capita per day) as per CPHEEO manual.
5. To study elimination of tube wells for groundwater sustainability for future generations.

4 Methodology to Meet the Desired Goal of Achieving Continuous Water Supply

1. Hydraulic Modelling
2. Integration of DMA with SCADA
3. Calibration of Distribution network
4. Revision of Tariff System
5. NRW Reduction Measures
6. Reuse of treated water
7. IoT-based System for continuous supply.

4.1 Hydraulic Modelling

Hydraulic modelling is essential when converting or expanding an existing network. It is about recuperating as much as possible from the faults of the existing running process. The effects of increasing the required load on the already existing system can be estimated. With this, any lapse that can arise during the design of new models can be detected easily.

A few of the measurement devices required for the modelling are:-

- Flow measurement devices such as electromagnetic flowmeter, ultrasonic flowmeter, bulk water flowmeter, and differential flowmeter
- Sensor-based pressure transmitters
- Actuators
- Valves such as air valves, sluice valves, gate valves, and butterfly valves
- Electrical parameter controls—SCADA (Clifford et al. 2005).

4.2 Integration of DMA with SCADA

The recording of the operating parameters of the water supply system was carried out, and the results obtained have been discussed later in the paper. District metering area (DMA) has been an excellent method to individually monitor and examine the water parameters, such as flow rate and pipeline pressure. DMA amalgamated with SCADA is primarily suitable for real-time insights across water treatment plants. In contrast, DMA integrated with SCADA helps identify the area of the issue more quickly and accurately. SCADA solutions may contribute a great deal towards integrating leak detection and periodically implementing planned repair programs. The following relevant measures and practices can be implemented with the use of a SCADA system (Fig. 3).

- Estimating the level of water losses via undetectable small leaks (in unknown locations)
- Constant monitoring and regulating of the pressure in the network at critical locations
- Recording and analysing sudden changes in flow rates for detecting new leaks and bursts
- Reducing the actual response time to isolate the troubled section [2].

Current System as per data acquired from government agencies

- Remote Terminal Unit (RTU)—204 Nos
- Total Tube wells Covered—155 Nos
- Total Booster Station Covered—49 Nos
- Main Master Control Unit—2 Nos.

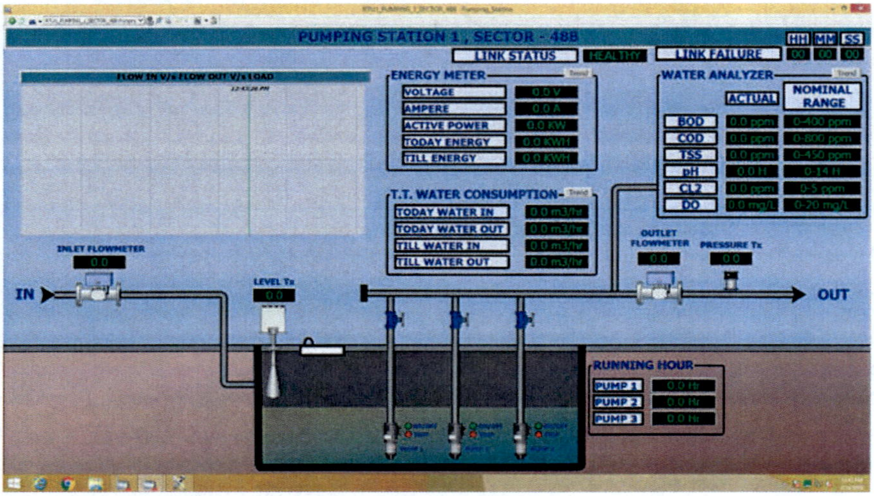

Fig. 3 Dashboard: SCADA control display

Wireless Water SCADA System

Entire process of monitoring takes place on the concept of SMS messaging system and artificial intelligence. The system provides fixed time data, alarms, and customized reports. In an alarming situation, the status is informed to the concerned officer by SMS. On pre-configured mobile sets or on the control room, P.C. provides present status of essential parameters on mobile instantaneously. In a new addition, artificial intelligence has been induced in the system and functions like auto cut, auto start-up are taken by the system itself [2].

Existing and proposed System for SCADA

The existing SCADA system was first introduced in the year 2007 at Municipal Corporation Chandigarh. There has been no significant up-gradation in that system since 2007. The software component of HMI (human–machine interface) should be made informative, and it should support the local alarm management system incorporated at Central Control Room. Screens should display the layout schematically as per actual configuration. Screens should have authentication and user login to access reports and manage the site efficiently. I.T. Friendly network across all layers and levels will help in improving the response time and speed of the entire system. The reporting systems should be enhanced with a secure reporting and analysis system. Detailed energy-related reports will identify the energy-intensive operations. System performance can be improved through a SITE-level database/data buffer. Asset management and innovative engineering document management software should be incorporated. The hardware component of SCADA will be operating under harsh local environmental conditions. The control room for SCADA has been made in Integrated Command Control Centre, proposed in sector 17. A complete

Table 2 Tube well data showing working and non-working no. of tube wells as per site visits done

Total No. of Tube wells	Average no. of tube wells working daily	Kept off by department	Tube well abandoned by department	Other daily issues faced by tube wells, such as motor faulty, low water level issues etc.
289	229	9	17	34

Table 3 Tube well data comparison as per site visit and data collected from government department

A1	A2	A3	A4	A5	A6	A7
Total No. of Tube wells working	As per pump specifications, the average flow in gallons/hr	Average working hours per day	Total average flow in gallons per day	Total average flow in MGD	Actual flow per day without losses in gallons (i.e. A1 * A2 * A3)	Difference between actual and observed flow data in M.G
229	9860	10	27,739,224	27	22,579,400	5.15

modern cum advanced SCADA program will be executed and operated in the project area. As required, approx. 252 Nos. of remote terminal unit (RTU) and 315 no. remote reading devices (RRD) have been proposed at various locations within the project area, Chandigarh.

4.3 Calibration of Distribution Network

The importance of the hydraulic model has been discussed above. However, a hydraulic model without calibration would not be of much or any use. The calibration of the models is based on either a demand-driven or pressure-driven approach. Due to the lack of details of the system and high degree of uncertainty, only with an accurate calibrated model, the model's credibility may be achieved [3] (Table 2).

As per the site visits done and comparing the data obtained with the data provided by the government (Table 3).

4.4 Revision of Tariff System

With the increase in daily water availability to the consumers, there is also a risk of consumers being slack in using water and therefore again leading to water wastage from their end. Keeping the cost revenues involved in the process and the risk of water wastage, it is best to maintain a practice such that water is utilized in the

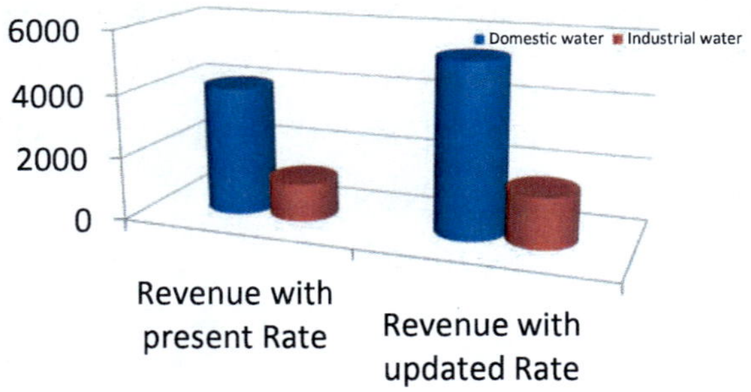

Fig. 4 Revenue comparison with updated tariff rates

most efficient manner possible. One such means to achieve this target is by raising the current water cost per unit paid by the consumer. This will lead to awareness for reducing water losses done by the consumer and, more importantly, add to the amount of revenue to be met for implementing the hydraulic model [3].

Below is a graph observation based on a rough estimation if the tariff is increased by 10% for domestic and industrial water (Fig. 4).

4.5 NRW Reduction Measures

As per data collected from the grievance cell of water supply department, a graph representing the percentage of non-revenue water is made. The graph clearly shows that most of the non-revenue water is from underground water leakages with as high as 43%. The other reasons of water wastage or non-revenue water being such as damaged pipes, slums and stand posts, unmetered and illegal connection, and UGR leakages. Also, a little margin (aprrox. 4%–5%) of NRW is generated from the water supplied to government buildings which is not billed as per the government law in U.T (Fig. 5).

As per research, water available—300 MG (Table 4).

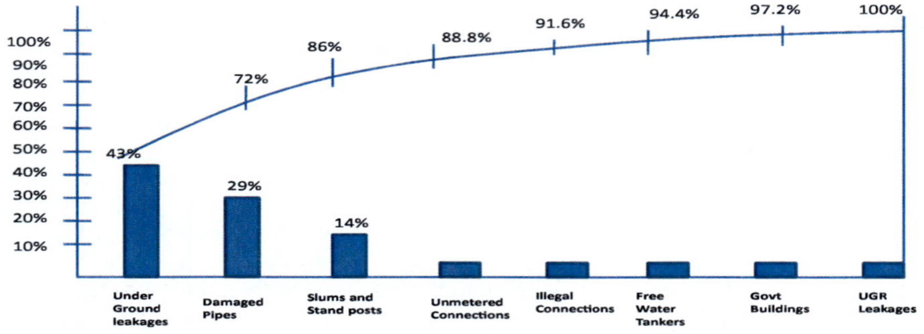

Fig. 5 NRW identification using Pareto chart

Table 4 Water availability in case of losses reduced by 20%

Water available—300 MG per day	Losses at present 35%
In case of losses reduced from present 35–15%, i.e., reduction in the loss by 20%	
Water availability—300 MGD + 20% of 300 MGD = 360 MGD	

Rate revenue: As per the present data, there are 1,60,000 households. By reducing the losses from 35 to 15%, as found above, there would be an increase in the daily water availability by 60 MGD as per above calculations [3].

4.6 *Reuse of Treated Water*

Water reuse offers promising opportunities to reduce the risk of non-availability of water resources for domestic and industrial purposes. Effective water reuse requires the integration of water and reclaimed water supply functions. Water reuse will become increasingly important in the years ahead to achieve sustainable use of the world's water. There is enough advancement in today's technology that makes it possible to recycle water for all domestic uses. However, there is still a widespread belief among people that reused water is a spoiled or unclean resource. To gain people's confidence in this matter and make them aware of the reuse of treated water, it is required that treated water parameters always be met under the designated value [3].

4.7 *IoT-Based System for Continuous Supply*

Various IoT resources shall be used for different applications in a vast range. As per the schematic diagram shown above, the authorized users shall have a direct accessibility either through mobile phones or by logging in the computer in main control station using SCADA. As shown in the schematic diagram, the main control station receiving the information from various iRTUs shall be forwarding ahead the information to certain users authorized by the department for keeping a proper monitoring on the running project [5] (Fig. 6).

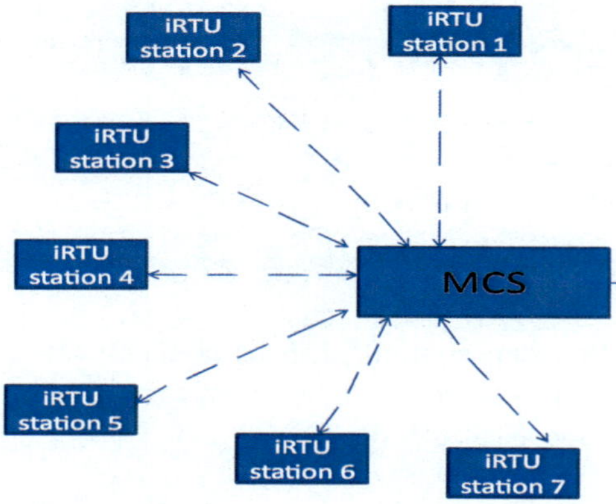

Fig. 6 A schematic diagram showing the proposed architecture of the IoT system to work upon

5 Other Proposed Algorithm

Another method that can be proposed for a continuous water supply is the use of overhead tanks. However, the drawback being that using overhead tanks is efficient only if the targeted city is having population not more than 1 million. The continuous water supply to a pan city with over 1 million population would not be achievable by using overhead tanks. The required pressure to be maintained in pipes using overhead tanks is achievable only for cities having very less population. Moreover, the cost and infrastructure required to construct new overhead tanks or renovate old existing head tanks is quite high and will be a further disadvantage. Overall, the use of head tanks would require a large amount for its construction and infrastructure, and still, it would be feasible only for cities having population less than 1 million.

6 Literature Review

A brief description of relevant literature for design for conversion of intermittent water supply to continuous water supply of Chandigarh: A case study.

Author(s)	Title of the paper	Description and findings

(continued)

(continued)

Author(s)	Title of the paper	Description and findings
Ramón Martínez, Nuria Vela, Abderrazak el Aatik, Eoin Murray, Patrick Roche, and Juan M. Navarro	On the Use of an IoT Integrated System for Water Quality Monitoring and Management in Wastewater Treatment Plants (2020)	This paper explains on how the analytical device such as nitrite analyser based on novel ion chromatography detection method, and other such devices can be beneficial for the detection method in wastewater treatment plants. The devices discussed here and integrated using an Internet of things software platform and are also tested under real conditions. The purpose of the paper is to create a decentralized smart water quality monitoring system for better water quality monitoring and management
Elad Salomons 1, Uri Shamir 2, and Mashor Housh	Optimization Methodology for Estimating Pump Curves Using SCADA Data	Supervisory control and data acquisition (SCADA) acquired a lot of data. This paper explains that this data collected by SCADA, if used and analysed properly can increase the pumps performance, efficiency, and longevity. Pump characteristic curves are derived using this data which further assisted in determining the fixed and variable speed pump curves. The methodology discussed in the paper has been demonstrated in a real-world case study, and the practical data obtained have been shared in the paper

(continued)

(continued)

Author(s)	Title of the paper	Description and findings
Allen M, Mudasser I, Srirangarajan S, Lim B L, Girod L, Whittle A	Water Distribution System SCADA Survey Report	This paper discusses on the importance of hydraulic modelling and provide statistics based on the projects on which the survey was done. This paper explains that hydraulic modelling was used in more than half of the projects under the survey. Six or more applications of SCADA which are used in water distribution system are discussed, and the survey was developed based on these applications. The majority of the plants using SCADA have been monitoring the system using SCADA for more than 15 years. Apart from the survey reports, this paper discusses on how the data collected from various SCADA sensors can be used for various related purposes, such as bill generation, generating water regulatory report, and create reports on trending of water quality

7 Pilot Study and Experimental Analysis

7.1 Pilot Study 1

A pilot study was also conducted with conventional pressure gauges to discover chronic water contamination problems at sectors 28 C and Sec 28 D Chandigarh. The pressure gauges were installed at various locations for a distance of 165 m with six testing points.

The pilot study's objective was to analyse the effect of measurement devices to find underground leakages. Six pressure transmitters were used in the water pipeline at unequal intervals (Fig. 7).

Up to 175 m, the pressure decrease was under the norm parameters, from 35 to 24 Psi. However, after 175 m, from point D onwards, a rapid decrease in pressure was observed in the following testing point, which was kept at a distance of 125 m. The pressure observed was 10 Psi. This indicated a sure sign for leakage between

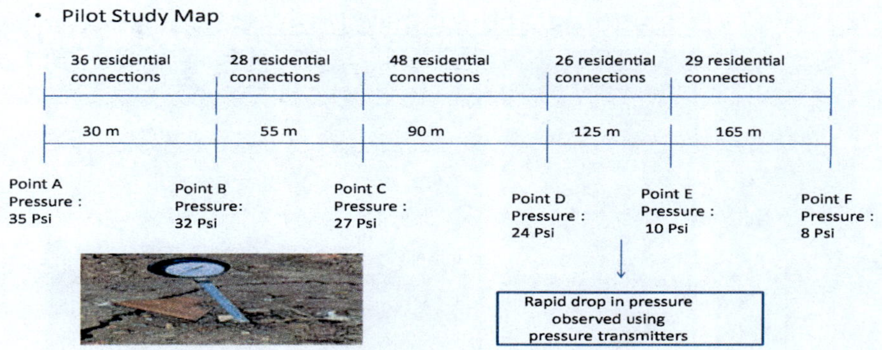

Fig. 7 Pressure transmitter used in the pilot experiment

two points, i.e., D and E. A further Site investigation was done, and it was found that the leaked water was being discharged into the stormwater line and was producing loss to the government.

7.2 Pilot Study 2

Below is the architecture of the pilot project carried out for obtaining the importance of RTU in SCADA (Fig. 8).

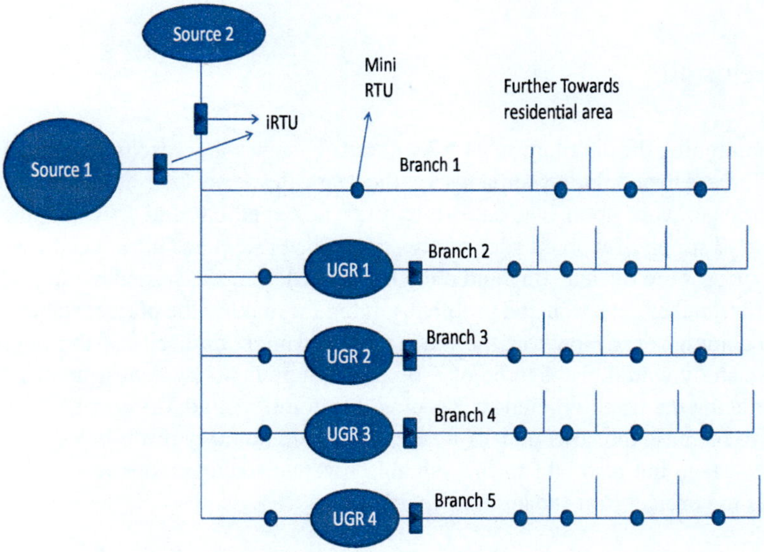

Fig. 8 Pictorial representation of working of RTU in SCADA

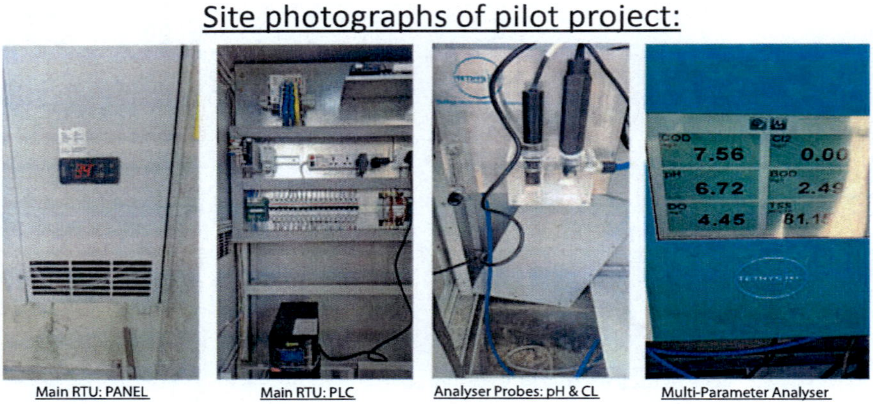

Fig. 9 Pictures taken during pilot project of main RTU panel, PLC, analyser probes, and multi-parameter analyser

The recycled water is drawn from two primary sources, i.e., STP 3BRD and STP Diggian. This recycled water is distributed into five branches, including four supported by UGRs at Sector 29B, 29C, Sec 48, and Sec 28. The main iRTUs with BOD COD sensors, TSS sensors, flowmeters, pressure transmitters, and electrical sensors were installed. Whereas, branch iRTUs known as mini iRTU were provided with flowmeters, pH sensors, and pressure transmitters at 13 different locations. This resulted in practical and real-time monitoring of qualitative and quantitative parameters of recycled water (Fig. 9).

8 Conclusion

To keep continuing the distribution 24 * 7 without any new water resources available, plus the increase in population demand over the years, the project seems very difficult and ludicrous to work upon. The case study finds new methods and checks that can lead to 24 * 7 supply if worked adequately upon. The basic plans introduced for the project's success are by detecting and reducing water leakages assisted by upgraded SCADA for quicker detection and problem solving for making the project stable and robust for coming years. However, the above proposed methods discussed thoroughly in the case study would prove to be of little use if there is no awareness among the public for using the water efficiently. Keeping this point in mind, the case study also proposes to increase the tariff plan of water, which will not only result in a decrease in water wastage but also add to the valuable revenue required for the successful running of the project over the coming years.

References

1. Massingham. (2016). The Lifecycle of an AWS IoT Thing from Manufacture to Retirement. https://nest.com/works-with-nest
2. Salomons, E., Shamir, U., & Housh, M. (2021). Optimization Methodology for Estimating Pump Curves Using SCADA Data.
3. Kojima, S., & Okaga, T. (2005). Basic design study report on the project for improvement of water supply system
4. Rahman, L. F., Ozcelebi, T., & Lukkien, J. J. (2016). Choosing your IoT programming framework: architectural aspects. In *2016 IEEE 4th International Conference on Future Internet of Things and Cloud (FiCloud)* (pp. 293–300). https://doi.org/10.1109/FiCloud.2016.49
5. Martínez, R., Vela, N., el Aatik, A., Murray, E., Roche, P., & Navarro, J. M. (2020). On the Use of an IoT Integrated System for Water Quality Monitoring and Management in Wastewater Treatment Plants
6. Final reference architecture of Open AIS system. Tech. Rep. d2.7, v1.0; OpenAIS: Open Architectures for the Intelligent Solid State Lighting Systems in year 2017 with the URL as: http://www.openais.eu/user/file/openais_final_reference_architecture_(d2.7)_v1.0-pub.pdf
7. Garcia-Carrillo, D., & Marin-Lopez, R. (2016). Lightweight CoAP-based bootstrapping service for the internet of things. *Sensors, 16*(3).
8. Stolikj, M. (2015). *Building blocks for the internet of things*. Ph.D. dissertation; Technische Universiteit Eindhoven.
9. Gielen, F. (2018). Software architecture for the internet of things: Q.A. Interoperability. Coursera Lecture. https://www.coursera.org/learn/iot-software-architecture/lecture/KXaEr/qa-interoperability. Accessed: 1 Feb 2018.
10. Albee, S., & Byrne, R. (2007). Report 1: A global vision for driving infrastructure asset management improvement. In: H. Alegre and M. Almeida (Eds.), *Strategic asset management of water supply and wastewater infrastructures*. International Institute for Environment and Development.
11. Carter, R., & Rwamwanja, R. (2006). *Functional sustainability in community water and sanitation: a case study from South-West Uganda*. Tearfund.
12. De la Harpe, J. (2011). *Lessons for Rural Water Supplies: Assessing progress towards sustainable service delivery, South Africa.* [online] The Hague: IRC International Water and Sanitation Centre. Available at: Accessed 10 Nov 2012
13. Fonseca, C., et al. (2011). *Lifecycle costs approach: costing sustainable services. (WASHCost Briefing Note 1a)* [online] The Hague: IRC International Water and Sanitation Centre. Available at: Accessed 17 Dec 2012
14. Kamienski, C., Soininen, J. P., Taumberger, M., Dantas, R., Toscano, A., Salmon Cinotti, T., Filev Maia, R., & Torre Neto, A. (2019). Smart water management platform, IoT-based precision irrigation for agriculture. *Sensors*, 276.
15. Dong, J., Wang, G., Yan, H., Xu, J., & Zhang, X. (2015). A survey of intelligent water quality monitoring system.
16. Geetha, S., & Gouthami, S., Internet of things enabled real-time water quality monitoring system bright water.
17. Bassi, A., & Horn, G. (2008). Internet of things in 2020. In *Proceedings of the Joint European Commission/EPoSS Expert Workshop on RFID/Internet-of-Things*.
18. Kamienski, C., Soininen, J. P., Taumberger, M., Dantas, R., Toscano, A., Salmon Cinotti, T., & Filev Maia, R. (2019). A Smart water management platform: IoT-based precision irrigation for agriculture. *Sensors*.
19. Dahasahasra, D. S., & Dahasahasra, D. V. (2010). A model transforming an intermittent into a 24x7 water supply system.
20. Garcia-Carrillo, D., & Marin-Lopez, R. (2016). Lightweight CoAP-based bootstrapping service for the internet of things. *Sensors*.

21. Gielen, F. (2018). Software architecture for the internet of things: Q.A.—Interoperability. Coursera Lecture. https://www.coursera.org/learn/iot-software-architecture/lecture/KXaEr/qa-interoperability
22. De la Harpe, J. (2011). *Lessons for Rural Water Supplies: Assessing progress towards sustainable service delivery, South Africa.* [online] The Hague: IRC International Water and Sanitation Centre.

A Novel Compression Method for Transmitting Multimedia Data in Wireless Multimedia Sensor Networks

Richa Tiwari and Rajesh Kumar

Abstract This paper discusses various compression methods used in wireless sensor networks. Compressed sensing is the emerging signal processing tool that makes the transmission of data easy via low-data rate links. In the wireless sensor network applications, a group of sensors is used to sense any events and make decisions, and the collaborated information sensed by different tiny sensing devices are used to give the decisions about the occurrence of the particular events. According to the different applications and data types, the quality of service parameters and designing parameters for nodes are different. For dealing with low bandwidth in a sensor network, it is most important to reduce the transmitted data bits between sensor nodes or from nodes to sink. In the case of multimedia data such as image signals, the compression is beneficial for the reduction of these bits because fewer bits required less transmission energy. In some situations of the multimedia sensor network, some loss is accepted without affecting the too much quality of results. Data collected by nodes are spatially correlated with each other, so the image samples collected over time by the nodes are also correlated with each other. If only some samples are transmitted, then these samples are sufficient to give the knowledge about the suspected object inside the monitoring area, so the transformation-based compression technique is the good solution for the compression in the case of the multimedia sensor network. In this paper, a Hadamard transform-based compression technique is discussed for image compression with the consideration of different designing parameters of an image signal. In that manner, this work helps us to select the transform and source coding schemes for the compression of image data inside the wireless multimedia sensor network.

R. Tiwari (✉)
Krishna Engineering College, Ghaziabad, UP, India
e-mail: richavgi@gmail.com

R. Kumar
North Eastern Regional Institute of Science and Technology, Nirjuli, Arunachal Pradesh, India
e-mail: rk@nerist.ac.in

P. K. Singh et al. (eds.), *Proceedings of Third International Conference on Computing, Communications, and Cyber-Security*, Lecture Notes in Networks and Systems 421,
https://doi.org/10.1007/978-981-19-1142-2_3

Keywords Data compression · Data grouping · Wireless multimedia sensor networks · Hadamard transform · Discrete cosine transform · Mean square error · Peak signal to noise ratio · Similarity index · Optimization

1 Introduction

In wireless sensor network applications, the data sensed by sensor nodes have been shared with the base station. To efficiently perform that task, sensor networks are used with many emerging computer network fields such as cloud computing and the Internet of things. These technologies make long-distance transmission easy and provide a high amount of virtual storage space. Internet of things (IoT) is a useful technology in many aspects when it is used in conjunction with sensor network technology. It is widely used in the transmission of multimedia data, because this data require high speed in transmission. The block diagram for WMSN is shown below in Fig. 1.

However, some major challenges have been resolved, such as data compression and energy management inside tiny sensing devices. A large amount of data has been sensed by sensors that need to be transferred inside the IoT layers. The sensors consumed a large amount of energy in the transmission of image signals in the case of the multimedia sensor network. If the total amount of transmitted frames bits is reduced from the perception layer of sensors to the network layer of IoT network architecture, then the network lifetime has been improved. For achieving this task, sensors are required to perform several signal processing operations on the aggregated data and make decisions about the events happening in the sensor network field. In

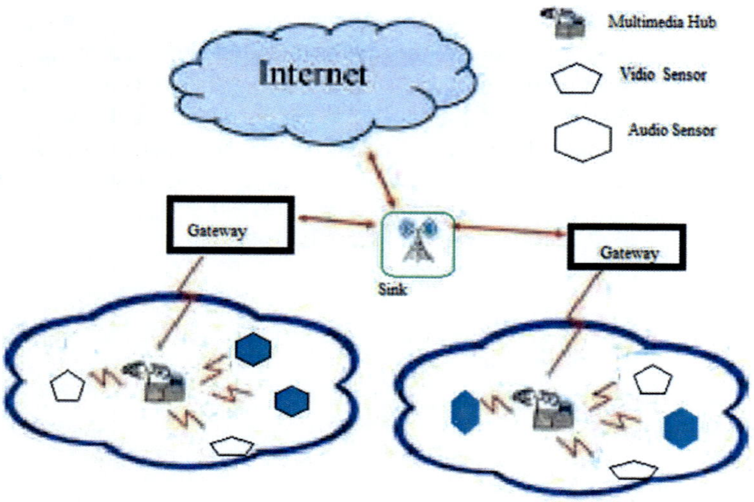

Fig. 1 Wireless multimedia sensor network

this task, sensor nodes are required to communicate in two manners: First, locally between the neighborhood nodes and second, globally between the long-distance sensor nodes and/or to base stations. Sensor nodes in sensor networks communicate using radio links that operate at low power and have short-range communication capability and also have low bandwidth for the transmission of data. Typical data rates of these radio links are from tens to a few hundred Kbps, so it is a very critical task to ensure the transfer of multimedia data on that sensor network links because these applications not suitable for low-data rate sensor link, and in practice, around only 20% of sensor power is consumed in sensing, rest power is consumed in data transmission. It will be very useful in these practical scenarios to compress the data communicated among sensor nodes and sink for reducing the bandwidth required for transmission and to achieve the high-data rate fast requirements. Since sensors operate with portable batteries, the benefit from that data compression is reduced energy consumption and improved network lifetime. The motivational aspect behind this approach is the availability of the hardware for Hadamard transform that is very low cost and consume very little energy. There are many methods of data compression in the case of lossless data compression high degree of accuracy has been obtained. The recovered signal at the destination is the same as the originally transmitted signal. In this case, the compression ratio is very small. The hardware complexity and storage requirements are very high in those cases. In the sensor network applications generally, the considered application area is the place that has been monitored for security purposes, in these situations, for the information about the existence of the suspected object, the good quality of an image has not been required. In lossy compression, some information content is lost in the data compression and decompression process. The quality of the decompressed signal is not very good but the hardware required and the requirement of the complex algorithm has been removed or in other words, the transformation techniques available in signal processing can perform this task. The sensor nodes have a small storage buffer that stores the compressed image of the area in form of bits. The sensor nodes sensed the image and compressed that image by using the transformation technique, if there is no difference in the number of bits up to a predefined threshold level, the sensor not transferred the sensed image signal, otherwise the sensor nodes send these signals. In that manner, the transmission energy should be saved.

In the transmission of multimedia data, it is useful to compress data by using various transforms. The number of transforms is available for compression in this paper, various techniques are discussed, and after that Hadamard transform-based compression technique is proposed for the compression of the image signal. There are many reasons behind considering that technique for image processing purposes. In the Hadamard transform, the multiplication is performed with an identity matrix called Hadamard matrices. Hentati et al. [1] proposed a hierarchical implementation of Hadamard transform modules by using RVC-CAL data flow programming. This FPGA-based implementation consumed less power and updated up to 32 input–output data buses. This type of implementation is easily possible inside the sensor module. That provides compression as well as low-power consumption that is the reason behind to the selection of this transformation technique.

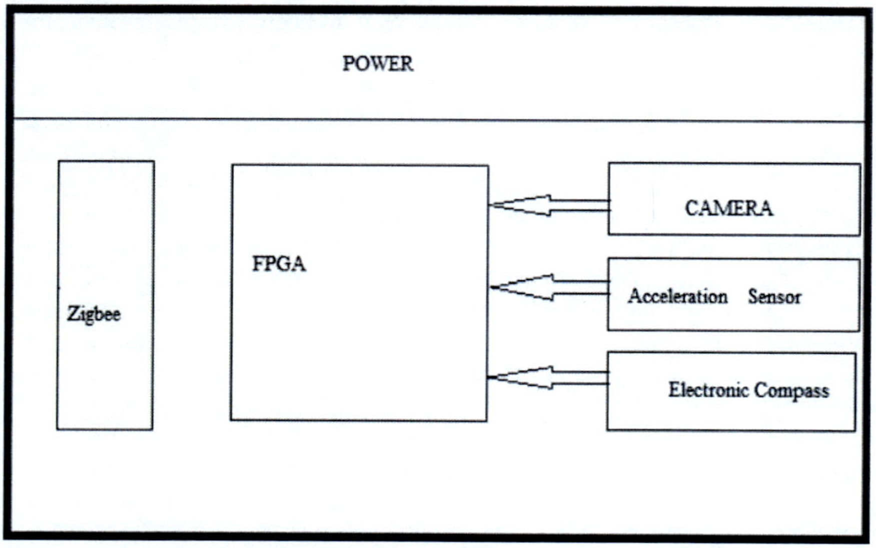

Fig. 2 FPGA-based design of a multimedia sensor node

The general architecture for designing of FPGA-based nodes used for sensor networks is shown in Fig. 2.

This paper proposes an image compression method that is based on the comparison of similar repeated data. This whole process is done on the images sensed by sensors. The similarity index of the image is calculated by root to mean square criteria.

Following the image, parameters are important to evaluate like compression ratio (CR), mean square error (MSE), and peak signal to noise ratio (PSNR). These parameters descriptions are following:

Mean Square Error: MSE is the cumulative square of the error between the original image and the received image. It is given as

$$\text{MSE} = \frac{1}{MN}\sum_{m=1}^{M}\sum_{n=1}^{N}[(S(i,j) - (R(i,j))]^2 \tag{1}$$

where $S(i,j)$ is the original image signal sensed by the sensor node, $R(i,j)$ is the received image signal.

Peak Signal to Noise Ratio: In this case, the signal represents an original image that is originated from the source, and the noise represents the compression and transmission losses. Thus, PSNR gives the overall quality of the received signal.

$$\text{PSNR} = 10\log 10\frac{\text{MAXO}}{\text{MSE}} \tag{2}$$

where MAXO is the maximum value of pixels in the given image, each pixel is represented by 8 bits hence MAXO $= 2n - 1 = 255$, n is bits per pixel.

Compression Ratio: This ratio is the ratio of the difference between original size and compressed size to original image size. The formula for compression ratio is

$$\text{CR}(\%) = \left[(\text{Original size} - \text{Compressed size})/(\text{Original size})\right] * 100 \quad (3)$$

The rest paper is organized as follows: In Sect. 2, we describe the various types of compression methods that are used in multimedia sensor networks. Section 3, the compression plan with Hadamard transforms method. In Sect. 4, MATLAB simulation results are obtained in terms of compression ratio, MSE, and PSNR. The conclusion remark on the paper is presented in Sect. 5.

2 Related Work

In sensor networks, the research has been focused on resource utilization with low-energy consumption. In a paper presented by El Gamal et al. [2], various scheduling algorithms with energy efficiency have been discussed for minimizing the energy used in data transmission, these algorithms used the variable packet-transmission time. In Sankar Subramaniam et al. [3], an energy-efficient frame-size optimization has been presented, and the practicality of forward error correction (FEC) in WSNs has also been considered. Pal et al. [4] presented balance in cluster-size clustering algorithm to extend the lifetime of a sensor network. This algorithm upgrades the cluster standard and provided a lower number of nodes dead inside the network. Lee et al. [5] designed the layout for the energy-efficient wireless video sensor networks. In this algorithm, "capture rate and network control" are developed for multimedia camera networks. This paper reveals that the quality of service is the function of many parameters like energy, captures rate, and transmitted power and has been improved by maintaining the optimal values of these parameters. Zhou Wei et al. [6], the distributed image compression has been analyzed in this paper for pass on the multimedia data over WSNs. The processing model inside the network is used for the development of distributed multiple-node cooperative network model; this model upgrades the performance of this network. Further, for achieving the improvement in image compression ratio, image quality, and SNR, a new algorithm NDICPCA algorithm is developed by Shikang Kong et al. [7], a non-negative factorization method for matrix is suggested, and this method is based on NMF for multimedia WSN enhances image compression ratio, data recovery rate with reduce amount of overall energy consumed by the sensor nodes.

For removing the small battery limitation, energy harvesting is the best solution in WSN [8, 9], and these schemes utilized the energy gathering from the environment for improving the life span of network and the connectivity between nodes. The energy harvesting via a RF signal in a relay network is analyzed in Nasir et al. [8], this work gives an sensors-relays-destination node architecture, a relay node having the energy harvesting capability, this relay utilizes the radio signal transmitted by the sensor node for energy harvesting, that energy is used to send the sensed data to the sink node. Gunduz et al. [9] proposed tools and analytical model, and this mathematical tool is used for designing reliable energy harvesting communication systems. The energy in WSNs is optimized by various methods; out of these methods, some methods are duty cycling between the resources and various signal processing inside the network. In the schemes based on the duty cycling approach, the energy saving is achieved by the scheduling of wake-up/sleep time at sensor nodes, and various MAC layer protocols are utilized for this purpose. Various network processing schemes are data aggregation, compression, etc. The main concept behind these methods is to reduce the amount of data to be transmitted. Some effective methods to reduce the amount of data to be transmitted, data compression, have been actively proposed by many researchers [10–13]. Using these methods, small-size data packets are transmitted inside the sensor network. By this compression process, the consumed energy is reduced in data transmission inside the network. Razzaque et al. [14] give a survey about various data compression methods used for energy-efficient WSNs. Data compression algorithms for WSNs are divided into several categories such as distributed source modeling, transform coding, source coding, and compressive sensing, various authors compared them for various performance parameters such as the compression ratio and power consumption. Application of data compression in multi-hop WSNs is to be considered, where packets originated from a source node are relayed toward the destination node through multiple wireless paths. These nodes throughout this multiple paths are capable of performing data compression for reducing the energy consumption in transmission. When packets are relayed toward the sink from the source node, the size of these packets is reduced by various data compression methods, which automatically decrease the overall energy consumption inside the wireless multimedia sensor network. In this paper, the energy harvesting via compression has been consider for increasing the lifetime of a multimedia sensor network. Via compression sensor nodes are transmitting and receiving comparatively smaller-size packets. This method reduces the overall energy consumed inside the network. We also consider battery-operated sensor nodes, which are rechargeable by energy harvesting. To reduce the average energy consumption of sensor nodes and extend the lifetime of the networks, we propose a compression algorithm for image signal by considering comparison between the upcoming image signals bits generated after compression with the previously stored image bits.

3 Materials and Methods for Data Compression

Wireless multimedia sensor network (WMSN) is based on the transmission of multimedia signals that needs high-data rate. For the processing and transferring of that data within limited storage space, the compression of every frame becomes the basic requirement. Network has less bandwidth, so the transmission of compressed data becomes easy. The compression is grouped into two parts such as lossy compression and lossless compression. In the case of lossless compression, transmitted image and recovered image differences are very small, but in lossy compression, distorted image is received. The compressing of the image frame is to achieve a large compression ratio with a smaller loss. For achieving the compression in the sensor network, two types of terminologies are used compression inside a node and cooperative compression between nodes. The data compression method employed in one node based on the compression process occurred inside the single sensor node, the exchanging of information between other nodes is not happen inside the network. The main object of these schemes is the achievement of a high-compression ratio of data, for reducing the packet size at node level before sending to another node or destination. In cooperative compression, the data packets are sending over the network by sharing the parameters such as the relative position of each node or the corresponding relation between sensed data at various nodes. Each sensor node will do the compression of raw data. The computation complexity is increased in this type of compression, but network lifetime is increased by proper energy balance in this situation. Khedker et al. [15] proposed that a transformation is a powerful tool for achieving a high-compression ratio. This can be achieved by using many two-dimensional image transforms like DCT, KLT, Slant, Hartley, Hadamard, and DST. By using transform techniques, the images are converted linearly from pixel to bits. In a sensor network, many applications accept lossy image compression up to a significant level. So this is beneficial because the high number of insignificant coefficients is to be discarded, and a high-compression ratio is achieved. Network energy is also saved by using these transform approaches. Hence, the transform domain-based compression methods are very suitable for WMSN. The block size of transform coefficients is of the same order as the block size of original image pixels. The original signal is reconstructed by applying inverse transform on the reduced coefficient block. Numbers of transformation techniques are used in signal processing. In this section, the effect of various 2-dimensional image transformation techniques like KLT, DCT, Slant, Hartley, Hadamard, and DST is discussed according to their energy requirement and compression properties. Karhunen–Loeve transform (KLT) is reconstructing the original image from less coefficients present in the transformation domain. This transform provides the minimum MSE between the transmitted image signal and the received image. The main limitation of this transform is its complexity as the kernel needs to be calculated each time an image signal is changed for compression. In discrete cosine transform (DCT), the kernel is fixed for the input image, thus computational cost is less as compared to KLT and both JPEG and MPEG compression standards used with this transform. It is a block transform approach that is applied over non-overlapping blocks of the image signal. In general, DCT block sizes are 8 by 8 or 16 by 16. Hartley transform is having a real linear operator, and it is symmetric transform method. Hadamard transform is just like the Walsh transform which represents the signal by using the set of orthonormal square wave functions. This transform is computationally simple because Hadamard functions are real and will take values either $+1$ or -1. Discrete sine transform is Fourier-based transform in which an odd symmetrical portion of the real data is extracted from original data.

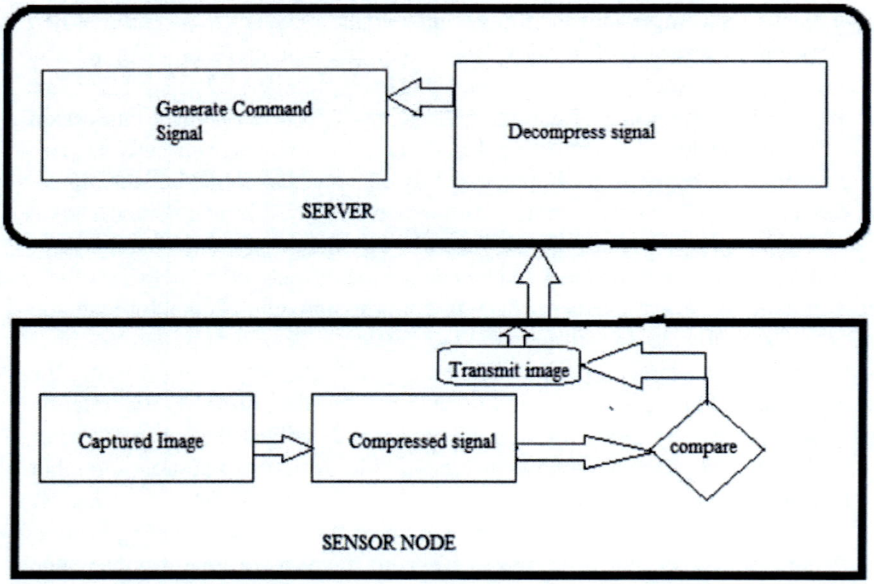

Fig. 3 Data flow from node to server in MSNs

Its operation is based on the fact that the function used by that transform operates on a finite number of discrete data points. The basic difference between the DCT and DFT is that DCT contains real and even parts of DFT, whereas DST contains real and odd parts of DFT.

In this section, we present a cooperative compression technique for sensor networks for the compression of the image signal. The energy consumption is using data compression for multi-hop delivery of data packets. As a data packet is forwarded along multiple paths, Kim et al. [16] propose a scheme that determines the compression level status at each node for minimizing the energy consumption and to maximize the node lifetime (Fig. 3).

In future aspect, the combination of the above-proposed technique with Hadamard transform for the compression of an image signal is very much useful in energy balancing. In this approach for every cycle, the base station provides the compression ratio information to all nodes. Nodes compress the image signal according to their required compression ratio. Now in this work, the Hadamard transform-based node has been designed that compresses the signal using this and after compression, comparator compares the total number of bits between the generated data and stored data if difference increases above the predefined limit then only data transfer occurs. The server generates security signal for all nodes, the inclusion of that signal inside that algorithm is now not taken into consideration. The algorithm steps are discussed below,

Algorithm

Step 1: Take input image
Step 2: Extract frames of that image
Step 3: Apply Hadamard transform-based compression on each extracted frame
Step 4: Convert pixel matrix of frames into serial data for transmission
Step 5: Compare these bits with stored bits
Step 6: If bit differences are large then transmit data, otherwise no transmission
Step 7: Set the baud rate of Tx and Rx unit at 19,000
Step 8: Open port at the transmitter for transmitting these data
Step 9: Fix output buffer sizes that are suited for serial data transmission
Step 10: Output serial data
Step 11: Fix buffer size at receiver for both input and output
Step 12: Open receiver port
Step 13: Receive serial data coming from sensor nodes
Step 14: Separate serial data into parts for reconstruction
Step 15: Reshape serial data into the matrix
Step 16: Repeat steps 3 to 15
Step 17: End.

4 Simulation Results

For determine the performance of the above algorithm, a multi-hop WSN is to be selected, in which N sensor nodes are deployed inside the observation area. The sensor node type is a MICAz device powered by rechargeable batteries and transmitted data serially via Zigbee at 19,200 bps. Each nodes are having 5 kJ as the initial energy. The transmission range of the RF transceiver is between 65 and 90 m. All sensors are within the range of each other. The energy consumed for compression, transmission, and reception is 3.5, 600, and 670 nJ/bit. Initially, the image signal is transformed into frames, and each frame is compressed by transformation. Each of these compressed frames are resized into 72 × 72 and then transmitted serially over Zigbee at a 19,000 bps baud rate. At the receiver, these frames are recovered back. After getting received frames, image parameters are calculated on both transmitted and received frames, by using MATLAB software. The comparative parameters are shown in Table 1, and this table shows the image parameters such as MSE, PSNR, and CR for seven transmitted and received frames at 19,000 bps.

4.1 Performance Parameters for Image Signal

Table 1 shows the calculation of various parameters related to the compressed image in the sensor network. These parameters are related to the quality of the image, for the reduction in the number of transmitted frames and energy saving, and it is

Table 1 Performance parameters

Frame no.	Original frame size (KB)	Compressed frame size (KB)	Compression ratio (%)	MSE	PSNR
Frame 1	16.5	10.70	35.1515	12.7072	26.0694
Frame 2	16.8	9.80	39.8809	13.7073	26.7271
Frame 3	16.0	9.50	40.6250	12.9899	26.7555
Frame 4	16.7	10.00	40.1197	13.7072	26.0232
Frame 5	16.5	10.20	38.1818	12.7077	26.3232
Frame 6	16.8	10.50	37.5000	12.7075	26.3134
Frame 7	16.3	10.60	34.9693	13.7271	26.3535

necessary to compromise among these values, the best way is to design an algorithm in such a manner that adjusts the compression ratio of nodes according to application requirement.

The performance measures of image signal are collected in Table 1.

5 Conclusions

This paper, a compression method based on Hadamard transform, is discussed for image compression. Network lifetime improvement has been done by comparing the compressed signal bits, so the conditional transmission reduces the transmission time and saves the transmission energy. In that manner, this technique gives advantage of duty cycling without the in-cooperation of any complex MAC protocol. The comparison is between the previous stored bits, so there is no requirement of any additional storage inside the node. This concept is beneficial in energy harvesting in wireless multimedia sensor network applications. The future aspect of this paper is to generate a command from server to sensor nodes for transmitting all images without comparison when any suspected article comes inside the monitoring area.

References

1. Hentati, M., Aoudni, Y., Nezan, J. F., & Abid, M. (2012). A hierarchical implementation of Hadamard transform using RVC-CAL dataflow programming and dynamic partial reconfiguration. In *Conference on Design and Architectures for Signal and Image Processing (DASIP)*, Oct 2012, Karlsruhe, Germany (pp. NC. hal-00763876).
2. El Gamal, A., Nair, C., Prabhakar, B., et al. (2002). Energy-efficient scheduling of packet transmissions over wireless networks. In *Proceedings of IEEE INFOCOM*, June 2002 (pp. 23–27), New York.
3. Sankarasubramaniam, Y., Akyildiz, I., & McLaughlin S. (2003). Energy efficiency based packet size optimization in wireless sensor networks. In *Proceedings of the IEEE Sensor Network Protocols and Applications (SNPA)*, 11 May 2003, Anchorage, AK.

4. Pal, V., Singh, G., & Yadav, R. (2015). Balanced cluster size solution to extend lifetime of wireless sensor networks. *IEEE Internet Things Journal, 2*(5), 399–401.
5. Lee, S., Lee, I., Kim, S., Lee, S., & Bovik, A. C. (2014). A Pervasive network control algorithm for multi-camera networks. *IEEE Sensors Journal, 14*(4), 1280–1294.
6. Wei, Z., Lijuan, S., Jian, G., & Linfeng, L. (2016). Image compression scheme based on PCA for wireless multimedia sensor networks. *The Journal of China Universities of Posts and Telecommunications, Science Direct, 23*(1), 22–30.
7. Kong, S., Sun, L., Han, C., & Guo, J. (2017). An image compression scheme in wireless multimedia sensor networks based on NMF. *Information Journal MDPI., 8*, 1–26.
8. Nasir, A., Zhou, X., Durrani, S., et al. (2013). Relaying protocols for wireless energy harvesting and information processing. *IEEE Transaction Wireless Communication, 12*(7), 3622–3636.
9. Gunduz, D., Stamatiou, K., Michelusi, N., et al.: Designing intelligent energy harvesting communication systems. *IEEE Communications Magazine, 52*(1), 210–216.
10. Medeiros, H. P., Maciel, M. C., Demo Souza, R., et al. (2014). Lightweight data compression in wireless sensor networks using Huffman coding. *International Journal of Distributed Sensor, 2014*, 1–11.
11. Alsalaet, J. K., & Ali, A. A. (2015). Data compression in wireless sensors network using MDCT and embedded harmonic coding. *ISA Transactions, 56*, 261–267.
12. Incebacak, D., Zilan, R., Tavli, B., et al. (2015). Optimal data compression for lifetime maximization in wireless sensor networks operating in stealth mode. *Elsevier Ad Hoc Network, 24*, 134–147.
13. Ma, N. (2019). Distributed video coding scheme of multimedia data compression algorithm for wireless sensor network. *EURASIP Journal on Wireless Communications and Networking*. https://doi.org/10.1186/s13638-019-1571-5
14. Razzaque, M. A.., Bleakley, C., & Dobson, S. (2013). Compression in wireless sensor networks: A survey and comparative evaluation. *ACM Transaction Sensor Network (TOSN)*, 10(1), 5.
15. Khedkar, M., Asutkar, G. M., Hariprakash, R. (2019). Evaluation of data compression techniques for video transmission over wireless sensor networks. *International Journal of Engineering and Advanced Technology (IJEAT), 8*(6), 5328–5335.
16. Kim, S., Cho, C., Park, K.-J., & Lim, H. (2017). Increasing network life time using data compression in wireless sensor networks with energy harvesting. *International Journal of Distributed Sensor Network, 13*(1), 1–10.

Live Temperature Monitoring: IoT-Based Automatic Sanitizer Dispenser and Temperature Detection Machine

Rudresh V. Kurhe, Anirban Sur, and Sharnil Pandiya

Abstract This study presents a live monitoring non-contact temperature detection and sanitizer dispensing system. The method is designed for preventing infection of COVID19 viruses. It maintains and improves community health and reduces the infection's adverse economic and social effects. The temperature detection (TD) and sanitizer dispenser (SD) subsystems of the LTM are controlled by a single microcontroller. In this study, the TD was created to operate similarly to existing and commercially available handheld infrared thermometers in terms of accuracy, show the temperature read to the user, and provide visual and audible alarms when the sensed temperature exceeds the average body temperature. Furthermore, the SD is designed to efficiently distribute sanitizer by dispensing only once and at the required amount. The experimentation study suggests that the final test findings are satisfactory, demonstrating that the LTM contributes to temperature monitoring and hand disinfection. In the end, we have discussed the limitations and future directions.

Keywords IoT · Sensors · Automated detection system · Sanitizer · Temperature · COVID-19 protection

1 Introduction

COVID-19 infections are currently posing a global danger. Many individuals have been affected due to the virus's rapid spread, which has infected 84% of countries worldwide. Spreading of severe acute respiratory syndrome coronavirus 2 (SARS-Cov-2) virus throughout the world (almost in 84% of counties in the world) developed a global pandemic, affecting the global economy very badly. Colossal health care infrastructure developed across the globe for fighting the coronavirus as the virus changes its symptoms and constantly mutations occurs.

R. V. Kurhe · A. Sur (✉) · S. Pandiya
Symbiosis Institute of Technology, Symbiosis International (Deemed University), Lavale, Pune, Maharashtra, India
e-mail: anirban.sur@sitpune.edu.in

P. K. Singh et al. (eds.), *Proceedings of Third International Conference on Computing, Communications, and Cyber-Security*, Lecture Notes in Networks and Systems 421,
https://doi.org/10.1007/978-981-19-1142-2_4

The world has been in a state of high tension since December 2019. Unfortunately, the numbers were growing by day by day. COVID-19 came to know after the report flashed from Wuhan, China. The vaccine was developed in the early stage, but due to the constant mutation, researchers, microbiologists, and pharmaceutical companies cannot conclude the vaccine's effectiveness. The World Health Organization (WHO) released a notice on how to protect yourself against the virus and assist people in preventing its spread. Frequently washing hands using soap or alcohol-based solutions show some excellent result against fighting the spreading virus. It has a symptom like influenza or pneumonia, so checking human body temperature frequently in public places for finding COVID infected people was a good idea. Although an infected individual may not have a high fever, it is one technique to identify a person with a COVID-19 infection.

Policies implemented around the world have helped mitigate its impact, but they have not eliminated it. Due to the lockdown, most countries weakened their economic growth, and the testing of many medications was not adequate. The debate now is between life and livelihood. The weaker sections of society are suffering as a result of the tightening of security across the globe. In India, during the fast wave of COVID-19, scenario was very worst; people were rushing for a small piece of grain. The pain is visible in the famished faces. Due to lockdown, industries were in decline mode; employees were losing jobs. Nations economic growth was in reversed gear. So lockdown was not a proper remedy for stopping the virus from spreading. Then, researchers realized that frequent body temperature monitoring and hand sanitization could be a good solution for spreading the virus. Current global situation keep in mind, proper sanitization kiosk in every public place with the temperature monitoring facility should be a good option, be it a manufacturing unit, information technology corporate office, educational institute or health monitoring unit, or a retail mall.

To resolve the discussed issues, we have presented a sensor fusion-based automatic sanitizer dispenser and temperature detection machine in the undertaken study. The presented IoT-based sensor fusion approach is designed to detect an individual with or without mask conditions and provide appropriate notification to the security personnel by raising the alarm. Moreover, the intelligent tunnel is also equipped with a thermal sensing unit embedded with a camera, which can detect the real-time body temperature of an individual concerning the prescribed body temperature limits as defined by WHO reports.

The presented article is organized as follows: Sect. 2 discusses the presented system's necessity and uniqueness from currently available systems. Section 3 discussed the methodology used for the proposed unit. Sections 4 and 5 discuss the system's detailed architecture design, software and hardware used, connections, sensing arrangements, deployments, and the step-wise workflow. In the end, expected outcome and future scope of the product have been discussed in Sect. 6.

2 System Design

This work has designed and implemented a live temperature monitoring automated hand sanitizer dispenser unit's prototype. Before designing the system, we have deeply studied the following works done by different researchers.

Joshi et al. [1] have proposed a portable kiosk model based on the CFD simulations. However, the proposed research is based on simulation, which is extremely difficult to execute in real-world settings. Murthy et al. [2] have proposed a small sanitizer system for passenger luggage disinfection. The technology, however, was not intended to disinfect humans. It also made recording timestamps and calculating the number of persons difficult. Pandya et al. [3] have proposed a smart sanitizer tunnel. However, the proposed system cannot detect the thermal temperature of individuals and notify them. Wan et al. [4] proposed a hand sanitization unit for food and process industries. They have used the hands motion detection approach for monitoring hand-hygiene compliance. Sarkar et al. [5] have proposed an automatic sanitizer dispenser system used in public places to disinfect humans. Ghayvat and the team [6] have proposed a location-based system to detect the possibility of COVID-19 in individuals. However, the system cannot detect and notify thermal temperature of the body-related variations. Lippi et al. [7] reviewed several bio-safety strategies to prevent COVID-19 infections in clinical laboratories. However, no disinfection-based techniques for humans were discussed or presented in the suggested system. Pandya and fellow researchers have proposed an intelligent home antitheft system to protect family members from theft possibilities. However, the proposed system cannot detect humans without face masks and notify human body temperature variations [8]. Marques et al. [9] have proposed an IoT-based non-contact infrared temperature acquisition system based for laboratory work. Shah and their team [10] have proposed a smart cardiac system for detecting the possibility of cardiac arrest and risks. However, the proposed system is not capable of detecting variations in body temperatures and detecting heartbeats. Pandya and their team [11] have developed an audio-based system to detect intruders and detect the lifestyles of individuals. However, the proposed method cannot detect variations in body temperatures. Recent researchers have made genuine attempts to design and develop intelligent systems for pandemics; however, the issue of seeing real-time body temperature and getting an emergency notification is an open research problem for fellow researchers [12–16].

2.1 Novelties of Proposed Work:

The following are the novelties of work flow of proposed system:

- Within 5 s, the given LTM can prevent or disinfect an outsider accessing a specific facility or property from becoming infected with COVID-19.
- The offered LTM can assist you in keeping track of people by allowing you to watch them and keep entrance timings.

- Finally, the Web and mobile interfaces were created to deliver live daily, weekly, and monthly updates of individual counts.

3 Detail Design for the Setup

3.1 The Layered Design

This section described the layered design of the proposed IoT-based automatic sanitizer dispenser and temperature detection machine. Figure 1 represents the graphical representation of the layered design of the proposed system; whereas, Fig. 2 represents the circuit design of the proposed system.

- Sensor Layer: Several sensing modules, including an infrared sensor module, an infrared temperature detection sensor, an Arduino microcontroller, a Wi-Fi module, and a submersible motor, make up the detecting layer. The detecting layer is also in charge of recognizing persons in front of the machine and disinfecting them for 5 s with a sanitizer spray on its palm.

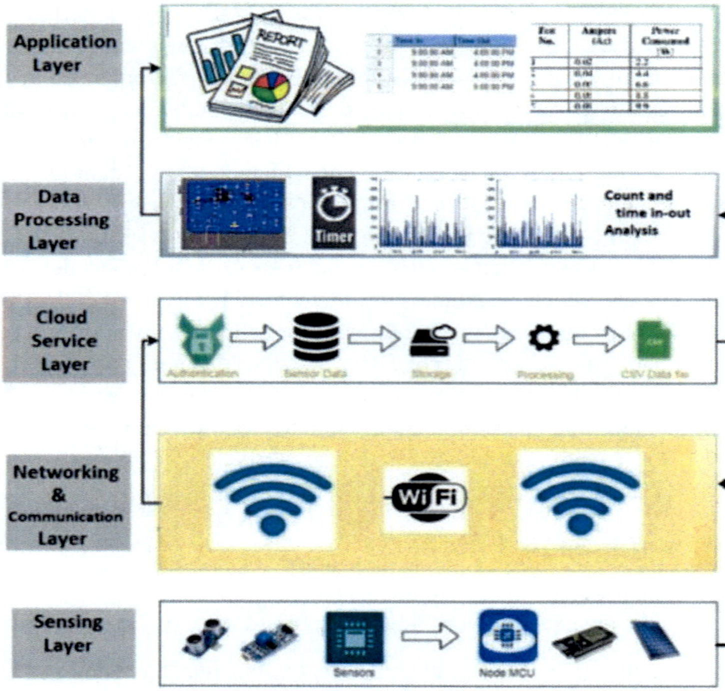

Fig. 1 Layered design for the system

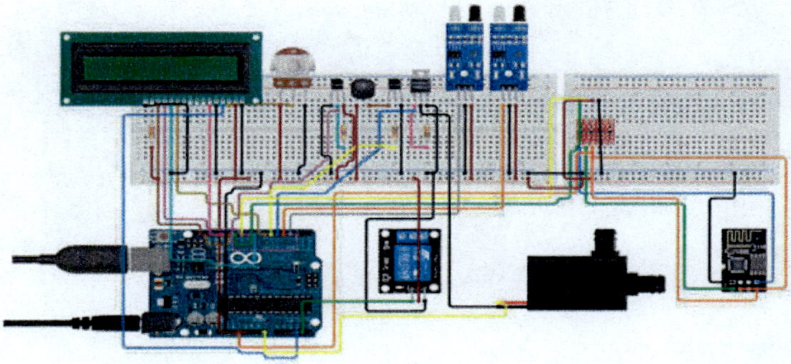

Fig. 2 Circuit design representation of the proposed system

- Networking Layer: The networking and communication layer connect a sensing layer with a networking and communication layer, a MQTT broker architecture, Google Firebase, and Web and mobile interfaces.
- Cloud Layer: The in and out timestamp values of all the people who stand in front of the machine and take timings are stored in the cloud service layer. It also keeps track of the number of people who have used machine. The described system's cloud platform is an open-source Google Firebase database. Real-time Data: The data have been processed and passed to the application layer for further examination.
- Processing Layer: The processing layer is in charge of receiving data from a Google Firebase cloud computing platform via a MQTT broker architecture, as well as processing the incoming in and out time stamp values, as well as the number of people in front of LTM. This layer also produces.
- Application Layer: The application layer comprises of a Web and mobile interface with a graphical user interface (GUI) that gives real-time statistics, daily, weekly, and monthly updates on the number of people who have accessed the LTM, as well as timings. The application also offers a number of graphical representations that may be submitted to the security control center for further study

3.2 List of Components Used in the Undertaken Study

Figure 3 represents the prototype implementation of the proposed system. A list of software's and hardware used in the undertaken study is described below:

- For system input and output data temperature sensors, LED, LCD, motor, pump, and infrared sensors are used.
- An Arduino Uno microcontroller is used for controlling the input and output devices, which have Microchip ATmega328P microprocessor.

Fig. 3 Prototype implementation

- For remort, contact measurement temperature sensor (MLX90614) is used, which is operated by infrared.
- For measuring surrounding infrared radiation, an infrared (IR) sensor is used.
- A DC 9 V motor pump is used for sucking and spraying sanitizer
- For displaying the temperature of the user, liquid crystal display (LCD) is used.
- For Wi-Fi module, Node MCU ESP8266 is used. It provides a self-contained SOC with an integrated TCP/IP protocol stack that can give any microcontroller access to the user Wi-Fi network.

4 Materials and Method

The (LTM) sanitizer dispenser with temperature detection is a unique solution designed to fight against the current pandemic and help people make them properly sanitized and protected from COVID in common places. This system can be installed outside public places, such as an ICU, OT of the hospital, vegetable markets, shopping malls, bus stops, railway stations, airports, and the entrance of housing societies.

System Requirements: India ministry of health has already requested all industries (retail/health, manufacturing/transportation/construction, etc.) to monitor their workers' body temperature and proper and sanitization during entering and leaving their industry premises. If any symptoms regarding fever detect, he/she should be isolated and observed with the further clinical investigation. The proposed system can achieve these requirements. System item components are selected in such way so that the system's overall cost should remain less than the other such kind of system available in the market. Working of the proposed IoT-based automatic sanitizer dispenser and temperature detection machine.

An IoT-based temperature monitoring machine has been proposed for the real-time. Detection of persons in the study.

Step 1: Working

- When a person approaches the machine, the IR sensor sends data to Arduino, which activates the IR contactless thermometer.
- The IR thermometer will detect a person's temperature, and Arduino displays the data on a 16X2 LCD.
- When you place your palm beneath the second IR sensor, Arduino receives the information, activating the submersible motor.
- The sanitizer is dispensed on your palm by the submersible motor.
- Wi-Fi modules help to monitor the live data using Web-based mobile interference.

Step 2: Specification

- IR sensor module:
- Range: The module detects a person from distance of 2 ~ 15 cm
- Detection angle: 35°
- Power supply: 3–5 V DC power supply module can be used.
- IR contactless temperature detection sensor:
- Range: 2–7 cm (approx.)
- Temperature measuring range −40 to 380 °C)
- Ambient temperature measuring range: −40 to 100 °C
- Accuracy: 0.02 °C
- Power supply: 3.6–5 V DC power supply module can be used.
- Submersible motor:
- Flow Rate: 80–120 L/H
- Power supply: 3–5 V DC power supply module can be used.
- Relay is used to activate submersible motor and its only for 4 s.
- Wi-Fi module:
- Mobile application is used to display the live data given by Arduino through Wi-Fi module.

5 Results and Discussion

Temperature detection machine and LTM prototype tests all had comparable results. Also, tests were conducted at various times throughout the day to establish system reliability in various environments and compare the results to those obtained in produced portable temperature detection devices. It is observed that the system is performing admirably, and that all of the sensors are functioning appropriately and delivering the desired results. The sample size the data collected for the conducted experiment of 1000 per parameter. Table 1 lists the data readings collected and used in the conducted experiments.

As shown in Fig. 4, the LTM and handheld device readings are nearly identical. It demonstrates that the LTM is operational and producing the desired results. Readings were taken at various times during the day. It demonstrates that the gadget eliminates the greatest external effect in temperature variance while producing the desired result.

Table 1 A representation of data readings of the proposed LTM system

No.	Time	Temp (*X*)	Hand Temp (*Y*)	LTM (Y)
1	7.00	1	36.3	36.4
2	7.03	2	36	36.2
3	7.07	3	35.9	35.5
4	7.27	4	36.1	36
5	7.40	5	36.4	36.5
6	12.05	6	36.5	36.6
7	12.15	7	35.8	35.9
8	1.10	8	36	36.2

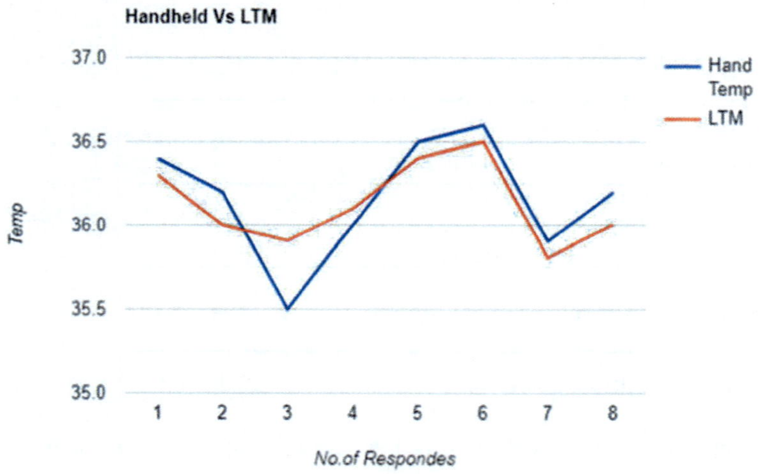

Fig. 4 Performance evaluation of the handheld versus LTM system

It demonstrates that the LTM is effective over time and that its values are not greatly influenced by external factors. Figure 4 represents the performance comparison of the handheld and the LTM system.

6 Limitations of the Proposed Work

The proposed approach is an experimental study to measure the body temperature of humans using thermal scanning approach. The system cannot claim to display the 100% accurate body temperature of humans due to changes in the weather conditions and complex nature of the human body.

7 Conclusions and Future Enhancements

As previously indicated, the device circuit is created in software and then simulated. As previously indicated, the device circuit is created in software and then simulated. Some power distribution to each module can be a hindrance while prototyping the hardware; to overcome this problem, relays must be installed to drive the spray pumps/submersible pumps, ensuring that the sensors, LCD, and other minute modules receive enough power from the inbuilt 5 and 3.3 V ports of the controller. Web mobile interface used to display live data. The major finding of the study is as follow:

- Within 5 s, the given LTM can prevent or disinfect an outsider accessing a specific facility or property from becoming infected with COVID-19.
- The offered LTM can assist you in keeping track of people by allowing you to watch them and keep entrance timings.
- Finally, the Web and mobile interfaces were created to deliver live daily, weekly, and monthly updates of individual counts.

In any organization, the system assists frontline workers in checking the temperature and administering alcohol to employees. The breakthrough of the sanitizer dispenser is that viruses will be easily removed because no one will contact the pump, and this device will distribute only a small amount of alcohol every motion activation, resulting in minimal waste. In future, oximeter integrated automatic hand sanitizer and temperature detection AI-based machine will can be designed to protect and fight against COVID-19.

References

1. Door-to-door temperature screening launched in Ethiopia. https://www.aa.com.tr/en/africa/door-to-door-temperature-screeninglaunched-in-ethiopia/1803391. Accessed August 05, 2020.
2. Non-contact Temperature Assessment Devices During the COVID-19 Pandemic|FDA. https://www.fda.gov/medical-devices/coronavirus-covid19-and-medical-devices/non-contact-temperature-assessment-devicesduring-covid-19-pandemic. Accessed August 05, 2020.
3. Pandya, S., Sur, A., & Kotecha, K. (2020). Smart epidemic tunnel: IoT-based sensor-fusion assistive technology for COVID-19 disinfection. *International Journal of Pervasive Computing and Communications.* https://doi.org/10.1108/IJPCC-07-2020-0091
4. Thermal Imaging Systems (Infrared Thermographic Systems/Thermal Imaging Cameras)|FDA. https://www.fda.gov/medical-devices/generalhospital-devices-and-supplies/thermal-imaging-systems-infrared
5. Public Health Response to the Initiation and Spread of Pandemic COVID-19 in the United States, February 24–April 21, 2020, Weekly/May 8, 2020/69(18);551–556 On May 1, 2020, this report was posted online as an MMWR Early Release. Anne Schuchat, MD; CDC COVID-19 Response Team. (https://www.cdc.gov/mmwr/volumes/69/wr/mm6918e2.htm)
6. Ghayvat, H., Awais, M., Gope, P., Pandya, S., & Majumdar, S. (2021). Recognizing suspect and predicting the spread of contagion based on mobile phone location data (counteract): a system of identifying covid-19 infectious and hazardous sites, detecting disease outbreaks

based on the internet of things, edge computing, and artificial intelligence. *Sustainable Cities and Society, 69*, 102798.
7. Ahmad, S. (2020). A review of COVID-19 (Coronavirus Disease-2019) diagnosis, treatments, and prevention. *Eurasian Journal of Medicine and Oncology, 4*(2), 116–125.
8. Pandya, S., Ghayvat, H., Kotecha, K., Awais, M., Akbarzadeh, S., Gope, P., Mukhopadhyay, S. C., & Chen, W. (2018). Smart home anti-theft system: A novel approach for near real-time monitoring and smart home security for wellness protocol. *Applied System Innovation., 1*(4), 42. https://doi.org/10.3390/asi1040042
9. Dabh, M. D. (2016). Geofencing: a generic approach to real-time location based tracking system. In: *IRACST—International Journal of Computer Networks and Wireless Communications, 6*(6).
10. Shah, A., Ahirrao, S., Pandya, S., Kotecha, K., & Rathod, S. (2021). Smart cardiac framework for an early detection of cardiac arrest condition and risk. *Frontiers in Public Health, 9*, 762303. https://doi.org/10.3389/fpubh.2021.762303
11. Pandya, S., & Ghayvat, H. (2021). Ambient acoustic event assistive framework for identification, detection, and recognition of unknown acoustic events of a residence. *Advanced Engineering Informatics, 47*, 101238.
12. El Majid, B., Motahhir, S., El Hammoumi, A., Lebbadi, A., & El Ghzizal, A. (2020). Preliminary design of a smart wristband disinfectant to help in the covid-19 fight. *Inventions, 5*(3).
13. Gupta, M., Abdelsalam, M., & Mittal, S. (2020). Enabling and enforcing social distancing measures using smart city and its infrastructures: A COVID-19 use case. arXiv:2004.09246 [cs.CY].
14. Joshi, J. R. (2020). COVSACK: An innovative portable isolated and safe COVID-19 sample collection kiosk with automatic disinfection. *Transactions on Indian National Academy and Engineering, 5*(2). https://doi.org/10.1007/s41403-020-00139-1
15. Kim, S. I., & Lee, J. Y. (2020). Walk-through screening center for COVID-19: an accessible and efficient screening system in a pandemic situation. *Journal of Korean Medical Science, 35*(15). https://doi.org/10.3346/jkms.2020.35.e154
16. Kwon, K.T., Ko, J.H., Shin, H., Sung, M., & Kim, J. Y. (2020). Drive-through screening center for COVID-19: A safe and efficient screening system against massive community outbreak. *Journal of Korean Medical Science, 35*(11).

A Comparative Study of Security Issues and Attacks on Underwater Sensor Network

Samiksha Kumari, Karan Kumar Singh, Parma Nand, Gouri Sankar Mishra, and Rani Astya

Abstract UWSNs are susceptible due to the unprotected acoustic path, extreme underwater atmosphere, and unique characteristics. UWSNs are subject to a broad range of security risks and malicious assaults due to their open auditory channel, hostile underwater atmosphere, and inherent characteristics. So, we outline several possible assaults at several stages of a typical UWSN communication protocol stack and discuss viable defenses. This article presents an overview of UWSN attacks, difficulties, and security and privacy issues. Also shown and addressed are contemporary security research and techniques.

Keywords UWSN · Security · Threat · Attacks · DOS attack · Intrusion detection

1 Introduction

Underwater sensor networks (UWSNs) have demonstrated their effectiveness in a variety of marine scenarios such as ocean tracking, mineral exploitation, spying, and combat activities in a hostile environment. To observe the undersea ecosystem, it consists of hundreds and hundreds of tiny sensors, each having sensing, computing, and connectivity capabilities. UWSN technologies are primarily concerned with detecting and sending real-time sense data from a specified tracking environment

S. Kumari · K. K. Singh (✉) · P. Nand · G. S. Mishra · R. Astya
Department of Computer Science and Engineering, Sharda University, Greater Noida, India
e-mail: 2018016287.karan@ug.sharda.ac.in

S. Kumari
e-mail: 2018016170.samiksha@ug.sharda.ac.in

P. Nand
e-mail: parma.nand@sharda.ac.in

G. S. Mishra
e-mail: gourisankar.mishra@sharda.ac.in

R. Astya
e-mail: rani.astya@sharda.ac.in

P. K. Singh et al. (eds.), *Proceedings of Third International Conference on Computing, Communications, and Cyber-Security*, Lecture Notes in Networks and Systems 421,
https://doi.org/10.1007/978-981-19-1142-2_5

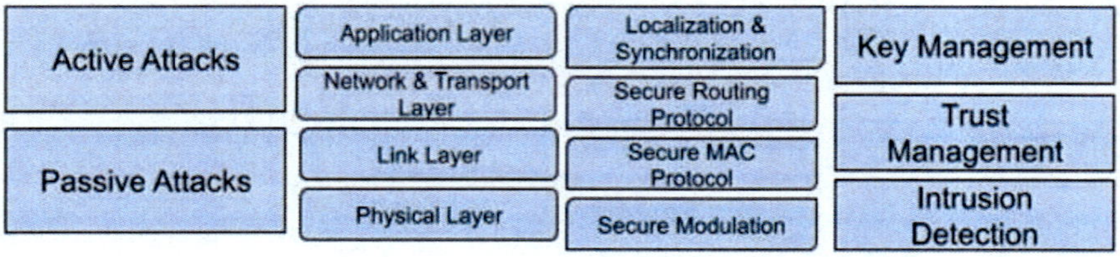

Fig. 1 Security architecture

to the base station for more processing and interpretation. However, because of their limitation of computing power, storage, and battery, the sensor nodes in UWSNs have significant resource restrictions.

Because these sensor networks are typically placed in remote locations and are left unchecked, they need to be provided with security features to protect them from threats such as black hole, Sybil attack, physical tampering, eavesdropping, and denial of service, and other threats are all possible. Yet, these studies are limited in their ability to combat security risks in UWSNs therefore to resource constraints and the fact that the security issue is more server based [1]. The security architecture of UWSNs is shown in Fig. 1, we are going to discuss each block in detail, and we will give a detailed review of numerous security problems in UWSNs in this paper.

2 Constraints and Distinctive Features

A UWSN is made up of a huge number of resource-limited sensor nodes. These sensor nodes have minimal computing power, a small amount of storage, and restricted connection bandwidth. These limitations are a result of the sensor nodes' restricted energy, actual size, and mobility. Because of these limitations, it is impossible to use traditional security methods in UWSNs. Some characteristics of UWSNs are comparable to those of wireless sensor networks (WSNs) [2, 3]. So, the difference between the constraints of both of them is given in Fig. 2. However, owing to the challenging work situation, there are certain unique characteristics and restrictions, which are listed below.

- *Hardware components are quite limited*: Hardware components, such as energy, computing capabilities, and memory size, are severely restricted in underwater sensor nodes. The energy required for underwater acoustic transmission is significantly greater than for terrestrial radio transmission due to greater ranges and highly complicated data computation at the receiver's end to adjust for signal degradation. Sensor nodes are used in shallower or deeper water when charging or replacing the node's battery is difficult. Computing capabilities and memory capacity are limited to extend the network's lifespan.
- *Storage constraints*: A sensor is a small gadget with limited brain and storage capacity. A sensor brain generally consists of internal storage and RAM. Installed

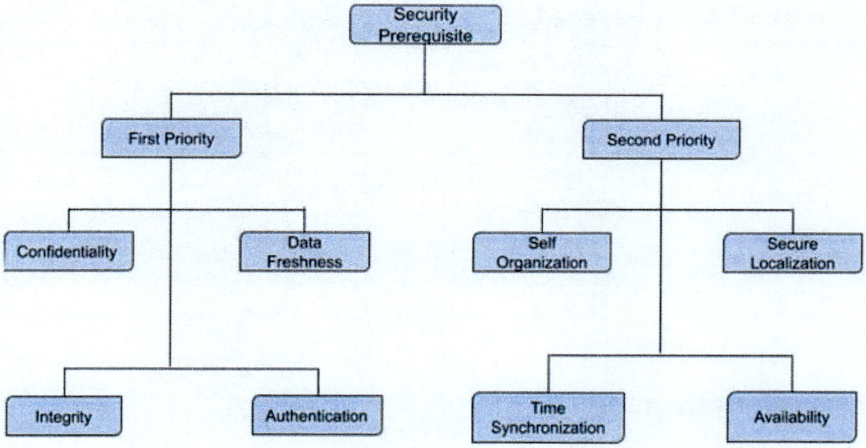

Fig. 2 Requirements of security based on priority

application software is stored in flash memory, whereas application programs, sensor data, and intermediary outputs of operations are stored in RAM. After installing the OS and program code, there is generally insufficient space to perform complex algorithms. Sensors, for example, TelosB, are equipped with a 16-bit, 8-MHz RISC processor. As a result, conventional security methods are ineffective in these sensors [4].

- *Transmission Channel That Is Not Efficient*: Underwater channels are spatially and geographically changeable, have a finite bandwidth, and are highly dependent on communication range and speed. Temperature of water, propagation loss, distortion, multipath impact, and Doppler dispersion all have an impact on the channel [24]. All of these variables contribute to significant bit mistake and latency variation, which leads to signal loss and a significant node rejection rate. Furthermore, because nodes inside the transmission range maintain the same undersea channel, an attacker can quietly collect and analyze data, or worse, aggressively damage network functionality. As a result, protecting UWSNs against snooping and other unauthorized assaults is a significant issue.
- *Architecture of a Changing Network*: While land sensor nodes are extensively distributed, underwater sensors are seldom distributed owing to the expense and difficulties of installation. Resulting from water movement, most underwater sensors are mobile. In a normal underwater environment, underwater items may travel at a velocity of 2–3 knots, according to empirical evidence [25]. As a consequence, the network architecture is very dynamic.
- *Risk with an Unsafe Environment*: The functional conditions of UWSNs may not be protected in particular specific domains of operation, such as underwater safety surveillance and target detection. The sensor nodes are used to keep an eye on unfriendly objects in the open seas or in hostile sea areas. As a result, these nodes may be extremely susceptible to cyber attacks and malicious assaults (Table 1).

Table 1 Difference between UWSNs and WSN

Characteristic	UWSNs	WSN
Energy absorption	High	Low
Latency in propagation	High	Low
Bandwidth	Low	High
Dynamic topological function	High	Low
Efficiency	Low	High
Rate of data transfer	Low	High
Storage capacity	Low	High
Mode of communication	Acoustic based	RF based

3 Security Prerequisite

A UWSN is a unique kind of network. It has some similarities to a normal computer network, but it also has a number of distinguishing features. The safety functions in a UWSN should safeguard the data and capabilities exchanged over the network against assaults and node misbehavior. The security constraints of UWSNs, being a subset of WSNs, are comparable to those of terrestrial WSNs [2]. So, we divided the security prerequisite into first priority and second priority as shown in Fig. 2.

3.1 *First Priority*

It is divided based on first we need to secure these four categories. The categories are as under:

- *Confidentiality*: No information in the connection should be interpreted by anybody other than the intended receiver, according to the security system. It is about stopping unauthorized nodes from deciphering the confidential data's contents. Confidentiality refers not just to the recipient's personal information, yet also to the Ip, routing information, and other data. Unauthorized attackers should not be able to view or meddle with this sensitive information. Confidentiality could be obtained by the use of a low-power cryptography method.
- *Data Freshness*: It denotes current information and guarantees that no attacker may repeat old information. Whenever the UWSN sensor nodes employ shared-keys for data transmission, this criterion is very critical; while the new key has been updated and transmitted to all sensor nodes, a prospective opponent can start a repeat attack by utilizing the old key.
- *Integrity*: Data integrity guarantees that incoming information is not updated, deleted, or distorted by unauthorized nodes throughout the transmission process,

whether due to node failure or malicious assault. This is especially important in situations like tactical operations and technical controls, when little modifications might result in catastrophic consequences.

- *Authentication*: In acoustic channels, without the use of encryption, a hostile attacker may simply collect messages and alter their content. As a result, in order to prevent malicious attacks, the collecting node must verify the origin of the data. To connect and manage channels, resources, applications, and information on such a network, nodes must be authorized. These techniques guarantee that only authorized sensor nodes have access to the network's resources.

3.2 Second Priority

This is also divided into four parts but the secondary security requirements after first priority. The categories are as under:

- *Self-organization*: Self-organization guarantees that networks can recover from threats independently and without assistance in real time. However, if the intruder stays in the network, if a sensor node is self-stabilizing in the face of malware activity, it can restore its natural condition on its own. So, every sensor node in a UWSN should be able to self-organize and heal.
- *Secure Localization*: In many cases, it is important to find individual sensor nodes inside a UWSN precisely and autonomously. Because of the mobility in water, nodes change their position from one place to another place, so it is difficult to locate.
- *Time Synchronization*: The majority of sensor network operations necessitate time synchronization. Each and every UWSN security method must be time-synchronized as well. A set of sensor nodes in a coordinated UWSN may need to be synchronized. A collection of secure synchronization techniques was presented in [26].
- *Availability*: Availability guarantees that the connection channel is sufficiently stable. Although if certain nodes collapse or the channel is assaulted, functionalities will always be available. UWSNs can have availability if they use the right robustness and self-adaptive strategies.

4 Threats on UWSN

As previously stated, UWSNs have several limitations. As a result, UWSNs are susceptible to a wide range of threats and harmful actions. Threats could be passive or active depending mostly on the malicious attacker's actions. In Fig. 3, we categorize the threats in parts.

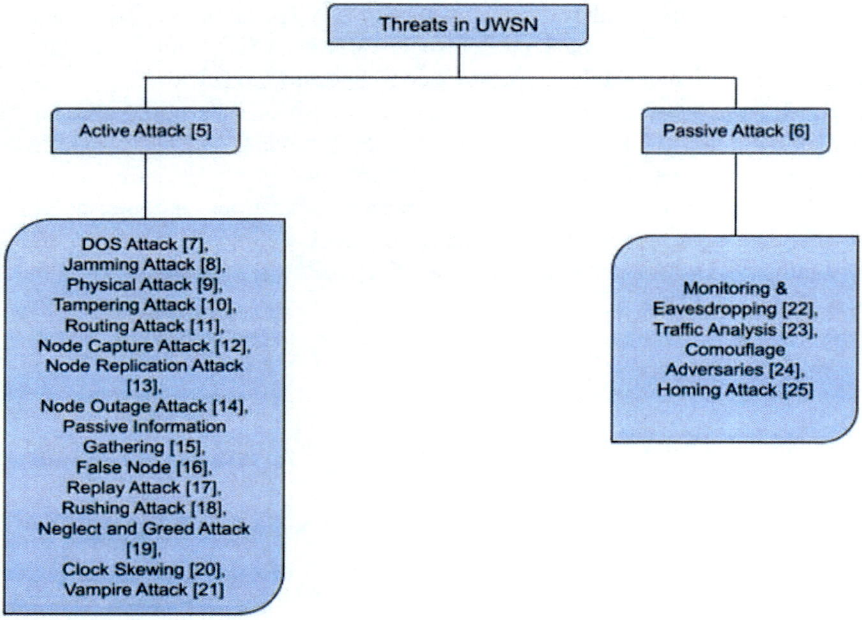

Fig. 3 Different types of threat

4.1 Active Attack

In active attack [5], attackers try to change, insert, remove, or corrupt data sent across the network. Active assaults may monitor data and seek to influence or discard packets that can be performed by malicious attackers on the inside or outside. External assaults, which are simpler to identify and counter, are launched out by nodes that are not part of the network. Internal assaults are carried out by inner nodes and have the potential to inflict significant harm. Internal assaults are thus more complex to identify and may result in more serious consequences. The easiest way to avoid this issue is to use security methods like cryptography, authentication, and trust management. Some of the active attacks are given below:

- *DOS Attack*: In DOS attack [7], by supplying a rogue node to mislead them, the attackers limit network operations. The major concern is an attack mostly on accessibility of data transmission facilities and services. The attackers may damage or divert the node network's channel settings. By altering the protocols of the transmitting media, wasting resources, and damaging physical parts, a DoS attack can arise in any tier in the OSI model.
- *Jamming Attack*: In jamming attack [8], an attacker can start the assault both outside and inwardly. The attacker blocks the host from sending data packets or denies valid packets from being sent. There are a variety of jammers and jamming methods which can disrupt network functionality.

- *Physical Attack*: In physical attack [6], wireless networks are susceptible to physical assaults because of their dispersed and unsupervised nature. In the physical world, sensor nodes are damaged indefinitely, resulting in a lifelong destruction of nodes. One technique of avoiding physical assaults is tamper protection.
- *Tempering Attack*: In tempering attack [9], the intruder changes or destroys the sensor node's capabilities before gaining total control of the compromised node. The physical equipment is harmed in this attack, resulting in a shortage of capabilities. The attack is thwarted by constantly altering the encrypted key and using suitable key management methods.
- *Routing Attack*: In routing attack [10], routing and data transmission are critical functions in nodes, and also the protocols must be energy-efficient and resistant to attacks. The approved receiver must get the exact message, as well as the message's authenticity and sender's identification, that the transmitter offered in the network. Some examples of routing attack are Sybil attack, blackhole attack, selective forwarding, sinkhole attack, wormhole attack, HELLO flood attack.
- *Node Capture Attack*: In this attack [11], the invader carries out a variety of tasks and threatens the entire network. The node acquisition assault takes portions of the sensor nodes and repurposes them for numerous strikes. The data in the network connection is modified by a rogue node that has been redeployed.
- *Node Replication Attack*: We can also call it clone attack [12], and the UWSN is vulnerable to an unprotected system in which rogue nodes can be copied into many copies. The duplicated nodes will have valid IDs and keys, allowing them to connect with neighboring nodes in the operating network like regular nodes. The solution is to create a one-of-a-kind pair-wise key that enables safe transmission among neighbors.
- *Node Outage Attack*: In this attack [13], the sensor components, such as sensor nodes, transmission links, and parent nodes, are entirely disabled. As a result, connectivity with nearby clustered nodes in various regions is disrupted.
- *Passive Information Gathering*: In this attack [14], the attacker intercepts information with sophisticated algorithms, allowing them to locate and disable nodes unencrypted data, including the actual position of the sensor nodes, was given. In addition, the attacker has access to the information of application-specific messages. A very well antenna with encryption keys and strong analytics is utilized to fight this assault.
- *False Node*: In this attack [14], the invader adds a malicious node with incorrect information or obstructs the right information channel. As a result, the rogue node transmits erroneous data to every node in the operating network, and it has the ability to either acquire the network services and place it under the control of the invader or completely destroy the network.
- *Replay Attack*: In this attack [15], for energy usage, the intruder repeats the corrupt node several times and also controls the communication routes. Because the sensor nodes are mobile in this assault, the network architecture is constantly changing. Deploying session keys and adding timestamps with data is an efficient approach to counter this assault.

- *Rushing Attack*: In this attack [16], the rogue node attempts to transfer messages out from the neighboring node to another final node in a separate tunnel. A rushing assault occurs when packets are tunneled over a fast communication path among the wormhole attack's endpoints. The safeguard to this threat is to incorporate a node list in the path records.
- *Neglect and Greed Attack*: In this attack [17], by routing data to the incorrect node, the rogue node chooses the shortest way to convey them. The major objective is to reduce data loss and increase network node performance. The data is received by the attacker, but he denies sending it to the adjacent nodes.
- *Clock Skewing*: In this attack [18], by changing the timing of the transmitting messages, the attacker imitates the specific skew. All peripherals in the sensor network have a different clock skew depending on the function.
- *Vampire Attack*: In this attack [19], it is a type of DoS attack that drains the nodes' energy and totally destroys the network. The AODV routing system uses Bcast id as an extra column inside the routing table through each node and in information packets to identify oriented antenna assaults.

4.2 Passive Attack

In passive attack [6], rogue nodes seek to discern the kinds of actions and collect data transferred in the network without interfering with the functioning of the network. Furthermore, the attacker may record data and then evaluate the traffic to anticipate the patterns of transmission, monitor the data transfer, determine the interacting hosts, and pinpoint the origin. These passive assaults are hard to identify since they have no effect on the network's functionality. Here, some of the passive attacks are as under:

- *Monitoring and Eavesdropping*: In this attack [20], eavesdropping threats have no effect on the channel's integrity. In the functional network, the rogue node detects the information. The intruders spy on the information to figure out the transmission pathway and compromise the network's security. This is the more prevalent data security breach.
- *Traffic Analysis*: In this attack [21], the intruder examines the transmission patterns that were followed. To injure and assist destruction to the sensor network by any form of active assault, the intruder reveals the sequence to the enemy. To avoid this threat, the network is continually checked.
- *Camouflage adversaries*: In this attack [22], the intruder impersonates an ordinary node in the sensor network to camouflage the required amount of nodes. As a result, transmissions are misrouted to various connectivity channels.
- *Homing Attack*: In this attack [23], the intruder does not really change or change the messages; instead, he or she seeks for the network's insight sources, which are then utilized to conduct any active assaults. To avoid this sort of attack, cluster chiefs, often called as cryptographic key administrators, utilize network behavior

monitoring and header encoding techniques to detect and strike nodes. In Table 2, we summarize the attack based on their description.

5 Layered-Wise Attack

As previously stated, sensors are sensitive to security attacks due to the unique properties of the fundamental networking protocols. Physical, link, network, transport, and application layers are all vulnerable to attacks. The majority of all these routing techniques lack security measures, making it much simpler as an attacker to breach security. The layer-wise diagram is shown in Fig. 4, and the defense against these attacks is given in Table 3.

5.1 *Physical Layer Attacks*

- *Tampering*: This is also known as node capture, and it involves compromising a node. It is simple to do and may be rather dangerous. Physically altering and damaging nodes is referred to as tampering.
- *Jamming*: It is produced by interference with the channel's components' radio transmissions that is an intrusion on the sensor channel's functionality. It differs from standard radio transmission because it is undesired and invasive, leading to denial-of-service situations.

5.2 *Link Layer Attacks*

- *Exhaustion*: An interrogation assault might cause a channel's storage power to be depleted. A hacked node might send frequently, utilizing higher battery capacity than necessary.
- *Collision*: It occurs in the link layer, which is responsible for nearest communication as well as connection arbitration. If an attacker is able to induce conflicts of even a small portion of a communication, the whole packet can be interrupted, and a single bit mistake can create a CRC mismatch, which may need repetition.

5.3 *Network Layer Attacks*

- *Hello Flood Attack*: When an intruder with a sufficient transmission capacity can transmit or replay hello messages that are utilized for neighbor finding, this is what happens. As a result, the intruder gives the appearance of being a neighbor to

Table 2 Summarization of attacks

Type of threat	Active or passive	Summary	How to defend
Jamming attack	Active	Restricts the source from sending messages or denies valid messages from being sent	Jamming methods, region mapping
DoS attack	Active	The facilities and services that are available. Disable or divert the infrastructure's architecture	Message prioritization, monitoring, and encryption techniques
Physical attack	Active	The physical assault destroys the sensor nodes indefinitely	Proofing against tamper
Tampering attack	Active	The node's capabilities are harmed, and the seized node is taken over completely	Updating the key on a regular basis, as well as suitable key management systems
Neglect and greed	Active	By packet forwarding to the incorrect node, it chooses the shortest way to convey them	Authentication methods, redundancy
Homing attack	Active	The goal is to discover the network's understanding capabilities, which are then utilized to conduct any active assaults	Cryptography
Node capture attack	Active	Several sensor nodes are removed and redeployed to execute numerous assaults	LEAP protocol
Node outage attack	Active	Sensor nodes, transmission links, and parental nodes are all entirely disabled	Powerful computations, time protocols
Monitoring and eavesdropping	Passive	In the functional network, it detects the data	Directional antenna
Traffic analysis	Passive	To hurt and assist destruction, the attacker reveals the sequence to the defender	Surveillance of the network
Camouflage adversaries	Passive	The messages are being misrouted to other communication channels as a result of this	Privacy analysis

(continued)

Table 2 (continued)

Type of threat	Active or passive	Summary	How to defend
Rushing attack	Active	The information from the neighboring node is rushed to the other target node through a separate tunnel	Adding a node list to a page
Clock skewing	Active	By changing the timing of the relaying messages, the attacker imitates the targeting skew	FTSP and the interval of variable time synchronization
Vampire attack	Active	This is a type of DoS attack that drains the sensor battery and totally destroys the network	Methods for validation

Fig. 4 For layer-wise attack

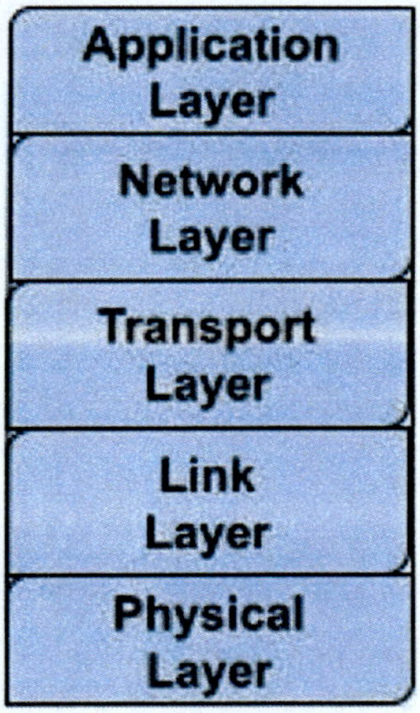

adjacent nodes, and the fundamental routing protocol could be disturbed, allowing for more sorts of assaults.

- *Wormhole Attack*: It occurs as a result of the establishment of a reduced link, which allows messages to go from one side to another quicker than they would via a multi-hop path. Wormhole attacks are a danger to routing protocols that are difficult to identify and avoid. An attacker can persuade remote nodes which are only one or two steps away via the wormhole, leading network routing algorithms to get confused.
- *Sybil Attack*: It occurs when a rogue equipment is used by an intruder to generate a huge number of participants in order to acquire control over channel traffic. Fake networking connections or replication of existing genuine identities may

Table 3 Layer-wise attacks and their defense

Layer	Types of attacks	Defense against such attacks
Physical layer	Tempering, jamming	Detect and sleep, route around the congested area, tamper-evident packaging
Link layer	Exhaustion, collision	Anti-replay protection and authentication
Network layer	Hello flood attack, wormhole attack, Sybil attack, sinkhole attack	Creation of a safe cluster, anti-replay protection and authentication, authenticate pair-wise, header encryption
Transport layer	Flooding attack	SYN cookies
Application layer	DOS attack, cloning attack	Anti-replay protection and authentication, tuning sensor, authentication streams

have resulted in the identification of these rogue nodes. The Sybil attack is most commonly used against responsibility to fix systems such as shared storage, topology management, and multi-hop networking.

- *Sinkhole Attack*: It occurs whenever an intruder disables a network's central server from receiving correct and comprehensive sensing data, posing a severe danger to higher-layer operations. An offender can use a sinkhole attack to capture virtually all of the activity from a given location. Sinkhole attacks operate by creating a rogue node that appears unusually appealing to other nearby nodes in terms of routing protocols and the underlying optimization algorithm.

5.4 *Transport Layer Attacks*

- *Flooding Attack*: This is a (DoS) attack that floods a connection or service with enormous quantities of traffic in order to knock it down. Whenever a channel or service gets overburdened with messages, flood assaults occur; as a result, it starts sending out partial connectivity attempts that it can no further handle. By flooding a system with links that cannot be completed, you might cause it to crash, the server's storage capacity is ultimately filled by a flood attack, because once the capacity is filled, the server crashes.

5.5 *Application Layer Attacks*

- *DOS Attack*: This assault is sometimes referred to as an adversary's intentional attack with the goal of damaging or breaking the sensor connectivity. The sensor connectivity functioning may be limited or eliminated as a result of a DoS attack.

A DoS attack on a UWSN can happen anywhere at layer of the OSI levels. DoS degrades the performance of targeting connections by interfering with their protocols and draining resources.

- *Cloning Attack*: It occurs when opponents may simply acquire and hack nodes, and then install an endless quantity of copies in the captured nodes in the sensor network. Because these copies have legal entry toward the sensor connection, they can do anything they want with it. They may readily take part in sensor connectivity activities as if they were a genuine node, leading in a wide range of insider assaults, taking over its entire network. When these copies in the sensor connection go unnoticed, the sensor connectivity becomes very susceptible to attackers.

6 Privacy Concern in UWSN

As discussed earlier, threats and assaults against UWSNs are many. To meet the privacy criteria, a set of procedures and security solutions to protect UWSNs against assaults must be suggested. Key management, intrusion detection, trust management, secure localization, secure synchronization, and routing security are the primary privacy concerns which are shown in Fig. 1.

- *Key Management*: Secrecy, authenticity, validity, and message integrity are the basic aims of encryption and key management. Unapproved users cannot see or modify sensitive data or send in insecure channels like the underwater channel thanks to cryptography. However, current encryption and key management systems have certain flaws, such as ciphertext enlargement and high computation. After applying encryption, data padding and codes lengthen the text and raise energy usage during transfer and processing [18]. Message verification is generally done with a digital signature. A verified message is added with a signature, which results in message enlargement and connection latency [27].
- *Intrusion Detection*: Inner and outside intruders are detected, identified, and isolated from the network using intrusion detection techniques. Intrusion detection methods, on the other hand, generally operate after malicious assaults have taken place and been identified. Suspicious intruders are tough to identify in the early stages of an assault. As a result, authentic detection techniques must be investigated and enhanced. Intrusion tolerance techniques, on the other hand, may be used to secure infrastructure while permitting harmful attackers to exist. Furthermore, techniques and intrusion detection systems (IDSs) have already been suggested to increase the privacy of UWSNs.
- *Trust Management*: The trust management method, as a crucial supplement to cryptographic security defense, provides substantial benefits in intrusion detection. The study on trust management methods in UWSNs confronts greater problems due to the unique characteristics and restrictions of UWSNs [28]. There are three types of trust management systems now in use: centralized schemes,

distributed schemes, and hierarchical schemes. A root node or a ground station provides trust management for every node in the channel in a centralized system. For UWSNs, centralized systems are ineffective. Because transferring trust values among nodes and the base station consumes a lot of power, it is a costly burden. Every node in a distributed system is responsible for computing and maintaining the channel's trust level. The assessment and propagation of trust level are executed hierarchically in hierarchical systems. From the bottom level to the top layer, the trust levels are transferred and combined.

- *Localization Security*: In origin identification and monitoring applications, location estimate is critical. Even during the localization phase, underwater sensors obtain information on the position and movement of moving nodes, which is utilized to pick the optimal intermediary node to transfer data. The base station is unable to determine the source of the collected signal without the position data. Because of the peculiarities of underwater channels, suggested WSN localization methods cannot be used in underwater operations [29].
- *Synchronization Security*: Several underwater operations and MAC protocol sequencing need synchronization. Furthermore, achieving accurate time synchronization in underwater situations is very challenging. Despite the importance of security in UWSN problems, neither of the current time synchronization techniques [30, 31] took it into account. However, the recommended solutions for latency attacks for WSNs [27, 32, 33] do not apply to UWSNs.
- *Routing Security*: Routing security is made up of fundamental transit and connections security methods that are implemented to routing protocols and sensors individually. Furthermore, in addition to using any of the routing protocols, nodes should communicate data from their neighbors in order to build the network architecture. There are two elements to routing security: safe routing and safe data transmission. Safe routing requires nodes to work together in order to communicate accurate routing data and preserve the network linked. Information transmissions should be secured against manipulation, dropping, and alteration by any unauthorized person in safe data forwarding.

7 Conclusion

The constraints, threats, and security concerns in UWSNs are covered in this paper. The unique characteristics and constraints of UWSNs and the underwater atmosphere are investigated. UWSNs are subject to a variety of security risks including malicious assaults that disrupt the network's connectivity and collaboration. The security measures of UWSNs are introduced to prevent these attacks. Finally, we will go through several particular security technologies and privacy concerns. Furthermore, various operations may have varied security needs, and redundant security methods would be a significant energy drain. As a result, future study will focus on how to incorporate these characteristics into security system design.

References

1. Heidemann, J., Ye, W., Wills, J., et al. (2006). Research challenges and applications for underwater sensor networkin. In *Wireless Communi-cationsand Networking Conference, 2006 (WCNC 2006)* (pp. 228–235). IEEE.
2. Lopez, J., Roman, R., & Alcaraz, C. (2009). Analysis of security threats, requirements, technologies and standards in wireless sensor networks. In *Foundations of Security Analysis and Design V* (pp. 289–338). Springer.
3. Perrig, A., Stankovic, J., & Wagner, D. (2004). Security in wireless sensor networks. *Communications of the ACM, 47*(6), 53–57.
4. Perrig, A., Szewczyk, R., Wen, V., Culler, D. E., & Tygar, J. D. (2002). SPINS: Security protocols for sensor networks. *Wireless Networks, 8*(5), 521–534.
5. Verma, R., & Bharti, S. (2020). A survey of network attacks in wireless sensor networks information. *Communication and Computing Technology*, 50–63.
6. Butun, I., & Osterberg, P., & Song, H. (2019). Security of the internet of things: vulnerabilities, attacks and countermeasures. *IEEE Communications Surveys & Tutorials*, 1–1.
7. Shi, L., Liu, Q., Shao, J., & Cheng, J. (2021). Distributed localization in wireless sensor networks under denial-of-service attacks. *IEEE Control Systems Letters, 5*(2), 493–498.
8. Verma, R., Darak, S. J., Tikkiwal, V., Joshi, H., & Kumar, R. (2019) Countermeasures against jamming attack in sensor networks with timing and power constraints. In *11th International Conference on Communication Systems & Networks (COMSNETS).*
9. Dewal, P., & Narula, G. S., Jain, V., Baliyan, A. (2018). Security attacks in wireless sensor networks: A survey.
10. Raoof, A., Matrawy, A., & Lung, C. (2019). Routing Attacks and Mitigation Methods for RPL-Based Internet of Things. *IEEE Communications Surveys & Tutorials, 21*, 1582–1606.
11. Butun, I., Osterberg, P., & Song, H. (2009). Security of the internet of things: vulnerabilities, attacks and countermeasures. *IEEE Communications Surveys & Tutorials*, 1–1
12. Xie, H., Yan, Z., Yao, Z., & Atiquzzaman, M. (2019). Data collection for security measurement in wireless sensor networks: a survey. *IEEE Internet of Things Journal, 6*(2), 2205–2224.
13. Butun, I., Osterberg, P., & Song, H. (2019). Security of the internet of things: vulnerabilities, attacks and countermeasures. *IEEE Communications Surveys & Tutorials* 1–1.
14. Riaz, M. N., Buriro, A., & Mahboob, A, (2018) Classification of attacks on wireless sensor networks: a survey. *International Journal of Wireless and Microwave Technologies (IJWMT), 8*(6), 15–39.
15. Jadhav, R., & Vatsala, V. (2017). Security issues and solutions in wireless sensor networks. *International Journal of Computer Applications*, 14–19.
16. Sinha, P., Jha, V. K., Rai, A. K., & Bhushan, B. (2017). Security vulnerabilities, attacks and countermeasures in wireless sensor networks at various layers of OSI reference model: A survey. In *International Conference on Signal Processing and Communication (ICSPC).*
17. Verma, R., & Bharti, S. (2020) A survey of network attacks in wireless sensor networks information. *Communication and Computing Technology*, 50–63.
18. Rajendran, G. B., Kumarasamy, U. M., Zarro, C., Divakarachari, P. B., & Ullo, S. L. (2020). Land-use and land-cover classification using a human group-based particle swarm optimization algorithm with an LSTM Classifier on hybrid pre-processing remote-sensing images. *Remote Sensing, 12*(24), 4135.
19. Sharma, M.K., & Joshi, B. K. (2017). Detection & prevention of vampire attack in wireless sensor networks. In *International Conference on Information, Communication, Instrumentation and Control (ICICIC)* (pp. 1–5)
20. Vu, D. L., Nguyen, T. K., Nguyen, T. V., Nguyen, T.N., Massacci, F., & Phung, P. H. (2019). A convolutional transformation network for malware classification. In *2019 6th NAFOSTED conference on information and computer science (NICS)* (pp. 234–239). IEEE
21. Do, D. T., Le, T. A., Nguyen, T.N. Li, X., & Rabie, K. M. (2020). Joint impacts of imperfect CSI and imperfect SIC in cognitive radio-assisted NOMA-V2X communications. *IEEE Access, 8* (pp. 128629–128645).

22. Zhen, L., Bashir, A. K., Yu, K., Al-Otaibi, Y. D., Foh, C. H., & Xiao, P. (2020) Energy-efficient random access for leo satellite-assisted 6G internet of remote things. *IEEE Internet of Things Journal*. https://doi.org/10.1109/JIOT.2020.3030856
23. Verma, R., & Bharti, S.: *A survey of network attacks in wireless sensor networks information, communication and computing technology* (pp. 50–63).
24. Akyildiz, I. F., Pompili, D., & Melodia, T. (2005). Underwater acoustic sensor networks: Research challenges. *Ad hoc networks, 3*(3), 257–279.
25. Cui, J. H., Kong, J., Gerla, M., et al. (2005). Challenges: building scalable and distributed underwater wireless sensor networks (UWSNs) for aquatic applications. *Channels, 45*(4), 22–35.
26. Ganeriwal, S., Capkun, S., Han, C.- C., Srivastava, M. B. (2005). Secure time synchronization service for sensor networks. In *Proceedings of the 4th ACM Workshop on Wireless Security* (pp. 97–106). ACM Press.
27. Boukerche, A., & Turgut, D. (2007). Secure time synchronization protocols for wireless sensor networks. *IEEE Wireless Communications, 14*(5).
28. Goyal, N., Dave, M., & Verma, A.K. (2017). Trust model for cluster head validation in underwater wireless sensor networks. *Underwater Technology, 34*(3).
29. Chen, K., Zhou, Y., & He, J. (2009). A localization scheme for underwater wireless sensor networks. *International Journal of Advanced Science and Technology, 4*.
30. Liu, J., Wang, Z., Zuba, M., et al. (2014). DA-Sync: A Doppler-assisted time-synchronization scheme for mobile underwater sensor networks. *IEEE Transactions on Mobile Computing, 13*(3), 582–595.
31. Liu, J., Wang, Z., Peng, Z., et al. (2011). TSMU: a time synchronization scheme for mobile underwater sensor networks. In *Global Telecommunications Conference (GLOBECOM 2011)* (pp. 1–6). IEEE.
32. Song, H., Zhu, S., & Cao, G. (2007). Attack-resilient time synchronization for wireless sensor networks. *Ad Hoc Networks, 5*(1), 112–125.
33. Du, X., Guizani, M., Xiao, Y., et al. (2008). Secure and efficient time synchronization in heterogeneous sensor networks. *IEEE Trans. Vehicular Technol, 57*(4), 2387–2394.

Discrete Event Driven Routing in SHIP Network using CupCarbon Simulation Tool

Himanshu Duseja, Ashok Kumar, Rahul Johari, and Deo Prakash Vidyarthi

Abstract In the world of data communication and data sharing, reliable routing and subsequent delivery of the message with minimum latency is a challenge. Various academicians and researchers around the world have devised various techniques, algorithms, and methodologies to ensure safe, secure and timely delivery of the message between source and destination nodes in the network. In the proposed work, effort has been made to demonstrate the broadcasting of messages in smart, hybrid, intermittent, and partitioned (SHIP) network to the sink node with the help of multiple intermediate neighboring node(s) with minimum delay using CupCarbon simulation tool.

Keywords Sensor · Latency · Routing · CupCarbon

1 Introduction

As well known, the last decade has seen the emergence of incredible and amazing technologies such as Internet of things (IoT), blockchain technology, and artificial intelligence including machine learning and deep learning, Mist, Internet of things (IoT), Cloud, Edge Fog computing (MICEF), quantum computing, big data tools, and technologies et al. which are helping the programmers and researchers to design, develop, and deploy world class innovative products for the benefit of mankind.

Progress and development in the field of wireless communication and electronics have given rise to class of heterogeneous network popularly called as Smart, Hybrid, Intermittent, and Partitioned network (SHIP). These are made up of a large number

H. Duseja · A. Kumar · R. Johari (✉)
SWINGER: Security, Wireless, IoT Network Group of Engineering and Research, University School of Information, Communication and Technology (USICT), Guru Gobind Singh Indraprastha University, Sector-16C, Dwarka, Delhi, India
e-mail: rahul@ipu.ac.in

D. P. Vidyarthi
School of Computer and System Sciences, Parallel and Distributed System Lab, JNU, New Delhi, India

P. K. Singh et al. (eds.), *Proceedings of Third International Conference on Computing, Communications, and Cyber-Security*, Lecture Notes in Networks and Systems 421,
https://doi.org/10.1007/978-981-19-1142-2_6

of sensors that can be positioned at precise latitude–longitude coordinates or at any random location in the network. A sensor is a small electronic device that can collect data from its environment and send it to a base station. The type of data collected varies depending on the application and the type of the sensors.

Sensors are electronic components that operate in networks with autonomy. The term sensor is usually used to refer to a sensor node (having dimensions of a few centimeters). It has components which are given below:

1. A microcontroller (programmable integrated circuit)
2. A radio antenna (for wireless communication)
3. A battery
4. Set of sensors that we will call capture unit in order to avoid confusion. A capture unit (having a dimension of a few millimeters) captures or intercepts environmental information (motion, temperature, humidity, gas, etc.). For communication (read or receive information), a sensor must be programmed.

A SHIP network is an ad hoc network that is composed of a number of sensor nodes that are able to collect, send and receive autonomously environmental data via wireless communications. SHIP network is one of the most important technologies that has changed the world and facilitates many daily tasks. They offer the possibility of observing and controlling physical and biological phenomena in several areas; industrial, scientific (temperature, pressure, humidity, light etc.), the environment (pollution, CO_2 etc.), health (patient monitoring, epidemiological studies etc.), transport (accident prevention etc.), home automation and so on. Sensor networks have many applications in different fields such as health, environment, agriculture, geology, and military. Most applications of SHIP network are challenging for designers, and this is due to the limited capacity of the nodes in terms of energy (battery), buffer space, computing power, and their deployment in tough inaccessible terrains. This makes the design of algorithms and programs for SHIP network too constrained. Therefore, performance evaluation tools such as CupCarbon simulator become very essential in the process of designing a SHIP network. The architectural framework of the CupCarbon simulator is shown in Fig. 1, and the simulator interface is shown in Fig. 2, respectively.

2 Literature Survey

Lopez et al. [1] present the evaluation of CupCarbon simulator and WSN evolution along with its application. Author(s) of the paper suggests that before implementation and deployment of any IoT application, network must be designed and simulate the operation because of expensive WSN equipments. This paper also proposes the modified version of Dijkastra algorithm for calculating cost for best available route.

Lopez et al. [2] present initial version of the CupCarbon simulator and describe that network can be designed by user-friendly interface and OSM framework by deploying sensors on the map. OpenStreetMap (OSM) is a framework which provides

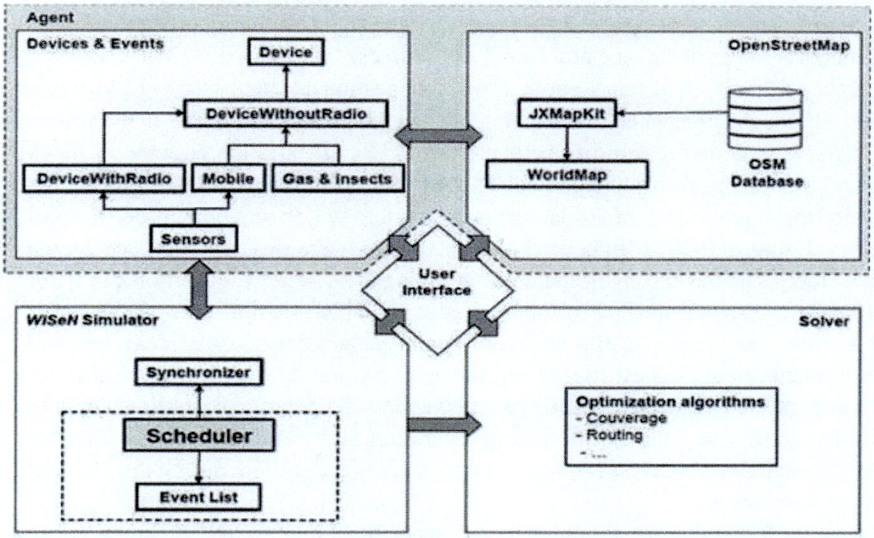

Fig. 1 CupCarbon simulator architectural framework

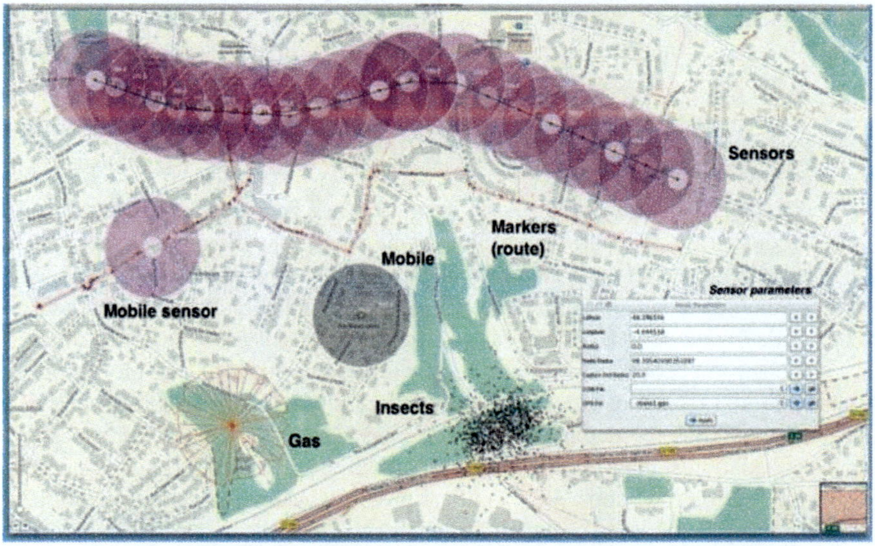

Fig. 2 CupCarbon simulator interface

satellite view by using which network or prototype can be designed with the help of real-life object such as buildings and routes of particular area. This paper also presents how sensors work and power diagram in the user interface of CupCarbon describes the battery level consumption by each sensors used in the network. This paper also presents that how a small delay in the SenScript (communication script) can increase the performance and life of the sensors.

Alzyoud et al. [3] discuss about traffic congestion problem around the world especially for the congested cities which creates the problem of traffic management and its safety. Broad reason of death in the road accident is the absence of quick medical treatment or mismanagement of the traffic. CupCarbon simulator tool was used for thorough testing of Jordan transport system which was composed of smart sensors. Using the OSM framework, smart sensors were directly deployed on the map of the particular testing domain or deploying area. Here, CupCarbon simulator was used to test and design various situation(s) which demonstrate actual roads and vehicle movements in the scenario. A particular scenario was designed and tested to determine the best route from the accident area to the nearest hospital for rescue operation enhancement. This paper considers the use of IoT intelligent system of transportation to enhance the rescue operation and hospital response time for handling emergency situation and controlling traffic light system during the accident management system.

Imran et al. [4] discuss about smart city with the help of WSN and smart nodes in real life like smart parking, healthcare center, smart transportation system, and banking. This paper proposed intelligent traffic monitoring system using WSN for smart city and implements different types of nodes in the particular area of the city which includes roads, objects, and signals of the traffic when each nodes are connected with each other and share the data of objects such as vehicles and rate of traffic. This paper suggests the approach of wireless detector for tracking the city traffic. All the sensors would be deployed in the city area like in the building, roads, schools, etc. Xbee wireless technology was effectively used between the communications of all sensors.

Bounceur et al. [5] presented CupCarbon simulator as a new architecture platform for WSN dedicated to IoT and smart cities. This paper also says that connected device is increasing, and in near future, it is expected to grow exponentially in the large cities. This paper explains about how 2D/3D visualization realistic environment helps in simulating the service of WSN and performance of sensors. Visual of 2D/3D environment helps to debug and validate the developed algorithm. 2D/3D city model represents a digital format of cities. It depicts different buildings, roads, areas, etc., and this model allows the deployment of different sensors in different areas and places for calculating signals propagation and interference. Saoudi et al. [6] discuss about event detection in various WSN applications such as environmental pollution and forest fire and natural calamity, and in such scenario, event must be detected early to reduce the threats and damages due to mishaps, for example: In forest fire, detection mechanism uses deployed sensors in the forest that would detect the fire and particular node alert the sink in order to inform the fire fighters well in time. In this paper, the proposed application is tested with the help of CupCarbon simulator.

John et al. [7] describe about green computing, and how this technique is used for protecting the environment as its main focus is power management and reuse of resources in the application of smart cities. IoT smart road network energy management is implemented with the help of CupCarbon simulator and road side unit (RSU),

Initial Sensor	Broadcasting to all sensor(s)	Sink

Loop
send "Message_A" *
delay 1000
send "Message_B" *
delay 1000
	Loop
	receive v
	send v *
	mark 1
	print v
	delay 1000
		//Receiver
		loop
		receive v
		if (v== "Message_A")
		print "Received Message A"
		mark 1
		Else
		print "Received Message B"
		mark 0
		End

Fig. 3 SenScript for pseudocode in CupCarbon

and MQTT IoT protocol [8] was used for managing power in street light in the network of roads.

Johari et al. [9] discussed the importance of routing in IoT and the requirement of studying and developing protocols for routing in IoT. Some of the advantages of IoT, as discussed in the paper, include enhanced customer satisfaction, digital optimization, and waste minimization. Tracking and reducing energy consumption, healthcare industry, education purpose, and government projects are some of the domains where IoT can prove to be useful (Fig. 3).

3 Methodology Adopted

In the proposed approach, effort to determine the most optimized route among all the best possible path (as determined using OpenStreetMap) is made. The flowchart depicting behavior of a simulated agent in CupCarbon simulation tool has been shown in Fig. 4. The pseudocode for the proposed approach is detailed as:

(1) Create a new project from project menu
(2) Give particular project name
(3) Open the SenScript window
(4) Add multiple sensor on the OSM GUI as per requirement
(5) Open the SenScript window
(6) Write the script
(7) Assign the SenScript file to the sensor node
(8) loop
(9) send "Message A" to the next node
(10) add delay of 1000 ms
(11) send "Message B" to the next node

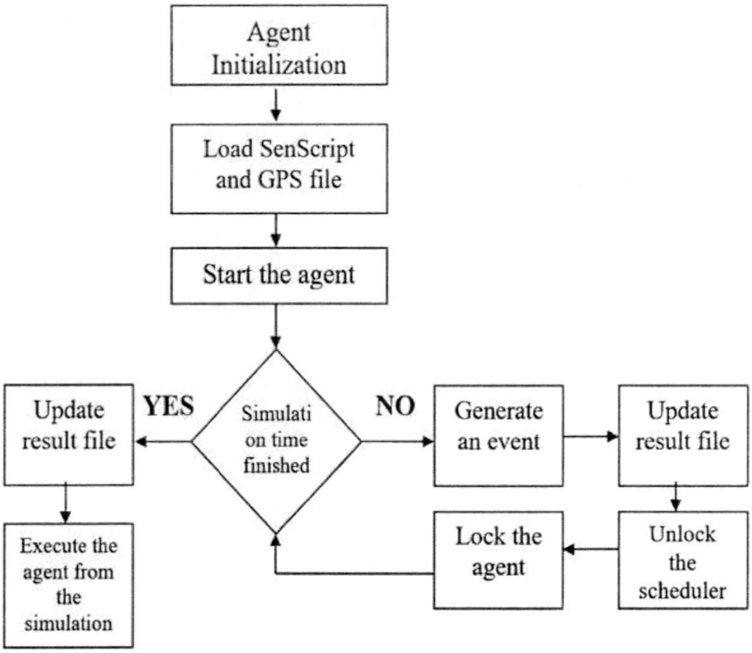

Fig. 4 Flowchart depicting behavior of a simulated agent in CupCarbon

```
      add delay of 100 ms
    loop
      receive the value in the variable "v"
      broadcast received value "v" using command "send v*"
      mark node as 1
      print received message from its neighbor by variable "v"
      add delay of 1000 ms
    end loop
  loop
  receive value in variable "v"
   If v = = "Message A"
      print "Received Message A"
      mark the node as 1
        else
        print "Received Message B"
      mark the node as 0
        end
      end loop
  end
```

Table 1 Hardware and software used in simulation

S. No.	Hardware and software requirements	Description
1	Operating system	Windows
2	CupCarbon simulator	Version 5.0
3	JDK	Version 1.8.0
4	CPU processor	Intel i5

4 Experimental Setup

The CupCarbon simulator [10, 11] which is based on multiple agents (sensors) and geo-location was used for carrying out simulation work. It enables designing and simulation of SHIP network for the proposed application on user-friendly interface with OpenStreetMap (OSM) framework. CupCarbon interface has a set of configurable and easy-to-use objects (usually sensors). Figure 1 shows the graphical interface of this simulator. The use of this system provides better optimization of sensors because it gives simulation time, performance, and energy consumption by agents. CupCarbon [12, 13] has three main components:

- Simulation environment
- Mobile simulation
- WSN simulator (WiSen).

Working steps are detailed as follows:

1. Initialize the first sensor with broadcasting message(s) for example in proposed case "Message A"* or "Message B" *. Here, * means it will send the particular message to all its neighbors.
2. Other intermediate sensor except the sensor in step 1 will be associated with the broadcasting communication script, which will receive the previous value and forward it to other associated nodes in the network.
3. Sink node will wait for the incoming message from its associated nodes, and finally, sink will check whether the intended message is received or not.
4. Process will continue or stop as per the requirement. Table 1 shows the hardware and software requirement to install and configure the CupCarbon simulation tool.

5 Result

The SenScript code for routing scenario has been showcased in Figure 3, and snapshot to demonstrate the broadcasting of messages to sink node with help of intermediate nodes in CupCarbon simulator has been shown in Figure 5.

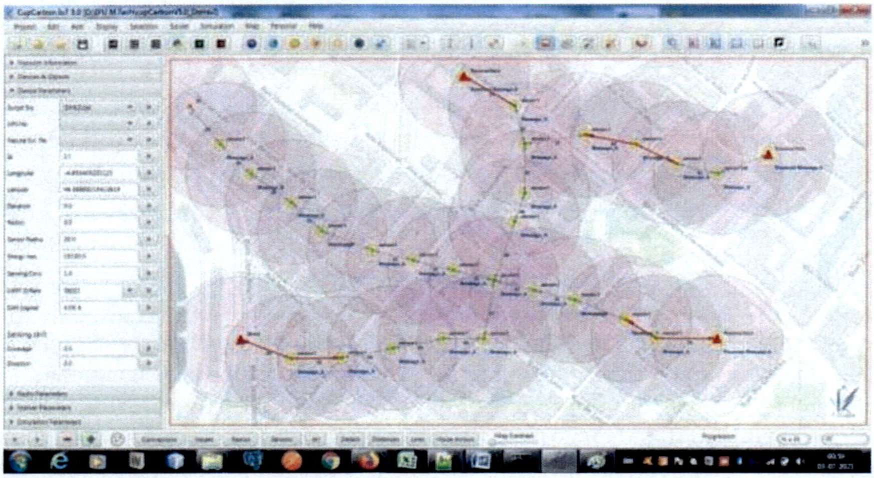

Fig. 5 Snapshot to demonstrate the broadcasting of messages to all the sink via all the neighbor node in CupCarbon simulator

6 Conclusion

In the work presented in the current research paper, a SenScript code to demonstrate the broadcasting of messages to the sink node through all multiple intermediate neighbor node(s) with minimum delay using CupCarbon simulation tool has been showcased. In future, it is proposed to introduce the concept of secure routing of messages, by encrypting them using symmetric and asymmetric cryptographic algorithms like advanced encryption standard (AES) and RSA, respectively.

References

1. Lopez-Pavón, C., Sendra, S., & Valenzuela-Valdés, J.F., 2018. Evaluation of CupCarbon network simulator for wireless sensor networks. *Network Protocols and Algorithms, 10*(2), 1–27.
2. Lounis, M., Mehdi, K., & Bounceur, A. (2014). A cupcarbon tool for simulating destructive insect movements. In *1st IEEE International Conference on Information and Communication Technologies for Disaster Management (ICT- DM'14), Algiers, Algeria.*
3. Alzyoud, F., AL Sharman, N., Al-Roosan, T., & Alsalah, Y. (2019). Smart accident management in jordan using cup carbon simulation. *European Journal of Scientific Research,* 128–135.
4. Imran, L. B., Amir Latif, R. M., Farhan, M., & Tariq, T. (2019). Real-time simulation of smart lighting system in smart city. *International Journal of Space-Based and Situated Computing, 9*(2), 90–98.
5. Bounceur, A., Clavier, L., Combeau, P., Marc, O., Vauzelle, R., Masserann, A., Soler, J., Euler, R., Alwajeeh, T., Devendra, V., & Noreen, U. (2018). CupCarbon: A new platform for the design, simulation and 2D/3D visualization of radio propagation and interferences in

IoT networks. In *2018 15th IEEE Annual Consumer Communications Networking Conference (CCNC)* (pp. 1–4). IEEE.

6. Saoudi, M., Bounceur, A., Euler, R., Kechadi, T., Cuzzocrea, A. (2016). Energy-efficient data mining techniques for emergency de-tection in wireless sensor networks. In *2016 International IEEE Conferences on Ubiquitous Intelligence Computing, Advanced and Trusted Computing, Scalable Computing and Communications, Cloud and Big Data Computing* (pp. 766–771). Internet of People, and Smart World Congress (UIC/ATC/ScalCom/CBDCom/IoP/SmartWorld). IEEE.
7. John, A., Ananth Kumar, T., Adimoolam, M., & Blessy, A. (2021). Energy management and monitoring using IoT with CupCarbon platform. In *Green Computing in Smart Cities: Simulation and Techniques* (pp. 189–206). Springer.
8. Johari, R., Bansal, S., & Gupta, K. (2020). Routing in IoT using MQTT protocol. In *2020 12th International Conference on Computational Intelligence and Communication Networks (CICN)* (pp. 1–5). IEEE.
9. Johari, R., & Adhikari, S. (2020). Routing in IoT network using CupCarbon simulator. In *2020 7th International Conference on Signal Processing and Integrated Networks (SPIN)* (pp. 301–306). IEEE.
10. http://www.cupcarbon.com/
11. http://labsticc.univ-brest.fr/bounceur/cupcarbon/examples/ [for cupcarbon tutorials]
12. https://github.com/bounceur/CupCarbon [for source code of cupcarbon simulator]
13. http://cupcarbon.com/cupcarbonug.html (for user guide)

Multiband Dual-Layer Microstrip Patch Antenna for 5G Wireless Applications

Vineet Vishnoi, Pramod Singh, Ishan Budhiraja, and Praveen Kumar Malik

Abstract A microstrip patch antenna (MPA) is compatible with 5G wireless applications due to its lightweight, compact and conformal shape, small volume, and minimal susceptibility to manufacturing tolerances. In the past few years, communication systems have frequently required multiband antennas to prevent the use of several antennas. Here, a stacked microstrip antenna with coaxial probe feed is proposed, and its multiband characteristics are studied for different 5G applications. For 5G applications, the proposed antenna works effectively in the frequency range of 0–10GHz. The variations in the shapes of the stacked microstrip patch antenna, such as circle, pentagon, hexagon, and octagon, are investigated, and it is observed that there is multi-resonance with decreasing lower resonance frequency as the shape varies. Here, two-layer geometry is used with one driving patch and another parasitic patch. This design covers a frequency range from 2.7 to 9.6 GHz with a gain of 9 dB, directivity of 13 dBi, and radiation efficiency of up to 75%. The proposed structure is simulated on IE3D Zealand software and return loss, directivity, gain, and radiation efficiency have been analyzed and measured.

Keywords Multiband antenna · Returns loss · Directivity · Gain · 5G Wireless applications

V. Vishnoi (✉) · P. Singh
Meerut Institute of Engineering and Technology, Meerut, Uttar Pradesh, India
e-mail: vineet.vishnoi@miet.ac.in

P. Singh
e-mail: pramod.singh@miet.ac.in

I. Budhiraja
Bennett University, Greater Noida, Uttar Pradesh, India

P. K. Malik
Lovely Professional University, Jalandhar, Punjab, India

P. K. Singh et al. (eds.), *Proceedings of Third International Conference on Computing, Communications, and Cyber-Security*, Lecture Notes in Networks and Systems 421,
https://doi.org/10.1007/978-981-19-1142-2_7

1 Introduction

The fifth generation (5G) of connectivity has been widely debated as a potential means of providing high data rate communications. The knowledge of the transmission channels is critical to the design and testing of the 5G communication system. Below 6 GHz, the 5G candidate frequency bands have been widely debated, with the following frequency ranges being proposed: 4500–4990 MHz, 470–694, 1427–1518 MHz, 3300–3800 MHz, and 1427–1518 MHz. Because of its lightweight, low cost, small size, and simlicity of interfacing with other microwave systems, the need for microstrip antennas in different types of communications networks has been rapidly growing [1–4]. As a result, the MPA has grown in popularity and has become a significant research topic in both theoretical and experimental settings. Microstrip antennas, on the other hand, have a narrow band range, which is one of their biggest drawbacks. There are several methods to enhance the bandwidth of the patch antenna like use of high dielectric, introduction of shorting walls [5], use of slots and notches [6], and insertion of defect in the ground plane [7]. The multilayer structure is well known as a valuable tool for resolving these issues [8]. The researchers looked into their basic characteristics and put in a lot of work into designing an electromagnetically coupled two-layer elliptical microstrip stacked antenna [9], a stacked square patch antenna for Bluetooth, and an analysis of stacked microstrip rectangular microstrip antenna [10]. Several approaches, including a thicker substrate reactive matching network and stacked patches, have been proposed in recent years to improve it [11, 12]. When high dielectric constant materials are used, single-layer patch antennas do not perform well. It has been stated that a HI-LO configuration, which combines high and low dielectric constant laminates, can produce impedance bandwidths and radiation efficiency comparable to that of a traditional stacked patch. Unlike single-layer cases with a low dielectric constant substrate, the HI-LO stacked patch has a very high surface wave efficiency across the impedance bandwidth. This stunning discovery opens up a new area of research in the domain of integration science. HI-LO stacked patches have a number of advantages that make them ideal for use with MMICs and OEICs: (1) good bandwidth can be achieved; (2) due to the ground plane, there is minimal back radiation; (3) they are relatively easy to design; (4) low cross-polarization levels are radiated, implying that good quality CP can be easily produced. As a result, it is obvious that piling a parasitic patch on top of the fed patch will increase bandwidth. Therefore, we observed on an electromagnetically stacked rectangular microstrip antenna with a large number of parasitic elements by using two stacked patches of different shapes in this paper. The effect of stacking on various antenna parameters is being investigated through experiments.

2 Related Work

Ghosal et al. proposed a rectangular microstrip antenna with an array of L-shaped slots. This structure is capable to provide multiband operation with five resonant frequencies:1.48, 1.25, 1.8, 2.25, and 2.9 GHz [13]. Shelar et al. proposed modified nicrostrip antenna for 5G application with better reflection coefficient, improved matching, and enhanced frequency coverage. This structure is fabricated on FR4 substrate with dimension of 18 mm × 16 mm × 1.6 mm [14]. Siri Chandana et.al. proposed single band MSA for 5G application using 2 E and 1 H-shaped slot. This structure is fabricated on Roger's RT5880 substrate. This structure covers bandwidth of 6.25 GHz with a center frequency of 59 GHz [15]. Yusnita Rahayu et al. proposed a triangular-shaped microstrip antenna array with triangular-shaped slot on the ground plane on Duroid 5880 substrate. In this structure, the gain of antenna is increased by increasing the number of array elements. It shows gain of 7.47 dBi at 28 GHz and 12.1 dBi at 38 GHz with six elements [16]. Kaur et al. proposed a truncated edge microstrip antenna on Duroid substrate for bandwidth enhancements. This structure has dimension of 12 mm × 35 mm and simulated on HFSS software [17].

3 Organization

The remainder of the paper is arranged in the following manner. The design considerations are described in Sect. 4. In Sect. 5, design parameters related to proposed antenna are presented. Section 6 discusses the performance of proposed antenna. Section 7 contains the conclusion.

4 Design Consideration

The microstrip antenna can be shaped in any way. It may be any of the following shapes: circle, pentagon, hexagon, or octagon. A total of five patches make up the proposed microstrip stacking antenna, and it is intended for use at 10 GHz. Table 1 shows the designing specification of dual-layer microstrip antenna. Patches for various shapes such as circle, pentagon, hexagon, and octagon are designed separately and shown in Fig. 1. They are stacked in the pattern depicted in Fig. 2. The lower patch is parasitic and the outer patch is coaxial feed. The outer patch is parasitic and feeds in a coaxial manner. The inner and outer patches are both parasitic, and the outer patch is feed. All the designed patches have been stacked one over the driving patch. The cross-sectional view of the antenna is shown in Fig. 3.

Table 1 Designing specification of dual-layer microstrip antenna

Parameters	Values
Thickness of the dielectric substrate ($h1 = h2$)	1.6 mm
Relative permittivity of the dielectr substrate 2	4.5
Relative permittivity of the dielectr substrate 1	2.2
Radius of upper patch (driven patch)	12.5 mm
Radius of lower patch (parasitic patch)	25.0 mm

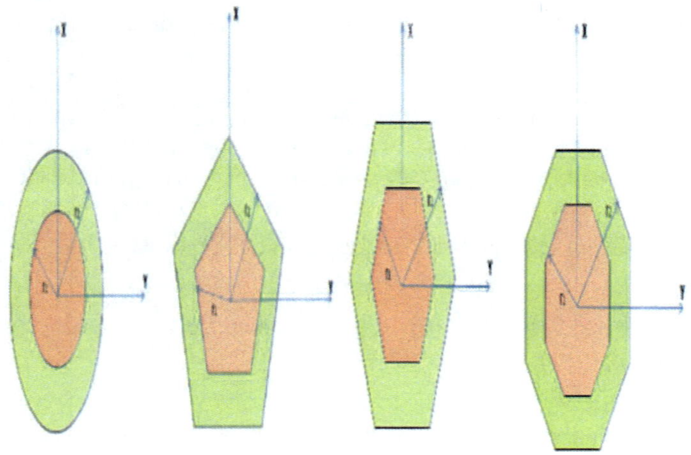

Fig. 1 Variation in patch shape for dual-layer antenna geometry

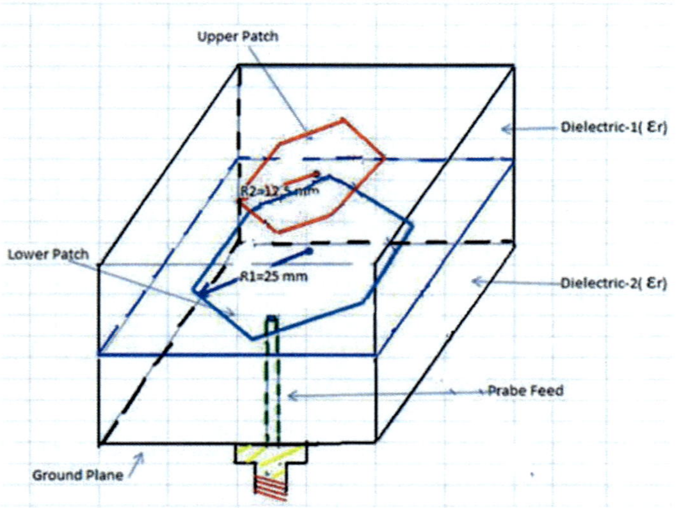

Fig. 2 3-Dimensional geometry of dual-layer antenna for hexagon

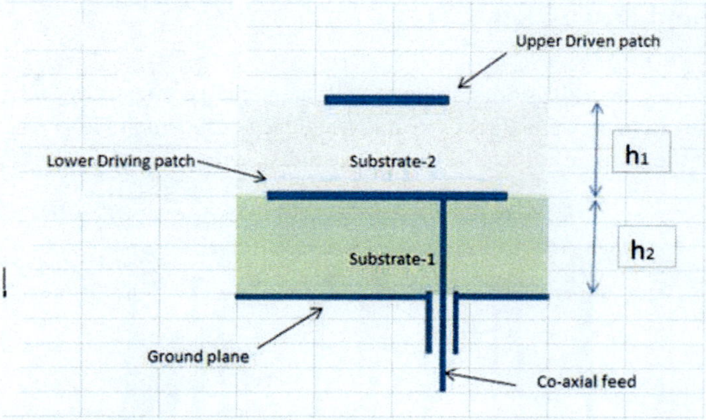

Fig. 3 Cross-section view of antenna

Table 2 Comparison table of return loss for antenna having variation in shapes (circle, pentagon, hexagon, and octagon)

Circular		Pentagon		Hexagon		Octogon	
Freq. (GHz)	S11 (dB)	Freq. (GHz)	S11 (dB)	Freq. (GHz)	S11 (dB)	Freq. (GHz)	S11 (dB)
2.33	−12.5	1.305	−12.19	1.274	−13.09	1.233	−17.5
9.55	−20.43	4.894	−12.83	5.75	−11.15	6.67	−10.9
10	−17.5	9.369	−10.05	8.57	−10.25	7.79	−9.68
		10	−14	9.01	−27.75	8.14	−15.15
				10	−12	9.01	−20.44
						9.98	−18.8

5 Design Parameters

The antenna's different design parameters are as follows:

Table 2 has the values of return loss for different frequency points. From the table values, it is quite clear that the antenna suits various commercially available frequency range applications, such as for GSM/UMTS (1.9 and 2.1 GHz), WiFi and IEEE802.11std. (3.6 GHz) for WLAN and Wi-Max, and for the ISM band (2.4/5.8 GHz.). This shows that the proposed antenna has a broad application range for commercial applications.

6 Result and Discussion

Antenna efficiency, as well as other important parameters like directivity and gain, is discussed. The curve between directivity and frequency is shown in Fig. 5, and it can be shown that the average value of directivity is 5 dBi, with a maximum value of 12

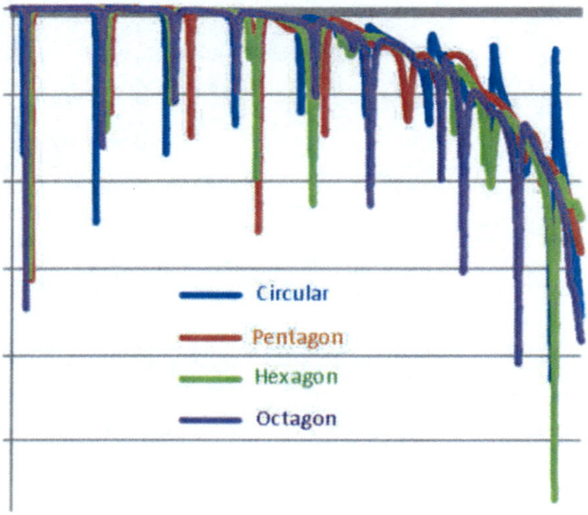

Fig. 4 Comparision curve (simulated) of S11 (dB) for all the geometries of dual-layer antenna for variation in geometries

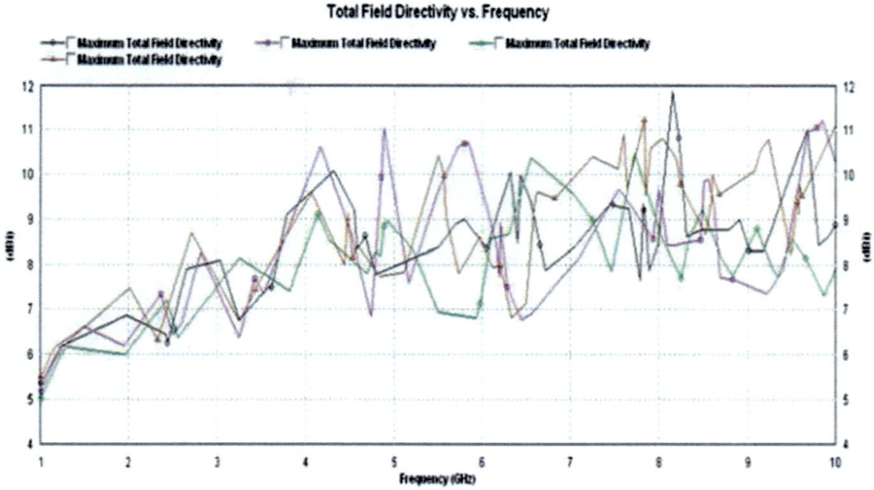

Fig. 5 Antenna directivity comparison curve: with, variation in shape for circle (purple), pentagon (green), hexagon (black), and octagon (red)

dBi for antenna designs with hexagonal shapes. Similarly, other parameters such as gain are compared for all of the proposed antenna's design geometries, which vary in their geometrical shapes. As shown in Fig. 6, the average value of gain is 5 dBi, with some frequency points approaching 9.0 dBi. Although the gain of patch antennas is considered to be poor, the gain of the proposed antenna is found to be satisfactory. Figure 7 shows antenna efficiency w.r.t frequency variation. It is clearly shown that the antenna efficiency approaches up to 65% value in its respective frequency domain.

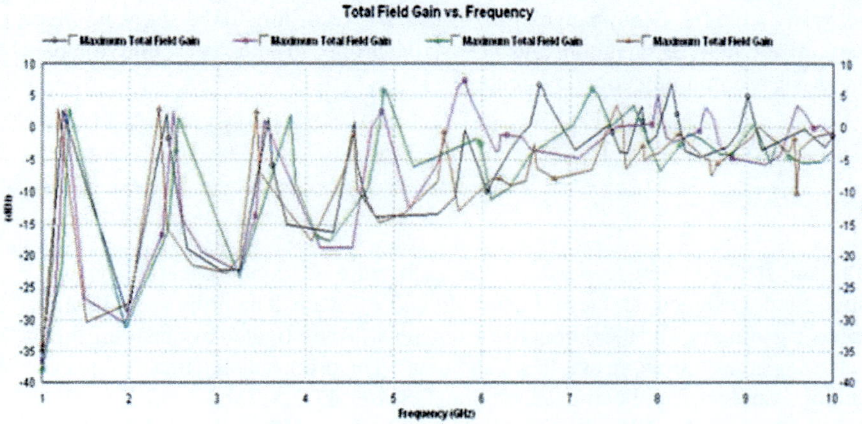

Fig. 6 Antenna gain versus frequency comparison curve: with, variation in shape for circle (purple), pentagon (green), hexagon (black), and octagon (red)

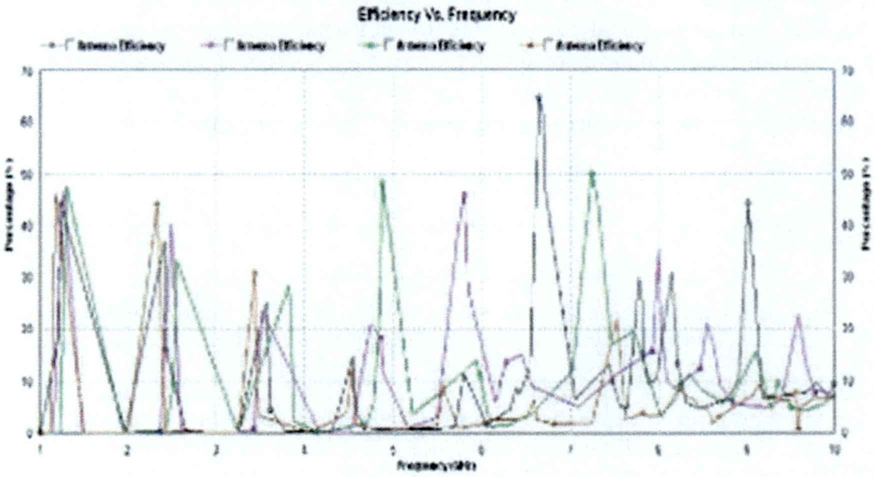

Fig. 7 Antenna efficiency versus frequency comparison curve: with, variation in shape for circle (purple), pentagon (green), hexagon (black), and octagon (red)

All of the four geometries of the proposed antenna for multiband stacked antenna are analyzed on the IE3D simulator tool for variations in geometrical shapes. Figure 4 depicts a combined comparative graph, with Table 2 providing details in tabular fashion. Table 2 and Fig. 4 compare the return loss of all four shapes and show the effect of changing their geometrical shapes. Furthermore, it is clear from Fig. 5 that the suggested antenna aims for significant directivity ranging from 5 to 13 dBi over a large frequency range. The overall gain versus frequency plot is shown in Fig. 6. It is worth noting that overall gain approaches 9 dB on the curve. The antenna efficiency versus frequency curve is shown in Fig. 7, and it can be seen that the antenna efficiency reaches 65% in the frequency range of 2.7–9.6 GHz. According to the above figures, a successful antenna design is conceivable, and the proposed antenna is quite suitable for 5G wireless communication application.

7 Conclusion

A multiband stacked microstrip patch antenna for wireless communication systems was suggested, designed, and tested using Zeland software. The proposed antennas have a basic geometry, and the output of the antennas is investigated by changing their geometrical shapes (circle, penta, hexa, and octa). The proposed antenna design can be used for a variety of applications, including GSM/UMTS (1.9 and 2.1 GHz), WiFi, and IEEE802.11std. (3.6 GHz) for WLAN and Wi-Max, and ISM band (2.4/5.8 GHz), because it is suitable for a wide range of frequencies from 2.7 to 9.6 GHz and has characteristics such as 75% radiation efficiency and highly allowable gain up to 9.0 dB, directivity up to 13 dBi. These promising characteristics are found to be more prominent than in conventional wireless antennas of the past. The stacked microstrip antenna exhibits excellent directional-radiation properties. This antenna also offers several benefits, including ease of construction, low cost, and small size. As a result, it has been found to be very suitable for wireless/WLAN/ applications in 5G.

References

1. Budhiraja, I., Kumar, R., & Pal, M. K. (2012). Slotted dual arrowhead multiband rectangular microstrip patch antenna. *International Journal of Scientific and Engineering Research, 3*(9), 758–762.
2. Budhiraja, I., Khan, A. A., Farooqi, M., & Pal, M. K. (2012). Multiband stacked microstrip patch antenna for wireless applications. *Journal of Telecommunications, 16*(2), 925–931.
3. Budhiraja, I., Khan, A. A., Farooqi, M., & Pal, M. K. (2012). Performance analysis of SFBC-OFDM system with frequency domain equalization. *International Journal of Scientific and Engineering Research, 3*(11), 678–682.
4. Bahl, J., & Bhartia, P. (1981). *Microstrip antennas*. Dedham, MA: Artech House.
5. Singh, P., & Aggarwal, R. (2016). Design of ultra-wideband antenna with triple band notch for minimum EMI. *Microwave and Optical Technology Letter, 7*(58), 75–83.

6. Singh, P., & Aggarwal, R. (2016). Comparative study of UWB microstrip antennas with different defected ground structures. In *Proceedings of ICMETE92016* (pp.42–46). IEEE Computer Society, SRM University.
7. Vishnoi, V., Pal, M. K, & Kannujia, B. K. (2012). Slotted octagonal shaped antenna for wireless applications. *International Journal of Scientific and Engineering Research, 3*(9), 678–682 (2012)
8. Vishnoi, V., Malik, P. K., & Pal, M. K. (2020). Horseshoe-shaped multiband antenna for wireless application. In *Proceedings of First International Conference on Computing,Communications, and Cyber-Security (IC4S 2019)* (LNCS, Vol. 121, pp. 33–45). Springer.
9. James, J. R., Hall, P. S., & Wood, C. (1981). *Microstrip antenna theory and design*. IEE Electromagnetic, Series 12, London, U.K., Peter Peregrinus.
10. Das, P. K., Malik, P. K., Singh, R., Gehlot, A., Gupta, K. V., & Singh, A. (2019). Industrial hazard prevention using raspberry Pi. In *International Conference on Intelligent Computing and Smart Communication* (pp. 1487–1499).
11. Rahim, A., Malik, P. K., & Sankar Ponnapalli, V. A. (2019). State of the art: A review on vehicular communications, impact of 5G, fractal antennas for future communication. In *Proceedings of First International Conference on Computing, Communications, and Cyber-Security (IC4S 2019)*. Lecture Notes in Networks and Systems (Vol. 121). Singapore: Springer.
12. Shaik, N., & Malik, P. K. (2020). A retrospection of channel estimation techniques for 5G wireless communications: Opportunities and challenges. *International Journal of Advanced Science and Technology, 29*(5), 8469–8479.
13. Ghosal, A., Das, S. K., & Das, A. (2019) Multifrequency rectangular microstrip antenna with array of L-slots. *AEU—International Journal of Electronics and Communications, 111*.
14. Shelar, S., Kasambe, P. V., & Kumbhare, R. (2021). Microstrip patch antenna with partial ground plane and parasitic patch for K band application in 5G. In *2021 International Conference on Communication information and Computing Technology (ICCICT)* (pp. 1–6).
15. Siri Chandana, R., Sai Deepthi, P., Sriram Teja, D., Veera Jaya Krishna, N., & Sujatha, M. (2020). Design of a single band microstrip patch antenna for 5G applications. *International Journal of Engineering and Technology, 10*(2).
16. Rahayu, Y., & Hidayat, M. I. (2018). Design of 28/38 GHz dual-band triangular-shaped slot microstrip antenna array for 5G applications. In *2018 2nd International Conference on Telematics and Future Generation Networks (TAFGEN)* (pp. 93–97).
17. Kaur, A., & Malik, P. K. (2020). Tri state, T shaped circular cut ground antenna for higher 'X' band frequencies. In *2020 International Conference on Computation, Automation and Knowledge Management (ICCAKM)* (pp. 90-94).

Distance-based Energy-Efficient Clustering Approach for Wireless Sensor Networks

Bhawnesh Kumar, Naveen Kumar, Harendra Singh Negi, and Rakesh Kumar Saini

Abstract Wireless sensor network (WSN) is a vast field for the research and development in many types of applications. In WSN, multiples nodes deployed into the environment; each node has energy level. Optimized energy consumption is main concern for any kind of applications like as military/battlefield, smart farming, medical science, vehicular ad hoc networks (VANET). Thousands of nodes deployment in sensor area become a typical task and later maintain the energy consumption as well. The level of consumption of energy consumption of network needs to be focused to prolong the life time. This research work increases the energy level of sensor network using distance-based technique for WSN. It elects the cluster head on the bases of distances between sensor nodes and from base station also considered. The implementation shows the graphical representation of sensor nodes and calculates the energy consumption of each node along with cluster head and also gives the comparison between clustering and quad clustering. This technique represents total energy which is transmission energy, receiving energy, and data aggregation energy through a graph. This proposed work examines that enhanced distance-based technique increases the life time of sensor network for the advancements in the WSN applications.

Keywords Clustering · Network life time · Energy efficiency · Data aggregation · Quad clustering

B. Kumar (✉) · N. Kumar · H. S. Negi
Graphic Era Deemed to be University, Dehradun, India
e-mail: bhawneshmca@gmail.com

R. K. Saini
DIT University, Dehradun, India

P. K. Singh et al. (eds.), *Proceedings of Third International Conference on Computing, Communications, and Cyber-Security*, Lecture Notes in Networks and Systems 421,
https://doi.org/10.1007/978-981-19-1142-2_8

1 Introduction

Wireless sensor network (WSN) is composed of collection of distinguish node which is small in size with limited energy level. Main controller of network is sink node which is responsible to collect the sensed data from all nodes known as base station (BS). In so many years, energy-efficient techniques [1] of WSN are the growing field for researchers' community. Deployment of all the nodes depends on the type of WSN application used [2]. And, input from environment and output of sensor node in the form of electrical pulses further transmit to the BS for ahead WSN processing.

The network can be decided to spread the node inside the network, and placement of each node depends on application specific. In WSN, sensor node is having battery backup which charges by solar energy panel board, and that remaining energy level of node is known as residual energy (RE) and helps to transmit the data from the environment to the BS or main computer system.

Different topologies [3] are used to generate sensor networks. And, the number of nodes used depends on the area or size of space, where battery backup performs main role to maintain the network's whole life cycle. If each node has enough battery backup that is able to transmit the enough data, that data transmit the distance from some to destination with usage of energy. Now, energy is main concern of networks which should be in consideration. So need to emphasis on residual energy usage of node to improve the efficiency of whole sensor networks. When deployment of nodes into hostile environment [4] is really very typical, so it is not easy to reboot or recharge or regenerate the replacement of battery of each and every node which is named as mote or hub seen in researches, but it needs to focus on energy usage of each node.

Figure 1 shows that WSN is a network that has so many scattered nodes to work collectively and just receive input from field (information gathering), and nodes are linked together wirelessly to form a network topologies. WSN has two things: firstly, cluster of nodes connected to each other, and second one is base station. Each node has a main device to be there for all working, i.e., battery; electric battery is a device

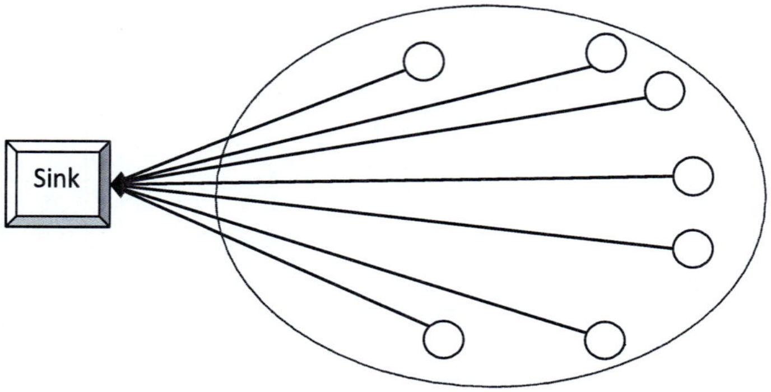

Fig. 1 WSN architecture

which consists one or more electro-chemical cells. There are some previous efforts for literature review: the characteristics, properties, applications, and communication routing protocols in WSNs [5–8]. To design the clustering technique is a challenging task in network area due to different factors [7] to be considered first. Main goal of WSN is to increase the life time of network. Clustering improves the life time of sensor networks supported by all types of applications. In case of clustering, two things are considered first: formation of cluster and selection of cluster head. Moreover, all the cluster members are responsible to send the sensed data to respective cluster head. Apart from this, in non-clustering, all the nodes directly communicated with base station (BS). Clustering is far better than non-clustering in the consumption of energy of sensor nodes.

The parts of the paper are divided into five sections which are followed: Sect. 2 covers the literature work. System model and proposed work are mentioned in Sect. 3. The implementation of work and simulation graph of work are described in Sect. 4. Section 5 has the conclusion and future work finally.

2 Literature Review

The most emerging field in the WSN-based application is energy usage and consumption of sensor nodes. It is needed to make a clustering approach which enhances the energy level of whole network. With the aid of clustering, energy usage can be minimized to prolong the long distant node energy level [9]. In WSN, two types of clustering approaches are there: homogeneous where all nodes have same energy amount and heterogeneous where each node has different energy amount. To support the homogeneous clustering, multiple protocols are available such as LEACH [10], HEED [11], PEGASIS [12], whereas heterogeneous clustering protocols are SEP [13], LEACH-E [14], DEEC [15]. The following Table 1 described the summary of energy-based clustering approaches research papers in WSN:

3 System Model

3.1 Energy Consumption Model

Energy consumption formula calculates the consumed energy for each node or whole network as well. This section describes the transmission, receiving, and data aggregation of energy consumption. As proposed work depends on the distance (d), energy also depends on the distance. So, consideration of distance to calculate the energy is focused on how far away the nodes are from BS location. The following formulas presented in Eqs. (1), (2), and (3) are used to calculate the energy consumption for k bit data.

Table 1 Summary of research papers

Authors	Objective	Features	Limitations
Liaqat et al. [16]	Proposed approach enhances the life time of sensor networks and also maintains the stability	It works for mobile network, throughput and minimizes the delay time and packet loss	Efficient energy utilization remains an open issue
Mahboub et al. [17]	Need to have a approach which minimizes the energy consumption and improves the network life time	It uses K-means clustering algorithm which optimized utilization of energy of the nodes for whole network	Mostly energy consumption in transmission phase specifically in large distant nodes
Shanmukhi et al. [18]	Aim for this paper is to maximize the network life time	It is weighted compressing sensing technique to maintain the suitability of network	Try to solve the load balancing problem and throughput as well
Liu et al. [19]	To achieve optimal energy consumption with the use of power line connection method	This approach balances the energy between nodes and uses energy optimal path to transmit	Increases the system, cable, and installation cost
Behera et al. [20]	Focused on an efficient and rotational selection of cluster head scheme which improved the energy level of network	It enhanced the throughput by 60%, network life by 66%, and residual energy by 64%	This work can be extended to other parameters for CH selection
Zhu et al. [21]	İt is a double cluster head strategy to minimize the energy level for unequal clustering	It improved the different parameters such as life time, throughput, stability, and energy consumption. It uses the best hop count method	Base station location can be at centered to improve the energy level
Singh et al. [23]	This new protocol objective enhanced the stable lifetime of the WSN and ultimately to smoothness transmission	Introduces an independent node, which is nearer to the sink, and independent nodes can transmit their own sensed data directly to the sink	Base station can be at center position and rotate the duty of CH as well
Wei et al. [22]	This paper focused on the energy consumption and data delay parameters	Examined the approach for mobile sink optimized cluster-based energy technique	Extends to have multiple mobile sink or base station

$$E_{Tx}(k, d) = E_{elec} * k + \varepsilon_{fs} * k * d^2 d < d_{thres} \tag{1}$$

$$E_{Tx}(k, d) = E_{elec} * k + \varepsilon_{amp} * k * d^4 d > d_{thres} \tag{2}$$

where distance threshold is $d_{thres} = \sqrt{\frac{\varepsilon_{fs}}{\varepsilon_{amp}}}$.

To receive the k bit data

$$E_{Rx}(k) = E_{elec} * k \tag{3}$$

3.2 Proposed Work

This proposed work developed a technique to design an efficient cluster head approach in wireless sensor network that can consume the less energy and save the energy level of each node. In WSN, cluster head selection process should enhance the energy level of each node and comprise the whole network in consideration. Here distance parameter is used to select the cluster head for each cluster for sensor network. The main concern to find the cluster head to pay the role for whole cluster is to collect the data from cluster member.

Enhanced distance-based cluster head selection technique applied the following steps:

1. Initially, sensor nodes are deployed into the environment, and each node has the co-ordinate position as x and y.
2. Set the center location of BS with co-ordinate position.
3. Divide the whole area by equal size into four partitions. Draw a vertical and horizontal line. (BS location will be worked as origin position).
4. Each quadrant calculates the distance between nodes using Euclidean distance formula as from BS as well.
5. Calculate the net distance of each node by the addition of BS distance of each node for each quadrant.
6. Node with minimum net distance is the node elected as cluster head (CH) for corresponding quadrant.

Figure 2 represents the single cluster formation of network where triangle symbol is the location of BS at center of sensor network, square represents the CH, and circle represents the nodes.

Single cluster considered in [24] needs the whole network partitioned into four clusters which formed quadrants. Quad clustering is also implemented by [25] to form the four clusters for energy efficiency. WSN rectangular window area for whole cluster lower-left corner is (X_L, Y_L) and upper-right corner is (X_R, Y_R) as shown in Fig. 3. The whole window is divided into four clusters. Further cluster head selection

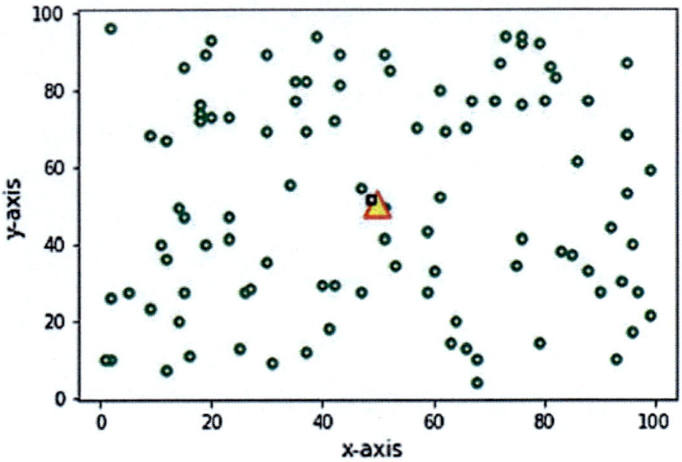

Fig. 2 Single cluster formation of 100 nodes

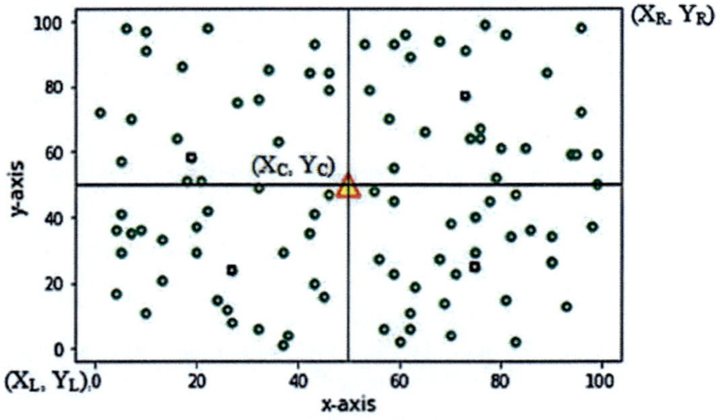

Fig. 3 Single cluster converts into four clusters

procedure of four clusters as per distance between nodes and BS takes place. So, quadrant positions are described as:

Quadrant cluster 1—starts from (X_L, Y_L) and ends with (X_C, Y_C)
Quadrant cluster 2—starts from (X_C, Y_L) and ends with (X_R, Y_C)
Quadrant cluster 3—starts from (X_L, Y_C) and ends with (X_C, Y_R)
Quadrant cluster 4—starts from (X_C, Y_C) and ends with (X_R, Y_R).

The methodology here to elect the CH depends on the two parameters.

i. Energy level
ii. Distance.

In case of energy level, energy is main aspect of WSN, because if a node is more capable to transmit, the data completely depend on the power or remaining energy of node. So, consider the energy of each node as main concern. If any approach can save the energy, that can also improve and enhance the energy level of whole network. So, methodology which is based on energy is going to be famous. Here, the CH selection is based on the energy of each node which can travel the minimum distance to save the energy.

Another parameter is distance which is also a main consideration to overcome the problem of energy usage. Because, a node travels the small distance obviously uses the less energy. The sending energy formula is as follows. As per distance consideration, initially, distance is consider between nodes and later on the distance from each to base station location. In sensor network area, there is deployment of the sensor nodes into two-dimensional space area axes like x, y-axis. So that, it calculates the distance using Euclidean distance formula as mentioned below.

$$\text{Distance} = \sqrt{(x_2 - x_1)^2 + (y_2 - y_1)^2}$$

whereas (x_1, y_1) and (x_2, y_2) are co-ordinate positions of node and base station as well.

4 Simulation Result and Analysis

4.1 Simulation Environment

Here is the Python implementation. For simulation, consider hundred to five hundred nodes for 100 m by 100 m, field network area for WSN. Parameters used to evaluate the energy consumption are shown in Table 2.

Table 2 Parameters of simulation

Parameters	Value
Network field area (in meters)	(100, 100)
Number of nodes	100–500
E_{elec}	50 nj/bit
E_{fs}	10 pJ/bit/m2
E_{amp}	0.0013 pJ/bit/m4
D_{thres}	87 m
E_{da}	5 nj/bit/signal

4.2 Simulation Outcome

Simulation of proposed work distributes 100 to 500 nodes in various field sizes, and BS location is at center of the area which is fixed. Initially, Fig. 4 shows hundred nodes' deployment in 100 m by 100 m sensor area, and Fig. 5 is for 100 m by 100 m with 500 nodes deployment, and further cluster has, respectively, 26, 30, 26, and 18 number of nodes. Location of first quadrant CH position is at (33, 30), second quadrant CH position is at (76, 24), third quadrant CH position is at (31, 72), and fourth quadrant CH position is at (72, 82) as (x, y) co-ordinates.

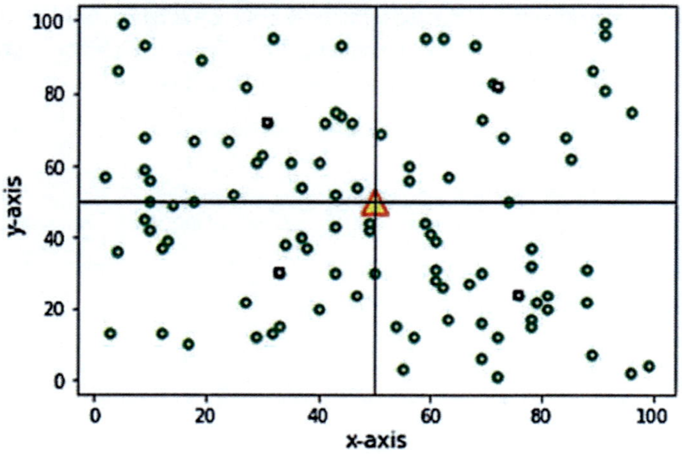

Fig. 4 100 nodes deployment

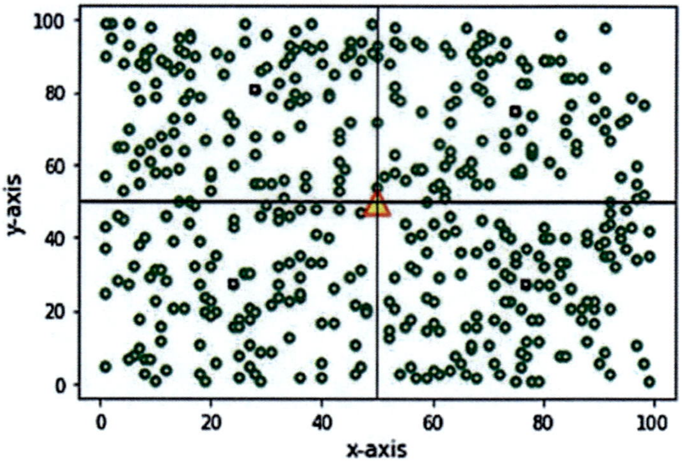

Fig. 5 500 nodes deployment

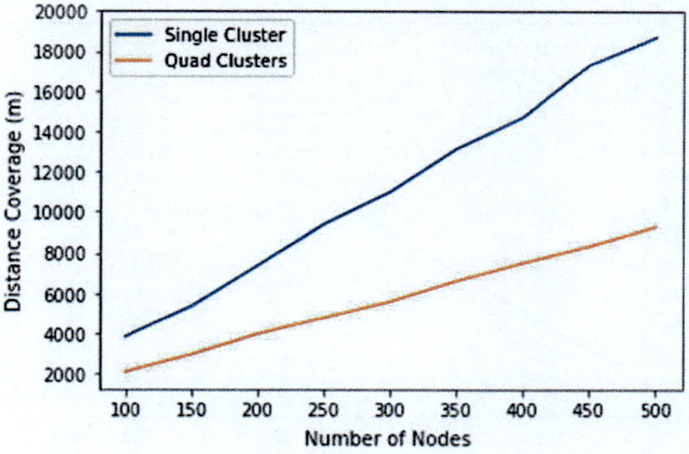

Fig. 6 Single cluster and quad clusters distance coverage

As comparison between single clustering and quad clustering, quad clustering is better than single cluster. In single cluster, all nodes transmit the data to CH and then aggregate the data sent to BS directly, whereas in quad clustering, four CHs are responsible to collect the data from corresponding cluster members and then to BS.

In Fig. 6, the graph of distance shows the comparison of distance coverage by single cluster and quad clusters, and the proposed work covered less distance in comparison of single cluster. As per energy formulas mentioned in Eqs. (1), (2), if the covered distance is less then energy consumption is also less. Distance coverage graph shows hundred to five hundreds nodes represented in 100 by 100 m area.

As per energy consumption of network, Fig. 7 shows energy usage comparative graph of 100 by 100 m with deployed 100–500 nodes for single cluster and quad clusters. This comparison shows that the proposed work gives the use of energy consumption which is lower than the single cluster. Different WSN applications can be beneficial to implement the distance-based clustering approach. Proposed work gives the balanced and utilization of energy level of whole network. Distance-based quad clustering gives the enhancement to improve the life time of WSN. It is observed that our proposed work is 40% to 50% better than single cluster and direct transmission in case of energy consumption and distance coverage. It also supports the balanced clusters which are near to equality of number of nodes.

5 Conclusion

Mostly, energy-based approaches consumed more energy which leads to degrade the network life time. Cluster formation may improve the network life time for many applications. The proposed work is energy-efficient approach based on distance

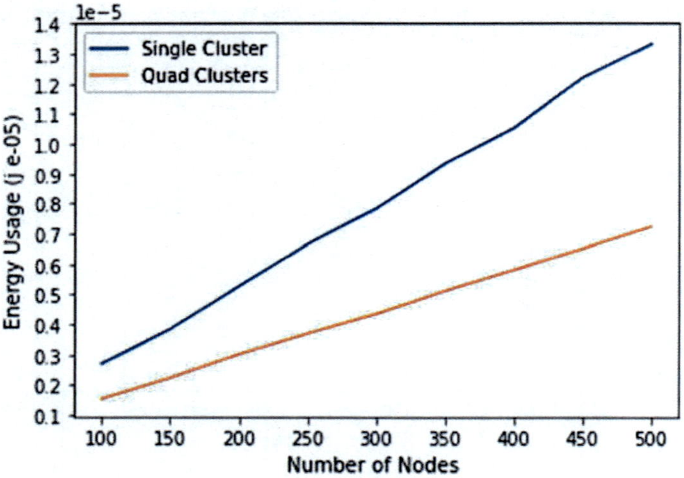

Fig. 7 Energy usage by single cluster and quad clusters

and energy level of node. In this paper, area of network is divided into four partitions to see the improvement in wireless sensor network life time. Each partition represents a quadrant and assigned a CH to collect the sensor data from respective members. Assignment of CH is based on distance between nodes and BS as well. The proposed implemented work used the energy of each node efficiently of wireless sensor network. The implementation shows that proposed work gives better result in case of performance than the single cluster and direct transmission. It is observed that our proposed work shows 40–50% increment in the energy level of network. Multi-level quad clustering can be further explored to improve the energy level of whole network. This proposed work is only simulation-based and later on can also implement on real-time application.

References

1. More, A., & Raisinghani, V. (2017). A survey on energy efficient coverage protocols in wireless sensor networks. *Journal of King Saudi University—Computer and Information Science, 29*(4), 428–448. https://doi.org/10.1016/j.jksuci.2016.08.001.
2. Mainwaring, A., Polastre, J., Szewczyk, R., & Culler, D. (2002). Wireless sensor network for habitat monitoring.pdf. *IEEE Communication Magazine,* 102–114
3. Santha Meena, S., & Manikandan, J. (2018). Study and evaluation of different topologies in wireless sensor network. In *Proceedings of 2017 International Conference on Wireless Communiation and Signal Processing and Networking (WiSPNET 2017)* (Vol. 2018, pp. 107–111). https://doi.org/10.1109/WiSPNET.2017.8299729
4. Sharma, V., Patel, R. B., Bhadauria, H. S., & Prasad, D. (2016). Deployment schemes in wireless sensor network to achieve blanket coverage in large-scale open area: A review. *Egyptian Informatics Journal, 17*(1), 45–56. https://doi.org/10.1016/j.eij.2015.08.003

5. Tilak, S., Abu-Ghazaleh, N. B., & Heinzelman, W. (2002). A taxonomy of wireless micro-sensor network models. *ACM SIGMOBILE Mobile Computing and Communication Review, 6*(2), 28–36. https://doi.org/10.1145/565702.565708
6. Al-Karaki, J. N., & Kamal, A. E. (2004). Routing techniques in wireless sensor networks: A survey. *IEEE Wireless Communications, 11*(6), 6–27. https://doi.org/10.1109/MWC.2004.1368893
7. Ketshabetswe, L. K., Zungeru, A. M., Mangwala, M., Chuma, J. M., Sigweni, B. (2019). Communication protocols for wireless sensor networks: A survey and comparison. Heliyon **5**(5), e01591. https://doi.org/10.1016/j.heliyon.2019.e01591
8. Akyildiz, I. F., Su, W., Sankarasubramaniam, Y., & Cayirci, E. (2002). Wireless sensor networks: A survey. *Computer Networks, 38*(4), 393–422. https://doi.org/10.1016/S1389-1286(01)00302-4
9. Abbasi, A. A., & Younis, M. (2007). A survey on clustering algorithms for wireless sensor networks. *Computer Communications, 30*(14–15), 2826–2841. https://doi.org/10.1016/j.comcom.2007.05.024
10. Heinzelman, W. R., Chandrakasan, A., & Balakrishnan, H. (2000). Energy-efficient communication protocol for wireless microsensor networks, pp. 1–10
11. Younis, O., & Fahmy, S. (2004) HEED: a hybrid, energy-efficient, distributed clustering approach for ad-hoc sensor networks. *IEEE Transactions on Mobile Computing, 3*(4), 366–379.
12. Raghavendra, C. S., & Lindsey, S. (2002). PEGASIS: Power-efficient gathering in sensor information systems.
13. Smaragdakis, G., Matta, I., & Bestavros, A. (2004). SEP: a stable election protocol for clustered heterogeneous wireless sensor networks, pp. 1–11.
14. Heinzelman, W. B., Chandrakasan, A. P., & Balakrishnan, H. (2002). An application-specific protocol architecture for wireless microsensor networks. *IEEE Transactions on Wireless Communications, 1*(4), 660–670. https://doi.org/10.1109/TWC.2002.804190
15. Qing, L., Zhu, Q., & Wang, M. (2006). Design of a distributed energy-efficient clustering algorithm for heterogeneous wireless sensor networks. *Computer Communications, 29*(12), 2230–2237. https://doi.org/10.1016/j.comcom.2006.02.017
16. Liaqat, M., Gani, A., Anisi, M. H., Ab Hamid, S. H., Akhunzada, A., Khan, M. K., Ali, R. L. (2016). Distance-based and low energy adaptive clustering protocol for wireless sensor networks, 1–29. https://doi.org/10.1371/journal.pone.0161340
17. Mahboub, A., Arioua, M., & En-Naimi, E. M. (2017). Energy-efficient hybrid K-means algorithm for clustered wireless sensor networks. *International Journal of Electrical and Computer Engineering, 7*(4), 2054–2060. https://doi.org/10.11591/ijece.v7i4.pp2054-2060
18. Shanmukhi, M., Patil, D. A., Amudahavel, Sathish, G. N. (2018) "Weighted compressive sensing with k-means algorithm ın wireless sensor networks. *International Journal of Pure and Applied Mathematics, 120*(6), 3681–3706.
19. Liu, X., & Wu, J. (2019). A method for energy balance and data transmission, 1–14. https://doi.org/10.3390/s19133017
20. Behera, T.M., Mohapatra, S.K., Samal, U.C., Khan, M.S., Daneshmand, M., & Gandomi, A.H. (2019). Residual energy based cluster-head selection in WSNs for IoT application. *IEEE Internet Thıngs J.*
21. Zhu, F., & Wei, J. (2019). An energy-efficient unequal clustering routing protocol for wireless sensor networks. *International Journal of Distributed Sensor Networks, 15*(9). https://doi.org/10.1177/1550147719879384
22. Wei, Q., Bai, K., Zhou, L., Hu, Z., & Jin, Y. (2021). A cluster-based energy optimization algorithm in wireless, 1–18. https://doi.org/10.3390/s21072523
23. Singh, T. S., & Khan, A. K. (2020). Distance-based clustering protocol (DBCP) in wireless sensor network. In *Progress in Computing, Analytics and Networking. Advances in Intelligent Systems and Computing* (Vol. 1119). Springer, Singapore. https://doi.org/10.1007/978-981-15-2414-1_42

24. Park, G. Y., Kim, H., Jeong, H. W., & Youn, H. Y. (2013). A novel cluster head selection method based on K-means algorithm for energy efficient wireless sensor network
25. Sharath, S. T., & Veena, N. (2014). Quad clustering routing protocol to enhance the stability in WSN, 3982–3988

Emerging Communication Technologies for Industrial Internet of Things: Industry 5.0 Perspective

Nagesh Kumar, Bhisham Sharma, and Sushil Narang

Abstract The Internet of things (IoT) has emerged into various application areas like agriculture, healthcare, defense, transportation, and manufacturing. The transformation of real things in the physical world to the Internet of things given a rise to industrial IoT (IIoT). IIoT applications are intended for the automation of the manufacturing industry, called Industry 4.0. Also, due to needs of end-user personalization, Industry 5.0 is becoming popular nowadays. Industry 5.0 is intended to inject artificial intelligence (AI) into human lives to improve capabilities and productivity. To make Industry 5.0 a successful revolution, IIoT must provide better efficiency, improved productivity, and better asset management. In this context, device-to-device communication plays an important role. IoT devices must be enabled with seamless communication technologies over heterogeneous networks. In this paper, communication standards, technologies, and various published research contributions are reviewed. Further, an analysis is presented to formulate challenges and opportunities for designing communication methods for IIoT. The paper also provides general directions for developing communication techniques in perspective to Industry 5.0.

Keywords Industrial Internet of things · Cyber physical systems · Device-to-device communication · Industry 4.0 · Industry 5.0 · Fog computing

1 Introduction

Industrial Internet of things (IIoT) is a promising application of IoT, resulted due to sensor and Internet interface leading to better human lives. IIoT involves machines in communication process though Internet and process data produced, by using advanced analytics techniques. IIoT technology has a broader range of applications

---

N. Kumar (✉) · B. Sharma · S. Narang
School of Engineering and Technology, Chitkara University, Chitkara University, Kallujhanda, Himachal Pradesh, India
e-mail: nagesh.kumar@chitkara.edu.in

P. K. Singh et al. (eds.), *Proceedings of Third International Conference on Computing, Communications, and Cyber-Security*, Lecture Notes in Networks and Systems 421,
https://doi.org/10.1007/978-981-19-1142-2_9

including healthcare, defense, finance, agriculture, manufacturing, retail, and advertising. In context to machine-to-machine communication, IIoT provides the advantage of semi-independent machine operations or operations with minimal human intervention [1]. These IIoT-based systems will be able to behave intelligently and can change course of action based on feedback established within framework.

Today's manufacturing industry is moving toward digital and intelligent operations, thinking about the customized production according to end-user needs. In this context, development of new computational techniques, smart devices, and other electronic technologies resulted in a new revolution called as Industry 4.0 [2]. Manufacturing process is digitally transformed in Industry 4.0, also called as 4th industrial revolution. The basis for Industry 4.0 is the cyber physical systems like smart machines. These machines are using modern control and software systems connected via Internet. Internetworking of machines and processes intelligently is possible due to IIoT and advancements in information and communication technologies [2, 3]. The process of manufacturing and actual product is networked, and communication technologies enable novel production, good values, and real-time optimization. Cyber physical systems able to fulfill needs for smart factories like remote monitoring or track and trace [3–7]. Industry 4.0 is basically intended to transform conventional machines to interactive machine. These interactive machines can communicate the issues related to health, production, and upgradation, so that there will be overall performance improvement. In simple words, Industry 4.0 is an open, smart platform for manufacturing, and generating industrial networked information [8–11] (Fig. 1).

With Industry 4.0, as artificial intelligence and big data analytics techniques are evolving, and there is new term coined, i.e., Industry 5.0 [12, 13]. Industry 5.0 can be thought of as intelligent machines or robots coworking with human beings. This vision of human–robot coworking [13, 14], humans will focus on innovations and creativity and rest of the work will be performed by robots. This will increase performance and industry profit as well. In another way, Industry 5.0 can be thought of

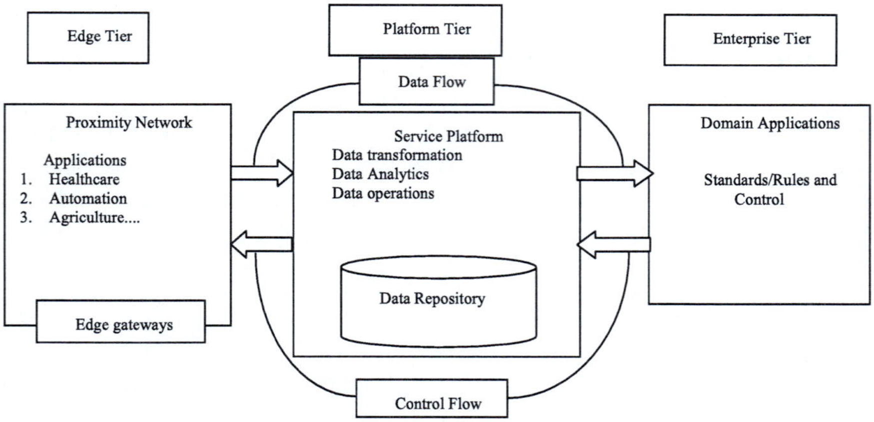

Fig. 1 Three-tier industrial IoT architecture [12]

Table 1 Industry 4.0 and Industry 5.0 perspectives

	Motivation	Energy sources	Technologies	Area(s) of research
Smart manufacturing (Industry 4.0)	Increasing production	Electricity, fossil fuel, renewable power sources	IoT, big data, cloud technologies	Improvement of methods of production, business analytics, organizational research
Human–Robot coworking (Industry 5.0)	Improving human lives (smart social life)	Electricity, renewable power sources	IoT, robotics and artificial intelligence, big data, cloud technologies	smart cities and environments, improvement of methods of production, business analytics, organizational research
Bioeconomy (Industry 5.0)	Sustainable development	Electricity, renewable power sources	Renewable resources, sustainable environments, bionics, and agriculture	Waste engineering, agriculture, healthcare, improvement of methods of production, business analytics

as a bioeconomy [14]. In this, the focus is on better utilization of biological resources for industrial use. This intends to balance the ecology, industry, and economy for better lifestyle of human beings. Table 1 provides overview of perspectives followed in Industry 4.0 and 5.0.

Internet of things can be thought of as large number of devices internetworked together to perform application specific tasks. If we want these devices to perform well, the communication techniques play an important role. IoT network protocols are designed to connect medium to high-power devices over the network. IoT network protocols allow data communication within the scope of the network. In this paper, IoT communication protocols and the specific protocol design requirements for Industry 4.0 will be discussed. The paper presents a detailed literature review on machine-to-machine communication protocols with their pros and cons. Few of popular protocols discussed over here are hypertext transfer protocol (HTTP), message queuing telemetry transport (MQTT), Zigbee, LoRaWAN, 6LowPAN, and advanced message queuing protocol (AMQP). After completing the literature review, an analysis will be performed to fetch out some challenges and opportunities in context to development of new protocols. Also, there will be a discussion carried out about the opportunities and directions toward using these technologies in Industry 5.0.

2 Literature Review Process

The research or review articles considered in this paper are mostly from last five years, and the old papers are only considered when it seems to be important to include. Before starting this review article, the research questions are formulated, and the research libraries are searched accordingly.

2.1 *Research Questions*

The focus in this article is to answer questions associated with communication technologies in IIoT, Industry 4.0, and Industry 5.0. Major research questions are

RQ1: What type of communication technologies and standards are available for Industry 4.0?
RQ2: How the performance of communication protocols is affected by constraints in IoT devices?
RQ3: What are the challenges, opportunities, and future directions toward Industry 5.0 in context to communication technologies?

2.2 *Information Source*

Due to wide scope, it was advised that various research libraries must be searched to find out research articles on Industry 4.0 and Industry 5.0. According to popularity, the focus was to consider IEEE Xplore, Science Direct, ACM digital library, Google scholar, and Springer Nature to formulate the state of the art in use of communication protocols for IIoT, Industry 4.0, and Industry 5.0. The searching keywords were related to industrial revolution, IoT, industrial IoT, applications of IIoT, Industry 4.0, Industry 5.0, artificial intelligence in manufacturing, routing protocols, fog computing in industries, communication technologies for IoT, Industry 4.0 and 5.0, and other related words.

3 Literature Survey

Due to advancements in information and communication technologies, connecting large number of IoT devices became easy in last few years. IoT technology mostly focused on device-to-device communication to fulfill the requirements of Industry 4.0 and Industry 5.0 [15–18]. As thousands of heterogeneous devices are connected, there is a production of large amount of data. Due to this, there is a requirement of good IoT architecture to store and process this many data. Cloud computing architectures are mainly focused to deal with it and new paradigm, fog computing focused to provide optimized and scaled up communication architecture for IoT [19–21].

In computer networking, basic communication protocols can be differentiated based on device interactions. Device interactions can be of two types, i.e., request-reply model and publish-subscribe model [22, 23]. The most commonly used model is request-reply model, in which a client requests server for data and server after processing request respond with the same. This type of model is generally centralized, and server will be responsible for most of the data communication. For IoT communication, the most common examples of request-reply communication models are representational state transfer HTTP (REST HTTP) and constrained application protocol (CoAP). On the contrary, publish-subscribe communication model is a distributed, loosely coupled, asynchronous type, which focus on the requirements of data for a specific end-user. Several research has been carried out on publish-subscribe to handle problems like data security, integrity, and availability. MQTT, AMQP, and data distribution service (DDS) are common examples of communication protocols which are applying publish-subscribe models.

Besides utilizing conventional communication protocols, some of new standards and protocols are also proposed for IoTs. These standards and protocols are mainly focused to various applications of IoT in Industry 4.0 and Industry 5.0. As most of the manufacturing applications of IIoT require short-range communication, IEEE standard for wireless personal area network, i.e., 802.15 is of higher impact since 2003. The communication range specified in 802.15 is in between 1 and 100 m. Bluetooth compatible standard 802.15.1 is a point-to-point communication technology, while ZigBee compatible standard 802.15.4 is a star topology-based peer-to-peer communication technology. These both technologies support unicast and broadcast communications. Near-field communication (NFC) is another short-range communication protocol applied in many applications of IoT which can communicate within few centimeters. In addition to these protocols, 6LoWPAN communication protocol, based on IPv6, supports power constrained IoT devices by doing header compression to save energy [24]. These protocols are most widely used in IoT devices and are very common in most of the applications. Apart from short-range communication, IoT devices are also utilizing cellular networks for long-range communication. Although, short-range communication technologies are more effective in Industry 4.0 and Industry 5.0. The researchers around the globe try to improvise commonly used protocols according to industry application demands and introducing more efficient protocols as alternatives.

Kim et al. [24] proposed improvised ZigBee protocol named as UPnP-ZigBee communication architecture which mirrored ZigBee topology by utilizing proxies. The protocol utilized UPnP gateways to improve communication capabilities of IIoT networks. Song et al. [25] proposed a communication system by taking smart home application into consideration. Authors worked on energy efficiency, security, and privacy preservation. In their proposed approach, they have used symmetric key encryption and message authentication codes (MAC) to ensure data integrity and data authentication. The comparison of results showed better performance in terms of complexity, memory consumption, and communication cost. The application scenario considered by authors was smart home networks and may be applicable to other industry applications of IoT.

As most of the IoT devices are energy constrained, Li et al. [26] proposed an improvement to ZigBee protocol named as passive-ZigBee. This proposed protocol offers low-duty cycle and ultra-low-energy consumption. The authors have developed new gateway that can generate combined ZigBee and Wi-Fi signals. Authors have claimed that passive-ZigBee can reduce power consumption by a factor of 1440 times to conventional ZigBee. Authors have developed a hardware-based demonstration and tested performance of passive-ZigBee-based devices on ZigBee, Wi-Fi, and FPGA platforms. As per the authors, the experimental evaluation showed the energy consumption around 25 μW up to 55 m. Author suggested that the applications of proposed protocol are intended to IIoT and smart factories.

Currently, as IoT networks are growing day by day, the researchers are motivated to explore more opportunities in energy efficiency, security, and high-communication throughput. Working in this direction, Li et al. [27] have proposed a new communication protocol to get high throughput named as Chiron. Chiron focused to support physical layer component in heterogeneous IoT network including ZigBee and Wi-Fi. Authors ensured concurrent transmissions of Wi-Fi and ZigBee packets in Chiron, which resulted in utilizing spectrum 16 times better than conventional gateways. For sharing spectrum in heterogeneous IoT, Liu et al. [28] have proposed a new multichannel framework for IoT applications. Authors proposed the framework to share spectrum with 5G devices to ensure simultaneous 5G and IoT communication. Another contribution of authors includes an optimization algorithm using Lagrange dual composition. After simulation, it was observed that proposed IoT framework has improved the throughput of 5G and IIoT communication, when used simultaneously in Industry 5.0.

Considering IIoT as focus Petrenko et al. [29] raised the issues of creating single integrated internetwork platform for data generators and data centers. The authors have proposed a solution for IIoT devices' interactions. The study was focused to Russian Federation only but may be implemented in many IIoT applications. Piyare et al. [30] addressed the disadvantages of LoRAWAN technology, i.e., latency, low control, and higher packet collisions. Authors have proposed an energy efficient IoT architecture by utilizing time-division multiple access (TDMA) protocol. The authors claimed in the paper that they have merged the abilities of short-range radios, asynchronous communication, and long-range connectivity of LoRAWAN. After simulation, the results depicted that there were no packet collisions, and energy consumption is below 46 mJ with a latency of few milliseconds.

Due to energy constraints in IoT devices, there is always a tradeoff between energy efficiency and other performance parameters like latency, packet delivery ratio, and data fragmentation. Considering data fragmentation as major issue, Khaled and Helal [31] have proposed a framework called as Atlas IoT communication framework. The framework addressed the problem of heterogeneous IoT networks and introduces translator for communication protocols. The authors believed that Atlas framework can provide flawless communication between different devices in an IoT. Authors have discussed about the use of framework in IIoT and recognized Atlas as a major step toward smart factories and Industry 4.0. Another communication technique, i.e., visible light communication (VLC) was discussed by Liu et al. [32] with respect to industrial environments. Authors have addressed the problem of power line connection and battery replacement in IoT devices and proposed an energy harvesting solution. As per the authors, VLC can be powered using EM interference signals by introducing advanced signal processing systems with new energy efficient technique. The simulation results depicted good results and promising applications of VLC in IIoT.

Drone technologies are emerging as unmanned aerial vehicles (UAVs), which require seamless wireless communication to be utilized in most of the applications. Keeping these points in view, Zhang et al. [33] proposed time slot computation-based strategies for UAVs and IoTs. Terminal devices are taken into consideration for assigning time slots to complete latency-critical tasks. Authors have proposed three different computation aspects for terminal devices in which devices can do self-computation, assign tasks to UAVs, and third was assigning task bits to access points. In paper, there was an optimization algorithm proposed to reduce energy consumption. The analysis was done based on numerical results only which showed good results as compared to benchmarked cases.

Debroy et al. [34] have proposed another protocol for communication in IoT devices named as SpEED-IoT (spectrum aware energy-efficient multi-hop multi-channel routing scheme for D2D communication in IoT mesh network). The authors have introduced a dedicated spectrum sensor for generating the radio environment map (REM). REMs are then utilized to calculate best path, channel allocation, and minimum energy consumption. According to authors, SpEED-IoT preserves energy, protect licenses, converge fast enough, and improve data rate. The paper by Rathee et al. [35] illustrated the use case of IIoT and proposed a communication method for wireless technology using blockchain. Blockchain was used to improve security, privacy, and transparency in wireless sensor systems and IoT. Proposed framework was simulated against various attack metrics and analyzed for authenticity in different setups. Simulation results showed good results in terms of energy efficiency and security analysis, but end-to-end delay was high. Zezulka et al. [36] reviewed open platform communication-unified architecture (OPC-UA) protocol for Industry 4.0. Authors have divided the article into two parts, first part has explained and analyzed OPC-UA. In the direction toward machine-to-machine communication, OPC-UA protocol was developed in Europe as a standard by OPC foundation [37]. For communication in Industry 4.0, OPC-UA is now adopted by various other countries like USA and other developed countries.

Researchers are more focused toward introducing IoT in manufacturing technologies and Industry 4.0. Most of the research articles discussed here are focused to Industry 4.0. Although there are other research articles [38–51] also, which focus on different applications like smart agriculture, smart grid, defense technology, and other promising areas. A discussion and analysis are presented in upcoming section to find out the promising applications of these protocols in Industry 4.0 and Industry 5.0.

Discussion and Analysis

The analysis and comparative performance of IoT communication protocols were published by many authors e.g. [52] in last few years. Most of the article are focused for open-source available standards and protocols in IoT communication. Few review articles like [52, 53] discussed IoT communication protocols for specific application like smart agriculture, smart grid, and smart homes. In this article, industrial IoT has been focused, and the protocols and published articles which are focused to IIoT were discussed in literature survey. This section carries out a discussion and analysis on the basis literature survey and specify the applicability of these protocols in manufacturing process. In the 1st step, Table 2 presents the analysis of various IIoT communication protocols discussed above, in terms of key characteristics. In the second step, a ranking of IoT protocols is completed and recommendation according to key performance factor is done. Table 3 presents the recommendation of discussed protocols with key performance metrics. The suggestions given over here depend on application scenario too, like some applications may require better bandwidth utilization while in some other applications, energy efficiency is an important aspect.

4 Open Research Challenges and Opportunities

Although most of the issues and challenges researchers have addressed for Industry 4.0, but still, most of the businesses in many countries are not adopting IIoT. Nowadays, we are moving toward Industry 5.0 and inclusion of artificial intelligence has been started already. By keeping this fact in view, major open research challenges for development of communication protocols are listed and discussed in this section.

Heterogeneity: IoT devices to be distributed to smart manufacturing in an organization need not be of homogeneous in nature. Heterogeneous devices may have different type of protocols to follow. This impose a great challenge toward data communication as data formats and data rates will be different. Hence, whenever a communication protocol needs to be developed, protocol translation must be considered as a major factor.

Interoperability: When two or more devices communicate with each other and perform certain common tasks, then there will be the requirement of interoperability. Interoperability depends on types of devices, platform of devices and networks, and protocols a particular device are following. Due to heterogeneous environment and devices interoperability in communication protocols becomes a challenge and should be addresses at the time of protocol development.

Energy efficiency: This issue remains the focus of researchers from the time of sensor system development. Whenever a researcher tries to develop a new protocol for wireless sensor networks or IoT, energy efficiency is always one of the challenges. Most of the researchers discussed in literature survey here have focused on this issue and tried to reduce power consumption.

Table 2 Comparative analysis based on key features

Communication protocol	Type (request-reply/publish-subscribe/Both)	Security	Bandwidth utilization	QoS support	Service type
REST HTTP [15]	Request-reply	Secure socket layer (SSL) in HTTPS	Fair	No	TCP
CoAP [15]	Both	Optional datagram transport layer security (DTLS)	Fair	Yes	UDP
MQTT [16]	Publish-subscribe	Security is optional (TLS/SSL)	Fair	Yes	TCP
AMQP [17]	Both	TLS/SSL	Fair	Yes	TCP
DDS [17]	Request-reply	TLS/DTLS	Fair	Yes	TCP/UDP
NFC [18]	Request-reply	Secure element identifier (SEID)	Good	Yes	TCP
6LowPAN [18]	Request-reply	End-to-end (E2E) security	Good	Yes	TCP
ZigBee [18]	Publish-subscribe	Encryption using network key	Good	Yes	UDP
UPnP-ZigBee [24]	Both	Encryption using network key	Good	Yes	UDP
Song et al. [25]	Request-reply	End-to-end (E2E) security	Low	No	TCP
Passive-ZigBee [26]	Request-reply	Encryption using network key	Good	Yes	UDP
Chiron [27]	Publish-subscribe	TLS/DTLS	Good	Yes	UDP

(continued)

Table 2 (continued)

Communication protocol	Type (request-reply/publish-subscribe/Both)	Security	Bandwidth utilization	QoS support	Service type
Liu et al. [28]	Request-reply	End-to-end (E2E) security	Low	Yes	TCP/UDP
Petrenko et al. [29]	Publish-subscribe	Not addressed	Low	Yes	TCP/UDP
Piyare et al. [30]	Request-reply	End-to-end (E2E) security	Fair	Yes	TCP
Atlas IoT [31]	Request-reply	End-to-end (E2E) security	Fair	Yes	TCP
Zhang et al. [33]	Both	TLS/DTLS	Fair	No	TCP
SpEED-IoT [34]	Both	End-to-end (E2E) security	Good	Yes	TCP
Rathee et al. [35]	Request-reply	Blockchain	Low	No	UDP

Table 3 Recommendations based on key performance indicators

Performance indicator	Bandwidth utilization	Energy efficiency	Throughput	Reliability	Preference in manufacturing applications of IoT
Most suitable protocol	CoAP, MQTT, AMQP NFC	CoAP, MQTT, AMQP NFC	MQTT, DDS, CoAP, passive-ZigBee, UPnP-ZigBee	MQTT, AMQP, Passive-ZigBee, 6LowPAN, NFC	Passive-ZigBee, 6LowPAN, NFC, Song et al., Petrenko et al.
Least suitable protocol	Song et al., Liu et al., Rathee et al.	REST HTTP, Atlas, Rathee et al.	AMQP, Zhang et al., Liu et al.	Piyare et al., SpEED-IoT	Rathee et al., SpEED-IoT

Security: As IoT devices generates lots of data, security will be another challenge and can cost organization's important assets. Data required to be secured in storage as well as in communication. Insecure communication is the biggest challenge in IIoT and required to be addressed at the time of protocol design. The researchers need to develop secure mechanisms of communication within an organization.

Scalability: To work smoothly, IIoT protocols need to scalable because of large number of devices connected. As number of devices keep on changing time-to-time, there will an issue of data variability, which needs to be addressed at the time of protocol development.

Reliability: Reliability of IoT in an organization depends mostly on the communication between the devices. There can be huge losses in terms of finance and human lives if communication gap is there between IIoT devices. The communication protocols must ensure fault tolerant capabilities to handle failures. Failures in IIoT can be of any type like device failure and failure of links. There are different solutions proposed for this problem, and some promising and popular ones are caching and dynamic spectrum access capabilities.

Quality ofService: QoS is the biggest challenge for any type of communication protocol, whether it a conventional one or specifically designed for IoT. Packet loss, delays, bandwidth, resource allocation, and packet loss are major QoS factors. To optimize all these performance factors is the biggest challenge and needs to be addressed according to application requirements. In IIoT manufacturing applications, these factors very crucial, and researchers must pay attention for QoS design in a communication protocol.

Despite these challenges, there may be other hurdles also in developing a communication protocol like legal challenges, replacement of old machinery, or updating machines for modern manufacturing, etc. Researchers around the globe are working for the improvement of communication protocols for future technologies. Intelligent

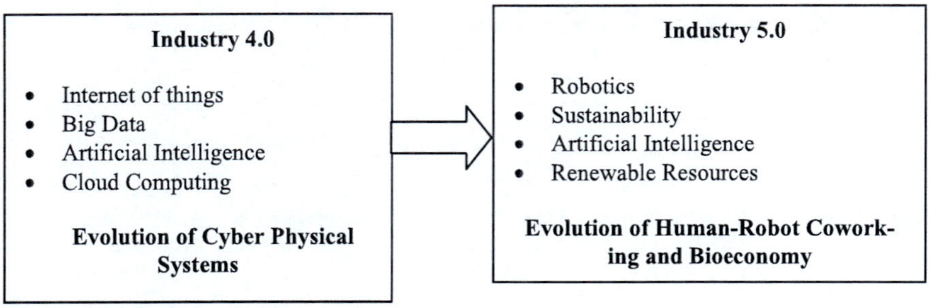

Fig. 2 Moving from Industry 4.0 to Industry 5.0

communication protocols may be the biggest revolution for Industry 5.0 as we are thinking of human–robot coworking.

5 Directions toward Industry 5.0

Industry 5.0 concepts are still emerging and require a lot of research on various aspects. Human–robot coworking may turn out the biggest revolution in future and require very precise and reliable communication techniques. Existing communication protocols like ZigBee, MQTT, Bluetooth, and other protocols discussed in this paper may not cope up with the intelligent robots. The researchers must pay attention toward these technical advancements and develop efficient protocols to improve organization and production environments. Robots working with human will make a change to human lives and replace human workers in few years. Other aspect of industry 5.0, i.e., bioeconomy development involves sustainable environments and of course better human lives. In this paper, communication protocols were discussed that are proposed by various researchers to tackle the issues raised during manufacturing process. Manufacturing process and methods are going to be changed if Industry 5.0 is evolved in few years. Communication protocols are going to play major role in improving the same and require lot of attention. The challenges and opportunities listed over here are important aspects to be focused on by a researcher developing new communication protocol for Industry 5.0 (Fig. 2).

6 Conclusion

In this paper, communication protocols have been discussed, and certain research questions have been answered which were formulated before starting of this review article. In context to RQ1, multiple communication protocols were discussed, and in Table 2, the types of those protocols are listed out. Identified types are request-reply and publish-subscribe type of communication protocols. RQ2 was about looking at the performance of communication protocols survey in this article. Tables 2 and 3 provide analysis and performance comparison based on the results given by researchers in their respective articles. Standardized protocols need an improvement as per the analysis, and some researchers have done this in context to various manufacturing applications. The last question RQ3 was answered by listing out various challenges and giving views on how communication protocols are to be developed and improved if we must move toward industry 5.0. Overall, improvement to any networking technology mostly depends on the communication system available for it. Certain characteristics need to be addressed for manufacturing applications like range, fault tolerance, reliability, capacity, scalability, and QoS parameters like delay, energy efficiency, mobility, and resource utilization. The researchers working on communication protocols for new revolution Industry 5.0 may find this paper helpful and address various challenges.

References

1. Tayeb, S., Latifi, S., & Kim, Y. (2017, January). A survey on IoT communication and computation frameworks: An industrial perspective. In *2017 IEEE 7th annual Computing and Communication Workshop and Conference (CCWC)* (pp. 1–6). IEEE.
2. Marcon, P., Zezulka, F., Vesely, I., Szabo, Z., Roubal, Z., Sajdl, O., Gescheidtova, E., & Dohnal, P. (2017, May). Communication technology for industry 4.0. In *2017 Progress in Electromagnetics Research Symposium-Spring (PIERS)* (pp. 1694–1697). IEEE.
3. Hedi, I., Speh, I., & Sarabok, A. (2017, May). IoT network protocols comparison for the purpose of IoT constrained networks. In *2017 40th International Convention on Information and Communication Technology, Electronics and Microelectronics (MIPRO)* (pp. 501–505). IEEE.
4. Mukherjee, S., & Biswas, G. P. (2018). Networking for IoT and applications using existing communication technology. *Egyptian Informatics Journal, 19*(2), 107–127.
5. Anusha, M., Babu, E. S., Reddy, L. S. M., Krishna, A. V., & Bhagyasree, B. (2017). Performance analysis of data protocols of internet of things: A qualitative review. *International Journal of Pure and Applied Mathematics, 115*(6), 37–47.
6. Bhoyar, P., Sahare, P., Dhok, S. B., & Deshmukh, R. B. (2019). Communication technologies and security challenges for internet of things: A comprehensive review. *AEU-International Journal of Electronics and Communications, 99*, 81–99.
7. Vaidya, S., Ambad, P., & Bhosle, S. (2018). Industry 4.0–a glimpse. *Procedia Manufacturing, 20*, 233–238.
8. Souri, A., Hussien, A., Hoseyninezhad, M., & Norouzi, M. (2019). A systematic review of IoT communication strategies for an efficient smart environment. *Transactions on Emerging Telecommunications Technologies*, e3736.
9. Dhanda, S. S., Singh, B., & Jindal, P. (2019). Wireless technologies in IoT: Research challenges. In *Engineering vibration, communication and information processing* (pp. 229–239). Springer, Singapore.
10. Kozma, D., Soás, G., Ficzere, D., & Varga, P. (2019, October). Communication challenges and solutions between heterogeneous Industrial IoT systems. In *2019 15th International Conference on Network and Service Management (CNSM)* (pp. 1–6). IEEE.
11. Wang, W., Capitaneanu, S. L., Marinca, D., & Lohan, E. S. (2019). Comparative analysis of channel models for industrial IoT wireless communication. *IEEE Access, 7*, 91627–91640.

12. Demir, K. A., Döven, G., & Sezen, B. (2019). Industry 5.0 and human-robot co-working. *Procedia Computer Science, 158*, 688–695.
13. Nahavandi, S. (2019). Industry 5.0—A human-centric solution. *Sustainability, 11*(16), 4371.
14. Martynov, V. V., Shavaleeva, D. N., & Zaytseva, A. A. (2019, September). information technology as the basis for transformation into a digital society and Industry 5.0. In *2019 International Conference on Quality Management, Transport and Information Security, Information Technologies (IT&QM&IS)* (pp. 539–543). IEEE.
15. Moraes, T., Nogueira, B., Lira, V., & Tavares, E. (2019, October). Performance comparison of IoT communication protocols. In *2019 IEEE International Conference on Systems, Man and Cybernetics (SMC)* (pp. 3249–3254). IEEE.
16. Glaroudis, D., Iossifides, A., & Chatzimisios, P. (2020). Survey, comparison and research challenges of IoT application protocols for smart farming. *Computer Networks, 168*, 107037.
17. Butun, I. (2020). *Industrial IoT*. Springer International Publishing.
18. Bryndin, E. (2020). Formation and management of Industry 5.0 by systems with artificial intelligence and technological singularity. *American Journal of Mechanical and Industrial Engineering, 5*(2), 24–30.
19. Malik, P. K., Sharma, R., Singh, R., Gehlot, A., Satapathy, S. C., Alnumay, W. S., Pelusi, D., Ghosh, U., & Nayak, J. (2020). Industrial internet of things and its applications in Industry 4.0: State of the art. *Computer Communications, 166*, 125–139.
20. Khan, W. Z., Rehman, M. H., Zangoti, H. M., Afzal, M. K., Armi, N., & Salah, K. (2020). Industrial internet of things: Recent advances, enabling technologies and open challenges. *Computers & Electrical Engineering, 81*, 106522.
21. Salimova, T., Vukovic, N., & Guskova, N. (2020). Towards sustainability through Industry 4.0 and Society 5.0. *International Review, 3–4*, 48–54.
22. Alvarez-Aros, E. L., & Bernal-Torres, C. A. (2021). Technological competitiveness and emerging technologies in industry 4.0 and industry 5.0. *Anais da Academia Brasileira de Ciencias, 93.*
23. Sari, A., Lekidis, A., & Butun, I. (2020). Industrial networks and IIoT: Now and future trends. In *Industrial IoT* (pp. 3–55). Springer, Cham.
24. Kim, S. H., Kang, J. S., Park, H. S., Kim, D., & Kim, Y. J. (2009). UPnP-ZigBee internetworking architecture mirroring a multi-hop ZigBee network topology. *IEEE Transactions on Consumer Electronics, 55*(3), 1286–1294.
25. Song, T., Li, R., Mei, B., Yu, J., Xing, X., & Cheng, X. (2017). A privacy preserving communication protocol for IoT applications in smart homes. *IEEE Internet of Things Journal, 4*(6), 1844–1852.
26. Li, Y., Chi, Z., Liu, X., & Zhu, T. (2018, November). Passive-zigbee: Enabling zigbee communication in iot networks with 1000x+ less power consumption. In *Proceedings of the 16th ACM Conference on Embedded Networked Sensor Systems* (pp. 159–171).
27. Li, Y., Chi, Z., Liu, X., & Zhu, T. (2018, June). Chiron: Concurrent high throughput communication for IoT devices. In *Proceedings of the 16th Annual International Conference on Mobile Systems, Applications, and Services* (pp. 204–216).
28. Liu, X., Jia, M., Zhang, X., & Lu, W. (2018). A novel multichannel Internet of things based on dynamic spectrum sharing in 5G communication. *IEEE Internet of Things Journal, 6*(4), 5962–5970.
29. Petrenko, A. S., Petrenko, S. A., Makoveichuk, K. A., & Chetyrbok, P. V. (2018, March). The IIoT/IoT device control model based on narrow-band IoT (NB-IoT). In *2018 IEEE Conference of Russian Young Researchers in Electrical and Electronic Engineering (EIConRus)* (pp. 950–953). IEEE.
30. Piyare, R., Murphy, A. L., Magno, M., & Benini, L. (2018). On-demand LoRa: Asynchronous TDMA for energy efficient and low latency communication in IoT. *Sensors, 18*(11), 3718.
31. Khaled, A. E., & Helal, S. (2019). Interoperable communication framework for bridging RESTful and topic-based communication in IoT. *Future Generation Computer Systems, 92*, 628–643.

32. Liu, X., Wei, X., Guo, L., Liu, Y., Song, Q., & Jamalipour, A. (2019). Turning the signal interference into benefits: Towards indoor self-powered visible light communication for IoT devices in industrial radio-hostile environments. *IEEE Access, 7*, 24978–24989.
33. Zhang, T., Xu, Y., Loo, J., Yang, D., & Xiao, L. (2019). Joint computation and communication design for UAV-assisted mobile edge computing in IoT. *IEEE Transactions on Industrial Informatics, 16*(8), 5505–5516.
34. Debroy, S., Samanta, P., Bashir, A., & Chatterjee, M. (2019). SpEED-IoT: Spectrum aware energy efficient routing for device-to-device IoT communication. *Future Generation Computer Systems, 93*, 833–848.
35. Rathee, G., Balasaraswathi, M., Chandran, K. P., Gupta, S. D., & Boopathi, C. S. (2021). A secure IoT sensors communication in Industry 4.0 using blockchain technology. *Journal of Ambient Intelligence and Humanized Computing, 12*(1), 533–545.
36. Zezulka, F., Marcon, P., Bradac, Z., Arm, J., Benesl, T., & Vesely, I. (2018). Communication systems for Industry 4.0 and the IIoT. *IFAC-PapersOnLine, 51*(6), 150–155.
37. Zhilenkov, A. A., Gilyazov, D. D., Matveev, I. I., & Krishtal, Y. V. (2017, February). Power line communication in IoT-systems. In *2017 IEEE Conference of Russian Young Researchers in Electrical and Electronic Engineering (EIConRus)* (pp. 242–245). IEEE.
38. Rahimi, H., Zibaeenejad, A., & Safavi, A. A. (2018, November). A novel IoT architecture based on 5G-IoT and next generation technologies. In *2018 IEEE 9th Annual Information Technology, Electronics and Mobile Communication Conference (IEMCON)* (pp. 81–88). IEEE.
39. Alam, T. (2018). A reliable communication framework and its use in internet of things (IoT). CSEIT1835111, Received, 10, 450–456.
40. Lavric, A., & Petrariu, A. I. (2018, May). LoRaWAN communication protocol: The new era of IoT. In *2018 International Conference on Development and Application Systems (DAS)* (pp. 74–77). IEEE.
41. Cruz-Piris, L., Rivera, D., Marsa-Maestre, I., De La Hoz, E., & Velasco, J. R. (2018). Access control mechanism for IoT environments based on modelling communication procedures as resources. *Sensors, 18*(3), 917.
42. Durand, A., Gremaud, P., Pasquier, J., & Gerber, U. (2019, October). Trusted lightweight communication for IoT systems using hardware security. In *Proceedings of the 9th International Conference on the Internet of Things* (pp. 1–4).
43. Aceto, G., Persico, V., & Pescape, A. (2019). A survey on information and communication technologies for industry 4.0: State-of-the-art, taxonomies, perspectives, and challenges. *IEEE Communications Surveys & Tutorials, 21*(4), 3467–3501.
44. Hayajneh, A., Zaidi, S. A. R., Hafeez, M., McLernon, D., & Win, M. (2019, May). Coverage analysis of drone-assisted backscatter communication for IoT sensor network. In *2019 15th International Conference on Distributed Computing in Sensor Systems (DCOSS)* (pp. 584–590). IEEE.
45. Salim, M. M., Wang, D., Elsayed, H. A. E. A., Liu, Y., & Abd Elaziz, M. (2020). Joint optimization of energy-harvesting-powered two-way relaying D2D communication for IoT: a rate-energy efficiency tradeoff. *IEEE Internet of Things Journal, 7*(12), 11735–11752.
46. Gorodetsky, V., Larukchin, V., & Skobelev, P. (2019, October). Conceptual model of digital platform for enterprises of Industry 5.0. In *International Symposium on Intelligent and Distributed Computing* (pp. 35–40). Springer, Cham.
47. Liu, Z., Liu, J., Zeng, Y., & Ma, J. (2020). Covert wireless communication in IoT network: From AWGN channel to THz band. *IEEE Internet of Things Journal, 7*(4), 3378–3388.
48. Maddikunta, P. K. R., Gadekallu, T. R., Kaluri, R., Srivastava, G., Parizi, R. M., & Khan, M. S. (2020). Green communication in IoT networks using a hybrid optimization algorithm. *Computer Communications, 159*, 97–107.
49. Longo, F., Padovano, A., & Umbrello, S. (2020). Value-oriented and ethical technology engineering in Industry 5.0: a human-centric perspective for the design of the factory of the future. *Applied Sciences, 10*(12), 4182.
50. Kumar, V. V., Devi, M., Raja, P. V., Kanmani, P., Priya, V., Sudhakar, S., & Sujatha, K. (2020). Design of peer-to-peer protocol with sensible and secure IoT communication for future internet architecture. *Microprocessors and Microsystems, 78*, 103216.

51. Çorak, B. H., Okay, F. Y., Güzel, M., Murt, Ş., & Ozdemir, S. (2018, June). Comparative analysis of IoT communication protocols. In *2018 International Symposium on Networks, Computers and Communications (ISNCC)* (pp. 1–6). IEEE.
52. Al-Sarawi, S., Anbar, M., Alieyan, K., & Alzubaidi, M. (2017, May). Internet of Things (IoT) communication protocols. In *2017 8th International conference on information technology (ICIT)* (pp. 685–690). IEEE.
53. Dizdarevic, J., Carpio, F., Jukan, A., & Masip-Bruin, X. (2019). A survey of communication protocols for internet of things and related challenges of fog and cloud computing integration. *ACM Computing Surveys (CSUR), 51*(6), 1–29.

Explainable Artificial Intelligence (XAI): Connecting Artificial Decision-Making and Human Trust in Autonomous Vehicles

A. V. Shreyas Madhav and Amit Kumar Tyagi

Abstract Automated navigation technology has established itself as an integral facet of intelligent transportation and smart city systems. Several international technological organizations have realized the immense potential of autonomous vehicular systems and are currently working towards their complete development for mainstream application. From deep learning algorithms for road object detection to intrusion detection systems for CAN bus monitoring, the functioning of a self-driving vehicle is powered by the simultaneous working of multiple inner vehicle module systems that perform proper vehicle navigation while ensuring the physical safety and digital privacy of the user. Transparency of the vehicle's thought processes can assure the user of its credibility and reliability. This paper introduces explainable artificial intelligence, which aims to converge the decision-making processes of Autonomous Vehicle Systems (AVS). Here, the domain of Explainable AI (XAI) provides clear insights into the role of explainable AI in autonomous vehicles and increase human trust for AI based solutions in the same sector. This paper exhibits the trajectories of transportation advancements and the current scenario of the industry. A comparative quantitative and qualitative analysis is performed to compare the simulations of XAI and vehicular smart systems to showcase the significant developments achieved. Visual explanatory methods and an intrusion detection classifier were created as part of this research and achieved significant results over extant works.

Keywords Intelligent transportation systems · Autonomous navigation · Explainable artificial intelligence (XAI) · Smart vehicle vision · Vehicle security · Intrusion detection

A. V. S. Madhav (✉) · A. K. Tyagi
School of Computer Science and Engineering, Vellore Institute of Technology, Chennai, Tamil Nadu 600127, India
e-mail: shreyas.madhav@gmail.com

A. K. Tyagi
Centre for Advanced Data Science, Vellore Institute of Technology, Chennai, Tamil Nadu 600127, India

P. K. Singh et al. (eds.), *Proceedings of Third International Conference on Computing, Communications, and Cyber-Security*, Lecture Notes in Networks and Systems 421,
https://doi.org/10.1007/978-981-19-1142-2_10

1 Introduction

Smart urban mobility opportunities are currently on a rapid rise due to the recent advances of autonomous driving systems. The autonomy of vehicles and transportation systems is now becoming an integral part of every smart city's propaganda and is set to serve as a more efficient replacement to traditional transportation infrastructure. Urban administrators, policymakers, politicians and legislators are unnerved by the increase in machine autonomy due to the different disruptions that they may inflict upon existing policies and urban strategies. The evolution of artificial intelligence-based systems is now leading towards a point where humans are expected to accept the decisions of the system as it possesses higher insights and the computational power to produce accurate results in real time. The entry of AI systems into critical and essential domains such as finance, defence, education, etc., has now instigated people to look towards a solution to the completely opaque decision-making process of the system. A typical AI system provides very little explanation or inferences, which is understandable by the layman, on how a particular decision was. Hence, the concepts of explainability and interpretability have now established themselves imperative for producing mainstream AI-powered applications. The extent to which the explanation provided by a system can be understood by a person is defined to be its interpretability. The biggest challenge is to produce a viable explainability solution to different neural networks to encourage transparency in machine decision-making to make humans understand and trust the conclusions provided by the learning networks. This generates the ability for the human to rationalize with the decision made by the system. Explainable AI (XAI) is interdomain of AI [1] created for the aforementioned purpose. Self-driving vehicles perform several computations ranging from vision recognition to network intrusion detection, and the ability to explain these computational decisions made by the vehicle will propagate human trust from an ethical, moral and legal standpoint.

Hence, the organization of the work is as follows: Sect. 3 of the article focuses on the concepts behind autonomous and semi-autonomous vehicles, followed by Sect. 4 which dives into the details of vehicle autonomy and related processes. Section 5 provides the principles behind explainable AI, and Sect. 6 describes how autonomous vehicles use their vision to make decisions. The focus is now shifted towards the ethics and morals surrounding self-driving cars followed by the exploration of methods to establish trust in connected vehicles and VANETS in Sect. 7. The simulation results of vision-based explanations and novel intrusion detection algorithm are provided in Sect. 8. Section 9 discusses and concludes this work in brief (with including some interesting future work).

2 Autonomous and Semi-Autonomous Vehicles

The intertwining research advances in embedded systems, wireless communication, vision networks, data analytics, sensors and ad hoc networks have led to the widespread emergence of autonomous vehicle and intelligent transportation systems [2]. Emerging in the 1920s, the origin of vehicle anonymity is rooted in the remote-controlled phantom autos which were showcased to prove the limitless potential of modern science and initiate the concept of driverless cars. Other significant developments in the path towards completely driverless vehicles were documented in the 1980s by the invention of the autonomous land vehicle (ALV) by Carnegie Mellon University and the introduction of Mercedes's Prometheus project. These developments, however, required human intervention at certain levels of their operational process. The increase in high-performance hardware and low-cost implementations in the twenty-first century has aggregated quite a lot of interest towards autonomous cars. Figure 1 showcases the different levels of automation in driving.

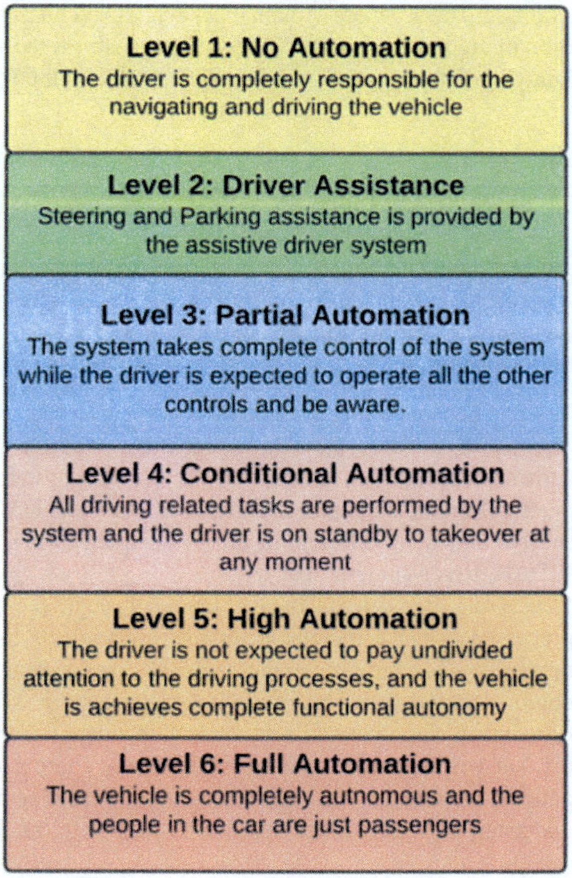

Fig. 1 Levels of driving automation

Several steps need to be executed in sync for the seamless functioning of self-driving vehicles. The vehicle must always be aware of its surroundings, constantly learn its environment, plan the route to achieve the lowest time for the travel and make well-defined manoeuvers in the street [3]. The aforementioned process can be split into three different operational segments such as environmental awareness, navigational planning and movement control. Apart from this, the QoS and safety of the passengers are also accentuated by accommodating assistive machine systems like stability control, assistive brakes, sensors of different modalities and GPS positioning [4]. Autonomous cars not only ease the burden of driving for existing drivers but also expand the market vastly to accommodate people with different disabilities. Mental health is also promoted in this process as this relieves the stress of driving and prevents most cases of road rage.

With the decrease in health risks, passengers may experience improved quality and life and higher productivity. The implementation of connected autonomous cars in smart regions can intercommunicate to make sure minimal congestions and perform optimal route optimizations. It is essential that the passengers completely understand their self-driving cars to yield the aforementioned benefits to completely trust their functioning. This is where explainable navigational intelligence comes into play. The fruition of autonomous cars will certainly transform the world and the human race, and only the future can tell whether the outcome will be positive or negative.

3 Current Processes in Vehicle Autonomy

Several different components function in unison for the smart vehicles to understand their environment and make optimized decisions while ensuring the safety of the passengers. The most critical components for the functioning of the vehicle are as mentioned below.

- *Sensors*: Sensors form the initial perception layer that interacts with the immediate surroundings. These hardware components observe and record data about the surroundings that are utilized by the learning systems of the vehicle to make decisions on the manoeuvers and navigational changes that need to be made. The sensors may be of different modalities ranging from simple IR sensors to radars, LIDAR, stereo cameras, etc.
- *Structuring inputs*: The data obtained through the sensors of the vehicle is processed to make it suitable for interpretation by the decision-making neural networks. This pre-processing stage involves accentuating the features of the obtained visual data employing image processing, segmentation, object detection, image classification, etc., to provide a detailed analysis of the environment [5] that will be used by the network to take appropriate decisions. The combinational information obtained through both mediated perception (multiple visual understandings interpreted together to formulate the data) and direct perception

systems (visual affordance extractions through scalar indicators) of the vehicle is utilized to understand the street environment.
- *Output representations*: The outputs of the main internal vehicle governing system are to produce and initiate vehicle controls to navigate the environment safely and efficiently. Dual methodologies are employed for this purpose. End-to-end strategies obtain the final output directly by feeding deep learning networks with the sensory inferences obtained; whereas, end-to-mid strategies tend to predict the future path of the vehicle, which is to be followed by a PID or similar controller.
- *Learning*: The learning algorithms [6] employed for propelling the vehicle to function autonomously fall into two major categories, namely reinforcement learning and behavioural cloning [7]. Reinforcement learning depends upon the trial-and-error learning of the system where it is exposed to an unknown environment with no prior knowledge and is expected to iterate through multiple attempts until it achieves its goal, finding a unique balance between reinforcement and self-exploration [8]. Behavioural cloning on the other hand is similar to supervised learning tasks where prior knowledge is fed into the networks in the form of well-defined data sets.

Many useful improvement details with respect to vehicle automation can be found in [9, 10].

4 Explainable Artificial Intelligence (XAI) and Its Principles

Explainable artificial intelligence (XAI) deals with eliminating the black box feeling associated with artificial decision-making and improving the transparency of the process to make the final decisions understandable by non-expert humans [11]. It propagates the social right for an explanation that can be exercised by humans to interpret why a particularly critical decision was made so that they can accept it through reasoning and rationalization. XAI extends beyond any legal or ethical obligation and provides improved service in AI applications. The users feel more comfortable as they can now trust their AI system wholeheartedly and understand its thinking methodologies. Hence, XAI plays an integral part in human–AI trust building. The ability of XAI to document its proceedings by demonstrating its past performed, present executing and future predicted actions helps confirm extant information, generate new assumptions and challenge the available information. Figure 2 depicts the process of explainable passenger—AI process.

Explainability, transparency and interpretability are the three major principles that drive the development of XAI algorithms. The concept of explainability should provide an array of interpretable feature representations that can educate humans on the exact background processes that yield a particular decision. Interpretability of a model is often interchanged with comprehension of the model's underlying basis [12]. Transparency is about providing descriptions of the classification or regression

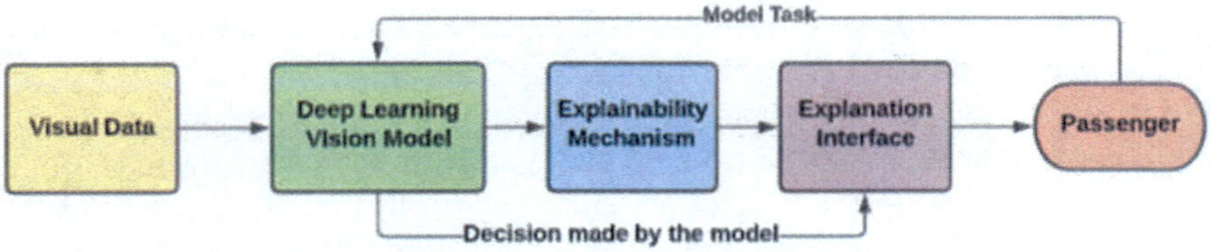

Fig. 2 Visual explanatory process in a vehicle

processes carried out by the machine learning models. There are several different approaches towards solving the problem of AI explainability.

In Fig. 3, interrelation of explainability concepts is depicted. Layer-wise Relevance Propagation (LRP) is a commonly used methodology which functions define set of propagation axioms and rules. This approach is highly appropriate for explaining deep and complex neural networks by propagating the prediction backward. Counterfactual algorithms carry out impact evaluation to recognize the factors contributing to the impact of the intervention by analysing the different parts of the observed actual improvement. Local interpretable model agnostic explanations (LIME) [13] as the name suggests focuses on the behaviour of the model towards the immediate prediction also known as local fidelity. Several other visual methods like Grad-CAM and its variations have been known to provide viable qualitative representations of the vehicle's vision [14–18]. Generalized additive modelling (GAM) works by combining the characteristics of additive models and generalized linear models. The linearized model relates a response univariate attribute to a selected number of predictor attributes. Let us consider Y to be the former and Xi be the latter. Equation (1) below showcases a link function g of a Y exponential distribution family relating the predictor variables and the Y expected value. The functions f_1, f_2, etc., are referred to as the smooth functions. GAM is quite flexible in assuming and establishing relationships between predictors and response variables

$$g(\mathbf{E}(Y)) = \beta_0 + f_1(x_1) + f_2(x_2) + \cdots + f_m(x_m) \tag{1}$$

Rationalization in artificial intelligence is geared towards mimicking human explanatory behaviour for autonomous systems. These algorithms translate internal state-action representations of an autonomous agent into human natural language

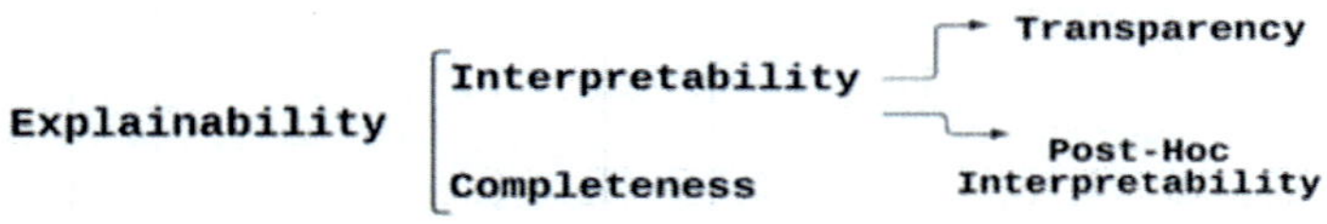

Fig. 3 Interrelation of explainability concepts

through the process of neural machine translations. The effectiveness of rationalization techniques is highly regarded as it provides more satisfaction to non-expert humans who can interpret the process much more easily through natural language [19].

5 Motivation

Human trust is a major factor in determining the future of autonomous systems, and XAI [20] has been positioned in the forefront of providing interpretable solutions to customers in order for them to gain understandings of the underlying processes [20]. Several experiments in the past have confirmed that providing information of the self-driving cars [21] and educating customers on its operational decision-making process made sure that the riders experienced higher trust and lower anxiety levels, according to the works of Koo et al. [22]. Peterson et al. [23] confirmed that the situational awareness among passengers propagated the impact of trust in driver assistance application and autonomous cars. Studies have also been conducted to explore the passengers' preferences among the four divisions of vehicle assistance: zero assistance traditional vehicles, semi-autonomous vehicles, autonomous vehicles without inferences/explanations and autonomous vehicles with interpretable results [24]. A diverse group of people ranging through different age groups were involved in the study. In autonomous cars, 88% of the people felt comfortable in the driver's seat [25] rather than the other seats even when the driving was autonomous. The stress and worry of the passengers motivated them to make sure that they were in a position of control so that they could take over if anything unexpected occurred. Close to 83% of the participating population felt much more comfortable in the autonomous vehicles with the interpretable systems. These works portray the importance of stress reduction and its interdependency towards development of human trust. The success of vehicle autonomy is tightly wound around the complete development of vehicle system explanations. This paper hopes to accentuate the scope of research in explainable autonomous vehicle development and inch closer towards a utopian future of complete human-AI trust.

6 Vision Learning for Autonomous Vehicles

Multiple streams of observational data are fed into the vehicle decision-making systems of autonomous vehicles for analytical processing. A majority of these data streams are obtained from the vehicle's environment through sensory devices of different modalities such as radars, cameras, GPS, ultrasonic sensors and LiDAR sensors. These play a vital role in the manoeuverability and navigational functioning of the vehicle. The decisions taken by the vehicle's driving system are computed in an end-to-end learning fashion (direct mapping to output controls) or through a perception planning-action pipeline, where the inferential system is built upon deep learning networks or non-learning-based conventional planning algorithms [26]. Combinational execution of such algorithms is also facilitated in the real-world driving scenarios where an object detection network feeds its outputs to an A-star algorithm for path planning. Most common components are localization and perception, complex path planning, behaviour arbitration and manoeuver control. The entire system can be demarcated as four different components that accommodate different strategies including classical methods and AI-based algorithms [9]. The safety of each component is always monitored with appropriate safety monitors. Deep learning methodologies [27] always grab the spotlight when it comes to vehicle vision systems. These network architectures are utilized for learning to detect and identify different common objects during the vehicle's travel. The identification is done on the 2D images obtained through cameras or 3D point clouds obtained through LiDAR-based sensors.

7 Self-Driving Ethics/Morality and Trust in Connected and Autonomous Vehicles

The philosophical branch of ethics provides a collection of principles and morals that help define positive and justified outcomes for both the person and society in general. For the digital era and the concept of self-driving cars, the most appropriate ethical frameworks were selected based upon different premises. The frameworks include pluralism, absolutism, relativism, deontology and utilitarianism. Self-driving cars employ a unique combinational approach of ethical frameworks that have a standard set of ethical standards (absolutism) and a separate adaptable set of rules and policies based upon the consequences of the final situation (Utilitarianism). The primary goal of all ethical notions in vehicle autonomy focuses on minimizing human injury, preventing casualties and non-discrimination based on gender, age, race or other factors.

7.1 Trust in Connected and Autonomous Vehicles

The trust of vehicles in VANET networks/intertransmissions is based upon the vehicle's reputation and trustworthiness that is developed based upon observing its previous activities as part of the network. The broadcasts and messages passed on by the vehicle system are evaluated to ensure a minimum trust score in order to prevent spoofing or imitation attacks. The establishment of such trust algorithms in vehicular networks helps protect the functionality of the network and ensures that no customer data is accessed by unauthorized intruders. The prevention of cyberattacks is crucial

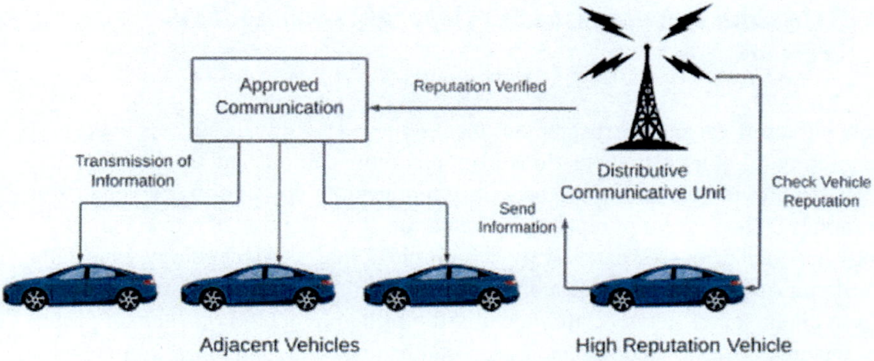

Fig. 4 Reputation-based communication system for intervehicle network trust

to the success of intelligent transportation systems. This has motivated the development of trust-based models for connected vehicles, namely data-centric, entity-centric and combinational trust approaches. Each transmitted message is linked with the vehicle's reputation score to determine its validity. The communications between the vehicles may be of three different types: beacon, alert and disclosure. Beacon messages are transmitted periodically with simple driving status information. Alert messages are sent in the case of emergencies, and disclosure messages are sent by witnessing vehicles and those with conflicting information. Figure 4 depicts an entity-centric trust system that works primarily based upon vehicle reputation as its assessment criteria. The former actions of each vehicle are considered to build a weighted past behavioural analysis in terms of reliability. Another highly regarded approach to enhance the security of connected vehicles and gain consumer trust is a blockchain backbone as a means to decentralize the system. Figure 4 showcases a reputation-based communication system for intervehicle network trust (RCSINT).

7.2 *Privacy in Connected and Autonomous Vehicles*

Privacy is a fundamental right in India and other countries [28]. Privacy is taken care by user during accessing services by connected vehicles or autonomous vehicles (or both), but here this is our responsibility or service provider's responsibility (as ethics) not to use their passenger information with another un-authorized user. Privacy is essential in such intelligent vehicles, also the detail description about leaking of privacy/personal information of users in such vehicles can be found in [9].

7.3 Trust and Privacy Issues in General Vehicle Adhoc Networks

Lightweight self-organized trust models have achieved significant success in the past in extracting reliable trust evaluations from recommendations and trust certificates. These also provide the additional benefit of not needing any third-party vendors or supernodes in the process of reputation evaluation. The adoption of blockchain technology has been discussed extensively in multiple domains of technology, especially in the domain of autonomous transportation. While most application systems function under a client server network architecture, the principle behind blockchain propagates a peer-to-peer form of transmission establishing intercommunication between multiple entities on the network. The employment of this strategy in transportation as a service application system will eliminate the controlling entities and enable transport operators to moderate its use. This distributes the responsibilities/dependencies from a singular entity, eliminates the singular node of failure risk, thus improving public trust. These aforementioned methodologies can help promote the secureness and robustness of the CAV applications while improving consumer awareness and their technological literacy. The widely tested and reliable history of blockchain technology will certainly gain the approval of both the public and related officials as a viable means of opening up the system's operational process.

Moreover, privacy is a serious issue in VANETs [26, 27] and its related applications like location-based services, navigation, carpooling, parking, etc. In the past decade, many solutions towards privacy issue in VANETs have been recommended by various experts which can be found in [10] in detail. Further, the role of software or software-based solutions for VANETs can be found in [29] to improve VANET's component efficiency.

8 Simulation Results

It has been determined that the key factors in enhancing human trust are explainable visual solutions of how the vehicle is interacting with the environment and the reassurance of complete security to the customer. Hence, we create an explainable simulation of Grad-CAM and LIME for autonomous vehicle RGB camera vision. We also explore a novel model for the results were obtained in a Windows 64 bit × 64-based processor Intel Core™ i7-8550U CPU@1.80 GHz 1.99 GHZ and 16 GB RAM. The simulations were carried out in Python on the Google Colab platform. Figure 5 depicts the LIME explanations for predictions made by ImageNet pretrained InceptionV3 weights for the rear end image of a moving vehicle. The image depicts the image regions that have determined the label prediction process of the vehicle's classification system and helps passengers attain a qualitative understanding.

Another viable visual explanation methodology is Grad-CAM, which analyses the last convolutional layer of the networks to utilize the flowing gradient information. An

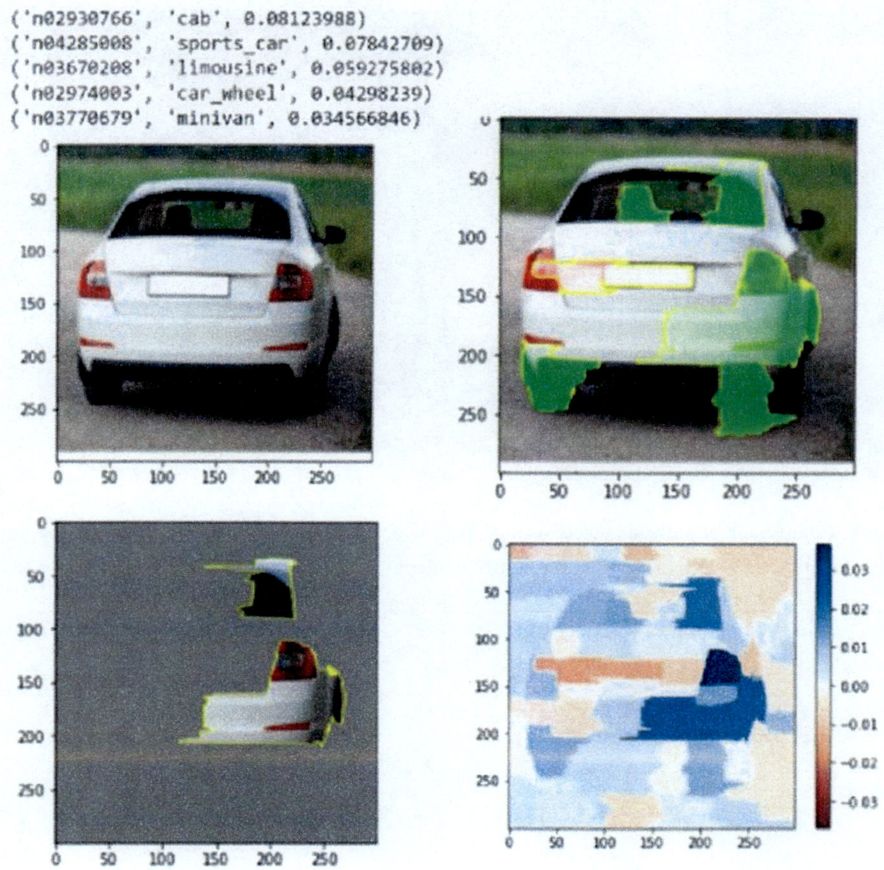

Fig. 5 LIME explanations for vehicle rear image

understanding of each neuron helps in deriving a decision of interest. The importance weights of the neurons are extracted by global average pooling for predicting the target label. Figure 6 showcases the Grad-CAM visualization in a driving scenario for recognizing nearby passing vehicles. The heatmap ranging from red to blue depicts the hotspots in the image responsible for the classification outputs of the InceptionV3 model.

A network intrusion detection system has also been proposed with a unique recursive feature selection algorithm that focuses on a score-based strategy to remove irrelevant features and keep the most imperative features. A decision tree model was employed as a classifier, and the system was tested on the NSL-KDD data set. The NSL-KDD data set provides extensive network traffic records with both normal authorized connections and intrusion connections for benchmarking cyber-attack detection models. Table 1 showcases a comparative analysis of our model coupled with the feature selection algorithm with other extant works.

Fig. 6 Grad-CAM visualization for vehicle vision

Table 1 Comparison of intrusion detection models

Methodology	DOS accuracy (%)	PROBE accuracy (%)	R2L accuracy (%)	U2R accuracy (%)
Naïve Bayes	99.3	97.5	95	60
Correlation-based J48	99.1	99	97.8	98.7
SVM with genetic optimal selection	99.1	99	96	97
Proposed model	99.8	99.88	99.7	98.9

In the last, several privacy-preserving techniques for vehicle ad hoc network (including future vehicles) have been included in detail. The researchers are recommended to refer these articles for enhancing their knowledge towards preserving of privacy of users in this smart era with emerging technologies/modern tools.

9 Conclusion

Autonomous vehicles have positioned themselves as crucial contributors to the widespread adoption of smart city infrastructures. The ability to explain the decision-making processes and help passengers interpret why the vehicle acted plays an important role in determining the level of trust and reliability placed on it by humans. Visual explanations and the assurance of security to the passengers are the two most influential factors in reducing stress and anxiety levels of the passengers. This paper introduced explainable navigational intelligence, which aims to converge the decision-making processes of autonomous vehicle systems and the domain of explainable AI (XAI) to provide clear insights into the role explainability plays in increasing human trust on AI solutions. Visual explanatory methods were tested out along with a novel intrusion detection system for security and proved more accurate than other extant works.

References

1. Arrieta, A. B., Díaz-Rodríguez, N., Ser, J. D., Bennetot, A., Tabik, S., Barbado, A., García, S., et al. (2020). Explainable artificial intelligence (XAI): Concepts, taxonomies, opportunities and challenges toward responsible AI. *Information Fusion, 58*, 82–115.
2. Yoganandhan, A., Subhash, S. D., Hebinson Jothi, J., & Mohanavel, V. (2020) Fundamentals and development of self-driving cars. *Materials Today: Proceedings, 33*, 3303–3310.
3. Cysneiros, L. M., Raffi, M., & Sampaio do Prado Leite, J. C. (2018). Software transparency as a key requirement for self-driving cars. In *2018 IEEE 26th International Requirements Engineering Conference (RE)* (pp. 382–387). IEEE.
4. Hilgarter, K., & Granig, P. (2020). Public perception of autonomous vehicles: A qualitative study based on interviews after riding an autonomous shuttle. *Transportation Research Part F: Traffic Psychology and Behaviour, 72*, 226–243.
5. Hussain, R., & Zeadally, S. (2018). Autonomous cars: Research results, issues, and future challenges. *IEEE Communications Surveys & Tutorials, 21*(2), 1275–1313.
6. Ras, G., van Gerven, M., & Haselager, P. (2018). Explanation methods in deep learning: Users, values, concerns and challenges. In *Explainable and Interpretable Models in Computer Vision and Machine Learning* (pp. 19–36). Springer.
7. Czubenko, M., Kowalczuk, Z., & Ordys, A. (2015). Autonomous driver based on an intelligent system of decision-making. *Cognitive Computation, 7*(5), 569–581.
8. Rödel, C., Stadler, S., Meschtscherjakov, A., & Tscheligi, M. (2014). Towards autonomous cars: The effect of autonomy levels on acceptance and user experience. In *Proceedings of the 6th International Conference on Automotive User Interfaces and Interactive Vehicular Applications* (pp. 1–8).
9. Tyagi, A. K., & Aswathy, S. U. (2021). Autonomous intelligent vehicles (AIV): research statements, open issues, challenges and road for future. *International Journal of Intelligent Networks, 2*, 83–102. ISSN 2666-6030. https://doi.org/10.1016/j.ijin.2021.07.002
10. Varsha, R., et al. (2020). Deep learning based blockchain solution for preserving privacy in future vehicles. *International Journal of Hybrid Intelligent System, 16*(4), 223–236.
11. Adadi, A., & Berrada, M. (2018). Peeking inside the black-box: A survey on explainable artificial intelligence (XAI). *IEEE Access, 6*, 52138–52160.
12. Samek, W., & Müller, K.-R. (2019). Towards explainable artificial intelligence. In *Explainable AI: interpreting, explaining and visualizing deep learning* (pp. 5–22). Springer.
13. Lee, E., Braines, D., Stiffler, M., Hudler, A., & Harborne, D. (2019). Developing the sensitivity of LIME for better machine learning explanation. In *Artificial intelligence and machine learning for multi-domain operations applications* (vol. 11006, pp. 1100610). International Society for Optics and Photonics.
14. Selvaraju, R. R., Cogswell, M., Das, A., Vedantam, R., Parikh, D., & Batra, D. (2017). Grad-cam: Visual explanations from deep networks via gradientbased localization. In *Proceedings of the IEEE International Conference on Computer Vision* (pp. 618–626).
15. Chattopadhay, A., Sarkar, A., Howlader, P., & Balasubramanian, V. N. (2018). Grad-cam++: generalized gradient-based visual explanations for deep convolutional networks. In *2018 IEEE Winter Conference on Applications of Computer Vision (WACV)* (pp. 839–847). IEEE.
16. Song, W., Dai, S., Huang, D., Song, J., & Antonio, L. (2021). Median-pooling grad-CAM: An efficient inference level visual explanation for CNN networks in remote sensing image classification. In *International Conference on Multimedia Modeling* (pp. 134146). Cham: Springer.

17. Ramaswamy, H. G. (2020). Ablation-cam: Visual explanations for deep convolutional network via gradient-free localization. In *Proceedings of the IEEE/CVF Winter Conference on Applications of Computer Vision* (pp. 983–991).
18. Shrikumar, A., Greenside, P., & Kundaje, A. (2017). Learning important features through propagating activation differences. In *International Conference on Machine Learning* (pp. 3145–3153). PMLR.
19. Arya, V., Bellamy, R.K.E., Chen, P.-Y., Dhurandhar, A., Hind, M., Hoffman, S. C., Houde, S., et al. (2020). AI explainability 360: hands-on tutorial. In *Proceedings of the 2020 Conference on Fairness, Accountability, and Transparency* (pp. 696–696).
20. Wiegand, G., Eiband, M., Haubelt, M., & Hussmann, H. (2020). I'd like an explanation for that!" Exploring reactions to unexpected autonomous driving. In *22nd International Conference on Human-Computer Interaction with Mobile Devices and Services* (pp. 1–11).
21. Omeiza, D., Webb, H., Jirotka, M., & Kunze, L. (2021). Explanations in autonomous driving: a survey. arXiv preprint arXiv:2103.05154
22. Koo, J., Kwac, J., Ju, W., Steinert, M., Leifer, L., & Nass, C. (2015). Why did my car just do that? Explaining semi-autonomous driving actions to improve driver understanding, trust, and performance. *International Journal on Interactive Design and Manufacturing (IJIDeM), 9*(4), 269–275.
23. Petersen, L., Robert, Jessie Yang, X., Tilbury, & D. M. (2019). Situational awareness, drivers trust in automated driving systems and secondary task performance.
24. Shen, Y., Jiang, S., Chen, Y., Yang, E., Jin, X., Fan, Y., & Campbell. K. D. (2020). To explain or not to explain: a study on the necessity of explanations for autonomous vehicles.
25. Wiegand, G., Schmidmaier, M., Weber, T., Liu, Y., & Hussmann, H. (2019). I drive-you trust: Explaining driving behavior of autonomous cars. In *Extended Abstracts of the 2019 Chi Conference on Human Factors in Computing Systems* (pp. 1–6).
26. Tyagi, A. K., & Sreenath, N. (2015). A comparative study on privacy preserving techniques for location based services. *British Journal of Mathematics and Computer Science, 10*(4), 1–25. ISSN: 2231-0851
27. Tyagi, A. K., & Sreenath, N. (2015). Location privacy preserving techniques for location based services over road networks, 2–4 April 2015. In *Proceeding of IEEE/International Conference on Communication and Signal Processing (ICCSP), Tamil Nadu, India* (pp. 1319–1326). ISBN: 978-1-4799-8080-2
28. Nair, M. M., & Tyagi, A. K. (2021). Privacy: History, statistics, policy, laws, preservation and threat analysis. *Journal of Information Assurance & Security, 16*(1), 24–34.
29. Midha, S., Tripathi, K., & Sharma, M. K. (2022). Software defined network horizons and embracing its security challenges: From theory to practice. In *Cloud and IOT Based Vehicular Ad hoc Networks*, Chap. 9. Wiley. https://doi.org/10.1002/9781119761846.ch9

Advanced Computing Technologies

An Empirical Study of Design Techniques of Chatbot, a Review

Akanksha Yadav and Namrata Dhanda

Abstract In recent times, evaluation of the informal coordination in the form of communication between the human being and the electronic brain is making the good progress. Human being or the electronic brain protected system is being used extensively for logical language/terminology development procedure. Chatbot acts as electronic brain which permits human being through the electronic brain applying logical terminology. Chatbot coordination is being used in the different areas like travel, e commerce, customer service, etc. Representation of the chatbot requires different procedures. Hence, by this study or in such work, authors or we introduce summary of the procedures which are used to layout the chatbot. Some steps of the chatbot layout are shown by generally reviewed knowledge as in what way chatbot layout works and which forms of methods are useful for the evolution of chatbot. By fast evolution of the chatbot technology, we can expect chatbot which can enhance human being limitations as well as maximize efficiency.

Keywords Chatbots · Artificial intelligence · Cleverbot · Machine learning · User interface · Natural language processing

1 Introduction

A chatbot is a software that interacts with human regarding a particular domain by natural conversation either by text or by the help of voice. These chatbots are prepared for many different purposes, and they are used all over the world. These chatbots are used over a wide range of domain, for example, in customer service, in library, educational institute, in industrial sector, in research and development, technical support, etc., development in the field of chatbot recommends that the interaction

A. Yadav (✉) · N. Dhanda
Department of CSE, ASET, Amity University, Lucknow, Uttar Pradesh, India
e-mail: itsakanksha9@gmail.com

N. Dhanda
e-mail: ndhanda@lko.amity.edu

P. K. Singh et al. (eds.), *Proceedings of Third International Conference on Computing, Communications, and Cyber-Security*, Lecture Notes in Networks and Systems 421,
https://doi.org/10.1007/978-981-19-1142-2_11

with technologies either by natural language or by voice is only possible because of development of technologies and human being become more interactive to digital world. Developing a chatbot is far better than making a human like smart machine and performing the same task for collecting information.

Commonly, bot is computing device which executes automatic function and helps in messenger stage, termed as chatbot. Chatbot is like usual texting application, and the unusual thing is that another receiving device is a robotic device. Another way to explain this circumstances same as person is doing chit chat at that electronic machine, exchange communication, for example, speak control, check chitchat, and the graphic interfaces.

Currently, chatbot is highly prevalent method that can help a person to do numerous work [1]. He proposed numerous benefits to create chatbots such as it can support human being. Questions as well as provide response 24 h basis and further enhance capability using finished works wherefore people have not been necessary. Greater benefit about chatbot is it has power to attain broad public of texting method or for customized communication capability [2]. Chatbot is using into the different sectors for providing details, example, weather forecasting, airline ticket booking, etc., [1] or buying commodity. Well-known software is also using this techniques for applications like Twitter, WhatsApp, Instagram, Facebook Messenger, Google Assistant, etc. According to [3], layout method generally selected through creators could be pattern matching, cleverscript, chatscript, AIML. Common method is pattern matching thus bot shall paired idioms from the given key phrase in a particular word list [3]. Hence, study focuses on examine some forms of the chatbot layout. In the steps, chatbot structure normally introduced as a part of this study. Outcome is considered, and consequences will draw finally.

At present, the chatbot area is wide. Chatbot is not only belongs to a single field or single category chatbot can be used in various categories. Here, a proper classification was done for both voice input and messaging channel.

By input.

1. Button based
2. Keyword recognition
3. Contextual
4. Voice enabled
5. Standalone application: done by the help of desktop and mobile
6. Web-based service: done on both integrated and individual basis
7. Integrated: done on instant messaging app as well as communication and collaboration through platform

2 Background

2.1 Chatbot System

Another name of chatbot is chapter robots or chatterbots, [4] which are electronic device which co-ordinate into people through texting applications [5]. It recognize various queries which are raised through person. It has capability to specify among particularity of the term as well as smiles. For the purpose of obtaining enhanced features of the chatbot communication, which require plenty of the terminology communication between persons [6].

Chatbot resembles general testing applications. It works on application layer, database, and the APIs which functioning on the background. Front-end illustrate interact create connection from the client chatbot is user friendly; in framework, its performance is complicated. So many chatbots have records about communication also designers utilize record for understanding the client enquiries. Records can be use enhance chatbot communication [7]. Working of chatbot is to match query from the person by means of neutral network. Such as, the query of any person-" let me take a look of the institute checklist of curriculum or show curriculum, both the sentence represents the same context. It is the responsibility of the creator, instruct chatbot for understanding queries giving similar result. As reported by [3], chatbot has instructed using reviews of the record of the people communication. Records are greater, and features make much smarter [7].

2.2 Applications of Chatbots

Applications of chatbot devices are using in different sectors. Because of resilience, it is using in sectors like educational sector, medical management, etc. Here is the name of some companies which have built in chatbot in its devices such as Messenger, Alexa, Google Assistance, Siri, and Cortana. Facebook has built in Messenger that assists chatbot devices. Chatbot is also helping other companies through providing automated client communicator.

Chatbot is also using in the learning sector. According to [8], chatbot works as smarter trainer to support the Internet trainee. Chatbot has capacity of analyzing human law also considers certainty of the communication. While communication rate is correct, it is create chatbot to using add a device for learners such as chatbot can resolve queries also help correspondence of 200 learners on particular base. While in medical management, chatbot is also using to help medical specialist for supporting sufferer using electronic device and the program moderate such as AI chatbot [9] serves on communicative subordinate for assisting prolonged-period to uphold healthy lifestyle, arbitration. On such events, chatbot works as duplex with medical specialist and a person who are taking advice like getting fat through providing consultation on the good nutrition routine, exercise, etc.

Although currently analyzing presents that chatbot is also using in the businesses such as collect. Chat is responsive chatbot which is created for recording client's details on the companies' Websites. Chat can be used for fetching data like commodity, working poll, respond to queries, etc.

3 Review of Chatbot Design

In this part, review about the multiple uses of chatbot like companies, marketing, education, and day-to-day life.

3.1 Chat.io

Chatbot devices can be used to assist business for communicating along with clients through number of facilities in groups. It is merged along with Facebook Messenger for assisting administrator for communicating from the Facebook client. Chatbot is developed through extensible structure which may easily communicate through Websites, apps, etc. AI serves for developing chatbot, thus chatbot may foresee message advice and then deliver good reply which are based on the analysis of discussion record.

Figure 1 primary page of Chat.io. Chatbot initiate working by asking personal details such as name and email id of the user. User 1 needs to click on agree for further processing.

Figure 2 represents dashboard for Chat.io.

Three sections are on the left side in the dashboard such as 1. customer name, 2. text, and 3. other details. Main functions of this Chat.io are to integrated with the messages, automatic chat assignments, intelligent can responses or answer, customer can message and chat. This can be use as messenger in mobile apps.

3.2 Collect.chat

Collect.chat is mostly used in marketing. It is also illustration of the chatbot devices, in this function is depend on the gadget interact of persons rather AI. Benefits of chatbot are different, and it transform new comer as a person and conduct communication unless fill any kind of forms.

Figure 3 represents Collect.chat primary design of front end. Front-end communication begins from "HOLA" then chatbot will ask some queries from the clients. Objective questions will display on the screen, and it was asked by the system. It is necessary for the client to submit response of the questions which are displaying on the screen.

Fig. 1 Primary page of Chat.io

Communication acts in the new style, and it is different from the previous type. It can further make communication with another resources such as sales force, Pax sheets, and slack. It can automatically complete the work.

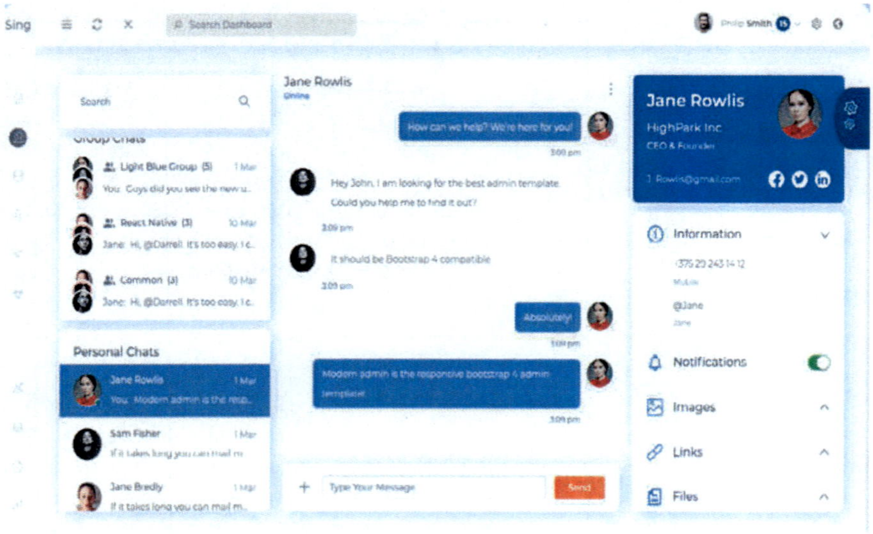

Fig. 2 Instance of Chat.io admin dashboard

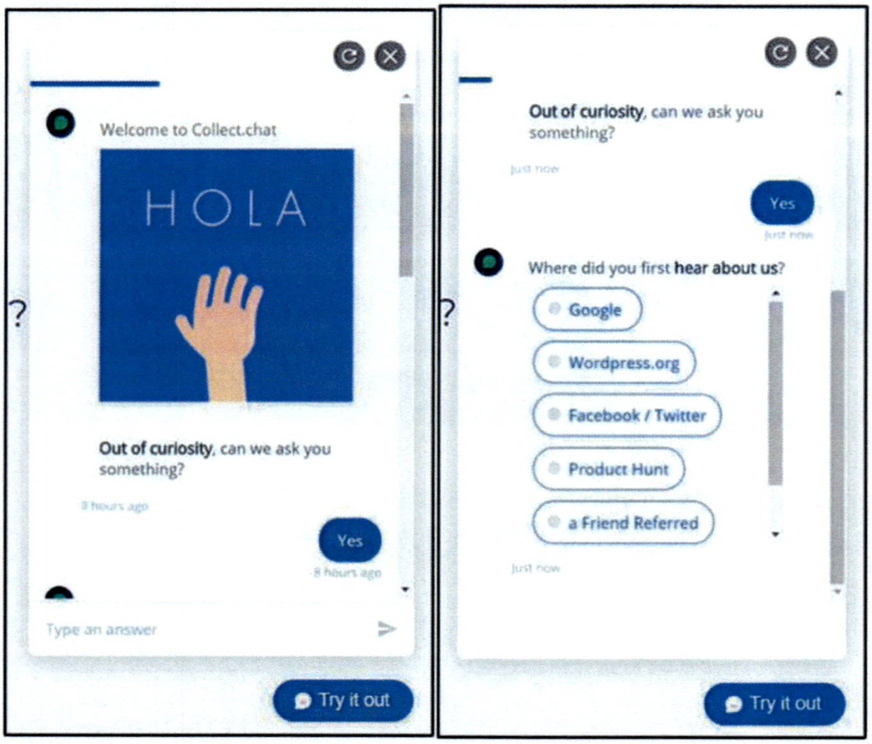

Fig. 3 Instance of Collect.chat interface

3.3 Cleverbot

Cleverbot is also one of the instances. It is an AI chatbot, namely cleverbot. British AI scientist Rollo Carpenter in 1997 creates a chatbot online-Web enable application cleverbot reply is easy. Cleverbot accepts the user input in the communication procedure while user inserts few information, device start to searching the related words which are similar to the insert information. It ensures to the information to the person through searching in process which can be reply as for the information.

This bot is available in mobile as Android app and on iOS platform too [10] (Fig. 4).

Specifically, cleverbot answer to people queries through pass study of the person's replies. People would insert his questions within entry field and device shall search for the words which are related to the query. Then, searching from already stored communication cleverbot shares answer to the people through searching by which person answer insert information earlier.

(GPU) working method: Graphical processing unit is a technical unique electronic switch design on speedy and modify storage for the purpose of developing photos and graphics memory which will appear. Currently, one of the mechanisms in the backside cleverbot or its API is currently provides commercially for the every creator in public.

Fig. 4 Instance of cleverbot interface

4 Discussion

4.1 Working of Chatbot

The Architecture

The modeling of chatbot initiates by planning the desired aims, the procedure that was followed and the requirement of the user. Developing a chatbot by the help of programming language and after that their testing which was done locally; furthermore, it was then published or uploaded to a online Website or a data center as the chatbot is connected with a network in order to send and receive the response or message. For all this, the proper implementation and integration are very important for the desired functioning of chatbot, and there are three integration methods to do this, they are integration using API, manual integration, or third-party integration [11]. Deciding of algorithm in order develop chatbot basically depends on the response that is chatbot going to make against the user response and what category it is falling in [12].

The proper selection can provide a good benefits for connectivity and functioning as well as easy updating with minimum effort by developer. That chatbots that can be used directly or connect directly to the user without downloading and installing are consider be the more efficient as compare to other chatbot.

The chatbots that were develop by the help of programming languages such as Java and Python or the development platform for chatbot (commercial or open source) [13]. Open-source platform includes RASA, Botkit, Chatterbot About, Pandorabots, and Microsoft Bot Framework and commercial platform includes Botsify, Chatfuel, Manychat, and FlowXO. Some NLU cloud platforms [14] powered by machine learning are Google Dialogflow [15], IBM Watson Conversation (IBM Watson), Amazon Lex [16], and SAP Conversation AI (SAP Conversational AI |Automate Customer Service with AI Chatbots 2019).

Diagrams are example of procedures of the chatbot system. Initially, person should have a system to access chatbot artificial intelligence to use the chatbot. A text will be shown on the interface where one has to input in form of text through that interface.

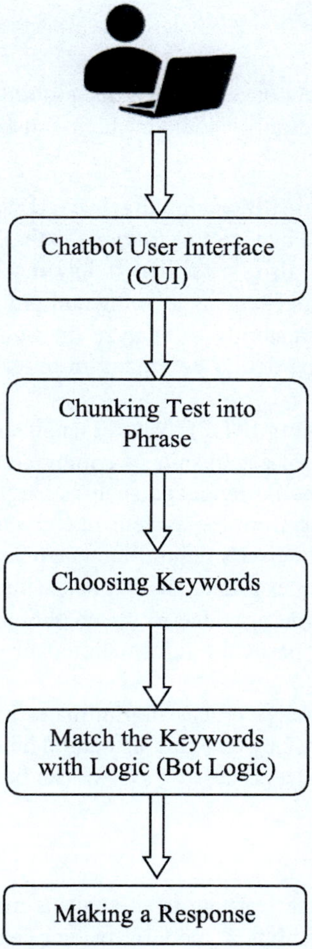

Then, message information which will be insert through a person as words so make it fragment. Fragment is a method of pieces of the message in divided phrase for tagging [17] Outcome after fragment method is numerous significant words which will used in corresponding method process of aligning. Words work as list in the matching process.

At the end, words outcome through fragmentation methods will pair at the pattern within chatbot system. Procedure identical along with pattern are known as BOT LOGIC. Outcome of the chatbot is programmable answer, and another message template Web form is shown in diagram 3.

4.2 Chatbots Design Techniques

After analyzing, set of document have included this plotting of chatbot needs some method and techniques. Commonly, method which can be used to creator is as follows:

- **Artificial Intelligence (AI)**: Artificial intelligence (AI) is the simulation of intelligence of human in machine by various programs so that it can think and respond like human. The term may also be applied to any machine that exhibits traits associated with a human mind such as learning and problem-solving. The basic characteristics of AI it is capability to analyze the human command and take proper action against that particular command in order to achieve a particular goal.
- **Natural Language Processing (NLP)**: Natural language processing (NLP) is a artificial intelligent method of establishing a communication between a system and human or user in a specific language such as English. NLP is used when user wants any specific task from the system like chatbot will follow the user commands by processing the user's natural language and providing the desired result in the same context. For example, a user wants to find some information over Web, then a command will be provided to the chatbot and chatbot will process that command and find that particular information over Web and then deliver to the user.

  This area of natural language processing intimates in developing automated chatbot to perform desired task by recognizing natural human language. The input and output of natural language processing system can be:

  Written text
  Speech

- **Machine Learning**: Machine learning is an analysis method that helps in data analysis which automates analytical model. Machine learning is a branch of AI by which a system can learn from the set of data provided by the user, and on the basis of them, the machine or the automated chatbot makes decision with the least human intervention. Machine learning is the most interesting technology of current times. As by its name, it makes machine more similar to humans. It is the ability to learn from human for the human. Machine learning is quite oftenly used today, and it can be employed to many more places that anyone can expect [18].

  Chatbot that used machine learning approach rather than pattern matching basically acquires the information from the user by the help of natural language processing (NPL) and persuades the learning ability from the conversation between the user and chatbot. It consider the conversation as a dialog context and don not consider as a term which don not require a any specific reply. Basically, it require a training and the finding from it and difficulty that was faced are collected as datasets. For example, movie scripts corpus may be too broad, or an IT helpline may be too specific [19]. Frequently, ANNs are required for the execution of these chatbots. Retrieval models use the neural network for the working and to provide the proper set of response against the user's response. We can say that the generative model incorporates the response by using deep learning technique.

- **AIML**: It is the essential method using markup language developed by Dr. Richard S. Wallace [20], rarely use through the developers. Aim of AIML language consists directly process the communication modeling into a stimulus response process. This method is also called as frequent tags. Since AIML obviating expertise in specific programming language, therefore, this method is maximum ease for the evaluation of chatbots.
- **Pattern Matching**: Pattern matching method is considered by various chatbots. Commonly, this method used for matching pattern for initiate suitable answer from the persons queries, depend upon matching forms like plain word expressive implication of the questions
- **Language Tricks**: The language ticks basically comprise of four tricks that are basically used for the modeling, and they are canned responses, no logical conclusion, typing errors, and stimulating key strokes. The language tricks basically assess paragraph, sentence, and phrases in chatbot in order to increase the data base and knowledge base which will further help chatbot to respond in a more convenient way. The use of four tricks makes the intelligent machine behave more like human. These tricks replicate the user behavior made while a conversation such as mistakes made during conversation, repetition of a particular word, typing error, the personality of the user, and the irritated response made by the user [21].
- **Chatscript:** Developing script like cleverscript in the development of chatbot. Chatscript is also a method which can be used when there is no matches appear in AIMl. Method focusing on providing good syntax to develop intelligent revert response
- **Parsing**: In this method, it is used to examine the message or series of images as well as using natural language instructions. Additionally, in computational linguistic, parsing is a method which is used to examine set of strings into its elements which can restrain linguistics or data. This method used NLP features like trees in Phyton NLTK.
- **SQL and relational database**: SQL and relational database are the current method which are using in chatbots to assure chatbots recall past communications. The algorithm from SQL-based chatbot can be used to improve the capacity of chatbot's word list and pattern matching through giving an expand way of data storage also enhancing the method execution.
- ***Markov Chain:*** Markov is also a method through developing replies which can easily implement and good. Markov method assists through find probolistic or terms incident text form set of information.

5 Conclusions

Almost everybody focuses on a system which are identical to the person. Most people I have no idea that chatbot will too as well as provide response true message and voice command, while at present, chatbot gives responsive assist data through diagrams. Advantage makes use capable to attain wide public also good range through predecessor software. Beyond those, automatic person- computer communication program act completely gives effective benefit in different sectors to help people in different manners.

By this study, analysis covers some studies which concentrate on the chatbot framework. At the start, clarified with chatbot system and their utilization in some essential sectors like learning, medical, etc. Then, we will explain on future bot framework in current market. Analysis depends on functions, how communicative with people and as well as interaction.

References

1. Naveen Kumar, M., Linga Chandar, P. C., Venkatesh Prasad, A., Sumangali, K. (2017). Android based educational chatbot for visually impaired people. In *2016 IEEE International Conference on Computational Intelligence and Computing Research (ICCIC 2016)* (pp. 0–3).
2. Harris, R. (2016, October). The advantages and disadvantages of chatbots. *App Developer Magazine, 3*.
3. Masche, J., Le, N. (2018). A review of technologies for conversational systems. In *Advances in Intelligent Systems and Computing* (Vol. 629, pp. 212–225).
4. Hatwar, N., Patil, A., Gondane, D. (2016). AI based chatbot. *International journal of Emerging Trends in Engineering and Basic Sciences*, 3(2), 2349–696785.
5. Ciechanowski, L., Przegalinska, A., Magnuski, M., Gloor, P. (2018). In the shades of the uncanny valley: An experimental study of human–chatbot interaction. *Future Generation Computer Systems*, 1–10.
6. Hill, J., Randolph Ford, W., Farreras, I. G. (2015). Real conversations with artificial intelligence: A comparison between human–human online conversations and human–chatbot conversations. *Computers in Human Behavior*, 49, 245–250.
7. Garrigós, I., Eds, M. W., Hutchison, D. (2018). Case study: building a serverless messenger chatbot. In *Current Trends in Web Engineering* (Vol. 10544, pp. 75–86).
8. Doshi, S. V., Pawar, S. B., Shelar, A. G., Kulkarni, S. S. (2017). Artificial intelligence chatbot in android system using open source program-O. *International Journal of Advanced Research in Computer and Communication Engineering*, 6(4), 816–821.
9. Fadhil, A., Gabrielli, S. (2017). Addressing challenges in promoting healthy lifestyles: the AI-chatbot approach. In *PervasiveHealth '17*.
10. Cingillioglu, N. (2017). *Neural logic framework for digital assistants*.
11. Trivedi, A., Gor, V., & Thakkar, Z. (2019). Chatbot generation and integration: A review. *International Journal of Advance Research, Ideas and Innovations in Technology, 5*(2), 1308–1311.
12. Nimavat, K., & Champaneria, T. (2017). Chatbots: An overview types, architecture, tools and future possibilities. *International Journal for Scientific Research and Development, 5*(7), 1019–1024, Retrieved from http://www.ijsrd.com/.
13. Nayyar, D. A. (2019). Chatbots and the open source tools you can use to develop them. Open Source For You website: https://opensourceforu.com/2019/01/chatbots-andthe-open-source-tools-you-can-use-to-develop-them/.

14. Braun, D., Hernandez-Mendez, A., Matthes, F., & Langen, M. (2017). Evaluating natural language understanding services for conversational question answering systems. Undefined website:/paper/Evaluating-Natural-Language-Understanding-Services- 336 Braun-Hernandez Mendez/ab8c725e04fc25dc03e96332e4490573cd87abd8.
15. Dialogflow. (2019). Dialogflow website: https://dialogflow.com/.
16. Amazon lex—build conversation bots. (2019). Amazon Web Services, Inc. website: https://aws.amazon.com/lex/.
17. Abdul-Kader, S. A., & Woods, J. C. (2015). Survey on chatbot design techniques in speech conversation systems. *International Journal of Advanced Computer Science and Applications*, 6(7), 72–80.
18. Bickmore, T., & Cassell, J. (2005). Social Dialogue with Embodied Conversational Agents. In J. C. J. Kuppevelt, N. O. Bernsen, & L. Dybkjær (Eds.), *Advances in Natural Multimodal Dialogue Systems* (Vol. 30, pp. 23–54). Springer.
19. Lin, L., D'Haro, L. F., & Banchs, R. (2016, October). A web-based platform for collection of human-chatbot interactions. In *Proceedings of the Fourth International Conference on Human Agent Interaction (HAI '16)* (pp. 363–366). Association for Computing Machinery . https://doi.org/10.1145/2974804.2980500.
20. Krantz, A., & Lindblom, P. (2017). Generating topic-based chatbot responses. *Blekinge Institute of Technology.*
21. Aza, M., Ahmad, Z., & Akma, N. (2018). Review of Chatbots design techniques. *International Journal of Computer Applications, 181*, 7–10.
22. Shaikh, A., More, D., Puttoo, R., Shrivastav, S., & Swati Shinde (2019). A survey paper on Chatbots. *International Research Journal of Engineering and Technology (IRJET), 06*(04). e-ISSN: 2395-0056. www.irjet.net p-ISSN: 2395–0072.

An Approach for Cloud Security Using TPA- and Role-Based Hybrid Concept

Pooja Singh, Manish Kumar Mukhija, and Satish Kumar Alaria

Abstract In the cloud environment, enormous amount of the data is shared on the server for the availability of access to the employees or customers related to the organization. Two main issues which are generally faced—when data is shared in cloud environment, first is authenticating the user who can access the data, and secondly, to secure the data itself. Seeing the concern, we proposed the hybrid concept which involves the role-based security as well as TPA-based security. By making use of role-based security, first we have authenticated the users using the graphical authentication in which first the image requires to be selected, then the image gets segmented into image blocks, which when selected then the pattern is formed which is used for the authentication purpose, after that when the user shares the file, then he/she specifies the role who can access the file.

Keywords Cloud environment · Cloud security · TPA · Role-based access

1 Introduction

Cloud computing is a significantly versatile and savvy foundation for running HPC and try Web applications. Regardless, the creating interest of the cloud foundation has profoundly extended the energy use of the data centers, which has transformed into a fundamental issue. High energy usage not simply implies high functional expense, which lessens the net income of cloud providers, yet moreover prompts high-carbon outpourings which are not naturally neighborly.

The commercialization of these advancements is portrayed as of now as cloud computing [2], where computing is passed on as utility on the pay-as-you-go reason. By and large, business affiliations are used to contribute enormous proportion of

P. Singh (✉) · M. K. Mukhija
Department of Computer Science & Engineering, AIET, Jaipur, India
e-mail: Poojasinghsfdc@gmail.com

S. K. Alaria
Department of Electronics & Communication, AIET, Jaipur, India

P. K. Singh et al. (eds.), *Proceedings of Third International Conference on Computing, Communications, and Cyber-Security*, Lecture Notes in Networks and Systems 421,
https://doi.org/10.1007/978-981-19-1142-2_12

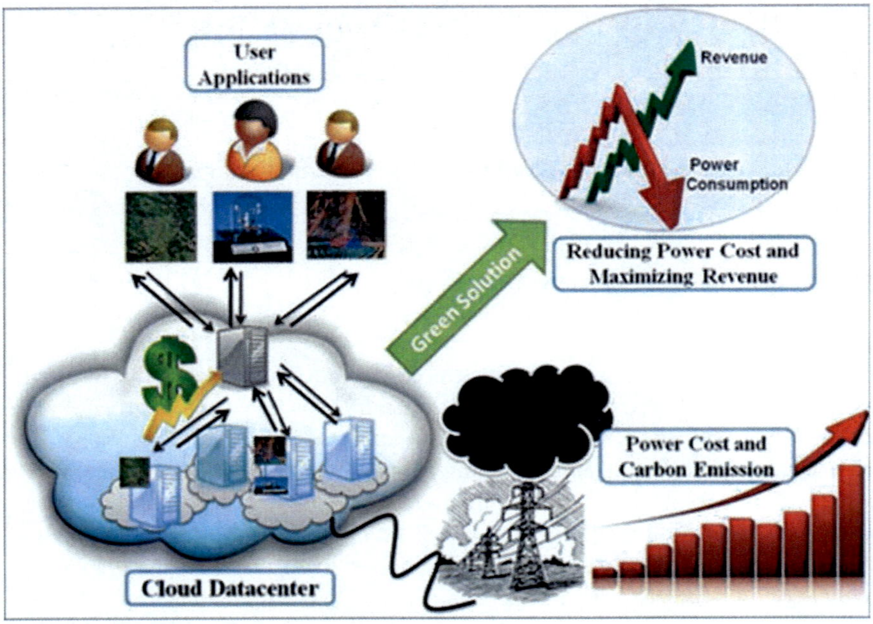

Fig. 1 Cloud and environmental sustainability

capital and time in acquiring and backing of the computational-related resources. The improvement of cloud computing is rapidly changing this belonging-based approach to manage enrollment arranged methodology by offering admittance to versatile foundation and organizations on-demand. Customers can then store, also access, and offer any proportion of data in cloud.

As such, numerous associations not simply see clouds as a significant on-demand advantage, yet moreover a potential market opportunity. As demonstrated by International Data Corporation (IDC) report [1], the overall IT cloud organizations' spending is evaluated to augment from the $16 billion of each 2008 to $42 billion of each 2012, addressing an accumulate yearly advancement rate (CAGR) of 27% (Fig. 1).

2 Literature Survey

Shree et al. [1] Optimization accepts a huge part in various issues that expect the specific yield. Security of the dataset away in far off specialists forth plainly reliant upon secret key one which is then used for the purpose of encryption and also then for the interpreting reason. Various strange key age estimations, for instance, the RSA, AES, are available to make the key. The key made by such estimations are ought to be smoothed out to give more noteworthy security to your data from unapproved customers similarly as from the TPA who will check their data for trustworthiness reason. In this paper, a procedure to redesign the baffling key by then making use of cuckoo search computation (CSA) is proposed.

Thakare et al. [2] Duties are separated inside a gathering by using the work-based induction control (RBAC) in the Azure Internet of Things (IoT) framework, and simply an appropriate level of access is surrendered to customers to perform unequivocal tasks, dependent upon a given situation. In any case, a comparative confirmation and endorsement framework is used for "sort of customer," which grows the movement over-trouble on the cloud laborer. Also, in view of its RBAC nature, the IoT structure is inefficient in dealing with an amazing situation where various customers request tantamount kinds of resources, by making a couple of repeated positions.

Belkhiria, et al. [3] In the domain of advanced real structures, the improvement of Smart Living Spaces (SLS) design offers people the opportunity to benefit with better methodologies for living. Such mechanical example that incorporates a couple of parts of step-by-step life allows the occupants of the space to all the more probable modify and control their present situation. SLS stresses, essentially, energy conservation, convenience, and comfort similarly as clinical consideration concerns.

Suganthy and Prasanna Venkatesan [4] With the happening to Web and its associated propels, the affiliations are more stressed in offering security to their resources. Access control models help in giving quite far to the customers when the resources are being gotten by them, and role-based induction control (RBAC) model is one such access control instrument in which the customers access the resources subject to the positions they obtained in the system.

Mu and Liu [5] According to the arrangement considered RBAC model, the SSM framework subject to JAVA language is obtained together with Maven to execute the establishment work module. The front-end advancement, for instance, AJAX and Bootstrap, is used to design the web page.

Liu et al. [6] to the extent people's lifestyles, direct data can be adequately accumulated and taken apart. Directions to get and utilize these data is a charming issue today. There are various direct affirmation models in insightful world, yet the technique for recognizing conduct is at this point considering standard organizing in industry. The major clarification is the impact of work circles. Makers propose a procedure for dynamically creating RBAC models. Then, makers segment the "min-advantage social events" and join the results of the get-togethers to get their RBAC-outfit model. In the assessments, makers examine the effects of a couple of models already, then, at that point afterward using "min-advantage social events". The results show that approvals do influence rehearses. Considering "min-advantage social occasions", makers take a gander at the effects of the other three models. The model makers propose achieves the best affirmation.

Ghafoorian et al. [7] Cloud computing is an extensive development, which has attracted a lot of thoughts nowadays. Accordingly, in this paper, makers at first present the security goals that should be considered in a powerful trust-based structure. Second, makers propose a sharp trust and reputation-based RBAC model that not only can true to form withstand the security risks of trust-based RBAC models, yet moreover is versatile as it has reasonable execution time. Third, makers survey the proposed model using the notable trust association of advogato dataset. At last, makers contrast the proposed model and actually appropriated ones to the extent mean absolute bungle, execution period of variant trust computation, and gave features. The refined results are illustrative of the need of the proposed model to be used in certified cloud conditions.

Zou et al. [8] Access control expects a huge part in binding the passageway of unmistakable advantages, keeping from assault of unlawful customers or the damages achieved by legal customers' unintentional undertakings.

3 Proposed Work

The proposed concept is divided into two main segments,

1. Role-based security
   a. According to the role-based security, the files are assigned for access for the particular role.
2. TPA-based security
   a. TPA is the third-party auditor which will cross-check the user requests to access the file and generate OTP and access key.

3.1 New User Creation

[In order to access the services related to cloud sharing platform, first the user is required to be registered using the platform, then the user can access the services.]

Step 1: First the user name, name of employee, and role of employee are specified.
Step 2: Grid of the pictures is available for selection, and the user has to select single image from it.
Step 3: The image is partitioned in small segments, and the image is organized in the second grid.
Step 4: The size of the image selected is calculated and displayed in bytes.
Step 5: To use, then select the segments of image, the selected segment turns gray, and after all the desired image blocks are selected, click on the generate button.
Step 6: The pattern will be formed, on the basis of the selection of image block, and the basis of pattern is,

Image(ImageNumber)_part(partnumber1)_sizeofimage_
Image(ImageNumber)_part(partnumber2)_sizeofimage_
::::
Image(ImageNumber)_part(partnumberN)_sizeofimage_

Step 6: Save the pattern with the other details of the user in the database table meminfo, which contains the details of registered users.
Step 7: END.

3.2 Existing User Login

[Now, after the registration is done, the registered user can login using the procedure adopted for registration].

Step 1: Enter the user name.
Step 2: Grid of the pictures is available for selection, and the user has to select single image from it.
Step 3: The image is partitioned into small segments, and the image is organized in the second grid.
Step 4: The size of the image selected is calculated and displayed in bytes.
Step 5: To use, then select the segments of image, the selected segment turns gray, and after all the desired image blocks are selected, click on the generate button.
Step 6: The pattern will be formed, on the basis of the selection of image block, and the basis of pattern is,

Image(ImageNumber)_part(partnumber1)_sizeofimage_
Image(ImageNumber)_part(partnumber2)_sizeofimage_
::::
Image(ImageNumber)_part(partnumberN)_sizeofimage_

Step 6: If Details Correct then

Login Successful

Else.
Login Failed

[End of If structure.]

Step 7: END.

3.3 File Upload

[This algorithm is used to upload the file on the server, according to role.]

Step 1: Access username according to session variable.
Step 2: Select the file to share.
Step 3: Select the role for which the file is to be shared.
Step 4: Store the details in the database table fuploads.
Step 5: Stop.

3.4 File Request

[This algorithm is used to request the file access from TPA].

Step 1: Select the name of file for which access to be requested.
Step 2: Save details in the Reqdata which is table for user requests.
Step 3: Autoincrement-based request ID is generated.
Step 4: Stop.

3.5 TPA Grant Access

[This algorithm is used to grant the access to the user requesting the file.]

Step 1: Select the Request ID.
Step 2: Fetch the file details and user details.
Step 3: Generate OTP which generated using 10 random numbers, each ranges from 30 to 126, and the character corresponding to these number will be combined in order to form the OTP.
Step 4: Generate Hash using SHA-512 algorithm for file which is requested and Hash of user name who requested the file. Now, extract 20 characters from Hash of File and 20 characters of Hash of Username to generate the access key.
Step 5: Save the details in the database table Reqdata for the requested ID.
Step 6: End.

3.6 User Accessing File

[This algorithm is used access the file requested].

Step 1: Select the Request ID.
Step 2: Fetch the file details.
Step 3: Enter OTP and access key.
Step 4: If Details Correct then:

File Download

Else
Invalid Details
[End of If structure]
Step 5: End.

4 Implementation and Result Analysis

The implementation of the algorithms is graphically created using the VS 2010 and using the database system of MS SQL server. The system has the general requirement of the 2 GB RAM, and the processor required for the execution of the implementation is Intel I3 or more (Figs. 2 and 3).

Now, to start up the system flow, first we require the new user to be registered. In the registration process, we have made the conde of the graphical authentication system, in which we have the picture from grid and then that picture segmented and size of picture gets listed up. Now, to start up the system flow, first we require the new user to be registered. In the registration process, we have made the conde of the graphical authentication system, in which we have the picture from grid and then that picture segmented and size of picture gets listed up. The image-segmented blocks are then clicked to form the pattern as,

Image(ImageNumber)_part(partnumber1)

_sizeofimage_Iage(ImageNumber)_part(partnumber2)_sizeofimage_
::::Image(ImageNumber)_part(partnumberN)_sizeofimage_

Now, the user once registered then required to login in order to access the system for the purpose of requesting the file over the cloud server or to upload file over server (Fig. 4).

In the case of the File Access Module, now, the TPA will first select: Select the Request ID. Then, Fetch the file details and user details. After that, Generate OTP which generated using 10 random numbers and each range from 30 to 126, and the character corresponding to these number will be combined in order to form the OTP. Generate Hash using SHA-512 algorithm for file which is requested and Hash of user name who requested the file. Now, extract 20 characters from Hash of File and 20 characters of Hash of Username to generate the access key.

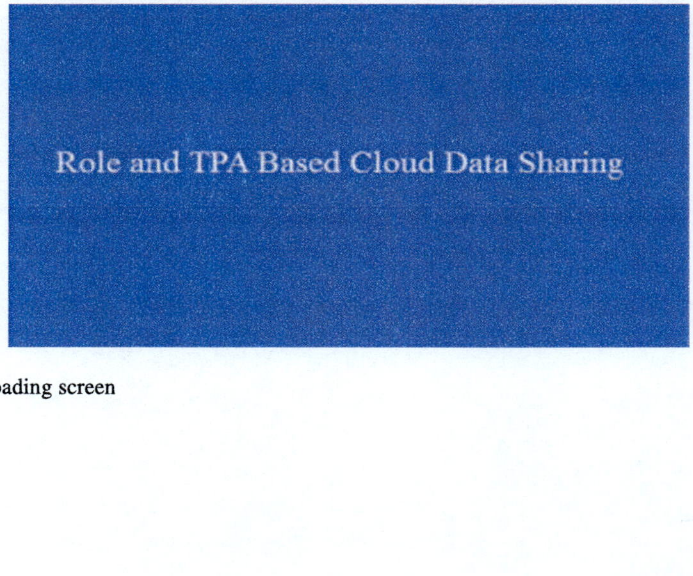

Fig. 2 Loading screen

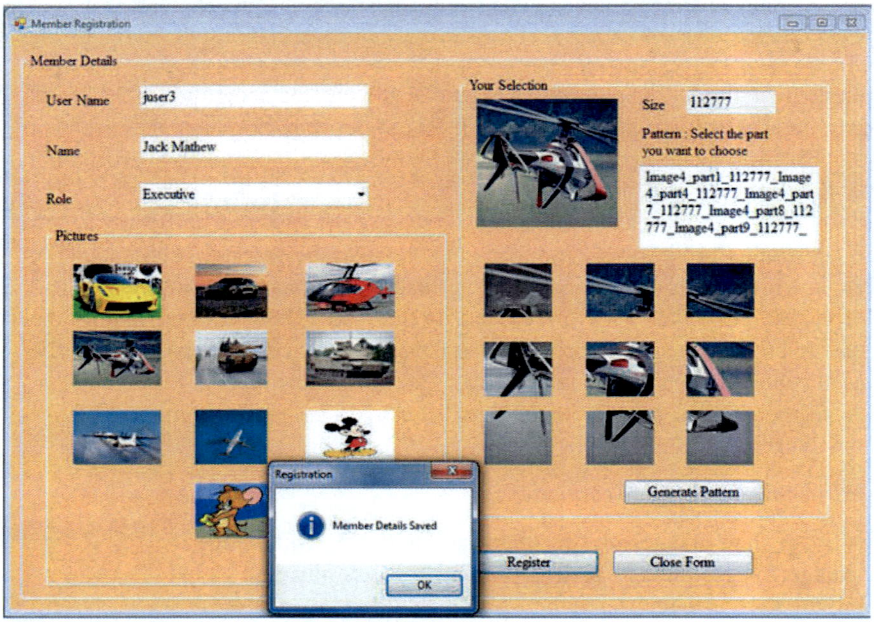

Fig. 3 User registration

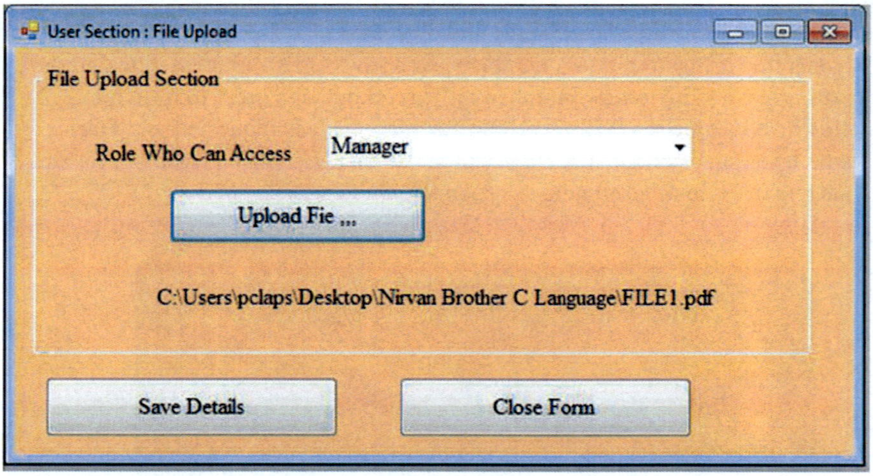

Fig. 4 File upload module

Table 1 Analysis of keys test 1

	Authentication key	Access key
Proposed approach	445.1	168.8

4.1 Device 1: Rumkin Test

This secret word checker will measure your secret key and give it a score dependent on how great of a secret phrase it is.

Authentication Key:

Image4_part1_112777_Image4_part4_112777_Image4_part7_112777_Image4 _part8_112777_Image4_part9_112777_

Access key:

e026ea23e40021c0086fB0BE79A29AB853C20553 (Table 1).

4.2 Device 2: Cryptool

CrypTool is an open-source tool for examining password strength (Figs. 5, 6, and 7; Tables 2 and 3).

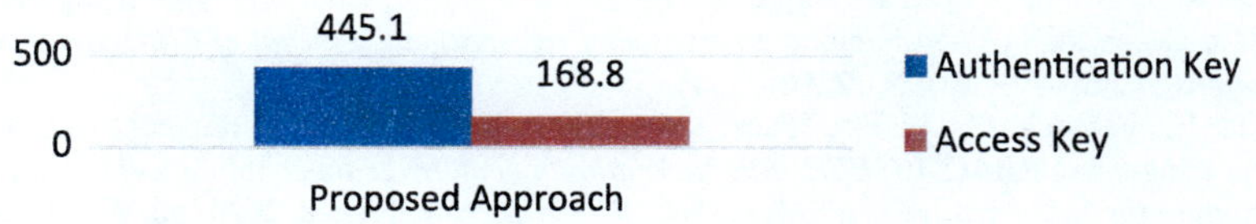

Fig. 5 Key Analysis 1 graph

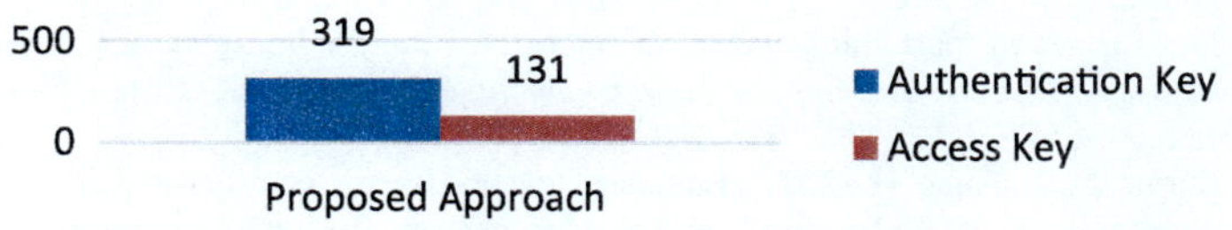

Fig. 6 Key Analysis 2 graph

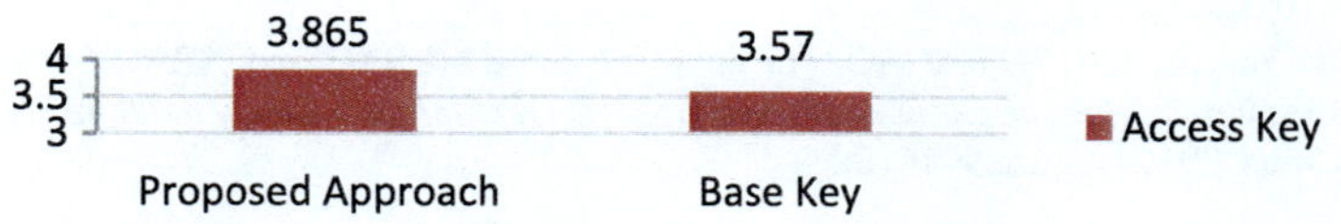

Fig. 7 Analysis 2 base and proposed graph

Table. 2 Analysis of keys test 2

	Authentication key	Access key
Proposed approach	3.621	3.865

Table. 3 Analysis of base keys test 2

	Access key
Base approach	3.57

5 Conclusion

Two main issues which are generally faced—when data is shared in cloud environment, first is authenticating the user who can access the data, and secondly, to secure the data itself. Seeing the concern, we proposed the hybrid concept which involves the role-based security as well as TPA-based security. The authentication pattern and access key are tested over various platforms and tools for testing password strength, and results are quite satisfactory.

References

1. Akkaoui, R., Hei, X., Guo, C., & Cheng, W. (2019). RBAC-HDE: On the design of a role-based access control with Smart contract for healthcare data exchange. In *IEEE international conference on consumer electronics—Taiwan (ICCE-TW)*, Yilan, Taiwan, 2019 (pp. 1–2).
2. Alaria, S. K., & Kumar, A. (2018). Implementation of new cryptographic encryption approach for trust as & service (TAAS) in cloud environment. *International Journal of Computers and Applications, 4*(July–August)(8), (2250–1797).
3. Belkhiria, H., Fakhfakh, F., & Rodriguez, I. B. (2020). Resolving multi-user conflicts in a Smart building using RBAC. In *IEEE 29th international conference on enabling technologies: infrastructure for collaborative enterprises (WETICE)*, Bayonne, France, 2020 (pp. 181–186).
4. Ghafoorian, M., Abbasinezhad-Mood, D., & Shakeri, H. (2019). A thorough trust and reputation based RBAC model for secure data storage in the cloud. *IEEE Transactions on Parallel and Distributed Systems, 30*(4), 778–788. https://doi.org/10.1109/TPDS.2018.2870652
5. Jadaun, A., Alaria, S. K., & Saini, Y. (2021). Comparative study and design light weight data security system for secure data transmission in Internet of things. *International Journal on Recent and Innovation Trends in Computing and Communication, 9*(3), 28–32. https://doi.org/10.17762/ijritcc.v9i3.5476
6. Kumar, M., Kumar, S., & Nagar, H. (2021). Enhanced text and image security using combination of DCT steganography, XOR Embedding and Arnold transform. *Design Engineering, 2021*(3), 732–739. ISSN: 0011-9342.
7. Liu, C., Cheng, S., Luo, Y., & Jiang, F. (2019). Behavior recognition based on RBAC-ensemble model. In *11th International conference on knowledge and smart technology (KST)*, Phuket, Thailand, 2019 (pp. 29–34).
8. Long, S., & Yan, L. (2019). RACAC: An approach toward RBAC and ABAC combining access control. In *IEEE 5th international conference on computer and communications (ICCC)*, Chengdu, China, 2019 (pp. 1609–1616).

Decision Tree Algorithm for Diagnosis and Severity Analysis of COVID-19 at Outpatient Clinic

Ritika Rathore, Piyush Kumar, and Rushina Singhi

Abstract This study investigates the feasibility of decision tree algorithm like CART recursive method for classifying participants into test-based positive cases and negative cases to detect COVID-19 in the outpatient and suggest admission or home isolation according to the evaluated parameters. It also evaluates the severity of the outpatients using the values of RTPCR test and Chest X-Ray imaging results. A theoretical and predicted decision tree is proposed in the study after focus group interview with a clinical physician. Primary data was collected from the survey of patients visiting a physician for treatment of COVID-19 during the first wave. CART algorithm was applied for predicting the required decision tree. According to the predicted decision tree, it was determined that the most important feature while treating a COVID-19 patient is their history of contact with the positive coronavirus patient. Based on the valuation of dataset, the predicted decision tree provided similar results to that of the conceptual tree. Thus, comparing both trees, it can be evidently said that the predicted decision tree is a subset of conceptual decision tree and can be used by physicians for diagnosis and severity analysis of COVID-19.

Keywords Decision tree · COVID-19 · Diagnosis and severity analysis · Healthcare · Coronavirus · Decision making · CART algorithm

---

R. Rathore (✉) · R. Singhi
Amity Business School, Amity University, Noida, India
e-mail: rathoreritika81@gmail.com

R. Singhi
e-mail: rsinghi@amity.edu

P. Kumar
Department of Computer Science and Engineering, ASET, Amity University, Noida, India

P. K. Singh et al. (eds.), *Proceedings of Third International Conference on Computing, Communications, and Cyber-Security*, Lecture Notes in Networks and Systems 421,
https://doi.org/10.1007/978-981-19-1142-2_13

1 Introduction

COVID-19 or coronavirus is a kind of virus that causes contagious disease in the human respiratory system. It is also termed as severe acute respiratory syndrome coronavirus-2 (SARS-CoV-2). The disease was unheard to the world until 2019 when Wuhan, China reported its first case of SARS-CoV-2. The humankind has been facing an ongoing pandemic due to this outburst of the coronavirus. The COVID-19 virus is similar in genetics to the SAR-CoV virus which was accountable for the epidemic back in 2002–2003 [1] which originated in Guangdong. The Middle East respiratory syndrome originated in the year 2012 [2].

After an individual is infected by the coronavirus, he or she may have symptoms of cold, fever, pneumonia, bronchiolitis, etc. [2]. Transmission of the disease is one of the biggest aspects due to which the world is suffering the pandemic. The transmission of the virus can be done from a sick patient. Individuals who have mild and asymptomatic infections can also lead to the distribution of virus [2]. Other than these, two main reasons for spreading the virus are respiratory droplet transmission and close contact transmission [2]. The growth of the respiratory droplet transmission is mainly due to coughing, sneezing, or talking. The extended distance for droplets more than 5 μm in diameter is limited and generally less than 1 m. Hence, in case of close contact, the virus can easily contact with person or droplets can contaminate the surface of the objects [2]. If the patient's hands make contact with the polluted environment or object, the hands will also be polluted. The contaminated hands can further make contact with the adenoidal cavity, face, oral cavity, etc. which might lead to communicating of virus via close contact [2]. The virus can spread via aerosol present in the confined space, with aerosol exposed for a long period. The aged pupils are excessively prone toward the virus while the children and infants experience milder effects during the first wave.

According to World Health Organization (WHO) by January 21 2021, there have been over 95 million cases of SAR-CoV-2 worldwide and over 20 lakh people have died due to this infectious virus. 224 counties, territories have been affected by this virus (Coronavirus disease (COVID-19) pandemic 2021). In India, there have been over 10 million confirmed cases and over 15 thousand people have died due to the coronavirus by January 21 2021. India ranks 2nd highest in confirmed cases after the USA [3].

With no antecedent of COVID-19, any aid with respect to diagnosis of disease and its progression has been welcomed in healthcare sector. Decision making through algorithms have always been utilized by clinicians. It has been seen that the role of technology in the field of health care sector has been the biggest factor for the advancement of vaccines, detecting and analyzing the disease at an early stage, conducting complicated surgeries, etc. More prominently, with the development of machine learning and artificial intelligence, a novel model has been established for the healthcare sector. Diverse machine learning algorithms such as support vector machine (SVM), convolutional neural network (CNN), Fuzzy C-means, KNN, decision tree, regression models, have been widely and still a portion of research in current years [4].

The decision tree is a unique, powerful as well as well-known tools for classification and prediction purposes [5]. It is generally utilized for explicit and visual representation of decisions/decision making [6]. A decision tree, as the name goes, is a flow chart forming a structure of the tree, normally upside down, where every inner node signifies a test on the attribute, each branch provides an outcome on the test and each leaf node holds a class label [5]. The decision tree has its huge impact on the health care sector where hospitals use some kind of decision tree classification to pre-determine the seriousness of illness [7]. Whether it is heart disease, thyroid disease prediction, false alarm prediction in ICU, etc. decision tree is very helpful. The emergency departments in the hospital ask series of questions like age, symptoms, etc. before admitting the patients. Yoo et al. [8] proposed deep learning-based decision tree classifier for pre-screening patients to conduct triage and fast-track decision making before RTPCR results are available in COVID-19 diagnosis. A decision tree model can be developed which could assist the department to take vital decisions [7]. Hence, a decision tree is very crucial in medicinal analysis. Tanner et al. [9] showed a proof-of-concept that decision algorithms based on simple clinical and hematological criteria can accurately predict dengue disease diagnosis and prognosis, a finding that could help with disease management and surveillance. Shouman et al. [10] evaluated the execution of the other decision trees. Their research recommends a framework that outclasses J4.8 decision tree as well as bagging algorithm during the analysis of heart ailment patients.

There are majorly five different decision tree algorithms [11]:

- Iterative Dichotomiser 3 (ID3)
- C4.5 (Successor of ID3)
- Classification and Regression Tree (CART)
- Chi-Square automatic interaction detection (CHAID)
- MARS.

1.1 Objective

In this paper the authors will be restricting themselves to use CART recursive partitioning procedure for the analysis and severity of COVID-19 which is the core purpose of the paper. Since developments of patients' symptoms may vary from the better-known inpatient disease, a collection and analysis of data set of clinical indicators for coronavirus will assist in identifying patients' symptoms and those who would benefit mostly in circumstances of restricted testing availability [12].

2 Literature Review

Van Pelt et al. [13] signifies an approach toward identifying the coronavirus infection among college students returning to campus. The tests recommended for students vary from zero to two tests per student. This paper further signifies the strategy that adjusts the number of positive and negative cases identified and the RTPCR test needed. Five different schemes of tree diagram analysis were evaluated. The 1st scheme was to classify students with clear indications of coronavirus. Secondly, the RTPCR test was recommended to all symptomatic students. The third strategy consists of the RTPCR test for all students. The fourth strategy was to conduct an RTPCR test for all students and all other symptomatic students whose first test was negative. The fifth strategy was to test all students and retesting all students whose first test was negative. The tests were conducted for 20,000 students returning to the campus. The fifth strategy resulted in the truest positive but also requires more tests. From strategy 1 to 5, the percentage of correctly identifying the infection was 40.6%, 29%, 53.7%, 72.5% and 86.9%, respectively. The examination of true positive proved that the reappearance of the RTPCR test dominated the single RTPCR test strategy and the emergence of more intensive RTPCR test decrease with the widespread increase of infection. As a conclusion, based on true positives, the single RTPCR test is never preferred.

Wiguna et al. [14] observed and responded to activities to identify PDP, ODP, OTG, or confirmed cases of coronavirus. C4.5 algorithms was applied for SARS-CoV-2 surveillance. Secondary data was extracted from state administration publications or some official websites of health agencies. The testing method in a three-class confusion matrix yielded an accuracy rate of 92.86%, placing it in the excellent classification group.

The [15] paper signifies major concerns over increased thyroid cases in women over the age of 30 years. In the paper, the investigators have applied the information excavation techniques like the random forest, decision tree, and Classification and Regression tree (CART) on the thyroid illness dataset. The results were further enhanced using the bagging ensemble technique. The thyroid disease dataset consists of 3710 cases and 29 features of thyroid patients. The dataset was distilled from the UCI-machine learning repository. The consequences of each of the classifiers were 98%, 99%, and 93% for the decision tree, random forest, and CART, respectively. This accuracy was obtained based on distinct num-fold and seed values. The bagging ensemble method gave better precision of 100% after combining all three trees. The seed value 35 num-fold value 10 was used in the bagging ensemble method. According to Yadav et al., this aimed model can be used for better forecasting of thyroid disease.

Tanaka and Voigt [16] focuses on malignancy in the liver transplant. Non-melanoma skin tumor like basal cell carcinoma or squamous cell cancer is believed to be the malignancy in the liver transplant. A tree diagram analysis was used to formulate an instrument to distinguish and evaluate the risk of non-melanoma skin cancer on a liver transplant. The data was extracted from the Organ Procurement Transplant Network (OPTN) star files of September 2016. The Cox regression analysis constituted to identify variables for the tree diagram analysis. The dataset consists of around 105,948 patients out of which 4556 patients tolerated from non-melanoma skin tumor. The mean of these patients suffering from skin cancer after transplantation is 5.6 years. The cox regression extracted features like gender, age, body-mass index (BMI), Caucasian race, and sirolimus of the patients during liver transplantation. The resulted tree showed that the non-Caucasians a low risk of about 0.8% whereas Caucasians males above 47 years of age with a body-mass index (BMI) below forty who did not get sirolimus are at a high chance of suffering from non-malignant skin tumor. This constitutes about 7.3% of the patients. According to Tanaka et al., the proposed decision tree framework precisely predicts and distinguishes the danger of arising non-melanoma skin tumor in long term after liver transplantation.

Manna et al. [17] focuses on proposing a novel approach in detecting the false alarm in the Intensive Care Unit (ICU) during arrhythmia. To detect the false alarm various feature inputs like electrocardiogram (ECG) signals, atrial blood pressure photoplethysmogram signals (PLETH), and respiration were used. Decision tree predictive learner is used to predict the false alarm. Since the dataset was small so it was divided in the ratio of 70/30. After the primary tree was achieved, it was pruned to achieve better accuracy. Agreeing to the method proposed in the paper the precision of forecasting false alarms is 97%.

Table1 shows the comparison between the review works already done in this background.

Table 1 Comparison table of review work

S. No.	Author	Accuracy	Model
1	Tanaka & Voigt [16]	99.2	A tree diagram analysis was used to formulate an instrument to distinguish and evaluate the risk of non-melanoma skin cancer on a liver transplant
2	Manna et al. [17]	97%	Detecting the false alarm in the Intensive Care Unit (ICU) during arrhythmia
3	Yadav et al.[15]	98%, 99%, and 93% for the decision tree, random forest, and CART, respectively	Random forest, decision tree, and Classification and Regression tree (CART)
4	Wiguna et al. [14]	92.68%	C4.5 algorithm
5	Van Pelt et al. [13]	Accuracy of 5 different strategies: 0.6%, 29%, 53.7%, 72.5%, and 86.9%, respectively	5 different strategies were used

3 Methodology

The anticipated procedure is a conceptual representation of how a COVID-19 confirmed case is detected. A consultation was done with the doctors treating COVID-19 patients and based on their diagnosing patterns for certain clinical indicators presented to them, a decision tree flowchart was proposed. Later on, we will be using primary data to evaluate similar kind of decision tree.

Figure 1 represents the portrayal of a decision tree, representing a classification model for COVID-19 analysis that can be used to forecast whether a person is a patient with COVID-19 infection or not, based on the COVID-19 detection traits which was obtained from the doctors. The same tree can also can be referred to find out the severity of the infection from the analytical outcomes with a number of trial indicators. The model will assist the medicinal staffs in analyzing the above-mentioned indicators of the analysis evaluation quickly and judge the severity of the infection.

During the pandemic, when a patient visits the doctor, he/she may or may not be a patient of coronavirus. Thus, the root node, which consists of symptoms like fever, chills, sore throat, headache, etc. along with coronavirus exposure, can be broadly classified into four categories.

The root node contains the basic clinical indicators measured for the early diagnosis of the COVID-19 infection and these diagnostic parameters can be tabulated as shown below in Table 2.

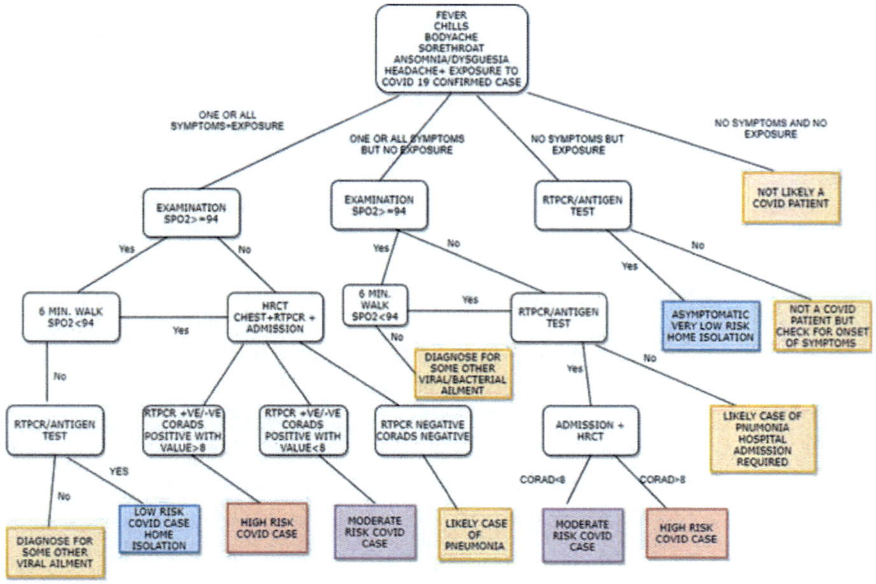

Fig. 1 Decision tree depicting detection of COVID-19

Table 2 Diagnostic parameters for initial screening

S. No.	Questions asked	Parameter assessed	Possibility
1	Have you come in contact with a confirmed COVID-19 patient?	Exposure	Yes/No
2	Have you felt a recent loss of taste or smell?	Ansomnia/Dysguesia	Yes/No
3	Do you have fever with body temperature more than 100 degree Celsius?	Fever	Yes/No
4	Do you have a cold with running nose?	Cold	Yes/No
5	Do you have cough?	Cough	Yes/No
6	Do you have chills?	Chills	Yes/No
7	Do you have a persistent headache?	Headache	Yes/No

For testing of people with severe symptoms as a probable coronavirus situation. It is significant to note the order of the enquiries as mentioned in the Table 1. If a single person 1st reports of been exposed to a coronavirus situation, the person is likely to have COVID-19 infection, and thus testing is suggested. If the person denies exposure to a coronavirus case, then the next question by the doctor should be, "Have you felt a current loss of taste or smell?" and the related symptoms questions. If the response is objectionable, then the necessity for testing is low, as maximum percentage of these pupils' test outcomes will be undesirable for coronavirus, thus no test is recommended to those people. A disagreeable outcome to these two questions can be clearly visible in the left side of the Fig. 1 would allow considerable number of the people to be eliminated as probable negative cases without admission and inpatient testing/X-ray imaging. All four categories are discussed in detail below:

1. *Either one or all symptoms along with coronavirus exposure*

In this case, a person may visit to doctor with one symptom, few symptoms or all symptoms along with exposure of coronavirus. When visited, and the basic diagnostic parameters all positive, the first step the doctor take is to examine the oxygen saturation level (SpO_2) of the patient. A normal body has a SpO_2 ranging from 95–100%. Hence, a visiting patient may or may not have oxygen saturation above 94%.

1.1 *Case-1*

If the SpO_2 level is greater than 94%, the patient is advised to take a six-minute walk. After a walk, the SpO_2 level is estimated again. Again, it may happen that the SpO_2 level either lowers down or stays above 94%. When the SpO_2 level does not decrease below 94%, the risk of respiratory illness due to SARS-CoV-2 is minimized but the outpatient is still recommended to go through an RTPCR test. If the test results positive, the patient is at low risk of coronavirus and is recommended for home isolation. If the test results are negative, the patient has visited the doctor for some other viral ailment and may be treated for the same.

1.2 *Case-2*

In this case, if the SpO_2 level is either less than 94% or if the oxygen saturation level decreases below 94% after taking a six-minute walk (as discussed in *Case-1*), the patient is immediately admitted in the hospital and undergoes HRCT test and RTPCR test. The RTPCR test is most likely to be positive, but might give either positive or negative results due to the space of the sickness in which the test was conducted; hence, a high-resolution chest computed tomography scan (CT scan) is done. A HRCT scan especially for COVID-19 detection provide the chest imaging wherein a score is given in terms of CORAD value which indicates the number of chest segments involved or damaged due to coronavirus. If a patient's CORAD value is greater than 8, in that case patient has high severity of COVID-19 Infection. If the CORADS value is less than eight, the patient is moderately severe COVID-19 infection. If both RTPCR and CORADS result is negative, then the patient is not likely a COVID-19 patient but may be suffering from other respiratory illness like pneumonia.

2. *Either one or all symptoms but no exposure to coronavirus*

In this case, a person may visit to doctor with either one symptom or all symptoms similar to that of coronavirus without actually coming in connection with a confirmed COVID-19 case or being unaware of it. During visit of the patient, the first step the doctor take is to examine the oxygen saturation level (SpO_2) of the patient. The visiting patient may or may not have oxygen saturation above 94%.

2.1 *Case-1*

If the SpO_2 level is greater than 94%, the patient is advised to take a six-minute walk. If the SpO_2 level does not decreases, then the patient is suffering from any other viral ailment or bacterial ailment.

2.2 *Case-2*

If the SpO_2 level decreases below 94% during testing or it decreases after a 6-min walk, then the patient is hospitalized and recommended for an RTPCR test as well as antigen test. If the tests are positive, then the inpatient undergoes an HRCT test. During the HRCT test, if CORADS is greater than 8 then the inpatient is at high risk of coronavirus and if the CORADS is lower than 8, the patient is at mild risk of COVID-19 infection. If the RTPCR test and antigen test come out to be negative, in that case, the patient is suffering from pneumonia, and hence the patient should be admitted immediately.

3. *No symptom but exposure to coronavirus*

If a patient has been exposed to coronavirus but does not reflect any symptoms, then they are recommended for an RTPCR test and antigen test. If the test results positive, the patient has a less danger of the virus and they are recommended for home isolation. If the report is negative then the patient is not declared as a coronavirus patient rather, clinicians advise the patient to be vigilant for the commencement of symptoms.

4. *No symptoms and no exposure*

If a patient does not have any of the symptoms and no contact to coronavirus, then the patient is probably not a COVID-19 patient and must have visited the clinic in concern for some other illnesses not including the diagnostic parameters as discussed in Table 2.

3.1 Data Collection

The data used during the research work was the primary data. It was fetched from general physician of a private clinic who is examining coronavirus patients. The dataset consists of 51 entries and 10 attributes. The various attributes consist of *"age", "sex", "SpO_2 at initial level", "RTPCR", "HRCT test"*, etc. The dataset provides broad scope to predict the severity of coronavirus.

3.2 SPSS Modeler

The software used for the prediction of decision tree was SPSS Modeler. SPSS Modeler is one of the leading software provided by IBM for data analytics and machine learning. The software assists in accelerating the operational tasks for analysis of data for data mining enthusiasts. In Fig. 2 we can see the flow diagram of evaluation of dataset in SPSS Modeler.

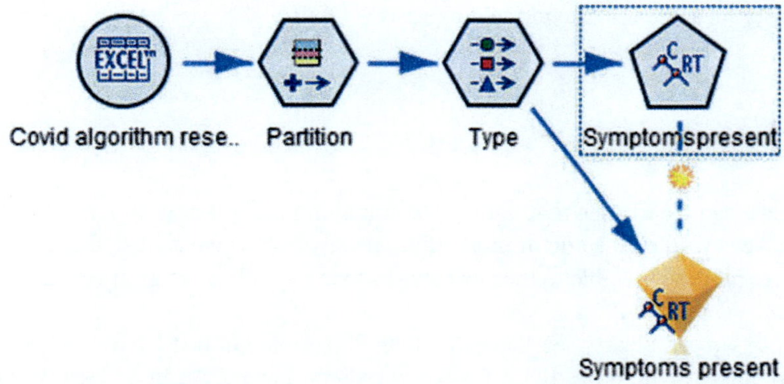

Fig. 2 Represents the flow diagram of evaluation of dataset in SPSS modeler

3.3 Partitioning and Feature Selection

The dataset was partitioned in the ratio of 60:40 where, 60% of the data was employed for training purposes and 40% was employed for the testing purposes. The dataset consisted of ten diverse features. Out of these ten features, four features were removed. Attribute *"Symptoms present"* was used as a target variable and *"SpO_2 at initial checkup", "RTPCR", "HRCT test", "hospitalized or isolated at home"*, and *"history of contact with COVID-19 patient"* were utilized as an input feature of the decision tree.

3.4 CART—Classification and Regression Tree Algorithm

In this research work, we will be using CART algorithm for prediction of confirmed cases of coronavirus. CART, as the term recommend, works for classification as well as regression problems. In our study, we have used classification feature to predict the desired outcome.

4 Results

The results shown below represents the predictor importance and predicted decision tree. Maximum tree depth was set as 5. Minimum change in impurity was 0.0001 and impurity measure for categorical targets was GINI.

4.1 Data Analysis

In Fig. 3, we can clearly see that the people belonging to the age group of 60 years or more have consulted the doctor more often. It can clearly give an idea that the old people were more vulnerable to the coronavirus than any other age group during the first wave.

In Fig. 4, we can clearly see that more than 30 people out of 51 have history of coming in contact with a positive COVID-19 patient. Hence, it can be clearly said that this feature is the most important feature for detection of coronavirus.

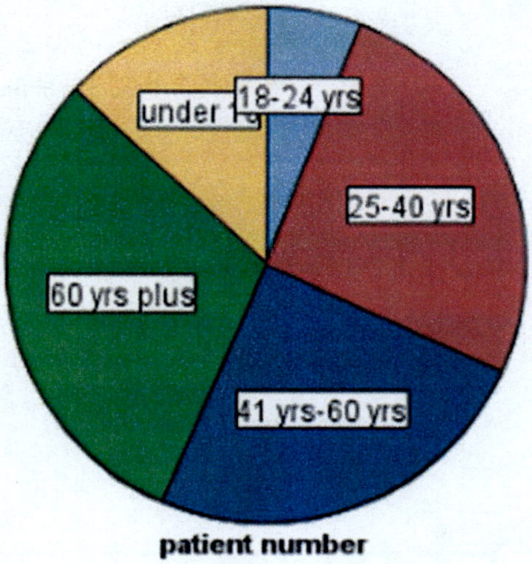

Fig. 3 Pie chart showing different age groups of patients visited doctor

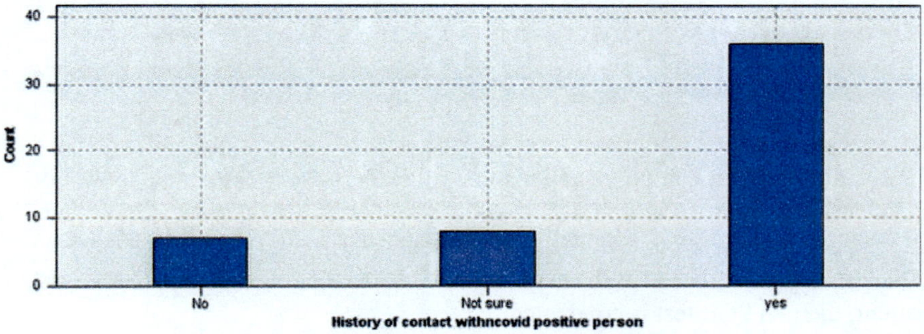

Fig. 4 Represent the count of people who have history of contact with COVID-19 patient

4.2 Predicted Decision Tree

Below is graphical representation of Predictor Importance and flow diagram of decision tree.

Figure 5 represents the predictor importance of the input attributes used for the evaluation of decision tree. According to the graph, *"History of contact with COVID-19 positive person"* is most important attribute whereas the *"HRCT test"* was found to be least important. The predictor importance is generated when Tree is created using SPSS Modeler. In making of the tree shown in Fig. 4, the predictor importance plays significant role along with attribute. In other words, it can be said that the above three features are shown in predictor importance are key attributes of the tree.

Figure 6 represents the predicted decision tree using the primary data. Now, observing at the decision tree, following predicted outcomes can be seen. These are:

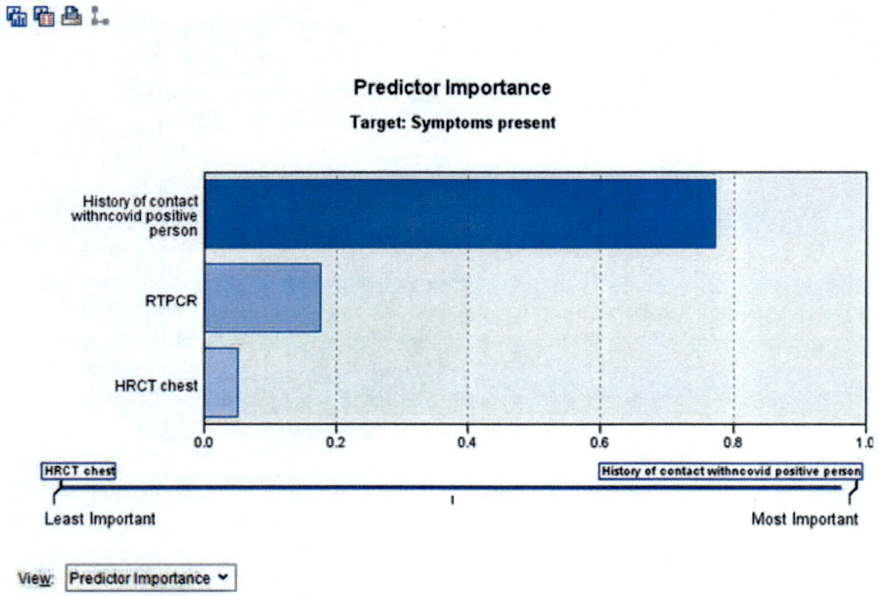

Fig. 5 Graphical representation of predictor importance

- Firstly, at node 0, it can be visualized that after the patient explained about his sickness to the doctor, the first doctor ask is whether or not the patient has come in contact with any positive COVID-19 patient.
- If the patient is sure that he or she has come in contact with the positive coronavirus patient, then RTPCR test is conducted.
- If the test outcomes are positive, then keeping in mind that the patient came in contact with positive coronavirus person, HRCT test is conducted.
- If HRCT test is positive with a CORADS value less than 8, then the person has moderate risk of COVID-19 virus.

5 Discussion

Referring to Fig. 1, high risk COVID-19 confirmed cases occur only at two times. First, when a patient has all symptoms along with coronavirus exposure, and second when a patient has all symptoms but has not been exposed to the virus in his awareness. In either case, there remains a probability of having a high risk confirmed case. In case 4 of Fig. 1, when a patient does not have any kind of symptoms or exposure to the coronavirus, the patient might have visited the doctor for other illness. In all the other situation the patient suffers from pneumonia. Moreover, the main focus relies on the 1st situation where the patient suffers from all symptoms and has exposure to

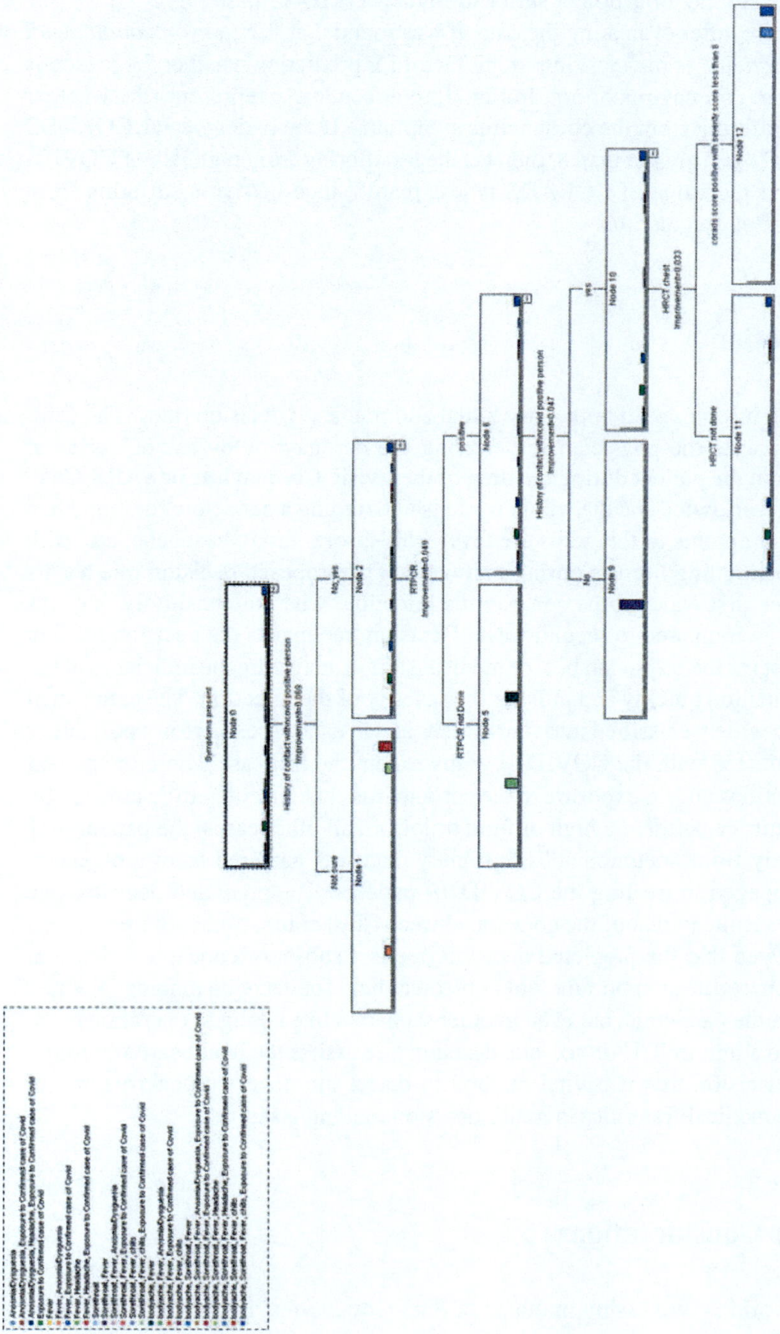

Fig. 6 Predicted Decision Tree

the coronavirus because theoretically there are few cases where a patient, even after being exposed to the virus do not suffer from the COVID-19 case.

Furthermore, after evaluating the data, it was found that *"History of contact with COVID-19 patient"* is the most important factor for predicting whether the person is suffering from coronavirus or not. In Fig. 4, the tree clearly represents the whether the person suffering from the coronavirus symptoms. If the patients have CORADS value of HRCT test greater than 8, then he/she is suffering from high risk of COVID-19 case but if the value of CORADS is less than 8, then he/she is suffering from moderate risk of coronavirus.

6 Conclusion

This research has proposed both, conceptual and predicted decision trees. The decision tree supports the physician in detecting the coronavirus by asking series of questions from the patient during the time of their visit. Coronavirus or SARS-CoV-2 has spread worldwide and the whole world is undergoing a pandemic for more than a year. The symptoms of this virus are fever, chills, sore throat, headache, etc. with few cases getting ling fibrosis during first wave. The proposed decision tree has its validity in the first wave of the coronavirus infection, when the positivity rate was much lower as compared to second wave. Decision tree makes the decision making easier not just for the physician but for medical staff in analyzing the indicators of the analysis evaluation quickly and judging the severity of the infection. The conceptual tree has been widely classified into four classes and in each case, there is a possibility of getting infected with the COVID-19 virus except the last case where the patient has no indications and no exposure to the coronavirus. The risk of getting affected by the coronavirus can either be high or mild or low. In all other cases, the patient may suffer severely from pneumonia. The primary data was gathered from the general physician engaged in treating the COVID-19 patients. The predicted decision tree gave similar results to that of the conceptual tree. Thus, comparing both trees, it can be clearly stated that the predicted decision tree is a subset of conceptual decision tree. Decision tree is an important tool in medical field for decision making. Not just in the time of the pandemic, but even in other scenarios like taking to operation room or whether to admit in ICU or not, etc. decision tree assists the healthcare fraternity. Hence, the decision tree is a vital method to detect the novel coronavirus and in assisting the medical fraternity in major decision making tasks.

7 Future Considerations

The entire world is witnessing, mutants of Sars-Cov-2 wreaking havoc throughout the world through second wave and third wave. Although the spread of the virus is steadily declining in India, the decision tree proposed in this paper can assist the healthcare fraternity to triage the flu like symptoms which might possibly be coronavirus infection and thus help in diagnosis and severity analysis of the disease. The paper may give way to find out predictor importance in diagnosis of COVID-19 in future waves of different mutants. Also, it would help in the vaccination drive. In India still more than 50% of the population is not vaccinated. People who are filtered out from testing might be less probable to have been infected by the coronavirus thus may be line-up for vaccination when supplies are limited. Besides, recognizing those who are less expected to have been unprotected to the coronavirus possibly will instantaneously recognize those who might respond well to a possible inoculation.

Limitations: The present study took inputs from medical practitioner/ general physician with a small number of databases. The same need to be checked on larger databases for assessing the validity of the decision tree.

Acknowledgements The authors thank Dr. Abhishek Rathore for his insights on the current situation of COVID-19 pandemic and sharing his patient database for research work.

References

1. Koley, T. K., & Dhole, M. (2020). The COVID-19 pandemic: The deadly coronavirus outbreak. The COVID-19 Pandemic: The Deadly Coronavirus Outbreak (pp. 1–157). *Taylor and Francis*. https://doi.org/10.4324/9781003095590
2. Qu, J.-M., Cao, B., & Chen, R.-C. (2021). Chapter 1—Respiratory virus and COVID-19. In J.-M. Qu, B. Cao, & R.-C. Chen (Eds.), *COVID-19* (pp. 1–6). Elsevier. https://doi.org/10.1016/B978-0-12-824003-8.00001-2
3. WHO Coronavirus Disease (COVID-19) Dashboard. (2021, January 21). Retrieved from World Health Organization: https://covid19.who.int
4. Gao, W., Bao, W., & Zhou, X. (2019). Analysis of cough detection index based on decision tree and support vector machine. *Journal of Combinatorial Optimization, 37*(1), 375–384. https://doi.org/10.1007/s10878-017-0236-8
5. Gupta, S. (2019, April 17). Decision tree. Retrieved from GeeksforGeeks: https://www.geeksforgeeks.org/decision-tree/
6. Gupta, P. (2017, May 18). Decision trees in machine learning. Retrieved from Towards Data Science: https://towardsdatascience.com/decision-trees-in-machine-learning-641b9c4e8052
7. Qiu, S. (2020, March 3). Why decision trees could help save lives. Retrieved from Towards Data Science: https://towardsdatascience.com/why-decision-trees-could-help-save-lives-a9fbaf15e7ef
8. Yoo, S. H., Geng, H., Chiu, T. L., Yu, S. K., Cho, D. C., Heo, J., Choi, M. S., Choi, I. H., Van Cung, C., Nhung, N. V., Min, B. J., & Lee, H. (2020). Deep learning-based decision-tree classifier for COVID-19 diagnosis from chest X-ray imaging. *Frontiers in Medicine, 7*(July), 1–8. https://doi.org/10.3389/fmed.2020.00427
9. Tanner, L., Schreiber, M., Low, J. G. H., Ong, A., Tolfvenstam, T., Lai, Y. L., Ng, L. C., Leo, Y. S., Puong, L. T., Vasudevan, S. G., Simmons, C. P., Hibberd, M. L., & Ooi, E. E. (2008). Decision tree algorithms predict the diagnosis and outcome of dengue fever in the early phase of illness. *PLoS Neglected Tropical Diseases*, *2*(3). https://doi.org/10.1371/journal.pntd.0000196
10. Shouman, M., Turner, T., & Stocker, R. (2010). Using decision tree for diagnosing heart disease patients. *Conferences in Research and Practice in Information Technology Series, 121*, 23–30.

11. Decision tree learning. (2021, January 10). Retrieved from Wikipedia: https://en.wikipedia.org/wiki/Decision_tree_learning
12. Zimmerman, R. K., Nowalk, M. P., Bear, T., Taber, R., Clarke, K. S., Sax, T. M., Eng, H., Clarke, L. G., & Balasubramani, G. K. (2020). Proposed clinical indicators for efficient screening and testing for COVID-19 infection using classification and regression trees (CART) analysis. *Human Vaccines and Immunotherapeutics*. https://doi.org/10.1080/21645515.2020.1822135
13. Van Pelt, A., Glick, H. A., Yang, W., Rubin, D., Feldman, M., & Kimmel, S. E. (2021). Evaluation of COVID-19 testing strategies for repopulating college and university campuses: a decision tree analysis. *Journal of Adolescent Health, 68*(1), 28–34. https://doi.org/10.1016/j.jadohealth.2020.09.038
14. Wiguna, W., & Riana, D. (2020). Diagnosis of coronavirus disease 2019 (COVID-19) surveillance using C4.5 algorithm. *Jurnal Pilar Nusa Mandiri, 16*(1), 71–80. https://doi.org/10.33480/pilar.v16i1.1293
15. Yadav, D. C., & Pal, S. (2020). Prediction of thyroid disease using decision tree ensemble method. *Human-Intelligent Systems Integration, 2*(1–4), 89–95. https://doi.org/10.1007/s42454-020-00006-y
16. Tanaka, T., & Voigt, M. D. (2018). Decision tree analysis to stratify risk of de novo non-melanoma skin cancer following liver transplantation. *Journal of Cancer Research and Clinical Oncology*. https://doi.org/10.1007/s00432-018-2589-5
17. Manna, T., Swetapadma, A., & Abdar, M. (2019). *Decision tree predictive learner-based approach for false alarm detection in ICU*.

CSBRCA: Cloud Security Breaches and Its Root Cause Analysis

Vivek Kumar Prasad, Vipul Chudasama, Akshay Mewada, Madhuri Bhavsar, and Asheesh Shah

Abstract Over the last three decades, the computing world has shifted from centralized to distributed systems, and we are now returning to virtual centralization like Cloud Computing (CC) systems. An individual user has complete control over the data and operations in his or her machine. On the other hand, there is CC, in which the operation and data preservation are given by a vendor, leaving the client/customer oblivious of where the activities are operating or where the information is saved. As a result, the customer has no management control over it. The Internet is used as a communication medium in CC. When it comes to data security in the cloud, the vendor must provide some guarantee in service level agreements (SLA) to persuade the user on privacy concerns. As a result, the SLA must specify several degrees of security. Their complexities are based on the services in order for the customer to comprehend the security measures that are in operation. Regardless of the supplier, there must be a defined method for preparing the SLA. The results obtained here indicates management of the SLAs using the Vitrage plugin of the Open stack public cloud eco-system. The results readings were based on the CPU utility of the cloud IaaS resources. The proposed approach shows how the prediction of the resources are managed in real-time when the occurrence of the attack will be identified.

Keywords Cloud computing · Security · Service level agreement · CPU Utility · Computing

V. K. Prasad · V. Chudasama (✉) · A. Mewada · M. Bhavsar
Institute of Technology, Nirma University, Ahmedabad, India
e-mail: vipul.chudasama@nirmauni.ac.in

V. K. Prasad
e-mail: vivek.prasad@nirmauni.ac.in

A. Mewada
e-mail: akshay.mewada_jrf@nirmauni.ac.in

M. Bhavsar
e-mail: madhuri.bhavsar@nirmauni.ac.in

A. Shah
Mewar University, Chittorgarh, Rajasthan, India

P. K. Singh et al. (eds.), *Proceedings of Third International Conference on Computing, Communications, and Cyber-Security*, Lecture Notes in Networks and Systems 421,
https://doi.org/10.1007/978-981-19-1142-2_14

1 Introduction to Cloud Security

CC has grown through many implementations such as application service provision (ASP), utility computing, and grid computing. The architecture of clouds varies depending on the services they supply [1]. The data is housed in centralized locations known as data centres, which have vast amounts of data storage. Both the data and the processing are stored on servers. As a result, clients must rely on the provider for reliability as well as data protection. The Service Level Agreement (SLA) is the only legal agreement between the client and vendor. The SLA is the only way for the supplier to win the client's trust, hence it must have a standard and always be followed. It should: identify and clarify the customer's demands, therefore provide framework for understanding, simplify complex problems, reducing overall areas of conflict, encourage discussion in the event of disagreements, eliminate unreasonable expectations if done appropriately [2]. A security-based resource management technique that manages cloud resources automatically and delivers safe cloud services is required to provide a secure cloud service [3]. We present Cloud Security Breaches and its RooT Cause analysis (CSBRCA) framework, a self-protection approach in cloud resource management that enables self-protection against security assaults and ensures continuing availability of services to authorized users in this work. The vitrage methodology was used to assess CSBRCA's performance. To accomplish this, a security-based resource allocation method that controls cloud resources and provides secure cloud services is necessary. A computing system's ability to defend itself against attacks and intrusions is known as self-protection. A self-protection component assists in recognizing and distinguishing threatening conduct and reacts independently to protect itself from harmful attack. These systems protect themselves against attackers by distinguishing between legitimate and illegitimate activities and taking the necessary procedures to prevent such attacks without the user's knowledge.

Figure 1 shows the growing cyber-attacks in the field of CC [4]. This also indicates that in the future this may increase, and these are issues that need to be resolved. Motivated from this, we have come up with three layer architecture and named this as a CSBRCA. Figure 2 describes the CC's risk areas that need to be solved out to maintain the SLAs [5].

1.1 Contribution

- Maintaining the resources in the dynamic environment of the cloud computing is difficult, and this requires the monitoring of the current resources of the cloud infrastructure.
- The SLA will be managed in autonomous ways based on the current demand and availability of the IaaS infrastructure.
- The cloud service provider will be intimated about every rise in the resource demand so that the appropriate actions need to be taken to manage the resources.

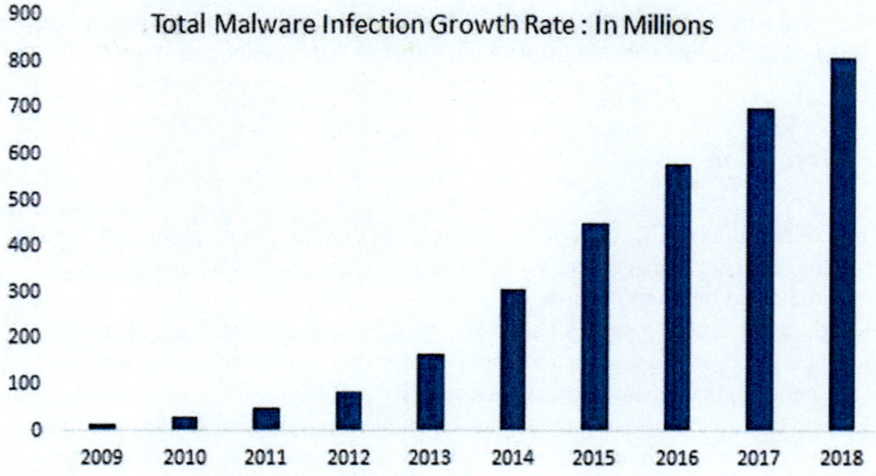

Fig. 1 Growing rate of the cyber-attacks in the cloud computing

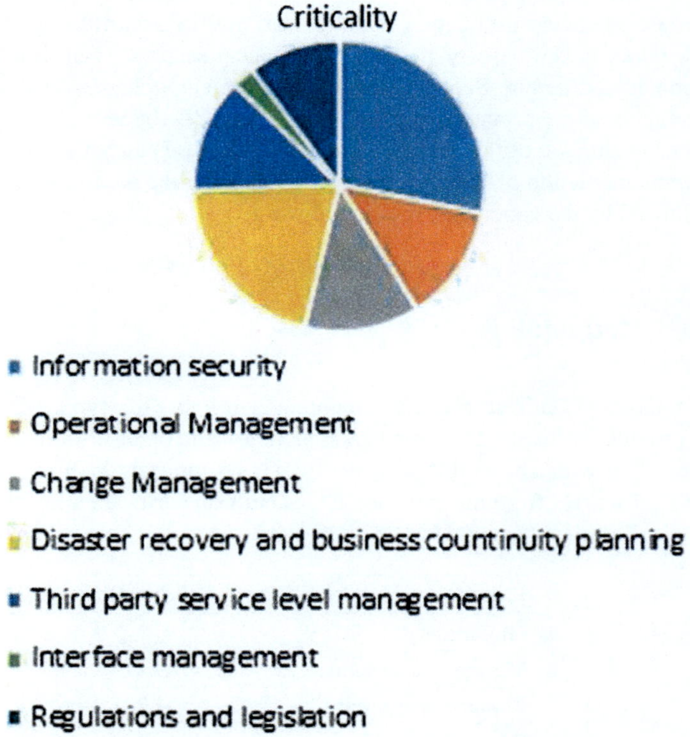

Fig. 2 CC and virtualization risk areas

Every security breach and its affect on the cloud resources will be notified or identified and can be used to mitigate the affect of the security breach.

1.2 Motivation

- The most significant CC concern, according to the literature study, is security. Data is no longer under management's control and is vulnerable when applications and data are hosted by a service provider.
- Unauthorized access to applications and data hosted on shared infrastructures increases the risk of unauthorized access and raises issues such as data privacy and identity management and network security and physical security concerns. SLA and third-party provider management, vendor lock-in, quality of service and viability of vendors are amongst other concerns. Data and application management, workload control (including performance), change control and availability of service are also concerns (Table 1).

Section 1 describes the introduction of the Cloud and its security importance. As the cloud works in the base of the Internet. Hence, security is an important factor to be considered herein. Section 2 mentions the methodologies used to full fill the requirements of the proposed approach, i.e. CSBRCA. In Sect. 3, describes the performance evaluation of the approach and Sect. 4, a case study, describes and explains the implementation of the proposed approach using the parameter as CPU utilization followed by the conclusion and future work.

2 Proposed Methodology

Applications and other IT infrastructure are maintained in-house in a typical IT environment. CC provides software, IT frameworks, memory, and other resources in the cloud. Services are provided by a third-party provider, who conceals the fundamental infrastructure's intricacies from the end user. CC is built on hardware and software

Table 1 CC risk areas

Risk areas	Criticality (%)
Information SECURITY	91.7
Management operations	41.7
Change management	41.7
Recovery from disaster	66.7
SLA management	41.7
Management of interface	8.3
Legislation and regulations	33.3

designs that enable infrastructure expansion and virtualization for multitenant environment. In today's CC environment, diverse resources are distributed around the globe, necessitating security-aware resource management to combat security risks. Existing solutions, on the other hand, are unable to safeguard systems from cyber-attacks. Recently, researchers worked on developing new strategies for intrusion detection and prevention in computing systems and determined that the Intrusion Detection System (IDS) is an excellent tool to safeguard networks from threats. It investigates security flaws to help avoid such issues in the future. CC architecture consists of cloud services (measured services) given across a networked infrastructure by cloud service providers (third parties, suppliers, or brokers) to cloud consumers (end users, businesses, or IT staff) (i.e. the Internet or a virtual private network). Consumer needs and the provider's commitment to them are specified in contractual agreements (SLAs) that regulate cloud computing services. Recently, researchers worked on developing new strategies for intrusion detection and prevention in computing systems and determined that the Intrusion Detection System (IDS) is an excellent tool to safeguard networks from threats. It investigates security flaws to help avoid such issues in the future. CC governance, risk, and control are so crucial in the execution of any assurance management process. The application of policies and procedures is how governance is enforced. These rules and procedures should be based on industry best practices and connected with business and IT goals. Prioritizing the installation (amount and timeline) of governance and controls, as well as establishing scope for monitoring or auditing cloud computing systems, requires risk identification and analysis. Controls should be created based on risk identification and analysis to guarantee that essential actions are made to address risks and accomplish commercial and IT targets [6].

Figure 3a depicts the proposed architecture of the system. The architecture is classified as three layered architectures, and in the top layer there is information about the present status of the cloud system. This indicates the current utility of the cloud infrastructure systems. There are various parameters which the cloud will be providing. Here in this research paper we have considered the CPU utility of the system to verify the results obtained herein. The second layer manages to create the alarm as per the predefined threshold and passes the data to the vitrage so that it will take the preventive measure in reactive ways. This will result in the proper management of the SLA and thus the management of the IaaS cloud resources [7]. The systems collect the present utility of the cloud CPU usages, then this validated the same with the threshold values [8]. If this does not cross the threshold, the systems monitor the cloud status [9]. In case, the values crosses the threshold, then the information is transferred to the vitrage to take appropriate action to maintain the resources. The cloud service provider will be notified about the same, and then proper actions have to be taken to resolve this issue, and this has been discussed in the result section.

Figure 3b shows the benefits of the proposed scheme. The CPU utility of the system is analysed and compared with the predefined threshold calculated based upon the base line approach. If the threshold is crossed, then the vitrage will generate the

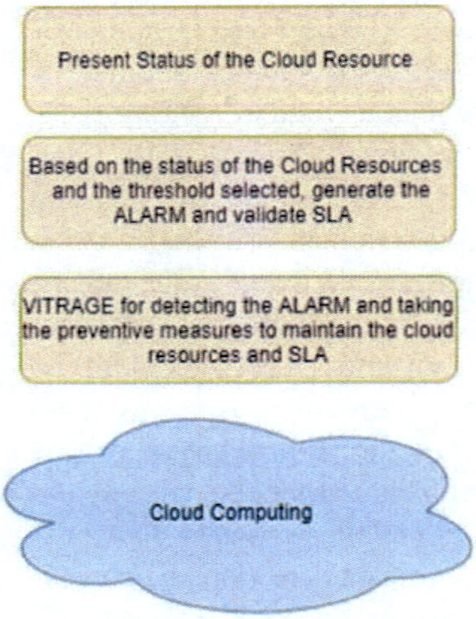

(a) CSBRCA Architecture

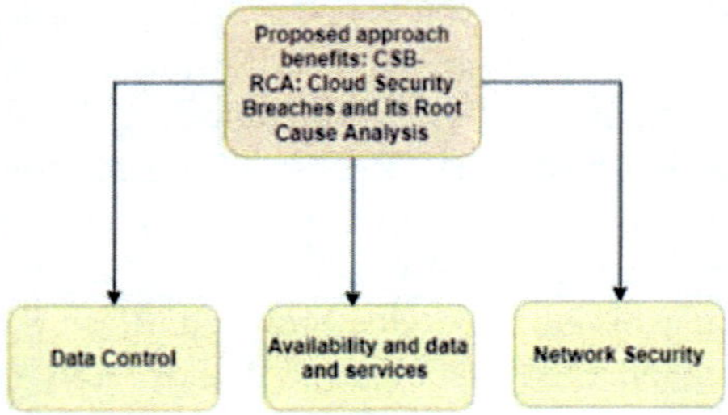

(b) Benifits of CSBRCA

Fig. 3 The architecture and its benefits

alarm and will notify the CSP to take appropriate action to maintain the resources or SLA of the cloud.

Algorithm 1 describes the scenario where the alarm is created using the threshold of 70% utilization of the mentioned resources, i.e. CPU and memory. When the threshold will cross the 70% utilization, the alarm will be generated. In Algorithm 2, the root because analysis of the system will be invoked as the threshold value is crossed and is identified through the continuous monitoring of the system. The present status of the system plays an important role to identify the status of the IaaS cloud. Hence, monitoring of the system is important.

Algorithm 1 CSBRCA : algorithm for root cause analysis in Cloud Computing

Require: $T\,CPU = 0.70, T\,mem = 0.70$

Ensure: $Alarm(A_{CPU}, A_{mem})$

$CV\ CPU \leftarrow currnet.CPUload()$. Initializing the current value of CPU $CV\ mem \leftarrow current.memload()$. Initializing the current value of memory **if** $CV\ CPU >= T\ CPU$ and $CV\ mem >= T\ mem$ **then**

Initialization of an alarm for CPU

if $a_{CPU} == True$ or $a_{mem} == True$ **then**

Notify vitrage for A_{CPU}

$A_{CPU} \leftarrow A_{CPU1}, A_{CPU2}, A_{CPU}, \ldots A_{CPUn}$. Deducing the alarms for CPU.

$A_{mem} \leftarrow A_{mem1}, A_{mem2}, A_{mem}, \ldots A_{memn}$. Deducing the alarms for mem.

Check SLA Status

Apply Solution $S_{CPU} \leftarrow A_{CPU}$ Apply

Solution $S_{mem} \leftarrow A_{mem}$ **end if**

end if Stop

Algorithm 2 CSBRCA : algorithm for root cause analysis in Cloud Computing

Require: $T\,CPU = 0.70, Current_{C}PUload$

Ensure: $State(S_{Optimal}, S_{SubOptmial})$

$CV\,CPU \leftarrow currnet.CPUload()$. Initializing the current value of CPU **if** $CV\,CPU >= T\,CPU$ and $CV\,mem >= T\,mem$ **then**

Update State $S_{Current} \leftarrow S_{SubOptimal}$ **else**

Update State $S_{Current} \leftarrow S_{Optimal}$

Continue to Monitoring

end if Stop

3 Performance Analysis

To test the performance of the proposed approach (CSBRCA), 9 VMs have been created in the open stack environment using the following infrastructure and is displayed in Sect. 3.1.

3.1 Experimental Setup

The experiment is carried out by the following system configuration: The system configuration where the experiments are carried out is stated below: PDL380 10th generation sff rackmount server, OS: MS windows server standard core sngl olp 16 lic-ae, CPU: intel xeon silver 4110 (2 nos.), RAM: 128 GN (32GB*4) DDR4-2666 MHz, ODD: HP 9.5mm SATA DVD Writer, HDD: 2.4 TB SAS 12G 10k SFF HDD (3 nos.) and, RAID: HPE Smart Arrayp 8161-a SR 10th Gen CNTRLR.

3.2 Methodology Used

Vitrage is an OpenStack Root Cause Analysis (RCA) service that organizes, analyses, and expands OpenStack alarms and events, providing insights into the root cause of problems and determining their presence before they are explicitly discovered. Vitrage has some requirements that should be satisfying when one should use it. The Fig. 4 shows the modules that needs to be integrated whilst installing vitrage plugins for OpenStack RCA. Vitrage has three major part that needs to be taken care whilst integrating it.

- **Vitrage Data Source(s)**: The main objective of the data source is to obtain the information from the different resources. The information is mainly about the physical and virtual sources along with the alarm data. Vitrage Graph then get all the information as input and produces the graph of current status of the system on the OpenStack vitrage dashboard. In our proposed system, we have integrated

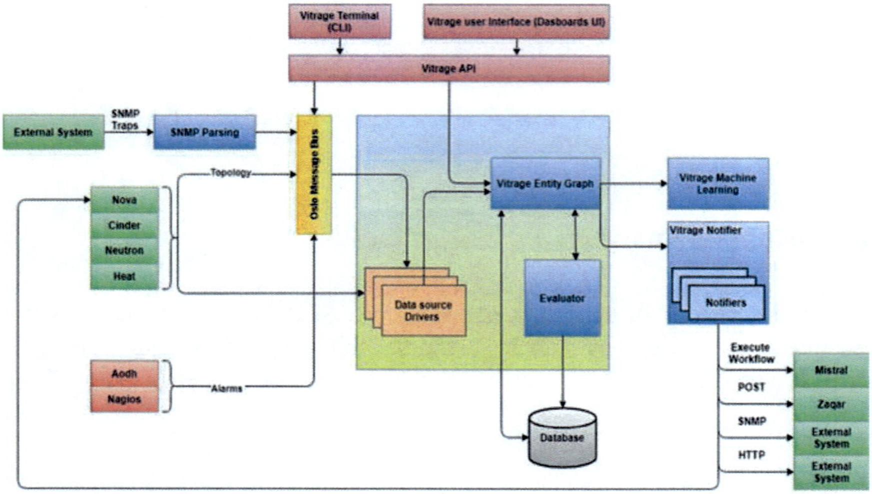

Fig. 4 Vitrage architecture diagram

default data sources like Nova, Cinder, Neutron, Heat including Nagios for alarms and Static physical data sources.

- **Vitrage Graph**: It captures the information gathered by the Data Sources with their inner relations. With the captured data, it generates primary graph algorithms that will further used by the vitrage evaluator as input.
- **Vitrage Evaluator**: It synchronizes the survey of the vitrage graphs and undertaking the outcomes of that survey. Primarily it works for executing various type of template-based action in vitrage. Similarly this also used to capture the alarm state and produces the deduced alarm or similar set of a deduce state.

To evaluate the performance of vitrage for OpenStack private cloud Root cause analysis (RCA), we have setup OpenStack private cloud using DevStack which is a sequence of a script to instal complete OpenStack environment by few clicks. Before executing DevStack scripts, we have added the vitrage plugins in the local.conf file.

In the Fig. 4, The Aodh and Nagious are the alarm services that are integrated with the vitrage RCA services. There are two types of alarm services provided by the vitrage and OpenStack cloud, firstly the RCA services provide internal alarms for the cloud system, and in addition to that it also provides external alarms like Nagios for monitoring purposes. To create threshold alarm, we have to integrate gnocchi as a back-end service. In Fig. 5, through 5a–e shows the OpenStack resource utilization

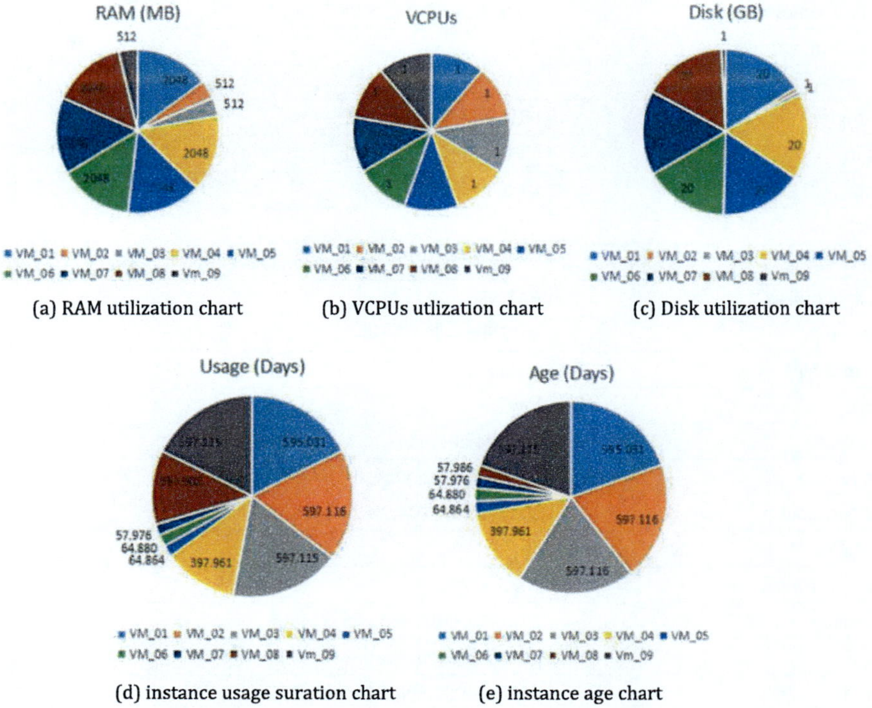

(a) RAM utilization chart (b) VCPUs utlization chart (c) Disk utilization chart

(d) instance usage suration chart (e) instance age chart

Fig. 5 Openstack resource utilization chart

from where Cloud Service Provider (CSP) can manages all the resources of a cloud. The command that used to create an alarm for CPU utilization threshold value it triggers an alarm when the utilization is goes beyond the threshold value is created.

If the systems satisfies all the requirements which will be needed for threshold alarm generation, then it generates the output including all the parameters mentioned in Table 2. In mentioned Table, we can see all the parameters of threshold alarm like *aggregation method, alarm actions, alarm id, comparison operator,* etc. Now whenever the system breaks the threshold value that is mentioned in alarm, it will immediately triggers the alarm in the form of logs.

Table 2 Aodh alarm generation information for RCA

Aodh alarm parameters	
Field	Values
aggregation method	mean
alarm actions	['log:']
alarm id	1b1af13c-4e71-463e-9c47-39fb272f0314
comparison operator	gt
description	instance running hot
enabled	TRUE
evaluate timestamp	2021–09-01T04:23:28.385285
evaluation periods	3
granularity	600
insufficient data actions	NIL
metric	CPU util
name	CPU hi
ok actions	NIL
project id	5fds5c5456as276nb72ew35sa434gh92
repeat actions	FALSE
resource id	ab125694-d658sds4-5d2g5as-1ad1sg21
resource type	instance
severity	low
state	insufficient data
state reason	Not evaluated yet
state timestamp	2021–09-01T04:23:28.333863
threshold	70
time constraints	NIL
timestamp	2021–09-01T04:23:28.333863
type	gnocchi resources threshold
user id	6016aex9825cvd545r1d1h256d0b1a15

When any threshold alarm is created for monitoring the resource utilization in OpenStack cloud, they will work on the alarm deduction method. For example, we have created an alarm to detect higher CPU utilization for the individual system. To check the functionality of an alarm, we have used python load generator library to create load so that CPU utilization increases beyond threshold value, an Aodh alarm at host level to be activated. The Fig. 6a, b represent the logical relationship between the resources in the OpenStack system. Here we can see that all the instances from VM 01 to VM 09 are connected with the Host 001. At the last level one alarm named alarm 001X is triggered on the host level which shows that amongst all, some of the instances are using CPU resource beyond threshold value as mentioned in Fig. 7. Now problem on the Host as per the alarm leave negative impacts on the other currently working resources. Hence, all the instances that are going to be diagnosed will be marked and change state to "ERROR". By following deduction process, the stage is identified and will get the exact location from where the original problem is generated. At last the resources can be reached and is shown in the Fig. 10. We will take further actions to find out the causes of higher CPU usage and if any unauthorized

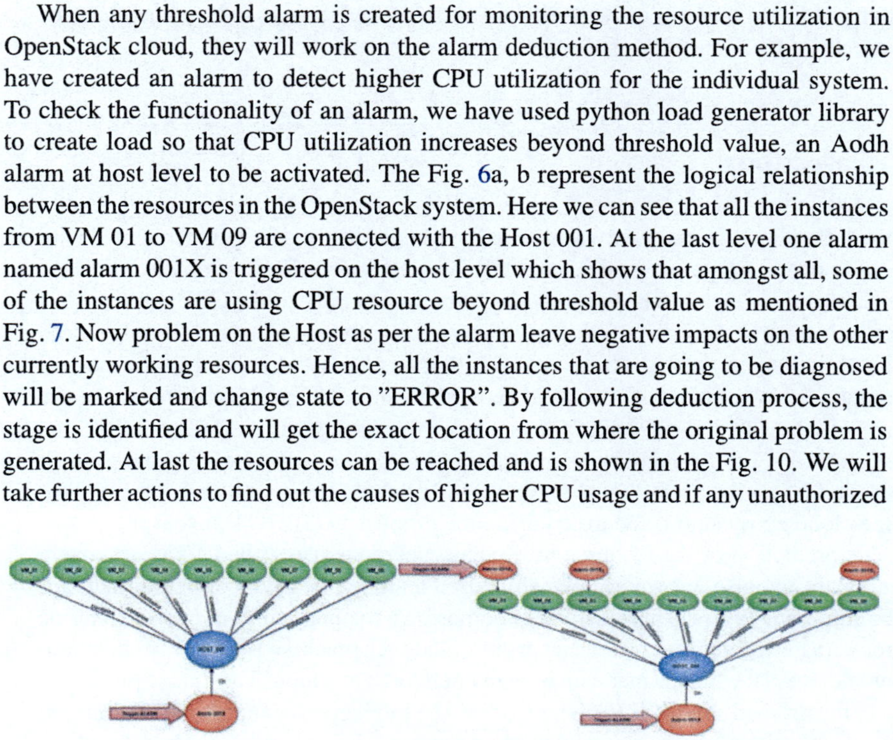

(a) Vitrage baseline chart (b) Vitrage deduce alarm and state chart

Fig. 6 Vitrage baseline chart and its alarm generation

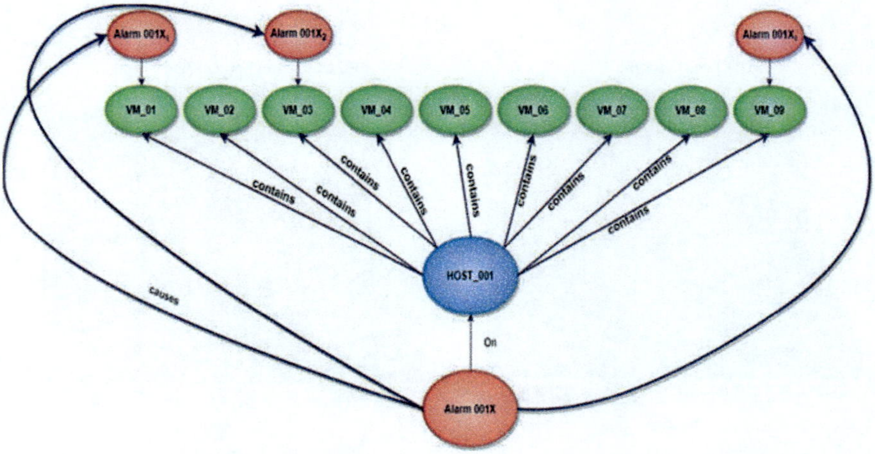

Fig. 7 Vitrage root cause indicator chart

activity is found then we can shut down that resources immediately to maintain the resources of the cloud ecosystem.

4 Case Study

In the following case study, two scenarios have been analysed and experimented with the help of the proposed model, i.e. CSBRCA approach. For example, in Fig. 8a, the case of the attack has been showcased where the utility of the CPU resource has been exceeded due to the attack, and it crosses the threshold value. The threshold is identified using the base line approach that has been carried out in the normal operational environment of the cloud ecosystem and without the occurrence of the attack. Figure 8b indicates the points where the change in the behaviour of the predicted CPU utility and the present status of the CPU mismatches due to the attack happened (the attack scenario was explained in the previous sections). This change in the cloud ecosystem is identified using the proposed CSBRCA approach.

The prediction of the resource utility has been done using the LSTM approach. Deep learning, also known as deep structured learning, is a type of neural network with numerous layers. These networks outperform regular neural networks in terms of retaining information from earlier events. One such machine is the recurrent neural network (RNN), which is made up of many networks in a loop. The data is preserved due to the networks in loop. Each network in the loop receives input and information from the previous network, conducts the requested action, and outputs data whilst transferring it to the next network. Some applications just demand recent information, whilst others may require more information from the past. The common recurrent neural networks lag in learning when the gap between the required past knowledge and the moment of necessity widens. However, long short-term memory (LSTM)

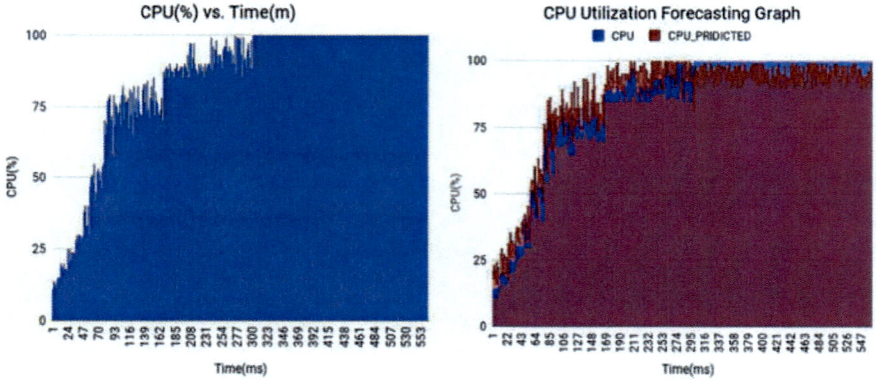

(a) CPU Utilization Graph (b) CPU Utilization Graph

Fig. 8 Prediction and utilization of the CPU

networks, a type of RNN, are capable of understanding such events [10]. These networks are specifically designed to avoid the long-term reliance problem associated with recurrent networks. LSTMs are excellent at memorizing for extended periods of time [11]. Because more prior knowledge may affect the performance of the model, LSTMs are a perfect choice for resource forecasting in the cloud ecosystem [12]

5 Conclusion and Future Work

As a general guideline to assist management in the adoption of cloud computing processes, procedures, and controls, we gave an overview of cloud computing benefits and security threats in this article. To assure the completeness, integrity, and availability of applications and data in the cloud, risks should be considered. We also identified a number of controls that may be explored for reducing the security concerns associated with cloud computing. Data security, quality control, logical access, network monitoring, data security, conformance, and virtualization were amongst the controls. The results readings were based on the CPU utility of the cloud IaaS resources. The proposed approach shows how the prediction of the resources are managed in real-time when the occurrence of the attack will be identified. Furthermore, the study will concentrate on the creation of a comprehensive risk and control framework for cloud computing and virtualization, which will offer management with rules and control guidelines for dealing with CC and virtualization issues in a consistent manner.

Acknowledgements This research paper is a part of the project approved under DST (Department of Science and Technology), Government of India, New Delhi (Grant: DST/ICPS/CPS-Individual/2018/). We thank DST for funding the project.

References

1. Prasad, V. K., Bhavsar, M. D., & Tanwar, S. (2019). Influence of montoring: Fog andedge computing. *Scalable Computing: Practice and Experience, 20*(2), 365–376.
2. Choi, Y., Kim, Y., & Rhu, M. (2021). Lazy batching: An SLA-aware batching system for cloud machine learning inference. In *2021 IEEE International Symposium on High-Performance Computer Architecture (HPCA)* (p. 493506). IEEE. https://doi.org/10.10007/1234567890
3. Prasad, V. K., Shah, M., Patel, N., & Bhavsar, M. (2018). Inspection of trust basedcloud using security and capacity management at an IaaS level. *Procedia Computer Science, 132*, 1280–1289.
4. Home page: Cyber security statistics https://purplesec.us/resources/cybersecurity-statistics/. Last. Accessed 4 Septe 2021
5. Mthunzi, S. N., Benkhelifa, E., Bosakowski, T., Guegan, C. G., & Barhamgi, M. (2020) Cloud computing security taxonomy: From an atomistic to a holistic view. *Future Generation Computer Systems, 107*(2020), 620–644.

6. Li, Q., Lv, P., Wang, M., Zhang, Z., Wang, S., Fang, P, & Gao, F. (2020). A risk assessment method of smart grid in cloud computing environment based on game theory. In *2020 IEEE 5th International Conference on Cloud Computing and Big Data Analytics (ICCCBDA)* (pp. 67–72). IEEE.
7. Prasad, V. K., & Bhavsar, M. (2020). Preserving SLA parameters for trusted IaaS cloud: An intelligent monitoring approach. *Recent Patents on Engineering, 14*(4), 530–540.
8. Mewada, A., Gujaran, R., Prasad, V. K., Chudasama, V., Shah, A., & Bhavsar, M. (2020). Establishing trust in the cloud using machine learning methods. In *Proceedings of first international conference on computing, communications, and cyber-security* (IC4S 2019) (pp. 791–805). Springer, Singapore.
9. Chudasama, V., Mewada, A., Prasad, V. K., Shah, A., Bhavasar, M. (2021). CS2M: Cloud security and SLA management. *Annals of the Romanian Society for Cell Biology*, 4459–4465.
10. Prasad, V. K., & Bhavsar, M. D. (2020). Monitoring IaaS cloud for healthcare systems: Healthcare information management and cloud resources utilization. *International Journal of E-Health and Medical Communications (IJEHMC), 11*(3), 54–70.
11. Prasad, V. K., & Bhavsar, M. D. (2020). Monitoring and prediction of SLA for IoT based cloud. *Scalable Computing: Practice and Experience, 21*(3), 349–358.
12. Prasad, V. K., Tanwar, S., Bhavsar, M. (2021). C2B-SCHMS: Cloud computing and bots security for COVID-19 data and healthcare management systems. In *Proceedings of second international conference on computing, communications, and cyber-security* (pp. 787–797). Springer, Singapore.

A Mobile-Based Patient Surgical Appointment System Using Fuzzy Logic

Femi Emmanuel Ayo, Sanjay Misra, Joseph Bamidele Awotunde, Ranjan Kumar Behera, Jonathan Oluranti, and Ravin Ahuja

Abstract The advent of artificial intelligence in medical field is playing a significant role in improving healthcare services. In healthcare, there is always need for an intelligent method to schedule resources and patients in order to reduce patient waiting time. The treatment process of patients from their arrival to the starting time of consultation is accompanied by uncertainties. Therefore, this study developed a fuzzy and a mobile-based solution for patient surgical appointment system based on some relevant input variables. The proposed system was simulated using MATLAB fuzzy inference system with a triangular member function. The range of the fuzzy inputs was then fed into the developed mobile-based application for an optimal patient surgical appointment system. The evaluation findings revealed that the proposed framework is efficient in terms of scheduling patient surgical consultations.

Keywords Patient appointment system · Fuzzy logic · Expert system · Mobile application

---

F. E. Ayo
Department of Computer Science, McPherson University, Seriki-Sotayo, Abeokuta, Nigeria
e-mail: ayofe@mcu.edu.ng

S. Misra (✉)
Department of Computer Science and Communication, Østfold University College, Halden, Norway
e-mail: sanjay.misra@hiof.no

J. B. Awotunde
Department of Computer Science, University of Ilorin, Ilorin, Nigeria
e-mail: awotunde.jb@unilorin.edu.ng

R. K. Behera
Birla Institute of Technology, Mesra, India
e-mail: ranjan.behera@bitmesra.ac.in

J. Oluranti
Center of ICT/ICE, CUCRID, Covenant University, Ota, Nigeria
e-mail: Jonathan.oluranti@covenantuniversity.edu.ng

R. Ahuja
Delhi Skill and Entrepreneurship University, New Delhi, India

P. K. Singh et al. (eds.), *Proceedings of Third International Conference on Computing, Communications, and Cyber-Security*, Lecture Notes in Networks and Systems 421,
https://doi.org/10.1007/978-981-19-1142-2_15

1 Introduction

In human lives, healthcare systems are an essential and pivot part of providing access to medical and non-medical services that are directed at refining the quality of life of individuals. However, patients in many healthcare facilities suffer from long waiting time and waste of money in taking appointments due to the complexities involved. The amount of time that patients spend waiting for services to be given in clinics is a big issue for several health centers [1–3]. To measure the quality of services of a good hospitals, the waiting time of patients in clinics is very useful [4, 5].

A hospital's appointment scheduling status is the first part of treatment, and it determines whether the treatment will be possible or efficient. Hospitals are constantly battling a scheduling issue that results in dissatisfaction with the time spent between patient arrival and actual consultation start time, particularly for patients with urgent surgical needs. The case of large number of patients occur in many large hospitals where some patients come way earlier than required in order to register for patient number. Patients encounter difficulties going to the hospital to take appointments, only to hear that the doctor is not available, this delay in treatment could result in adverse health effects. The quality of the waiting time of a patient is strongly related to the degree of stratified care received within a clinic in healthcare systems [6–8].

The indirect form of waiting has to do with patients contacting the healthcare administrator to schedule an appointment and waiting for reply of confirmation which may take days or even months. This raises the likelihood of patients not showing up, which has an impact on healthcare use because patients see extended wait times as a barrier to actually receiving services. Ambulatory care waiting time can be classified into two categories: waiting before appointment and waiting after appointment [9]. The time spent waiting for consultation can be separated into two categories: registering time and consulting time [10]. The current procedure in outpatient department has to do with the first-come, first-served basis. Another problem apart from the high-waiting time is the referring of patients to different doctors as it is not possible for the new doctor to know all about the previous records of the patient in the short checkup time. The efficiency of healthcare services is based on the use of patient scheduling and appointment system which reduces the number of unsatisfied patients. The central purpose in ideal patient arranging is to choose a course of action technique for which a particular extent of execution is upgraded under questionable conditions [11–13]. Appointment and booking frameworks are frameworks for arranging arrangements between assets like patients, offices, and suppliers. It is likewise utilized to limit holding up occasions, make a smooth patient stream, focus on arrangements, and advance the usage of assets.

The purpose of this study is to create a surgical appointment management system that can be used on a mobile device using fuzzy logic where patients can take the appointment of any doctor through a smart phone device and make the best use of their time. The system possesses the resource of the list of appointments on the android phone of the patients. The procedure for scheduling is based on the severity and the availability of the surgeon for a particular surgical operation. It also depends on the date and session for the surgical operation.

The significance of this study is to (1) enable patients with severe surgical level to take appointment of doctors without having to waste time waiting in hospitals and (2) resolve uncertainties surrounding the scheduling of appointments.

The contributions of this study include

- design of a mobile-based surgical appointment system for patients
- design of a fuzzy logic model to provide grading levels for scheduling of patients for surgical operations.

The second section contains a review of relevant work. The methodology is presented in Sect. 3. The fourth section focuses on the implementation and discussion. The work comes to a close with Sect. 5.

2 Related Work

Authors in [14] proposed hybrid appointment system (HAS) that can schedule patients in Dubai for doctors based on preferences. The study conducted a survey to get feedback from healthcare providers and patients on the need to implement HAS. The study's findings provide useful feedback for the development of an automated appointment system. In [15], the authors proposed an online appointment scheduling system to facilitate access to primary care in Brazil. The proposed system allows patients to schedule, view available time slots, and book and cancel appointments. Reference [16] modeled a multi-agent system and proposed an intelligent algorithm that schedules patients and resources in real-time based on the actual status of available resources. The proposed algorithm can also reorganize the resources in the outpatient clinics in real-time in order to match demands with resources. The results showed improved waiting time and resources utilization compared to isolated methods.

In [17], the authors presented a Web-based appointment system to investigate the efficacy of Web-based appointment system in the registration service for outpatients. The proposed method provides an increase in the satisfaction of patients with registration. Reference [18] proposed an automated patient appointment reminder for cross-platform mobile application. The authors proposed this model so that doctors and patients can observe their appointments via mobile application. The authors in [19] presented an Internet-based expert system used in outpatient department. The method provides reduction of the queues in hospitals, number of consultation, and it saves time of doctors and patients. In a comparable work in [20], the authors proposed an appointment keeping system to help in booking an appointment with a consultant in a clinic. The proposed system given the patients the opportunity of changing or reschedule an appointment that are not convenient for them so as to meet up with an arrangement with the clinical administrations system by means of an Internet browser.

In [21], the authors designed an appointment system using Android app. The proposed system created an interaction between the patient and the clinic specialist to be able to create a convenient time for both the patient and the specialist. The patient personal phone can be used to book an appointment without going out of the environment, and the specialist can take track of the patients he/she need to attend to daily, weekly, or monthly. The author in [22] created an online dental appointment and booking system. The main purpose of the system is to help in saving time and resources of patients from going to clinic just because of booking an appointment with a consultant. The system helps the patients to book and check the time frame for their dental clinic with other-related services. The creators proposed a dental specialist online reservation framework. This help the patients and specialist to check their appointment and booking online without any stress, use for appointment plan, and can be used to check the patients details in real-time. Reference [23] proposed a structure for arrangement frameworks with patient inclinations. They established a model for arrangement mechanisms that automatically update patients' preferences to increase booking options. The creators in [24] fostered a Bilevel fluffy possibility compelled model to tackle the test of clinic short-term arrangement booking fixated on income the board. Every branch of the emergency clinic settles on the choice of the arrangement booking to expand its benefit.

2.1 Patient Waiting Time

The time a patient waits in the clinic before being seen for consultation and treatment is referred to as patient waiting time. It is the whole time from enrollment to a doctor's appointment [25, 26]. The length of a hospital's waiting time is a function of the patient's happiness and an important component in determining the quality of healthcare. Waiting time is an essential factor that predicts access of health services, utilization, and patients' retention [27]. Patients become unhappy the longer they wait because waiting can be frustrating, annoying and can cause vast ambiguity. Sometimes patients are faced with the problem of arriving early to the hospital for outpatient care only to hear that the doctor will be late for a while or would not be available for the day. The factors that lead to long waiting time include high-patient load and poor communication.

2.2 Appointment and Scheduling Management System

An important element of healthcare is a good appointment and scheduling system that encourages patient satisfaction [28]. In hospitals, scheduling appointments are a critical administrative task. Outpatients' requests for appointments are based on various priorities, and it is critical to evaluate the priority of patients when scheduling outpatients. A scheduler carries out the scheduling activity, and in turn, it allows multiple users to share system resources effectively. Real-time scheduling involves very little intervention of schedulers and therefore can assist in reducing the waiting time caused by human factors. The time slots available are visible to patients through the Web interface [29]. The appointment and scheduling system are used to minimize waiting time, optimize the utilization of resources, and prioritize appointments [30].

2.3 Types of Patient Appointment Scheduling

Time-specified Scheduling

It is also known as stream scheduling, and it entails making appointments at precise times. The goal of time-specified appointments is to reduce patient wait times while maintaining a consistent flow of patients through the facility (like a stream of water). The length of time allotted for a timed appointment is determined by the reason for the visit.

Wave Scheduling

The wave scheduling is very useful if more than one patients are planned for the same time by splitting the available time within them in other to be able to take care of them concurrently. This means that patients show up in a way to wave other patient that is booking for the same time, and there is consistently a patient holding on to be seen.

Double Booking

This is used to attend to an emergency cases within a clinic for two or more patient to concurrently see a consultant for an injury or acute illness has to be added to already booked schedule.

Open Booking

This is used to give patients the opportunity to book within a range of time and the consultant attended to them in the order they arrive the clinic. Here, patients with serious illness are attended to before those with less significant complaints. This type of patient appointment scheduling works best when the practice is not busy.

3 Methodology

The fuzzy logic paradigm describing the methodology for the system is shown in this section. The developed model consists of the user interface where details are collected from the patient. The system next translates the details to fuzzy values and schedules the patient's appointment using inference engine-based rules stored in the fuzzy knowledge base.

3.1 Fuzzy Set

Fuzzy logic depicts a multi-valued logic that captures its true value between the intermediate values of 0 and 1 inclusive [31]. A fuzzy set represents the input to the fuzzy logic with various grades of membership between the intervals of (0–1). The following list formed the fuzzy set for this study:

$$A = \{\text{Surgical level, Specialist status, Date, Session}\} \quad (1)$$

- Surgical level is used to denote the severity grade for a surgical operation.
- Specialist status is used to denote the availability of a specialist for a selected surgical operation.
- Date is used to denote the week of the month that the surgical operation may be schedule.
- Session is used to denote the time of the day (morning, afternoon, evening, or night) the surgery will be performed.

3.2 Membership Function

A membership function is used to define the degree of membership in a fuzzy set between 0 and 1. The membership function converts the crisps values of each fuzzy set element to their equivalent fuzzy values between 0 and 1 as depicted in Tables 1, 2, 3, and 4. Table 5 depicts the membership function of the expected output. This process of turning a crisp input into a fuzzy input using the membership function is known as fuzzification. Because of its simplicity, the triangle membership function given in Eq. 2 was used in this investigation.

Table 1 Fuzzy-based decision input variables for surgical level

Linguistic value	Value range
Mildly	$0.1 \le x < 0.3$
Moderately	$0.3 \le x < 0.6$
Moderately severe	$0.6 \le x < 0.8$
Severely	$0.8 \le x \le 1.0$

Table 2 Fuzzy-based decision input variables for specialist status

Linguistic value	Value range
Mildly busy	$0.1 \le x < 0.3$
Moderately busy	$0.3 \le x < 0.6$
Moderately severe busy	$0.6 \le x < 0.8$
Severely busy	$0.8 \le x \le 1.0$

Table 3 Fuzzy-based decision input variables for date

Linguistic value (W)	Value range
First week (1)	$0.1 \leq x < 0.3$
Second week (2)	$0.3 \leq x < 0.6$
Third week (3)	$0.6 \leq x < 0.8$
Fourth week (4)	$0.8 \leq x \leq 1.0$

Table 4 Fuzzy-based decision input variables for session

Linguistic value	Value range
Morning session	$0.1 \leq x < 0.3$
Afternoon session	$0.3 \leq x < 0.6$
Evening session	$0.6 \leq x < 0.8$
Night session	$0.8 \leq x \leq 1.0$

Table 5 Fuzzy-based decision input variables for scheduling

Linguistic value	Value range
Double	$0.1 \leq x < 0.3$
Wave	$0.3 \leq x < 0.6$
Open	$0.6 \leq x < 0.8$
Time-specific	$0.8 \leq x \leq 1.0$

$$\mu_A(x; [a, b, c]) = \begin{cases} 0, & \text{if } x = a \\ \frac{x-a}{c-a}, & \text{if } x \in [a, c] \\ \frac{b-x}{c-b}, & \text{if } x \in [b, c] \\ 0, & \text{if } x \geq c \end{cases} \tag{2}$$

where a, b, and c are the y-coordinate range between 0 and 1, and x represents the x- coordinate of real values.

Fuzzy rules

A sum of 16 guidelines was characterized utilizing the rule of thumb. Since the semantic factors utilized were 4, then, at that point, we have 24 = 16 guidelines. The AND work was utilized for rules mix and assessment. Table 6 shows the standards dependent on specialists' information.

Inference engine

For reasoning, the constructed fuzzy model uses Eq. 3 for the root mean square (RMS) formula.

$$\sqrt{\sum R^2} = \sqrt{R_1^2 + R_2^2 + R_3^2 + \cdots + R_n^2} \tag{3}$$

Table 6 Fuzzy rule base for patient surgical scheduling

S. No.	Surgical level	Specialist status	Date	Session	Scheduling (conclude)
1	Mildly	Mildly busy	First week	Morning	Wave
2	Mildly	Moderately busy	Second week	Afternoon	Wave
3	Mildly	Moderately severe busy	Third week	Evening session	Double
4	Mildly	Severely busy	Fourth week	Night session	Double
5	Moderately	Mildly busy	First week	Morning	Wave
6	Moderately	Moderately busy	Second week	Afternoon	Wave
7	Moderately	Moderately severe busy	Third week	Evening session	Double
8	Moderately	Severely busy	Fourth week	Night session	Double
9	Moderately severe	Mildly busy	First week	Morning	Open
10	Moderately severe	Moderately busy	Second week	Afternoon	Open
11	Moderately severe	Moderately severe busy	Third week	Evening session	Time-specific
12	Moderately severe	Severely busy	Fourth week	Night session	Time-specific
13	Severely	Mildly busy	First week	Morning	Time-specific
14	Severely	Moderately busy	Second week	Afternoon	Time-specific
15	Severely	Moderately severe busy	Third week	Evening session	Time-specific
16	Severely	Severely busy	Fourth week	Night session	Time-specific

$R_1^2 + R_2^2 + R_3^2 + \cdots + R_n^2$ are the values of different rules in the fuzzy rule base that have the same result. The center of gravity is calculated by adding all the resultants of the same firing rules.

3.3 Defuzzification

Defuzzification involves the transformation of the fuzzy values back to their crisp values for easy interpretation. The centroid model was adapted as the defuzzification method. This method determines the centroid value and uses that value for the conclusion of the fuzzy logic system. It is represented as

$$\mathrm{CoG}(Y^*) = \frac{\sum \mu y(X_i)x_i}{\sum \mu y(X_i)} \tag{4}$$

where x_i is the mid-point of fuzzy value ranges between 0 and 1 and the $\mu y(X_i)$ *RMS* for rules with the same result.

4 Implementation and Discussion

Flutter, PHP, MySQL, and MATLAB were used to implement the system. The front end of the proposed system, which is a mobile app, was built with Flutter. The front end of the system was built with PHP, and the back end, which included the patient's login and information, was built with MySQL. MATLAB was used for the rules generation.

The mobile-based patient surgical appointment system using fuzzy logic was built on MATLAB for easy integration, visualization, and programming. The rules generated from MATLAB were integrated into the Flutter application program for accurate and quick patient scheduling for surgical operations. In this study, the various fuzzy input ranges as shown in Tables 1, 2, 3, and 4 were used to convert the patient inputs to their equivalent fuzzy sets.

4.1 *Fuzzy Logic Membership Function Plots and Rule Editor*

Figures 1 and 2 depict the input variable membership function plots and fuzzy rule editor, respectively. The input variable membership function plots are the graphical display of the fuzzy inputs and their respective linguistic variables. The values 0 and 1 denote fuzzy inputs that are only substantially part of the fuzzy set. A fuzzy input with a value of 0 is not a member of the fuzzy set; a fuzzy input with a value of 1 is completely a member of the fuzzy set. A fuzzy rule editor is an interface in fuzzy logic for creating, adding, deleting, and modifying rules, which are used by the fuzzy inference system for patients scheduling. The fuzzy logic system produces the output based on the rule combination of the linguistic variables assigned to the fuzzy inputs. These rules are formulated in the form of "IF-THEN" statements required by the system for intelligent decisions.

In addition, the output of the fuzzy system depicts the scheduling decisions described by the linguistic variables assigned to the output variable. In order to avoid difficulty in result interpretation, the generated fuzzy rules were fed into a mobile-based application program developed for easy interaction and patient appointments. The system used icons for the purpose of easy operation and interaction.

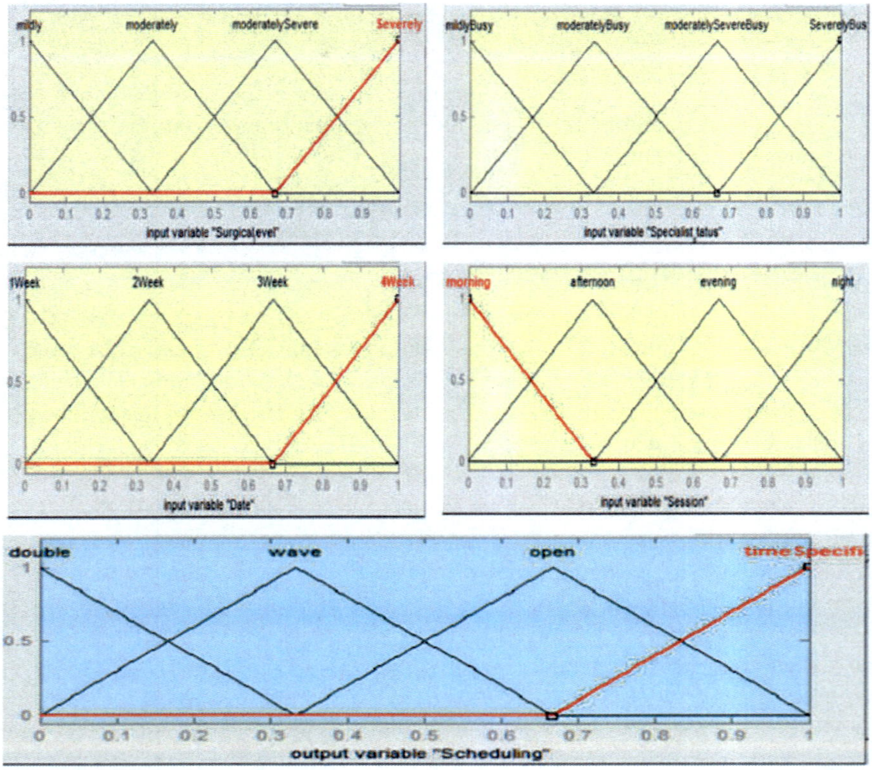

Fig. 1 Input variable membership function plots

5 Discussion

The input variables for the developed fuzzy logic model include surgical level, specialist status, date, and session. The output variable is the scheduling grades for patients based on the rule combination of the linguistic variables assigned to the fuzzy inputs. The fuzzy rules generated from the MATLAB fuzzy inference system were fed into the developed mobile-based application for patient surgical appointment system.

A screenshot of the patient's login into the system is shown in Fig. 3a. Patients must use the login button to log into the system and enter their card number as their login details. The patient then selects the book appointment button in Fig. 3b, which leads to the selection of the specialist he or she wants to see. After selecting the specialist in Fig. 4a, the patient has to pick the date in Fig. 4b. Once the date is entered, the appointment is set using the fuzzy logic process, and the details of the appointment are displayed as in Fig. 5. The performance of the developed fuzzy system was measured repeatedly with human operator based on the time required to schedule various number of patients (See Table 7). The goal of this test is to evaluate

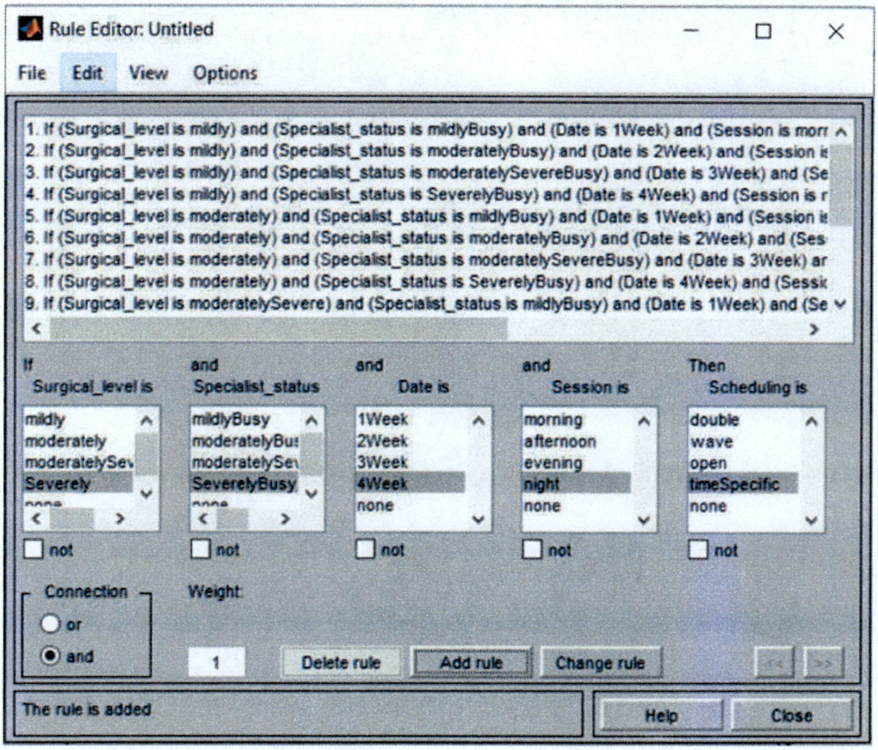

Fig. 2 Fuzzy model rule editor

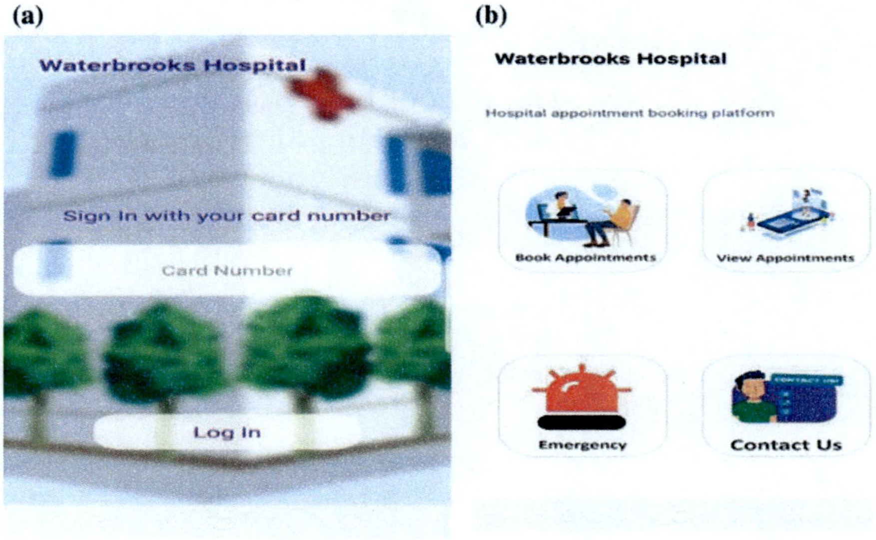

Fig. 3 **a** Patient login pane; **b** activity selection pane

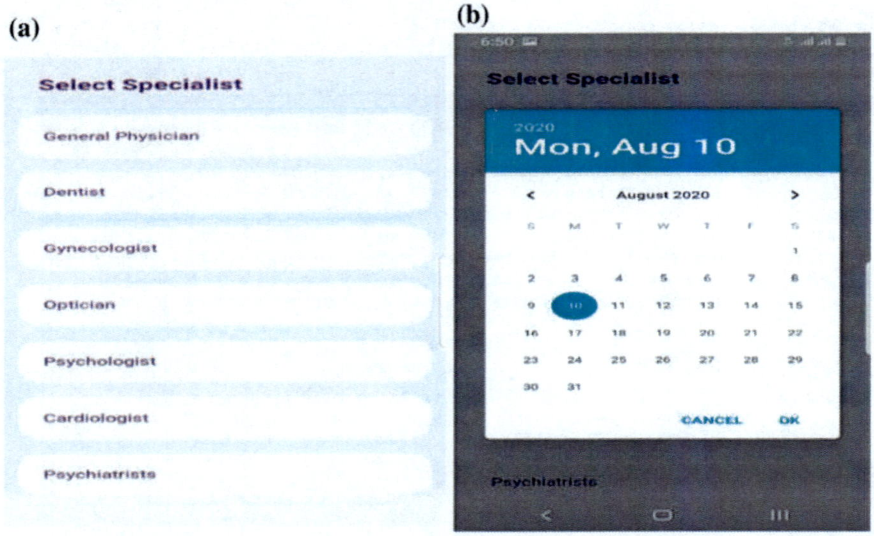

Fig. 4 **a** Specialist selection pane; **b** date selection pane

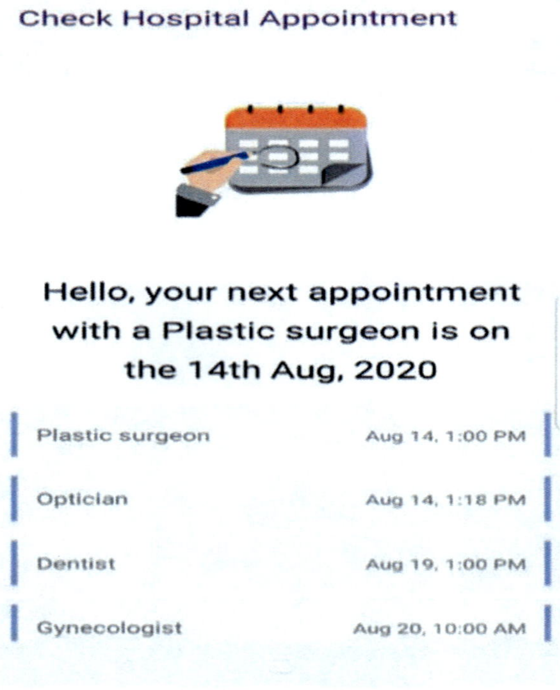

Fig. 5 View appointment panel

Table 7 Performance evaluation

No. of patient	Time required by human to schedule (seconds)	Time required by developed fuzzy-based system to schedule (seconds)
1	60	10
2	120	15
3	180	20
4	240	25
5	300	30

that the produced fuzzy system is right and accurate. The results of the study revealed that the designed fuzzy system performed well and can provide patient appointment and scheduling at reduced waiting time.

6 Conclusion

The mobile-based fuzzy patient appointment system was developed to be a user-friendly and intuitive interface in which scheduling of appointments is made possible through a smart phone device with the ability to access it anywhere and anytime. Healthcare providers can make changes and updates in real-time, and patients can make use of the designed fuzzy expert system to book appointments to reduce waiting times and save cost. The adoption of the developed system in hospitals will reduce the long waiting lines and also assist in saving lives. This study was able to deploy artificial intelligence in scheduling patients for surgery sessions based on the fuzzy input variables. The study also provides grading levels for scheduling patients for surgery operations. The evaluation results of the developed fuzzy inference system showed that it can provide appointment for patients seeking for surgery sessions.

References

1. Mahdi, F., Saeid, J. G., Marzieh, K., & Shadi, D. (2020). Patient waiting time management through fuzzy based failure mode and effect analysis. *38*(6), 1–12.
2. Ayeni, F., Omogbadegun, Z., Omoregbe, N. A., Misra, S., & Garg, L. (2018). Overcoming barriers to healthcare access and delivery. *EAI Endorsed Trans. Pervasive Health Technol.*, *4*(15), e2.
3. Dorosti, S., Fathi, M., Ghoushchi, S. J., Khakifirooz, M., & Khazaeili, M. (2020). Patient waiting time management through fuzzy based failure mode and effect analysis. *Journal of Intelligent & Fuzzy Systems, 38*(2), 2069–2080.
4. Rossiter, C. E., & Raynolds, F. A. (1968). Automatic monitoring of the time waited in an outpatient clinic. *Journal Storage: Medical Care., 1*, 218–225.
5. Adeloye, D., Adigun, T., Misra, S., & Omoregbe, N. (2017). Assessing the coverage of e-health services in sub-saharan Africa. *Methods of information in medicine, 56*(03), 189–199.

6. Ayeni, F., & Misra, S. (2014, September). Overcoming barriers of effective health care delivery and electronic health records in Nigeria using socialized medicine. In *2014 11th International Conference on Electronics, Computer and Computation (ICECCO)* (pp. 1–4). IEEE.
7. Ayo, F. E., Awotunde, J. B., Ogundokun, R. O., Folorunso, S. O., & Adekunle, A. O. (2020). A decision support system for multi-target disease diagnosis: a bioinformatics approach. *Heliyon, 6*(3), e03657.
8. Soyemi, J., Misra, S., & Nicholas, O. (2015, March). Towards e-Healthcare deployment in Nigeria: The open issues. In *International Conference on Soft Computing, Intelligence Systems, and Information Technology* (pp. 588–599). Springer, Berlin, Heidelberg.
9. Chen, B. L., Li, E. D., Yamawuchi, K., Kato, K., Naganawa, S., & Miao, W. J. (2010). Impact of adjustment measures on reducing outpatient waiting time in a community hospital: Application of a computer simulation. *China Medical Journal., 123*, 574–580.
10. Yu, Q., & Yang, K. (2008). Hospital registration waiting time reduction through process redesign. *International Journal of Six Stigma and Competitive Advantage., 4*, 240–253.
11. Cayirli, T., & Veral, E. (2003). Outpatient Scheduling in health care: A review of literature. *Production Operation Manager, 12*(4), 519–549.
12. Elleuch, M. A., Hassena, A. B., Abdelhedi, M., & Pinto, F. S. (2021). Real-time prediction of COVID-19 patients health situations using artificial neural networks and fuzzy interval mathematical modeling. *Applied soft computing, 110*, 107643.
13. Braune, R., Gutjahr, W. J., & Vogl, P. (2021). Stochastic radiotherapy appointment scheduling. *Central European Journal of Operations Research*, 1–39.
14. Aburayya, A., Al Marzouqi, A., Ayadeh, I., Albqaeen, A., & Mubarak, S. (2020). Evolving a hybrid appointment system for patient scheduling in primary healthcare centres in Dubai: Perceptions of patients and healthcare provider. *International Journal on Emerging Technologies, 11*(2), 251–260.
15. Celuppi, I. C., Lima, G. D. S., Felisberto, M., Lacerda, T. C., Wazlawick, R. S., & Dalmarco, E. M. (2021). PEC e-SUS APS online appointment scheduling system: A tool to facilitate access to Primary Care in Brazil. *Ciência & Saúde Coletiva, 26*, 2023–2034.
16. Munavalli, J. R., Rao, S. V., Srinivasan, A., & van Merode, G. G. (2020). An intelligent real-time scheduler for out-patient clinics: A multi-agent system model. *Health Informatics Journal, 26*(4), 2383–2406.
17. Cao, W., Wan, Y., Tu, H., Shang, F., Liu, D., Tan, Z., & Xu, Y. (2011). A web-based appointment system to reduce waiting for outpatients: A retrospective study. *BMC health services research, 11*(1), 1–5.
18. Chaiwongsai, J., Preecha, P., & Intem, S. (2016). Automated patient appointment reminder for cross-platform mobile application. In *2016 International Symposium on Intelligent Signal Processing and Communication Systems (ISPACS)* (pp. 1–6). IEEE.
19. Aktepe, A., Turker, A. K., & Ersoz, S. (2015). Internet based intelligent hospital appointment system. *Intelligent Automation & Soft Computing, 21*(2), 135–146.
20. Gowthem, S. S., & Kaliyamurthie, K. P. (2015). Smart appointment reservation system. *International Journal of Innovative Research in Science, Engineering and Technology, 4*(6).
21. Choudhari, S. B., Kusurkar, C., Sonje, R., Mahajan, P., & Vaz, J. (2014). Android application for doctor's appointment. *International Journal of Innovative Research in Computer and Communication Engineering, 2*(1), 2472–2474.
22. Kerdvibulvech, C., & Win, N. N. (2012). The dentist online reservation system design and implementation web based application and database management system project. In *International Conference on Education Technology and Computer (ICETC2012) IPCSIT vol* (Vol. 43).
23. Wang, W. Y., & Gupta, D. (2011). Adaptive appointment systems with patient preferences. *Manufacturing & Service Operations Management, 13*(3), 373–389.
24. Zhou, X., Luo, R., Zhao, C., Xia, X., Lev, B., Chai, J., & Li, R. (2016). Bilevel fuzzy chance constrained hospital outpatient appointment scheduling model. Scientific Programming.
25. Belayneh, M., Woldie, M., Berhanu, N., & Tamiru, M. (2017). The determinants of patient waiting time in the general outpatient department of Debre Markos and Felege Hiwot hospitals

in Amhara regional state, North West Ethiopia. *Global Journal of Medicine Public Health, 6*(5), 2277–9604.

26. Awotunde, J. B., Ogundokun, R. O., & Misra, S. (2021). Cloud and IoMT-based big data analytics system during COVID-19 pandemic. *Internet of Things, 2021*, 181–201.
27. Umar, I., Oche, M. O., & Umar, A. S. (2011). Patient waiting time in a tertiary health institution in Northern Nigeria. *Journal of Public Health and Epidemiology, 3*(2), 78–82.
28. Appleman, E. R., O'Connor, M. K., Boucher, S. J., Rostami, R., Sullivan, S. K., Migliorini, R., & Kraft, M. (2021). Teleneuropsychology clinic development and patient satisfaction. *The Clinical Neuropsychologist, 35*(4), 819–837.
29. Zhao, P., Yoo, I., Lavoie, J., Lavoie, B. J., & Simoes, E. (2017). Web-based medical appointment systems: A systematic review. *Journal of medical Internet research, 19*(4), e134.
30. Li, N., Li, X., Zhang, C., & Kong, N. (2021). Integrated optimization of appointment allocation and access prioritization in patient-centred outpatient scheduling. *Computers & Industrial Engineering, 154*, 107125.
31. Awotunde, J. B., Ajagbe, S. A., Oladipupo, M. A., Awokola, J. A., Afolabi, O. S., Mathew, T. O., & Oguns, Y. J. (2021). An Improved Machine Learnings Diagnosis Technique for COVID-19 Pandemic Using Chest X-ray Images. *Communications in Computer and Information Science*, 2021, 1455 CCIS, pp. 319–330.

Implementation of Green Technology in Cloud Computing

Soha Bhatia, Anushka Shrivastava, Radhika Nigam, and Punit Gupta

Abstract The increase in the use of modern technology worldwide in various sectors like business, software development, automation, etc., has made human life considerably easy, comfortable, and far from any danger. For this reason, there has been a rapid growth in technologies and upcoming trends. Cloud computing is one such paradigm which has rapidly developed in recent years. It has a range of applications in various domains due to the features it offers, like scalability, elasticity, and cost saving with low maintenance, security, and reliability. However, the production and consumption of these advanced technologies have a negative impact on the environment causing massive energy consumption and generating carbon footprints. Due to this, the concept of green technology has gained a lot of attention in order to make positive changes to the environment. Green computing is the implementation of environmentally friendly concepts in computing to increase power and energy and reduce carbon content. In this paper, we review that how adopting a cloud-based architecture has reduced the energy consumptions levels, and we further survey various methods and algorithms which can make clouds greener and more energy efficient.

Keywords Green computing · Cloud computing · Green cloud computing · Cloud data centers · Task scheduling · Load balancing · Green computing algorithms

1 Introduction

In today's world, the increased use of modern technologies has not only led to the misuse of resources but has also intensified global warming. An estimate indicates that only around 20% of the waste is recycled from the 50 million tons of e-waste

S. Bhatia · A. Shrivastava · P. Gupta (✉)
Department of Computer and Communication Engineering, Manipal University Jaipur, Jaipur, Rajasthan 303007, India
e-mail: punitg07@gmail.com

R. Nigam
Department of Information Technology Engineering, Manipal University Jaipur, Jaipur, Rajasthan 303007, India

P. K. Singh et al. (eds.), *Proceedings of Third International Conference on Computing, Communications, and Cyber-Security*, Lecture Notes in Networks and Systems 421,
https://doi.org/10.1007/978-981-19-1142-2_16

which is generated worldwide and thus leading to landfills and water contamination. A large number of resources are utilized in making modern technological devices which have led us to focus on more sustainable methods and approaches. This drives the need for a green computing solution which refers to the environmentally friendly use of computers and their resources, that is, the study of engineering, design, manufacturing, and disposal in a sustainable manner intending to maximize energy efficiency and service life while reducing the use of toxic materials and increasing the recyclability or biodegradability of discarded products and factory waste. Green technology or green IT has quickly become the most effective means of using technology, pioneered by the U.S. Environmental Protection Agency (EPA) in 1992.

This paper focuses mainly on the area of cloud computing and surveys the various solutions and methods which would present a more energy-efficient solution in today's world. But before that it is important to understand how the coming up of cloud altogether affected the aspect of green computing and improved the past scenario. There is no doubt that shifting to cloud would lead to a lot of improvement in overall energy reduction, as derived from a study conducted by Microsoft in 2018 [1], that cloud computing can increase the efficiency of a business by 22 to 93%. In the study, Microsoft compared the greenhouse gas emissions of its cloud-based products to traditional on-site data storage solutions and found an evident reduction in the amount of carbon footprints which came from the operational, infrastructural, and equipment efficiency of the data centers [2]. In today's time, most organizations are moving from traditional data centers to cloud-based resources because these modern networks provide economic and technological advantages. These technologies provide energy-saving ways to reduce carbon footprint and e-waste [3]. If we talk about green computing, then cloud computing has helped a lot in minimizing the electricity and power consumption of highly efficient cloud networks. Due to virtualizations and outsourcing itself, there has been a considerable amount of reduction in the capital expenditure, hardware-related requirements, and power usage. According to a report by Accenture, there are a lot of factors which have made shifting to a cloud-based architecture very energy efficient.

- Due to dynamic provisioning, most IT companies do not end up deploying a lot more infrastructure for the data requirements than actually required, but data centers always maintain only the active servers as per the demand of the company and hence lower energy consumption avoiding over provisioning.
- Multi-tenancy approach helps reduce the carbon emission in cloud infrastructure.
- Cloud provides the ability to host different applications on the same server but in isolation with the help of virtualization techniques and hence reduces the power consumed.
- Cloud providers can most importantly do the amendment of the power usage effectiveness (PUE) of their data centers, using the most energy-efficient technology [4].

Though there are a lot of benefits of cloud computing, the question about whether or not cloud computing is the most efficient way of saving energy and moving toward a greener solution and is still debatable. Andreas Berl states that cloud computing is an inherently and potentially efficient technology for ICT if, along with the focus on its hardware aspect, we explore more about its system operations and networking aspect [5]. It is very important to understand that data centers are a very wide terminology, and with the ever-increasing demand of cloud-based application, a lot of data centers have come into existence. A data center with 50,000 square feet of area requires about 5 MW electricity which is enough to support about 5000 households for an entire year. A cloud-based infrastructure is definitely more energy efficient than on premise data centers, but it still consumes a large amount of power, and due to the demand-based solution, the CDC needs to be functional 24/7 leading to increased energy costs, carbon emissions, and issues that degrade the environment [6].

2 An Overview of Cloud Computing

Cloud computing is an emerging paradigm which includes virtualization of resources and using them to run various applications and services on distributed networks. Cloud has really shaped the way of computing pay-as-you-use, making services universally available, and managing highly scalable systems which are customizable and reusable along with requiring a minimum manpower [7].

Cloud is extremely cost effective as it mitigates the installation and running cost of the software as there is a very minimalistic need for any sort of infrastructure. Along with that cloud provides expanded storage as compared to the traditional data storage systems. Hence, in today's competitive businesses and IT sector, it has become very crucial for companies to espouse cloud-based architecture. Deployment model and service model are two kinds of models for distributed cloud computing based upon

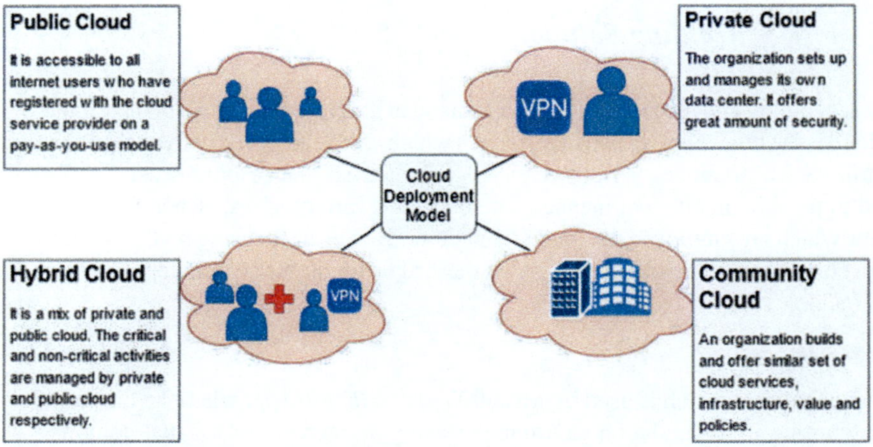

Fig. 1 Cloud deployment model

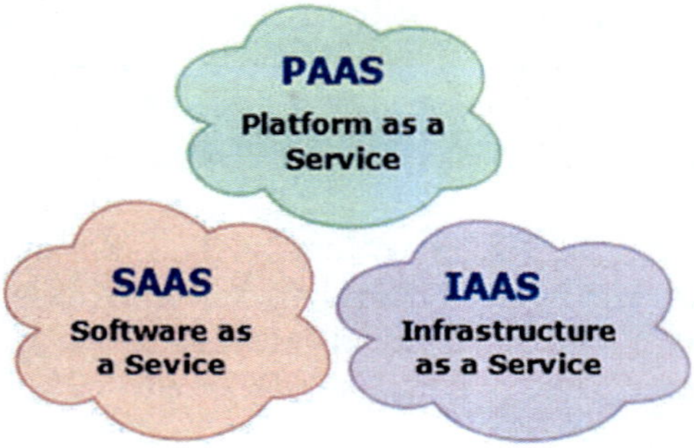

Fig. 2 Cloud service model

the accessibility and function of where the cloud is based. Figures 1 and 2 show these different types of models [8].

3 Survey

As mentioned above, cloud is proven to be environment friendly and has made the utilization of the resources quite efficient which in turn leads to lower power consumption and lower carbon emissions. However, there are various parameters and aspects of cloud computing where we can come up with more efficient solutions and further reduce the impact on the environment. Here, we survey the solutions proposed by various authors with respect to load balancing, task scheduling, and data centers.

3.1 Task Scheduling Solution

Task scheduling and resource allocation go hand in hand. One of the pivotal attributes of cloud is the proper utilization of resources which can be achieved with a scheduling algorithm that organizes various requests and tasks so that resources can be allocated to them in an efficient manner. There exists a variety of task scheduling algorithms which are employed in cloud architecture. However, focusing on the scope of green computing, there is a need to make these algorithms energy efficient and more sustainable.

Green scheduler

It is an algorithm which is used by Mridul Wadhwa in his paper where he has tested the algorithm in a simulation environment using a green simulator that combines 2 languages—C++ and the other TCL's core code which is written in C++. The basic principle of the green scheduler algorithm is that when the load increases the servers will turn on and when load decreases the servers will turn off. This leads to minimization of power consumption and load on data centers. In the paper, the author has compared the results on the basis of energy consumption among different, already existing, algorithms like DENS and round robin task scheduling. The goal of this was to find out the most optimal algorithm in terms of the amount of the energy consumed and distribution of the tasks among all the servers ideally. The green scheduler consumes the minimum amount of energy because it distributes the task among all the servers and is able to send large amounts of tasks to fewer servers. It was also observed that green scheduler had an increased data center load as compared to DENS algorithm. If we look into the energy efficiency of the green scheduler algorithm, the following results are obtained in the study [9] (Table 1).

Another simulation of the green scheduler was also presented by Jagadeeswara Rao. G in his paper using a green cloud simulator and Ubuntu 12.04, 32-bit on Intel Core i5 system. To set up the configuration of data centers, 144 servers were used, 3 VMs per machine, 327,529 of average submitted tasks. Along with this, the simulation time was 62.5 s with 10 cloud users. A 3-tier debug DC topology was employed, and system load was 30%. (Further details are mentioned in the paper). The total energy consumed by the green scheduler was found to be about 637.2 Wh and simulation time of 1320 s [10]. Looking at the results and comparisons with other algorithms like round robin, HEROS scheduler, and random scheduler, we can conclude that green scheduler is one of the greener options we can develop and practice.

Genetic-ACO task scheduling

A dynamic fusion between ant colony algorithm and genetic algorithm proposed by Zhong Zong would also provide a minimum energy utilization and processing time realizing the goal of green cloud computing. The core idea is to combine the global optimization abilities of GAs and local optimization abilities of ACOs to fasten the convergence, reduce task execution time, improve task scheduling efficiency, and mitigating energy consumptions. Cloud computing simulation software—CloudSim. The main initial parameters for both testing the algorithms are all set and illustrated in the paper. The genetic-ACO task scheduling performance is compared with that of individual algorithms taking around 200 execution tasks. At about 190 tasks, the GCA consumed about 80 units of energy consumption which was considerably less than that of GA and ACO which was about 100 and 105 units, respectively. Hence, we can say that the combined algorithm can achieve greater energy optimization [11].

Table 1 Energy efficiency of green scheduler algorithm

Parameters (energy efficiency)	Green scheduler
Data center	203.3 K (48%)
Servers	161.8 K (45%)
Network switches	41.5 K (65%)

Clonal Selection Algorithm (TSCSA)

Another algorithm is the task scheduling clonal selection algorithm (TSCSA) put forward by R. K. Jena, which aims at optimizing the energy consideration and processing time. Cloud simulator used—CloudSim-3.0.1, which is an open-source environment to test the algorithm. Clonal selection algorithm (CSA) is a based upon clonal selection theory and was first put forward by de Castro and Von Zuben. In the TSCSA algorithm, jobs are added to the arrival queue, and based on the available data center, the job is selected for execution using FCFS principle. The details of the algorithm are mentioned in the paper. If we look into the energy consumption statistics of TSCSA model, it states that the TSCSA reduces the energy consumption by 10–30% and the time by 5–25% when compared to other algorithms like various genetic algorithms-based TSGA, maximum applications scheduling algorithm (MASA), and random scheduling algorithm (RSA) [12].

3.2 *Data Centers*

Energy consumption by data centers

With the emerging digitized economy, an ever-increasing amount of data is generated and stored in data centers worldwide. As analyzed by the 2019 paper by Ralph Hintemann, the energy consumption of servers and data centers has grown from 225 billion kWh per year in 2010 to around 350 billion kWh per year in 2017. The Fig. 3 illustrates this development [13].

And the rate at which the energy consumption by data centers is increasing is definitely not ameliorating. It is concluded that the gains that we are getting from the innovation in technology do not compensate for the growth in the usage and

Fig. 3 Energy consumption of data centers from 2010 to 2017

massive consumption of electricity. If all the other growth factors remain the same, it is expected that the electricity consumption by the data centers would reach up to 321 TWh by 2030 [14].

Green data centers

Green data centers are a service which facilitates storage in the cloud while utilizing energy-efficient technologies and avoiding obsolete systems. With the exponential increase in the application of the Internet, the power consumption has also increased proportionally. Due to this, there has been a creation of sustainable green data centers that have become essential in an environmental sense. Industry estimates suggest that data centers consume three to five percent of the world's global energy [15] which has a similar footing with the aviation industry. Due to this, various methods have been introduced to implement green computing in data centers and henceforth making them green data centers. The methods are as follows:

AI for sustainable data centers: The use of AI to cut down on power consumption is being explored by tech giants like Google, which has created a machine learning system that can optimize the cooling of data centers and hence lead to the reduction of energy consumption by them. The cooling bill for Google's data centers has reduced by about 40% by using deep mind AI [16]. The software uses the micro-variation technique wherein components can be memory caches or various searching and sorting algorithms which assemble themselves according to the type of work that is being done by a server [17].

Safe haven for hyper scale cloud firms: While AI is being used to control the energy consumption of data centers, Stockholm's data center hubs aim at using the waste heat released from them to provide the local areas with heated homes [15]. The city has formed a consortium of several companies working as a data park that will provide heating for homes in the local areas and allow the country to become a viable location for firms looking to build their European data center footprint [18]. In the current scenario with the consumption of large amounts of energy by the data centers, Greenpeace [19] has been pressing companies like Amazon Web Services (AWS), Facebook, and Google to ensure that their growth does not come at the expense of the environment. This allows Sweden to not only be an attractive hub for companies targeting Europe but also provide for the country's residents. Setting the minimum data center load to 10 MW allows operators to use the waste heat for heating around 20,000 modern apartments [20].

Underwater subsea data facilities: Underwater data facility concentrates on the longevity of the machinery. Currently, Microsoft has been working on underwater data center [21] trials to see if the pros out way the cons with respect to the impact on the environment. Microsoft's Natik team deployed a data center 117 feet deep to the seafloor in 2018 to test the servers in the center. The aim of this experiment was to see if the submerged data center is feasible or not. Oftentimes-due to contact with oxygen and humidity on land, data centers get corroded along with bumps and jostles from component repairs by people-equipment failure can occur. According to Microsoft, the underwater data center avoids the need for a mechanical cooling system. Using Nitrogen decreases the corrosion of the materials greatly and allows for a sustainable data center. Though the underwater data center facility has yet to be deployed as a commercial product, the research has shown great results in the short term and hope to show similar results in the long term as well.

3.3 Load Balancing Solutions

Volkova stated in his paper [22] that load balancing focuses on improving system performance by distributing more load to smaller processing nodes and aims at fair allocation of computing resources in order to deliver higher levels of user satisfaction. As efficient load balancing helped resources and networks to achieve maximum throughput with least response time. Improved performance results in less consumption of resources and reduced costs. The key factors that affect cloud performance are storage utilization and download speed for the user. Some efficient load balancing techniques [23] can be employed to improve cloud performance and efficiency. A good load balancing strategy should consider load estimation, load comparison, stability of system performance, interaction between the nodes, and selection of nodes [24].

Load balancing algorithms are mainly divided into two types and further into subcategories, depending on the system state and depending on who initiated the process, shown in Fig. 4 [25].

As mentioned above, energy efficient load balancing can be achieved through high-resource utilization and can further lead to an optimal utilization of resources, hence mitigating their consumptions. Along with this, it can also help in preventing over-provisioning, reducing the response time, achieving high scalability, and avoiding bottlenecks. Here, we present some load balancing solutions which could adhere to the mentioned goal and contribute toward green computing [26].

In their paper [27], Mallikarjuna has proposed one such energy-efficient load balancing technique which uses a mathematical model to prove the efficiency of servers and effective transmission of data. The author proposes an algorithm which can provide a solution from not only the perspective of cloud provider, service provider but also the end user. The main idea is to predict a balancing factor of efficiency, availability, reliability, and cost in order to maintain balance in the large customers and abate the load continuously. The various steps followed by the algorithm are edified properly in the paper. In order to test and implement, the proposed algorithm the simulator used is -The CloudSim 3.0.3 (Calheiros et al. 2011, 2009; Buyya et al. 2009), as this provides a lot of flexibility in designing the solution. The algorithm showed the following results in terms of power and data utilization. Therefore, it provides a methodical load balancing solution to maximize the energy efficiency during product life in the cloud environment providing energy management, server consolidation and virtual machine migration, etc., temperature-based distribution of the load by the scheduler to the virtual machines which is very far

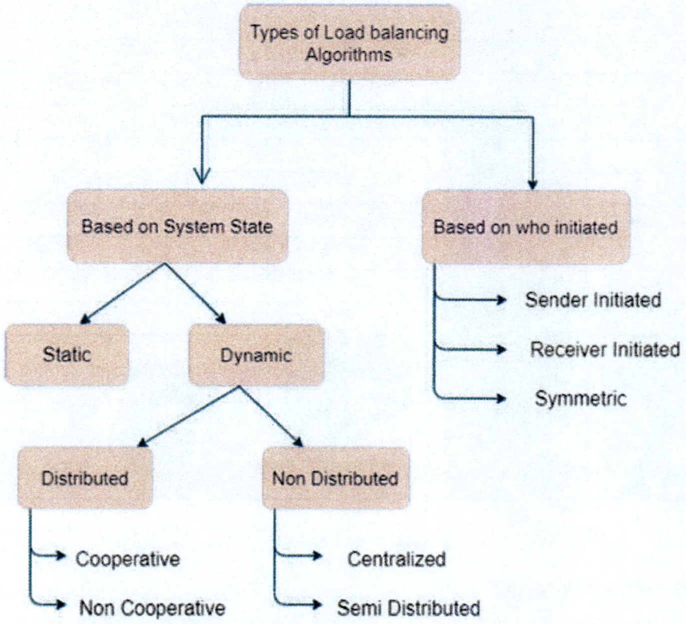

Fig. 4 Types of load balancing algorithms

from its critical temperature and focuses on less power consumption. The basic functioning of this algorithm is that it makes decisions based on the critical temperature of the system and thereafter the power consumption of that system. This is done by a scheduler which maintains a queue for the same. The 3 important specifications of the algorithm are critical temperature, minimum temperature, and power consumed. The experiment is performed on a cloud simulator and developed in Java. The energy and the temperature consumed by using this algorithm is between 200 and 250 units. So, we can say that this thermal and power-based scheduling policy are not only able to distribute workload in an efficient manner but also reduce the temperature of the nodes (Fig. 5).

The next algorithm put forward by the author, TanuShree [28], involves.

Another algorithm mentioned by Yatendra Sahu in their paper is honey bee foraging, and this algorithm is a type of distributed load balancing technique. Distributed load balancing techniques focus on distributing their workload to different host machines connected by host machines that share resources at a global level. Honey bee foraging is inspired from the behavior of honey bees as they follow a procedure for finding and reaping food. This algorithm is ideal when the multiform types of services are preferred [29]. The compare and balance algorithm maintain an equilibrium condition of the cloud server to manage the workload of the cloud system. On the basis of no. of virtual machines running on, the current host, and the whole cloud system, the current host randomly selects a host and compares their workload. If the current host has more workload than the selected host, then the

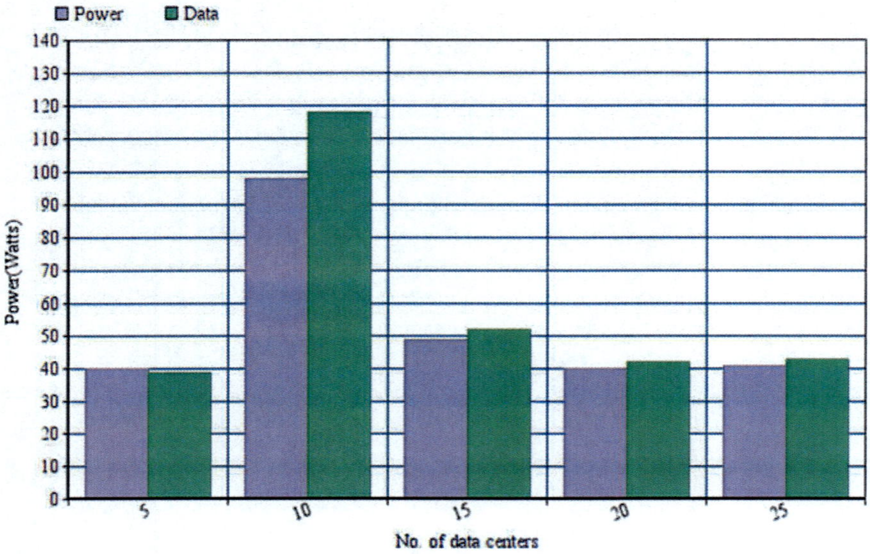

Fig. 5 Effective energy utilization

difference of the workload is transferred to that node. This algorithm decreases the migration time using live migration technique.

4 Future Scope

Green cloud computing that it is today is the result of much technological advancement that happened in the last many decades such as distributed systems, grid computing, virtualization, utility computing, and other advancements. To further improve green computing, it is necessary to improve existing technologies to minimize resource usage while meeting the quality-of-service requirements and robustness. Also, new approaches and techniques should be analyzed and used. In addition to this, large number of servers and data centers work toward providing pay-per-use services to the consumers. These resources occupy a huge area and require a large amount of power. If this energy is derived from green sources like solar energy and wind energy, then it will further help reduce the carbon footprint of green computing.

5 Conclusion

This paper focuses on highlighting the effect of cloud computing on the environment and thereafter presents a survey of various methods and solutions, primarily concentrating on 3 aspects—task scheduling, load balancing, and data centers, for making the cloud more sustainable and environment friendly. We see many effective strategies which can be implemented on the mentioned aspects in order to make the cloud greener. However, we can only see a considerable amount of difference if solutions like these are practiced consistently worldwide for a certain amount of time. Cloud computing is burgeoning at an exponential rate and the harmful effects, and it will have on the environment will be visible pretty soon. So, it is high time that we work toward optimizing such methods and making them as cost efficient as possible. Nonetheless, green cloud computing comes with a bunch of challenges like:

- A good amount of awareness in the area of green computing is needed among the stakeholders of computing ranging from manufacturer user to organization.
- Green computing requires a standardized policy from the government of each country [30].
- Cost-efficiency is also required while considering green solutions by building less costly computation. Performance deprivation as a degrading performance of servers will ultimately increase the consumption of power and energy throughput [31].

There is still a lot of research happening up with more solutions to green cloud computing taking into account all the challenges and attempts to mitigate them. Despite the challenges, there are a lot of benefits of a green cloud and hence should definitely be practiced as much as possible.

References

1. Study: Carbon, energy efficiency benefits of the Microsoft cloud. Microsoft. (n.d.). Retrieved September 28, 2021, from https://www.microsoft.com/en-us/download/details.aspx?id=56950.2.
2. The cloud paradox: Less energy used, more energy wasted by michaelmh.Engineering.com. (n.d.). Retrieved September 28, 2021, from https://www.engineering.com/story/the-cloud-paradox-less-energy-used-more-energy-wasted.
3. Radu, L.-D. (2017). Green cloud computing: A literature survey. *Symmetry, 9*(12), 295.
4. Sandhu, S. S., Rawal, A., Kaur, P., & Gupta, N. (2012). Major components associated with green networking in information communication technology systems. In *2012 International conference on computing, communication and applications* (pp. 1–6). IEEE.
5. Berl, A., Gelenbe, E., Di Girolamo, M., Giuliani, G., De Meer, H., Dang, M. Q., & Pentikousis, K. (2010). Energy-efficient cloud computing. *The Computer Journal, 53*(7), 1045–1051.
6. Garg, A. K., M. L., & Ritika, R. (2019). The issues of energy efficiency in cloud computing based data centers. *Bioscience Biotechnology Research Communications, 12*(2), 485–490.
7. Kaur, M., & Singh, P. (2013). Energy efficient green cloud: Underlying structure. In *2013 International conference on energy efficient technologies for sustainability* (pp. 207–212). IEEE.
8. Malik, M. I. (2018). Cloud computing-technologies. *International Journal of Advanced Research in Computer Science, 9*(2), 379–384.
9. Wadhwa, M., Goel, A., Choudhury, A., & Mishra, V. P. (2019). green cloud computing-a greener approach to IT. In *2019 International conference on computational intelligence and knowledge economy (ICCIKE)* (pp. 760–764). IEEE.

10. Rao, G. J., & Stalin Babu, G. (2017). Energy analysis of task scheduling algorithms in green cloud. In *2017 International conference on innovative mechanisms for industry applications (ICIMIA)* (pp. 302–305). IEEE.
11. Zong, Z. (2020) An Improvement of Task Scheduling Algorithms for Green Cloud Computing. In *2020 15th International conference on computer science & education (ICCSE)* (pp. 654–657). IEEE.
12. Jena, R. K. (2017). Energy efficient task scheduling in cloud environment. *Energy Procedia, 141*, 222–227.
13. Hintemann, R., & Hinterholzer, S. (n.d.). Innovation alliances for sustainable ICT—good practices and success factors, using the example of initiatives to improve the energy efficiency of data centers. https://doi.org/10.29007/k8d7
14. Koot, M., & Wijnhoven, F. (2021). Usage impact on data center electricity needs: A system dynamic forecasting model. *Applied Energy, 291*, 116798.
15. Focus green datacenter—bitpipe. (n.d.). Retrieved September 28, 2021, from https://docs.media.bitpipe.com/io_10x/io_102267/item_1306461/Focus_Green_datacentre.pdf.
16. DeepMind, A. I. (2016). Reduces google data center cooling bill by 40%. 2018-08-20. https://goo.gl/QTdU2T
17. Donnelly, C. (2016, November 1). Lancaster University hails potential of AI software to cut Datacenter power consumption.ComputerWeekly.com. Retrieved September 28, 2021, from https://www.computerweekly.com/news/450402095/Lancaster-University-hails-potential-of-AI-software-to-cut-datacentre-power-consumption.
18. Stadigs, J. (2015, March 18). Interview: Building the 'world's greenest Datacentre' in falun, sweden.ComputerWeekly.com. Retrieved September 28, 2021, from https://www.computerweekly.com/news/2240242517/Interview-Building-the-worlds-greenest-datacentre-in-Falun-Sweden.
19. Greenpeace. (n.d.). Retrieved September 28, 2021, from https://www.greenpeace.org/.
20. Build your data center in the Silicon valley of Scandinavia- welcome to stockholm DATA PARKS. Stockholm Data Parks. (2020, April 20). Retrieved September 28, 2021, from https://stockholmdataparks.com/benefits-of-green-computing-in-stockholm/.
21. Roach, J. (2020). Microsoft finds underwater datacenters are reliable, practical and use energy sustainably.
22. Volkova, V. N., Chemenkaya, L. V., Desyatirikova, E. N., Hajali, M., Khodar, A., Osama, A. (2018). Load balancing in cloud computing. In *2018 IEEE conference of russian young researchers in electrical and electronic engineering (EIConRus)* (pp. 387–390). IEEE.
23. Al Nuaimi, K., Mohamed, N., Al Nuaimi, M., & Al-Jaroodi, J. (2012). A survey of load balancing in cloud computing: Challenges and algorithms. In *2012 second symposium on network cloud computing and applications* (pp. 137–142). IEEE.
24. Mondal, B., Dasgupta, K., & Dutta, P. (2012). Load balancing in cloud computing using stochastic hill climbing-a soft computing approach. *Procedia Technology, 4*, 783–789.
25. Kaur, R., & Luthra, P. (2012). Load balancing in cloud computing. In *Proceedings of international conference on recent trends in information, telecommunication and computing, ITC*.
26. Rawat, P. S., Gupta, P., Dimri, P., & Saroha, G. P. (2020). Power efficient resource provisioning for cloud infrastructure using bio-inspired artificial neural network model. *Sustainable Computing: Informatics and Systems 28*, 100431.
27. Kumar, N. (2018). Green computing: Efficient energy load balancing technique in cloud computing. *SSRN Electronic Journal*.
28. Shree, T., & Badal, N. (2016). Energy efficient load balancing algorithm for green cloud. *International Journal of Engineering Research, 5*(03).
29. Sahu, Y., Pateriya, R. K., & Gupta, R. K. (2013). Cloud server optimization with load balancing and green computing techniques using dynamic compare and balance algorithm. In *2013 5th International conference and computational intelligence and communication Networks* (pp. 527–531). IEEE.

30. Paul, P. K., & Chaterjee, D. (2012). Cloud computing and green computing: challenges & issues in Indian perspective. *Asian Journal of Computer Science and Technology, 1*(2), 50–54.
31. Rawat, S., Kumar, P., Sagar, S., Singh, I, & Garg, K. (2017). An analytical evaluation of challenges in Green cloud computing. In *2017 International conference on infocom technologies and unmanned systems (trends and future directions) (ICTUS)* (pp. 351–355). IEEE.

Concurrency Control in Distributed Database Systems: An In-Depth Analysis

Husen Saifibhai Nalawala, Jaymin Shah, Smita Agrawal, and Parita Oza

Abstract Distributed databases are databases that spread across multiple locations, often crossing geographical boundaries. It has been a popular research topic because of the novel set of problems it brings to the table. One of the problems is maintaining consistency in the database. Concurrent access to the database gives rise to consistency and integrity issues that need to be resolved. Various methods have been put forward, and this paper explores some of those methods, particularly on-lock and timestamp-based techniques. It also analyzes all these methods based on various parameters.

1 Introduction

A database is used to store an organization's data, which is further used for various activities. Traditionally, the database has been centralized in nature, i.e., all the data stored in a single site. However, as businesses grows, centralized databases tend to show certain limitations, so distributed or decentralized databases are preferred [1]. Distributed databases have their own set of problems, but the advantages outgrow them. This paper focuses on one such problem, which is concurrency control in distributed databases. A distributed database (DDB) is spread across multiple sites and interconnected using a medium, usually a network. Thus, it spread across multiple locations rather than keeping all the data in a centralized location. There

H. S. Nalawala · J. Shah · S. Agrawal (✉) · P. Oza
Computer Science and Engineering Department, Nirma University, Ahmedabad, India
e-mail: smita.agrwal@nirmauni.ac.in

H. S. Nalawala
e-mail: smita.agrwal@nirmauni.ac.in

J. Shah
e-mail: 19mca035@nirmauni.ac.in

P. Oza
e-mail: parita.prajapati@nirmauni.ac.in

P. K. Singh et al. (eds.), *Proceedings of Third International Conference on Computing, Communications, and Cyber-Security*, Lecture Notes in Networks and Systems 421,
https://doi.org/10.1007/978-981-19-1142-2_17

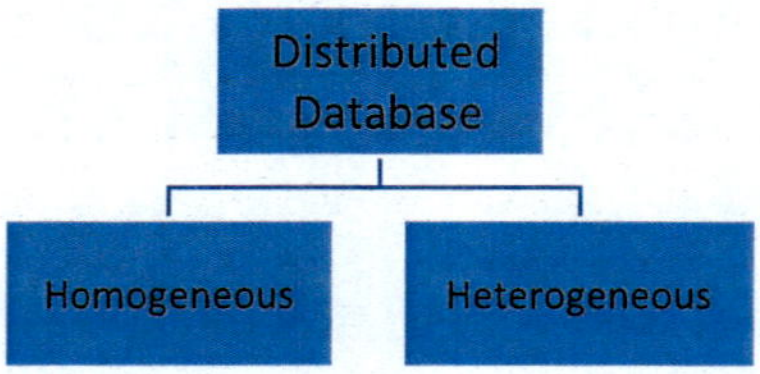

Fig. 1 Types of distributed systems

are various reasons an organization might want to prefer implementing DDB [2]. These include valuable data protection, making data highly available by maintaining multiple copies, promoting autonomy among various geographical regions, increasing database performance, etc. Even though DDBs are widely used, it is a difficult task to design them properly. A chief characteristic every DDB should possess is that the end-user of the system should not distinguish between a DDB and non-DDB.

Distributed databases can be classified into two types as per Fig. 1. In homogeneous systems, all the sites use the same type of database software and heterogeneous systems where at least one site uses a different database software.

Numerous problems exist which are distinct to DDBs [3, 4]. Some of them are discussed below:

Concurrency Control—In DDBs, the integrity and the consistency of the database as well as its copies have to be taken care of. It is one of the most researched areas and is the central topic of this paper.

Replication Control—In a distributed database, multiple copies of the data and its schema objects like tables, views, etc., will be copied in different locations. The real issue lies in maintaining the same data across all the copies and thus maintaining consistency.

Deadlock Management—This is similar to deadlock management in operating systems. Multiple users can try to access specific data, leading to deadlock if locking techniques are implemented. Like operating systems, the solution is to apply mechanisms like deadlock prevention, avoidance, and detection.

Operating System Support—To support distributed systems, special operating systems are required to support such a mechanism in the first place. It is a challenge to maintain such an operating environment.

Resource Handling—Resources are located in different locations, and routing through the network is one of the issues.

The paper is structured in the following manner—Sect. 2 gives a detailed introduction to concurrency control in distributed databases. Section 3 is about various concurrency control methods in depth. Section 4 analyzes all the methods discussed in Sect. 3 along with various parameters. Finally, the paper concluded in Sect. 4.

2 Concurrency in Distributed Database

In a multiuser database, multiple users can access the same block of data concurrently. This will not be an issue if they perform only a read operation on the same block. But reading and writing on the same data block simultaneously is where the issue is created. This can lead to consistency and integrity problems. Concurrency control solves the problem of managing concurrent access to the database and maintaining it consistently. It gives an illusion to the users that they are the only ones interacting with the database, all while maintaining the atomicity, consistency, isolation, durability (ACID) properties of the database system [5]. Concurrency control is challenging in the case of DDBs. Problems exclusive to DDBs need to be tackled, like a failure of a particular site, deadlocks across multiple sites, network failures, and managing multiple copies of a data item. Some of the problems or anomalies that we are bound to face without any concurrency control are discussed in subsequent subsections.

a. **Lost Update Anomaly**

Consider that two customers are accessing the same bank account and depositing money into it at the same time. Then, the final account balance will be inconsistent without concurrency control as shown in Fig. 2 with example.

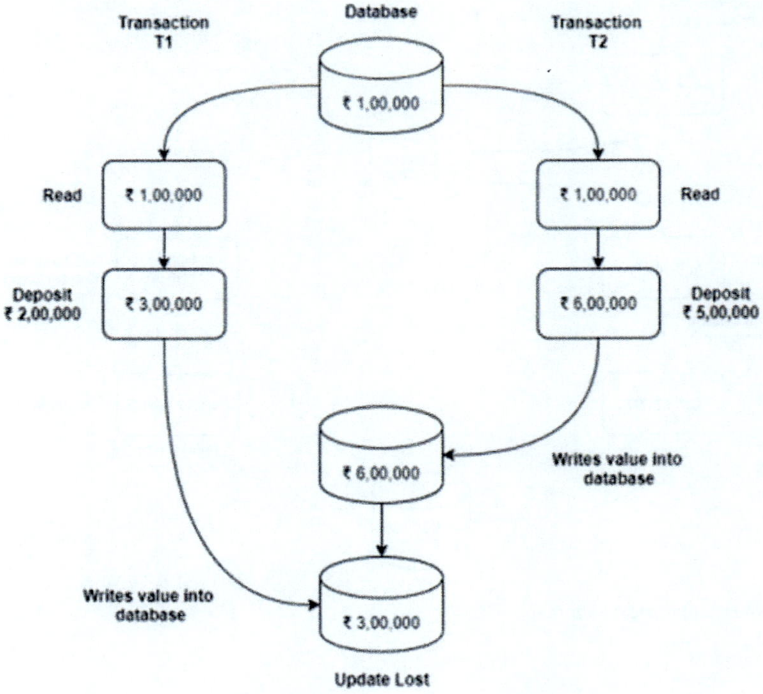

Fig. 2 Lost update anomaly with bank example

b. **Inconsistent Retrieval Anomaly**

This type of anomaly does not put inconsistent values into the database, but the values retrieved by the transaction could be invalid [6], e.g., continuing with our bank example, let's bifurcate the bank account into savings and current account. Suppose a transaction T1 is transferring some funds from savings to the current account, and another transaction T2 reads the account balance in between. In that case, the transaction T2 will be left with an inconsistent value. A graphical representation of this anomaly can be observed in Fig. 3.

The described problems are the two most common problems which are found in DDBs. Of course, this list is not exhaustive. Problems like phantom read, temporary update, etc., can also take place.

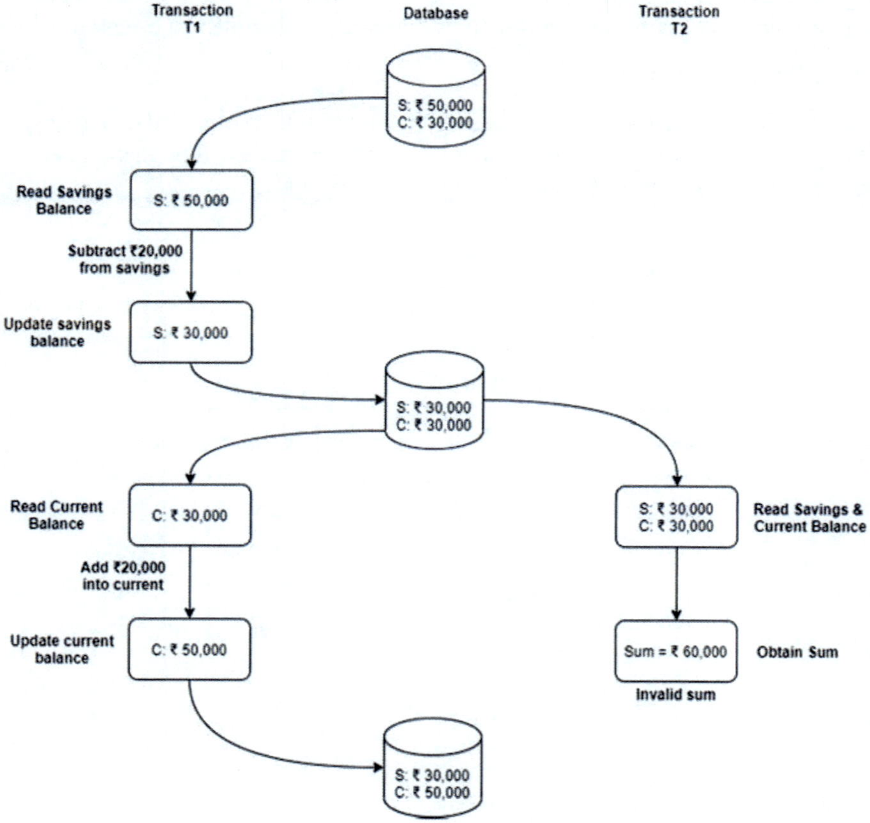

Fig. 3 Inconsistent retrieval anomaly

3 Different Methods of Concurrency Control

Concurrency control is challenging in the case of DDBs. Problems exclusive to DDBs need to be tackled, like a failure of a particular site, deadlocks across multiple sites, network failures, and managing multiple copies of a data item [7–9]. The various concurrency control methods discussed in this section are some of the problems or anomalies that we are bound to face without any concurrency control.

3.1 Two-Phase Locking (2PL)

Two-phase locking (or simply 2PL) is an algorithm that works on the principle "read any, write all," i.e., a read transaction requires only the lock to be placed on any copy of the data block, and that version will be locked. On the other hand, a write (update) operation requires a lock to be placed on all the versions of the data block. The lock can be either shared (S—read) or exclusive (X—write). This algorithm works in two phases:

- **Growing Phase**

  This phase can also be called the "lock acquisition" phase, wherein a transaction acquires all the locks on the data block it needs for successful completion. No locks are released in this phase. The transaction reaches "lock pint" after securing all the locks. The transaction has to restart if it fails to acquire all the locks and wait for some time before it can start acquiring again, as shown in Fig. 4.

- **Shrinking Phase**

  After releasing the first lock, the transaction enters the shrinking phase [1]. Here, all the locks previously acquired are released, and no further locks are secured.

Fig. 4 Two-phase locking

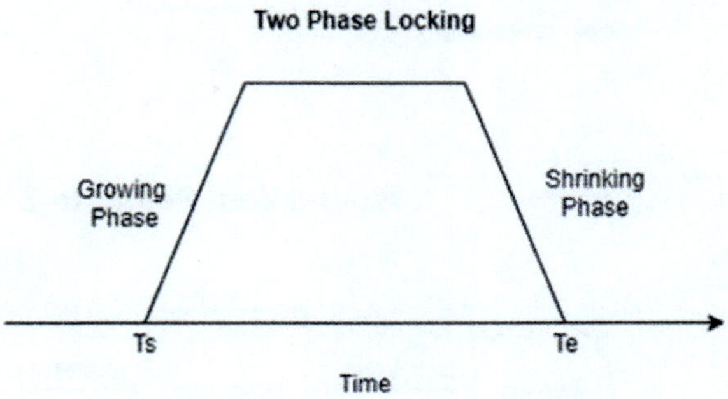

3.2 *Strict Two-Phase Locking (S2PL)*

Similar to 2PL, S2PL varies only in the fact that it releases all the locks at once in the shrinking phase, unlike one at a time in 2PL that is represented in Fig. 5. Thus, no locks are released unless the transaction performs commit or aborts.

- **Wound-Wait** [1, 3]

This is a preemptive technique where a transaction says Tn can wait for a data item held by a conflicting transaction Tc, only if Tn has an older timestamp than Tc; otherwise, Tc is killed. Two scenarios of wound-wait are represented in Fig. 6. In other words:

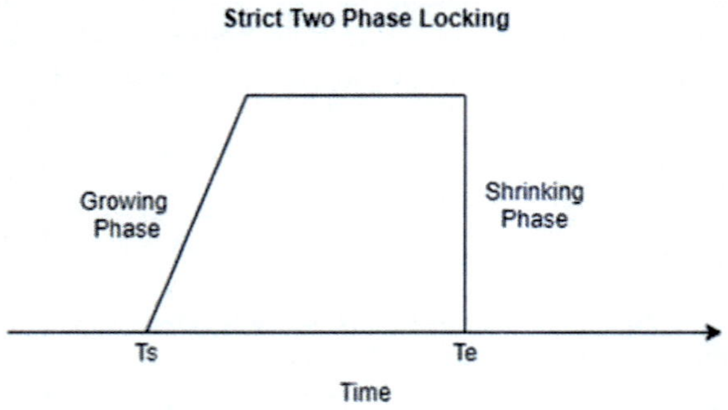

Fig. 5 Strict two-phase locking

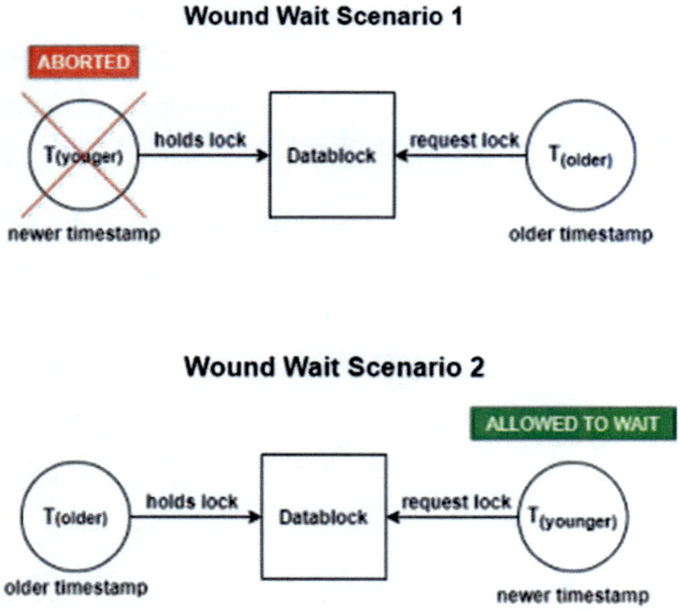

Fig. 6 Wound-wait scenario

Timestamp (Tn) < Timestamp (Tc): It means that Tn is older than Tc, so Tc is "wounded" or "killed" and is restarted later with the same timestamp.
Timestamp (Tn) > Timestamp (Tc): It means that Tn is younger than Tc, so it has to wait until Tc finishes.

- Wait-Die

Wait-die is a non-preemptive technique where a transaction, say Tn is allowed to wait for a data item held by a conflicting transaction Tc, only if Tn has an older timestamp than Tc. Otherwise, it is killed. The two scenarios for the wait-die are represented in Fig. 7. In other words:

Timestamp (Tn) < Timestamp (Tc): It means that Tn is older than Tc, so it is allowed to wait.
Timestamp (Tn) > Timestamp (Tc): It means that Tn is younger than Tc, so it is aborted and restarted later on with the same timestamp.

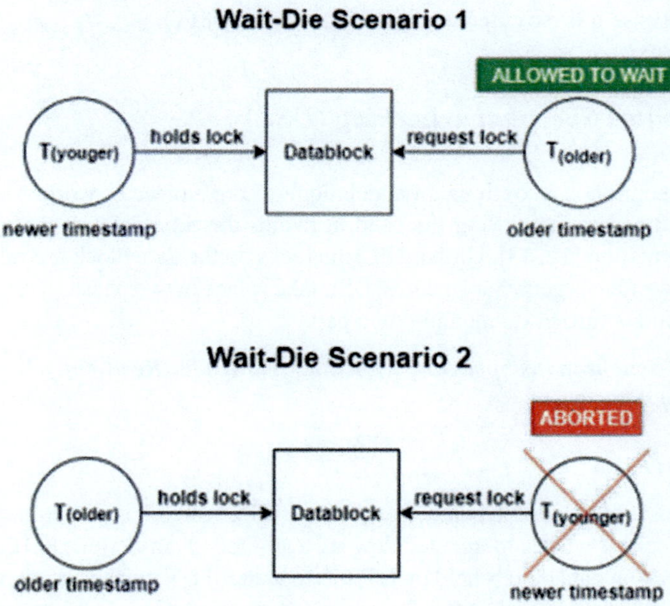

Fig. 7 Wait-Die scenario

3.3 Basic Timestamp Ordering (BTO)

The database assigns each transaction a timestamp. This timestamp indicates the order in which the transaction should be executed in the database. In other words, the timestamp indicates the starting time of the transaction [10, 11]. For two conflicting transactions, the timestamps determine the execution order and provide serializability. Additionally, BTO uses two timestamps associated with each data item, namely Last Successful Read—LSR (T) and Last Successful Write—LSW (T) on that particular data item.

Suppose a transaction T with timestamp T(S) wants to perform a write operation on a data item X. In that case, it is aborted if LSR(X) > TS(T) or LSW(X) > TS(T) (data was changed since the last fetch); otherwise, it is executed, and LSR(X) and LSW(X) is set to TS(T).
If a transaction T with timestamp T(S) wants to perform read operation on a data item X, then it is aborted if LSW(X) > TS(T) (data was changed since the last fetch), otherwise it is executed, and LSR(X) is set to TS(T).

3.4 Distributed Speculative Locking (DSL)

Speculative execution is an optimization technique where a piece of work or task is performed even before identifying the need. It avoids the time delay if the system waits for confirmation [12, 13]. Unlike 2PL, the locks on the data block are released as soon as a new value is generated in DSL. DSL takes place in two phases, Execution and Commit, and is further divided into two parts:

- ***Distributed Synchronous Speculative Locking (DSSL) for Read-Only Transactions (ROT)***

Execution Phase

Consider a transaction called Update Transaction (UT), which has placed locks on several data blocks it wishes to update. Now, a Read-Only Transaction (ROT), who wants a read lock on data blocks held by UT, has to wait till UT finishes its execution and produces an updated value. After UT completes its execution, both the original and updated values are sent to ROT as a response. The ROT does not wait for UT to commit the transaction and carries out a speculative execution with the help of both original and updated values. ROT enters the commit phase after acquiring all the locks and completing its execution.

Commit Phase

To select appropriate speculative execution, ROT communicates with the UT home site to know the commit status of UT [14]. If UT commits the transaction, the ROT commits the speculative execution with updated value; otherwise, the speculative execution with old value is committed.

- ***Distributed Asynchronous Speculative Locking (DASL) for Read-Only Transactions (ROT)***

Execution Phase

Here, the ROT does not wait for the conflicting UT to produce an updated value. Instead, it carries out speculative execution with the original value. After the UT finishes its execution, the updated value is sent to ROT. The ROT does not wait for UT to commit the transaction and carries out a speculative execution with the help of updated value.

Commit Phase

To select appropriate speculative execution, ROT communicates with the UT home site to know the commit status of UT [14]. If UT commits the transaction, the ROT commits the speculative execution with updated value; otherwise, the speculative execution with old value is committed.

4 Analysis

The comparative analysis of concurrency control methods is discussed in Table 1 by concerning parameters like performance, serializability, and deadlock. In 2PL, because of the "read all, write any" rule, it ensures serializability. Also, this algorithm performs wells in centralized environments rather than distributed environments. On the other hand, there is a possibility of deadlock and starvation in the case of 2PL. Also, cascading rollbacks are possible. In the case of strict 2PL, the transaction will not read data until it is committed, and thus prevents cascading rollbacks, but deadlocks are still possible. An easy way to resolve deadlocks is to put a wait time or a timeout on the transaction—if a transaction T1 is still waiting for locks after time t, abort it and start over.

Wound-wait and wait-die are two deadlock prevention mechanisms. We can theoretically say that older transactions would have acquired more locks and performed more tasks compared to newer transactions. Thus, it is expensive to abort the older transactions. In wound-wait, the abort rate is lower compared to wait-die, while in the case of wait-die, the younger transaction will die more times, as they need to acquire many locks. On the other hand, the amount of work done by a younger transaction might be significant in wound-wait, and aborting it will result in all the work being rolled back. In wait-die, the aborted transaction would not have done any work, and so nothing is wasted. Thus, wound-wait has fewer rollbacks, but the wastage of work can be more. Wait-die has more rollbacks, but there is no wastage of work.

In BTO, better performance is obtained if some additional information related to the database or transaction is available and used. It also ensures serializability in case

Table 1 .

Methods	Performance	Serializability	Deadlock
2PL	• Suitable for databases with lots of frequent updates • Increased overhead because of maintaining locks	Ensures Serializability for both read-only and update transactions	Possible transactions are restarted to resolve it. Cascading rollbacks are also a possibility
Strict-2PL	• Suitable for databases with lots of frequent updates • Increased overhead because of maintaining locks	Just like 2PL, Serializability is ensured	Deadlock is still possible, but cascading rollbacks are avoided
Wound-wait	• Less aborts, more wastage of work • Aborts transaction even if there is no deadlock	Serializability is ensured	It is a deadlock prevention technique, so deadlock is avoided No starvation
Wait-die	• More aborts, less wastage of work • Aborts transaction even if there is no deadlock	Serializability is guaranteed	It is also a deadlock prevention technique, so deadlock is avoided No starvation
BTO	• Provides better performance if additional information is available about the database or transaction is used • Better concurrency than 2PL, as no transactions are blocked	As transactions are processed as per the timestamp, it ensures Serializability	Free of deadlocks Possibility of starvation and cascading deadlocks
DSSL	• Better than DASL for transactions <60 • Overall less effective than DASL	It uses parallelism and maintains Serializability	As it is a locking-based approach, deadlock is possible and additional mechanisms are required to handle it. For example, wait-for graphs can be used to detect deadlock, and in the case of one, the waiting / requesting transaction is aborted
DASL	• Better than DSSL as there is less waiting time and data contention	It uses parallelism and maintains serializability	As it is a locking-based approach, deadlock is possible and additional mechanisms are required to handle it. For example, wait-for graphs can be used to detect deadlock, and in the case of one, the waiting/requesting transaction is aborted

of conflict, as the conflicting transaction will be restarted with the same timestamp. It is also deadlock-free. However, maintaining timestamps is expensive and puts a load on the memory. Another problem with BTO is that starvation is possible, particularly a more extended transaction getting restarted by a shorter transaction.

In the case of DSSL, ROT has to wait for the UT to produce the updated value, so there is more wait time, while in DASL, there is relatively less wait time. Thus, because of this, the performance of DASL is more than DSSL. The validation in both methods is dependent on the commit status of the UT.

5 Conclusion

This paper explored concurrency in distributed databases. First, we understood in detail what distributed databases are and how they differ from traditional centralized databases. We also explored various problems unique to distributed databases. We then focused on concurrency control which is a popular research topic in distributed systems. Various methods were studied, like 2PL, Strict-2PL, wound-wait, wait-die, basic timestamp ordering (BTO), and speculative locking, which provide concurrency control. We also analyzed the pros and cons of all the methods. A summary table has been provided, which gives a bird's eye view of the analysis.

References

1. Abbas, Q., Shafiq, H., Ahmad, I., & Tharanidharan, S. (2016). Concurrency control in distributed database system. In *2016 International conference on computer communication and informatics (ICCCI)* (pp. 1–4).https://doi.org/10.1109/ICCCI.2016.7479987
2. Haroon, M. (2019). Challenges of concurrency control in object oriented distributed database systems.
3. Kanungo, S., & Morena, R. D. (2016) Comparison of concurrency control and deadlock handing in different OODBMS. *International Journal of Engineering Research & Technology (IJERT), 05*(05). https://doi.org/10.17577/IJERTV5IS050615
4. Philip, A. B., & Goodman, N. (1981). Concurrency control in distributed database systems. *ACM Computer. Survey, 13*(2), 185–221. https://doi.org/10.1145/356842.356846.
5. Nasseri, M., & Jameii, S. M. (2017). Concurrency control methods in distributed database: A review and comparison. 200–205. https://doi.org/10.1109/COMPTELIX.2017.8003964.
6. Cheung, S. Y. (2020) Comparing the wait-die and wound-wait schemes. http://www.mathcs.emory.edu/~cheung/Courses/554/Syllabus/8-recv+serial/deadlock-compare.html
7. Bernstein, P., & Goodman, N. (1981). Concurrency control in distributed database systems. *ACM Computing Surveys, 13*, 185–221. https://doi.org/10.1145/356842.356846.
8. Liut, M. Michael liut deadlock prevention. https://www.michaelliut.ca/uploads/2/1/0/3/21032302/week_11_c.pdf
9. Wakefield, R. (2016) Locking Colorado State University. https://www.cs.colostate.edu/~cs430dl/yr2016sp/more_examples/Ch14/Locking.pdf
10. Wang, C., & Qian, X. (2021). RDMA-enabled concurrency control protocols for transactions in the cloud Era. *IEEE Transactions on Cloud Computing*.

11. Guo, H., Zhou, X., & Cai, L. (2021). Lock violation for fault-tolerant distributed database system. In: *2021 IEEE 37th international conference on data engineering (ICDE)*. IEEE.
12. Gbaranwi, P. B., & Asagba, P. O. (2021). Distributed transactions and distributed concurrency control.
13. Domdouzis, K., Lake, P., & Crowther, P. (2021). Distributed databases. In *Concise guide to databases* (pp. 213–222). Springer, Cham.
14. Mhatre, A., & Shedge, R. (2014). Comparative study of concurrency control techniques in distributed databases. In *2014 Fourth international conference on communication systems and network technologies* (pp. 378–382).https://doi.org/10.1109/CSNT.2014.81

House Pricing Prediction Based on Composite Facility Score Using Machine Learning Algorithms

Santosh Kumar and Mohammad Haider Syed

Abstract Various features of a house play some role to determine its price. Out of these, location is the dominant feature to determine the price. Besides location, there are some other features which affect the price of a house like area, sports facility, hospital, 24 × 7 security, etc. In this paper, 40 features, available in dataset of houses, are taken from Kaggle platform and have been considered for prediction of house prices. The data of six different cities of India has been included, and these are Delhi, Bangalore, Hyderabad, Kolkata, Mumbai, and Chennai. Here, we endeavored to develop a predictive model for anticipating the price dependent on a specific number of highlights that influence the price. Six machine learning algorithms are used to develop models and compared based on their accuracy of prediction, and the most accurate model is used to determine the price of houses.

Keywords Linear regression · Random forest · Decision tree · KNN · XGBoost regressor · Support vector machine

1 Introduction

House is a fundamental necessity, and the price of the house is a critical factor for house buyers as well sellers. To buy a house is the most fruitful and cost-effective investment for all the time. Delhi is among one of the cities with a high number of house owners. A survey conducted in Delhi showed that there are 66.63% families that live in their own houses, whereas 32.38% families live as a tenant. Price of the house varies depending upon different residential and geographical factors. Even different countries have their own way of analyzing and giving preference to certain

S. Kumar (✉)
Department of Computer Science and Engineering, ABES Engineering College, Ghaziabad, Uttar Pradesh, India
e-mail: santoshg25@gmail.com

M. H. Syed
College of Computing and Informatics, Saudi Electronic University, Riyadh, Saudi Arabia
e-mail: m.haider@seu.edu.sa

P. K. Singh et al. (eds.), *Proceedings of Third International Conference on Computing, Communications, and Cyber-Security*, Lecture Notes in Networks and Systems 421,
https://doi.org/10.1007/978-981-19-1142-2_18

factors which might not be common to every country. In the phase of transactions related to the sale and purchase of houses, one should consider all the criteria that affect the cost of the house and choose the best deal. Looking at Lancarter's (1966) Consumer Theory, how a dwelling is valued by a customer is dependent on certain factors. Such factors include the number of bedrooms, area of land, number of floors, and various other geographical factors, including the surrounding area. Sometimes price variation is dealt with postcode dummies, which is understood as a description of locations also referred to as absolute by the geographers. In this paper, a house price prediction model has been implemented using different machine learning algorithms and procedures to give a clear idea and analysis of different algorithms to get a perfect and efficient way to find out the correct price of the house. This would be beneficial to many sectors of society, including real estate, banks, and common people who want to purchase the house.

There are a significant large number of houses that are sold on a day-to-day basis. There are several questions in a buyer's mind while purchasing a house. Some of these questions include what would be the best cost for this house, am I paying too much, does the features of this house properly synchronize with the price. Thus, it becomes necessary to accurately determine the cost of the property. We are going to solve this problem using a house price prediction system based on machine learning techniques. The price of the house will be determined by considering various parameters on which the price depends. Some features affect the cost quite significantly, while a change in some features has a little impact on the value. These features include the region that the house occupies, the state and the shape in which the house is in, the year of establishment of the house, etc. A model for predicting the house price can prove to be a very important tool for the buyers as well as sellers and help them in making well-informed decisions. The sellers can use this tool to determine the average price at which the house can be sold. On the other hand, the buyers can use this tool to find out the fair price while purchasing the house. This would do nothing but will make it easy for buyers and sellers. They don't have to rush out and meet people to get an estimated cost to their house, and everything would be automated.

The proposed model would be able to predict the house price according to the area; to calculate house price depending upon surrounding environment like railway station, hospital area, ATM, college, banks so a customer can purchase flat/house with full facilities; to suggest builder price prediction for the new constructions and to provide a comparison of house pricing to customers. The comparison of sale and price forms the basis of the traditional house price prediction model lacking an accepted standard and a certification process. The model helps to reduce the gap of information and make the real estate market more structured.

2 Related Work

In [1], using the city of Oslo as their case to find the difference in the explanatory power of absolute versus relative location in a hedonic model. It was observed that the post code dummies power declined to a large stand when certain factors such as distance to commercial places were taken into consideration. The distance to places like parks, ATMs have very high explanatory power as they are important to house buyers. In the end, this analysis showed that the location of nearby places plays an important part in determining the price of the house in Oslo. If these factors are not considered, then it would reduce the value of neighborhoods and affect the price of the houses gravely. In [2], an innovative solution is used for analyzing and mapping real estate. The data is collected systematically, and then the data is analyzed and assessed about the real estate market's changes by the software. It is mentioned that the software assembled over 650,000 price quotations about the sale of houses, building lots, and business properties. All the information is stored and analyzed by the software for its credibility. The analysis of comparisons based on statistics helped the researchers and other market professionals to observe deeply the changes in the price of real estates of Czech Republic. Both normal people and large multinational companies can use this output to make smart business decisions. The second quarter of the year 2008 saw a long-term decrease in the cost of real estate properties. It seems that this trend would likely remain, resulting in no increase in real estate prices in Czech Republic with the exception of Prague and Middle Bohemia. In [3], the predictive power of the artificial neural network model and hedonic model is compared. It involves a randomly selected sample from the Harcourt website containing data of 200 houses in the city of Christchurch, New Zealand. It was seen that R^2 score of the neural network model outperformed the hedonic model. The performance of hedonic models was much poorer for out-of-sample data in comparison to neural network. This paper thus focused on the potential of neural network model when it comes to predicting the price of houses. In [4], several techniques such as artificial neural networks, fuzzy logic, and KNN are compared and analyzed. This analyzed data is then used to find the most efficient method for determining the price of the houses. The actual cost and the prediction determined using the MAPE formula are matched to examine the methods. The result showed that the fuzzy method outperformed all other methods and resulted in the most accurate model to predict the price of the houses. In future, for improving the prediction accuracy, optimization of fuzzy rule and increasing data training are required. In [5], the solution to "House Prices: Advanced Regression Techniques" contest which took place on Kaggle platform has been depicted. The motive of the contest is to predict the property cost based on the aspects given in the training data, some of which are lot area, lot size, house type, many classic algorithms were used to find the solution and finally got 18th position in the competition. It has also been depicted that the finest approach to enhance the outcome is to teach the model for an ensemble that needs more calculative assets. The issue with several model ensembling is how to make their errors unrelated failing, which would not lead to any significant improvement. Learning the same models on different feature sets is also one way to achieve models decorrelation. In [6], numerous procedures like neural networks, multiple linear regression, and simple linear regression are utilized to track down the best procedure which has the lowest mean square error, and afterward, that calculation is picked for anticipating the house cost. From the outcome, it is inferred from the result that the neural network method performed better prediction among all others. In [7], the use of predictive analytics in the area of real estate has been depicted. Numerous procedures are utilized to estimate property costs. A wide range of examination work done in the real estate area is referenced, and furthermore, in-depth analysis of methods like decision tree and neural networks that are used to anticipate the deal cost of the house is portrayed. It concluded that the neural networks give the most noteworthy precision among all with the least error although using this is basically a trial and error procedure. Decision tree is the second method that provides high precision after neural networks. The decision tree gave output that was binary in nature, and this was further partitioned into several classes, and the size of the class was decreased. This led to the output, which was very near to the real cost and thus, the accuracy of this method increased significantly, which concluded that regression problems could be resolved using classification. In [8], it is mentioned about real estate customers that the cost of some property is based on how much can customer spend and how early customer wants to buy. So, the cost of some property in the time to come will always be predicted by calculating the previous trends in the real estate market and also considering the new developments and the cost range. Now the problem mentioned in this paper is that all the data is needed to be stored and extracted on the basis of one's requirements is becoming tough. By using the linear regression technique on the extracted information should be utilized. The setup uses a very efficient way so as to utilize such huge data, and a linear regression algorithm helps to fulfill customers by reducing the risk of investing in real estate and increasing the accuracy of estate choice. One of the major future scopes of this system is to add the estate database of more cities which will provide users to explore more real estates and reach an appropriate and accurate decision. In [9], predictive model is created by considering various aspects of the property that affects the cost. Several regression techniques are applied based on the performance of all these techniques the most efficient predictive model is built. The paper sheds some light on the fact that an optimal model cannot always guarantee a robust model as sometimes data can be noisy or might contain very few records, which can make the model remain fit. After observing the estimation metrics for advanced regression models, it is concluded that both behaved in the same manner. Either model can be chosen for the prediction of the price of the house compared to the primary model. Also with the help of many techniques such as box plots, outliers can be found and removed, after which model's performance can be checked for improvement. In [10], different methods to find the assessment in different conditions on the basis of area dependence of the property are compared. A relatively fluid approach is used; this method may allow the smaller market to differ from property to property. It is concluded that the accuracy is gained from different methods is lower than when adding the small markets in the specification. In [11], the current sale price of modern property is estimated by inflating the previous selling price. Modeling how the asking price influences the time and price of the sale is possible with prediction. This study looked at 105 pairs of new and old sale prices for properties constructed in the UK since January 1999. A tool for changing the price of a home by using data from previous sales and a method that modifies the house price using observed sales of houses is analyzed by using the best published index, which gives an average error of 10.9%, while the published index adjusted by the sales information gives an error of 8.4%. In [12], NJOP is used with regression analysis and particle swarm optimization to predict the property cost in Malang city. This study finds that combining regression and PSO is efficient and yields the lowest predicted error. To forecast house prices, several experiments were conducted using particle swarm optimization and linear regression. The system divided the NJOP data of nine houses into seven models, each representing a different region. Kelurahan Karang Besuki, Tunggul Wulung, Lowokwaru, Puncak, Trikora, Sumbersari, Dinoyo, and Manggar are all part of the modeled region. It can be concluded that M-1 represents the Karang Besuki region and obtains the best parameter for optimal prediction based on the results of the iteration test, particle test, and inertia weight test. The prediction error values for other models are still very high. To fit the time-series data that will be used in future studies, various different methods have to be used.

3 Data Set Description

Data Description: The dataset is taken from Kaggle platform, and it consists of various features that affect the price of houses in a city. There are around 40 features containing 4898 records (Table 1).

Table 1 Feature set in the dataset

Feature set			
Price	Rainwater Harvesting	Car parking	Lift available
Area	Indoor games	Staff quarter	BED
Location	Shopping mall	Cafeteria	Vaastu Compliant
No. of bedrooms	Intercom	Multipurpose Room	Microwave
Resale	Sports facility	Hospital	Golf course
Maintenance staff	ATM	Washing machine	TV
Gymnasium	Clubhouse	Gas connection	Dining table
Swimming pool	School	AC	Sofa
Landscaped gardens	24 × 7 security	Wifi	Wardrobe
Jogging track	Power backup	Children's play area	Refrigerator

4 Process Flow Diagram

Step-1: Importing all the libraries needed for the implementation.

Step-2: Loading the dataset into pandas dataframe and exploring the dataset using various Python functions and heat map.

Step-3: After loading the dataset, split the dataset into a test and train set.

Step-4: To find the predicted values, different models have been developed.

The proposed model is based on the following algorithms.

4.1 Linear Regression

Linear regression is supervised in nature, and it is one of the machine learning algorithms. This algorithm's expected output is continuous in nature, with a constant slope. Rather than being grouped into groups, the expected values are found in a continuous range.

4.2 Random Forest

The random forest algorithm is a simple supervise algorithm. This algorithm is adaptable and produces good results even when the hyper-parameters aren't tuned. Because of its simplicity, this algorithm is one of the most widely used in machine learning.

4.3 Decision Tree

For predictive modeling, the decision tree learning algorithm is widely used in data mining, machine learning, and statistics. A decision tree is used in this algorithm to evaluate an item's observation by branching and decide the item's target value, which is defined by leaves in the decision tree.

4.4 KNN

Evelyn Fix and Joseph Hodges created a nonparametric machine learning algorithm in 1951. The KNN algorithm is also known as K-nearest neighbors algorithm. This

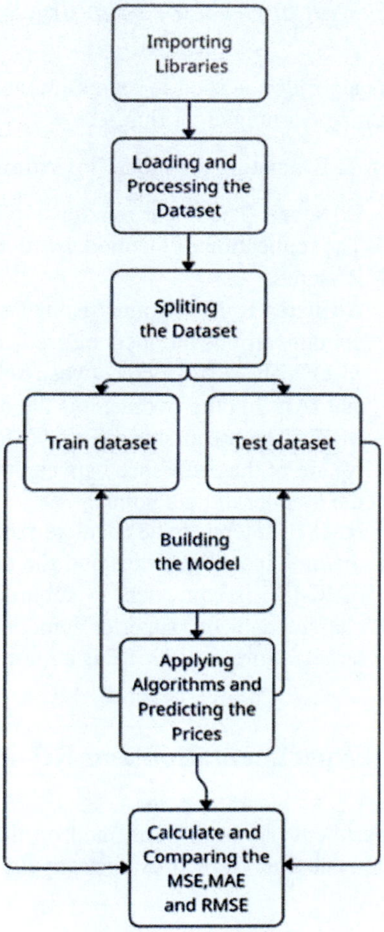

Fig. 1 Flow diagram for determining the cost of houses

algorithm is capable of both regression and classification. This algorithm's input is usually k closest training examples (Fig. 1).

4.5 XGBoost Regressor

It is an algorithm that has won several Kaggle competitions. XGBoost is a high-speed and high-performance implementation of gradient boosted decision trees.

4.6 Support Vector Machine

This algorithm is used in regression and classification studies. This is a supervised machine learning algorithm.

Step-5: Evaluating the model's performance on the following metrics:

1. **R^2 score:** The linear relationship between X and Y is referred to as R^2 score. The replication of the model with the actual result is another way of describing R^2 score.
2. **Adjusted R^2:** The comparison of a regression model, explanatory powers which contain various numbers of predictors, is referred to as adjusted R-squared.
3. **MAE:** The difference between actual and predicted values of *Y* is referred to as MAE. In other words, it is the mean of the absolute value of the errors.
4. **MSE:** The mean square error (MSE) is the sum of overall data points of the square of the difference between the actual and predicted variables, divided by the number of data points.
5. **RMSE:** It tends to be characterized as the standard example deviation between the anticipated qualities and the noticed ones. It is to be noticed that the unit of RMSE is equivalent to subordinate variable y. The lower RMSE esteems characteristic of a superior fit model. On the off chance that the model's essential target is forecast, RMSE is a more grounded measure.

5 Experimentation and Result

Experiment has been performed on the dataset of five metropolitan cities having various number of records respectively. The procedure used is given as following:

5.1 Procedure

1. The Python library used for experimental setup is pandas, NumPy, and Sklearn.
2. Convert the features from string to numeric values.
3. Check for all features to be numeric values.
4. Check for null values in feature set. If more null values in any feature, then ignore that feature.
5. Find the heat map to get the correlation degree between any two features.
6. X = FS //Feature Set
7. Y = output variable (Price)
8. $X' = X - Y$
9. Predict *Y*from *X*' i.e. *X*'*Y*
10. By utilizing the train_test_split which is present in sklearn.model_selection to split the dataset into training and testing.

Table 2 Number of records present in dataset of different cities

City	Chennai	Delhi	Hyderabad	Kolkata	Mumbai	Bangalore
Sample data size	5014	4998	2518	6507	7719	6207

11. Select the classification model (linear regression and other one by one) after importing linear regression and other models one by one in each iteration.
12. Apply the classification model.
13. Perform the prediction using training dataset.
14. Calculate the R^2, mean absolute error, etc. to get the relationship between X' and Y.
15. Now the prediction on testing dataset using the same procedure (i.e., step 10 to step 14) as used in training dataset.

Facility Score $= \sum$ Score of individual facilty ('No. of Bedrooms', 'Resale', 'Maintenance Staff', 'Gymnasium', 'Swimming Pool', 'Landscaped Gardens', 'Jogging Track','RainWater Harvesting', 'Indoor Games', 'Shopping Mall', 'Intercom', 'ATM', 'School', '24X7Security', 'Power Backup', 'Car Parking', 'Staff Quarter', 'Cafeteria', 'Multipurpose Room', 'Hospital', 'Washing Machine', 'Gas connection', 'AC', 'Wifi', 'Children's play area', 'Lift Available', 'BED', 'Vaastu Compliant', 'Microwave', 'Golf Course', 'TV', 'Dining Table', 'Sofa', 'Wardrobe', 'Refrigerator',)

(Location, Area, Facility Score)Price (Table 2).

5.2 *Location, Area, and Facilitywise Price Prediction*

In Table 3, the different evaluation scores such as mean absolute error, R^2 score, etc. have been calculated for different cities in India using different mechanisms.

In Table 4, mean absolute error of different cities is evaluated on the basis of different procedures.

Figure 2 shows the pictorial representation of comparison of various evaluation parameters on different cities in India as obtained using several machine learning mechanisms.

In Table 5, R^2 score of different cities is evaluated on the basis of different procedures.

In Table 6, mean squared error of different cities is evaluated on the basis of different procedures.

Figure 3 shows the pictorial representation of comparison of mean squared error obtained for different cities in India against several machine learning algorithms.

In Table 6, root mean squared error of different cities is evaluated on the basis of different procedures.

Figure 4 Comparison of mean squared error obtained for different cities in India against several machine learning algorithms

Table 3 Analysis of different cities based on different parameters using different machine learning mechanisms

Index	Model	MAS	RSS	MSE	RMSE	Index	Model	MAS	RSS	MSE	RMSE
Chennai						Delhi					
0	Linear regression	3.27E+06	47.63981	3.30E+13	5.75E+06	0	Linear regression	7.61E+06	38.34434	5.94E+14	2.44E+07
1	Random forest	1.60E+06	80.30135	1.24E+13	3.53E+06	1	Random Forest	5.83E+06	39.94554	5.79E+14	2.41E+07
2	SVM	3.76E+06	−7.306807	6.77E+13	8.23E+06	2	SVM	9.37E+06	-3.962984	1.00E+15	3.17E+07
3	XGBoost	1.40E+06	81.62519	1.16E+13	3.40E+06	3	XGBoost	5.45E+06	23.72509	7.35E+14	2.71E+07
4	Decision Tree	2.37E+06	60.18298	2.51E+13	5.01E+06	4	Decision Tree	5.93E+06	29.14715	6.83E+14	2.61E+07
5	KNN	1.99E+06	70.06742	1.89E+13	4.35E+06	5	KNN	5.64E+06	52.43665	4.59E+14	2.14E+07
Hydrabad						Kolkata					
0	Linear regression	2.43E+06	57.66533	3.95E+13	6.28E+06	0	Linear regression	2.43E+06	57.66533	3.95E+13	6.28E+06
1	Random forest	1.82E+06	55.48589	4.15E+13	6.44E+06	1	Random forest	1.82E+06	55.48589	4.15E+13	6.44E+06
2	SVM	4.70E+06	−4.984846	9.79E+13	9.89E+06	2	SVM	4.70E+06	-4.984846	9.79E+13	9.89E+06
3	XGBoost	1.62E+06	57.41894	3.97E+13	6.30E+06	3	XGBoost	1.62E+06	57.41894	3.97E+13	6.30E+06
4	Decision Tree	2.51E+06	48.7827	4.78E+13	6.91E+06	4	Decision Tree	2.51E+06	48.7827	4.78E+13	6.91E+06
5	KNN	2.02E+O6	56.16325	4.09E+13	6.39E+06	5	KNN	2.02E+06	56.16325	4.09E+13	6.39E+06
Mumbai						Bangalore					
0	Linear regression	1.17E+07	19.93145	9.26E+14	3.04E+07	0	Linear regression	3.24E+06	67.16284	3.36E+13	5.80E+06
1	Random forest	1.01E+07	28.6477	8.25E+14	2.87E+07	1	Random forest	1.77E+06	82.59728	1.78E+13	4.22E+06
2	SVM	1.28E+07	−6.979198	1.24E+15	3.52E+07	2	SVM	4.73E+06	−6.881138	1.09E+14	1.05E+07
3	XGBoost	8.99E+06	43.82023	6.50E+14	2.55E+07	3	XGBoost	1.62E+06	83.92624	1.64E+13	4.06E+06
4	Decision tree	1.10E+07	24.84923	8.69E+14	2.95E+07	4	Decision Tree	2.62E+06	76.85242	2.37E+13	4.87E+06
5	KNN	1.07E+07	28.71096	8.24E+14	2.87E+07	5	KNN	2.12E+06	80.38785	2.01E+13	4.48E+06

Table 4 Mean absolute error for different cities using different algorithms

	Mean absolute error					
Model	Chennai	Delhi	Hyderabad	Kolkata	Mumbai	Bangalore
Linear regression	4.67E+06	1.36E+07	2.21E+06	5.99E+06	1.01E+07	5.71E+06
Random forest	3.87E+06	1.20E+07	1.65E+06	6.83E+06	9.18E+06	4.99E+06
SVM	4.70E+06	1.37E+07	4.71E+06	5.09E+06	9.48E+06	5.85E+06
XGBoost	3.80E+06	1.32E+07	1.51E+06	6.32E+06	8.76E+06	4.95E+06
Decision tree	4.21E+06	1.37E+07	2.34E+06	5.66E+06	9.69E+06	5.32E+06
KNN	4.15E+06	1.28E+07	1.96E+06	6.20E+06	9.39E+06	5.40E+06

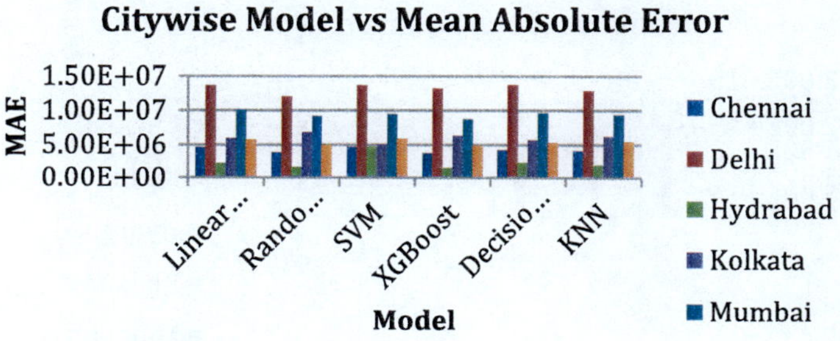

Fig. 2 Citywise model versus mean absolute error

Table 5 R-squared score for different cities using different algorithms

	R-squared score					
Model	Chennai	Delhi	Hyderabad	Kolkata	Mumbai	Bangalore
Linear regression	20.381074	19.82289	56.988162	6.172705	6.886441	14.198445
Random forest	31.217351	20.867331	58.77837	−38.947982	11.157949	15.644588
SVM	−9.479318	−5.06804	−5.345433	−7.158341	−8.478965	−5.674243
XGBoost	32.236071	−1.550947	59.040843	−4.877587	17.273554	15.689431
Decision tree	30.145754	−4.849819	55.595841	12.10711	5.792261	15.404721
KNN	31.056368	16.90262	54.854087	2.649249	5.772098	14.736855

The results are sorted on the basis of the mean absolute error in ascending order. And it can be clearly seen from Table 7 that XGBoost has the lowest value which denotes that XGBoost is performing a more accurate classification for predicting house prices.

Table 6 Mean squared error for different cities using different algorithms

	Mean squared error					
Model	Chennai	Delhi	Hyderabad	Kolkata	Mumbai	Bangalore
Linear regression	7.96E+13	2.05E+15	4.04E+13	1.58E+14	3.29E+14	1.87E+14
Random forest	6.88E+13	2.02E+15	3.88E+13	2.34E+14	3.14E+14	1.84E+14
SVM	1.09E+14	2.68E+15	9.90E+13	1.81E+14	3.83E+14	2.30E+14
XGBoost	6.78E+13	2.60E+15	3.85E+13	1.77E+14	2.92E+14	1.84E+14
Decision tree	6.99E+13	2.68E+15	4.17E+13	1.48E+14	3.33E+14	1.84E+14
KNN	6.89E+13	2.12E+15	4.24E+13	1.64E+14	3.33E+14	1.86E+14

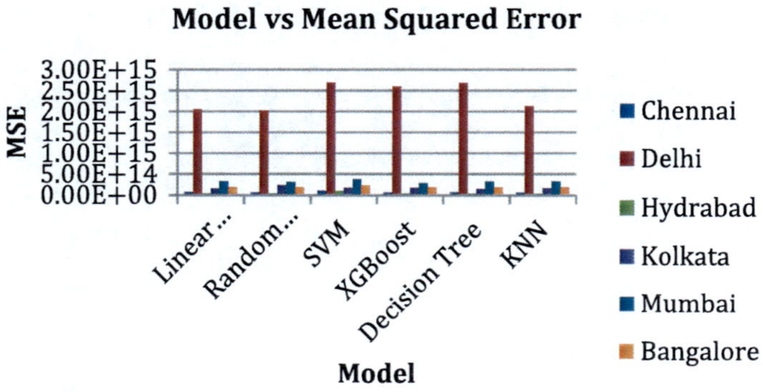

Fig. 3 Model versus mean squared error

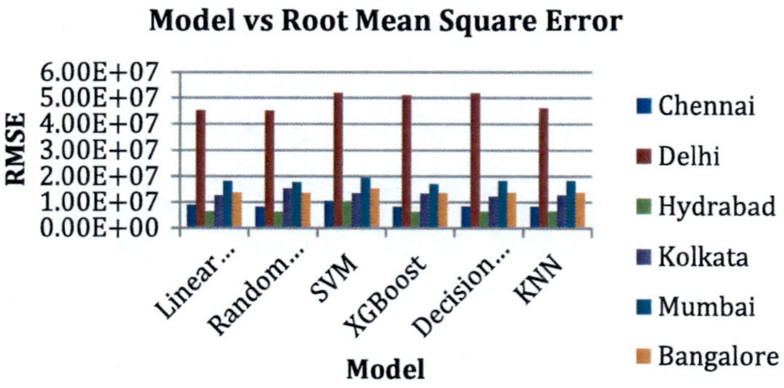

Fig. 4 Citywise model vs root mean square error

6 Conclusion

A model for predicting the house price can prove to be a very important tool for the buyers as well as sellers and help them in making well-informed decisions. The

Table 7 Root mean squared error for different cities using different algorithms

	RMSE					
Model	Chennai	Delhi	Hyderabad	Kolkata	Mumbai	Bangalore
Linear regression	8.92E+06	4.53E+07	6.36E+06	1.26E+07	1.81E+07	1.37E+07
Random forest	8.29E+06	4.50E+07	6.23E+06	1.53E+07	1.77E+07	1.36E+07
SVM	1.05E+07	5.18E+07	9.95E+06	1.34E+07	1.96E+07	1.52E+07
XGBoost	8.23E+06	5.09E+07	6.21E+06	1.33E+07	1.71E+07	1.36E+07
Decision tree	8.36E+06	5.18E+07	6.46E+06	1.22E+07	1.82E+07	1.36E+07
KNN	8.30E+06	4.61E+07	6.52E+06	1.28E+07	1.82E+07	1.36E+07

sellers can use this tool to determine the average price at which the house can be sold. On the other hand the buyers can use this tool to find out the fair price while purchasing the house. After observing the resultant metrics for various models, it can be concluded that XGBoost has the highest R^2 score. We can also check for outliers with the help of box plots and remove them if they exist and then analyze the improvement in the performance of the model. XGBoost has the lowest value, which denotes that XGBoost is performing a more accurate classification for predicting house prices.

7 Future Scope

Several other models can be constructed using advanced mechanisms such as particle swarm optimization or neural network, which can also improve the accuracy of predictions.

References

1. Heyman, A., & Sommervoll, D. (2019). House prices and relative location. Cities.
2. Hromada, E. (2015). *Mapping of real estate prices using data mining techniques.* Czech Technical University.
3. Limsonbunchai, V., Gan, C., & Lee, M. (2004). House price prediction: Hedonic price model vs. artificial neural network. *American Journal of Applied Sciences, 1*, 193–201.
4. Mukhlishin, M. F., Saputra, R., & Wibowo, A. (2017). Predicting house sale price using fuzzy logic, artificial neural network and k-nearest neighbour. In *1st International conference on informatics and computational sciences (ICICoS)* (vol. 1, pp. 171–176).
5. Aleksandrovich, P.V., Leopoldovich, K.I., & Viktorovich, P.A. (2018). Predicting sales prices of the houses using regression methods of machine learning. In *2018 3rd Russian-Pacific conference on computer technology and applications (RPC)* (pp. 1–5).
6. Vineeth, N., Ayyappa, M., & Bharathi, B. (2018). House price prediction using machine learning algorithms. In I. Zelinka, R. Senkerik, G. Panda, & P. S. LekshmiKanthan (Eds.), *Soft computing systems* (pp. 425–433). Singapore: Springer Singapore.

7. N. Shinde, and K. Gawande. Survey on predicting property price. In 2018 International Conference on Automation and Computational Engineering (ICACE) (pp. 1–7). IEEE. October 2018.
8. Nihar Bhagat, Ankit Mohorkar and Shreyas Mane (2016). House price forecasting using data mining. *International Journal of Computer Applications.*
9. Manasa, J., Gupta, R., & Narahari, N. S. (2020). Machine learning based predicting house prices using regression techniques. In *2020 2nd International conference on innovative mechanisms for industry applications (ICIMIA)* (pp. 624–630).
10. Bourassa, S.C., Cantoni, E., & Hoesli, M. (2007). Spatial dependence, housing submarkets, and house price prediction. *Journal of Real Estate Finance and Economics, 35*(2), 143–160.
11. Brint, A. (2009). Predicting a house's selling price through inflating its previous selling price. *Journal of Operational Research Society, 60*, 339–347.
12. Alfiyatin, A. N., & Febrita, R. E. (2017). Modeling house price prediction using regression analysis and particle swarm optimization. *International Journal of Advanced Computer Science and Applications.*

Malicious Website Detection Based on URL Classification: A Comparative Analysis

Swati Maurya and Anurag Jain

Abstract Phishing has been one of the most frequent cyber threats in the recent decade, prompting an increase in anti-phishing research and the development of numerous solutions for detecting and preventing phishing assaults. This paper identifies the system's vulnerabilities and adversaries' tactics to deceive Internet users into trusting the malicious email or website and providing sensitive information and credentials. For this study, the relevant URL features are retrieved from the collected dataset that includes phishing and legitimate URLs of websites. The correlation among different features is studied that can help users to identify fake web URLs by scanning phishing specific properties. This paper also analyzes the performance outcome of the machine learning, ensemble, and deep learning techniques on the collected dataset. Each model's performance is compared and measured, and random forest and gradient boosting with XGBoost are found to be the best optimal model for phishing binary classification problem in terms of accuracy (97.3%).

Keywords Anti-phishing · Phishing detection · Machine learning · Deep learning · URL-based classification

1 Introduction

Phishing is a technique adversaries used to gain personal and financial information such as login credentials and payment card details by impersonating the user or by tricking them into trusting fake websites or emails. It is a social engineering act in which an attacker uses a specially crafted message to random people in the hopes of obtaining sensitive information or utilizing the vulnerability of the user system for deploying and executing malicious software on the victim's infrastructure, such as

S. Maurya (✉) · A. Jain
Guru Gobind Singh Indraprastha University, New Delhi, New Delhi, India
e-mail: swatimaurya@hotmail.com

A. Jain
e-mail: anurag@ipu.ac.in

P. K. Singh et al. (eds.), *Proceedings of Third International Conference on Computing, Communications, and Cyber-Security*, Lecture Notes in Networks and Systems 421,
https://doi.org/10.1007/978-981-19-1142-2_19

viruses, Trojans, and ransomware [1]. Fake emails have increased where the scammer pretends to be a reliable and legitimate government or bank official, and facts are added to support their claim [2]. The scammer may notify the user that their account has been used illegally or questionably. The email contains details to convince the user that a significant purchase was made in another region and ask if the user has approved the payment. The fraudsters are prompt to confirm the credit card or bank account information so the 'bank' can examine the whole scenario. This way, they steal sensitive information from users.

Every day, cybercriminals conduct thousands of phishing campaigns like this, and most of them are successful. Attackers often modify their techniques, but specific characteristics might help the user identify a malicious email, text message, or website. User education to follow best practices while browsing the Internet and following safety guidelines can assist users to avoid getting trapped in attempts to steal sensitive information [3]. Advanced attack methods and assault tactics, which cannot be diagnosed by primary education and training, necessitate automated detection and prevention approaches. Software-based defense mechanisms help in detecting malicious emails, fake text messages, and phishing websites.

Phishing Tactics: A typical phishing assault might use various tactics, such as exploiting browser vulnerabilities or executing man-in-the-middle attacks. However, the most basic and often used technique is to create a webpage that looks identical to the one that the user is familiar with or craft emails or text messages in a way that appears to be genuine and helps them gain the trust of the user [4]. They hide URLs of fake external websites which replicate the appearance and user interface of the original website to trick the user into entering their details and credentials. The most popular phishing attacks are shown in Fig. 1.

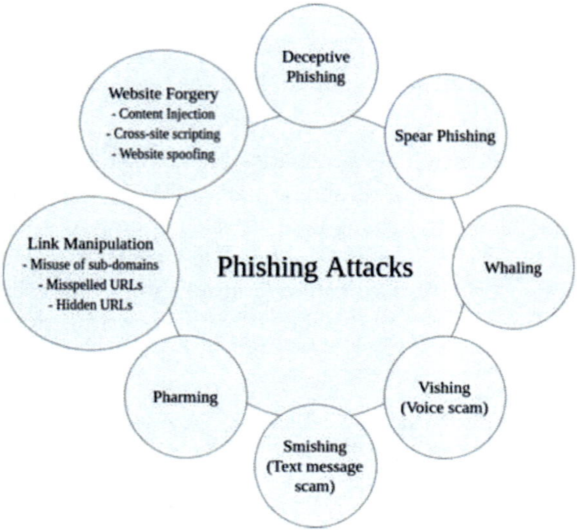

Fig. 1 Common phishing attacks used by adversaries for stealing sensitive information from the users

Phishing remains a severe security issue, and many Internet users are still victims of this deception. Furthermore, such attacks cause severe problems for Internet users and organizations that offer financial services over the Internet.

Filter-based phishing detection [5] techniques are incorporated these days into most email service providers to transfer suspicious emails to a 'Junk' or 'Spam' folder. When email filtering is enabled, incoming emails are scanned independently for features that indicate malicious content and transfer those emails to a different folder. Browser extension [3] and webpage content analysis [6]-based phishing detection and prevention solutions provide security from phishing websites if they match the suspicious criteria. Despite numerous anti-phishing solutions available these days for prevention from phishing attacks, the adversaries always stay a step ahead which makes all anti-phishing solutions incapable of preventing zero-day attacks. Hence, the need to deeply understand the correlation among different features present in a phishing webpage arises. The appropriate models should be selected while drafting a new adaptable anti-phishing solution making it more suitable for zero-day attack prevention.

This paper will analyze the effectiveness of phishing detection techniques based on URL classifications. With the advancements in machine learning algorithms and their accurate predictions in less time, the phishing domain research in the last decade has shifted to the machine learning, ensemble, and deep learning-based solutions. Section 2 presents the research methodology followed during this study and analysis. Section 3 discusses the details for feature analysis from URLs and the correlation among them. Section 4 covers popular classification models for phishing detection, and their performance is comparatively analyzed when used on the same phishing URL dataset. It analyzes the performance metrics and evaluates the results obtained for classification models on the URL dataset to find the best classifier. Section 5 summarizes the paper.

2 Research Methodology

Various research studies have been conducted by authors earlier to analyze the phishing detection techniques, and the approaches followed are as follows. As discussed in [7], the authors compared software-based phishing detection techniques like blacklists, heuristic detection techniques, visual similarity detection techniques, and data mining detection techniques available in the literature. A detailed survey of literature available for phishing detection approaches is also presented in [8]. Most of the surveys or analyses are focused on techniques mentioned in the literature, but there is no comparative analysis available that should guide toward selecting the best optimal model while designing an anti-phishing solution. This paper's study will aid academics and industry in determining the optimum algorithm that can be used for anti-phishing solutions based on requirements and resources.

This section briefs the steps followed in this study for URL feature analysis and performance comparison of latest machine learning, ensemble, and deep learning-

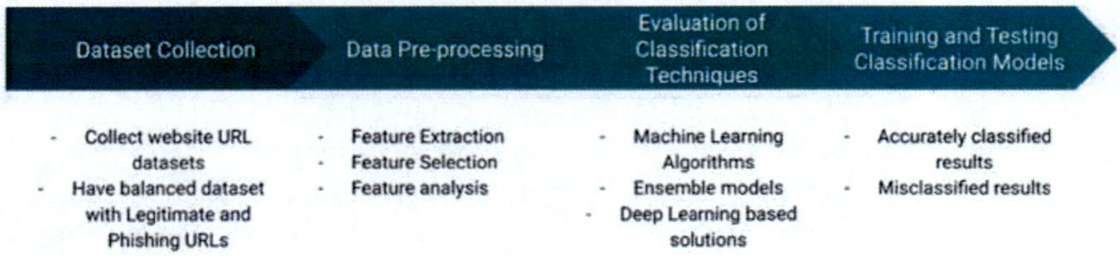

Fig. 2 Research methodology followed in the paper for URL feature analysis and performance comparison of latest classification models on phishing dataset

based classification models on phishing dataset provided by PhishTank [9], dmoz-tools.net [10]. Figure 2 presents the sequence of steps followed, and detailed explanation is given in following sections.

3 Analysis of Features Extracted from Web URLs

The URL and webpage content for fake websites are usually replicated to appear similar to the original website [11]. This research focuses on phishing website detection in real time by examining the features of the URL of the webpage. The malicious websites can be efficiently detected by thoroughly analyzing their URL. The attackers cannot utilize the exact URL of an original site, and they frequently misspell URL elements such as 'PrimaryDomain,' 'SubDomain,' and 'PathDomain' [12]. Identifying these phishing URL alteration tactics will undoubtedly assist in educating individuals and organizations about phishing attacks and ensuring prevention from them.

Features are extracted from the collected dataset of URLs and classified based on categories defined in [3] and are shown in Fig. 3. The analysis is done to understand the importance of each feature in classification for phishing or legitimate URL and the correlation among different features. Lexical characteristics of the URL [13] are analyzed in this study, and the efficacy for phishing prediction is studied. The observations from examining features of web URLs in the collected dataset are:

- Each data sample contains 30 features and a class label 'result' that indicates whether or not it is a phishing website (1 or −1).
- Size of URL: Long URLs are frequently associated with concealing the suspicious section of a fake website URL in the address bar to mislead the user.
- Number of dots: In comparison with legitimate websites, phishing pages frequently have more than 5–6 dots in their URLs.
- IP in the domain: Using an IP address instead of a domain name in a URL indicates an effort to steal personal information.

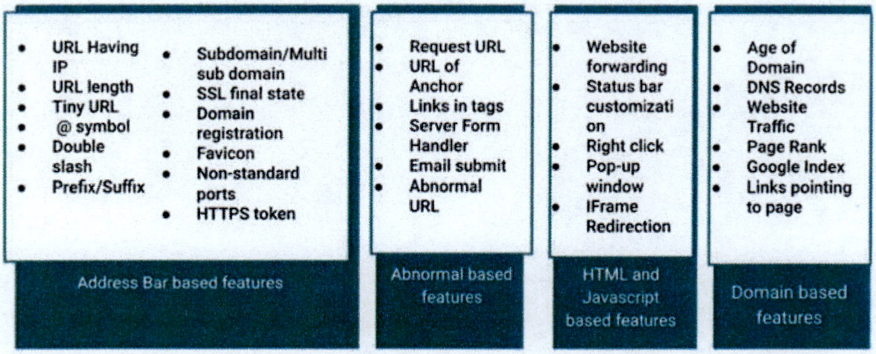

Fig. 3 Classification of website features based on address bar, abnormal, HTML and JavaScript and domain features

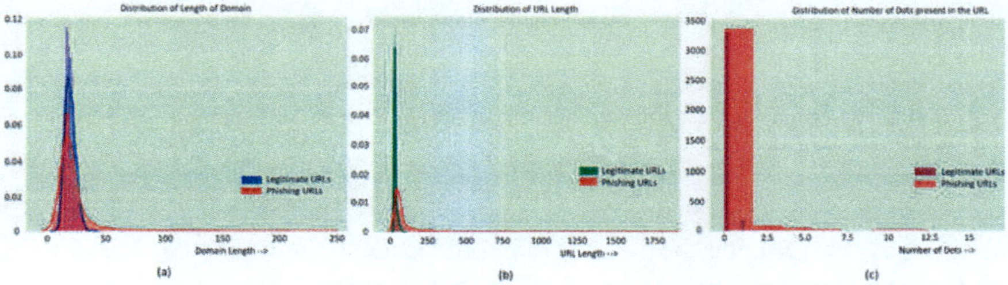

Fig. 4 Visual representation of distribution of **a** length of domain, URL length (**b**), and **c** number of dots present in legitimate and phishing URLs in the collected dataset

- Special characters in URL: The phishing link will perplex the user by adding a special character in the URL. "@" hides the phishing URL by commenting out the domain name that comes before it. The presence of the 'hyphen' and '@' symbol in the URL dominate in malicious URLs, whereas legitimate URLs avoid using them [12].
- Double slash ('//'): The presence of a double slash in a URL route indicates that the visitor will be redirected to a different website.
- Multiple sub-domains and a domain name mismatch: Phishers employ this type of technique to persuade victims that the message or email they received originated from a well-known organization. They use the genuine organization's domain name and append multiple sub-domains as a prefix to deceive users into thinking the crafted fake URL is genuine.
- Age of a URL: Phishing websites have been found to only exist for a short time, whereas trustworthy websites are registered and paid for several years in advance.

The distribution of domain length, URL length, and the number of dots present in legitimate and phishing web URLs is shown in Fig. 4a–c, respectively.

Correlation between URL features: The statistical measure of a linear relationship between two variables is known as correlation, and the visual representation is called

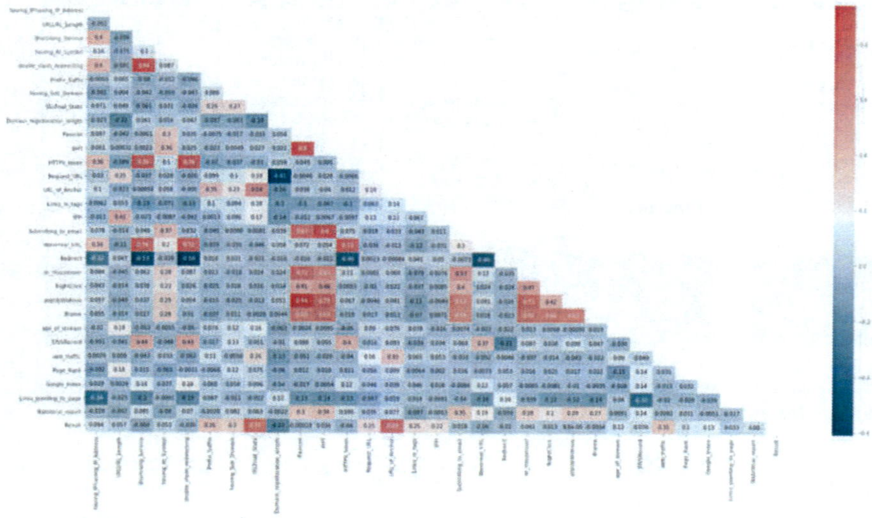

Fig. 5 Correlation heatmap representing the interdependence between URL features of the collected dataset

correlation heatmap [14]. It is the measure of interdependence between two variables. The matrix data format is utilized when there are numerous variables. Figure 5 shows a custom diverging colormap in the form of a matrix generated for the collected dataset and is drawn with the mask and correct aspect ratio. The correlation coefficient might have any value between −1 and 1 [15].

- Value = 1: The correlation between two variables is considered to be positive and indicates that while one variable rises, the other increases as well.
- Value = −1: Negative correlation between two variables is defined as a value of −1 and indicates that as one variable goes up, the other goes down.
- Value = 0: There is no connection between two variables if the value is 0 and indicates that the variables vary at random in relation to one another.

Identifying the characteristics of phishing URL alteration methods and their correlations can aid users in recognizing phishing attempts just by looking at them.

4 Classification Models for Phishing Detection Based on Web URLs

Phishing is a binary classification problem that classifies the given sample into two classes: legitimate or phishing. This research is focused on analyzing the performance of the latest classification algorithms provided by machine learning, ensemble techniques, and deep learning models that are best suited for binary classification.

Machine Learning (ML) Techniques: Data analysis is made easier and more efficient using machine learning. The capacity to construct adaptable models for specific tasks like phishing detection is a fundamental feature of machine learning. ML models might swiftly adapt to changes to identify patterns, which would aid in developing a learning-based identification system [3]. The following algorithms have been chosen because of their accurate prediction results for binary problems: decision tree, random forest, K-nearest neighbor, logistic regression, support vector machine.

Ensemble Classification Techniques: An ensemble is made up of several hypotheses that are created from training data using a primary learning method. Most ensemble methods generate homogeneous ensembles using a single base learning algorithm; however, other approaches employ several learning algorithms to produce heterogeneous ensembles. Several high-performance and advanced frameworks like AdaBoost, XGBoost, and a family of gradient boosting techniques [16] that focus on both speed and accuracy have recently been analyzed on the collected dataset for their performance.

Deep Learning (DL) Techniques: Artificial neural networks are used in deep learning models to conduct complex computations on large datasets [17]. It is a form of ML based on the human brain's structure and function. Algorithms extract features, organize objects, and find valuable data patterns during the training phase by using unknown elements in the input distribution. Machines are trained using examples. DL algorithms require high-end infrastructure to train in an acceptable amount of time. When there is a dearth of domain expertise for feature introspection, DL approaches shine since feature engineering is less of a concern [18].

4.1 Performance Comparison of Classification Models

The malicious and legitimate URLs are collected from PhishTank and dmoztools.net, and a dataset is formed. The collected dataset consists of a total of 65428 web URLs which have 29182 phishing URLs and 36246 legitimate URLs. The dataset is preprocessed, and features are extracted and chosen for analysis. The collected features from URLs are concatenated after random shuffling during the feature extraction step to prevent the overfitting [19] problem during model training. This also helps balance the distribution while splitting the data into training (75%) and testing (25%) sets. The implementation is done in Python with the help of machine learning libraries, and the pseudo-code is presented in Fig. 6.

Table 1 gives a comparison summary of performance metrics obtained as a result of applying classification algorithms for ML, ensemble, and DL techniques on the training set and validation set. Figure 7 shows visual representation of accuracies obtained from machine learning, ensemble, and deep learning techniques applied on collected dataset.

Pseudo-code for application of ML, Ensemble, and DL algorithms on the collected dataset for comparative analysis.	
Input:	Determine the training and validation datasets. - Import the collected dataset and pre-process to adjust missing values and balance.
Output:	Determine the training time (in sec), Validation accuracy, Recall, Precision, and F1 score.
Process:	• Scale the dataset. • Store/implement ML, Ensemble, and DL models. • Set scoring parameters to Accuray, Recall, Precision, and F1 score. • Set Name as name of the models. FOR Name, Model in models: • Store value of model selection using 10 splits in a variable. • Evaluate results using the cross-validation method for Training and Validation data. • Append results. RETURN scoring parameters.

Fig. 6 Pseudo-code used in experiments on the collected dataset for comparative analysis

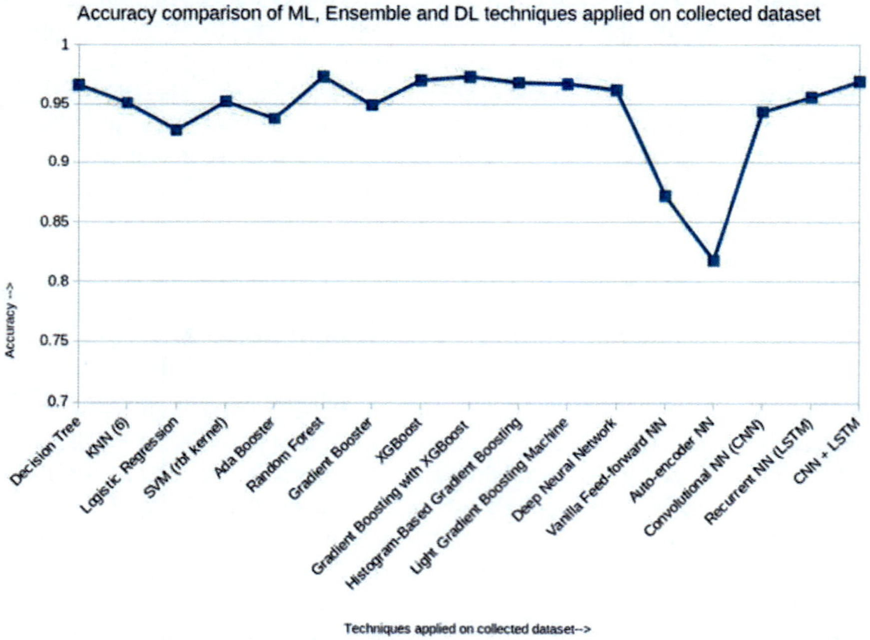

Fig. 7 Visual representation of accuracies obtained from machine learning, ensemble, and deep learning techniques applied on collected dataset

Table 1 Performance comparison of machine learning, ensemble, and deep learning techniques applied on collected dataset

Classifier	Training time (s)	Validation accuracy	Recall	Precision	$F1$ score
Machine learning classifiers					
Decision tree	0.017	0.966	0.971	0.968	0.969
KNN (5)	0.021	0.951	0.959	0.953	0.956
Logistic regression	0.095	0.927	0.944	0.926	0.935
SVM (rbf kernel)	1.509	0.952	0.969	0.947	0.957
Ensemble classifiers					
Ada booster	0.266	0.937	0.955	0.933	0.943
Random forest	0.399	0.973	0.981	0.967	0.974
Gradient booster	0.784	0.949	0.963	0.947	0.955
XGBoost	0.418	0.97	0.976	0.968	0.971
Gradient boosting with XGBoost	0.412	0.973	0.982	0.969	0.974
Histogram-based gradient boosting	0.396	0.968	0.975	0.967	0.971
Light gradient boosting machine	0.117	0.967	0.975	0.966	0.971
Deep learning classifiers					
Deep neural network	10.220	0.962	0.95	0.972	0.969
Vanilla feed-forward NN	12.54	0.872	0.868	0.869	0.867
Auto-encoder NN	8.657	0.818	0.825	0.802	0.821
Convolutional NN (CNN)	5.984	0.943	0.951	0.945	0.949
Recurrent NN (LSTM)	7.217	0.956	0.969	0.951	0.96
CNN + LSTM	5.844	0.969	0.979	0.965	0.972

4.2 Discussion and Results Analysis

This research evaluated the time taken for training different classification models (in seconds), validation accuracy obtained, recall, precision, and $F1$ score. Table 1 lists the results of the experiments mentioned above, and the following are the observations.

ML models get quickly trained, allowing them to make predictions and self-improve algorithms.

- Compared to other ML algorithms, the decision tree classifier (C4.5) predicts the phishing website accurately and fastest in training. It uses the 'Gini measure of impurity,' which helps the tree create 'pure nodes' with only one class label that does not need to be further divided, hence the fast execution [3].

- SVM has been tested with different kernels, but RBF kernel gave the best results compared to linear, poly, and sigmoid kernels.
- For testing KNN, there is no ideal number for setting k suitable for all types of datasets. Various experiments were conducted by altering the value of k to find the best-suited value for the collected dataset. A perfect balance needs to be found as noise has a greater influence on the outcome when the number of neighbors is small; moreover, a large number of neighbors makes obtaining the result computationally expensive [20].

Ensemble Models show better results as compared to standard ML models. Random forest (RF) and gradient boosting with XGBoost are the best in terms of accuracy (97.3%) and even take almost the same time for getting trained.

- RF adds randomness to the training and validation dataset and uses more trees, reducing variance, ultimately making the predictions fast and noise prune.
- The significant benefit of XGBoost over other algorithms is its rapid speed, as well as the 'regularization parameter,' which successfully lowers 'variance.' Usage of learning rate and subsamples from features like RF allows it to generalize even further. Hyperparameter tuning increases performance, and hybrid with gradient boosting algorithms reduces the training time and results in higher accurate predictions.

DL models take maximum time to get train due to the large number of hyperparameters. They incrementally learn high-level characteristics from data through the hidden layer architecture. The performance of DL models has been observed to increase with the amount of data [18].

- CNN combined with LSTM gave the best prediction results as compared to separately testing CNN or LSTM. The grid pattern analysis of CNN, when combined with feedback connections of LSTM, takes less time to train, and the performance increases.
- Although DL models take more time to train, once trained, their accuracy keeps on increasing with time as they can learn from past observations and utilize them for future predictions.
- DL models work best with a large amount of data, and the dataset used in this study was not that huge. With small datasets, the results are less accurate than ensemble models.

This study can be utilized before finalizing any model before drafting any anti-phishing solution based on URL characteristics. The comparative analysis highlights the best classifier in each section which can be chosen based on requirements and resources for the research. These results were retrieved using classifiers on a live dataset; thus, they are not theoretical, which enhances the study's reliability and dependability.

5 Conclusion

Recently with an increased emergence of phishing attacks through emails, fake websites, text messages, and phone calls, the need for awareness among Internet users has risen to identify a fake email or webpage. This research covered the basic phishing attack scenarios that deceive users into trusting scammers or adversaries. The URL feature analysis presents the correlation between web URL features which help any user scan the links provided in email before clicking them. A thorough understanding of these interdependencies among features prevents users from falling prey to fake websites that look similar to the original webpages but steal sensitive information. This study also analyzed the performance of the latest ML, ensemble, and DL algorithms. Their speed and accuracy are compared for the same dataset. Random forest and gradient boosting with XGBoost are found to be the best optimal solution for classifying URL-based phishing detection in terms of accuracy (97.3%). The limitation of this research is that comparative analysis was limited only to feature analysis of the web URLs. The classification techniques were tested on collected web URL datasets which can be extended to cover website content in the future.

References

1. Khan, F., Ncube, C., Ramasamy, L. K., Kadry, S., & Nam, Y. (2020). A digital DNA sequencing engine for ransomware detection using machine learning. *IEEE Access, 8,* 119710–119719.
2. Hoang, L. N., Faucon, L., Jungo, A., Volodin, S., Papuc, D., Liossatos, O., Crulis, B., Tighanimine, M., Constantin, I., Kucherenko, A., & Maurer, A. (2021). Tournesol: A quest for a large, secure and trustworthy database of reliable human judgments. arXiv preprint arXiv:2107.07334
3. Maurya, S., Singh, H., & Jain, A. (2019). Browser extension based hybrid anti-phishing framework using feature selection. *International Journal of Advanced Computer Science and Applications, 10*(11).
4. 12 Types of Phishing Attacks to watch out for. https://www.helixstorm.com/blog/x-types-of-phishing-attacks-to-watch-out-for/ Last Accessed August 9, 2021.
5. Surwade, A. U. (2020). Blocking Phishing e-mail by extracting header information of e-mails. In *2020 International Conference on Smart Innovations in Design, Environment, Management, Planning and Computing* (pp. 151–155).
6. Opara, C., Wei, B., & Chen, Y. (2020). HTMLPhish: Enabling phishing web page detection by applying deep learning techniques on HTML analysis. In *2020 International Joint Conference on Neural Networks* (pp. 1–8).
7. Khonji, M., Iraqi, Y., & Jones, A. (2013). Phishing detection: a literature survey. *IEEE Communications Surveys & Tutorials, 15*(4), 2091–2121.
8. Vijayalakshmi, M., Shalinie, S. M., & Yang, M. H. (2020). Web phishing detection techniques: A survey on the state-of-the-art, taxonomy and future directions. *IET Networks, 9*(5), 235–246.
9. PhishTank, https://phishtank.org/. Last accessed July 6, 2021.
10. DMOZ, https://dmoztools.net/. Last accessed July 20, 2021.
11. Tomaselli, J., Willoughby, A., Amezcua, J. V., Delehanty, E., Floyd, K., Wright, D., Lammers, M., & Vetter, R. (2021). Verifying phishmon: A framework for dynamic webpage classification. In *Proceedings of the 2021 ACM Southeast Conference* (pp. 185–189).
12. Mohammad, R. M., Thabtah, F., & McCluskey, L. (2015). *Phishing websites features.* School of Computing and Engineering, University of Huddersfield.

13. Khonji, M., Iraqi, Y., & Jones, A. (2011). Lexical URL analysis for discriminating phishing and legitimate websites. In *Proceedings of the 8th Annual Collaboration, Electronic messaging, Anti-Abuse and Spam Conference* (pp. 109–115).
14. Haarman, B. C. B., Riemersma-Van der Lek, R. F., Nolen, W. A., Mendes, R., Drexhage, H. A., & Burger, H. (2015). Feature-expression heat maps—A new visual method to explore complex associations between two variable sets. *Journal of Biomedical Informatics, 53*, 156–161.
15. Kumar, A. (2021). *Correlation concepts, matrix and heatmap using seaborn*. https://vitalflux.com/correlation-heatmap-with-seaborn-pandas/. Last accessed August 1, 2021.
16. Bentéjac, C., Csörgő, A., & Martínez-Muñoz, G. (2021). A comparative analysis of gradient boosting algorithms. *Artificial Intelligence Review, 54*, 1937–1967.
17. Maurya, S., & Jain, A. (2020). Deep learning to combat phishing. *Journal of Statistics and Management Systems, 23*(6), 945–957.
18. Mahapatra, S. (2021). *Why deep learning over traditional machine learning?*https://towardsdatascience.com/why-deep-learning-is-needed-over-traditional-machine-learning-1b6a99177063. Last accessed June 3, 2021.
19. Yeom, S., Giacomelli, I., Menaged, A., Fredrikson, M., & Jha, S. (2020). Overfitting, robustness, and malicious algorithms: A study of potential causes of privacy risk in machine learning. *Journal of Computer Security, 28*(1), 35–70.
20. Shahrivari, V., Darabi, M. M., & Izadi, M. (2020). *Phishing detection using machine learning techniques*. arXiv preprint arXiv:2009.11116

Attribute Selection, Sampling, and Classifier Methods to Address Class Imbalance Issues on Data Set Having Ratio Less Than Five

Aarchit Joshi, Kushal Kanwar, and Pankaj Vaidya

Abstract Many modern approaches to classification presume that the underlying training set is uniformly distributed. In a class unbalanced grouping, where the minority class is generally the more fascinating class, the majority class's training set greatly outweighed the minority class's training set. The authors of the current study discuss the difficulties that occur when training the machine as a result of the class imbalance. Research on current techniques checks the performance of the techniques on various parameters. Doing this will help us in getting idea how the standard techniques perform while minority class is suffering from one or more kind of issues. The purpose of this article is to present a comparative analysis of techniques for contemporary imbalance data analysis techniques, with a focus on data pre-processing, attribute selection, and algorithmic analysis, as well as a comparison of these techniques in the context of different data distributions.

Keywords Attribute selection · Sampling · Classifier · Class imbalance

1 Introduction

In many domain applications, learning with class imbalance distribution happens often. Imbalanced class distribution in datasets arises when one class, usually the one that is of more interest, i.e., the positive or minority class, is underrepresented. Simply said, the number of positive class (minority) instances is far lower than the number of negative class (majority). When unusual examples are seldom present, they are most considered to be rare events; unknown or ignored, or assumed to be noise or outliers, resulting in more positive class misclassifications (minority) relative to the majority class. Any diagnostic mistakes will stress patients and lead to further problems. The physicians couldn't afford to make a mistake since it may have a negative impact on the patients' health and possibly modify the therapy and pharmaceutical options available. A classification model must thus be capable of

A. Joshi · K. Kanwar · P. Vaidya (✉)
Shoolini University, Bajhol, HP, India
e-mail: pankaj.vaidya@shooliniuniversity.com

P. K. Singh et al. (eds.), *Proceedings of Third International Conference on Computing, Communications, and Cyber-Security*, Lecture Notes in Networks and Systems 421,
https://doi.org/10.1007/978-981-19-1142-2_20

achieving a better identification rate for uncommon events in datasets (minority class).

In recent years, research on class imbalance categorization has gotten a lot more attention [1]. Several domain applications, such as finding fault within [2, 3], exceptions findings(outliers) [4], medical examination [5, 6], detection of oil spills via satellite photos [7], recognition of people's faces [8], categorization of text [9], identification of protein sequences [10], as well as several others have been documented in classification works for class imbalance distribution. Many academics have been drawn to the topic of class imbalance because of its considerable problems and frequent occurrence pattern recognition, and data mining is used in real-world applications.

The intent behind this study is to know:

- The problems with unbalanced data classification in machine learning.
- When the imbalance ratio between the majority and minority classes is less than five, a comparative study of few approaches based on different assessment metrics has been undertaken to manage the issue of imbalance data.

This study has added to the literature about how different machine learning techniques work on different kind of class imbalance issues.

The following is the format of this review: Sect. 2 discusses the key difficulties and disadvantages of class imbalance categorization. Section 3 discusses sampling and attribute selection. The parameters on which the methods will be assessed are described in Sect. 4. A comparison chart of different approaches and strategies is provided for a better understanding. Section 5 defines the output metrics that are typically used to assess a classification algorithm's effectiveness in the categorization of datasets with imbalance characteristics. Finally, Sect. 6 concludes with a discussion of the frequent difficulties that arise when minorities are misclassified.

2 Class Issues

Class inequality exists because there are considerably fewer instances of preparation in one class than in another class. In two situations, the essence of imbalanced class distribution could occur; (1) when a class imbalance is the fundamental problem or when it occurs spontaneously. An intrinsically unbalanced class distribution arises in the event of credit card fraud or uncommon disease diagnosis. Another example is (2) when the information is not intrinsically unbalanced, but it is too difficult to gather such data for minority class learning owing to expense, secrecy, and substantial effort to identify a well-represented data set, such as a particularly uncommon occurrence of a space shuttle disaster. There are numerous learning difficulties associated with class imbalance, including unbalanced class distribution, training sample size, overlapping classes, and tiny disjuncts. The following sections explain all these variables in detail.

2.1 *Imbalanced Composition of the Class*

The distribution of the imbalanced class can be identified by the ratio of the number of minority class cases to that of the majority class [11–14]. The imbalance ratio may be as high as 1:10,000 [15] in some domain problems [14, 16–18].

2.2 *Shortage of Evidence Arising From a Limited Sample Size*

Recorded work has shown that the error rate of the imbalanced classification decreases as the training sample size increases [12]. This is also verified by [17], which used the fuzzy classifier to record-related findings. As a classifier creates improved coverage for classes with the more accessible testing sample, this is understandable since a larger variety of training sizes allows for more knowledge to be acquired from combinations of examples.

2.3 *Overlapping Groups or the Complexity of Classes*

Work in [13, 18] showed that the class imbalance aspect starts to impact the generalization power of a classifier as the degree of data complexity rises. The job from the job [19] proposed that in class imbalance classification, there is a correlation between overlap and imbalance, but the degree is not well established. Numerous research on class separability [20–27] offer proof that the issue of class overlap poses a significant impediment to the efficiency of a classifier relative to imbalanced class distribution. Standard classifiers that seek to optimize classification accuracy frequently fell into the pit of the overlapping issue as such classifiers typically categorized the overlapping area as belonging to while considering noise as the minority class, the dominant class [28].

2.4 *Minor Disjuncts Within the Imbalance of Class*

When a class consists of many subclusters containing numerous examples, the disparity within the class, also known as a minor disjunct, occurs [29–31]. [16, 32] examined the imbalance of the minority class within the class and claimed that by adding directed up sampling to the minority class, the underrepresented minority class generated by a little disjunction may be reinforced. Said that tiny disjuncts in class imbalance impact classification performance because (1) it strains a classifier during the minority class definition learning phase and (2) the instances of the query inside the class are often tacit.

3 Identification of Class Disparity Approaches

In general, there are two ways for addressing the issue of class imbalance [9, 33–38]. There are two approaches: (1) data-level approach and (2) algorithm-level approach. Traditional classification algorithms are fine-tuned at the algorithm-level approach to improve the learning task, particularly relative to the smaller class, while data-level approach approaches modify the class imbalance ratio to create a balanced distribution across classes.

3.1 Solution to the Data Stage for Managing the Class Imbalance Problem

A pre-processing stage is used in a data-level approach, also known as external techniques, to rebalance the class distribution. Undersampling removes fewer examples from the majority class, while oversampling duplicates examples from the minority class, narrowing the difference between the two races [39].

3.1.1 Sampling

In 2002, a paper [39] proposed an adaptive oversampling approach known as Synthetic Minority Over-sampling Technique (SMOTE), which has subsequently gained popularity in the classification of class imbalance. To model the smaller class, it uses a probability distribution, SMOTE adds additional examples to the minority class, thus widening the decision boundary to catch neighboring examples of the minority class. The study in [32, 40–42] that uses clustering to pick representative training examples to achieve improved accuracy estimation for minority groups proposes a cluster-based undersampling.

However, omitting examples from a class (down-sampling) may result in the loss of potentially useful class information, whereas repetition simply increases the number of examples in examples (oversampling) but does not include additional class information, so the problem is not solved by a lack of evidence [43–48].

3.1.2 Feature Selection

Feature selection, in addition to sample approaches, is another pre-processing procedure that is gaining traction in class imbalance classification. [49] presented a novel class decomposition-based feature selection method for correcting high-dimension class imbalance datasets, as well as a new Hellinger distance-based feature selection method for correcting smaller pseudo-subclasses produced by majority class partitioning [49–53]. The study [54] found that insignificant features do not dramatically boost the efficiency of the classification and indicated that more features slow down the process of induction. The collection of features excludes obsolete, repetitive, or noisy information [55] that represents the issue of class complexity or overlap in class imbalance. The wrapper approach wraps the mechanism of feature selection around thė induction algorithm. Compared to the former, while they are computationally costly, they are usually better at forecasting accuracy than filter methods [55–58].

4 Indicators of Success

Since the standard overall accuracy metric is no longer sufficient to define the output of a classifier [18, 44, 59], the confusion matrix and its derivatives will be utilized to explain the performance data. The confusion matrix is made up of four results from classification outputs: the number of true positives (TP), true negatives (TN), false positives (FP), and false negatives (FN) for a binary-class problem. The minority class is referred to as "positive," whereas the dominant class is referred to as "negative."

The confusion matrix's entries are labeled as,

- The number of positive cases that a classifier correctly predicts as positive is referred to as true positive (TP).
- The number of negative occurrences that a classifier properly identifies as negative is referred to as true negative (TN).
- The amount of negative samples that a classifier incorrectly identifies as positive is known as false positive (FP), also known as a false alarm.
- A classifier calculates false negative (FN), also known as miss, as the number of positive instances incorrectly classified as negative.

Examining the four elements in the uncertainty matrix is insufficient for evaluating a classifier's output. As a result, several derivatives based on the previously described uncertainty matrix are used in this study to evaluate a classifier. The matrix of uncertainty's output metrics is as follows:

Sensitivity (Sen) is the ability of a classifier to correctly categories a positive class as such. It ranges from 0 to 1, with 1 being the highest possible score. Sensitivity, also known as true positive rate or recall, is defined as.

$$\text{Sen} = \frac{\text{TP}}{\text{TP} + \text{FN}} \tag{1}$$

Specificity (Spy) signifies a classifier's ability to accurately classify negative groups as such. 1 is the optimal score, and 0 is the worst measure. Specificity or real adverse rate shall be calculated as

$$\text{Spy} = \frac{\text{TN}}{\text{TN} + \text{FP}} \tag{2}$$

Accuracy (Acy) is a percentage of the population's real outcomes (both true positives and true negatives). Accuracy shall be denoted as

$$\text{Acy} = \frac{\text{TN} + \text{TP}}{\text{FP} + \text{FN} + \text{TP} + \text{TN}} \tag{3}$$

Precision (Pre) is a precision metric that measures the proportion of positive class observations that are properly classified as such. For the positive class, that is the correct categorized number. This demonstrates how well a classifier prevents a negative class from being mistaken for a positive class.

$$\text{Pre} = \frac{\text{TP}}{\text{TP} + \text{FP}} \tag{4}$$

F-measure (F-msr).

$$\text{F}_{\beta}\text{measure} = \frac{\left(1 + \beta^2\right) * \text{Recall} * \text{Precision}}{\beta^2 * \text{Recall} + \text{Precision}} \tag{5}$$

A recall is a completeness test. It describes the percentage of positive class observations that should be returned, or how much the positive class learns from a classifier.

Recall (Rcl).

$$\text{Rcl} = \frac{\text{TP}}{\text{TP} + \text{FN}} \tag{6}$$

β is a coefficient that weighs the significance of recall and accuracy in relation to one another. F1 is often used for classification and is set to 1 according to standard practice [60]. When the F-measure is equal to 1, it indicates that recall and accuracy are equally weighted.

5 Research Work

The dataset was collected from the site UCI machine learning repository [61] and Kaggle [62]. **Data Set** is the name of datasets, **Total examples** number of examples, **Attributes** are characteristics of dataset, **No. of Classes** number of classes present in dataset, **Majority Examples** total examples present in majority class, **Minority examples** total examples present in minority class, * **symbolize** as the datasets are multiclass depending on selection of class no of example in majority class and minority class will vary.

The set of data (Glass) had six classes, “build wind float”, “vehic wind float”, tableware, “build wind non-float”, headlamps, containers. We tabulated classes: “build wind float” as 1, “vehic wind float” as 2, tableware as 3, “build wind non-float” as 4, headlamps as 5, and containers as 6 (Table 1).

There were two classifications in the KC3 dataset: N and Y. We tabulated classes: Y as 1, N as 0.

N and Y were the two classes in the MW1 dataset. We tabulated classes: Y as 1, N as 0.

The diabetes dataset had two classes, 1 and 0. No Need of any tabulation, 1 means person is diabetic else not.

The ionosphere dataset had two classes, g and b. We tabulated classes: g as 1, b as 0, g stands for good and b stands for bad.

The abalone dataset had twenty-nine classes, 1 to 29. No need of any tabulation.

Glass and Abalone datasets are multiclass datasets; we split them into two groups: a majority and a minority.

We have applied principal component analysis (PCA) to narrow down attributes to two components. As different dataset had different attributes, we normalized the dataset range between 0 and 1. Then we have used sampling techniques (1) SMOTE, (2) undersampling, and (3) oversampling followed by classification, i.e.,

Table 1 Dataset details

Dataset	Total examples	Attributes	No. of classes	Majority examples	Minority examples
Glass	214	9	6	*	*
KC3	200	21	2	164	36
MC2	127	39	2	83	44
Diabetes	768	8	2	500	268
Abalone	4177	8	29	*	*
Ionosphere	351	34	2	225	126

Table 2 Majority and minority classes

Dataset	Majority class	Minority class	Ratio minority: majority	Dataset	Majority class	Minority class	Ratio minority: majority
Glass (A)	4	5	1: 2.55	Abalone (A)	9	7	1: 1.76
Glass (B)	4	2	1: 4.35	Abalone (B)	9	11	1: 1.41
Glass (E)	1	5	1: 2.4	Abalone (C)	10	7	1: 1.62
Glass (F)	1	2	1: 4.11	Abalone (D)	10	11	1: 1.30
KC3	0	1	1: 4.55	Diabetes	0	1	1: 1.86
MC2	0	1	1: 1.88	Ionosphere	1	0	1: 1.7

(1) ADA BOOST (2) GRADIENT BOOST (3) KNN 4) RANDOMFOREST with each sampling technique (Table 2).

In Table 3, we can say all classifier applied with overs sampling method did well as compared with other two.

In Table 4, it can be evidenced SMOTE sampling when applied with different classifiers performed better than rest of the sampling methods. Undersampling when applied with KNN was unable to classify any attribute as true negative hence we are getting 0 value for specificity and rest follows. The sample size of the data was small.

Table 5 shows the performance of the classifiers with different sampling techniques. SMOTE and oversampling did equally good.

Table 3 Glass A

Dataset	Sampling	Classifier	Acy	Sen	Spy	Pre	Rcl	F-msr
Glass(A4/5)	SMOTE	ADA BOOST	0.97	1.00	0.93	1.00	0.93	0.97
		GRADIENT BOOST	0.97	1.00	0.93	1.00	0.93	0.97
		KNN	0.94	1.00	0.87	1.00	0.87	0.93
		RANDOMFOREST	1.00	1.00	1.00	1.00	1.00	1.00
	UNDER	ADA BOOST	0.92	1.00	0.86	1.00	0.86	0.92
		GRADIENT BOOST	0.92	1.00	0.86	1.00	0.86	0.92
		KNN	0.92	1.00	0.86	1.00	0.86	0.92
		RANDOMFOREST	0.92	1.00	0.86	1.00	0.86	0.92
	OVER	ADA BOOST	1.00	1.00	1.00	1.00	1.00	1.00
		GRADIENT BOOST	1.00	1.00	1.00	1.00	1.00	1.00
		KNN	0.94	1.00	0.87	1.00	0.87	0.93
		RANDOMFOREST	1.00	1.00	1.00	1.00	1.00	1.00

Table 4 Glass B

Dataset	Sampling	Classifier	Acy	Sen	Spy	Pre	Rcl	F-msr
Glass(B4/2)	SMOTE	ADA BOOST	0.97	1.00	0.94	1.00	0.94	0.97
		GRADIENT BOOST	0.94	1.00	0.88	1.00	0.88	0.93
		KNN	0.90	1.00	0.81	1.00	0.81	0.90
		RANDOMFOREST	0.94	1.00	0.88	1.00	0.88	0.93
	UNDER	ADA BOOST	0.86	0.67	1.00	0.80	1.00	0.89
		GRADIENT BOOST	0.57	0.67	0.50	0.67	0.50	0.57
		KNN	0.43	1.00	0.00	nan	0.00	#VALUE!
		RANDOMFOREST	0.71	0.67	0.75	0.75	0.75	0.75
	OVER	ADA BOOST	0.97	1.00	0.94	1.00	0.94	0.97
		GRADIENT BOOST	0.97	1.00	0.94	1.00	0.94	0.97
		KNN	0.84	0.87	0.81	0.87	0.81	0.84
		RANDOMFOREST	0.97	1.00	0.94	1.00	0.94	0.97

Table 5 Glass E

Dataset	Sampling	Classifier	Acy	Sen	Spy	Pre	Rcl	F-msr
Glass(E1/5)	SMOTE	ADA BOOST	0.96	0.93	1.00	0.93	1.00	0.96
		GRADIENT BOOST	0.96	0.93	1.00	0.93	1.00	0.96
		KNN	0.93	1.00	0.85	1.00	0.85	0.92
		RANDOMFOREST	0.96	0.93	1.00	0.93	1.00	0.96
	UNDER	ADA BOOST	0.92	1.00	0.86	1.00	0.86	0.92
		GRADIENT BOOST	0.92	1.00	0.86	1.00	0.86	0.92
		KNN	0.92	1.00	0.86	1.00	0.86	0.92
		RANDOMFOREST	0.92	1.00	0.86	1.00	0.86	0.92
	OVER	ADA BOOST	0.96	0.93	1.00	0.93	1.00	0.96
		GRADIENT BOOST	0.96	0.93	1.00	0.93	1.00	0.96
		KNN	0.93	1.00	0.85	1.00	0.85	0.92
		RANDOMFOREST	0.96	0.93	1.00	0.93	1.00	0.96

In dataset Glass F, applying oversampling helped to get better results, and we can see in Table 6 that oversampling and SMOTE sampling show almost same performances.

In KC3, we have overlapping issues in the dataset oversampling results are way better than under and SMOTE sampling as shown in Table 7.

Table 8 although overall oversampling performed well, but best combination was SMOTE and ADA BOOST.

Oversampling shows better results Table 9. Undersampling with RANDOMFOREST was consistent and best.

Table 6 Glass F

Dataset	Sampling	Classifier	Acy	Sen	Spy	Pre	Rcl	F-msr
Glass(F1/2)	SMOTE	ADA BOOST	0.93	0.87	1.00	0.87	1.00	0.93
		GRADIENT BOOST	0.96	1.00	0.92	1.00	0.92	0.96
		KNN	0.89	0.80	1.00	0.81	1.00	0.90
		RANDOMFOREST	0.96	0.93	1.00	0.93	1.00	0.96
	UNDER	ADA BOOST	0.86	0.67	1.00	0.80	1.00	0.89
		GRADIENT BOOST	0.86	0.67	1.00	0.80	1.00	0.89
		KNN	0.86	0.67	1.00	0.80	1.00	0.89
		RANDOMFOREST	0.86	0.67	1.00	0.80	1.00	0.89
	OVER	ADA BOOST	0.93	0.87	1.00	0.87	1.00	0.93
		GRADIENT BOOST	0.96	0.93	1.00	0.93	1.00	0.96
		KNN	0.93	0.87	1.00	0.87	1.00	0.93
		RANDOMFOREST	0.96	0.93	1.00	0.93	1.00	0.96

Table 7 KC3

Dataset	Sampling	Classifier	Acy	Sen	Spy	Pre	Rcl	F-msr
KC3(0/1)	SMOTE	ADA BOOST	0.77	0.81	0.73	0.76	0.73	0.74
		GRADIENT BOOST	0.73	0.69	0.77	0.68	0.77	0.72
		KNN	0.74	0.58	0.93	0.65	0.93	0.77
		RANDOMFOREST	0.85	0.81	0.90	0.79	0.90	0.84
	UNDER	ADA BOOST	0.67	0.86	0.50	0.80	0.50	0.62
		GRADIENT BOOST	0.73	0.71	0.75	0.75	0.75	0.75
		KNN	0.87	0.71	1.00	1.00	1.00	1.00
		RANDOMFOREST	0.73	0.57	0.88	0.70	0.88	0.78
	OVER	ADA BOOST	0.90	0.83	1.00	0.83	1.00	0.91
		GRADIENT BOOST	0.85	0.72	1.00	0.75	1.00	0.86
		KNN	0.85	0.75	0.97	0.76	0.97	0.85
		RANDOMFOREST	0.91	0.83	1.00	0.83	1.00	0.91

SMOTE sampling shows better results Table 10. SMOTE with RANDOMFOREST performed best.

Oversampling shows better results in Table 11.

Undersampling shows better results in Table 12.

Oversampling shows better results in Table 13.

Oversampling shows better results Table 14.

Table 8 MC2

Dataset	Sampling	Classifier	Acy	Sen	Spy	Pre	Rcl	F-msr
MC2(0/1)	SMOTE	ADA BOOST	0.79	0.81	0.77	0.71	0.77	0.74
		GRADIENT BOOST	0.71	0.71	0.69	0.60	0.60	0.60
		KNN	0.56	0.52	0.62	0.44	0.62	0.52
		RANDOMFOREST	0.74	0.67	0.85	0.61	0.85	0.71
	UNDER	ADA BOOST	0.61	0.67	0.50	0.43	0.50	0.46
		GRADIENT BOOST	0.50	0.75	0.50	0.50	0.50	0.50
		KNN	0.67	0.83	0.33	0.50	0.50	0.50
		RANDOMFOREST	0.67	0.75	0.50	0.50	0.50	0.50
	OVER	ADA BOOST	0.76	0.81	0.69	0.69	0.69	0.69
		GRADIENT BOOST	0.76	0.76	0.77	0.67	0.77	0.71
		KNN	0.71	0.76	0.62	0.62	0.62	0.62
		RANDOMFOREST	0.79	0.71	0.92	0.67	0.67	0.67

Table 9 Abalone A

Dataset	Sampling	Classifier	Acy	Sen	Spy	Pre	Rcl	F-msr
Abalone(A9/7)	SMOTE	ADA BOOST	0.86	0.85	0.86	0.84	0.86	0.85
		GRADIENT BOOST	0.99	0.99	0.98	0.98	0.98	0.98
		KNN	0.94	0.94	0.94	0.93	0.94	0.94
		RANDOMFOREST	0.99	1.00	0.98	1.00	0.98	0.99
	UNDER	ADA BOOST	0.86	0.82	0.90	0.84	0.90	0.87
		GRADIENT BOOST	0.93	0.94	0.93	0.94	0.93	0.93
		KNN	0.97	0.97	0.99	0.95	0.99	0.97
		RANDOMFOREST	0.99	0.99	0.99	0.99	0.99	0.99
	OVER	ADA BOOST	0.87	0.88	0.86	0.87	0.86	0.86
		GRADIENT BOOST	0.99	1.00	0.98	1.00	0.98	0.99
		KNN	0.95	0.94	0.95	0.94	0.95	0.95
		RANDOMFOREST	0.99	1.00	0.98	1.00	0.98	0.99

6 Conclusion

The article gives description of what are the issues present in the minority class. Problem is mainly classified in the category: Uneven distribution of class, size of sample is very small, overlapping of classes, minority disjuncts within the imbalance of class, and rare case and outliers. Tabulated data is also presented showing how different techniques works. This data might help researcher/user to select which

Table 10 Abalone B

Dataset	Sampling	Classifier	Acy	Sen	Spy	Pre	Rcl	F-msr
Abalone(B9/11)	SMOTE	ADA BOOST	0.79	0.77	0.80	0.79	0.80	0.80
		GRADIENT BOOST	0.99	0.98	1.00	0.99	1.00	0.99
		KNN	0.96	0.98	0.95	0.98	0.95	0.96
		RANDOMFOREST	1.00	0.99	1.00	0.99	1.00	1.00
	UNDER	ADA BOOST	0.80	0.81	0.79	0.81	0.79	0.80
		GRADIENT BOOST	0.97	0.96	0.99	0.96	0.99	0.97
		KNN	0.96	0.96	0.97	0.96	0.97	0.96
		RANDOMFOREST	0.97	0.97	0.99	0.96	0.96	0.96
	OVER	ADA BOOST	0.79	0.80	0.77	0.81	0.77	0.79
		GRADIENT BOOST	0.99	0.98	1.00	0.98	1.00	0.99
		KNN	0.93	0.97	0.90	0.97	0.90	0.93
		RANDOMFOREST	0.99	0.99	1.00	0.99	1.00	0.99

Table 11 Abalone C

Dataset	Sampling	Classifier	Acy	Sen	Spy	Pre	Rcl	F-msr
Abalone(C10/7)	SMOTE	ADA BOOST	0.89	0.90	0.88	0.90	0.88	0.89
		GRADIENT BOOST	0.94	0.91	0.98	0.92	0.98	0.95
		KNN	0.93	0.96	0.91	0.96	0.91	0.93
		RANDOMFOREST	0.98	0.98	0.99	0.98	0.99	0.98
	UNDER	ADA BOOST	0.80	0.81	0.79	0.81	0.79	0.80
		GRADIENT BOOST	0.97	0.96	0.99	0.96	0.99	0.97
		KNN	0.96	0.96	0.97	0.96	0.97	0.96
		RANDOMFOREST	0.97	0.96	0.99	0.96	0.99	0.97
	OVER	ADA BOOST	0.90	0.92	0.88	0.92	0.88	0.90
		GRADIENT BOOST	0.97	0.95	0.98	0.95	0.98	0.97
		KNN	0.96	0.97	0.95	0.97	0.95	0.96
		RANDOMFOREST	0.99	0.99	0.99	0.99	0.99	0.99

technique will suit the problem on which one person is working. As the ratio is near to 1, i.e., 1:1 oversampling performed good as the ratio increases toward 5, i.e., 1:4, SMOTE performed better.

Table 12 Abalone D

Dataset	Sampling	Classifier	Acy	Sen	Spy	Pre	Rcl	F-msr
Abalone(D10/11)	SMOTE	ADA BOOST	1.00	1.00	1.00	1.00	1.00	1.00
		GRADIENT BOOST	1.00	1.00	1.00	1.00	1.00	1.00
		KNN	0.94	0.92	0.97	0.92	0.97	0.94
		RANDOMFOREST	1.00	1.00	1.00	1.00	1.00	1.00
	UNDER	ADA BOOST	1.00	1.00	1.00	1.00	1.00	1.00
		GRADIENT BOOST	1.00	1.00	1.00	1.00	1.00	1.00
		KNN	0.96	0.95	0.98	0.95	0.98	0.96
		RANDOMFOREST	1.00	1.00	1.00	1.00	1.00	1.00
	OVER	ADA BOOST	1.00	1.00	1.00	1.00	1.00	1.00
		GRADIENT BOOST	0.96	1.00	1.00	1.00	1.00	1.00
		KNN	0.95	0.95	0.96	0.95	0.96	0.96
		RANDOMFOREST	1.00	0.99	1.00	0.99	1.00	1.00

Table 13 Diabetes

Dataset	Sampling	Classifier	Acy	Sen	Spy	Pre	Rcl	F-msr
Diabetes (0/1)	SMOTE	ADA BOOST	0.80	0.79	0.80	0.82	0.80	0.81
		GRADIENT BOOST	0.81	0.79	0.82	0.82	0.82	0.82
		KNN	0.81	0.79	0.83	0.83	0.83	0.83
		RANDOMFOREST	0.83	0.78	0.86	0.82	0.86	0.84
	UNDER	ADA BOOST	0.81	0.88	0.78	0.91	0.78	0.84
		GRADIENT BOOST	0.77	0.85	0.72	0.89	0.72	0.79
		KNN	0.84	0.90	0.81	0.93	0.81	0.86
		RANDOMFOREST	0.79	0.93	0.70	0.94	0.70	0.80
	OVER	ADA BOOST	0.82	0.80	0.83	0.83	0.83	0.83
		GRADIENT BOOST	0.86	0.81	0.90	0.85	0.90	0.87
		KNN	0.85	0.79	0.89	0.84	0.89	0.86
		RANDOMFOREST	0.80	0.76	0.93	0.82	0.93	0.87

Table 14 Ionosphere

Dataset	Sampling	Classifier	Acy	Sen	Spy	Pre	Rcl	F-msr
Ionosphere (1/0)	SMOTE	ADA BOOST	0.62	0.62	0.63	0.65	0.63	0.64
		GRADIENT BOOST	0.77	0.74	0.79	0.78	0.79	0.78
		KNN	0.82	0.69	0.94	0.78	0.94	0.85
		RANDOMFOREST	0.87	0.86	0.88	0.88	0.88	0.88
	UNDER	ADA BOOST	0.75	0.72	0.77	0.74	0.77	0.75
		GRADIENT BOOST	0.72	0.72	0.73	0.73	0.73	0.73
		KNN	0.78	0.76	0.81	0.78	0.81	0.79
		RANDOMFOREST	0.76	0.80	0.73	0.79	0.73	0.76
	OVER	ADA BOOST	0.76	0.79	0.73	0.80	0.73	0.76
		GRADIENT BOOST	0.89	0.98	0.81	0.98	0.81	0.89
		KNN	0.83	0.71	0.94	0.79	0.94	0.86
		RANDOMFOREST	0.88	0.93	0.83	0.93	0.83	0.88

References

1. Kotsiantis, S., Kanellopoulos, D., & Pintelas, P. (2006). Handling imbalanced datasets: A review. *GESTS International Transactions on Computer Science and Engineering, 30*, 25–36.
2. Yang, Z., Tang, W. H., Shintemirov, A., & Wu, Q. H. (2009). Association rule mining-based dissolved gas analysis for fault diagnosis of power transformers. *IEEE Transactions on Systems, Man, and Cybernetics, Part C (Applications and Reviews), 39*, 597–610.
3. Zhu, Z.-B., & Song, Z.-H. (2010). Fault diagnosis based on imbalance modified kernel Fisher discriminant analysis. *Chemical Engineering Research and Design, 88*, 936–951.
4. Tavallaee, M., Stakhanova, N. Ghorbani, A. A. (2010). Toward credible evaluation of anomaly-based intrusion-detection methods. *IEEE Transactions on Systems, Man, and Cybernetics, Part C (Applications and Reviews), 40*, 516–524.
5. Mazurowski, M. A., Habas, P. A., Zurada, J. M., Lo, J. Y., Baker, J. A., & Tourassi, G. D. (2008). Training neural network classifiers for medical decision making: The effects of imbalanced datasets on classification performance. *Neural networks, 21*, 427–436.
6. Soler, V., Cerquides, J., Sabria, J., Roig, J., & Prim, M. (2006). Imbalanced datasets classification by fuzzy rule extraction and genetic algorithms. In *Sixth IEEE international conference on data mining-workshops (ICDMW'06)*.
7. Kubat, M., & Matwin, S. (1997). Addressing the curse of imbalanced training sets: one-sided selection. In *Icml*.
8. Liu, Y.-H., & Chen, Y.-T. (2005). Total margin based adaptive fuzzy support vector machines for multiview face recognition. In *2005 IEEE international conference on systems, man and cybernetics*.
9. Li, Y., Sun, G., & Zhu, Y. (2010). Data imbalance problem in text classification. In *2010 Third international symposium on information processing*.
10. Al-Shahib, A., Breitling, R., & Gilbert, D. (2005). Feature selection and the class imbalance problem in predicting protein function from sequence. *Applied Bioinformatics, 4*, 195–203.
11. Kotsiantis, S., & Pintelas, P. (2004). Combining bagging and boosting. *International Journal of Computational Intelligence, 1*, 324–333.
12. Japkowicz, N. (2000). The class imbalance problem: Significance and strategies. In *Proceeding of the international conference on artificial intelligence*.
13. Nguyen, G. H., Bouzerdoum, A., & Phung S. L. (2009). Learning pattern classification tasks with imbalanced data sets. *Pattern Recognition*, 193–208.

14. Sun, Y., Wong, A. K. C., & Kamel, M. S. (2009). Classification of imbalanced data: A review. *International Journal of Pattern Recognition and Artificial Intelligence, 23*, 687–719.
15. Chawla, N. V., Japkowicz, N., & Kotcz, A. (2004). Special issue on learning from imbalanced data sets. *ACM SIGKDD Explorations Newsletter, 6*, 1–6.
16. Weiss, G. M., & Provost, F. (2003). Learning when training data are costly: The effect of class distribution on tree induction. *Journal of artificial intelligence research, 19*, 315–354.
17. S. Visa and A. Ralescu, "The effect of imbalanced data class distribution on fuzzy classifiers-experimental study," in *The 14th IEEE International Conference on Fuzzy Systems, 2005. FUZZ'05.*, 2005.
18. Japkowicz, N., & Stephen, S. (2002). The class imbalance problem: A systematic study. *Intelligent data analysis, 6*, 429–449.
19. Denil, M., & Trappenberg, T. (2010). Overlap versus imbalance. In *Canadian conference on artificial intelligence.*
20. Pedregosa, F., Varoquaux, G., Gramfort, A., Michel, V., Thirion, B., Grisel, O., Blondel, M., Prettenhofer, P., Weiss, R., Dubourg, V., Vanderplas, J., Passos, A., Cournapeau, D., Brucher, M., Perrot, M., & Duchesnay, E. (2011). Scikit-learn: Machine learning in python. *Journal of Machine Learning Research, 12*, 2825–2830.
21. García, V., Mollineda, R. A., Sánchez, J. S., Alejo, R., & Sotoca, J. M. (2007). When overlapping unexpectedly alters the class imbalance effects. In *Iberian conference on pattern recognition and image analysis.*
22. García, V., Sánchez, J., & Mollineda, R. (2007). An empirical study of the behavior of classifiers on imbalanced and overlapped data sets. In *Iberoamerican congress on pattern recognition.*
23. García, V., Mollineda, R. A., & Sánchez, J. S. (2008). On the k-NN performance in a challenging scenario of imbalance and overlapping. *Pattern Analysis and Applications, 11*, 269–280.
24. Xiong, H., Wu, J., & Liu, L. (2010). Classification with class overlapping: A systematic study. In *The 2010 international conference on e-business intelligence.*
25. Visa, S., & Ralescu, A. (2003). Learning imbalanced and overlapping classes using fuzzy sets. In *International conferences machine learning–workshop on learning from imbalanced datasets II.*
26. Batista, G. E. A. P. A., Prati, R. C., & Monard, M. C. (2005) Balancing strategies and class overlapping. In *International symposium on intelligent data analysis.*
27. Tomašev, N., & Mladenić, D. (2013). Class imbalance and the curse of minority hubs. *Knowledge-Based Systems, 53*, 157–172.
28. Weiss, G. M. (2004). Mining with rarity: A unifying framework. *ACM Sigkdd Explorations Newsletter, 6*, 7–19.
29. Japkowicz, N. (2001). Concept-learning in the presence of between-class and within-class imbalances. In *Conference of the Canadian society for computational studies of intelligence.*
30. Prati, R. C., Batista, G. E. A. P. A., & Monard, M. C. (2004). Class imbalances versus class overlapping: an analysis of a learning system behavior. In *Mexican international conference on artificial intelligence.*
31. Weiss, G. M. (2010) The impact of small disjuncts on classifier learning. In *Data Mining.*
32. Jo, T., & Japkowicz, N. (2004). Class imbalances versus small disjuncts. *ACM Sigkdd Explorations Newsletter, 6*, 40–49.
33. Ganganwar, V. (2012). An overview of classification algorithms for imbalanced datasets. *International Journal of Emerging Technology and Advanced Engineering, 2*, 42–47.
34. Guo, X., Yin, Y., Dong, C., Yang, G., Zhou, G. (2008). On the class imbalance problem, In *2008 Fourth international conference on natural computation.*
35. Alejo, R., Valdovinos, R. M., García, V., Pacheco-Sanchez, J. H. (2013). A hybrid method to face class overlap and class imbalance on neural networks and multi-class scenarios. *Pattern Recognition Letters, 34*, pp. 380–388, 2013.
36. Fatourechi, M., Ward, R. K., Mason, S. G., Huggins, J., Schloegl, A., & Birch, G. E. (2008). Comparison of evaluation metrics in classification applications with imbalanced datasets. In *2008 seventh international conference on machine learning and applications.*

37. Stefanowski, J., & Wilk, S. (2008). Selective pre-processing of imbalanced data for improving classification performance. In *International conference on data warehousing and knowledge discovery.*
38. Nunes, C., Silva, D., Guerreiro, M., Mendonça, A., Carvalho, A. M., & Madeira, S. C. (2013). Class imbalance in the prediction of dementia from neuropsychological data. In *Portuguese Conference on Artificial Intelligence.*
39. Chawla, N. V., Bowyer, K. W., Hall, L. O., & Kegelmeyer, W. P. (2002). SMOTE: Synthetic minority over-sampling technique. *Journal of Artificial Intelligence Research, 16*, 321–357.
40. Guo, H., & Viktor, H. L. (2004). Learning from imbalanced data sets with boosting and data generation. *ACM SIGKDD Explorations Newsletter, 6*, 30–39.
41. Yu, T., Jan, T., Simoff, S., & Debenham, J. (2007). A hierarchical VQSVM for imbalanced data sets. In *2007 International Joint Conference on Neural Networks.*
42. Yen, S.-J., & Lee, Y.-S. (2009). Cluster-based under-sampling approaches for imbalanced data distributions. *Expert Systems with Applications, 36*, 5718–5727.
43. Galar, M., Fernandez, A., Barrenechea, E., Bustince, H., & Herrera, F. (2011). A review on ensembles for the class imbalance problem: bagging-, boosting-, and hybrid-based approaches. *IEEE Transactions on Systems, Man, and Cybernetics, Part C (Applications and Reviews), 42*, 463–484.
44. Visa, S. (2007). Fuzzy classifiers for imbalanced data sets.
45. García, S., & Herrera, F. (2009). Evolutionary under sampling for classification with imbalanced datasets: Proposals and taxonomy. *Evolutionary Computation, 17*, 275–306.
46. Visa, S., & Ralescu, A. (2005). Issues in mining imbalanced data sets-a review paper. In *Proceedings of the sixteen Midwest artificial intelligence and cognitive science conference.*
47. Provost, F. (2000). Machine learning from imbalanced data sets 101. In *Proceedings of the AAAI'2000 workshop on imbalanced data sets.*
48. Maloof, M. A. (2003). Learning when data sets are imbalanced and when costs are unequal and unknown. In *ICML-2003 workshop on learning from imbalanced data sets II.*
49. Dash, M., & Liu, H. (1997). Feature selection for classification. *Intelligent Data Analysis, 1*, 131–156.
50. Zheng, Z., Wu, X., & Srihari, R. (2004). Feature selection for text categorization on imbalanced data. *ACM Sigkdd Explorations Newsletter, 6*, 80–89.
51. Chen, X. -W., & Wasikowski, M. (2008). Fast: a roc-based feature selection metric for small samples and imbalanced data classification problems. In *Proceedings of the 14th ACM SIGKDD international conference on knowledge discovery and data mining.*
52. Hall, M. A., & Smith, L. A. (1999). Feature selection for machine learning: comparing a correlation-based filter approach to the wrapper. In *FLAIRS conference.*
53. Yu, L., & Liu, H. (2003). Feature selection for high-dimensional data: A fast correlation-based filter solution. In *Proceedings of the 20th international conference on machine learning (ICML-03).*
54. Grobelnik, M. (1999). Feature selection for unbalanced class distribution and naive bayes. In *ICML '99: Proceedings of the sixteenth international conference on machine learning.*
55. Cuaya, G., Munoz-Meléndez, A., Morales, E. F. (2011). A minority class feature selection method. In *Iberoamerican congress on pattern recognition.*
56. Guyon, I., & Elisseeff, A. (2003). An introduction to variable and feature selection. *Journal of Machine Learning Research, 3*, 1157–1182.
57. Das, S. (2001). Filters, wrappers and a boosting-based hybrid for feature selection. In *Icml.*
58. Kamal, A. H. M., Zhu, X., Pandya, A., & Hsu, S. (2009). Feature selection with biased sample distributions. In *2009 IEEE international conference on information reuse & integration.*
59. Vapnik, V. (1998). *Statistical learning theory New York* (vol. 1, p. 2). Wiley.
60. Lewis, D. D., & Gale, W. A. (1994). A sequential algorithm for training text classifiers. In *SIGIR'94.*
61. Asuncion, A., & Newman, D. (2007). *UCI machine learning repository,* Irvine, CA, USA.
62. Kaggle, "kaggle," (Online). Available: https://www.kaggle.com/datasets.

Timely Prediction of Diabetes by Means of Machine Learning Practices

Rajan Prasad Tripathi, Punit Gupta, and Mayank Kumar Goyal

Abstract In the past few decades, the quality and quantity of medical data generated by digital devices have been significantly improved, which makes data generation cost-effective and simple, thereby increasing its leading position in the field of big data and machine learning. There is a huge application of machine leaning and artificial intelligence in health care sector. The use of machine learning to train the machine to classify the medical cases taking care of the historical data can be a boon in medical studies. In this paper, we have analyzed many machine learning algorithms and classifiers which are used to make prediction on the diabetes based on the chosen features and attributes of the dataset. The implementation of the algorithms and its performance are compared in terms of accuracy. The proposed model uses soft voting ensemble techniques to the standardized Pima diabetes data to best fit the data and high accuracy.

Keywords Machine learning · Diabetes · Ensemble · Soft voting · Data science

1 Introduction

The top ten causes of death in 2016 include diabetes. In 2016, 1.6 million people were affected by diabetes, up from fewer than 1,000,000 in 2000. HIV/AIDS was the seventh leading cause of death with this figure [1]. Diabetes figures grew from the number of diabetes people in the 1980s of 108 million to 422 million in 2014; global diabetes rose from 4.7% in 1980 to 8.5% in 2014 for adults aged over 18.

R. P. Tripathi
Amity University, Tashkent, Uzbekistan

P. Gupta (✉)
Department of Computer and Communication Engineering, Manipal University Jaipur, Jaipur, India
e-mail: punitg07@gmail.com

M. K. Goyal
Department of Computer Science & Engineering, School of Engineering & Technology, Sharda University, Greater Noida, India

P. K. Singh et al. (eds.), *Proceedings of Third International Conference on Computing, Communications, and Cyber-Security*, Lecture Notes in Networks and Systems 421,
https://doi.org/10.1007/978-981-19-1142-2_21

By 2040, diabetes is projected to be present in 642 million people (1 in 10 people). In addition, 46.5% of diabetes patients were not diagnosed [2]. It is important to develop strategies and procedures that aid early diagnosis of diabetes, since many deaths of diabetic patients are due to late diagnosis, to reduce diabetes-related deaths. We need advanced information technology to achieve state-of-the-art technologies for early diagnostics of diabetes, and the data mining sector is an important area for it. Data mining provides the ability to extract from a broad database repository and discover previously unknown, secret, yet interesting models. Such trends can help to diagnose and determine medically.

Diabetes mellitus is one of the diseases that affect a very large human population and is often called diabetes mellitus. Diabetes [2], a very large amount, affected more than 425 million people in 2017. In the same year, about 4 million people died of diabetes and associated complications. Though 74 million people in India have suffered from diabetes, India is recognized as the "World Capital for Diabetes." If this disease has not been taken seriously and there are no major steps to diagnose and prevent it, an estimated 629 million people worldwide will be affected by diabetes by 2045 [3]. Diabetes is a high blood glucose condition that is caused if the body cannot make the required quantity of the insulin or the body is unable to use the insulin that is produced effectively. Diabetes is most commonly caused by obesity, urbanization, physics inactivity, unhealthy diet, aging and diabetes family history. When diabetes is not rightly diagnosed or managed properly, it can cause many complications, such as cardiovascular problems, kidney diseases, blindness, neural complications such as stroke [4]. Early diagnosis is the most important fact for effective diabetes management and related complications. Early diagnosis and the recommended daily healthy lifestyle are the most important factors [5].

2 Literature Review

The following describes some of the various methods used on PIMA Indian Diabetes Datasets with their results. Bansal et al. used diabetes diagnosis KNN classifier; the attributes are selected using the PSO techniques. This method has proven to be 77% accurate [6]. In the case of the normalization and unconventional KNN algorithm model, i.e., the KNN class-specific classification algorithm, the preprocessing of the dataset is proposed as classwise KNN (CKNN) methodology for diabetes classification. The accuracy of this process is 78.16% [6]. Li et al. proposed one of the techniques known as weight-adjusted voting classification. This method is predictive of the accuracy of 77% following implementation of Pima Indian diabetes dataset [7]. The principle of modified extreme learning machines was used by Priyadarshini et al. to determine whether or not the patient is diabetic-dependent on the available data. In neural networks and extreme classifier learning, the authors draw comparative conclusions [8].

Prema et al. proposed to use ensemble technique on normalized Pima Indian diabetes dataset and got efficiency of 81% [9].

In its analysis, Iyer [10] indicated that a forecast for diabetes should be made with the use of the naïve Bayes algorithm. The study reported a 79.56% accuracy result. Throughout the classification of diabetic patients, Tarun [11] used a PCA and a support vector machine. Experimental tests have shown that while their accuracy is 93.66 percent, the previous amount can be enhanced. Kadhmi [12] suggested that after applying a nearest K algorithm to the elimination of unwanted data, a decision tree (DT) is used to assign each data sample to the appropriate class. Han et al. [13] developed a model using the diabetes prediction algorithm using the k-means algorithm. The model achieved an accuracy of 95.42 [14]. In [15], k-means clustering is used to identify and exclude outliers, genetic algorithms and CFS, extract relevant features and classify the nearest neighbors (ANN) of diabetic patients. Patil [16] proposed a hybrid prediction model that applies k-means to the original dataset and then uses the C4.5 algorithm to model the classifier. As a result, the classification accuracy rate was 92.38%. Anjali [17] suggested reducing the dimensionality of features extracted using neural networks (NN) as a classification technique dependent upon principal component analysis. The accuracy result was 92.2% [18].

3 Methodology

3.1 Pima Indian Diabetes Dataset

A list of different datasets is available for the research and implementation of ML algorithms in the UCI Machine Learning Repository. The data has been very regularly used as a primary source of machine learning datasets by researchers, students and educators. We took the Pima diabetes dataset [15] for our study from this repository. This dataset is made up of 768 patients' medical data (Fig. 1).

There are eight attributes in each data point, and they are:

- Number of pregnancies
- Body mass index
- Age
- Plasma glucose concentration
- Diastolic blood pressure
- Triceps skin fold thickness
- 2-h serum insulin
- Diabetes pedigree function.

The ninth attribute of each data point is the class variable. The outcome will be either 0 or 1 for positive or negative diabetes, respectively (Fig. 2).

```
df.describe().T
```

	count	mean	std	min	25%	50%	75%	max
Pregnancies	2000.0	3.70350	3.306063	0.000	1.000	3.000	6.000	17.00
Glucose	2000.0	121.18250	32.068636	0.000	99.000	117.000	141.000	199.00
BloodPressure	2000.0	69.14550	19.188315	0.000	63.500	72.000	80.000	122.00
SkinThickness	2000.0	20.93500	16.103243	0.000	0.000	23.000	32.000	110.00
Insulin	2000.0	80.25400	111.180534	0.000	0.000	40.000	130.000	744.00
BMI	2000.0	32.19300	8.149901	0.000	27.375	32.300	36.800	80.60
DiabetesPedigreeFunction	2000.0	0.47093	0.323553	0.078	0.244	0.376	0.624	2.42
Age	2000.0	33.09050	11.786423	21.000	24.000	29.000	40.000	81.00
Outcome	2000.0	0.34200	0.474498	0.000	0.000	0.000	1.000	1.00

Fig. 1 Description of dataset

3.2 Data Cleaning

The data was found to have many missing values. This missing values create a lot of problem in the analysis, and when we train the model with the help of original dataset having these missing values will not give good result, and hence, the missing values has to be taken care of. There are many methods available for cleaning the data like replacing the whole row or deleting the complete row, but that would result in less number of training data which we don't want, and hence, we have used the mean method we have replaced all the missing data with the mean of the values taken from other values, and hence, it has given the same kind of values and we can process further with the pipeline [16] (Figs. 3 and 4).

- We normally provide training and testing results. Only at the end of the measurement and the final performance assessment should we reach the test range. Then, we can set the train to train and check settings. We use the validation dataset to tune the model [17].
- High variance test issue with conventional train testing process. It means by dynamic change in input the results of the prediction changes. We tend to use the k-fold validation methodology in our train and validation set to unravel this problem [18] (Fig. 5).

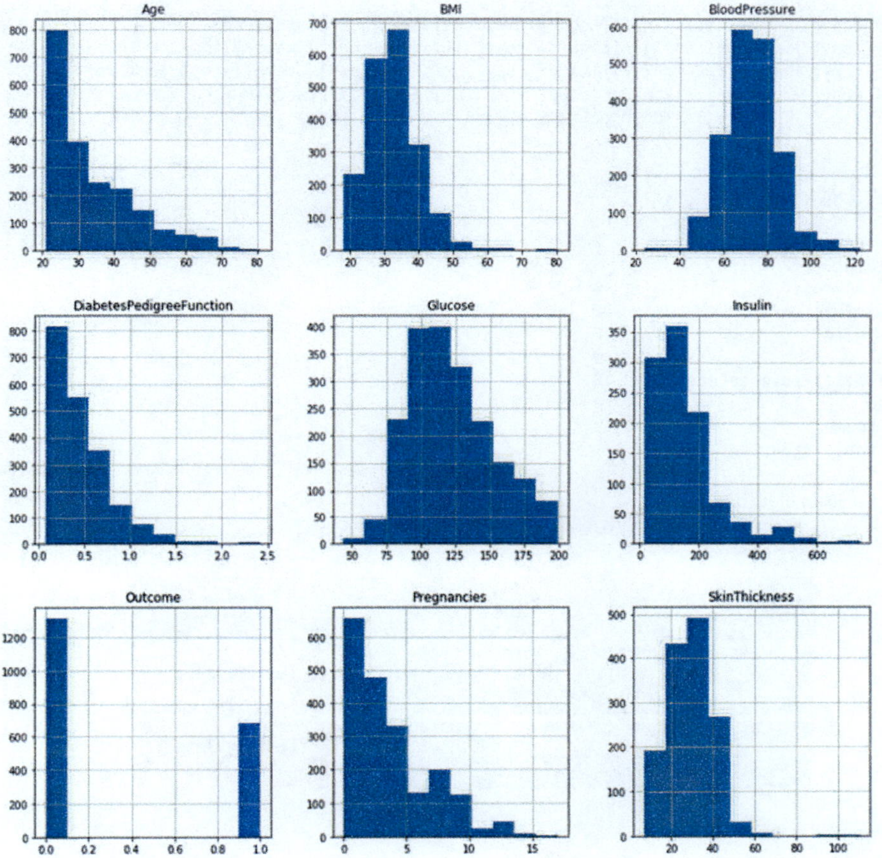

Fig.2 Histogram of features

We analyzed the data; after that, we visualized the data to understand the data better. We plotted a pair plot and found out there were lot of outliers in the data [19]. We investigated each feature distribution and checked its skewness and kurtosis. We followed this step with feature engineering which includes.

3.3 Data Preprocessing

The numerical preprocessing of tree model and non-tree model is different. Tree models are usually independent of scale. The tree-based model does not rely heavily on scaling. Where standard daviation is 1, std=1. Then, we delete outliers.

```
df_copy['Glucose'].fillna(df_copy['Glucose'].mean(),inplace=True)
df_copy['BloodPressure'].fillna(df_copy['BloodPressure'].mean(),inplace=True)
df_copy['SkinThickness'].fillna(df_copy['SkinThickness'].mean(),inplace=True)
df_copy['Insulin'].fillna(df_copy['Insulin'].mean(),inplace=True)
df_copy['BMI'].fillna(df_copy['BMI'].mean(),inplace=True)
```

```
df_copy.isnull().sum()
```

```
Pregnancies                 0
Glucose                     0
BloodPressure               0
SkinThickness               0
Insulin                     0
BMI                         0
DiabetesPedigreeFunction    0
Age                         0
Outcome                     0
dtype: int64
```

Fig. 3 Data cleaning

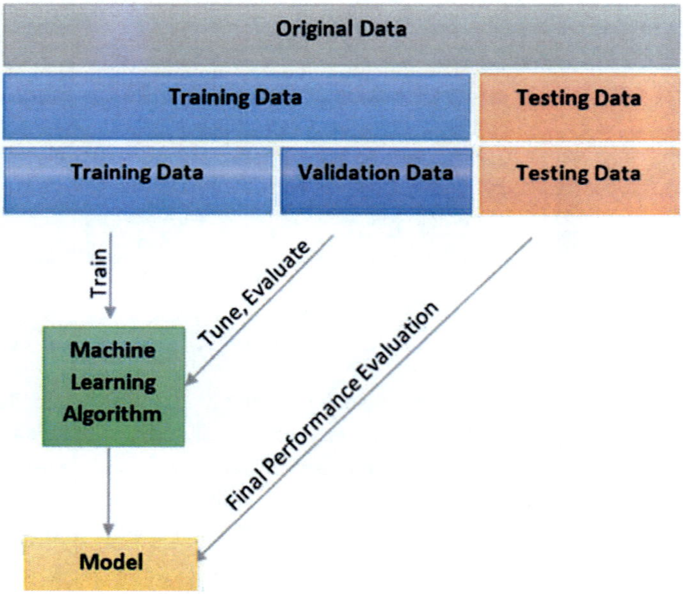

Fig. 4 The blueprint of algorithm

3.4 Feature Selection

Feature selection means that we will have to select those variable or features which will give very high dependency on our target variable which is whether diabetes is there or not in our case. In our data, the features or the attributes are automatically selected using the feature selection; the most relevant to the prediction of our test case variable will be taken up. Feature selection methods allow you to build a predictive model in our task. It allows us to choose those feature which will give very high dependency on the target class [20]. All the redundant and irrelevant features or the columns are deleted as they can have adverse effect on the prediction accuracy.

Models and chosen hyperparameters.

Algorithm	Parameter 1	Parameter 2
Logistic regression	C: Regulation parameter	
KNN	Count of neighbors	
SVC	Penalty parameter C	F(x): Linear, RBF, sigmoid
Decision Tree	Depth	sample
AdaBoost classifier	Learning rate	Count of trees
Gradient boosting	Learning rate	Count of trees

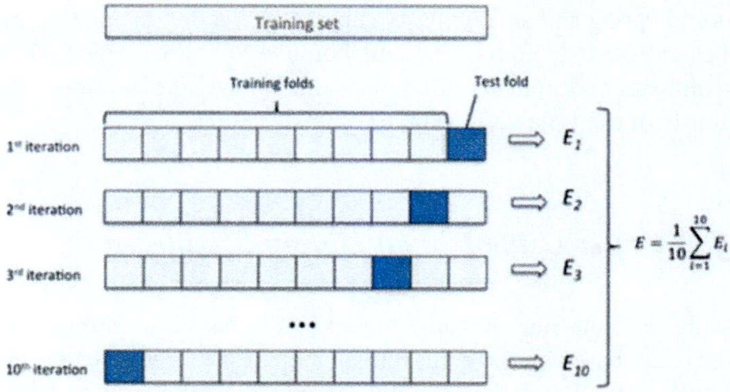

Fig. 5 Cross validation

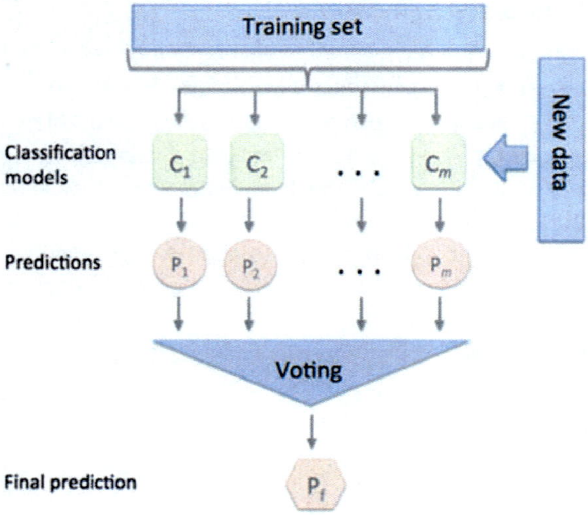

Fig. 6 Flow of ensemble method

Ensemble Methods

Ensemble is a machine learning technology that combines multiple machine learning technologies into an optimal predictive model. Reduce variance, bias or improve forecasting [21]. Compared with a single model, this method improves the prediction efficiency. There are multiple assembly methods, such as bagging, reinforcement, stacking, stacking, voting and averaging. We applied the voting integration method to PIMA diabetes records in India. The collaborative voting classifier is a meta-classifier that combines conceptually different or similar machine learning classifiers to classify majority or multiple votes (Fig. 6).

3.5 *Voting Classifier Using Python Library Scikitlearn*

A voting classifier is a machine learning model that is based on multiple models and predicts the result based on the highest likelihood of the selected category [22]. We pass the results of each classifier, and our voting classifier aggregates them and predicts the output category based on the vast majority of votes. Our idea is that instead of building different specialized models and calculating the accuracy of each model, it is better to build a single model to train the entire given model for machine learning. These models predict outcomes based on the cumulative majority of votes for each output category. Two types of votes are supported by voting classifier. Hard voting: The expected performance class in hard polling is a class which is most likely to be expected by each classifier, with the most number of votes. Suppose the output class (A, A, B) is foreseen by three classifiers, so that most predicted A as output. A is therefore the ultimate forecast.

Soft voting: The prediction in soft voting is based on the average probability given to this class. Assume the likelihood for class $A = (0.40, 0.57, 0.63)$ and $B = (0.30, 0.42, 0.50)$ given some inputs to three models. The average is 0.5333 for class *A* and 0.4067 for class *B*. The winner is clearly class *A*. In soft voting, class label is predicted on the predicted probabilities *p* for classifier.

$$y^{\wedge} = \text{argmax } i \sum j = 1mwjpij, \tag{1}$$

where *wj* is the weight that can be assigned to the *j*th classifier.

We assume as per our figure, a binary classification task with class labels $i \in \{0,1\}$, our ensemble could make the following prediction:

$$\begin{aligned} C1(x) &\to [0.8, 0.2] \\ C2(x) &\to [0.7, 0.3] \\ C3(x) &\to [0.3, 0.7] \end{aligned} \tag{2}$$

Using uniform weights, we compute the average probabilities:

$$p(i0|x) = (0.8 + 0.7 + 0.3)/3 = 0.6 \tag{3}$$

$$p(i1|x) = (0.2 + 0.3 + 0.7)/3 = 0.4 \tag{4}$$

$$y^{\wedge} = \text{argmax } i[p(i0|x), p(i1|x)] = 0 \tag{5}$$

4 Experiment and Results

We have applied different classification techniques for Pima Indian diabetes, and the results shown in Table 1. The data are sent to the classifier by dividing the data into 30% testing and 70% training. The accuracy of various models using cross validation technique is shown in Table 1, and the comparative analysis is shown in Fig. 1 as well.

5 Conclusion

Diabetes prediction is done using various machine learning model and classifiers. We have also used ensemble voting with a group Indian diabetes dataset for Pima classifiers compared highest consistency with different classification algorithms. We

Table 1 Various models with accuracy

Model	Accuracy	Parameters
Logistic regression	84.3	$C = 0.76$, penalty = l1
KNN	82.8	'n_neighbors': 15
SVC	84.3	'C': 1.7, 'kernel': 'linear'
Decision tree	76.5	criterion': 'gini', 'max_depth': 3, 'max_features': 2, 'min_samples_leaf': 2
AdaBoost classifier	81.2	'learning_rate': 0.05, 'n_estimators': 150
Gradient boosting	81.2	'learning_rate': 0.01, 'n_estimators': 100
Ensemble method	82.8	Soft voting

have used cross validation on dataset with tenfold CV data was distributed into 30% tests and training is 70%. Logistic regression performed surprisingly very well 84.3% and by using ensemble voting classifier with default soft voting, the accuracy came out to be 82.8%.

References

1. "IDF Diabetes Atlas—8th Edition. (2017). International Diabetes Federation. [Online]. Available: https://diabetesatlas.org/. Accessed: December 15, 2018.
2. http://www.who.int/news-room/fact-sheets/detail/diabetes. Retrieved July 27, 2018.
3. https://www.diabetesdaily.com/learn-about-diabetes/what-is-diabetes/how-many-people-have-diabetes/
4. Jhaldiyal, T., & Mishra, P. K. (2014). Analysis and prediction of diabetes mellitus using PCA, REP and SVM. *International Journal of Engineering and Technology Research (IJETR), 2(8)* (2014). ISSN: 2321-0869.
5. Prabhu, P., et al. (2011). Improving the performance of K-means clustering for high dimensional data set. *International Journal of Computer Science and Engineering, 3*(6) (2011).
6. Khandegar, A., & Pawar, K. (2017). Diagnosis of diabetes mellitus using PCA, neural Network and cultural algorithm. *International Journal of Digital Application & Contemporary Resarch, 5*(6) (2017).
7. Novakovic, J., & Rankov, S. (2011). Classification performance using principal component analysis and different value of the ratio R. *International Journal of Computers, Communications & Control, VI*(2), 317–327 (2011).
8. Prema, N. S., Varshith, V., & Yogeswar, J. (2019). Prediction of diabetes using ensemble techniques. *International Journal of Recent Technology and Engineering (IJRTE).*
9. Motka, R., Parmarl, V., Kumar, B., & Verma, A.R. (2013). Diabetes mellitus forecast using different data mining techniques. In *IEEE 4th International Conference on Computer and Communication Technology (ICCCT)* (pp. 99–103). IEEE (2013).
10. Global Report on Diabetes WHO Library Cataloguing-in-Publication Data Global report on diabetes (2016).
11. Seyed, S., Mohammad, G., & Kamran, S. (2015). Combination of feature selection and optimized fuzzy apriori rules: the case of credit scoring. *International Arab Journal of Information Technology, 12*(2) (2015).
12. Santhanam, T., & Padmavathi, M. S. (2015). Application of K-means and genetic algorithms for dimension reduction by integrating SVM for diabetes diagnosis. *Procedia Computer Science, 47*, 76–83 (2015).

13. Karegowda, A. G., Jayaram, M. A., & Manjunath, A. S. (2012). Cascading K-means clustering and K-nearest neighbor classifier for categorization of diabetic patients. *International Journal of Engineering and Advanced Technology, 1*(3) (2012).
14. PIMA Indian Diabetes Dataset, An open dataset. UCI Machine Learning Repository. [Online]. Available: http://ftp.ics.uci.edu/pub/machine-learningdatabases/pima-indians-diabetes/. Accessed: January 11, 2019.
15. Bansal, R., Kumar, S., & Mahajan, A. (2017). Diagnosis of diabetes mellitus using PSO and KNN classifier. In *2017 International Conference on Computing and Communication Technologies for Smart Nation (IC3TSN)* (pp. 32–38).
16. Christobel, Y. A., & Sivaprakasam, C. (2013). A new classwise K Nearest Neighbor (Cknn) method for the classification of diabetes dataset. *International Journal of Engineering and Advanced Technolology, 2*, 396–400.
17. Li, L. (2014). Diagnosis of diabetes using a weight-adjusted voting approach. In *2014 IEEE International Conference on Bioinformatics and Bioengineering*, pp. 320–324 (2014).
18. Priyadarshini, R., Dash, N., & Mishra, R. (2014). A Novel approach to predict diabetes mellitus using modified Extreme learning machine. In *2014 International Conference on Electronics and Communication Systems (ICECS)*, pp. 1–5 (2014).
19. Kotsiantis, S. B., Kanellopoulos, D., Pintelas, P. E. (2007). Data preprocessing for supervised leaning. *World Academy of Science, Engineering and Technology, International Journal of Computer, Electrical, Automation, Control and Information Engineering, 1*(12), 4091–4096 (2007)
20. Kavakiotis, I., Tsave, O., Salifoglou, A., & Maglaveras, N. (2017). machine learning and data mining methods in diabetes research. *Computational and Structural Biotechnology Journal, 15*, 104–116.
21. Ali, R., Siddiqi, M. H., & Idris, M., Kang, B. H., Lee, S. (2014). Prediction of diabetes mellitus based on boosting ensemble modeling. In *International Conference on Ubiquitous Computing and Ambient Intelligence* (pp. 25–28). Cham: Springer (2014).
22. Mohammadi M., Alizadeh, H., & Minaei-Bidgoli, B. (2008). Neural network ensembles using clustering ensemble and genetic algorithm. In *2008 Third International Conference on Convergence and Hybrid Information Technology* (Vol. 2, pp. 761–766). IEEE (2008).

Data Analytics and Intelligent Learning

Detection of Brain Tumor Using K-Means Clustering

Ravendra Singh and Bharat Bhushan Agarwal

Abstract Machine learning has been playing a vital role in the field of computer vision. It has many applications in the field of detection of diseases, especially brain tumor diagnosis. In brain tumor detection, segmentation has an important role. In this study, an efficient approach has been adopted. Segmentation has been done using the k-means clustering method. The main idea behind this color-based segmentation approach with K-means is to convert a gray-level MR image into a color space image and then use K-means clustering and histogram clustering to differentiate the position of tumor objects from other items in the MR image. Experiments reveal that the method can successfully achieve segmentation for MR brain images to help pathologists distinguish exactly lesion size and region.

Keywords Segmentation · Clustering · Cancer · Benign · Malignant

1 Introduction

Brain tumors are uncontrolled and abnormal growth of cells inside the brain. As normal, cells live and die in the brain after a short period, but sometimes, due to some complications cells live for a prolonged time and multiply the cells inside the brain and leads to death [1]. The diagnosis of these abnormal and uncontrolled growths of tissues in the brain is done at an early stage and accurately. The mass of cells grows rapidly and increases in size inside the brain which increases the pressure on brain cells that affects the normal operations of the brain, and if not treated on time leads to death. Tumors are categorized into two: benign and malignant. The most popular type of tumor is benign which originates in membranes around the brain and spinal cord. Benign tumors are not spread in the brain and treated easily. Malignant tumors are cancerous and spread to other parts of the body. Detection of the tumor at an early stage leads to life for the patient. MRI is an imaging procedure that is most commonly used for the diagnosis of brain tumors. MRI is based on the magnetic field

R. Singh (✉) · B. B. Agarwal
Department of Computer Science and Engineering, IFTM University, Moradabad, India
e-mail: ravendra85@gmail.com

P. K. Singh et al. (eds.), *Proceedings of Third International Conference on Computing, Communications, and Cyber-Security*, Lecture Notes in Networks and Systems 421,
https://doi.org/10.1007/978-981-19-1142-2_22

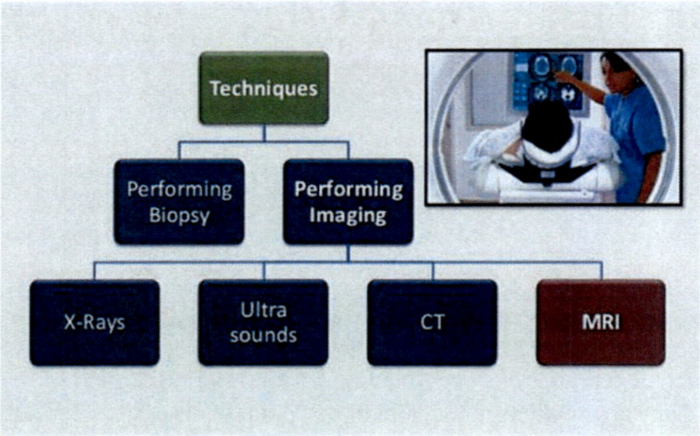

Fig. 1 Imaging techniques

principle. MRI system provides multidimensional nature of input data. Segmentation has an important role in the diagnosis of brain tumors. It differentiates the suspicious (tumor) region from the background MR image (Fig. 1).

To view the anatomical structures of the human body, imaging is an important element of medical science [2, 3]. Several new complicated medical imaging modalities, such as X-ray, MRI, and ultrasound, rely heavily on computer technology to generate or display digital images. Multidimensional digital images of physiological components can be processed and altered using computer techniques to aid in the visualization of hidden diagnostic features that would otherwise be difficult or impossible to detect using planar imaging methods. In most medical image analysis and classification for radiological evaluation or computer-aided [2] diagnosis, segmentation is a critical step. Image segmentation methods are divided into three types: edge-based methods, region-based methods [4], and pixel-based methods. In pixel-based approaches, K-means clustering is a crucial technique. The application is more practicable since pixel-based approaches based on K-means clustering are simple and have a low processing complexity compared to other region-based or edge-based methods. Furthermore, because the number of clusters is usually known for photographs of certain parts of the human anatomy, K-means clustering is excellent for biomedical image segmentation. Many academics have offered K-means clustering segmentation research [1, 5]. The advances made by [1, 5] are impressive, but greater computational complexity and software functionality are necessary. We carefully select appropriate features from brain images as clustering features in this research to produce good segmentation results while keeping the segmentation algorithm's low computation aspect. Because the color space transformation function in our proposed method is a basic operation in most image processing systems, the color space translation in the suggested scheme has no additional overhead. Therefore, by using color-based segmentation with K-means clustering to magnetic resonance (MR) brain tumors, the proposed image tracking method maintains efficiency.

The experimental results also confirm that the proposed method helps pathologists distinguish exact lesion sizes and regions.

2 Review of K-means Clustering and Histogram Statistics

In our suggested method, we use two pixel-based segmentation methods. Histogram statistics and k-means clustering are two examples. The histogram approach uses single or many thresholds to pixel-by-pixel classify an image. Analyzing the histogram for peak values and identifying the lowest point, which is often located between two successive peak values of the histogram, is a simple way to establish the gray value threshold t. The histogram statistics method can produce decent results if a histogram is clearly bi-modal. The k-means clustering algorithm divides data into k groupings and is commonly utilized. Clustering is the process of combining data points with comparable feature vectors into a single group cluster and for grouping data points with dissimilar feature vectors into different clusters.

3 Methodology

In the proposed method, we combine histogram statistics and K-means clustering to track the tumor objects in MR brain images. In K-means clustering segmentation, feature space selection is crucial. The original MR brain picture is presented as a gray-level image, which cannot handle fine details. The suggested technique uses pseudo-color transformation, a mapping function that converts a gray-level pixel to a color-level pixel using a lookup table in a preset color map, to get more usable characteristics and improve visual density. Each component in an RGB color map has R, G, and B values. Each gray value corresponds to an RGB value. The proposed method uses a conventional RGB color map, which converts gray-level values from 0 to 255 into blue-to-green-to-red color. The proposed method consists of various steps to get the final results. MRI image is acquired in the first step and preprocessing of the image is done. In the next step, segmentation of the image and feature extraction is done. All these steps are discussed in the next sections.

3.1 Preprocessing

In segmentation of brain MRI, bi-level thresholding technique uses an intensity value to differentiate between background and foreground object of brain MRI. This bi-level technique cannot distinguish b/w the background and region of interest; this technique is proved to be inefficient, and so we are moving to multilevel brain MRimaging thresholding technique. In initial consideration, an MRI image is taken

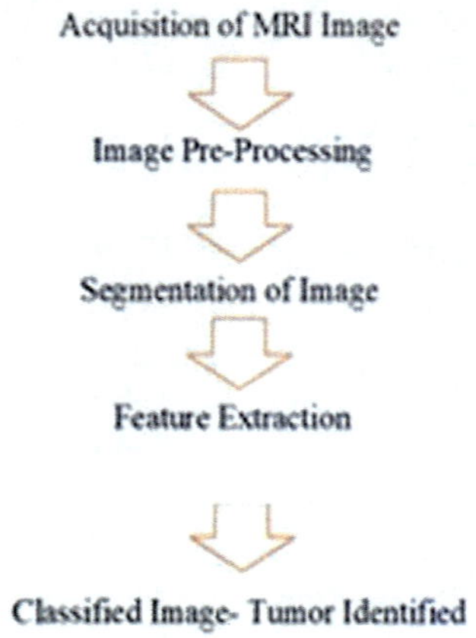

Fig. 2 Flowchart of proposed work

for making it ready to act as an input for the algorithm. After completing this step, the resizing [6] of the two-dimensional image is done. The purpose of the reshaping of an image is to make it uniform by converting it into the dimension of the same size. The next process which is involved here is a conversion of a two-dimensional image into a grayscale format [7]. The purpose of this conversion is to lessen the complexity of a typical RGB image. The flowchart of brain tumor detection (proposed) is illustrated in Fig. 2.

3.2 Segmentation Process

K-means is a widely used clustering algorithm to partition data into k clusters. Clustering is the process for grouping data points with similar feature vectors into a single cluster and for grouping data points with dissimilar feature vectors into different clusters. The procedure of partitioning an image into various regions that contain each pixel with the same attribute is called segmentation. Segmentation of an image could be thresholding, edge detection, statistical classification, or it may be any combination of these mentioned methods. The segmentation is categorized into two parts: edge-based and region-based. The edge-based methods are depending on the discontinuities in the values of intensities on regions of the border and this edge-based segmentation aims to find the boundary b/w the regions. Another technique of segmentation which is based on patterns of intensity values in the form of clustering of pixels neighbors. In these region-based techniques, the objective is to group the regions. The segmentation method aims to divide an image into various regions for further implementation and analysis. Segmentation of an image is essential to more focus on the parts of an image that are important. In the proposed work K-means method is adopted for the segmentation of brain MRI [8] which is a region-based technique. K-means is an unsupervised approach, in this unsupervised technique, the outcome is not known in advance, and it implements on the given data points by initialing labels.

Proposed Algorithm:

Input: D is a dataset containing n objects, k is the number of clusters.
Output: A set of k clusters.
Steps:

1. Randomly choose k objects from D as the initial cluster centroids.
2. For each of the objects in D do.

Compute the distance between the current objects and k cluster centroids.
Assign the current object to that cluster to which it is closest.

3. Compute the "cluster centers" of each cluster. These become the new cluster centroids.
4. Repeat steps 2–3 until the convergence criterion is satisfied
5. Stop.

In the k-means method initially, k centroid points are given that are $k = 3$ or $k = 5$ and so on. The Euclidean distance has been used to compare the other data points, and their clusters are formed that are closer to others. The mean of the clustered data is computed, and the new centroid is discovered. The formation of new clusters in the curiosity of the new centroid and continue this procedure until no new centroids are formed. The proposed method uses k-means that produce the no. of k-segments in MRI [1]. In MR image, there are three segments which are the outline of a skull, the suspected tumor area, and the normal brain area [8, 9]. The K-means approach for segmentation separates the tumor in an appropriate cluster and is given for further analysis and processing. To recompute the cluster centroids based on their group members and then regroup the feature vectors according to the new cluster centroids. The clustering procedure stops only when all cluster centroids tend to converge. The results of segmentation are depicted in Figs. 3 and 4.

The suggested technique uses pseudo-color transformation, a mapping function that converts a gray-level pixel to a color-level pixel using a lookup table in a preset color map to get more usable characteristics and improve visual density. Each component in an RGB color map has R, G, and B values. Each gray value corresponds to an RGB value. The proposed method uses a conventional RGB color map, which converts gray-level values from 0 to 255 into blue-to-green-to-red color [10–15].

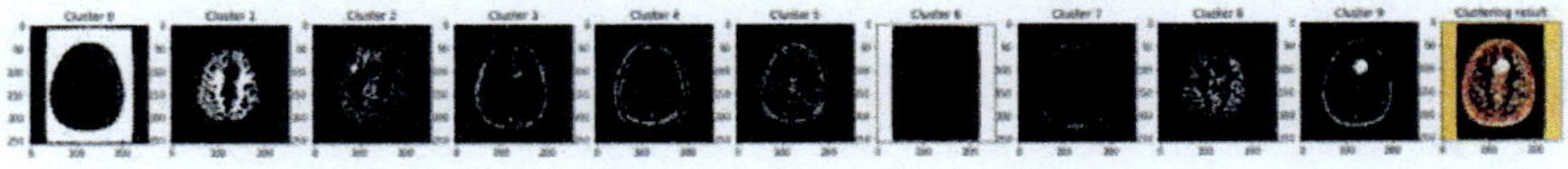

Fig. 3 Cluster result

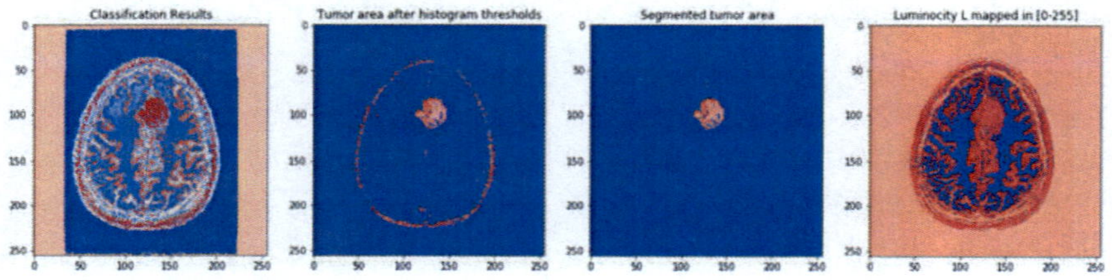

Fig. 4 Final segmented image

3.3 Feature Extraction

The MR image contains a large amount of information that is required so the feature extraction has been applied to the MR image, and relevant features have been retrieved from an image that described an image completely. The extracted features are described as follows:

Entropy: It provides information that varies from pixel to pixel in an MRI [16].

$$\text{Entropy} = \sum_{i=0}^{n} P_i \log P_i$$

RMS of an MR image is explained as it varies from the expected values and observed values [17].

Skewness represents an MR image surface information as the dark surfaces are having an upper value as compared to the lower surface [18, 19].

Kurtosis provides information about resolution and noise. The low value of kurtosis represents high resolution and high noise and vice-versa [20].

Energy provides the difference in intensity values of an image [21, 22].

Contrast represents the difference in luminance in an image [23].

Smoothness attempt to extract the vital sequences in data of an image [24].

4 Result and Discussion

In the proposed method, we convert a gray-level MR brain image into an RGB color image first and then convert the RBG color image into a color model. Therefore, colors in both the a* and b* spaces are feature vectors for K-means clustering. An MR brain picture containing the diseased alteration area depicted in Fig. 4 was used as a test image to demonstrate the proposed method's detection performance. Two independent datasets were created to demonstrate that the feature vectors offered by our method can indeed deliver better segmentation performance: the gray feature vectors of the original MR brain picture and the RGB features derived from the converted RGB color image. An MR brain picture, in general, is divided into areas that depict bone, soft tissue, fat, and background. In the test image presented in Fig. 3 where $k = 3$ show the image labeled by cluster index from the K-means procedure for different kinds of feature vectors, visual judgments from the gray and color test images propose three primary clusters. We can distinguish things in the brain image using index labels in three colors: white, gray, and black. Figure 3 depicts the final segmentation results obtained by histogram clustering. We can see that the white matter, cerebrospinal fluid, and ventricles are all recognized, in addition to the tumor (in the right part of the image). To put it another way, the segmentation result cannot pinpoint the specific location of the tumor in Fig. 4. The proposed method's segmentation result can ignore most of the white matter, cerebrospinal fluid, and ventricles while pinpointing the tumor's specific location.

5 Conclusion

In this paper, a color-based segmentation method based on K-means clustering for tracking tumor in the MRI brain image is proposed. A preliminary experiment conducted on the MRI brain image demonstrates encouraging results. The features derived can provide good segmentation performance with the proposed method, and the location of a tumor or lesion can be exactly separated from the colored image. Furthermore, the suggested method integrates color translation, K-means clustering, and histogram clustering in a single step, making it both efficient and simple to use.

References

1. Abiwinanda, N., Hanif, M., Hesaputra, S. T., Handayani, A., and Mengko, T.R.: Brain tumor classification using convolutional neural network. In *World congress on medical physics and biomedical engineering 2018* (pp. 183–189). Singapore: Springer.
2. Kumar, S., Dabas, C., & Godara, S. (2017). Classification of brain MRI tumor images: A hybrid approach. *Procedia computer science, 122*, 510–517.
3. Zhang, Y., Lu, S., Zhou, X., Yang, M., Wu, L., Liu, B., Phillips, P., & Wang, S. (2016). Comparison of machine learning methods for stationary wavelet entropy-based multiple sclerosis detection: Decision tree, k-nearest neighbors, and support vector machine. *SIMULATION, 92*(9), 861–871.
4. Zhang, Y. D., Chen, S., Wang, S. H., Yang, J. F., & Phillips, P. (2015). Magnetic resonance brain image classification based on weighted-type fractional Fourier transform and nonparallel support vector machine. *International Journal of Imaging Systems and Technology, 25*(4), 317–327.
5. Godara, S., Singh, R., & Kumar, S. (2016): A novel weighted class based clustering for medical diagnostic interface. *Indian Journal of Science and Technology, 9*(44).

6. Gomes, T. A., Prudêncio, R. B., Soares, C., Rossi, A. L., & Carvalho, A. (2012). Combining meta-learning and search techniques to select parameters for support vector machines. *Neurocomputing, 75*(1), 3–13.
7. Iqbal, A., & Jeoti, V. (2012). A novel wavelet-Galerkin method for modeling radio wave propagation in tropospheric ducts. *Progress in Electromagnetics Research, 36*, 35–52.
8. Mohsin, S. A. (2011). Concentration of the specific absorption rate around deep brain stimulation electrodes during MRI. *Progress in Electromagnetics Research, 121*, 469–484.
9. Kumari, M., & Godara, S. (2011). Comparative study of data mining classification methods in cardiovascular disease prediction.
10. Nasr-Esfahani, M., Mohrekesh, M., Akbari, M., Soroushmehr, S. R., Nasr-Esfahani, E., Karimi, N., Samavi, S., & Najarian, K. (2018). Left ventricle segmentation in cardiac MR images using fully convolutional network. In *2018 40th Annual International Conference of the IEEE Engineering in Medicine and Biology Society (EMBC)* (pp. 1275–1278).
11. Vijay, J., & Subhashini, J. (2013). An efficient brain tumor detection methodology using K-means clustering algoriftnn. In *2013 International Conference on Communication and Signal Processing* (pp. 653–657) (2013).
12. Ala, G., Francomano, E., & Viola, F. (2011). A wavelet operator on the interval in solving Maxwell's equations. *Progress in Electromagnetics Research, 27*, 133–140.
13. Chaturvedi, C. M., Singh, V. P., Singh, P., Basu, P., Singaravel, M., Shukla, R. K., Dhawan, A., Pati, A. K., Gangwar, R. K., & Singh, S. P. (2011). 2.45 GHz (CW) microwave irradiation alters circadian organization, spatial memory, DNA structure in the brain cells and blood cell counts of male mice, Mus musculus. *Progress in Electromagnetics Research, 29*, 23–42.
14. Wang, F. F., & Zhang, Y. R. (2011). The support vector machine for dielectric target detection through a wall. *Progress in Electromagnetics Research, 23*, 119–128.
15. Zhang, Y., & Wu, L. (2011). Crop classification by forward neural network with adaptive chaotic particle swarm optimization. *Sensors, 11*(5), 4721–4743.
16. Oikonomou, A., Karanasiou, I. S., & Uzunoglu, N. K. (2010). Phased-array near field radiometry for brain intracranial applications. *Progress in Electromagnetics Research, 109*, 345–360.
17. Tagluk, M. E., Akin, M., & Sezgin, N. (2010). Classıfication of sleep apnea by using wavelet transform and artificial neural networks. *Expert Systems with Applications, 37*(2), 1600–1607.
18. Li, D., Yang, W., & Wang, S. (2010). Classification of foreign fibers in cotton lint using machine vision and multi-class support vector machine. *Computers and Electronics in Agriculture, 74*(2), 274–279.
19. El-Dahshan, E. S. A., Hosny, T., & Salem, A. B. M. (2010). Hybrid intelligent techniques for MRI brain images classification. *Digital Signal Processing, 20*(2), 433–441.
20. Ghosh, A., Shankar, B. U., & Meher, S. K. (2009). A novel approach to neuro-fuzzy classification. *Neural Networks, 22*(1), 100–109.
21. Martiskainen, P., Järvinen, M., Skön, J. P., Tiirikainen, J., Kolehmainen, M., & Mononen, J. (2009). Cow behaviour pattern recognition using a three-dimensional accelerometer and support vector machines. *Applied Animal Behaviour Science, 119*(1–2), 32–38.
22. Zhang, Y. D., & Wu, L. (2008). Weights optimization of neural network via improved BCO approach. *Progress in Electromagnetics Research, 83*, 185–198.
23. Yeh, J. Y., & Fu, J. C. (2008). A hierarchical genetic algorithm for segmentation of multi-spectral human-brain MRI. *Expert Systems with Applications, 34*(2), 1285–1295.
24. Xu, Y., Guo, Y., Xia, L., & Wu, Y. (2008). A support vector regression based nonlinear modeling method for SiC MESFET. *Progress in Electromagnetics Research, 2*, 103–114.

On Efficient and Secure Multi-access Edge Computing for Internet of Things

Akshita, Yashwant Singh, and Zakir Ahmad Sheikh

Abstract The explosive growth of the Internet of Things (IoT) and smart devices currently has been drastically encouraging the development of edge computing. To improve the quality of service (QoS) with low latency in IoT applications, edge computing acts as a promising paradigm that transfers the data from cloud to edge nodes. The importance of minimal delay in critical IoT networks is being considered a highly prioritized task. The current review focuses on minimizing the gap between the resource allocation layers and resource consumer devices and nodes. Reduction in these gaps reduces the communication flow to external sources thereby involving real-time communication and reduction in delay. The remarkable development of edge computing leads to the ignorance of security threats in edge computing Therefore, keeping in view the security threats, we have addressed the security challenges on the edge of a network. The most recent security preserving mechanisms have also been taken into consideration for the secure transmission of data on edge.

Keywords Internet of Things · Multi-access edge computing · Edge computing · Security

1 Introduction

IoT is a network that is provided with unique identifiers (UIDs) and can exchange data without human-to-human interaction or human-to-computer interaction. The term IoT was proposed by Kevin Ashton in 1999. The devices of IoT interact with each other and do a lot of work without the efforts of humans. IoT is the network of physical things that are enclosed with sensors, software, and other applied technologies for the objective of transferring data with other devices over the Internet. These devices range from normal household things to sophisticated commercial instruments. Edge computing is empowering another age of revolutions that works near

Akshita · Y. Singh (✉) · Z. A. Sheikh
Department of Computer Science and Information Technology, Central University of Jammu, Bagla Suchani, J&K 181143, India
e-mail: yashwant.csit@cujammu.ac.in

P. K. Singh et al. (eds.), *Proceedings of Third International Conference on Computing, Communications, and Cyber-Security*, Lecture Notes in Networks and Systems 421,
https://doi.org/10.1007/978-981-19-1142-2_23

Fig. 1 Growth in mobile edge computing market and forecasts (2019–2025) [1]

end nodes. Edge computing decreases the amount of information that must proceed with the subsequent congestion to reduce the latency and lower the transmission cost. As edge computing is not all located within a secure data center, connectivity should be hardened with the use of VPNs and secure tunnels. The intruder can attack edge nodes that have security flaws and will use the edge nodes to hack the whole system. By encrypting the data, edge computing can upgrade the security and its features. With the use of various emerging technologies such as blockchain, machine learning, cryptography, and artificial intelligence, we can upgrade traditional encryption techniques to ensure security on edge (Fig. 1).

According to the statistical report from the research firm Mordor Intelligence research, it has been estimated that the global edge computing market was USD 93.35 million in 2020. In the coming year, its market value will be USD 2596.31 million by 2025. According to the report, North America is a hub for technological innovations such as 5G. In the future, when the new data produced by the market will increase, then, we will not be able to satisfy the customer requirements. Then, latency will be the critical component for the business. This will lead to radical changes in the business.

Multi-access edge computing is an advanced technology that provides computational resources and backhaul capacity for mobility support, low latency, location awareness to the edge of the network. To enhance the efficiency of end-user experience and IoT devices, security in MEC is a key challenge for the formation of the edge paradigm. To overcome the issues related to cloud computing, edge computing was developed. Edge computing enhances the quality of service, reduces latency, and provides high scalability. Edge computing is today's need for IoT. Rather than sending all data accumulated by IoT sensors to the cloud, the data are processed within the network by edge computing, and only, applicable data are sent by reducing latency. Therefore, in this study, we will focus on exploring the recent challenges and security mechanisms on the edge of a network.

The remainder of the paper is structured as follows. Section 2 provides the literature survey of IoT. Section 3 outlines the need for edge computing followed by its architecture. Section 4 discusses the challenges of multi-access edge computing.

Section 5 highlights the securing preserving mechanisms on edge; finally, Sect. 6 concludes the paper.

2 Literature Survey

This section briefly presents the work related to the existing technologies to secure edge computing in IoT. The use of multi-access edge computing in IoT networks brings cloud-like functionalities, but the technology's security risks must also be taken into consideration. The effectiveness of key IoT infrastructures depends on real-time connectivity. This goal can be achieved by implementing service requirements and facilities locally in networks or near the edge. However, due to limited storage, computational capacity, and low battery, local deployment in the IoT is not realistic. As a result, for these networks, it is suggested that these services be deployed at the edge node, also known as the multi-access edge. Real-time communication is important in an IoT network, and this can be achieved by ensuring that communication between devices and nodes occurs with little delay. The distance between communicating devices increases the delay and energy consumption. Many well-known services such as DHCP, DNS, HTTP, and so on are frequently utilized from cloud servers. The connectivity between IoT nodes and cloud servers causes significant delays that cannot be accepted in critical networks. As a result, we have switched the deployment of well-known services from the cloud to the edge to meet network attributes including reduced latency, efficiency [2], and secure communication. A multi-access edge is a network edge that contains a repository for all of these essential services. As a result, it will assist in the provision of services and the fulfillment of important network requirements.

Liu et al. [3] presented a use case that makes use of MEC to attain edge intelligence in the IoT framework. The main aim of this use case is to decrease the transmission rate caused by the interchange of data among the clients and the server. In this use case, the researcher has used five types of neural networks. The researcher has also proved that the networks that have maximum efficiency to perform best for resolving the problem which explains the strength of smart MEC-enhance proximity detection solutions are GRU neural network and LSTM.

Ranaweera et al. [4] analyzed the possible threat vectors in the main MEC formation framework that act with the ETSI standards. The author proposed a solution to reduce the identified threat vectors. The author has also given some approaches, for example, virtual machine introspection (VMI), trusted platform manager (TPM), network slicing (NS) to secure MEC by design.

Zhang et al. [5] proposed a strategy to upgrade the security of edge computing by virtualizing edge nodes. Firstly, the author proposed a technique of separating edge nodes, and then, the edge nodes are converted along with various kinds of things into different virtual networks that are installed in the edge nodes and the cloud server. Secondly, based on security level measurement, the author proposed a security technique. By conducting various demonstrations, the author has made a comparison between the existing algorithms to prove the efficiency in upgrading the security of edge computing.

Hassija et al. [6] presented security-related issues and security threats in IoT applications at different layers. The author has also discussed various technologies to secure IoT by using blockchain, fog computing, machine learning. The author has also discussed various open challenges that come from the solution itself.

Arfaoui et al. [7] proposed a 5G-security architecture by using the concept of domain and strata which were earlier used in 3G and 4G networks. Finally, the author has proposed a use case of a smart city that targets the two features of IoT gadgets. The first is to provide connectivity, and the second is to investigate software-defined networks in 5G. In this use case, a maximum number of IoT devices are installed to gather information for automatic control works. This use case focuses on network function virtualization (NFV) and software-defined network (SDN) technologies to give authority to the economic resources for isolating traffic for specific users. The issue in the network connection is the main warning in a mobile network.

Yu et al. [8] have proposed a structure for the security assessment of IoT networks with edge computing. This paper throws light on the interpretation of networks and has made a comparison in terms of response time, computation capacity, storage space, and analyzed the advantages of utilizing edge computing to support IoT. According to the researcher, edge computing transfers data computation to the edge of the network and near to the edge nodes. In this way, edge computing decreases the congestion to minimize the bandwidth needs in IoT.

3 Edge Computing

Edge computing is a distributed computing paradigm that aims to bring data storage closer to the site to save bandwidth and upgrade the response [9]. To mitigate the limitations [3] associated with cloud computing, edge computing was developed. Because of the fast expansion in the number of cell phones, ordinary unified cloud computing is trying to fulfill the quality of service (QoS) for certain operations. Edge computing will be the main reason to address this problem with 5G technology. Radio access network (RAN) is the main challenge that is connected with 5G technology. In this network, mobile edge computing offers ongoing RAN information. By utilizing the continuous RAN data, the network providers will upgrade the quality of experience for end clients. Consequently, the organization agents can apply RAN by outcast co-executives and rapidly developing the arrangement of modern operations. Also, the computational center points are performing their duties to transfer equivalent safety ideas to ensure the same degree of safety. Table 1 depicts the comparative analysis of secured edge frameworks in the Internet of Things (IoT).

Table 1 Comparative analysis of secured edge frameworks in the state-of-art IoT

Author's name	Year	Frameworks	Contributions	QoS metrics	Analysis	1	2	3	4	5	6
Ali et al. [10]	2021	Implementation of multiple layers of security control	The author has presented ETSI MEC reference architecture	Flexibility, latency	The researcher suggested that MEC must execute various layers of security control to alleviate targeted attacks	×	✓	✓	×	×	✓
Liu et al. [3]	2020	5G in IoT	MEC enhanced proximity detection architecture by using a connected neural network, convolutional neural network (CNN), recurrent neural network (RNN), long short-term memory (LSTM) neural network, and gated recurrent unit (GRU) neural network	High bandwidth and Latency	Latency was less than 1 ms. Here, low latency requirements can be met by MEC architecture	✓	✓	✓	×	×	✓
Medhane et al. [11]	2020	Security framework for IoT	A distributed security framework is proposed which integrates three technologies, namely blockchain, edge-cloud, and SDN	Less latency improve the quality of services	The proposed security framework will accurately and successfully encounter data confidentiality issues given by the method of blockchain, edge cloud, and SDN paradigm	×	✓	×	×	✓	×

(continued)

Table 1 (continued)

Author's name	Year	Frameworks	Contributions	QoS metrics	Analysis	1	2	3	4	5	6
Bhat et. al. [12]	2020	Edge computing	To upgrade the implementation of various technologies AI, blockchain for securing edge computing paradigm	Scalability, energy-efficiency, flexibility	To alleviate the security issues related to edge computing, the implementation of blockchain is necessary	×	✓	✓	×	✓	✓
Rahman et al. [13]	2019	Smart City	A blockchain-based framework is proposed to secure spatiotemporal smart contract services for IoT in a smart city	Security and privacy	The result shows that without the involvement of a central verification authority, the framework provides complex spatiotemporal services worldwide	×	✓	×	×	✓	×
Rahman et al. [14]	2018	To secure therapy applications	The author proposed a blockchain-based mobile edge computing framework to secure therapeutic data privacy	Low latency, security	The results show that without a substantial rise in mean processing time, this framework supports the maximum number of users	×	✓	×	×	✓	✓

1, IoT applications; 2, Security; 3, Challenges; 4, Machine learning; 5, Blockchain; 6, Edge computing. Notations: ✓, considered; ×, not considered.

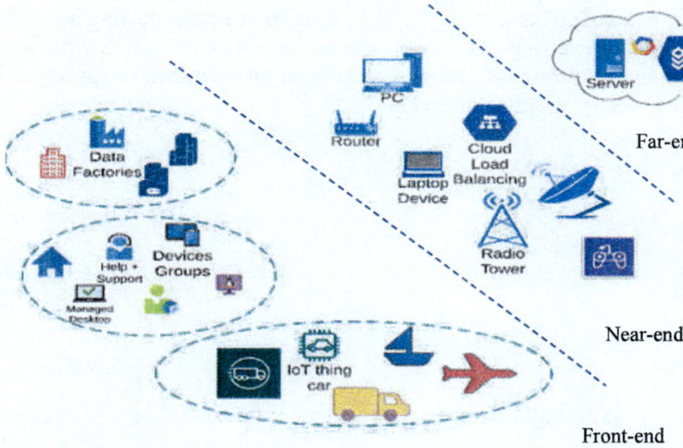

Fig. 2 Edge computing architecture [8]

3.1 Edge Computing Architecture

In Fig. 2, the cloud servers are far away from the edge computing servers than the end clients. As compared to the cloud servers, edge computing servers have lower computational ability to give superior quality of service (QoS) and lesser latency to the end users. Edge computing architecture has front end, close end, and far end.

- **Front end**. In edge computing architecture, the end gadgets (e.g., actuators, sensors) are situated at the front end. Edge computing can give immediate assistance to certain applications with calculation ability given by a large number of end gadgets. Because of the restricted limit of the end gadgets, the majority of the necessities can't be fulfilled in the front end. Subsequently, in this way, the servers will get the suitable necessities provided by the end gadgets [8].
- **Near end**. Most of the traffic flows in the network can be controlled by gateways that are in the near end. The maximum part of the processed information and capacity will be moved to the near end in edge computing. Due to this, the end clients can get a superior performance on processed information and capacity with a little expansion in the latency.
- **Far end**. The communication latency is important in the network when the cloud servers are installed far away from the end gadgets. However, maximum data capacity and computation power are provided by the cloud servers.

4 Multi-access Edge Computing Challenges

1. Security for MEC

Many researchers have focused on the security issues in edge computing, but limited work has been done in this field to provide the solution. The development pace of the security issues cannot fulfill the demand of advanced security challenges to meet the growing demand for security maintaining mobile services. Most edge devices are asset manageable. Therefore, it is difficult to apply ongoing information security techniques on edge devices. Consequently, it is a big challenge to secure a MEC system [3].

2. Site Selection for MEC Server

In an IoT network, more MEC servers are installed where maximum computational demands are required. Consequently, while installing MEC servers, a combined issue of reckoning resource arrangement and selection of the site is required to be resolved by the researchers [15].

3. Edge Intelligence

Intelligent offloading in IoT is still lacking in edge computing. To save cost, efficient energy, or to acquire quick calculation, conventional offloading generally carries calculated work to the MEC server at the edge. The issue in edge intelligence can be resolved by using advanced AI techniques such as reinforcement [16] and deep learning [17]. Thus, various parameters are needed to be examined by the researchers to acquire brilliant offloading and is a severe challenge.

4. Mobility Management

In a real-time application, mobility management technique cannot be used as VM migration acquire a heavy burden in the backhaul network which leads to a long delay. To mitigate long delays in the network, several delay-sensitive methods need the mobility management method [18]. Hence, this is a challenge for a researcher to give a better real-time method to pace up with the upcoming VM migration technique.

5. Pricing Models in Network Function Virtualization (NFV)

NFV technology encounters so many challenges. For example, when the resources are used in an unfair manner and obstruction on the physical infrastructure can be caused by the use of resources among many users. Appropriate pricing techniques can be utilized to help users to use the resources smartly.

6. Data Correlation

A large amount of data would be produced from the edge framework persistent-ly, where some of the data is tactful and put under strong safety but some of the data is not so tactful and revealed in front of the public without any safety in an edge-computing system. Consequently, invisible connections occur within the data that are not genuine. However, by utilizing these relationships and exploit-ing numerous models an intruder can reckon the data and even tamper [19]

7. Privacy in MEC

Data privacy is a crucial challenge as a large quantity of clients' personal information gathered from edge nodes is imparted to the MEC or MCC servers [3]. There is a need to pay attention among clients to use both privacy protection and other activities like data privacy (e.g, restoring, examining, and penetrating). Nowadays, location information is becoming a crucial challenge. The privacy of location can be secured by engaging cache proxies to accumulate the information regarding location rather than dispatching the actual location to location-based services.

5 Security Preserving Mechanisms on Edge

The various security preserving mechanisms on edge are blockchain, cryptography, machine learning, and artificial intelligence.

1. **Blockchain**: Blockchain is a method in which a record of transactions is made in a way that makes it impossible to hack the system. It is a digital ledger of transactions that are categorized in the whole computer system on the blockchain [20]. Various security issues already exist in an IoT network. Blockchain is a suitable technique to address the security issues in IoT. Blockchain technology brings radical change by tracking and keeping a record of every document. All the devices that are linked with the application must have authentication in the blockchain network. After registration, the devices can give the best performances according to their characteristics. In the same way, clients need to authenticate in the blockchain network. Consequently, in the network, client examines and observes the distinct objects.
2. **Cryptography**: Cryptography is a method of securing information by using codes to provide information to the intended user who processes and understands it. In this way, this method prevents the information from an unintended user. Before transmitting the data to the cloud servers, this method encrypts the context of the data. This method experiences excessive overhead and needs essential key management. To operate the data without disclosing the information, homomorphic encryption (HE) can assign the task to third parties [21]. HE technique produces a key pair that is based on numerical problems that are not able to give the solution by computers. There are two types of keys one is a public key, and the other is a private key. Data will be sent to the mediator by the public key; then, the mediator will do all the processes on the encrypted data and give feedback that will be decrypted by the private key. In the whole process, the data are confidential. RSA algorithm and the ECC algorithm are familiar homomorphic encryption algorithms.
3. **Machine Learning**: Machine learning is a subfield of artificial intelligence that makes a machine automatically learn from past experiences [22] without being programmed. The main aim of the machine learning technique is to enhance its efficiency and to provide the particular policies for security to implement in the data plane. The main goal of the ML technique is to alleviate a variety of attacks by defining access control policies or by labeling the network traffic. For example, a neural network can be used to find network intrusion, KNN, and DoS attacks in malware detection. Machine learning is used when humans are unable to use their experiences, for example, speech recognition, robotics, etc. Machine learning is also used in various smart systems, for example, Google is using machine learning to find out the threats against mobile applications and endpoints operating on android. Similarly, Amazon has introduced a Macie service [22] that uses machine learning to allocate information that is stored in the cloud. Machine learning is classified into three groups, namely supervised, unsupervised, and reinforcement learning [23].

- *Supervised Learning.* A machine learning technique in which training is given to models by using labeled data. In this type of learning, there is a need to find the mathematical function for the models to map the input variable (X) and the output variable (Y).

$$Y = F(X)$$

- *Unsupervised Learning.* In unsupervised learning, data need not be labeled. It tries to find hidden correlations on its own. With the help of an appropriate algorithm, the model can train itself, and only, input data are given to the model. It is less accurate as compared to supervised learning. For example, unsupervised learning was used by NASA for the formation of clusters of celestial bodies which were based on the identical characteristics of objects [24]
- *Reinforcement Learning.* It is used in many software and machine to monitor the result of its output and compute its value function [23]. It uses a try and error strategy which is distinct for the above two types. In reinforcement learning, particular results are not defined, and the user acquires knowledge from the feedback [22]. For example, autonomous parking.

4. **Artificial Intelligence**. AI is a technology that makes a computer system do tasks that are usually done by humans. It does not need any prior program. On the other hand, these systems use such algorithm which performs their job with their intelligence. It is used to solve complex problems. Artificial intelligence provides a solution for various IoT security threats [25]. The conventional solutions lack in terms of processing capability and have low real-time execution and less efficiency. Nowadays, the AI method provides new solutions to the problem such as DDoS attacks are becoming a global problem. On a cyber-platform, it is one of the most powerful weapons that use multiple servers and Internet connections to intrude on the targeted resource. Whenever a Web site is not working or crashing, it means it has been attacked by an intruder. To solve this problem, artificial intelligence approach is used. Different attackers can intrude in the different systems at the same time in different positions. Therefore, various AI-based detection methods, namely k-nearest neighbor (K-NN) algorithm, fuzzy logic, and support vector machine (SVM), are used [26].

6 Conclusion

Edge computing plays a pivotal role to mitigate the flaws related to cloud computing. Security and privacy are the crucial challenges limiting the acknowledgment of edge computing. Existing architectures are still relying on remote processing of data that increases the risk of data privacy and security. So, we emphasize on edge processing of data to reduce the security flaws and energy consumption issues. The most foreseeable challenges in multi-access edge computing are discussed that are the main barriers to the realistic view of edge computing. Lightweight machine learning and cryptographic algorithms must be developed for the edge devices in an IoT network because of the processing constraints of smart devices. In addition, blockchain is also an alternative solution to provide a distributed and more secure solution in an edge environment. The design and development of a secure edge computing model will be considered in our future work.

References

1. Edge computing market | growth, trends, forecasts (2021–2026). https://www.mordorintelligence.com/industry-reports/edge-computing-market-industry. Accessed May 30, 2021.
2. Nijim, M., & Albataineh, H. (2021). Secure-stor: A novel hybrid storage system architecture to enhance security and performance in edge computing. *IEEE Access, 9*, 92446–92459. https://doi.org/10.1109/access.2021.3092732
3. Liu, Y., Peng, M., Shou, G., Chen, Y., & Chen, S. (2020). Toward edge intelligence: multiaccess edge computing for 5G and Internet of Things. *IEEE Internet of Things Journal, 7*(8), 6722–6747. https://doi.org/10.1109/JIOT.2020.3004500
4. Ranaweera, P., Jurcut, A. D., & Liyanage, M. (2019). Realizing multi-access edge computing feasibility: security perspective. In *2019 IEEE Conference on Standard Communication Networking, CSCN 2019* (pp. 1–7). https://doi.org/10.1109/CSCN.2019.8931357
5. Zhang, P., Jiang, C., Pang, X., & Qian, Y. (2020). STEC-IoT: A security tactic by virtualizing edge computing on IoT. *IEEE Internet of Things Journal, 8*(4), 2459–2467. https://doi.org/10.1109/jiot.2020.3017742
6. Hassija, V., Chamola, V., Saxena, V., Jain, D., Goyal, P., & Sikdar, B. (2019). A survey on IoT security: application areas, security threats, and solution architectures. *IEEE Access, 7*, 82721–82743. https://doi.org/10.1109/ACCESS.2019.292404.5
7. Arfaoui, G., et al. (2018). A security architecture for 5G networks. *IEEE Access, 6*, 22466–22479. https://doi.org/10.1109/ACCESS.2018.2827419
8. Yu, W., et al. (2017). A survey on the edge computing for the Internet of Things. *IEEE Access 6*(c), 6900–6919. https://doi.org/10.1109/ACCESS.2017.2778504
9. Cao, K., Liu, Y., Meng, G., & Sun, Q. (2020). An overview on edge computing research. *IEEE Access, 8*, 85714–85728. https://doi.org/10.1109/ACCESS.2020.2991734
10. Ali, B., Gregory, M. A., & Li, S., Multi-access edge computing architecture, data security and privacy: a review. *IEEE Access 9*, https://doi.org/10.1109/ACCESS.2021.3053233
11. Medhane, D. V., Sangaiah, A. K., Hossain, M. S., Muhammad, G., & Wang, J. (2020). Blockchain-enabled distributed security framework for next-generation IoT: An edge cloud and software-defined network-integrated approach. *IEEE Internet of Things Journal, 7*(7), 6143–6149. https://doi.org/10.1109/JIOT.2020.2977196
12. Bhat, S. A., Sofi, I. B., & Chi, C. Y. (2020). Edge computing and its convergence with blockchain in 5G and beyond: Security, challenges, and opportunities. *IEEE Access, 8*, 205340–205373. https://doi.org/10.1109/ACCESS.2020.3037108
13. Rahman, M. A., Rashid, M. M., Shamim Hossain, M., Hassanain, E., & Alhamid, M. F., Guizani, M. (2019). Blockchain and IoT-based cognitive edge framework for sharing economy services in a smart city. *IEEE Access 7*, 18611–18621. https://doi.org/10.1109/ACCESS.2019.2896065
14. Rahman, M. A., et al. (2018). Blockchain-based mobile edge computing framework for secure therapy applications. *IEEE Access, 6*, 72469–72478. https://doi.org/10.1109/ACCESS.2018.2881246
15. Mao, Y., You, C., Zhang, J., Huang, K., & Letaief, K. B. (2017). A survey on mobile edge computing: The communication perspective (Vol. 19, No. 4, pp. 2322–2358).

16. Keneshloo, Y., Shi, T., Ramakrishnan, N., & Reddy, C. K. (2020). Deep reinforcement learning for sequence-to-sequence models. *IEEE Transactions on Neural Networks Learning Systems, 31*(7), 2469–2489. https://doi.org/10.1109/TNNLS.2019.2929141
17. Mao, Q., Hu, F., & Hao, Q. (2018). Deep learning for intelligent wireless networks: a comprehensive survey. *IEEE Communications Surveys and Tutorials, 20*(4), 2595–2621. https://doi.org/10.1109/COMST.2018.2846401
18. Mach, P., & Becvar, Z. (2017). Mobile edge computing: A survey on architecture and computation offloading. *IEEE Communications Surveys and Tutorials, 19*(3), 1628–1656. https://doi.org/10.1109/COMST.2017.2682318
19. Xiao, Y., Jia, Y., Liu, C., Cheng, X., Yu, J., & Lv, W. (2019). Edge computing security: state of the art and challenges. *Proceedings of IEEE 107*(8). https://doi.org/10.1109/JPROC.2019.2918437
20. Mohanta, B. K., Jena, D., Satapathy, U., & Patnaik, S. (2020). Survey on IoT security: Challenges and solution using machine learning, artificial intelligence and blockchain technology. *Internet of Things, 11*, 100227. https://doi.org/10.1016/j.iot.2020.100227
21. Xu, Z., Liu, W., Huang, J., Yang, C., Lu, J., & Tan, H.: Artificial intelligence for securing iot services in edge computing: A survey. *Security and Communication Networks, 2020.* https://doi.org/10.1155/2020/8872586
22. Hussain, F., Hussain, R., Hassan, S. A., & Hossain, E. (2020). Machine learning in IoT security: Current solutions and future challenges. *IEEE Communications Surveys and Tutorials, 22*(3), 1686–1721. https://doi.org/10.1109/COMST.2020.2986444
23. Bagaa, M., Taleb, T., Bernabe, J. B., & Skarmeta, A. (2020). A Machine learning security framework for IoT systems. *IEEE Access, 8*, 114066–114077. https://doi.org/10.1109/ACCESS.2020.2996214
24. Rana, B., Singh, Y., & Singh, P. K. (2020). A systematic survey on internet of things: Energy efficiency and interoperability perspective. *Transactions on Emerging Telecommunications Technologies,* 1–41. https://doi.org/10.1002/ett.4166.
25. Wu, H., Han, H., Wang, X., & Sun, S. (2020). Research on artificial intelligence enhancing Internet of Things security: A survey. *IEEE Access, 8*, 153826–153848. https://doi.org/10.1109/ACCESS.2020.3018170
26. Khalaf, B. A., Mostafa, S. A., Mustapha, A., Mohammed, M. A., & Abduallah, W. M. (2019). Comprehensive review of artificial intelligence and statistical approaches in distributed denial of service attack and defense methods. *IEEE Access, 7*, 51691–51713. https://doi.org/10.1109/ACCESS.2019.2908998

Execution Survey and State of the Art of Different ML-Based Ensemble Classifiers Approach Contextual Analysis of Spam Remark Location

Biswajit Mondal and Subir Gupta

Abstract The digital podium is proving as an increasingly important area for the contemporary development of civilization. However, it additionally engenders a rudimentary conundrum. Spamming is one of the most solemn quandaries that puts state-of-the-art security to the test. Spam wires, which send offensive messages to an immensely voluminous number of recipients, conventionally have become an apperceived security peril. There are various ways spam security issues can be addressed, including utilizing a machine learning (ML) complement system. Ensemble classifier is one of the most commonly used ML approximations. Ensemble methods use different models to amend execution. In various examination fields, like computational erudition, stats, and machine learning uses ensemble classifiers. This paper surveys traditional and verbally express-of-the-art ensemble approaches, accommodating a comprehensive overview for both practitioners and newcomers. In customary outfit strategies like Ada boost, Bagging classifier, extra trees sorts the ensemble techniques; gradient boost; logit boost; random forest; real Ada boost. This investigation is fixated on the ensemble frameworks to slant toward the spam (channel spam or ham remarks) security issue. Remark datasets are utilized for a fascinating judgment of over 41k comments and not for spam. We can split the experimental dataset into two parts. The first uses 30k for training, and the second utilizes the remaining 10k for testing. End-of-heuristics evaluation utilizing accuracy, precision, recall, $f1$ score, AUC score, model preparation time, and mean squared error reveals that Extra Trees outperforms numerous models in various exhibit metrics.

Keywords Ada boost · Bagging classifier · Extra trees · Gradient boost · Logit boost · Random forest · Real Ada boost

B. Mondal · S. Gupta (✉)
Department of Computer Science and Engineering, Dr. B. C. Roy Engineering College, Durgapur, West Bengal 713206, India
e-mail: subir2276@gmail.com

P. K. Singh et al. (eds.), *Proceedings of Third International Conference on Computing, Communications, and Cyber-Security*, Lecture Notes in Networks and Systems 421,
https://doi.org/10.1007/978-981-19-1142-2_24

1 Introduction

In the present scenario, spam has become a severe problem because of the increasing use of digital media. Various services like cordial value, web business, etc., have a prominent effect on spam [1, 2], which in turn has an impact on expressing customers and affiliations [3]. Regularly, the specialists use a unique method to investigate spam issues visually. In 1959, Arthur Samuel created ML, and to date, ML has become a well-known mainstream viewpoint [4]. Since ML has become ubiquitous, it is in high demand to cope with a slew of difficulties. It could be used in a variety of domains, including image processing, material science, and data visualization, to name a few [5–7]. ML is waiting for the likelihood of the boundary, where the pre-arranged dataset made the inhibition. Coordinated learning refers to a discrepancy if we pass the labeled dataset [8–11]. In the current years, because of the developing computational power which permits preparing tremendous ensemble learning in a reasonable period, the quantity of its applications has developed increasingly. One of the applications of ensemble classifiers is "Distributed denial of service"; it is possibly the most damaging digital attack that can befall a web access provider [12, 13]. By consolidating the yield of ensemble classifiers lessen the error of distinguishing and separating such assaults from natural glimmer swarms.

The next is "Malware Detection, " which characterizes malware codes, such as computerized viruses, worms, Trojans, ransomware, and spyware. The utilization of machine learning methods is enlivening the record classification problem [14]. Ensemble learning frameworks have shown legitimate adequacy around here. Another one is "Face Detection" [15]. It manages the distinguishing proof or confirmation of a person utilizing computerized photographs and has as of late become quite possibly the most mainstream research region in design acknowledgment. The following application is "Spam Detection," which manages recognizable proof of bank extortion, for example illegal tax avoidance, charge card misrepresentation, and media transmission misrepresentation, which have vast areas of examination and uses of machine learning. Ensemble learning works on the strength of the typical conduct display to propose as a productive strategy to identify such fake cases and exercises in banking and charge card frameworks [16, 17].

We can divide the main contributions of this study into various types. Classifier ensembles combine predictive models of different classifiers for feature selection in ML. This approach is known to improve classification performance, and there exists a wide range of ensemble methods and algorithms. However, up to our best knowledge, ensemble methods have never been analyzed extensively in the feature selection domain. Hence, there is a chance to find out the most fruitful ensemble method. This study investigates the effects of using different classifier ensemble methods in the feature selection domain. For this purpose, we analyze seven other classifier ensemble methods and compare their performances in seven metrics: accuracy, precision, recall, f1 score, AUC score, model preparation time, and mean squared error [18–20]. We also compare the results with existing ensemble algorithms and with the state-of-the-art algorithms. Moreover, we elaborate on your findings with statistical test results.

We can depict this review as follows. The covering-predicated part attestation procedure utilizes a classifier to assess the presence of every subset. Classifier gregarious events join canny models of sundry classifiers. Moreover, this system kens to cultivate demand execution. There exists a wide degree of outfit procedures and calculations. Regardless of our best information, pack procedures have never been reviewed broadly in the component cull space.

Consequently, there is no ultimate auxiliary outfit methodology existing. This review examines the impacts of utilizing concrete classifier outfit strategies in the component cull space. Therefore, we investigate seven exceptional classifier amassing techniques and cogitate their exordial in seven accuracies, precision, recall, f1 score, AUC score, model preparation time, and mean squared error.

The remnant of the primary copy is composed as follows. Section 2 presents the multiobjective substantiation calculation and ML techniques. In like way, model execution is depicted plenarily in Sect. 2. The primer climate and results have taken part in Sects. 3 and 4 autonomously. Wrapping up comments and conceivable future works are given in Sect. 5.

2 Background Knowledge

The primary target of this investigation is to recognize the state-of-art execution of the seven classifiers, namely (a) Ada boost; (b) Bagging classifier; (c) Extra Trees; (d) Gradient Boost; (e) Logit boost; (f) Random Forest, and (g) Real Ada boost [21–25]. For this purpose, we consider the seven estimation metrics, accuracy score, precision score, AUC score, recall score, $F1$ score, model training time, and mean square error. For the estimation of the exhibition, a typical dataset is compulsory. In such a manner, the common dataset represents, and we create, a 41k dataset. The dataset arrangement depicts in segment 2.1. Each of the seven classifiers follows diverse distinctive numerical models or consistent estimation that numerical clarification provides in 2.2 article as a pseudocode of the classifier.

2.1 Dataset Preparation

Our dataset derives from two open APIs. One is GitHub, and the other is Kaggle [26, 27], and the data processing is manual. After the data accumulation is over, the next phase investigates as the data storage phase. In this context, data storage refers to assiduously storing data and managing data amassments, including repositories such as databases and more specific storage types such as files. After preserving the data, it moves on to the next phase, kenned as data preprocessing, which has three components: data cleaning, feature extraction, and data fitting [28–30]. Here, data

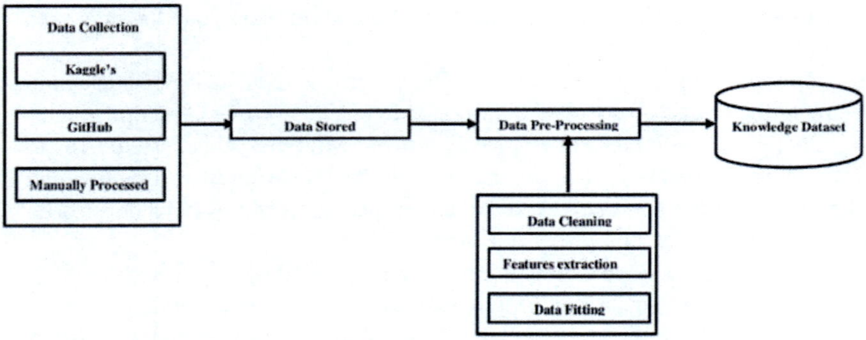

Fig. 1 Extraction of data for spam and ham comment identification

cleaning betokens identifying and redressing errors in the dataset that may negatively impact the predictive model. The feature extraction identifies the minimum use of required resources to describe an astronomically big data set. Lastly, data fitting is described as a process of fitting models to data and analyzing the precision of the fit. Determinately, after the preprocessing step is over, the processed data gets into the next phase, kenned as an erudition dataset, which denotes that there is lots of data amassed and stored together in a single place. Figure 1 shows the dataset preparation methodology.

The dataset utilized in this model has been acquired from the public dataset vaults of Kaggle's and GitHub and prepared physically. It has two pre-defined columns: Content, which contains comments, and Class, which is either 1 or 0, with one denoting spam and 0 representing ham. There are 41k comments in total [31].

2.2 *Pseudocode of the Different Ensemble Classifier*

2.2.1 Ada Boost Classifier

The Ada Boost classifier, short for Adaptive Boosting, is a boosting approach used in machine learning as an ensemble method. The weights are re-assigned to each instance, with consequences for incorrectly classified models, recognized as Adaptive Boosting.

Let the samples be $m_1 \ldots m_x$.
the outputs are $m_1 \ldots m_x, n\ \varepsilon \{-1, 1\}$,
the initial weights $i_{1.1} \ldots, i_{x-1}$ set to $\frac{1}{x}$,
the error function be Err $(f\ (m), n, j) = e_j^{-n} - f(mj)$.
the weak learners are g: $m \rightarrow \{-1, 1\}$.
For each iteration f in 1 … F:

We choose α $g_f(m)$ such that it minimizes $€_f$ which is the weighted sum error for the misclassified points present in the input where $€_f = \sum_{k=1}^{x} i_{k,f} g_k^{(m)_k} \neq n_k$.

Next, we choose α_f, which is the weighted error rate of the weak classifier where $\alpha_f \cdot \frac{1}{2}^{\ln}$.

Then, we did it to the ensemble in a way E no_f $(m) =$ Eno_{f-1} $(m) + \alpha_f g_f$ (m).

We update the weights using $ï_{1,f+1} = i_{1,f}\, e^{-n}{}_1\alpha_f g_f{}^{(m}{}_1{}^{)}$ For l in 1 … x and renormalize $I_{1,f+1}$ such that $\sum_l i_{1,f+1} = 1$

2.2.2 Bagging Classifier

Bagging is a machine learning ensemble meta-classifier meant to increase the stability and accuracy of machine learning classifiers used in statistical and regression. It also helps to avoid overfitting by reducing variance.

Let the input dataset = DT + $\{(m_1, n_1, \ldots\, m_1, m_z)\}$.

The base learners = BL, and, no of iterations = J.

We need to iterate for $j = 1 \ldots J$

Generation of bootstrap sample from DT is using as DT_j = Bootstrap (DT), where Bootstrap is a generic function.

Then, we need to train a base learner $\mathrm{BL_j}$ on the bootstrap sample DT_j, through BL. After execution of all the iterations, we finally obtain a strong learner in the form of BL (m) = argmax $\sum_{j=1}^{J} \mathrm{BL}(m = \mathrm{BL}_j(m))$, which can program any kind of classification $n \in N$.

2.2.3 Extra Trees Classifier

It involves overall three main steps:

Step 1: Splitting a node—Let L be the local input dataset corresponding to the node we want to split.

If we have reached a stage when no further splitting is possible, then abort.

Else, we select M attributes as $\{b_1, \ldots\, b_m\}$ among the non-constant candidate attributes present in L.

Then we draw M splits as $\{l_1, \ldots\, l_m\}$ where l_j involves selecting a random split using $\left(L \cdot b_j\right)$, $\forall\, j = 1 \ldots M$..

Finally, we return a split l_* such that the score_of $(l_*, L) = \max_j = 1 \ldots M$ score_of (l_j, L).

Step 2: Select a random split—Let L be a data subset and b be an attribute. Also, assume that $b_{\max}^{L}$ and $b_{\min}^{L}$ are the maximal and minimal values of b in L. Now, we draw a random cut-point b_t in $[b_{\min}^{L}, b_{\max}^{L}]$ uniformly.

Finally, we return a split $[b < b_t]$.

Step 3: Stopping the splitting of nodes—Let L = data subset from the input sample. Now, if $|L| < p_{\min}$, then we stop splitting.

If all the attributes in L are constant, then we stop splitting.

If the output in L is constant, then we stop splitting.

If the above three conditions are not satisfied, then we continue splitting.

Here, $p_{\min}$ is the stopping criterion.

Finally, be combining these steps together we generate an ensemble of extra trees denoted bt $T = \{r_1 \ldots r_2\}$ for $z = 1 \ldots z$.

2.2.4 Gradient Boost Classifier

The gradient boosting approach can use to forecast continuous and categorical target variables (as a regressor). And log loss is the cost function when used as a classifier.

Let the input dataset be $= \{(m_z, n_z\}^k{}_{z=1}$, differentiable loss function $=$ Loss $(n, \text{Ç}(m))$ and number of iterations y.

First, we need to initialize the classifier model with $\text{Ç}_0\,(m) = \text{argmin} \sum_{z=1}^{k} \text{loss}(n_z, \gamma)$

Repeat the following four steps for $y = 1 \ldots y$.

Complete the pseudo-residuals for $z = 1, 2, \ldots, k$ in the form of S_{zy}.

Fit the weak learner, i.e., the regression, which means that we use the training set $\{m_z, S_{zy}\}_{z=1}^{k}$.

Now, we need to compute γ_y using the formula $\gamma_y =$ argmin $\sum_{z=1}^{k} \text{loss}(n_z, \text{Ç}_{y-1}(m_z) + \text{Ç}_{wy}(m_z))$

Final step includes updating the model, i.e., $\text{Ç}_y(m) = \text{Ç}_{y-1}(m) + \gamma_y\, w_y(m)$

The final outcome is a strong classifier denoted by $\text{Ç}_y\,(m)$.

2.2.5 Logit Boost Classifier

Logit Boost is a classification classifier that uses boosting. It is very logical to perform an additive logistic regression.

At first, we initialize the decision function to $F_0\,(m, n)$.

Then, we repeat the following three steps for $x = 1, \ldots, X$.

Step 1: A base function g needs to be selected such that $g_x = \arg\min_\alpha \overline{\varepsilon}(g_\alpha i\, F_{x-1})$ where $\alpha =$ anyset—valued classifier and $\overline{\varepsilon} =$ weighted error rate.

Step 2: A ω efficient computed such that $\alpha_x = \arg\min_{\alpha \geq 0} M(F + \alpha g_x)$.

Step 3: Finally, the decision function needs to get updated in the way $F_x = F_{x--1} + \alpha_x g_x$.

The outcome of the iterative steps can combine into a single decision function $\text{F}_\text{x} = \text{F}_0 + F_x = F_0 + \sum_{x=1}^{x} \alpha_x g_x$.

2.2.6 Random Forest Classifier

Random Forest, also known as random decision forest, is an ensemble learning method for classification, regression, and other problems that work by training many decision trees.

This classifier involves two significant steps:

Step 1: Generation of bootstrap samples

For $c = 1$ to c repeat.

We randomly sample the input dataset T with replacements and produce T_c.

A root node R_c contains T_c.

Finally, a random decision tree can build by using R_c.

End for.

Step 2: Building random decision trees using R:

If R contains input data about one class only, then abort.

We perform a random selection of R % of all possible features splitting in R.

Then, select a feature U such that it has the highest gain from information on splitting.

We need to create u child nodes from R in a way $R_1 \ldots R_n$, where U: $(U_1 \ldots U_n)$.

Now, repeat the next few steps of R_z to T_z such that $T_z =$ all matching instances among R and U_z.

End for.

End if.

Finally, we build an ensemble of decision trees, i.e., random forest.

2.2.7 Real Ada Boost Classifier

Real Ada Boost is also a type of Adaptive Boosting classifier which comes under ensemble classification.

This boosting technique reaches the optimum result using much fewer trees than Ada Boost. The ultimate equation of this technique is as follows:

$F(p(\eta_\alpha = 1) = \text{Ç}(q_a) = \sum_{b=1}^{z} H_b(q_a)$.. At first,

We start with weights $i_m = x\frac{1}{x}, m = 1, 2, \ldots x$.

Then, we repeat the next three steps for $Y = 1, 2, \ldots, y$.

At first, we fit the weak classifier in order to obtain class probability estimation $P_y(q) = \text{þ}_i^{\wedge}(N = 1/q)\ \varepsilon\ [0, 1]$,, using weights $\ddot{I}_m$ on input data. Then, we set $H_y(q) \leftarrow \frac{1}{2}\log\left(\frac{\frac{p}{q}}{1} - p_y(q)\right) \in R - n_m H_y(q_m)$.

We update the weights $i_m \leftarrow i_m e$, $m = 1, 2, \ldots, x$ and renormalize it such that $\sum_m i_m = 1$.

Finally, the model's output is of form $\text{Ç}(q) = \sum_{y=1}^{y} H_y(q)$..

3 Methodology

Figure 2 shows the amelioration in connecting with the Ensemble classifier to expect the model. More than 41k of data is open in the encephalon dataset. Following this, the information is dissevering into two regions to a 3:1 degree that is unremarkable for training and testing. Where 75% of the learnedness dataset, which is around 30k data, pass on the orchestrating dataset, and the ensemble classifier supervises this organizing dataset, which is a coalescence of seven classifiers that are (a) Ada boost; (b) Bagging classifier; (c) Extra Trees; (d) Gradient Boost; (e) Logit boost; (f) Random Forest, and (g) Real Ada boost for spam ham remark revenue. Every ensemble classifier has its mathematical model that is the explication it will make different posit arrangements. The model relies on 25% of the dataset, around 10k data, utilized for model testing. Endeavored outcomes aggregates for heterogeneous examinations limits (i.e., accuracy, precision, recall, f1 score, AUC score, model preparation time, and mean squared error). It will benefit from working out the assessment execution and profit to find the best model in this substantial case. The mundane model directs this testing dataset, and the model's precision is unfaltering. The model has set something to the side for future change.

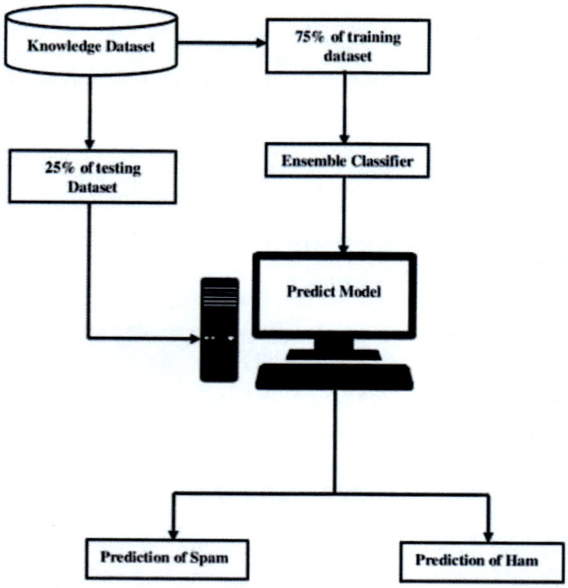

Fig. 2 Flow diagram for spam comment detection using ensemble classifier

4 Result Analysis

In this analysis, we use various cut-off points to survey the Ensemble classifier model presentation. Figure 3 shows the measurement of various cut-off points, namely AUC curve, precision curve, heatmap of seven different classifiers. Firstly, it serves in the AUC curve diagram. AUC curve includes the degree of recognizability in the model. It expresses how well equipped the model is for detaching between classes. The better the model is at culling 0 s as 0 s and 1 s as 1 s, the higher the AUC. The AUC scale ranges from 0 to 1. The AUC of a model with 100% inaccurate assumptions is 0.0; a model with 100% correct appraisal is 1.0. Specifically, we used precision-recall curves to identify data innovation settings, where the exactness review bend appears, which renders many recovered reports and crucial archives. Finally, the heat map used to survey the presentation of Extra Trees' detailed information in two-assessment is a concretely solid visual guide for a visual examiner, facilitating the rapid distribution of specific or information-driven data.

In Table 1, according to the initial investigation, the precision score, $F1$ score, and accuracy score of the Extra Trees have been calculated at 97.25, 96.49, and

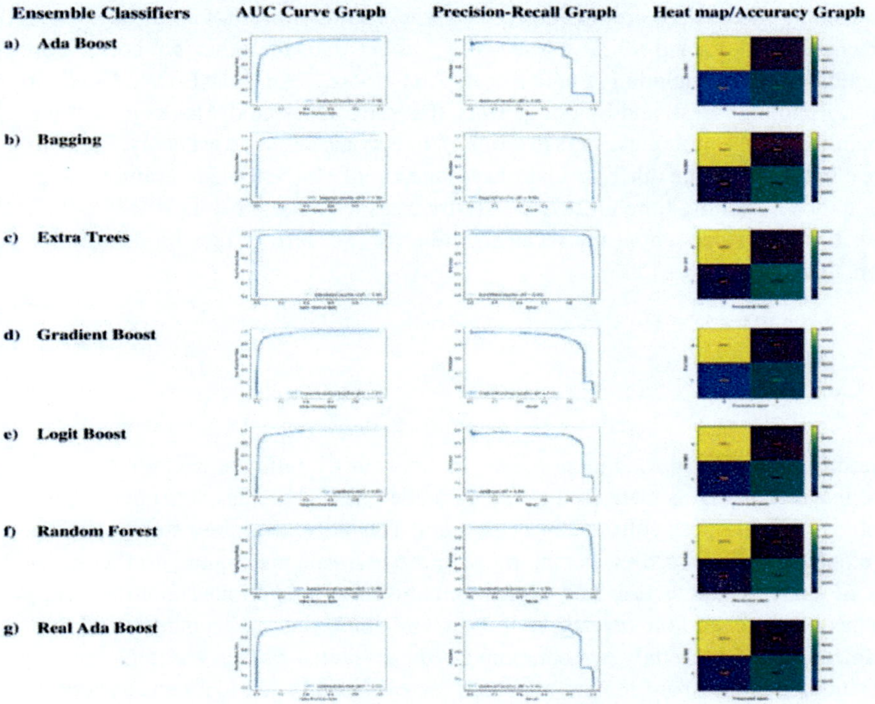

Fig. 3 Measurement of AUC curve, precision curve, heatmap of seven different classifiers **a** Ada boost; **b** bagging classifier; **c** extra trees; **d** gradient boost; **e** logit boost; **f** random forest, and **g** real ada boost

Table 1 Outcome score of seven models

Parameter	Ada boost	Bagging	Extra trees	Gradient boost	Logit boost	Random forest	Real ada boost
Accuracy score	87.92	95.54	97.28	89.82	91.97	97.09	92.32
Precision score	93.62	94.70	97.25	95.18	93.50	96.90	92.32
Recall score	74.08	93.81	95.73	77.84	85.36	95.60	86.39
$F1$ score	82.71	94.25	96.49	85.64	89.24	96.25	86.77
AUC score	94.15	98.65	99.32	96.89	96.90	99.22	97.26
Training time	376.35	4,690.95	572.52	866.03	396.57	267.62	375.27
MSE	0.12	0.044	0.027	0.10	0.080	0.03	0.08

97.28, respectively. These are higher than the other six correlation ensembles models, such as Logit Boost, Ada Boost, Real Ada Boost, Gradient Boost, Random Forest, and Bagging. Thus, it is essential to ensure that the Extra Trees' precision, $F1$, and accuracy scores are prominent. Extra Trees' recall and AUC scores have been calculated at 95.73 and 99.32, respectively, higher than the other six correlation ensemble models, including Logit Boost, Ada Boost, Real Ada Boost, Gradient Boost, Random Forest, and Bagging. Thus, the Extra Trees model looks to beat the remaining six ensemble classifiers in terms of precision, AUC, and accuracy. Random Forest outperforms the other six classifiers in terms of efficient model training time, with a model training time of 267.62 s. Extra trees have the most un-worth, 0.027, lower than the others about the mean square error. So, Extra Trees produce fewer errors than others.

5 Conclusion

Spammers and relaxed sodality assailers are becoming intelligent enough to befool the clients by acting as mannered as the bonafide ones. There has dependably been an objective to propose culls to supervise them. This paper examines sundry cutting edge outfit classifier studies, beginning with the assessments cognate to the thickness of misinformed records and exercises mundane in gregarious sodalities. And, it is passed on toward the revelation of spam and compromised accounts in relaxed organizations. In this study, we contemplate the exhibits of seven ensemble classifier methods in the spam remark apperceiving evidence of the IoT security locale, with affliction results. The tested ensemble methods are (a) Ada boost; (b) Bagging classifier; (c) Extra Trees; (d) Gradient Boost; (e) Logit boost; (f) Random Forest, and (g) Real Ada boost. We are footing the analysis on 41k well-known datasets and provide comparison results on seven metrics: accuracy score, precision score, AUC score, recall score, $F1$ score, model training time, mean square error. State of the art for performance evaluation divides into two parts: an essential condition and a sufficient condition. While a primary condition must be present for an event to occur, a satisfactory condition is a condition or group of factors that will cause the event to occur. Although an urgent circumstance should exist, it is insufficient to motivate the event's occurrence. The location raises the F1 score and cut-off (calculated using precision and recall). MSE is essential for every programmer, just as for any industry expert or subject matter expert. However, a required criterion is the AUC score lift and model planning time limit. Using this concept, it is unquestionably confident that in the case of binary labeling datasets, considering all conditions, different trees are superior to each of the seven techniques.

Acknowledgements We are acknowledging Priyanka Dhara, Shubham Bhattacharjee, and Sohom Bhattacharya for technical help.

References

1. Mewada, A., & Dewang, R. K. (2021). Research on false review detection methods: A state-of-the-art review. *Journal of King Saud University and Computer and Information Sciences*, (xxxx). https://doi.org/10.1016/j.jksuci.2021.07.021
2. Petschke, D., & Staab, T. E. M. (2019). A supervised machine learning approach using naive Gaussian Bayes classification for shape-sensitive detector pulse discrimination in positron annihilation lifetime spectroscopy (PALS). *Nuclear Instruments and Methods in Physics Research, Section A: Accelerators Spectrometers, Detectots and Associated Equipment, 947*, 162742. https://doi.org/10.1016/j.nima.2019.162742
3. Ning, B., Junwei, W., & Feng, H. (2019). Spam message classification based on the naïve Bayes classification algorithm. *IAENG International Journal of Computer Science, 46*(1).
4. Samuel, A. L. (1959). Eight-move opening utilizing generalization learning. (See Appendix B, Game G-43.1 Some Studies in Machine Learning Using the Game of Checkers) *IBM Journal*, 210–229.
5. Gupta, S., Sarkar, J., Kundu, M., Bandyopadhyay, N. R., & Ganguly, S. (2020). Automatic recognition of SEM microstructure and phases of steel using LBP and random decision forest operator. *Measurement, 151*(xxxx), 107224. https://doi.org/10.1016/j.measurement.2019.107224
6. Gupta, S. et al. (2020). Modelling the steel microstructure knowledge for in-silico recognition of phases using machine learning. *Materials Chemistry and Physics, 252*, 123286. https://doi.org/10.1016/j.matchemphys.2020.123286
7. Mondal, B. (2020). Artificial intelligence: State of the art. *Intelligent Systems Reference Library, 172*, 389–425.
8. Ligthart, A., Catal, C., & Tekinerdogan, B. (2020). Analyzing the effectiveness of semi-supervised learning approaches for opinion spam classification. *Applied Soft Computing, 101*, 107023. https://doi.org/10.1016/j.asoc.2020.107023
9. Padmanabha Reddy, Y. C. A., Viswanath, P., & Eswara Reddy, B. (2018). Semi-supervised learning: a brief review. *International Journal of Engineering and Technology, 7*(1.8), 81. https://doi.org/10.14419/ijet.v7i1.8.9977
10. Panahi, R., Ebrahimie, E., Niazi, A., & Afsharifar, A.(2021). Integration of meta-analysis and supervised machine learning for pattern recognition in breast cancer using epigenetic

data. *Informatics in Medicine Unlocked, 24*, 100629, 2021. https://doi.org/10.1016/j.imu.2021.100629

11. Wang, Y., et al. (2020). Unsupervised machine learning for the discovery of latent disease clusters and patient subgroups using electronic health records. *Journal of Biomedical Informatics, 102*, 103364. https://doi.org/10.1016/j.jbi.2019.103364
12. Reisach, U. (2021). The responsibility of social media in times of societal and political manipulation. *European Journal of Operational Research, 291*(3), 906–917. https://doi.org/10.1016/j.ejor.2020.09.020
13. Engström, E., & Strimling, P. (2020). Deep learning diffusion by infusion into preexisting technologies—Implications for users and society at large. *Technology in Society, 63*, 101396. https://doi.org/10.1016/j.techsoc.2020.101396
14. Gao, H., Cheng, S., & Zhang, W. (2021) GDroid: Android malware detection and classification with graph convolutional network. *Computers & Security, 106*. https://doi.org/10.1016/j.cose.2021.102264
15. Sharmila, V., Rejin Paul, N. R., Ezhumalai, P., Reetha, S., & Naresh Kumar, S. (2020). IOT enabled smart assistance system using face detection and recognition for visually challenged people. *Materials Today: Proceedings*. https://doi.org/10.1016/j.matpr.2020.10.198
16. Piryonesi, S. M., & El-Diraby, T. E. (2020). Role of data analytics in infrastructure asset management: Overcoming data size and quality problems. *Journal of Transportation Engineering: Part B Pavements, 146*(2), 04020022. https://doi.org/10.1061/jpeodx.0000175
17. Yang, S., Wu, J., Du, Y., He, Y., & Chen, X. (2017). Ensemble learning for short-term traffic prediction based on gradient boosting machine. *Journal of Sensors*. https://doi.org/10.1155/2017/7074143
18. Hasan, M., Islam, M. M., Zarif, M. I. I., & Hashem, M. M. A. (2019). Attack and anomaly detection in IoT sensors in IoT sites using machine learning approaches. *Internet of Things, 7*, 100059. https://doi.org/10.1016/j.iot.2019.100059
19. El-Dairi, M., & House, R. J. (2019). Optic nerve hypoplasia. In *Handbook of Pediatric Retinal OCT and the Eye-Brain Connection* (pp. 285–287). https://doi.org/10.1016/B978-0-323-60984-5.00062-7
20. Benussi, A., et al. (2021). Classification accuracy of TMS for the diagnosis of mild cognitive impairment. *Brain Stimulation, 14*(2), 241–249. https://doi.org/10.1016/j.brs.2021.01.004
21. Louzada, F., & Ara, A. (2012). Bagging k-dependence probabilistic networks: An alternative powerful fraud detection tool. *Expert Systems with Applications, 39*(14), 11583–11592. https://doi.org/10.1016/j.eswa.2012.04.024
22. Moral-García, S., Mantas, C. J., Castellano, J. G., Benítez, M. D., & Abellán,J. (2020). Bagging of credal decision trees for imprecise classification. *Expert Systems with Applications, 141*. https://doi.org/10.1016/j.eswa.2019.112944
23. Besharati, E., Naderan, M., & Namjoo, E. (2018). LR-HIDS: Logistic regression host-based intrusion detection system for cloud environments. *Journal of Ambient Intelligence and Humanized Computing*. https://doi.org/10.1007/s12652-018-1093-8
24. Padmaja, B., Prasad, V. V. R., & Sunitha, K. V. N. (2020). A novel random split point procedure using extremely randomized (Extra) trees ensemble method for human activity recognition. *EAI Endorsed Transactions on Pervasive Health and Technology, 6*(22), 1–10. https://doi.org/10.4108/eai.28-5-2020.164824
25. Sarker, I. H. (2021). CyberLearning: Effectiveness analysis of machine learning security modeling to detect cyber-anomalies and multi-attacks. *Internet of Things, 14*, 100393. https://doi.org/10.1016/j.iot.2021.100393
26. Mateen, M., Wen, J., Nasrullah, Song, S., & Huang, Z. (2019). Fundus image classification using VGG-19 architecture with PCA and SVD. *Symmetry (Basel), 11*(1). https://doi.org/10.3390/sym11010001
27. Thakur, S., Chakraborty, A., De, R., Kumar, N., & Sarkar, R. (2021). Intrusion detection in cyber-physical systems using a generic and domain specific deep autoencoder model. *Computers & Electrical Engineering, 91*. https://doi.org/10.1016/j.compeleceng.2021.107044

28. Sun, X. F., & Lin, X. G. (2017). Random-forest-ensemble-based classification of high-resolution remote sensing images and nDSM over urban areas. In *International Archives of the Photogrammetry, Remote Sensing and Spatial Information Sciences—ISPRS Archives, 42*(2W7), 887–892. https://doi.org/10.5194/isprs-archives-XLII-2-W7-887-2017
29. Wazarkar, S., & Keshavamurthy, B. N. (2018). A survey on image data analysis through clustering techniques for real world applications. *Journal of Visual Communication and Image Representation, 55*, 596–626. https://doi.org/10.1016/j.jvcir.2018.07.009
30. Maeder, M., McCann, N., Clifford, S., & Puxty, G. (2020). *Model-based data fitting* (2nd Ed.., Vol. 3). Elsevier.
31. Rao, S., Verma, A. K., & Bhatia, T. (2021). A review on social spam detection: Challenges, open issues, and future directions. *Expert Systems with Applications, 186*. https://doi.org/10.1016/j.eswa.2021.115742

Real-Time Eyesight Power Prediction Using Deep Learning Methods

Amit Saraswat, Abhijeet Negi, Kushagara Mittal, Brij Bhushan Sharma, and Nimish Kappal

Abstract This paper describes a real-time eyesight power prediction using deep learning methods. Artificial intelligence (AI) based on deep learning algorithmic methods has been widely adopted by researchers in health care for speech recognition, image processing, etc. Similarly, these deep learning methods can be useful in predicting the eyesight of a person. In order to check eyesight, we need refractometer, but due to its high cost, it is very rarely available in rural areas. Also, there are no online means to check your eyesight which leads to more difficulties. So, in this paper, we have proposed a model on the data we have collected from the survey which contained questions set by an ophthalmologist Vikas Saraswat for the prediction of axis, spherical power, cylindrical power, and addition power. These are the essential factors for the detection of eyesight power. Experimental results were shared with graphical representation comparing the traditional methods for the detection with the proposed models.

Keywords Medicine · Artificial intelligence · Eyesight · Deep learning methods · Real-time prediction · Smart health care

1 Introduction

People from both rural and urban area suffer from eyesight problems based on their age, sex, etc. Vision problems (hyperopia and myopia) will damage the optic nerve and also lead to more vision loss. If a person does not take proper initiative, then it may cause serious vision loss in early ages or even blindness in some serious cases [1].

A. Saraswat (✉) · A. Negi · K. Mittal · B. B. Sharma · N. Kappal
Shoolini University, Bhajol, Solan, Himachal Prasdesh 173229, India
e-mail: amitsaraswat71161@gmail.com

B. B. Sharma
e-mail: brijbhushan@shooliniuniversity.com

P. K. Singh et al. (eds.), *Proceedings of Third International Conference on Computing, Communications, and Cyber-Security*, Lecture Notes in Networks and Systems 421,
https://doi.org/10.1007/978-981-19-1142-2_25

Near-sightedness (**myopia**) is a common vision condition in which you can see objects near to you clearly, but objects farther away are blurry. It occurs when the shape of your eye causes light rays to bend (refract) incorrectly, focusing images in front of your retina instead of on your retina. Near-sightedness may develop gradually or rapidly, often worsening during childhood and adolescence. Near-sightedness tends to run in families [2].

Hyperopia (farsightedness) is a refractive error, which means that the eye does not bend or refract light properly to a single focus to see images clearly. In hyperopia, distant objects look somewhat clear, but close objects appear more blurred. People experience hyperopia differently. Some people may not notice any problems with their vision, especially when they are young. For people with significant hyperopia, vision can be blurry for objects at any distance, near or far. It is an eye focusing disorder [3].

Presbyopia is the gradual loss of your eyes' ability to focus on nearby objects. It is a natural, often annoying part of aging. Presbyopia usually becomes noticeable in your early to mid-40s and continues to worsen until around age 65. You may become aware of presbyopia when you start holding books and newspapers at arm's length to be able to read them. A basic eye exam can confirm presbyopia. You can correct the condition with eyeglasses or contact lenses. You might also consider surgery [4].

2 Literature Survey

Literature survey was done to gather the knowledge regarding various type of disease caused due to the weak eyesight problems, e.g., myopia, hyperopia, and presbyopia. In rural areas, people do not have the means to get their eyesight checked due to lack of eyesight clinics. In order to check eyesight, there are no online means to check your eyesight which leads to more difficulties (Table 1).

On the basis of above study, we found that from 1991 to April 2020, no one predict eyesight, but they have their own advantages like:

1. Most studies regarding intelligent diagnosis focused on binary classification for instance to detect AMD, or glaucoma or other retinal disease [11].
2. Out of 3002 children, 457 (15.22%) had defective vision. Myopia affected 418 (91.47%) students, while hyperopia was observed in 21 (4.60%) students; astigmatism was present in 18 (0.04%) [14].
3. Preschool children with vision problems age dynamic of their functional state permitted to establish that these indicators are improving, coinciding age [13].
4. Gabor filter has DC component, so it performs non-uniform coverage of edge segmentation, whereas Log-Gabor filter has no DC component, and it performs accurate edge segmentation [6].

Table 1 Literature survey

S. No.	Title	Authors	Published date	Major findings
1	A study of data mining techniques in glaucoma detection [5]	R. Gomathi et al.	Apr-20	1. To prevent the vision loss of the patient, early detection of glaucoma is very necessary and that can be done by data mining techniques 2. Optical coherence tomography (OCT) is an imaging technique that is used for medical imaging and industrial non-destructive testing
2	Non-invasive diagnosis of eye disease using image segmentation and neural networks [6]	L. Parvathvarthiny et. al	May-2014	1. Restorative determination of the retinal infection takes a long course of time, and at first comparative kind of treatment is carried out, this leads to vision misfortune from the retinal pictures of the locale show that injuries can be predicted 2. Gabor channel has DC component, so it performs non-uniform scope of edge division, while the Log-Gabor channel has no DC component, and it performs precise edge segmentation
3	A general regression neural network [7]	Donald F. Specht	Nov-2019	1. The general regression neural network (GRNN) learns in one pass through the information and can generalize from the case as before long as they are stored 2. The main disadvantage of GRNN relative to other technique is that it requires substantial computations to evaluate new points

(continued)

Table 1 (continued)

S. No.	Title	Authors	Published date	Major findings
4	Artificial intelligence and deep learning in ophthalmology [8]	Daniel Shu Wei Ting et. al	Oct-2018	1. To move forward a clinical acknowledgment of DL frameworks, it is imperative to disentangle the "Black Box" nature of DL utilizing existing and future methodology
5	Development and verification of ANN classifiers for eye diseases diagnosis [9]	Hossein Parsaei et. al	2008	1. Artificial neural network systems can learn distinctive designs of visual field misfortune. In common, the result shows that all these classifiers perform way better than the worldwide indices 2. ANN is utilized for eye infection determination and result approval on standard robotized perimetry information. The ANN in common beats the execution of the worldwide files of STATPAC files as measured by CCN and PCCN
6	Promising AI-ML- DL algorithms in ophthalmology [10]	Lokman Balyen	Aug-19	1. Robotized retinal imaging innovations may diminish the boundaries to get to wellbeing care frameworks and wellbeing screening

(continued)

Table 1 (continued)

S. No.	Title	Authors	Published date	Major findings
7	Applications of AI in ophthalmology: general overview [11]	Wei Lu et. al	Nov 2018	1. Most thinks about concerning cleverly determination centered on double classification for an occasion to identify AMD we come up short to consider a quiet with glaucoma. Choi and his colleagues carried out the work applying DL to consequently identify different distinctive retinal diseases 2. The Tall Reliance on the information quality ought to be considered; diverse imaging gadget, different imaging conventions, and inherent commotion of information can influence data's quality which may have colossal impacts to demonstrate performance
8	The Research of regression model in machine learning field [12]	Shen Rong et al	2018	1. The paper herein introduces a linear regression model used to analyze the sale of the iced product of the company
9	The Investigation of the level of preschool children with eye sights problem functional state [13]	Sedava O.O	April-2003	1. Preschool children with vision issues useful state concerning their sex; it was found that boy's pointers are progressing coinciding age 2. Introductory information of preschool children with vision issues age energetic of their utilitarian state allowed to set up that these markers are making strides, coinciding age

(continued)

Table 1 (continued)

S. No.	Title	Authors	Published date	Major findings
10	Spectrum of visual impairment among female school students of Surat [14]	Mausumi Basu et al.	Dec-2011	1. Authors have conducted the test on 3002 children, out of which 457 children are found to be suffering from the defective vision. 418 were found to have myopia, whereas the hyperopia was found in 21 children. Myopia had a prevalence of 13.93%, and this is noted during the test that astigmatism and myopia were higher the particular age group
11	The prevalence of eyesight deterioration in people aged over 50 years and its correlation with Type II Diabetes in Trinidad [15]	B. Shivnanda Nayak	Sep-2019	1. This paper authors have decided that there is, to a degree, a relationship between the frequency of vision weakening and sort II diabetes in individuals matured over 50 years
12	Physical inactivity in relation to self- rated eye sight [16]	Lee Smith et. al	Dec-2016	1. A add up to of 16.2, 35.6, 37.7, and 10.6% of the test evaluated their vision as fabulous, exceptionally great, great, and reasonable destitute, separately. Those with fair–poor self-rated vision were more seasoned (cruel 67.8 a long time) and more likely to be female (59.6%)

3 Proposed Model

Firstly, we split the data into two categories, left eye data and right eye data. After splitting, we took care of the null value data in our dataset by introducing value of the vision parameters as seven and non-vision parameters as 0. Then, the data was down casted from string to float. And the remaining data was encoded using one-hot encoder.

We created four models for the same.

1. Axis prediction model
2. Spherical power prediction model
3. Cylindrical power prediction model
4. Addition power prediction model.

4 Flow Diagram

See Fig. 1.

5 Experimental Study

Data Collection: We started our data collection by distributing hard copies of forms which contains questions related to lens power to the eye clinic centers and the other rural areas. This allowed us to get majority of the data which was used in machine learning. Another source for data collection was using online forms such as Google Forms. Google Form helped us to spread our data collection area.

Size of dataset—300.

6 Model for Axis

The model for axis is as follows: We applied sequential model on our axis model [17]; after that, we added densely connected with eight and 2670 units to the model with the activation function "ReLU" with one input unit.

The last layer is linear layer with one output unit (Fig. 2).

Testing of Model

In Fig. 3,

Pred* = reading from proposed model

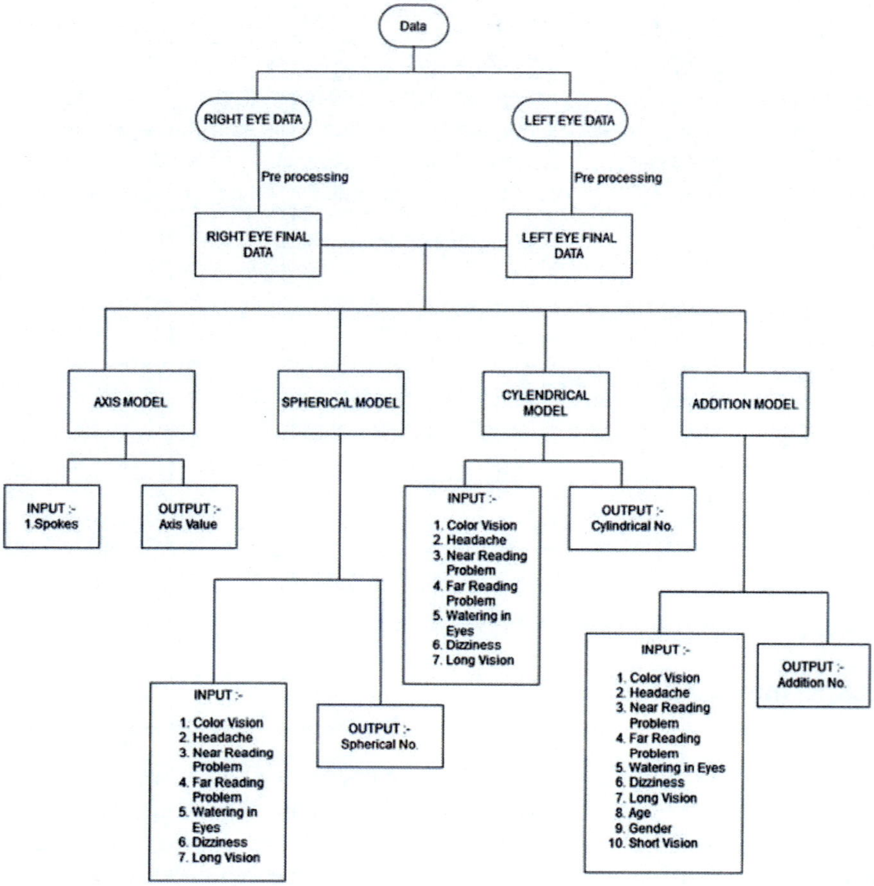

Fig. 1 Flow diagram for the proposed models

Real* = reading from the traditional method of testing.

Graphical Representation

See Fig. 4.

7 Model for Spherical Power

After axis next, we make a model for spherical power prediction. The model for spherical power is as follows:

```
In [24]: x = data.Spokes_Right
         y = data.Right_Eye_Prescription_Axis

In [25]: from sklearn.model_selection import train_test_split
         X_train, X_test, y_train, y_test = train_test_split(x, y, test_size = 0.2, random_state = 0)

In [26]: model_axis_r = Sequential()
         model_axis_r.add(Dense(8, input_dim=1, kernel_initializer='normal', activation='relu'))
         model_axis_r.add(Dense(2670, activation='relu'))
         model_axis_r.add(Dense(1, activation='linear'))
         model_axis_r.summary()

         Model: "sequential_1"
         _________________________________________________________________
         Layer (type)                 Output Shape              Param #
         =================================================================
         dense_3 (Dense)              (None, 8)                 16
         _________________________________________________________________
         dense_4 (Dense)              (None, 2670)              24030
         _________________________________________________________________
         dense_5 (Dense)              (None, 1)                 2671
         =================================================================
         Total params: 26,717
         Trainable params: 26,717
         Non-trainable params: 0
         _________________________________________________________________

In [27]: model_axis_r.compile(loss='mse', optimizer='adam', metrics=['accuracy','mse','mae'])

In [28]: model_axis_r.fit(X_train, y_train,epochs=900,batch_size=10)
```

Fig. 2 Proposed model for axis

	A	B	C	D	E
1	Name	R.Axis(Pred)	R.Axis (real)	L.Axis(Pred)	L.Axis(Real)
2	Vikas	89	90	120	120
3	Sahil	146	147	0	0
4	Rakesh	0	0	0	0
5	Ram Dhoc	170	170	27	27
6	Lata Saras	170	170	170	170
7	Parth	0	0	0	0
8	Kamal	0	0	0	0
9	Birender S	0	0	0	0

Fig. 3 Test results for the axis model with traditional methods

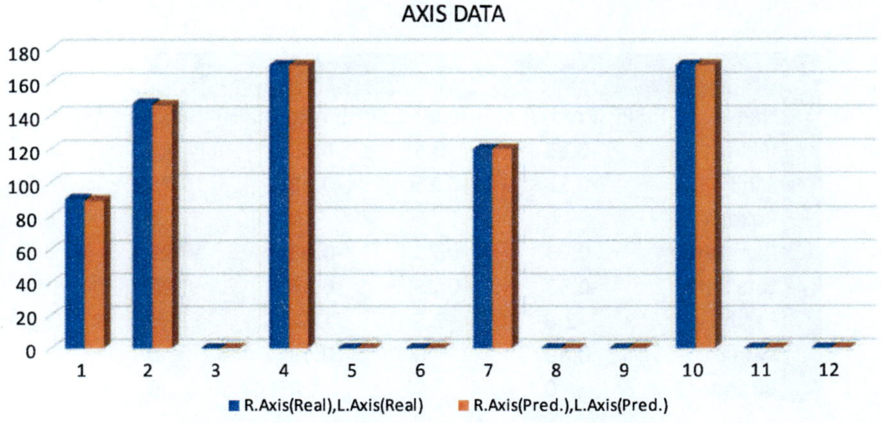

Fig. 4 Graphical comparison in between the proposed model and traditional testing methods

We applied sequential model on our spherical model; after that, we added densely connected with eight and 2670 units to the model with the activation function "ReLU" with nine input units.

The last layer is linear layer with one output unit (Fig. 5).

Testing of Model

In Fig. 6,

Pred* = reading from proposed model

Real* = Reading from the traditional method of testing.

Graphical Representation:

See Fig. 7

```
[3]: x = data[['Colour_Vision_Right_Both','Colour_Vision_Right_Green','Colour_Vision_Right_Red','Headache','Near Reading Problem','Fa
     y = data["Right_Eye_Prescription_Sph"]

In [30]: model = Sequential()
         model.add(Dense(8, input_dim=9, kernel_initializer='normal', activation='relu'))
         model.add(Dense(2670, activation='relu'))
         model.add(Dense(1, activation='linear'))
         model.summary()

         Model: "sequential_2"
         _________________________________________________________________
         Layer (type)                 Output Shape              Param #
         =================================================================
         dense_7 (Dense)              (None, 8)                 80
         _________________________________________________________________
         dense_8 (Dense)              (None, 2670)              24030
         _________________________________________________________________
         dense_9 (Dense)              (None, 1)                 2671
         =================================================================
         Total params: 26,781
         Trainable params: 26,781
         Non-trainable params: 0
         _________________________________________________________________

In [31]: model.compile(loss='mse', optimizer='adam', metrics=['accuracy','mse','mae'])

In [39]: model.fit(X_train, y_train,epochs=30000,batch_size=8)
```

Fig. 5 Proposed model for spherical power prediction

	A	B	C	D	E
1	Name	R.SPH(Pred)	R.SPH(Real)	L.SPH(Pred)	L.SPH(Real)
2	Vikas	-0.35	-0.5	-0.35	-0.5
3	Sahil	-0.18	-0.25	-0.18	-0.25
4	Rakesh	0	0	0	0
5	Ram Dhoo	-0.63	-0.75	-0.63	-0.75
6	Lata Saras	-0.33	-0.25	-0.33	-0.25
7	Parth	-2.6	-2.5	-1.41	-1.5
8	Kamal	0.49	0	0.49	0
9	Birender S	0	0	0	0

Fig. 6 Test results for the spherical power prediction with traditional methods

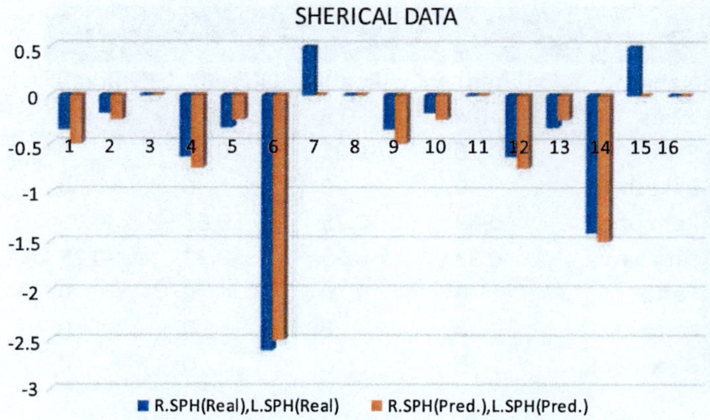

Fig. 7 Graphical comparison in between the proposed model and traditional testing methods

8 Model for Cylindrical Power

After spherical power, next step is to create a model for cylindrical power prediction. The model for cylindrical power is as follows:

We applied sequential model on our cylindrical model; after that, we added densely connected with eight and 2670 units to the model with the activation function "ReLU" with night input units.

The last layer is linear layer with one output unit (Fig. 8).

```
In [16]: from sklearn.model_selection import train_test_split
         X_train, X_test, y_train, y_test = train_test_split(x, y, test_size = 0.25, random_state = 42)

In [17]: import tensorflow
         from tensorflow.python.keras.layers import Dense
         from tensorflow.keras.layers import Dropout
         from tensorflow.python.keras.models import Sequential
         from sklearn.metrics import mean_absolute_error
         from sklearn.metrics import mean_squared_error
         from sklearn.model_selection import train_test_split

In [18]: model = Sequential()
         model.add(Dense(8, input_dim=9, kernel_initializer='normal', activation='relu'))
         model.add(Dense(2670, activation='relu'))
         model.add(Dense(1, activation='linear'))
         model.summary()

         Model: "sequential_1"
         _________________________________________________________________
         Layer (type)                 Output Shape              Param #
         =================================================================
         dense_3 (Dense)              (None, 8)                 80
         _________________________________________________________________
         dense_4 (Dense)              (None, 2670)              24030
         _________________________________________________________________
         dense_5 (Dense)              (None, 1)                 2671
         =================================================================
         Total params: 26,781
         Trainable params: 26,781
         Non-trainable params: 0
         _________________________________________________________________

In [19]: model.compile(loss='mse', optimizer='adam', metrics=['accuracy','mse','mae'])

In [24]: model.fit(X_train, y_train,epochs=15000,batch_size=5)
```

Fig. 8 Proposed model for cylindrical power prediction

	A	B	C	D	E
1	Name	R.Cyl(Pred)	R.Cyl(Real)	L.Cyl(Pred)	L.Cyl(Real)
2	Vikas	-0.4	-0.5	-0.4	-0.5
3	Sahil	-0.18	-0.25	-0.01	0
4	Rakesh	0	0	0	0
5	Ram Dhoc	-0.63	-0.75	-0.63	-0.75
6	Lata Saras	-0.33	-0.25	-0.33	-0.25
7	Parth	0	0	0	0
8	Kamal	0	0	0	0
9	Birender S	0	0	0	0

Fig. 9 Test results for the cylindrical power prediction with traditional methods

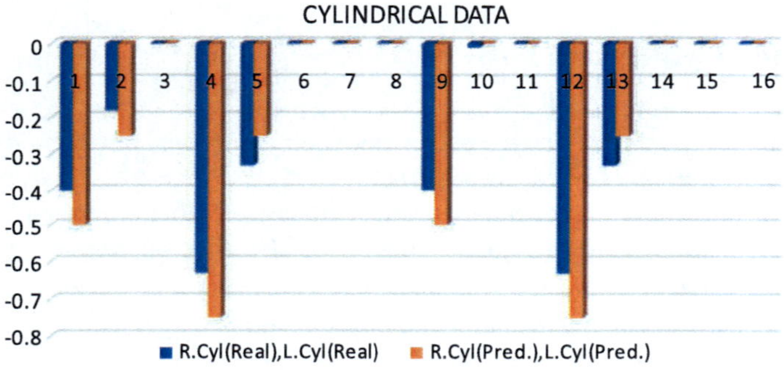

Fig. 10 Graphical comparison in between the proposed model and traditional testing methods

Testing of Model

In Fig. 9,

Pred* = reading from proposed model

Real* = reading from the traditional method of testing.

Graphical Representation

See Fig. 10.

9 Model for Addition Power (Presbyopia)

After cylindrical power, next step is to create a model for addition power. The model for addition power is as follows:

We applied sequential model on our addition model; after that, we added densely connected with eight and 2670 units to the model with the activation function "ReLU" with 12 input units.

The last layer is linear layer with one output unit (Fig. 11).

Testing of Model

In Fig. 12,

Pred* = reading from proposed model

Real* = reading from the traditional method of testing.

```
In [4]: from sklearn.model_selection import train_test_split
        X_train, X_test, y_train, y_test = train_test_split(x, y, test_size = 0.25, random_state = 42)

In [5]: import tensorflow
        from tensorflow.python.keras.layers import Dense
        from tensorflow.keras.layers import Dropout
        from tensorflow.python.keras.models import Sequential
        from sklearn.metrics import mean_absolute_error
        from sklearn.metrics import mean_squared_error
        from sklearn.model_selection import train_test_split

In [6]: model = Sequential()
        model.add(Dense(8, input_dim=12, kernel_initializer='normal', activation='relu'))
        model.add(Dense(2670, activation='relu'))
        model.add(Dense(1, activation='linear'))
        model.summary()

        Model: "sequential"
        _________________________________________________________________
        Layer (type)                 Output Shape              Param #
        =================================================================
        dense (Dense)                (None, 8)                 104
        _________________________________________________________________
        dense_1 (Dense)              (None, 2670)              24030
        _________________________________________________________________
        dense_2 (Dense)              (None, 1)                 2671
        =================================================================
        Total params: 26,805
        Trainable params: 26,805
        Non-trainable params: 0
        _________________________________________________________________

In [7]: model.compile(loss='mse', optimizer='adam', metrics=['accuracy','mse','mae'])

In [12]: model.fit(X_train, y_train,epochs=15000,batch_size=5)
```

Fig. 11 Proposed model for addition power prediction

	A	B	C	D	E
1	Name	R.Add(Pred)	R.Add(Pred)	L.Add(Pred)	L.Add(Pred)
2	Vikas	0	0	0	0
3	Sahil	0	0	0	0
4	Rakesh	1.3	1.5	1.3	1.5
5	Ram Dhoo	1.6	1.5	1.6	1.5
6	Lata Saras	0	0	0	0
7	Parth	0	0	0	0
8	Kamal	1.9	2	1.9	2
9	Birender S	1.35	1.25	1.35	1.25

Fig. 12 Test results for the addition power prediction with traditional methods

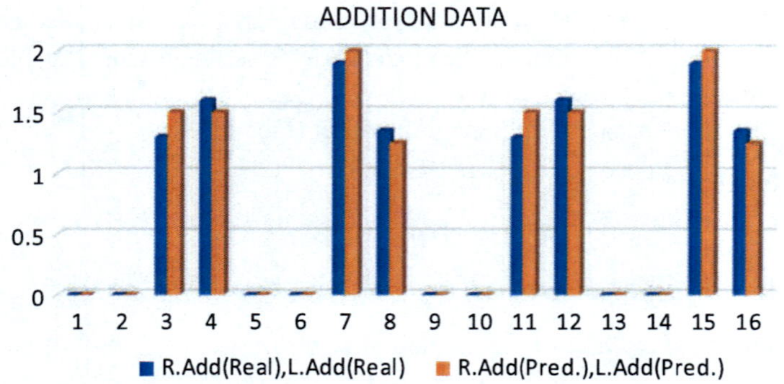

Fig. 13 Graphical comparison in between the proposed model and traditional testing methods

Graphical Representation

See Fig. 13.

10 Conclusion

To prevent the vision loss or serious vision problem of the patient, early detection of myopia, hyperopia, and presbyopia is very necessary in healthcare department. Our web application is a part of DL application, has been increased with the accuracy of predicting disease, and gives more importance to these techniques. DL techniques in the existing are to detect the disease as earlier, so it will avoid vision blindness. The mentioned attribute can conclude that a person can affect or affected or how much affected by these diseases or not. Using data and deep learning with effective results is major advantage to predict vision disease in early stage to reduce the chance of risky level.

Acknowledgements I cannot express my thanks to Dr. Vikas Saraswat and Dr. Lata Saraswat for their continued support and encouragement. My completion of this research cannot be done without their support. He has given me his clinic Parth Opticals (Pratap Vihar, Ghaziabad, UP) to test the model. This research cannot be done without his support.

References

1. Upadhyay, S. (2015). Myopia, hyperopia and astigmatism: A complete review with view of differentiation. *International Journal of Science and Research, 8*(4), 2319–7064 (Online Index Copernicus Value Impact Factor, vol. 4, no. 8, pp. 2319–7064.
2. Wolffsohnm, J. S. et al. (2019). IMI—Myopia control reports overview and introduction. *Investigative Ophthalmology and Visual Science, 60*, M1–M19.

3. Castagno, V. D., Fassa, A. G., Carret, M. L. V., Vilela, M. A. P., & Meucci, R. D. (2014). Hyperopia: A meta-analysis of prevalence and a review of associated factors among school-aged children. *BMC Ophthalmology, 14*(1) (2014).
4. Patel, I., & West, S. K. (2007). Presbyopia : prevalence , impact , and *Interventions. Commun Eye Health, 20*(63), 51–52.
5. Gomathi, R., Ramprashath, R, A. Gokulraja, A. Bharath, M., & Vishnu, K. S. H. (2020) A study of data mining techniques in glaucoma detection. *IJIRT, 6*(11), pp. 75–78.
6. Parvathavarthiny, L., & Batmavady, S. (2014). Non-invasive diagnosis of eye diseases using image segmentation and neural networks. *International Journal of Advanced research in Compuetr Engineering. Technology, 3*(5), 1651–1655.
7. Gholamrezaei, M., & Ghorbanian, K. (2007). Rotated general regression neural network.In *IEEE International Conference on Neural Networks* (Vol. 2, no. 6, pp. 1959–1964).
8. Ting, D. S. W., et al. (2019). Artificial intelligence and deep learning in ophthalmology. *British Journal of Ophthalmology, 103*(2), 167–175.
9. Parsaei, H., Moradi, M. H., & Parsaei, R. (2008). Development and verification of artificial neural network classifiers for eye diseases diagnosis (pp. 1–5).
10. Balyen, L., & Peto, T. (2019). Promising artificial intelligence–machine learning–deep learning algorithms in ophthalmology. *Asia-Pacific Journal of Ophthalmology, 8*(3), 264–272.
11. Lu, W., Tong, Y., Yu, Y., Xing, Y., Chen, C., & Shen, Y. (2018). Applications of artificial intelligence in ophthalmology: general overview. *Journal of Ophthalmology.*
12. Rong, S., & Bao-Wen, Z. (2018). The research of regression model in machine learning field. *MATEC Web Conference, 176*, 8–11.
13. S. O.O. (2013). The investigation of the level of preschool children with eyesight problems functional state. Physical Education of Students, 54–58.
14. Basu, M., Das, P., Pal, R., Kar, S., Desai, V. K., & Kavishwar, A. (2011). Spectrum of visual impairment among urban female school students of Surat. *Indian Journal of Ophthalmology, 59*(6), 475–479.
15. *Journal of Endocrinology and Metabolism, 9*(1–2), 9225.
16. Smith, L., Timmis, M. A., Pardhan, S., Latham, K., Johnstone, J., & Hamer, M. (2017). Physical inactivity in relation to self-rated eyesight: Cross-sectional analysis from the english longitudinal study of ageing. *BMJ Open Ophthalmology, 1*(1), 1–5.
17. Salehi, A. W. (2020). A CNN model: Earlier diagnosis and classification of Alzheimer Disease using MRI (pp. 156–161).

An Unsupervised Machine Learning Approach to Prediction of Price for Taxi Rides

Ankit Kumar, Kunal Jani, Abhishek Kumar Jishu, Visaj Nirav Shah, Kushagra Pathak, and Manish Khare

Abstract Taxi services are the primary method of transportation in urban areas. With the advent of technological sophistication and digital innovation used by companies like Uber and Ola, taxi businesses are undergoing a rapid transformation. Various methods have been developed by product engineers of software companies in the past, but they did not consider the demand for a customer's ride in a particular region. In this paper, a machine learning-based model has been proposed having the capability to automatically classify booking points into different areas based on optimizing the within-cluster sum of squared distances to estimate the taxi demand in different geographical zones of a city. A robust and accurate price prediction model has been developed which would assist in predicting the price of rides from one fixed location to another fixed location based on the time and location of booking.

Keywords Unsupervised learning · K-means clustering · Price prediction · Taxi rides

1 Introduction

Ride-hailing apps for everyday trips have become widely popular with the rise in demand and ease of technology of starting such an application [14]. Today, the market is saturated with large multinational companies like Uber and Ola to small start-ups which aim to provide service in a local region [23]. The technology to run such a service has been around for quite some time now, but it is still evolving rapidly

A. Kumar · A. K. Jishu · V. N. Shah (✉) · K. Pathak · M. Khare
DAIICT, Gandhinagar, India
e-mail: 201801016@daiict.ac.in

M. Khare
e-mail: mkharejk@gmail.com

K. Jani
Santa Clara University, Santa Clara, CA, USA
e-mail: kjani@scu.edu

P. K. Singh et al. (eds.), *Proceedings of Third International Conference on Computing, Communications, and Cyber-Security*, Lecture Notes in Networks and Systems 421,
https://doi.org/10.1007/978-981-19-1142-2_26

to accommodate the local factors and uncertainties [9]. One of the most important factors that these companies need to consider is the price determination of each ride and fluctuations in the same [18].

With the advent of machine learning, we have been able to mitigate the problems of predictions to some degree. The algorithms used by services like Ola and Uber for price prediction for any ride are fundamentally based on machine learning in one form or another [20]. Machine learning is a part of decision sciences that used past data to predict the future output by 'learning' the pattern (function) [6]. It helps ride-hailing apps predict the price based on various parameters like distance, time of the day, current demand, number of drivers in the vicinity. We need to understand how it is used so that we can further develop and improvise the system.

Usually, such an analysis leads us to find supervised machine learning used in research and practical applications [5]. We propose an unsupervised machine learning algorithm to predict the most profitable trips in terms of price [2]. In our problem, unsupervised learning has an advantage over supervised learning because the location of cab ride booking is not assigned to predetermined categories. Therefore, it is necessary to use unsupervised learning since our data is not labeled. Our proposed algorithm used K-means clustering to tackle this problem [13]. The main objective of this research is to use an efficient and robust method of unsupervised learning to predict the price of cab rides. The price of cab rides should be optimal and should be beneficial for both the customer and the driver.

2 Related Works

Altusher et al. [1] worked on the dynamics of ride sharing while traveling in cabs. The main idea was to see the stability of ride-sharing utilization over a long period of time. It involved modeling the ride-sharing utilization using the known. Using the New York Taxi dataset modeling, a ride-sharing utilization was concisely modeled, and it could be shown that the ride-sharing process is highly dynamic.

Ota et al. [19] worked on infrastructure for ride sharing that was simulated using a computer-based system. Numerous practical scenarios could be tracked and developed. It was also possible to monitor its success using a large number of parameters and stakeholders' constraints and demands with the use of a linear optimization algorithm.

Santi et al. [22] used a shareability network so that it was possible to measure the effectiveness of sharing of taxis using a GPS dataset. It was concluded computational traceability was possible when there was taxi sharing between two people, while sharing more than two people was not feasible.

Liu et al. [15] investigated hotspot taxi demand predictions and suggested CFM, RFM, and RRM. Taxi demand hotspots were identified, and a model was built to predict taxi demand based on the hotspots with the use of time, environmental, and meteorological factors.

A realistic method was proposed by Yang et al. [25] in this paper to describe taxi movements and search behavior for taxis. The taxi service models have also been validated and calibrated so that they are used for practical applications [18]. This paper is able to present an explanation of the research that the authors have carried out for the development of equilibrium models in networks, and solutions have been proposed for taxi services [2].

Markou et al. [16] worked on an algorithm that was based on knowledge about various activities in social life using numerous techniques like aggregation and natural language processing. The different stages of gathering of data, enrichment of data, and prediction for the purpose of generation of queries for searching were used to propose a framework [8].

Gholami et al. [10] conducted a study of bus and taxi services based on the varying demand of economic conditions and density. Since taxis can be used more frequently compared to buses, they are cost-friendly when used in the transit fleet [12]. Therefore, it would be possible in this case to direct taxis to areas having a low population density.

3 Motivation

Cab price prediction has been a machine learning problem for a long time now. One of the most popular problems is the New York City taxi fare price prediction [21]. Various attempts at tackling such problems have been made using different machine learning techniques and considering different types of data and parameters. The data collection methods are varied and diverse in terms of sources and consolidation. For example, some papers attempt to solve the problem using data collected via satellite monitoring, some directly use Google Maps API to calculate distances, or some use pre-existing databases for data collection [24].

The majority of the past work has been based around using supervised learning algorithms like regression algorithms, random forest, support vector machines, and others. Supervised learning is beneficial when the target variable is known. Regression algorithms are further subdivided into different approaches like linear regression, polynomial regression, and ridge regressions. Results of each approach are analyzed, and the best one is used depending on the data being considered. Selecting the right parameters are done using exploratory data analysis (EDA), correlation factors, the value of each parameter depending on the city or area [17].

Very few attempts have been made to study this problem using unsupervised learning algorithms. One approach of unsupervised learning relies on clusters of neighborhoods and the division of time hours into classes. Another important aspect found in past works is the lack of correlation factors in using unsupervised learning algorithms because there is no target value available. Since very little work has been done using unsupervised learning, this is an almost unexplored area. With an intention of exploring this domain, we are using K-means clustering to predict fares.

4 Methodology

K-means clustering is an unsupervised machine learning algorithm that classifies unlabeled data points into the most appropriate clusters [4]. These clusters denote the different self-generated categories that the objects can be classified into based on the parameters. The value of k is decided by the user. All the data points belonging to a particular cluster are supposed to denote similar qualities or can be grouped together because of their similar qualities. This is especially useful when we plan to make a collective decision for a particular cluster.

For selecting the centroids in K-means clustering, here, we have used the k++ algorithm. In this algorithm, the first centroid is randomly assigned to a data point. The next $k - 1$ centroids are selected from the remaining data points based on the probability proportional to the square of the distance of the point from the nearest centroid [7].

One of the issues that arise is the value of k which is to be chosen and the perfect number of categories to get the best accuracy [3]. One mathematical way of determining the best k is the elbow method. In the elbow method, we run the K-means clustering algorithm for a fixed range of k (here [1, 18]) and find the value of k depending on the change in slope of the curve.

For our dataset, from the elbow method graph, we select $k = 15$. Using the in-built K-means clustering module of the scikit-learn library, it is possible to obtain the required clusters. Each cluster denotes a unique category with points in a single cluster having similar characteristics.

The prices are dynamically calculated using the distance of the point from the centroid point and the time of the day. The exact formula is shown below. Based on these predictions, the drivers can decide which ride to accept to ensure that their profits are maximum [3].

5 Results

For visualizing the results, the locale.ai [11] dataset for data science was used for our experiments. It consists of 43,431 rows, where each row represents a booking done by a customer and 19 columns, where each columns represents a booking attribute. The attributes representing the location and the time of booking have been used for our experiments.

From Fig. 1, it can be observed that the required change in slope is obtained at $k = 15$. The graph shows the values of within-cluster sum of values (WCSS) on Y-axis and the number of cluster (k) on the X-axis. Off all the integral values of k in the range [1, 18], we find $k = 15$ to be the most suitable. This means that the city is divided into 15 areas for the purpose of determining which areas have a greater number of cab rides booked and which areas do not have a greater number of cab rides.

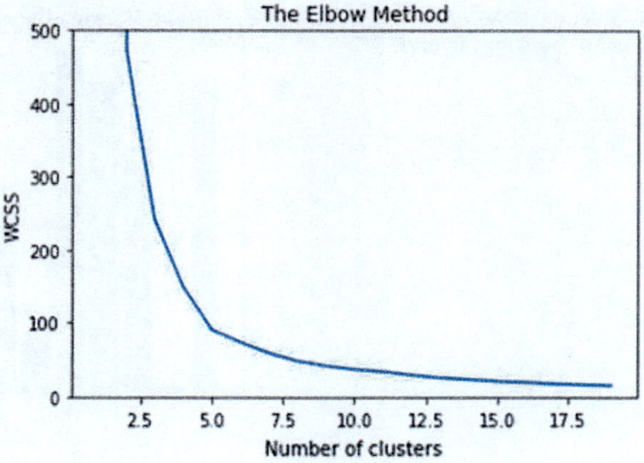

Fig. 1 Graph showing the variation of within-cluster sum of square distance with the number of clusters

The set of data points that are obtained from the dataset is divided into 15 clusters, which represent 15 different areas of the city that are generated from the K-means clustering model. Figure 2 shows each customer point and the cluster it belongs to. The graph is color-coded with each color representing a single cluster representing a unique area.

The *Y*-axis represents the number of bookings, and the *X*-axis denotes the area number. We calculate the number of bookings in each cluster to identify the popular regions. The areas with more bookings can be said to have more users and a higher frequency of trips which can mean more business. From Fig. 3, one can deduce that Area 4 has the maximum number of bookings, whereas Area 1 has the minimum number of bookings.

After considering the number of bookings areawise, we look at the number of bookings based on time of the day. There are four categories on the *X*-axis: morning, afternoon, evening, and night. Maximum bookings occur in the morning which makes sense primarily because the people are generally the busiest during this time. The

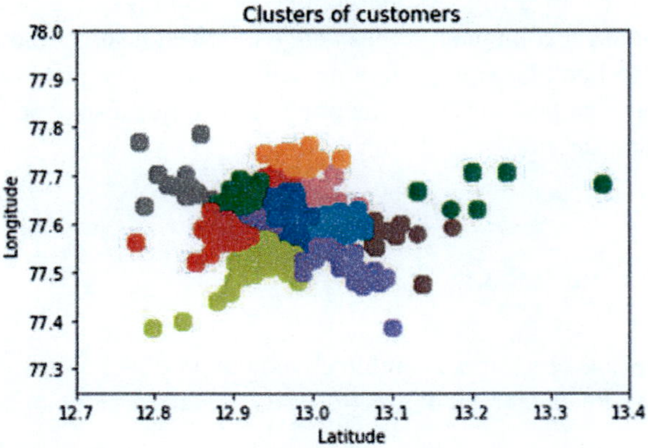

Fig. 2 Graph showing the location of bookings classified into clusters

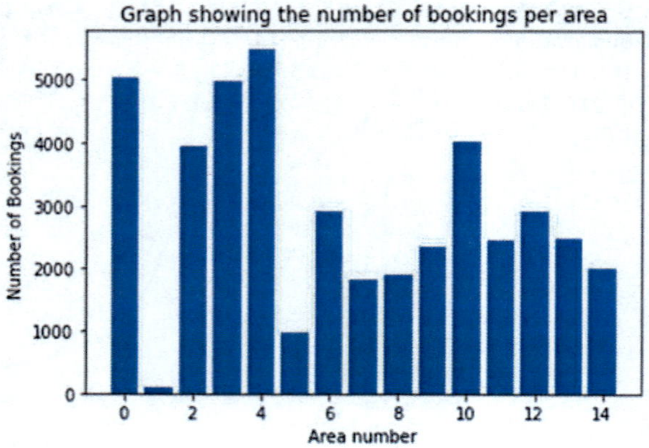

Fig. 3 Graph showing the number of bookings in each area

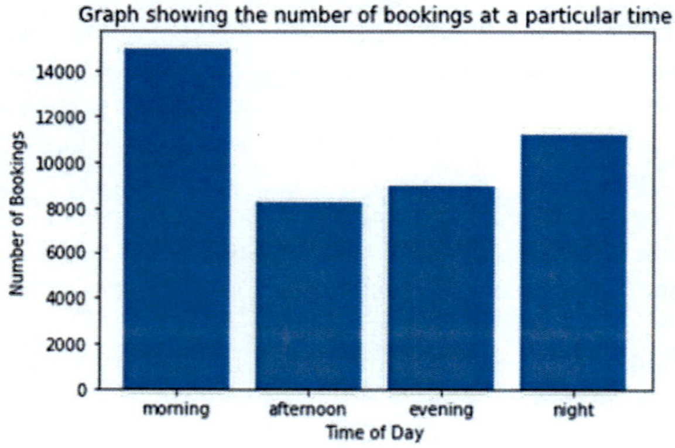

Fig. 4 Graph showing the number of bookings made within a particular time of the day

least takes place in the afternoon since people do not actively travel during this time of the day. This can be seen in Fig. 4.

The formula used to calculate the prices is accurate in determining dynamically the price of a trip. As the number of bookings in a particular area increases, our formula accommodates the change to reflect a surge in prices. Similarly, if the time of the booking is in a more demanding time, the prices shoot up, as shown by the dataset and the formula. The price of a cab ride also depends on the distance traveled during the ride.

6 Conclusion

K-means clustering is an effective method of using unsupervised machine learning to predict the fare. It helps us classify the different regions of a large city into clusters which can be used to make business decisions. Since the areas are classified on the most important parameters, we can make collective predictions for an entire area.

All the required data are available in the given dataset. Some pre-processing in calculating the time of the day is done for proper classification, which yields exactly the kind of data that is needed for such an algorithm.

The time of booking, area of booking, and distance of the trip are considered in determining the dynamic price. Our formula correctly accounts for the expected change with each of the factors. This shows that unsupervised machine learning can be effectively used to dynamically predict fare prices.

References

1. Altshuler, T., Katoshevski, R., & Shiftan, Y. (2017*). Ride sharing and dynamic networks analysis* (2017). arXiv preprint arXiv:1706.00581
2. Barlow, H. B. (1989). Unsupervised learning. *Neural Computation, 1*(3), 295–311.
3. Bholowalia, P., & Kumar, A. (2014). Ebk-means: A clustering technique based on elbow method and k-means in wsn. *International Journal of Computer Applications, 105*(9).
4. Bradley, P. S., & Fayyad, U. M. (1998). Refining initial points for k-means clustering. In *ICML* (Vol. 98, pp. 91–99). Citeseer.
5. Caruana, R., & Niculescu-Mizil, A. (2006). An empirical comparison of supervised learning algorithms. In *Proceedings of the 23rd International Conference on Machine Learning* (pp. 161–168).
6. Charbuty, B., & Abdulazeez, A. (2021). Classification based on decision tree algorithm for machine learning. *Journal of Applied Science and Technology Trends, 2*(01), 20–28.
7. Dahiya, A., & Saluja, K. (2014). Performance analysis of k++ and apriori algorithm in terms of their effectiveness against various diseases. *International Journal of Mechanical Engineering and Information Technology, 2*(06).
8. Duarte-Melo, E. J., & Liu, M. (2003). Data-gathering wireless sensor networks: Organization and capacity. *Computer Networks, 43*(4), 519–537.
9. Fielbaum, A., & Tirachini, A. (2021). The sharing economy and the job market: The case of ride-hailing drivers in Chile. *Transportation, 48*(5), 2235–2261.
10. Gholami, A., & Mohaymany, A. S. (2012). Analogy of fixed route shared taxi (taxi khattee) and bus services under various demand density and economical conditions. *Journal of Advanced Transportation, 46*(2), 177–187.
11. Kumar, H. (2020). Why you shouldn't use zip codes for your hyperlocal last-mile analysis.
12. Lewne, M., Nise, G., Lind, M. L., & Gustavsson, P. (2006). Exposure to particles and nitrogen dioxide among taxi, bus and lorry drivers. *International Archives of Occupational and Environmental Health, 79*(3), 220–226.
13. Likas, A., Vlassis, N., & Verbeek, J. J. (2003). The global k-means clustering algorithm. *Pattern recognition, 36*(2), 451–461.
14. Lim, K., Yeo, S., Goh, M., & Gan, J. (2018). A study on consumer adoption of ride-hailing apps in Malaysia. *Journal of Fundamental and Applied Sciences, 10*, 1132–1142.
15. Liu, Z., Chen, H., Li, Y., & Zhang, Q. (2020). Taxi demand prediction based on a combination forecasting model in hotspots. *Journal of Advanced Transportation.*
16. Markou, I., Kaiser, K., & Pereira, F. C. (2019). Predicting taxi demand hotspots using automated internet search queries. *Transportation Research Part C: Emerging Technologies, 102*, 73–86.
17. Mitchell, T. M., et al. (1997). Machine learning.
18. Nguyen-Phuoc, D. Q., Su, D. N., Tran, P. T. K., Le, D. T. T., & Johnson, L. W. (2020). Factors influencing customer's loyalty towards ride-hailing taxi services—A case study of Vietnam. *Transportation Research Part A: Policy and Practice, 134*, 96–112.

19. Ota, M., Vo, H., Silva, C., & Freire, J. (2016). Stars: Simulating taxi ride sharing at scale. *IEEE Transactions on Big Data, 3*(3), 349–361.
20. Park, B., & Bae, J. K. (2015). Using machine learning algorithms for housing price prediction: The case of fairfax county, Virginia housing data. *Expert Systems with Applications, 42*(6), 2928–2934.
21. Qasem, A. G., & Lam, S. S. (2020). Predicting taxi fare using multilayer perceptron and radial basis function networks: New York city as a case study. In IIE *Annual Conference. Proceedings* (pp. 1–6). Institute of Industrial and Systems Engineers (IISE).
22. Santi, P., Resta, G., Szell, M., Sobolevsky, S., Strogatz, S. H., & Ratti, C. (2014). Quantifying the benefits of vehicle pooling with shareability networks. *Proceedings of the National Academy of Sciences, 111*(37), 13290–13294.
23. Surie, A., & Koduganti, J. (2016). The emerging nature of work in platform economy companies in Bengaluru, India: The case of Uber and Ola cab drivers. *E-Journal of International and Comparative Labour Studies, 5*(3).
24. Svennerberg, G. (2010). *Beginning Google Maps API 3.* Apress.
25. Yang, H., & Wong, S. C. (1998). A network model of urban taxi services. *Transportation Research Part B: Methodological, 32*(4), 235–246.

Facial Landmark Features-Based Face Misclassification Detection System

Aditya Bakshi and Sunanda Gupta

Abstract Issues of face spoofing that can evade the verification system by placing the photo of real user on camera have been discussed a lot in the literature survey. By detecting the person through misclassification, the problem could be minimized. Therefore, in this paper, robust face misclassification detection system is proposed using ABT mechanism. The proposed system provides the additional level of security before face recognition module. Face landmark features such as eye, nose, and mouth movements are used for generating challenges for detecting fake users from genuine users using misclassification. The reliability of system is tested by placing photographs and videos from Replay-Attack database and live database. Proposed system gives good results under spoofing attacks such as eye imposter attack and mouth imposter attack. The results show that system detects the fake user when implemented on all types of attacks and confirms the 79.6% misclassification detection.

Keywords Face recognition · Face landmark features · Face spoofing · Replay-Attack database

1 Introduction

Authentication of any user can be possible using biometric authentication. This is same as humans authenticate the users in their life. For uniquely differentiating one user from another, characteristics like face, fingerprint, voice, gait, etc., and biometrics are used for authentication. But, face recognition system is most often used among all these biometric traits. In the year 1960, face recognition technology had started. The researchers worked a lot on various methods such as different facial

A. Bakshi (✉) · S. Gupta
Department of Computer Science and Engineering, Shri Mata Vaishno Devi University, Katra, Jammu & Kashmir, India
e-mail: addybakshi@gmail.com

S. Gupta
e-mail: sunanda.gupta@smvdu.ac.in

P. K. Singh et al. (eds.), *Proceedings of Third International Conference on Computing, Communications, and Cyber-Security*, Lecture Notes in Networks and Systems 421,
https://doi.org/10.1007/978-981-19-1142-2_27

terms, bad illumination, orientations, and even in fractional obstructions [1] that recognize the fake or genuine person. Also, in the last 50 years, the effort has been done to work in face recognition technology [2]. Access controls, human–robot interaction, surveillance, etc., are the areas that use the face recognition technology. Face recognition system is vulnerable to different kinds of attacks that motivate several researchers for increasing the integrity of system. Spoofing attack [3–5] is the most frequently used attack these days. The main problem of face recognition lies within its working principle. The major problem in face recognition system is the credentials submitted for accessing the system as the system does not check who is accessing the system. Earlier, for detection of spoofing attacks, challenge and response method has been proposed that throws some challenges using eye and mouth movement which is performed by real users not by photographs, and responses are analyzed by given challenges [6]. Most of the researchers use texture information such as edges and roughness for distinguishing between legitimate and illegitimate users. For dealing with spoofing attacks, multimodal approach is used by combining speech and face modalities for detection purposes.

In today's scenario, spoofing techniques are more complex as simple photograph, polymeric face, and fingers, etc., are used for detection purposes. Hence, various approaches have been proposed to deal with the problem of spoofing attacks. Also, the use of landmark features is the best to deal with spoofing attacks as misclassification gives robust results in detecting fake users. Misclassification checks only the enrolled user of the system, and there is no help for detecting the users if good resolution camera or even good classification mechanism can be used. Also, attacker attacks the system more accurately. If you think about the reliability of the system, the accuracy and efficacy are two problems that improve the detection. Detection of the imposter user using proper mechanism can solve this problem.

After covering the literature and various research techniques, a proper security mechanism is the need of an hour to differentiate between legitimate and illegitimate users. So, in this paper, different face landmark features such as eye, nose, and mouth movements are used for the detection of misclassification between the genuine and fake users. The main highlights of the proposed are explained in the paper as follows:

- As per the diverse taxonomy, detailed explanation on results has been done.
- The performance of system is improved in detecting the user under different illumination conditions and the size of the database.

2 Scope of Study

This section provides an insight into previous scientific work that has been done along with the consideration of current and future scenarios of face anti-spoofing detection model. Wang et al. [7] used physiological motion as an efficient and effective face live detection method. Their model worked by estimating an eye contour extraction algorithm from different sequence of videos. The result provided an optimizing fitting procedure using local match. The face recognition system proposed by them was very reliable and successfully distinguished a legitimate face from an illegitimate face. Kose et al. [8] presented a technique to countermeasure against mask attacks by analyzing the reflectance features of real and mask faces. Here, the authors use 2D and 3D face mask attack database for their research project. Using the texture images that were captured by 3D images, a countermeasure against the mask attack was calculated. The accuracy of 94.47% was achieved by implementing the proposed countermeasures. (Dhamecha et al. [9]) Here, authors presented multi-spectrum face images for face verification under disguise variations. Using thermal and visible face images, the proposed framework classifies between biometric and non-biometric classes. The proposed framework improved the performance of the system, but still there is a need for more research for getting better results.

Singh et al. [6] model was based on challenge and response technique for based robust liveness detection. Here, the eye and mouth face macro features were used to observe the user's response by generating random challenges. Except the eye and mouth imposter attack, the experimental results showed that the system could detect the person lively. But, there was a massive change in the facial structure that bypasses the liveness test. The result gave 75% misclassification after the test was conducted on 65 persons from University of Essex face database. Chen et al. [10] proposed an enhanced face authentication model, i.e., EchoFace, against media-based attacks without any need of hardware modification. The proposed model works well in differentiating the uneven structure of the face and the flat forged media as its uses acoustic sensing for robust liveness detection system. The model works robustly under various environmental conditions as efficient reflection profiles have been done for differentiating between forged media and live users. Sun et al. [11] proposed a fully convolutional network (FCN) depth-based face spoofing detection method. Here, two supervision schemes are comprehensively investigated, i.e., global and local label supervisions. An aggregation and FCN part is proposed on the basis of pixel-level local classifiers for spatial aggregation.

Table 1 provides the description of various biometric security approaches in terms of features type, database used, accuracy achieved as well as the highlights, or key finds of each of the approaches.

3 Biometric Anti-spoofing

The biometric community has still not reached a total arrangement on the basic terminologies used despite of current determinations and applications to reach a combined and uniform terminology for susceptibility concepts [12, 13]. Using an artificial fake version acquired through sensor of unique biometric trait, the process of detecting an illegal user as a genuine one that fools a biometric system is called as biometric spoofing. For such case, these types of attacks are called as direct attacks or presentation attacks [14]. Presentation attack is an attack that inhibits the properties of biometric subsystem in a style that inhibit with the planned procedure of the biometric system [12]. Human characteristics such as damaged fingers or traits

Table 1 Different face biometric security approaches

Ref No.	Author (year)	Feature type	Key findings	Database
[7]	Wang et. al. (2009)	Face + eye	(a) Physiological motion-based face live detection method (b) Estimating the results by using eye blinks from an eye contour extraction algorithm and captured video sequence	Yale face database
[8]	Kose et al. (2013)	Face	Detection of mask attacks based on reflectance characteristics of masks and real faces	TABULA RASA (EU Research Project)
[9]	Dhamecha et al. (2013)	Face	Face verification under disguise variations using multi-spectrum	Yale, AR database
[6]	Singh et al. (2014)	Face + Mouth	Challenge and response-based method for liveness detection	Face database from University of Essex
[10]	Chen et al. (2019)	Face	Face authentication using EchoFace	Live database
[11]	Sun et al. (2020)	Face	Local ternary label-based face spoofing detection supervision using convolutional networks	OULU-NPU and SiW datasets

and diverse living characteristic are included in the presentation attacks. Therefore, process of impersonating the new genuine identity using artificial trait is known as spoofing. Numerous methods that are typically considered for spoofing attacks are follows: (i) Verification system: In spoofing, a forged replica of the real user trait is presented to a sensor at the time of authentication. The real template of the genuine user is matched with the acquired artifacts. (ii) Verification system/identification system in closed set: In this, at enrollment stage, spoofing is performed by different user that accesses the system by generating a new identity with an artifact. (iii) Identification system in open set: Using spoofing artifacts, new identities have been created by lookup system which is not found in the watch list.

For automatically differentiating between actual biometric characteristics that have been accessed by sensor and falsely shaped objects comprising a biometric trait, anti-spoofing mechanisms are required for spoofing detection. The cues related to living features of biometric traits can be detected by anti-spoofing methods. Therefore, there is no difference between the liveness detection and all anti-spoofing methods. Even in certain anti-spoofing techniques, detection of other types of presentation attacks is also highly effective. The requirements for satisfying the anti-spoofing methods are follows: (i) Non-invasive: These techniques require an extreme interaction with the user and not harmful too. (ii) User friendly: For interaction purpose, users should not be reluctant in using such techniques. (iii) Fast: User interface with the sensor should be short as results generated in a very reduced lapse of time. (iv) Low cost: If cost is excessively high, use of such techniques cannot be expected. (v) Performance: Performance of the biometric system should not degrade the recognition procedure.

From a common viewpoint, anti-spoofing techniques are divided into three groups for biometric system module which are as follows:

(a) **Sensor-Level Techniques**: Referring the literature, the other name of sensor-level techniques is hardware-based techniques. Some methods, e.g., sweat of fingerprint, eye properties, etc., use sensor as a specific device for detection of living characteristics properties. The three major characteristics of hardware-based approaches are as follows: (i) Physical, electrical, and spectral properties are the intrinsic properties of a living body. (ii) The pulse, blood pressure, perspiration, pupillary unrest, brain wave signals (EEG), or electric heart signals, etc., are examples of involuntary signals of living body. (iii) Challenge–response methods, i.e., in response to an external signal, voluntary (behavioral), or involuntary signals are required based on user interaction. In recent work, it includes multibiometric techniques, although feature-level methods can also be classified. As the detection of fake traits is difficult to detect than an individual trait when it is generated, there is an increase in the robustness of direct attacks after combining different biometrics. Such increase can be detected by multibiometric anti-spoofing. Normally, complementary traits are used in different strategies for maintaining performance and finding vulnerabilities. For low recognition rates, traits robust to spoofing are combined between each other that are vulnerable to spoofing. Therefore, sensor-level group of anti-spoofing methods requires additional hardware acquisition devices. Therefore, an advanced level of security against deceiving attacks is not guaranteed using multi-biometry.

(b) **Feature-Level Techniques**: In literature, the other name of feature-level techniques is software-based techniques. With the help of standard sensor, the fake trait is detected from the sample. In case of sensor-level techniques, features are not extracted from the human body, but features are extracted from the biometric sample for distinguishing between real and fake traits. In feature extraction module, systems are joined after placing on sensor device. As per feature extractor module, all the approaches are combined across sensor devices. Over a period, arrangement of models or work on one case of the biometric trait depends on features-level techniques that can be classified as static and dynamic anti-spoofing methods. Static features are faster and less invasive as it requires less assistance too. Therefore, static features are preferred over dynamic techniques even if there is degradation in performance. In face recognition, the subdivision static and dynamic approaches are used for detection of facial images. Although in sensor-level type of anti-spoofing, multimodality will be considered [15–19]. Face and iris recognition is performed from one single high resolution image. There is no need of any detecting method for feature extractor level in multimodal strategy. Software-based techniques are proficient of sensing other types of illegitimate break-in attempts as they function directly on the sample. For avoiding the system against the inoculation of recreated or artificial models into the channel between the sensor and the feature extractor, feature-level methods are used.

(c) **Score-Level Techniques**: Apart from two types of classification, i.e., software- and hardware-based, a third group of protection method, i.e., score-level technique, has been used for analyzing the fingerprint anti-spoofing. In order to work on fusion approaches that rise the confrontation against fooling attempts, this protection technique uses the score level of biometric systems. Score-level techniques are usually integrated sensor-level and feature-level techniques in matching module as they are designed as additional measures. The combined scores can be calculated from: (i) unimodal biometric modules; (ii) anti-spoofing techniques and unimodal biometric modules; (iii) anti-spoofing modules result.

4 Proposed Approach

Security of the system is the major concern shown by the researchers in today's scenario. As all the work has been done virtually, so for running, a practical mechanism security possesses major problems in the system. To combat such security attacks, biometric classification is one such field that gives vigorous results. There is the number of types and applications in which biometric characteristics exist. The matcher module matches an application of a specific biometric by applying biometric characteristic properties and different application modes. Therefore, every biometric application has its own properties, strengths, and weaknesses.

In proposed approach, detection of fake user using misclassification mechanism is explained. In this, cascade classifier is used for extracting the features from the videos of standard database and live database. In this paper, features are extracted such as eye, nose, and mouth movement from the videos of the database. Here, movement means the eye, nose, and mouth extraction from the video. After that, these extracted features are applied for misclassification detection using threshold information of the valid user. In proposed model, different parameters have been considered for eye and nose extracted features such as eye blinking and eye closity. These eye parameters give an edge for detecting the user using threshold calculation mechanism. Also, mouth has been used as extracted feature for detection purpose. But, it is very difficult to find the displacement of the mouth when the user is smiling, laughing, shouting, and giggling. So, mouth movement does not give accurate results when the threshold has been calculated with the changes in mouth movement. Although, if combined features, i.e., eye, nose, and mouth movement, have been used, the model gives accurate results for misclassification detection.

4.1 ABT Concept

The ABT concept, i.e., average-based threshold, is used for generating probability between the genuine and fake user.

Let the probability be β_1 which can be written as follows:

$$\beta_1 = S(R_1, R_2, R_3 \ldots R_P|\alpha)$$

In another sequence of length R, we drop R_1 and append R_{P+1} in the sequence, generating $R_1, R_2, R_3 \ldots R_{P+1}$ as the new sequence. Let the new probability be β_2

$$\beta_2 = S(R_1, R_2, R_3 \ldots R_{P+1}|\alpha)$$

Let $\Delta\beta = \beta_1 - \beta_2$.

If $\Delta\beta > 0$, it means there is major modification in face spoofing detection and that results in misclassification.

The threshold information is calculated by calculating the genuine user with imposter user with genuine user under different illumination conditions. So, genuineness of the user is detected by misclassification. The proposed block diagram of misclassification detection is shown in Fig. 2.

5 Result and Discussion

The result shows the robustness of our method by comparing our method with different types of attack on the videos of Replay-Attack database. The experimental results have also been checked on the live videos of students of the university.

5.1 Replay-Attack Spoof Databases

In this, a renowned and frequently used public-domain face database, i.e., Replay-Attack database, is explained. The Idiap Replay-Attack database contains 1, 200 video clips of photo and video Replay-Attacks for 50 subjects [6] managed by Idiap research institute. Using the webcam of a MacBook, live face videos and images of subjects were captured. Each subject in Replay-Attack was captured using a Cannon

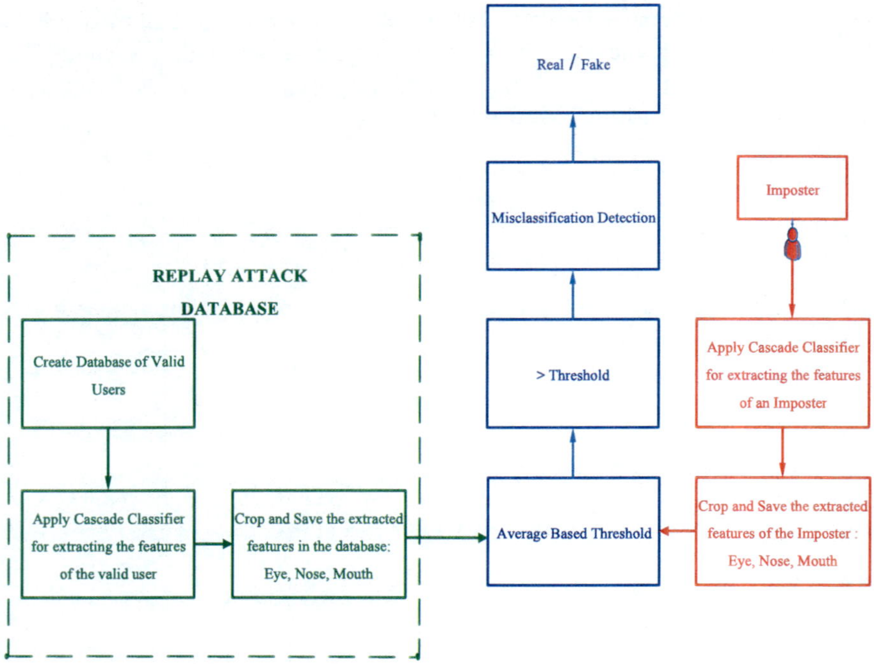

Fig. 2 Proposed block diagram for misclassification detection

PowerShot SX 150 IS camera that records 720p video clips. The high resolution camera of iPhone 3GS (480 × 320 resolution) and iPad 1 (1024 × 68 resolution) are used to capture Replay-Attacks.

In attack generation, different types of spoofing attacks have been considered in this paper.

(a) **Photo Imposter Attack**: Photo imposter attack is an attack in which photo of the real user evades the verification system.
(b) **Eye Imposter Attack**: If eye movement is detected by the system, it could be possible attackers bypass it.
(c) **Mouth Imposter Attack**: In this, attacker eliminates eye region and detects the mouth movement of the user's image. The attacker fools the system using mouth imposter attack.

In verification and recognition module, system blocks because of failure if attacker bypassed the system. There is major change in the result in face, if attacker has to falsify both the eye, nose, and mouth region of the genuine user result in misclassification. By calculating the threshold, system will verify the user from the database. If the value is greater than threshold, then misclassification is detected for real or fake user. The system shows successful authentication if person is already registered in the system. Figure 3 shows the misclassification detection percentage on attacks performed. In this, threshold is calculated on extracted features individually

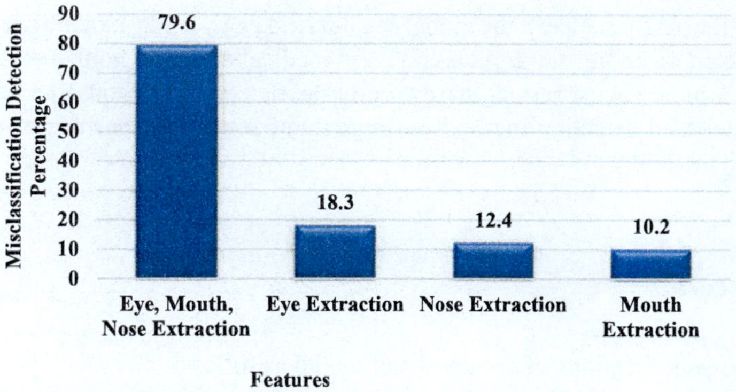

Fig. 3 Misclassification detection percentage

and combination of eye, nose, and mouth for detection purpose. Mouth as extracted feature gives 10.2% misclassification detection, whereas eye and nose features have 18.3 and 12.4%, respectively. But, the combined extracted features, i.e., eye, nose and mouth, give 79.6% misclassification rate after applying on different imposter attacks.

6 Conclusion

The next-generation system technologies have increased the role of biometric drastically. Therefore, securing these systems is the need for an hour. One of the steps toward this objective is fake biometric detection method. In this paper, a misclassification detection model is explained. The proposed method is using a threshold concept for detecting a spoof user. If there is a drastic change in the structure of genuine user's face, then misclassification is detected. We have tested our approach on the videos of Replay-Attack database. The system successfully identified all the fake users and has shown a good accuracy ratio.

7 Future Scope and Applications

After seeing the results and developments from proposed model, there are lot of future directions that can help the authors in further enhancing the work.

Firstly, even though an effective misclassification model is executed on live images samples and Replay-Attack live videos that produces good result of the 500 students that produce great experimental results. But results will be checked and compared by using other databases and approaches, respectively.

Secondly, the performance of the model can be checked by using authentic security mechanisms and machine learning classifiers in medical or health fields.

Finally, for more robust results, hardware-based detection methods can be used with software-based detection methods as large number of variations can be traced using sensing methods.

7.1 Applications

There are many applications of the proposed model as follows:

- Medical Field: During the insurance claims, the difficulties of fake bills can be resolved using facial recognition mechanisms. This will greatly help the patients for smooth functioning of whole process.
- Banking: For secure transactions, banks rely on the facial recognition system that helps the customers against fraud preventions.
- Security: Identifying management systems and video security are certain examples that help in detecting the fake or real users.

References

1. Zhao, W., Chellappa, R., Phillips, P. J., & Rosenfeld, A. (2003). Face recognition: A literature survey. *ACM Computing Surveys (CSUR), 35*(4), 399–458.
2. Abate, A. F., Nappi, M., Riccio, D., & Sabatino, G. (2007). 2D and 3D face recognition: A survey. *Pattern Recognition Letters, 28*(14), 1885–1906.
3. Schuckers, S. A. (2002). Spoofing and anti-spoofing measures. *Information Security Technical Report, 7*(4), 56–62.
4. Nixon, K. A., Aimale, V., & Rowe, R. K. (2008). Spoof detection schemes. In *Handbook of biometrics* (pp. 403–423). Springer.
5. Biggio, B., Akhtar, Z., Fumera, G., Marcialis, G. L., & Roli, F. (2012). Security evaluation of biometric authentication systems under real spoofing attacks. *IET Biometrics, 1*(1), 11–24.
6. Singh, A. K., Joshi, P., & Nandi, G. C. (2014, July). Face recognition with liveness detection using eye and mouth movement. In *2014 International Conference on Signal Propagation and Computer Technology (ICSPCT 2014)* (pp. 592–597). IEEE.
7. Wang, L., Ding, X., & Fang, C. (2009). Face live detection method based on physiological motion analysis. *Tsinghua Science & Technology, 14*(6), 685–690.
8. Kose, N., & Dugelay, J. L. (2013, July). Reflectance analysis based countermeasure technique to detect face mask attacks. In *2013 18th International Conference on Digital Signal Processing (DSP)* (pp. 1–6). IEEE.
9. Dhamecha, T. I., Nigam, A., Singh, R., & Vatsa, M. (2013, June). Disguise detection and face recognition in visible and thermal spectrums. In *2013 International Conference on Biometrics (ICB)* (pp. 1–8). IEEE.
10. Chen, H., Wang, W., Zhang, J., & Zhang, Q. (2019). Echoface: Acoustic sensor-based media attack detection for face authentication. *IEEE Internet of Things Journal, 7*(3), 2152–2159.
11. Sun, W., Song, Y., Chen, C., Huang, J., & Kot, A. C. (2020). Face spoofing detection based on local ternary label supervision in fully convolutional networks. *IEEE Transactions on Information Forensics and Security, 15*, 3181–3196.

12. Akhtar, Z., Fumera, G., Marcialis, G. L., & Roli, F. (2012, September). Evaluation of serial and parallel multibiometric systems under spoofing attacks. In *2012 IEEE Fifth International Conference on Biometrics: Theory, Applications and Systems (BTAS)* (pp. 283–288). IEEE.
13. Rodrigues, R. N., Kamat, N., & Govindaraju, V. (2010, September). Evaluation of biometric spoofing in a multimodal system. In *2010 Fourth IEEE International Conference on Biometrics: Theory, Applications and Systems (BTAS)* (pp. 1–5). IEEE.
14. Akhtar, Z., Rizwan, M., & Kale, S. (2011). Multimodal biometric fusion: Performance under spoof attacks.
15. Bakshi, A., Gupta, S., Gupta, A., Tanwar, S., & Hsiao, K. F. (2020). 3T-FASDM: Linear discriminant analysis-based three-tier face anti-spoofing detection model using support vector machine. *International Journal of Communication Systems, 33*(12), e4441.
16. Bal, R., Bakshi, A., & Gupta, S. (2019). Performance evaluation of optimization techniques with vector quantization used for image compression. In *Harmony Search and Nature Inspired Optimization Algorithms* (pp. 879–888). Springer.
17. Bakshi, A., & Gupta, S. (2020). An efficient face anti-spoofing and detection model using image quality assessment parameters. *Multimedia Tools and Applications*, 1–22.
18. Bakshi, A., & Gupta, S. (2021). A taxonomy on biometric security and its applications. In *Innovations in Information and Communication Technologies (IICT-2020)* (pp. 211–218). Springer.
19. Bakshi, A. (2018, July). A comparative analysis of different intrusion detection techniques in cloud computing. In *International Conference on Advanced Informatics for Computing Research* (pp. 358–378). Singapore: Springer.

Predictive Model for Agriculture Using Markov Model

Punit Gupta, Sumit Bharadwaj, Arjun Singh, and Dinesh Kumar Saini

Abstract With upcoming technologies in farming and new varieties of seeds, it is a new challenge to adopt new ways of farming with changing climate conditions. So IoT gives us a new way to evolve with upcoming challenges. Production of crops is always influenced by the weather conditions; climate change has drastically changed the scenario. In new generation, new seed is evolved to get better yield, but they come with their own climate and water requirements. So in this work, we have tried to train the machine with the life cycle of the crop and make suggestion to the farmer with automated maintenance to maintain the environmental requirement of the crop at various stages of life cycle of the crop. The highlighting feature of this smart farming project is to maintain and meet the dynamic requirement of the crop and maintain the environmental condition based on the life cycle of the crop rather than making static threshold-based system. The farmer can check for new suggestions based on the growth of the crop which is directly proportional to the condition and height of the crop. Controlling of all these operations can be handled through any computer connected to Internet and connecting sensors with Intel Galileo 2.

Keywords Microcontroller · IoT · Intelligent system · Smart agriculture

1 Introduction

The Internet of things (IoT) is an overall system to connect various computing devices, any object, people, or animals with wearable computing, and mechanical and electrical machines. IoT was coined by Kevin Ashton in 1999; he was the one who added the Internet with RFID tags. Since then, IoT has evolved from the last two decades creating huge impacts on human interactions with machines and objects. It

P. Gupta (✉) · A. Singh · D. K. Saini
Department of Computer and Communication Engineering, Manipal University Jaipur, Jaipur, India
e-mail: punitg07@gmail.com

S. Bharadwaj
Amity University, Noida, India

P. K. Singh et al. (eds.), *Proceedings of Third International Conference on Computing, Communications, and Cyber-Security*, Lecture Notes in Networks and Systems 421,
https://doi.org/10.1007/978-981-19-1142-2_28

created humans a different level of comforts by providing the Internet with various objects or things. Using IoT, any object in the system can transfer data to the network without a human-to-human interactions. IoT enables humans with a better quality of living and a well-managed environment. It also enhances services for enhancing the quality of life and utilization of resources. IoT has a very large-scale of applications, and these include smart cities, smart homes, healthcare services, irrigation, agriculture, etc. This evolving paradigm shows a new path to innovations that will build the novel type of interactions among things and humans beings.

As per McKinsey, IoT will be having a high impact in the global economy of 11 trillion dollars by 2025. It will be 11 percent of the global economy. As in today's environment, various IoT-enabled devices can be employed in the form to automate and improve productivity. Artificial intelligence shows promising results in predicting various outcomes in IoT based on historical data. This research article emphasizes the application of IoT in farming by applying AI predictive models to improve the farming outcome.

Farming is one of the important vital business where IoT can improve the thin margins of the farmers. The margin is thin as it involves a large supply chain in the distributions. Small farmers could be equipped with the IoT-enabled devices so that they can have optimized production. Presently, various IoT techniques are available to improve crop productions. They extensively use actuators, controllers, and sensors to automate various systems required for cultivations. These systems include pest control, irrigation, cultivations, etc. Due to the availability of the Internet in rural places, the possibilities and utilization of IoT have surged to the next level. The most important utilization of IoT in farming is to reduce the cost of cultivation and improve the level of simplification. Although lots of work have been done for smart farming, nevertheless there are still various gaps to be filled and more innovative ideas are required to be implemented.

2 Related Work

To improve the forming yield and productivity, precision forming is required. In various research articles, IoT-enabled forming is proposed to improve the efficiency. They have used various sensors to transmit the information through which some information can be sent and some action can be taken in response. This information can be weather conditions, soil quality, etc. Agritalk [1] proposed usage of IoT sensors by interoperating the message from environment to human understandable form. This paper particularly works on the turmeric cultivation by precising soil using IoT. Authors in [2] review the role of IoT, big data, and artificial intelligence jointly in agriculture and food industries. It covers the food assessment, drone-based image processing, intelligent forming, etc., to improve the productivity. Research has exploited machine learning-based analytics to predict the future of the crops on the basis of past data [3]. This will help to automate the yield with less or no human interventions. Some authors worked to improve water resources [4, 5] by applying smart devices and algorithm that improves irrigation system by providing suggestions on the basis of past soil conditions.

In the field of optimization are proposed using fuzzy logic, machine learning, and intelligent algorithm to optimize the system [6–11].

3 Proposed System

In this section, we have proposed an IoT-based smart system using Markov chain to predict the next current state of the farm and predict the steps to get maximum yield. The system uses Intel Galileo Gen 2 as the main processing unit. The proposed system is an active system, where the system uses live reading from moisture sensor, humidity sensor, pH sensor, and temperature sensor to get the live data from field. Galileo collects the live data and streams it to the web server where the data is stored for further prediction and actions. Figure 1 shows the proposed IoT architecture. The Markov model is used at web server to predict the actions.

The web server consists of model to evaluate the current input of sensor and provide the steps and future plan for the farmer with the current status of its farm healthy or not healthy (Fig. 2).

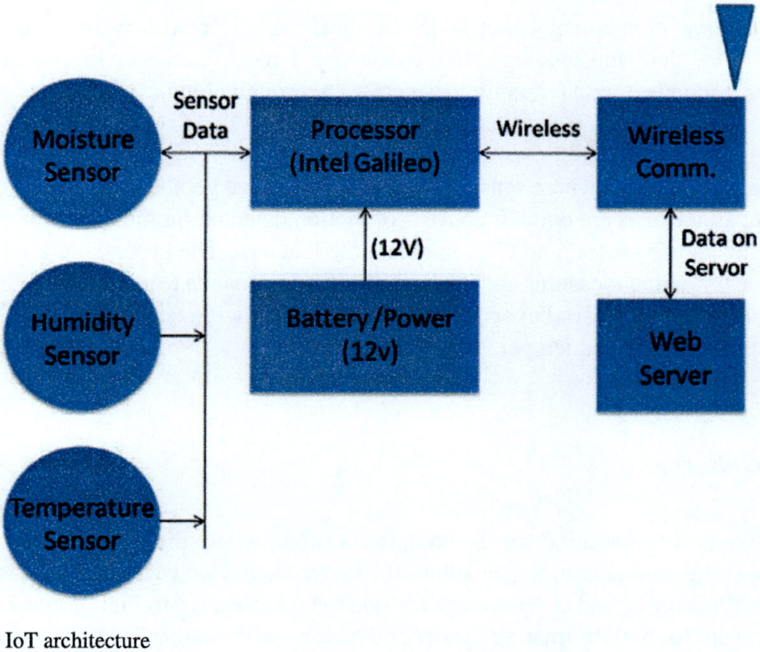

Fig. 1 IoT architecture

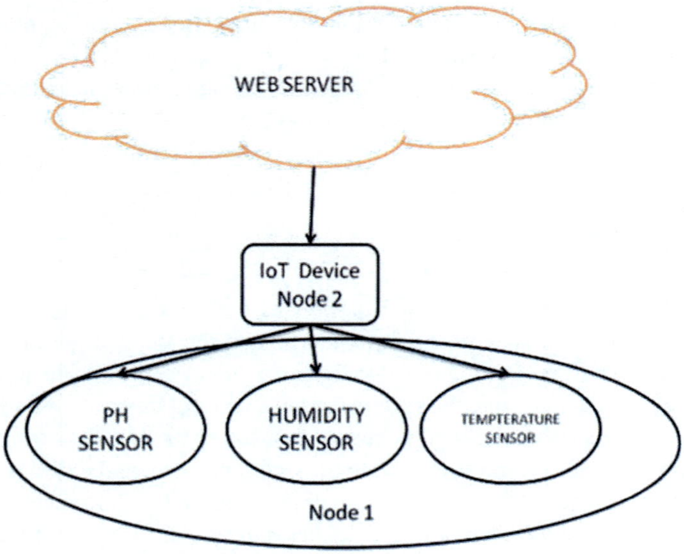

Fig. 2 Green farming model

Node 1

Node 1 is a server component which deals with receiving collected data from sensors deployed in the field and sending it to the database. Server interacts with requests, and server responds it with response to accept data through cloud.

Node 2

Node 2 consists of field, where sensors are deployed and we get the data. This is the area where all the tests are done. It consists of motion detector, humidity sensor, pH sensor, temperature sensor, and GPS to detect the location. The processing used is responsible for taking measures and taking action on commands from the server like if the temperature increases beyond threshold, the pump will start the sprinkler to maintain the humidity and temperature of the soil.

3.1 Web Server

This module acts as the mind for the complete system, where the server is trained with the specific seed dataset which allows the system to decide what are preferable conditions for farming and at what stage for good yield as shown in Fig. 3. The web server is meant for both intimating farmer and taking active actions for maintaining required temperature and humidity in the farm. Our research work started with data collection for various crops and various other related datasets. The dataset shown below is collected from Indian government agriculture research website. Dataset

MONTHLY ETO PENMAN-MONTEITH DATA
(File: KURNOOL.pen)

Country: Location 9529 Station: KURNOOL
Altitude: 281 m. Latitude: 15.80 °N Longitude: 78.06 °E

Month	Min Temp °C	Max Temp °C	Humidity %	Wind m/s	Sunshine hours	Radiation MJ/m²/day	ETo mm/day
January	17.0	31.3	47	1.2	8.8	18.7	3.79
February	19.3	34.3	37	1.3	9.3	21.2	4.66
March	22.5	37.5	30	1.4	9.7	23.5	5.67
April	26.0	39.3	34	1.6	9.2	23.6	6.42
May	27.2	40.0	37	2.6	8.3	22.2	7.60
June	25.0	35.6	54	4.1	5.8	18.3	6.85
July	23.8	32.5	64	4.2	4.4	16.2	5.52
August	23.5	32.1	63	3.5	4.9	16.9	5.28
September	23.3	31.9	65	2.4	5.5	17.3	4.69
October	22.4	32.4	61	1.1	8.7	20.7	4.49
November	19.2	31.0	56	0.9	7.7	17.6	3.62
December	16.6	30.3	51	0.8	8.4	17.7	3.27
Average	22.1	34.0	50	2.1	7.5	19.5	5.15

Cropwat 8.0 Bèta

Fig. 3 Atmospheric parameters

presents various atmospheric parameters affecting the crop yield at different stages of crop life cycle (Fig. 4).

We have proposed models for web server using learning-based model (Markov chain).

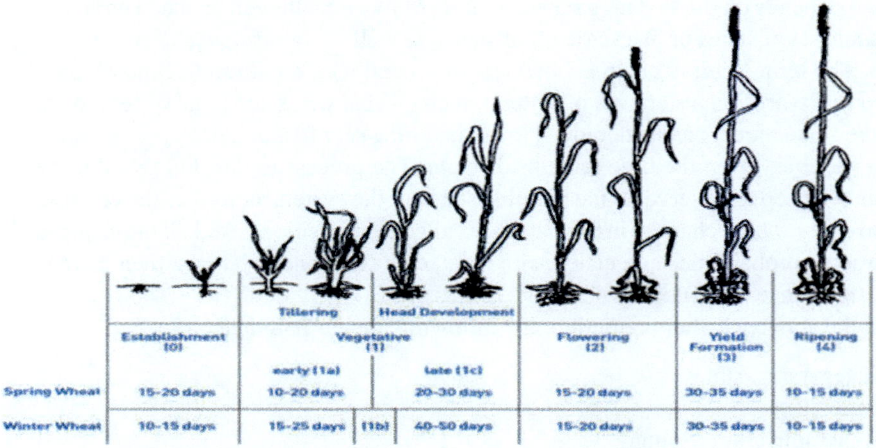

Fig. 4 Height of wheat in winter and spring

Table 1 Type of Markov model

	System state is fully observable	System state is partially observable
System autonomous	Markov chain	Hidden Markov model
System is controlled	Markov decision process	Partially observable Markov decision process

3.2 *Learning-Based Model (Markov Chain)*

A Markov model, named after mathematician Andrey Markov, is used to work on randomly changing systems. It considers all the possible states of the system and all the paths or transitions to that state to then model an output that simulates the behavior of the system. Markov model works on the *Markov property,* wherein the future states of the process depends only on the current state without considering whatsoever the preceding states were. This is the key assumption that helps the prediction and forecasting.

Markov models can be adjusted on the basis of observations made in the system, depends on the behaviour of variour states in the system. For this, there are mainly four types of models (Table 1).

For this work, our system state is fully observable, and the system is not controlled; therefore, we use the Markov chain. Markov chain is essentially the sequence of random states that a process yields, following the Markov property, of course. Therefore, it depends on the system's state and time. Markov chains can be made both on a dynamic set of states or in a static set of states, as well as for discrete and continuous time. The term Markov chain is however associated with the discrete-time Markov chain that works on a static set of states, which is what we assume our system to be, but the same model can be deployed to an incoming set of values/states, of the same state variable, given the time remains discrete. The process to develop the Markov chain, we need to observe all the possible states of the system as well as the changes in the states. These changes in the states are called the transitions. And all the changes have some probabilities associated with them, e.g., "if Monday it rains, then there is an 80% chance that Tuesday is sunny" (Fig. 5).

In the above weather prediction, the following are our parameters:

- States: rainy, sunny
- Time: days
- Transitions: rain to sunny
- Transition probability: 80%.

Fig. 5 Markov and temperature

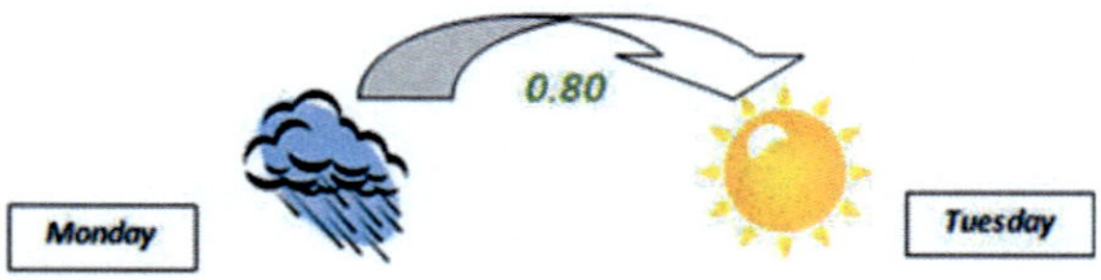

Once we have all the states and all the transition probabilities, we make a transition matrix that is a square matrix that beholds all the transition probabilities from every system state to every other system state and itself. All the states and this transition matrix amount to the definition of the process.

To calculate the transition probability pij, we have the formula:

$$\Pr(X_n + 1 = x_n + 1 | X_1 = x_1, X_2 = x_2, \ldots, X_n) = x_n$$
$$= \Pr(X_{n+1} = x_{n+1} | X_n = x_n) \quad (1)$$

Here, X_n is a state, which can have substates:

$$S = \{s_1, s_2, \ldots, s_n\}$$
$$pij = \Pr(X_1 = sj \mid X_0 = si) \quad (2)$$

Let us take an example data for 12 days: **R S S R S S R R S S R S**.

where R: Rainy and S: Sunny.

The transition probabilities are calculated as follows,

P(*R*|*R*) = 0.20, since after a rainy day, there is one rainy day.
P(*S*|*R*) = 0.80, and after a rainy day, there are two occurrences of a sunny day.
P(*S*|*S*) = 0.50, since after a sunny day, there are three occurrences of a sunny day.
P(*R*|*S*) = 0.50, and after a sunny day, there are three occurrences of a rainy day.

Say, now, you have for these two states, rainy and sunny, and the transition matrix is as follows (Table 2).

This is the following transition chart, for state 0 (Figs. 6 and 7).

For this work, we had the data for wheat's life cycle of 150 days, which gave us a transition matrix, with 17 unique states and a transition matrix like so (Fig. 8).

Table 2 Mapping of probability of various seasons

	Rainy	Sunny
Rainy	0.20	0.80
Sunny	0.50	0.50

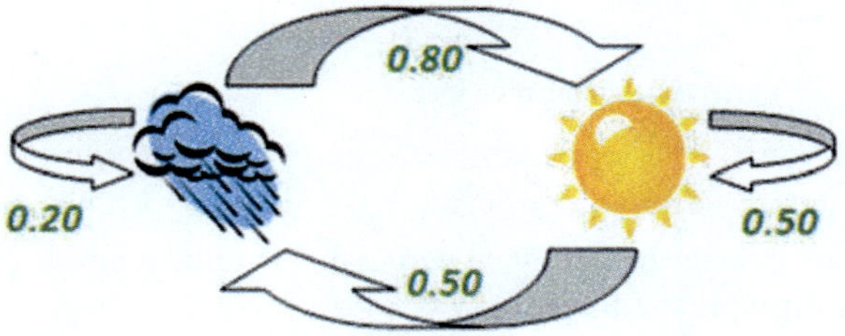

Fig. 6 Markov and temperature

Fig. 7 Machine learning model

```
> mcX
MLE Fit
 A  17 - dimensional discrete Markov Chain defined by the following states:
 10, 11, 12, 13, 14, 15, 16, 17, 18, 19, 20, 21, 22, 23, 24, 25, 26
 The transition matrix  (by rows)  is defined as follows:
   10 11 12 13 14 15 16 17 18 19 20 21 22     23    24    25  26
10  0  1  0  0  0  0  0  0  0  0  0  0  0 0.000 0.000 0.000 0.0
11  0  0  1  0  0  0  0  0  0  0  0  0  0 0.000 0.000 0.000 0.0
12  0  0  0  1  0  0  0  0  0  0  0  0  0 0.000 0.000 0.000 0.0
13  0  0  0  0  1  0  0  0  0  0  0  0  0 0.000 0.000 0.000 0.0
14  0  0  0  0  0  1  0  0  0  0  0  0  0 0.000 0.000 0.000 0.0
15  0  0  0  0  0  0  1  0  0  0  0  0  0 0.000 0.000 0.000 0.0
16  0  0  0  0  0  0  0  1  0  0  0  0  0 0.000 0.000 0.000 0.0
17  0  0  0  0  0  0  0  0  1  0  0  0  0 0.000 0.000 0.000 0.0
18  0  0  0  0  0  0  0  0  0  1  0  0  0 0.000 0.000 0.000 0.0
19  0  0  0  0  0  0  0  0  0  0  1  0  0 0.000 0.000 0.000 0.0
20  0  0  0  0  0  0  0  0  0  0  0  1  0 0.000 0.000 0.000 0.0
21  0  0  0  0  0  0  0  0  0  0  0  0  1 0.000 0.000 0.000 0.0
22  0  0  0  0  0  0  0  0  0  0  0  0  0 1.000 0.000 0.000 0.0
23  0  0  0  0  0  0  0  0  0  0  0  0  0 0.875 0.125 0.000 0.0
24  0  0  0  0  0  0  0  0  0  0  0  0  0 0.008 0.984 0.008 0.0
25  0  0  0  0  0  0  0  0  0  0  0  0  0 0.000 0.500 0.000 0.5
26  0  0  0  0  0  0  0  0  0  0  0  0  0 0.000 0.000 1.000 0.0

> |
```

Fig. 8 Markov chain transition matrix

To then apply the Markov process on which we used the Markov package in R, which gave us the following output for an initial temperature set at 11 degree Celsius, making a prediction (Fig. 9).

The aim of the work is to train a model so that the system will have dynamic behavior in farming, for example, high humidity in starting, then in second phase

```
> TempOfDays <- rmarkovchain(n =150, object = mcTemp, t0 = "11")
> TempOfDays[1:15]
 [1] "12" "13" "14" "15" "16" "17" "18" "19" "20" "21" "22" "23" "23" "23" "23"
> |
```

Fig. 9 Markov chain prediction

low humidity, and at last zero humidity is required, so static model fails in such scenario.

4 Experiment and Results

In this section, we have shown the results of complete system working using Markov chain model for data prediction to suggest better instructions and health of the crop to have better yield and decisions about cultivation of a field on a farm. Figure 10 shows an online dashboard for checking the current reading of the field and testing the health of the field using static model as shown.

Figure 11 showcases the web page which takes input from farmer, where the height of the crop is an input and the result will be the preferable condition prerequisite for the next months which is coming from trained model as shown in Fig. 12. Figure 13 shows the reading of the various sensors and varying field conditions, where Fig. 14 showcases the suggestion to the farmer.

Figures 15 and 16 showcase the accuracy of the proposed model as compared to the expected data using moisture and temperature as performance parameters. Similarly, Figs. 17, 18, and 19 display the variation in moisture, temperature, and soil pH for a crop of 90 over the period of time: days.

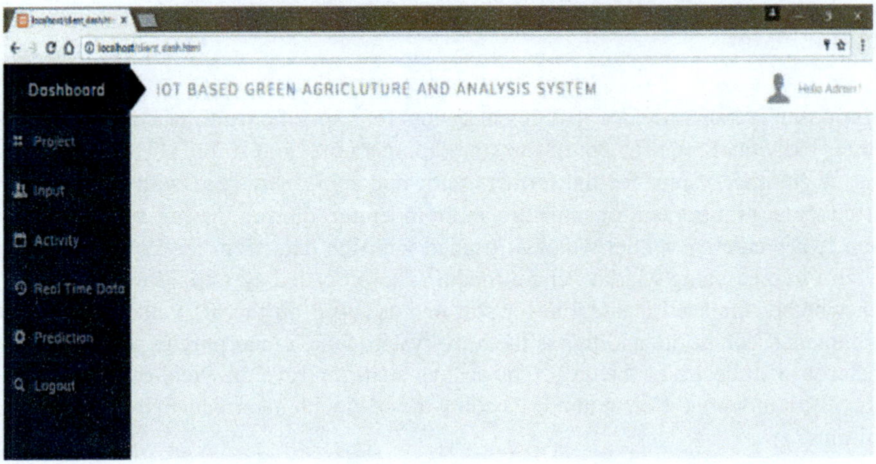

Fig. 10 Green farming dashboard

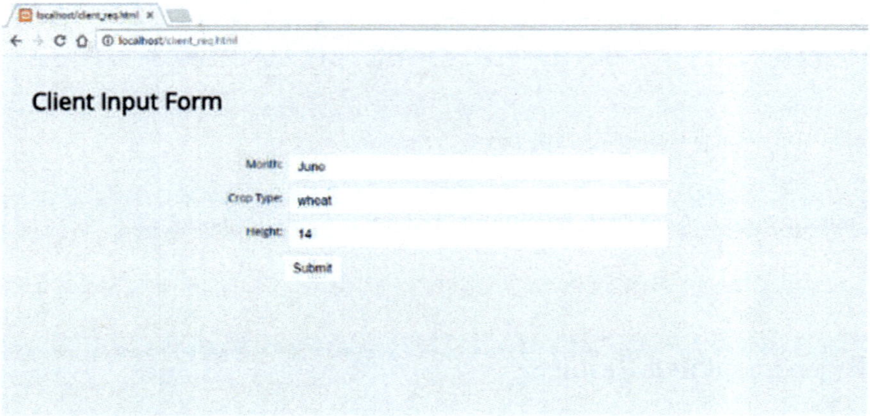

Fig. 11 Green farming input panel

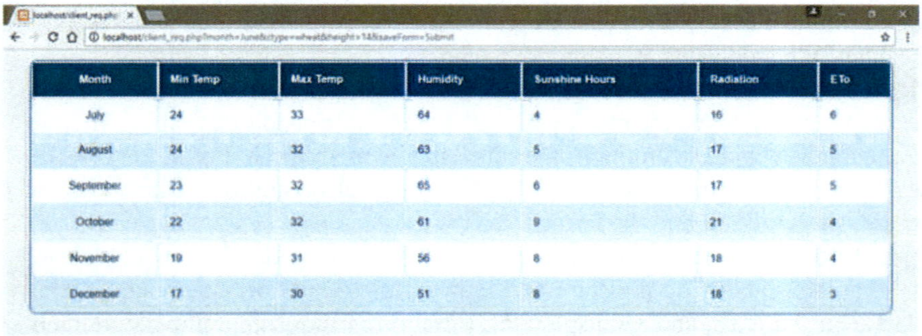

Month	Min Temp	Max Temp	Humidity	Sunshine Hours	Radiation	ETo
July	24	33	64	4	16	6
August	24	32	63	5	17	5
September	23	32	65	6	17	5
October	22	32	61	9	21	4
November	19	31	56	8	18	4
December	17	30	51	8	18	3

Fig. 12 Green farming sample dataset for static model

5 Conclusion

The proposed system provides an efficient system for a specific seed since every seed has their individual requirement of moisture, temperature, and soil pH. The system acts as a guiding system for the farmers with upcoming new seed with different environmental requirement of moisture and temperature during the life cycle. The system is a predictive model which is trained with the data from the government body and trained using Markov chain model. The proposed system allows you to check whether the seed is suitable for soil or not based on the pH value and the environmental condition, and this is the only system which incorporates the height of the crop with health of the crop. The system shall improve the yield of the crop drastically, removing the error in miss feeding the seed with incorrect environmental conditions.

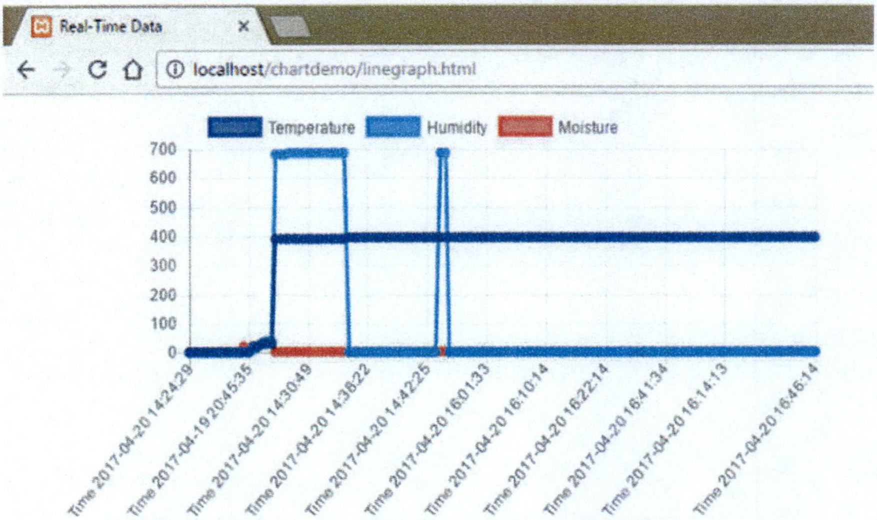

Fig. 13 Green farming log dataset plot

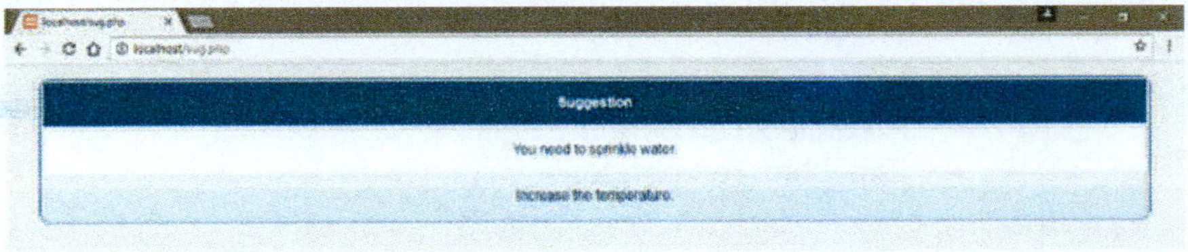

Fig. 14 Green farming predictive suggestions

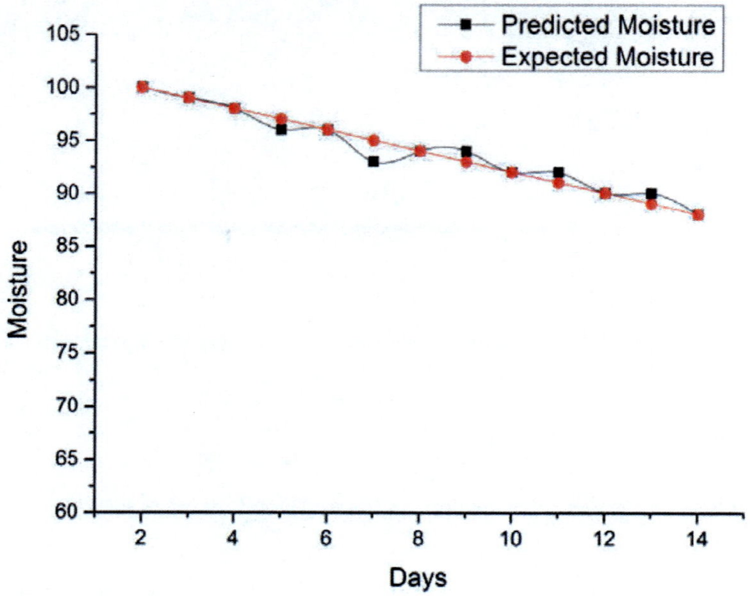

Fig. 15 Prediction of soil moisture

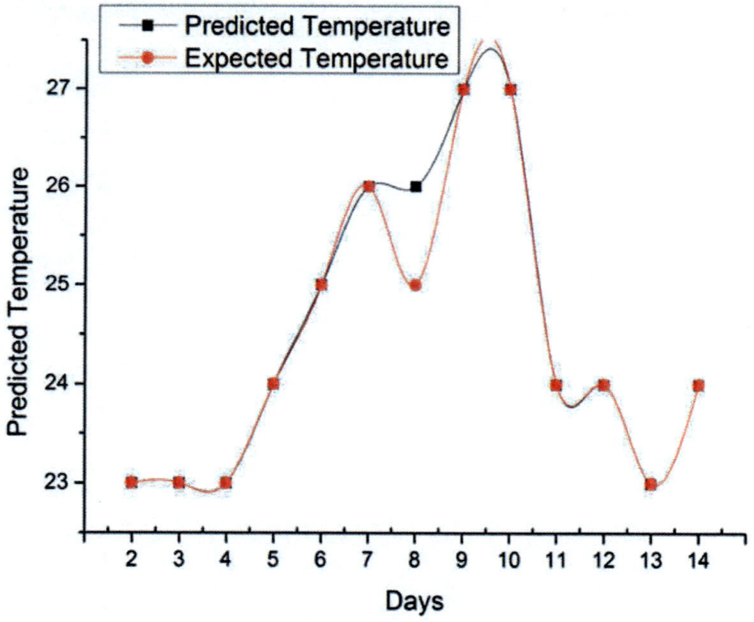

Fig. 16 prediction of soil temperature

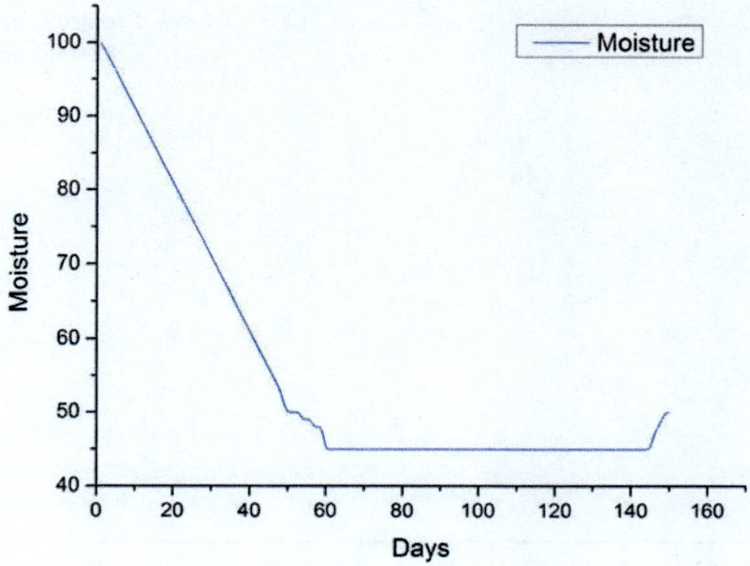

Fig. 17 Prediction of soil moisture using proposed model

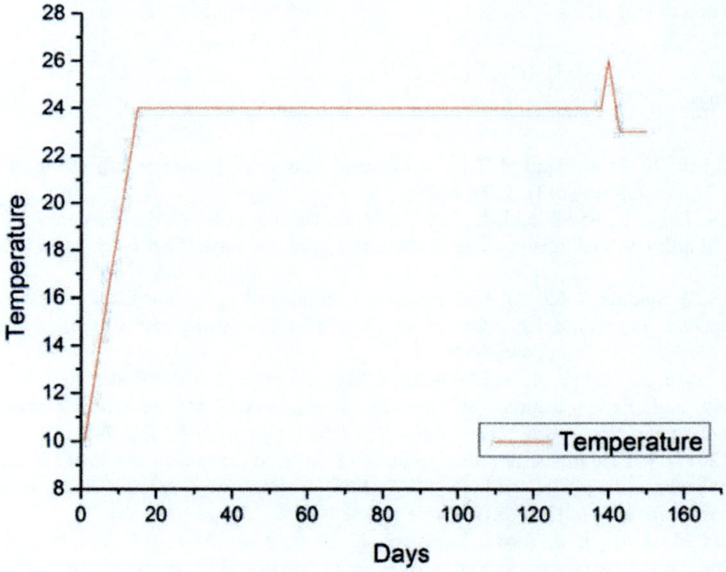

Fig. 18 Variation in soil temperature over time

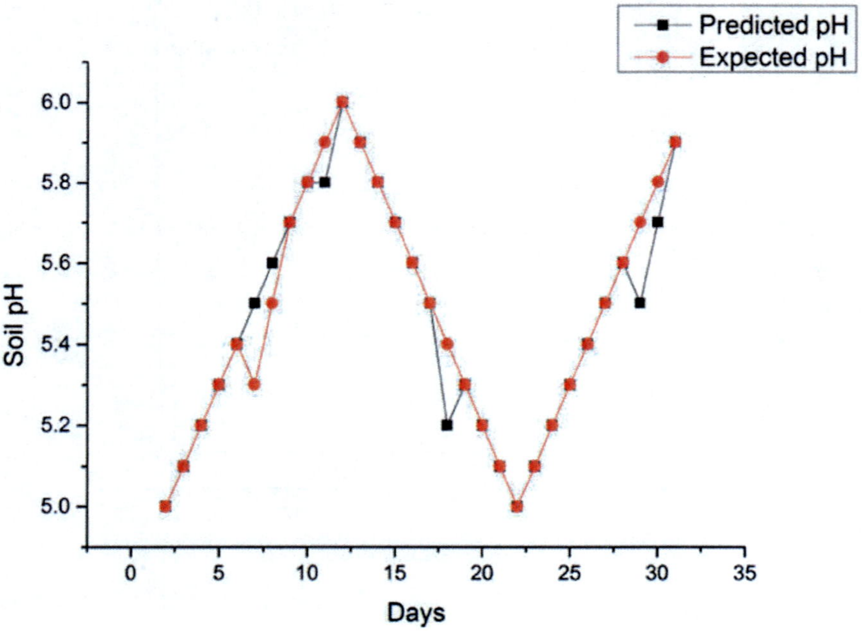

Fig. 19 Prediction of soil pH

References

1. Chen, W., et al. (2019). AgriTalk: IoT for precision soil farming of turmeric cultivation. *IEEE Internet of Things Journal, 6*(3), 5209–5223.
2. Misra, N. N., Dixit, Y., Al-Mallahi, A., Bhullar, M. S., Upadhyay, R., & Martynenko, A.: IoT, big data and artificial intelligence in agriculture and food industry. *IEEE Internet of Things Journal.*
3. Varghese, R., & Sharma, S. (2018). Affordable smart farming using IoT and machine learning. In *2018 Second International Conference on Intelligent Computing and Control Systems (ICICCS), Madurai, India* (pp. 645–650).
4. Syed, F. K., Paul, A., Kumar, A., & Cherukuri, J. (2019). Low-cost IoT+ML design for smart farming with multiple applications. In *2019 10th International Conference on Computing, Communication and Networking Technologies (ICCCNT), Kanpur, India* (pp. 1–5).
5. Dolci, R. (2017). IoT solutions for precision farming and food manufacturing: artificial intelligence applications in digital food. In: *2017 IEEE 41st Annual Computer Software and Applications Conference (COMPSAC), Turin* (pp. 384–385).
6. Suryanegara, M., Arifin, A. S., Asvial, M., Ramli, K., Nashiruddin, M. I., & Hayati, N. (2019). What are the Indonesian concerns about the internet of things (IoT)? Portraying the profile of the prospective market. IEEE Access **7**, 2957–2968.
7. Monteiro, J., Barata, J., Veloso, M., Veloso, L., & Nunes, J. (2018). Towards sustainable digital twins for vertical farming. In: *2018 Thirteenth International Conference on Digital Information Management (ICDIM)* (pp. 234–239).
8. Pini, M., Marucco, G., Falco, G., Nicola, M., & De Wilde, W. (2020). Experimental testbed and methodology for the assessment of RTK GNSS receivers used in precision agriculture. *Access IEEE, 8*, 14690–14703.

9. Singh, P., & Singh, N. (2020). Blockchain with IoT and AI: A review of agriculture and healthcare. *International Journal of Applied Evolutionary Computation (IJAEC), 11*(4), 13–27.
10. Sanjeevi, P., Prasanna, S., Siva Kumar, B., Gunasekaran, G., Alagiri, I., & Vijay Anand, R. (2020). Precision agriculture and farming using Internet of Things based on wireless sensor network. *Transactions on Emerging Telecommunications Technologies, 31*(12).
11. Khatri, N., Sharma, A., Khatri, K. K., & Sharma, G. D. (2018). An IoT-based innovative real-time pH monitoring and control of municipal wastewater for agriculture and gardening. In Proceedings of First International Conference on Smart System, Innovations and Computing (pp. 353–362).

A Comparative Analysis of Edge Detection Using Soft Computing Techniques

Ankush Verma, Namrata Dhanda, and Vibhash Yadav

Abstract Detecting edges is one of the most significant aspects of computer vision. Typical methods for edge detection like Sobel and Canny are robust and fast, but they are sensitive to noise. Soft computing techniques such as particle swarm optimization (PSO), ant colony optimization (ACO), genetic algorithms (GA) and fuzzy logic system (FLS) have extensive application in edge detection of images because of their adaptive behavior. Edge detection is identifying the discontinuities in intensity of the pixel and grouping the contour of edges. The quality of edges in ACO-based edge detection majorly depends on the choice of constants, pheromone evaporation rate, number of iterations etc. In PSO-based edge detection, the quality of images depends on the values of acceleration coefficients and inertia weight. However, thresholding is major stakeholder in determining the fitness of the chromosomes. The population contains 2-D chromosomes. Fuzzy systems are most suitable for designing edge detection hardware. This paper presents a thorough comparative study of soft-computing-based edge detection techniques and highlights their key features. The factors affecting quality of edges are compared, and the actual outcomes of the approaches are systematically arranged for better understanding.

Keywords PSO · ACO · Fuzzy logic · GA · Edge detection

A. Verma (✉)
Amity Institute of Information Technology, AUUP, Lucknow, India
e-mail: withankush@gmail.com

N. Dhanda
Amity School of Engineering and Technology, AUUP, Lucknow, India
e-mail: ndhanda@lko.amity.edu

V. Yadav
Rajkiya Engineering College, Banda, India
e-mail: vibhashds10@recbanda.ac.in

P. K. Singh et al. (eds.), *Proceedings of Third International Conference on Computing, Communications, and Cyber-Security*, Lecture Notes in Networks and Systems 421,
https://doi.org/10.1007/978-981-19-1142-2_30

1 Introduction to Edge Detection

1.1 Edge

"Edge in a gray level image is the boundary between two regions of different gray levels [1]." It can also be termed as discontinuities in the intensity of the image or first derivative of the image. There are two types of discontinuities in image intensity resulting in two types of edges: (1) Step discontinuity: Abrupt change in intensity. The resulting edge is called step edge. (2) Line discontinuity: Abrupt but returns back to original intensity at a certain distance. The resulting edge is called line edge. Ramp edges are distorted form of step edges and roof edges are distorted form of line edges [2].

1.2 Edge Detection

Edge detection is defined as finding boundaries of objects within image on the basis of texture and intensity [3]. The structure of the image depends on several parameters such as hardware specifications of sensing device, lighting conditions and noise [4]. Edge detection is generally done in three phases: (1) noise reduction by smoothening, (2) differentiation of image (computing magnitude of edge and its orientation) and (3) pixel labeling. Smoothening techniques reduce the noise in the image edges and prepare image for numerical computations. Poggio and Torre [5, 6] proposed significant techniques for image smoothening.

In digital images due to discrete quantification of pixels, discrete approximation of differentiation operators is required, and amplification of noise due to application of operator is inevitable [2]. Digitized image found from quantization of analogous image G is represented as $A \times B \rightarrow^{I} P$, where $A = \{0, 1, 2, \ldots, c\}$, $B = \{0,1,2, \ldots, r\}$ and $P = \{0, 1,2, \ldots, p\}$. c, r and p represent number of columns of image, number of rows of image and highest intensity of any pixel, respectively. First-order derivatives are represented as Eq. (1).

$$G_x(p,q) \cong I_x(p,q) = I(p,q) - I(p+1,q),$$
$$G_y(p,q) \cong I_y(p,q) = I(p,q) - I(p,q+1) \quad (1)$$

These operators are commonly represented as masks (Eq. (2)).

$$I_x = M_x \begin{bmatrix} I(p,q) \\ I(p+1,q) \end{bmatrix}, \quad I_y = [I(p,q) \quad I(p,q+1)]M_y,$$
$$M_x = \begin{bmatrix} 1 & -1 \end{bmatrix}, \quad M_y = \begin{bmatrix} 1 \\ -1 \end{bmatrix} \quad (2)$$

$$M_1 = \begin{bmatrix} 0 & +1 \\ -1 & 0 \end{bmatrix} \quad M_x = \begin{bmatrix} -1 & 0 & +1 \\ -1 & 0 & +1 \\ -1 & 0 & +1 \end{bmatrix} \quad M_x = \begin{bmatrix} -1 & 0 & +1 \\ -2 & 0 & +2 \\ -1 & 0 & +1 \end{bmatrix} \quad M_x = \begin{bmatrix} -1 & 0 & +1 \\ -\sqrt{2} & 0 & +\sqrt{2} \\ -1 & 0 & +1 \end{bmatrix}$$

$$M_2 = \begin{bmatrix} +1 & 0 \\ 0 & -1 \end{bmatrix} \quad M_y = \begin{bmatrix} +1 & +1 & +1 \\ 0 & 0 & 0 \\ -1 & -1 & -1 \end{bmatrix} \quad M_y = \begin{bmatrix} +1 & +2 & +1 \\ 0 & 0 & 0 \\ -1 & -2 & -1 \end{bmatrix} \quad M_y = \begin{bmatrix} +1 & +\sqrt{2} & +1 \\ 0 & 0 & 0 \\ -1 & -\sqrt{2} & -1 \end{bmatrix}$$

(a) (b) (c) (d)

Fig. 1 **a** Robberts, **b** Prewitts, **c** Sobel, **d** Frei-Chen

The above-mentioned masks are not symmetric; therefore, odd number of elements are used in masks as follows (Eq. (3)):

$$M_x = \begin{bmatrix} 1 \; 0 \; -1 \end{bmatrix}, \quad M_y = \begin{bmatrix} 1 \\ 0 \\ -1 \end{bmatrix} \tag{3}$$

Other significant approximations used are Roberts, Prewitt, Sobel and Frei-Chen [1, 7, 8] as shown in the Fig. 1.

Edge labeling is localization of edges. Localization using thresholding of gradient magnitude results in thick images. Another technique non-maximum suppression (NMS) proposed by Canny [9] is commonly used. It finds local maximum along the gradient's direction. The edge detection methods which are based on the first derivatives use above-mentioned derivatives—Robberts, Prewitt, Sobel etc. [10]. These algorithms are simple and fast, but they are highly sensitive to the noise [11]. Identified edges are thick as well. Second category of edge detection technique uses second-order derivatives. These are called zero crossing edge detectors. Marr-Hildreth [12] proposed edge detector based on second-order derivative in which localization is better than first-order derivative-based detectors. It uses approximation by Laplacian. But it is also sensitive to the noise. The techniques which use Gaussian filter for noise reduction are put in third category. Marr-Hildreth [12] had proposed Gaussian filter in 1980. Castan [13] used Laplacian-of-Gaussian (LoG) operator for the first time in 1993.

There is a wide range of other edge detection techniques, but in this paper, we are going to review soft computing technique-based edge detectors. Major soft computing techniques covered are ACO [14], PSO [15], GA [16] and fuzzy logic [17] in following sections.

2 Ant Colony Optimization

2.1 Ant Colony System (ACS)

Dorigo and Caro [14] presented an optimization technique which works on the patterns followed by ants to identify most suitable path. The technique is called ant colony system (ACS). ACS is suitable for optimization environments, hence the name ant colony optimization (ACO). ACO looks at the problem environment as collection of states in which one state is the initial state or starting state, and one state is goal state. Rest of the states are intermediate states which may or may not fall on one or more of several possible paths. The shortest path from initial state to final state is solution. ACO attempts to generate a sequence of state transfers from initial state to final state in discrete space of states. The ants start moving on adjacent neighboring randomly until they find the goal state. When an ant transfers from one state to another state, the choice of next state depends on the probability which is calculated using trial intensity (pheromone). The ants keep increasing the trial intensity as per the quality of solution they encounter on their path. The probability of migrating from state s_i to s_j is obtained using Eq. (4).

$$P_{ij}(t) = \begin{cases} \frac{[\tau_{ij}(t)]^{\alpha} \cdot [\eta_{ij}]^{\beta}}{\sum_{s_j \in \text{Allowed}} \left[[\tau_{ij}(t)]^{\alpha} \cdot [\eta_{ij}]^{\beta} \right]} & \text{if } s_j \in A \\ 0, & \text{Otherwise} \end{cases} \tag{4}$$

where $\tau_{ij}(t)$ represents trial intensity between s_i and s_j at time t. α and β are positive constants. η_{ij} is the heuristics, generally calculated as inverse of the distance between s_i and s_j. A contains unvisited states. The sequence of state transfer is captured in Tabu List. When all the ants finish traversing the states, the trial intensity of each state is updated using Eq. (5).

$$\tau_{ij}(t+1) = \sigma \cdot \tau_{ij}(t) + \Delta\tau_{ij}(t, t+1) \tag{5}$$

where $\tau_{ij}(t+1)$ and $\tau_{ij}(t)$ are new and old trial intensities of (s_i, s_j). σ is a constant s.t. $0 < \sigma < 1$. $\Delta\tau_{ij}(t, t+1)$ is calculated as shown in Eq. (6).

$$\Delta\tau_{ij}(t, t+1) = \sum_{k=1}^{n} \Delta\tau_{ij}^{k}(t, t+1) \tag{6}$$

where n represents number of ants. $\Delta\tau_{ij}^{k}(t, t+1)$ is trial intensity updated by the kth ant from s_i to s_j at tth iteration. It is calculated as follows (Eq. (7)).

$$\Delta\tau_{ij}^{k}(t, t+1) = \begin{cases} \frac{1}{L_k} & \text{if the } k \text{ - th ant goes from } s_i \text{ to } s_j \\ 0 & \text{otherwise} \end{cases} \tag{7}$$

where L_k is the length of the path covered by kth ant. ACS system is suitable for any optimization problem with discrete state space.

2.2 *Edge Detection Using Ant Colony Optimization*

Zhuang [18] used ACO to find edges in an image. The relationship among neighboring image points is represented by a weighted perceptual graph. An ACO-based layered computer vision model is developed. ACO helps in identifying the edge features in digital images. Ant colony system builds an evolving pheromone field corresponding to perceptual graph. Each point in the image which is not on the border has four neighbors, hence four connections to adjacent points. The example of perceptual graph is shown in the Fig. 2.

Three characteristics of perceptual graph are identified and examined in the experiments using following metrics of the edge point. (1) Maximum value, (2) length and (3) variance.

Ari et al. [19] developed an effective ACO-based edge detection technique which uses Fisher ratio (F ratio). F ratio is utilized to determine the most suitable threshold value from pheromone matrix. It is further used to extract binary edge map from the pheromone matrix. F ratio is defined as Eq. (8).

$$F \text{ ratio} = \frac{\text{Variance of means between the clusters}}{\text{Average Variance within the clusters}} \tag{8}$$

Liu and Fang [20] proposed a robust method for edge detection using ACO which uses user-defined threshold to update the pheromone using a new heuristic function. The new heuristic function is defined as Eq. (9).

$$\eta_{ij} = \frac{1}{X_{\max}} \cdot \max\left[\left|X_{(i-u,j-v)}, X_{(i+u,j+v)}\right|\right], k \tag{9}$$

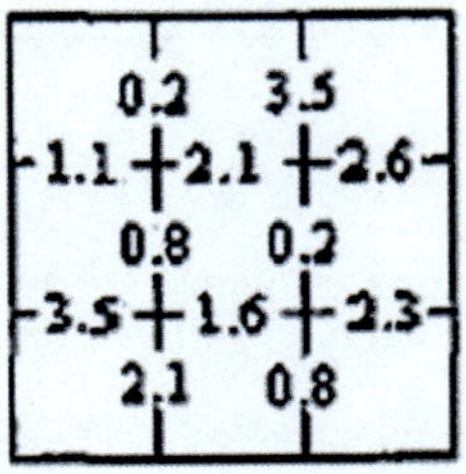

Fig. 2 Example of perceptual graph

where $X_{\max}$ is maximum value of gray-level image intensity. X_{pq} is intensity of pixel (p, q), and max[.] represents maximum difference of intensities of two pixels having same color. This approach gives upper hand as compared to traditional approaches, but its computational overheads a bit higher as compared to other traditional approaches which make it slow.

Kumar and Raheja [21] proposed an adaptive edge detection method based on ACO which takes weighted average of threshold as compared to other approaches which take simple average of threshold during pheromone update step. The weighted average of threshold is defined as Eq. (10).

$$Th^q = \frac{w_1 m_L^{(q)} + w_1 m_U^{(q)}}{2} \tag{10}$$

where $m_L^{(q)}$ and $m_U^{(q)}$ are means of intensities of pixels having intensity lower than threshold and greater than threshold, respectively, $w_1 + w_2 = 1$. The proposed algorithm outperforms recently proposed techniques in F-score. The overall efficiency is 87%.

Kumar and Raheja [22] in further research proposed a guided image filtering-based method for edge detection to enhance the edges; then, ACO is used to find the edges. The computational complexity does not increase in this approach and remains same as normal ACO-based edge detection method.

Table 1 displays parameter values taken by various techniques discussed in Sect. 2.2. However, recent researchers did not mention clearly the values of some or all parameters used during experiments. Such values are denoted by '–' in Table 1.

Figure 3 shows the sample of resultant edge detection by various techniques discussed in Sect. 2.

Table 1 Parameters comparison table for ACO-based edge detection techniques

Reference	Year of publication	K	τ_0	α	β	ρ	ψ	T	N	L
[18]	2004	–	–	–	–	–	–	–	–	–
[19]	2013	$\sqrt{M \times N}$	0.0001	4	0.2	0.05	0.05	–	2	250
[20]	2015	$\sqrt{M \times N}$	0.0001	2	2	0.02	-	–	3	$\sqrt[3]{M \times N}$
[21]	2020	Variable	0.0001	1	0.1	0.1	0.05	Adaptive	8	40
[22]	2020	$\sqrt{M \times N}$	0.0001	1	0.1	0.1	0.05	Adaptive	–	40

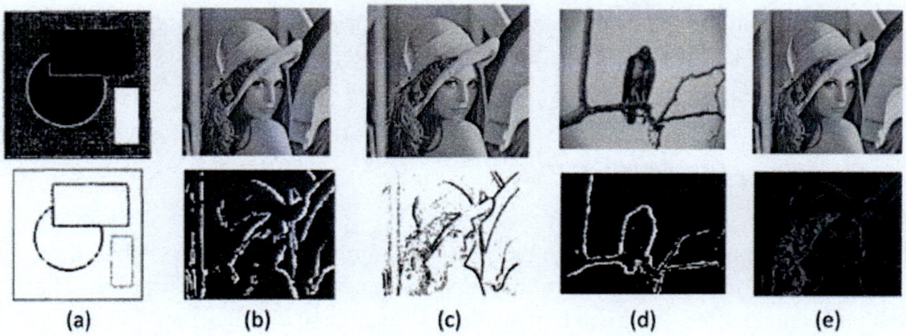

Fig. 3 First row—original image. And second row—output image of **a** [18], **b** [19], **c** [20], **d** [21], **e** [22]

3 Particle Swarm Optimization (PSO)

3.1 Introduction to PSO

Kennedy and Eberhart [15] presented PSO as a method of optimization of nonlinear functions. PSO has a direct derivation from swarming theory of fish schooling or bird flocking. Reynolds [23] and Heppener and Grenander [24] built a bird flocking simulators. Both models are based on the fact that the unpredictable dynamics of group of birds in order to maintain a significant distance among themselves. No requirement of the gradient information and simple implementation attracts researchers toward PSO [25]. Particles represent a population (swarm). Any of the possible swarms can be a potential solution. Local best of global best particles affects every individual particle's behavior to help them fly in the search space. The particles have their own memory to remember the unexplored positions in the search space.

Let the search space has D dimensions. Then, the i-th particle is represented as $K_i = (k_{i1}, k_{i2}, \ldots, k_{iD})$. The rate of position change of a particle is called velocity, which is denoted by $R_i = (r_{i1}, r_{i2}, \ldots, r_{iD})$. The most optimal position of any particle K_i is represented by $B_i = (b_{i1}, b_{i2}, \ldots, b_{iD})$. It is called pBest. Similarly, the best particle of swarm is denoted by g and called gBest. Manipulation of the particle is done as Eq. (11) and (12).

$$R_{id}(t+1) = w \cdot R_{id}(t) + c_1 * x_1 * (b_{id} - k_{id}) + c_2 * x_2 * \left(b_{gd} - x_{id}\right) \quad (11)$$

$$k_{id}(t+1) = k(t) + r_{id}(t+1) \quad (12)$$

where size of the swarm is N for $i = 1, 2, 3, 4, \ldots, N$. Inertia weight is w. c_1 and c_2 are acceleration coefficients in the range [0, 2], x_1 and x_2 are random constants in the range (0,1). Inertia weight maintains the balance between exploration and exploitation. Equation (11) calculates new velocity for each particle using previous velocity, and Eq. (12) finds new position of the particle.

3.2 Edge Detection Using PSO

Alipoor et al. proposed a novel edge detection technique in [26] which is based on PSO where a new filter-based on evolutionary computation has been proposed. The edge filter is constructed using a pair of synthetic images and its edge map. Generally, the edge pixel is determined using linear function shown in Eq. (13).

$$Y(i, j) = \sum_{p=i-1}^{i+1} \sum_{q=j-1}^{j+1} X(p, q) \times M\left(p^*, q^*\right) \tag{13}$$

where X and Y represent the original image and edge image, respectively. M represents edge filter. A window of size 3 × 3 of the image centered around pixel (p, q) is multiplied by M to determine whether pixel (p, q) is an edge pixel or not. The decision is taken on the basis of threshold T. The contents of M are directly optimized by using PSO with two objectives: (1) Sum of the members of M must be zero which leads to $Y(.) = 0$ for non-edge pixels. (2) All the edge pixels must result in $Y(.) > T$. Training and edge images are shown in Fig. 4.

Setayesh et al. [27] proposed a technique for improvement in broken edges detected in noisy environment. A new fitness function with updated encoding scheme has been developed to address the overhead of noise in images. The objective is to find best fitting curve. Generally, the curve separates two regions. So the intra-set distances are minimized and inter-set distances are maximized (Fig. 5a).

The curve is represented by particle. There are eight possible directions for each pixel. Figure 5b represents ways of dividing the neighborhood into two regions. The algorithm measures $\text{Max}_d(X)$ as shown in Eq. (14), where intra-set distance of pixels is minimum and inter-set distance of pixel is maximum.

Fig. 4 Training image and its edge map

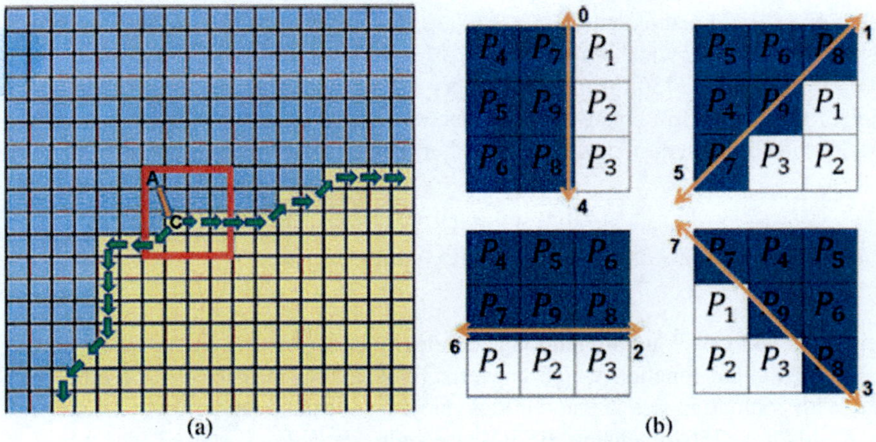

Fig. 5 **a** A is in neighborhood of pixel C, **b** ways to divide neighborhood into two regions

$$\mathrm{Max}_d(X) = \frac{\mathrm{Interset}_d(X)}{1 + \mathrm{Intraset}_d} \tag{14}$$

where X is targeted pixel on the curve and the direction of movement is d in the range 0–7. $\mathrm{Max}_d(X)$ represents the magnitude of change in the intensity in the direction d. Non-maxima suppression (NMS) value of each pixel is calculated. $\mathrm{NMS}_d(X)$ combines with $\mathrm{Max}_d(X)$ is utilized to calculate the probability of an edge pixel.

Uniformity factor [28] is used to measure pixel intensities along the curve. The overall process runs in two phases; the first phase finds edge probability of each pixel in all possible eight directions, and in second phase, best curve is identified that fits best to the edges using PSO technique.

Setayesh et al. [29] further extended the previous work by proposing a new fitness function along with two constraints for more smooth and accurate edge detection for a noisy environment. The previous algorithm [27] produced spiky edges. To overcome this shortcoming, curvature cost (CC) is introduced in the fitness function. It measures the local curvature of the edge pixel. Local curvature depends of the migration from one pixel to next pixel. The curvature cost of each particle is then considered to calculate final curvature in Eq. (15).

$$\mathrm{Curvature}(C) = \frac{1}{\max - 2}\left(\sum_{i=1}^{\frac{\max}{2}-1} CC(d_i, d_{i+1}) + \sum_{i=\frac{\max}{2}-1}^{\max-1} CC(d_i, d_{i+1})\right) \tag{15}$$

The new fitness function is given by Eq. (16).

$$\mathrm{Fitness}(C) = \mathrm{Score}(C) - \mathrm{Curvature}(C) \tag{16}$$

Score(C) $>$ T hreshold and $G(C) = 0$ are two constraints. $G(C)$ represents the frequency of curve C which crosses itself.

Dagar and Dahiya [30] proposed a binary PSO-based edge detection technique which minimizes multi-objective fitness function. BPSO operates in binary space. A particle's position kth dimension can be either 0 or 1 as shown in Eq. (17):

$$P_{id}^{t+1} = \begin{cases} 0 \text{ if rand()} \geq \text{Sig}(\text{vel}_{id}^{t+1}) \\ 1 \text{ if rand()} < \text{Sig}(\text{vel}_{id}^{t+1}) \end{cases} \tag{17}$$

where P_{id}^{t+1} and vel_{id}^{t+1} are population and velocity of ith particle in dth dimension. Sig(.) is sigmoidal function which converts velocity into probability in the range [0,1]. With population size 200 and initial velocities assigned randomly in the range $[-6, 6]$ in 256×256 dimension, BPSO-based edge detection is started. The fitness of each particle is calculated. Particle with smallest fitness is called *gbest*. The whole process is repeated again after updating velocities. Approximately, 500 iterations are done.

Figure 6 shows the sample of resultant edge detection by various techniques discussed in this section.

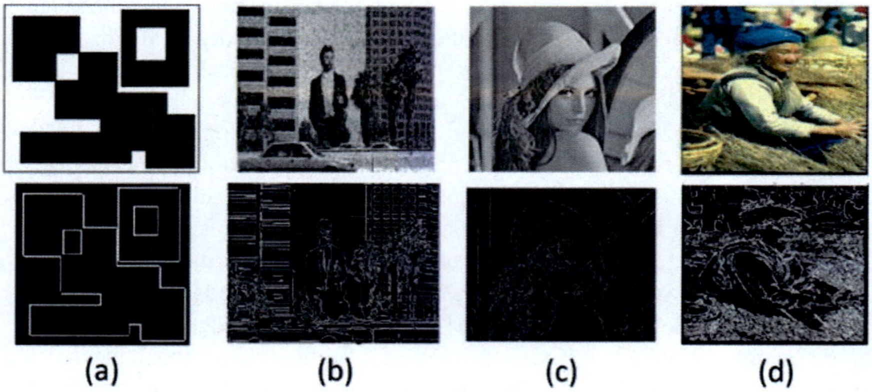

Fig. 6 First row—original image. And second row—output image of **a** [26], **b** [27], **c** [29], **d** [30]

4 Genetic Algorithm (GA)

4.1 Edge Detection Techniques Using Genetic Algorithms (GA)

Bhandarkar et al. [16] presented a GA-based technique for edge. The image is considered as 2-D chromosomes and fitness values are considered inversely proportional to the costs. Valid local edge structure is defined as shown in Fig. 7. Where $0 \leq C_i \leq 1$ and $w_i \geq 0$ are i-th cost factor and weight associated with i-th cost factor, respectively. Let us assume that there are two edge images I_i and I_j which are almost identical to each other except for the window W(x) centered around pixel x; then, comparative cost function is given in Eq. (18).

$$F\left(I_i, I_j, x\right) = \sum_{W(x)} \sum_{k} w_k\left[C_k(I_i, x) - C_k\left(I_j, x\right)\right] = \sum_{W(x)} \sum_{k} \Delta C_k\left(I_i, I_j, x\right) \quad (18)$$

The window size $W(x)$ can be a single pixel or a matrix of 3×3. If $F(I_i, \text{I}, x) < 0$ then I_i is a better configuration, if $F\left(I_i, I_j, x\right) > 0$ then I_j is better configuration, and if $F\left(I_i, I_j, x\right) = 0$, then the configurations are identical with respect to cost.

A randomly generated 2-D array of 0 s and 1 s is used to represent a chromosome in population where, 1 and 0 represent the edge pixel and non-edge pixel in the image, respectively. The cost associated with every pixel in the chromosome is calculated using decision tree technique. The fitness of the chromosome is calculated as Eq. (19).

$$\text{Fitness}(p) = (\text{cost}[\text{Worst}] - \text{cost}[p])^n \quad (19)$$

where n ranges from 2 to 5.

Fu et al. [31] proposed an edge detector based on genetic programming in which the sensitivity of the edge map to the threshold is handled by replacing linear transformation by S-shaped transformation.

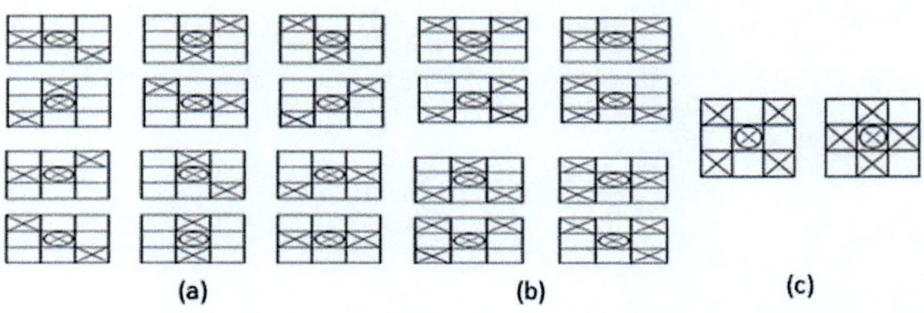

Fig. 7 **a** Valid structures of two neighbor edge, **b** valid structures of three neighbor edge, **c** valid structures of four neighbor edge

(a) X-direction

$(v^2-v)/6$	$(v^2-v)/6$	$(v^2-v)/6$
$-v/5$	$-v/5$	$-v/5$
$-v/5$	1	$-v/5$

(a) Y-direction

$(v^2-v)/6$	$-v/5$	$-v/5$
$(v^2-v)/6$	$-v/5$	1
$(v^2-v)/6$	$-v/5$	$-v/5$

(b) X-direction

0	$(v^2-v)/2$	0
0	$-v$	0
0	1	0

(b) Y-direction

0	0	0
$(v^2-v)/2$	$-v$	1
0	0	0

Fig. 8 **a** X-direction and Y-direction for fractional mask 1, **b** X and Y direction for fractional mask 2

Fu et al. [32] presented a technique for edge detection based on GA optimization. The technique maintains a balance between noise elimination and accuracy of localization. The pixels are selected automatically to avoid the problem of blurring edges and noise influence. The paper investigates that GP is able to systematically find an appropriate set of pixels for construction of subjective low-level edge detectors.

ElAraby et al. [33] presents four different edge detection algorithms. Two algorithms are based on fractional order differentiation and are used to detect edges. The other two algorithms are based on genetic programming which are used later to improve the quality of the edges. K-means principle is used to get the automatic threshold. Figure 8 shows the masks used by algorithm 1 and 2 in x and y directions. The fractional order $v = 0.2$.

Figure 9 shows the sample of resultant edge detection by various techniques discussed in this section.

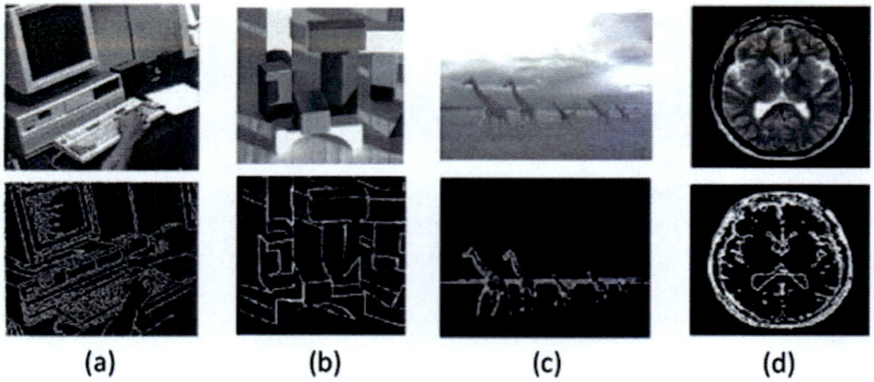

Fig. 9 First row—original image. And second row—output image of **a** [16], **b** [31], **c** [32], **d** [33]

5 Fuzzy Logic

5.1 *Fuzzy Logic System (FLS)*

The term fuzzy logic was given by Zadeh [17] for the first time in 1965. It provides ability to solve the problems with imprecise information. In recent span of 15 to 29 years, type-1 fuzzy logic system (T1-FLS) and type-2 fuzzy logic system (T2-FLS) have gained enormous popularity.

Type-1 FLS. Equation 20 shows a fuzzy set F. Type-1 FLS includes inference engine, defuzzifier, fuzzy rules and fuzzifier.

$$F = \{(x, \mu_F(x)) | x \in U\} \tag{20}$$

where U is universe of discourse, and $\mu_F(x)$ is membership function that accepts values in closed interval [0,1]. Degree of membership of each member of U is determined by the membership function. It also lies in the range [0,1]. Trapezoidal, triangular, Gaussian and singleton are few of the mostly used membership functions.

Interval Type-2 FLS. Interval T2 fuzzy set [34, 35] $\tilde{F}$ is represented as Eq. (21).

$$\tilde{F} = \left\{ \left((x, p), \mu_{\tilde{F}}(x, p) \right) | \forall p \in K_x \subseteq [0, 1] \right\} \tag{21}$$

where $0 \leq \mu_{\tilde{F}}(x, p) \leq 1$ is the membership function where, $x \in X$ and $p \in K_x \subseteq [0, 1]$. $\mu_{\tilde{F}}(x, p)$ is a type-1 fuzzy set, also called secondary set. $K_x \subseteq [0, 1]$ denotes primary membership of x.

5.2 *Fuzzy Logic-Based Edge Detection*

Gonzalez et al. [36] presented a technique for edge detection by combining type-2 fuzzy logic with Sobel technique. Fuzzy logic proves to be effective in dealing with real-world images with uncertainties. Metaheuristics are used for construction of optimal fuzzy system. The technique also uses cuckoo search (CS) and genetic algorithms (GA) inference system. The task of fuzzy inference engine is to map fuzzy set into fuzzy set. A Type-2 fuzzy system maps type-2 fuzzy set to type 1 fuzzy set for type reduction and then does defuzzification. This technique uses centroid type reduction method [34].

Bias et al. [37] developed a low-cost, easily accessible and efficient hardware for edge detection. The technique is specially focused on malaria thin blood smears. Histogram analysis-based dynamic thresholding is implemented which eliminates inter-cell interference. The proposed algorithm has three major parts, thresholding, unwanted artifacts removal and edge tracking. Dynamic threshold acquisition is

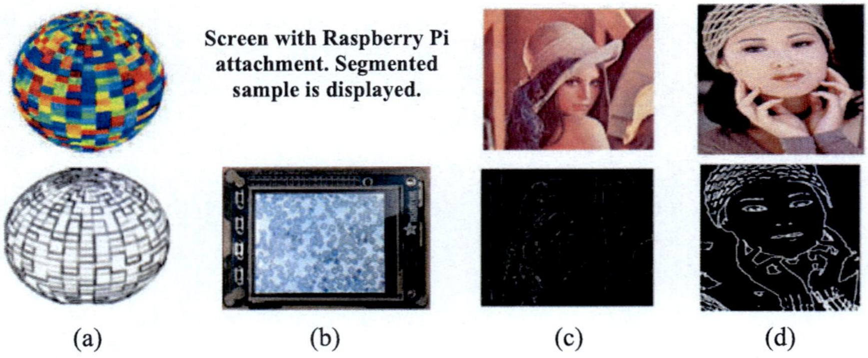

Fig. 10 First row—original image. And second row—output image of **a** [36], **b** [37], **c** [38], **d** [39]

done using histogram analysis. Unwanted artifacts are removed by constraining the minimum number of neighbors a background white pixel and a foreground black pixel to 500 and 150, respectively. Otherwise, the aforesaid cells are removed. Eight semi-ambiguous kernels are used to determine edges in binary image. Raspberry Pi3 is chosen as hardware implementation over field programmable gate array (FPGA) due to less cost and high compatibility with almost every possible IT device available in the market.

Raheja ad Kumar [38] developed a technique for edge detection which produces good-quality edges by using guided filter for sharpening of the edges. The proposed technique uses type-1 fuzzy logic.

Bozorgmehr et al. [39] used carbon nanotube field effect transistor (CNTFET) in a fuzzy edge detector. CNTFET realized with gate-all-around (GAA) structure offers outstanding electrical properties suitable for high performance fuzzy systems for edge detection. Implementing traditional edge detection methods like Sobel and Canny require complex electrical circuit, but fuzzy-based system requires simple implementation of inference rules. A 3×3 or 5×5 or wider than that mask can be used to detect the edges.

Figure 10 shows the sample of resultant edge detection by various techniques discussed in this section.

6 Conclusion

Edge detection is identifying image boundaries caused by discontinuity in intensity. Edge detection covers a major portion of computer vision. Traditional methods for edge detection mostly relied on first-order derivatives and second-order derivatives of image. Prewitts, Robberts, Sobel and Frei-Chen are few significant approximations deployed in for quantification of pixels. These are generally used in the techniques on the basis of first-order derivatives which are greatly sensitive toward the noise. Second-order derivative-based techniques like LoG are also sensitive to noise.

Due to adaptive behavior of optimization techniques used in soft computing like ACO, PSO, GA and fuzzy, these technique have substantial application in edge detection in digital images. ACO technique originally mimics the behavior of ants to find optimal path. Researchers have successfully applied ACO for detection of edges in digital images by converting the image into a perceptual graph. Number of ants initially put on the image, control parameters, pheromone evaporation rate, number of construction steps and ant movement steps majorly affect the quality of detected edges. Similarly, PSO is a direct derivation of birds flocking or fish schooling. Edge detection using PSO does not require gradient information. The population of swarms is called particle. Any swarm can be a potential solution. The edge generally separates two regions. The quality of edges detected using PSO majorly depends on constants acceleration coefficients, random constants, inertia weight and threshold. In contrast to ACO and PSO, GA-based edge detection technique majorly depend on various cost factors of edge image E corresponding to a pixel p. The randomly generated edge images are considered as 2-D chromosomes and reproduction is performed using crossover and mutation. Careful design of fitness function guarantees quality edge detection. Several edge detection algorithms use fuzzy logic system. Traditionally, FLS comprises of three modules, fuzzifier, inference engine and defuzzifier. Type-1 and type-2 FLS have prominent application in development of hardware for edge detection.

The results are presented comparatively which establishes the fact that relevant application of soft computing in edge detection, and computer vision is no less than other more popular approaches such as ML and deep learning.

References

1. Gonzalez, R. C., &Woods, R. E. (2007). *Digital image processing.* Prentice Hall.
2. Umbaugh, S. E. (2005). *Computer imaging: digital image analysis andprocessing.* CRC Press.
3. Lim, D. H. (2006). Robust edge detection in noisy images. *Computational Statistics and Data Analysis, 50*(3), 803–812.
4. Chidiac, H., & Ziou, D. (1999). Classification of image edges. In *Proceedings of the Conference on Vision Interface, Canada,* pp. 17–24.
5. Poggio, T., & Torre, V. (1984). Ill-Posed Problems and Regularization Analysis in Early Vision. Artificial Intelligence Lab. Memo, No. 773, Massachusetts Institut of Technology.
6. Poggio, T., & Torre, V. (1985). A Regularized Solution to Edge Detection. Artificial Intelligence Lab. Memo, No. 833, Massachusetts Institut of Technology.
7. Heath, M., Sarkar, S., Sanocki, T., & Bowyer, K. (1998). Comparison of edge detectors: A methodology and initial study. *Computer Vision and Image Understanding, 69*(1), 38–54.
8. Clavier, E., Clavier, S., Labiche, J.: Image sorting—image classification: A global approach. In *ICDAR '99: Proceedings of the Fifth International Conference on Document Analysis and Recognition. Washington, DC, USA* (pp. 123–129). IEEE Computer Society.
9. Canny, J. (1983). *Finding edges and lines in images.* Technical Report, Massachusetts Institute of Technology, Cambridge, MA, USA.

10. Roushdy, M. (2007). Comparative study of edge detection algorithms applying on the grayscale noisy image using morphological filter. *ICGST International Journal on Graphics, Vision and Image Processing, 6*, 17–23.
11. Sharifi, M., Fathy, M., & Mahmoudi, M. T. (2002). A classified and comparative study of edge detection algorithms. In *Proceedings of the International Conference on Information Technology: Coding and Computing* (pp. 117–120).
12. Marr, D., & Hildreth, E. (1980). Theory of edge detection. *Proceedings of the Royal Society of London Series B, 207*, 187–217.
13. Shen, J., & Castan, S. (1993). Towards the unification of band-limited derivative operators for edge detection. *Signal Processing, 31*(2), 103–119.
14. Dorigo, M., & Di Caro, G. (1999). Ant colony optimization: a new meta-heuristic. In: *Proceedings of the 1999 Congress on Evolutionary Computation-CEC99 (Cat. No. 99TH8406)* (Vol. 2, pp. 1470–1477). https://doi.org/10.1109/CEC.1999.782657
15. Kennedy, J., & Eberhart, R. (1995). Particle swarm optimization. In *Proceedings of ICNN'95-International Conference on Neural Networks* (Vol. 4). IEEE.
16. Bhandarkar, S. M., Zhang, Y., & Potter, W. D. (1994). An edge detection technique using genetic algorithm-based optimization. *Pattern Recognition, 27*(9), 1159–1180.
17. Zadeh, L. A. (1996). Fuzzy sets. In *Fuzzy sets, fuzzy logic, and fuzzy systems: selected papers by Lotfi A Zadeh* (pp. 394–432).
18. Zhuang, X. (2004). Edge feature extraction in digital images with the ant colony system. In *2004 IEEE International Conference onComputational Intelligence for Measurement Systems and Applications (CIMSA, 2004)* (pp. 133–136). https://doi.org/10.1109/CIMSA.2004.1397248
19. Ari, S., Ghosh, D., & Mohanty, P. (2014). Edge detection using ACO and F ratio. *Signal, Image and Video Processing, 8*. https://doi.org/10.1007/s11760-013-0569-4
20. Liu, X., & Fang, S. (2015). A convenient and robust edge detection method based on ant colony optimization. *Optics Communications, 353*, 147–157.
21. Raheja, S., & Kumar, A. (2020). Edge Detection using ant colony optimization under novel intensity mapping function and weighted adaptive threshold. *International Journal of Integrated Engineering, 12*(1), 13–26. Retrieved from https://publisher.uthm.edu.my/ojs/index.php/ijie/article/view/3278
22. Kumar, A., & Raheja, S. (2020). Edge detection using guided image filtering and enhanced ant colony optimization. *Procedia Computer Science, 173*, 8–17. https://doi.org/10.1016/j.procs.2020.06.003
23. Reynolds, C. W. (1987). Flocks, herds and schools: a distributed behavioral model. *Computer Graphics, 2*(4), 25–34.
24. Heppner, F., & Grenander, U. (1990). A stochastic nonlinear model for coordinated bird flocks. In S . Krasner (Ed.), *The ubiquity of chaos.* AAAS Publications.
25. Chen, Y., et al. (2004). A local linear wavelet neural network. In: *Proceedings of 5th world Congress on Intelligent Control and Automation, China* (pp. 15–19).
26. Alipoor, M., Imandoost, S., & Haddadnia, J. (2010) Designing edge detection filters using particle swarm optimization. In *2010 18th Iranian Conference on Electrical Engineering* (pp. 548–552). https://doi.org/10.1109/IRANIANCEE.2010.5507008
27. Setayesh, M., Zhang, M., & Johnston, M. (2010). Improving edge detection using particle swarm optimization. In *2010 25th International Conference of Image and Vision Computing, New Zealand* (pp. 1–8). https://doi.org/10.1109/IVCNZ.2010.6148810
28. Setayesh, M., Zhang, M., & Johnston, M. (2009). A new homogeneity-based approach to edge detection using PSO. In *Proceedings of the 24th International Conference on Image and Vision Computing, New Zealand* (pp. 231–236). IEEE Press.
29. Setayesh, M., Zhang, M., & Johnston, M. (2011). Edge detection using constrained discrete particle swarm optimisation in noisy images. *IEEE Congress of Evolutionary Computation (CEC), 2011*, 246–253. https://doi.org/10.1109/CEC.2011.5949625
30. Dahiya, P., & Singh, N. (2019). Edge detection technique using binary particle swarm optimization. *Procedia Computer Science, 167*. https://doi.org/10.1016/j.procs.2020.03.353

31. Fu, W., Johnston, M., & Zhang, M. (2012). Soft edge maps from edge detectors evolved by genetic programming. *IEEE Congress on Evolutionary Computation, 2012*, 1–8. https://doi.org/10.1109/CEC.2012.6256105
32. Fu, W., Johnston, M., & Zhang, M. (2014). Low-level feature extraction for edge detection using genetic programming. *IEEE Transactions on Cybernetics, 44*(8), 1459–1472. https://doi.org/10.1109/TCYB.2013.2286611
33. ElAraby, W. S., Madian, A. H., Ashour, M. A., Farag, I., & Nassef, M. (2017). Fractional edge detection based on genetic algorithm. In: *2017 29th International Conference on Microelectronics (ICM)* (pp. 1–4). https://doi.org/10.1109/ICM.2017.8268860
34. Liang, Q., & Mendel, J. (2000). Interval type-2 fuzzy logic systems: Theory and design. *IEEETrans. Fuzzy Syst., 8*, 535–550.
35. Mendel, J. (2001). *Uncertain rule-based fuzzy logic systems: Introduction and new directions*. Prentice-Hall.
36. Gonzalez, C. I., et al. (2016). Optimization of interval type-2 fuzzy systems for image edge detection. *Applied Soft Computing, 47*, 631–643 (2016).
37. Bias, S., Reni, S., & Kale, P. (2018). mobile hardware based implementation of a novel, efficient, fuzzy logic inspired edge detection technique for analysis of malaria infected microscopic thin blood images. *Procedia Computer Science, 141*, 374–381. https://doi.org/10.1016/j.procs.2018.10.187
38. Raheja, S., & Kumar, A. (2021). Edge detection based on type-1 fuzzy logic and guided smoothening. *Evolving Systems, 12*(2), 447–462.
39. Bozorgmehr, A., et al. (2020). A novel digital fuzzy system for image edge detection based on wrap-gate carbon nanotube transistors. *Computers & Electrical Engineering, 87*, 106811 (2020).

A Comprehensive Study of Pose Estimation in Human Fall Detection

Shikha Rastogi and Jaspreet Singh

Abstract According to a study, unexpected fall is one of the main causes of sudden demise in elder persons. Therefore, it is very important to take immediate safety measures for the people having age 65 or above, or the people who are physically or mentally disabled. A powerful fall detection system to identify and provide immediate assistance to senior citizens or the people who is prone to falls is needed. A medical alert system with fall detection allows the user to summon assistance without pressing the call button. This review paper identifies the comparison in the approaches used for fall detection based on machine learning algorithm. A brief discussion on the methods used in pose estimation like OpenPose and PoseNet, which are majorly used to detect the fall and non-fall of a person is done. Moreover, we have also discussed the privacy concern of a person while using camera-based technique for detecting fall.

Keywords Healthcare · Fall detection · Machine learning · Threshold-based · Vision-based

1 Introduction

According to a survey done by WHO, nearly 28% of population comes under the age group of 64, and this count has increased by 32–42% as the age increases over 70 [1]. Aging is the main factor of physical weakness among elderly people due to which they suffer a lot of fall-related health issues [2]. Fall is one of the major causes of death among elder people [3]. The results of such fall in elderly are fractures

S. Rastogi (✉) · J. Singh
GD Goenka University, Sohna, Gurugram, Haryana, India
e-mail: Shikha.bvcoe@gmail.com

J. Singh
e-mail: Jaspreet.singh@gdgu.org

S. Rastogi
Bharati Vidyapeeth's College of Engineering, New Delhi, India

P. K. Singh et al. (eds.), *Proceedings of Third International Conference on Computing, Communications, and Cyber-Security*, Lecture Notes in Networks and Systems 421,
https://doi.org/10.1007/978-981-19-1142-2_31

and long-lasting sicknesses, which will cause incapability of doing their own work, dependency on others, and mental pressure of falling another time [4]. So, it is very important to give extra care to these people, and prevent them from such conditions to happen by providing them proper surveillance. Moreover, if a person falls, then there should be some automated system which will provide information to their caretakers so that immediate action can be taken.

The major focus of this paper is on different pose estimation techniques used in detecting fall and non-fall of a person. There are many methods for detecting pose like OpenPose, PoseNet, AlphaPose, and MediaPipe Pose, but we majorly discussed the OpenPose and PoseNet, which are the most used methods in estimation the pose of a person for fall with high accuracy rate. We have also discussed the basic methods used for detecting fall and non-fall, like threshold and machine learning.

The rest of the paper is organized as: In Sect. 2, the related work in fall detection and pose estimation are discussed. In Sect. 3, post estimation methods are explained. In Sect. 4, how pose estimation is used in fall detection is discussed. Section 5 explains the issues related to privacy of a person while using camera for fall detection, and Sect. 6 provides the conclusion of the work.

2 Related Work

Over the years, this domain has seen significant and fruitful research. As we progress from one study to the next, the technology differs dramatically. Vallabh et al. explain that camera-based systems, though expensive, allow monitoring multiple people and require no effort on the part of the user [5]. Miaou et al. provide a novel method for detecting older people's falls; his detection system captures images with a MapCam (omni-camera) and conducts image processing on them. The processing work takes into account each individual's personal information [6]. Mobile phones are self-contained devices, i.e., they provide a sophisticated hardware and software environment which enables the construction of a widespread fall detection system. Dai et al. proposed a fall detection system based on mobile phones that can be used practically anywhere. With a few accessories, the embedded accelerometers are employed for fall detection [7]. Tao et al. presents a system-based infrared ceiling network for behavioral research and fall detection. The sensors generate numerous sequences from which they determine whether or not people are present under the sensors [8].

Rimminen et al. employs a floor sensor and pattern recognition to do near-field imaging (NFI). By identifying impedances with a framework of slight anodes underneath the floor, the floor sensor perceives individuals' areas and examples [9]. Putra et al. proposed an event-triggered machine learning (EvenT-ML) approach, which uses the distinguishing features of fall. KNN, logistic regression, and SVM were utilized for classification and regression [10]. In a semi-supervised context, Droghini et al. proposed a template-matching classifier which utilizes the one-class SVM (OCSVM) [10] to discriminate and classify between human falls and non-fall [11]. G. Sannino et al. explains mel-frequency cepstral coefficients and Gaussian mean

Supervectors (GMSs) for fall/non-fall classification using floor acoustic sensors that capture acoustic signals, which are subsequently processed to obtain the remote monitoring of old persons using wearable sensors [12]. To further the safety, comfort, and wellbeing of older individuals, Greene et al. proposed Internet of Things (IoT)-based fall detection systems were integrated into homes and cities that had preexisting "smart" infrastructure. Smart technology, such as digital AI-assisted cameras, wristbands (fitness trackers, watches etc.), and voice-controlled devices, can considerably improve the accuracy, promptness and effectiveness of a fall warning system [13]. In [14], Bosch-Jorge et al. proposed a camera-based low-cost detection system, in which wide-angle cameras are aimed to reduce the number of cameras deployed. A couple of novel features utilizing the gravity vector were introduced for fall detection. To detect fall occurrences, Cao et al. developed a new technique which used acceleration data as its basis and applied a hidden Markov model (HMM) [15].

The pose of human body shows the symmetrical structure of the humanoid movement. In the area of vision-based detection system, the pose of a body is defined by the keypoints [16]. In other words, on the basis of the movement of the joints in the skeleton, the position of the body is determined. A lot of work has been done in this field; Weiminng Chen et al. used the OpenPose technique for fall detection and with the help of three thresholds able to achieve high accuracy [16]. In [17], S. Jeong et al. used OpenPose with long short-term memory (LSTM) technique and determines the human fall with the help of coordinate and speed of human body. In [18], Y. K. Kang et al. detects the human fall with the help of PoseNet and gated recurrent unit (GRU) on the basis of velocity, height, and width of the body. In [19], Gibello Foglio et al. proposes that when PoseNet employs with convolutional neural network (CNN) and multilayer perceptrons (MLP) gives high accuracy. C. B. Lin et al. predicted that OpenPose with recurrent neural network (RNN) also works well in detecting fall with the use of skeleton map [20].

A comprehensive comparison of all the available devices is tabulated in Table 1.

3 Pose Estimation Methods

A computer vision-based system that recognizes and analyzes human posture is known as human pose estimation [27, 28]. The modeling of the human body is the most important aspect of human pose estimation. Skeleton-based, contour-based, and volume-based models are the most common types of pose estimation models. Pose estimation is an application in the field of computer vision which detects a subject's body pose (sitting, standing etc.) from an image or video. It can also be referred to as the task of determining a camera's angle or location relative to a subject. The system works by drawing/estimating the subject in 2D/3D space using up to 17 body keypoints on a person's body as shown in Fig. 1.

These keypoints represent significant joints in humans, such as the elbow and knee. Keypoints will be in different places compared to others if we bend our arms

Table 1 Quantitative comparison for related research in fall detection

Method for fall detection	Precision	Recall	Specificity	F-measure	Accuracy
Threshold analysis-based fall detection classifiers [21]	0.946	0.954	0.9	0.949	0.946
Camera-based detection using gravity vector [14]	0.93	0.906	0.94	0.918	0.969
Head tracking via RGB camera [22]	0.923	0.902	0.952	0.912	0.932
Threshold based on Omni-camera images [6]	0.789	0.909	0.86	0.844	0.81
Mobile phone sensor-based [7]	0.924	0.912	0.9	0.89	0.92
Infrared ceiling sensor network [8]	0.924	0.98	0.925	0.951	0.932
Floor sensor-based [9]	0.906	0.891	0.918	0.898	0.906
IoT-based [13]	0.88	–	0.735	–	0.88
Hidden Markov model (HMM) [15]	–	–	1	–	0.972
Support vector machine (SVM) [10]	0.931	0.863	–	0.882	0.909
One-class SVM [11]	–	–	–	0.899	–
Threshold-based [12]	0.85	–	0.92	–	0.93
Video-based approach [23]	0.833	0.882	–	0.857	–
3D head tracking [24]	0.85	0.667	0.9	0.747	0.789
Sensor-based approach [25]	0.976	0.817	0.987	0.889	0.921
Sensor-based [26]	0.967	–	0.966	–	0.983

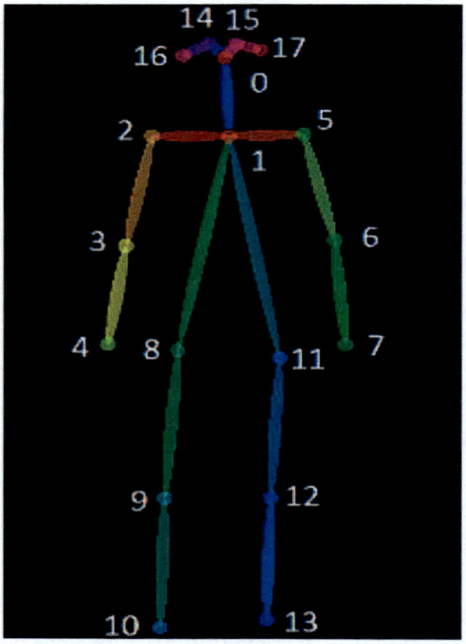

Fig. 1 Human pose estimation keypoints [29]

or legs. The majority of inanimate objects are inflexible. For example, regardless of the orientation of a brick, its corners are always the same distance apart.

A distinction must also be established between 2 and 3D pose estimation techniques. In 2D pose estimation, the position of the keypoints in 2D space in relation to an image is estimated. An X and Y coordinate is determined for each keypoint, thus we get the coordinates for the keypoint. It is when we add the third coordinate (Z), we step into 3D pose estimation.

An important classification criterion is whether the algorithm detects one thing/subject or can it work for multiple objects in a frame simultaneously. Based on this criterion, the techniques are classified into two categories: single-pose and multi-pose estimation. As the name suggests, multi-pose estimation approaches are capable of detecting and tracking body poses of multiple subjects at a time, whereas single-pose estimation approaches can detect and track only a single subject at a particular time. We can monitor an object or person in a familiar and 3D space at an extraordinarily detailed level using posture estimation. This tremendous capacity brings up a plethora of potential uses.

In some important aspects, pose estimation varies from other standard computer vision tasks which includes tasks like object detection, or recognition tasks. Object detection is a task that locates objects within an image frame or a video. Though important, object detection can only provide a rough estimate and a bounding box encompassing the object. Pose estimation, on the other hand, provides a much more detailed analysis such as the exact location of the subject's different body parts. When we analyze how pose estimation may be used to automatically detect human movement, we can see how powerful it is. Pose estimation has the potential to develop a new wave of automated systems, meant to quantify the precision of human movement, from virtual sports coaches and AI-powered personal trainer to tracking movements on factory floors to ensure worker's safety.

Convolutional neural networks are used in deep learning designs that are suited for pose estimation. There are two types of approaches—the top down approach and the bottom up approach. The model employs the bottom up approach to learn all the occurrences of a specific keypoint in an image and then create a grouping in the form of a skeleton. The top down approach works in the opposite way. An object detector is used to create bounding boxes around each instance of the object and then it tries to estimate the presence of the keypoints in the bounded area.

Many specific neural networks are used for this task, but here, we have highlighted a few reliable ones—PersonLab (PoseNet) and OpenPose.

3.1 PersonLab (PoseNet)

The PersonLab [30] employs a box free bottom up approach to pose estimation.

The method involves the occurrence of two steps. First is the detection of K keypoints, and then, the grouping of these points into personal instances like in Fig. 2. With each iteration, improvements in the form of short, mid-, and long-range

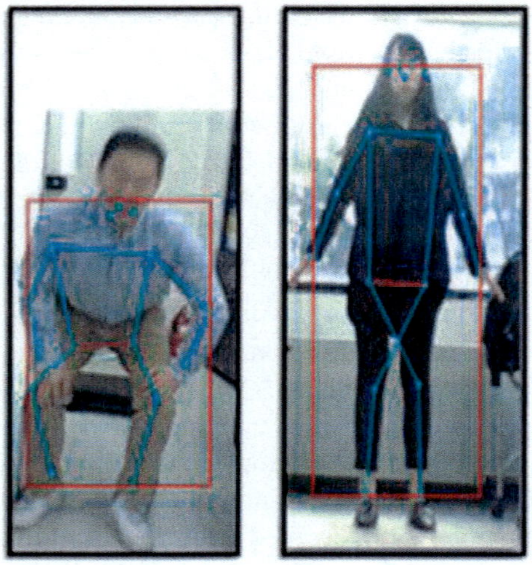

Fig. 2 Keypoint detection using PoseNet [18]

Table 2 Comparative analysis of ML algorithm on the basis of accuracy

Method	Features	Accuracy (%)
PoseNet + SVM [31]	Skeleton map	93.67
PoseNet + GRU [18]	Velocity, height, width	99.80
PoseNet + CNN [19]	Depth images	99.74
PoseNet + MLP [19]	Depth images	99.84

offsets are made to the heatmaps. The heatmap is used to gather any large area of the image where keypoint can be found, whereas the offsets help to fine tune the predictions and get a precise final result. Table 2 shows the comparative study of PoseNet with different machine learning algorithms. Figure 3 shows the graphical representation of the comparison done on PoseNet with ML algorithms, and on the basis of that, PoseNet with MLP gives highest accuracy of 99.84%.

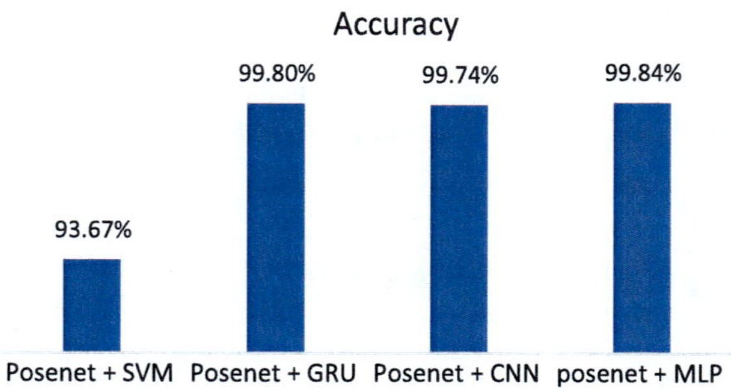

Fig. 3 Performance analysis of ML algorithms

Fig. 4 Skeleton detection using OpenPose [16]

An issue with using the PersonLab is its over-reliance on keypoint-level annotations for instance segmentation. The research intends to investigate approaches to address this constraint in the future, such as weakly supervised component finding.

3.2 OpenPose

OpenPose [32] is a real-time multiple-person detection library that is the first to demonstrate the ability to recognize human body, face, and foot keypoints simultaneously. The first step is to feed an RGB (red, green, blue) image into a two-pronged convolutional neural network (CNN), which will yield two separate outputs (Fig. 4).

The confidence maps of various body parts, such as eyes, elbows, knees, and nose are predicted by the first branch at the top. The affinity fields, which describe a degree of relationship between different body components in the input image, are predicted by the bottom branch. Last but not least, greedy inference is used to process the confidence maps and affinity fields. The 2D keypoint pose estimation outputs for all subjects in the image. We use multistage to improve the depth of the neural network technique in order to capture more fine outputs, which implies that the network is layered one on top of the other at each stage. Table 3 compares the machine learning algorithm which are used for fall detection with OpenPose. Figure 5 demonstrates the performance evaluation of ML algorithm on the basis of accuracy, and it is observed that when OpenPose is used with LSTM for fall detection, it gives high accuracy as compared to other ML algorithms.

Table 3 Comparative analysis of OpenPose with ML algorithm

Method	Features	Accuracy (%)
Open Pose + LSTM [17]	Coordinate, speed	98.70
OpenPose + convolutional neural network [33]	Skeleton map	91.70
OpenPose + three thresholds [16]	Velocity, Angle, Ratio	97
OpenPose + recurrent neural network [20]	Skeleton map	90.10
OpenPose + SSD MobileNet [34]	Skeleton map	93

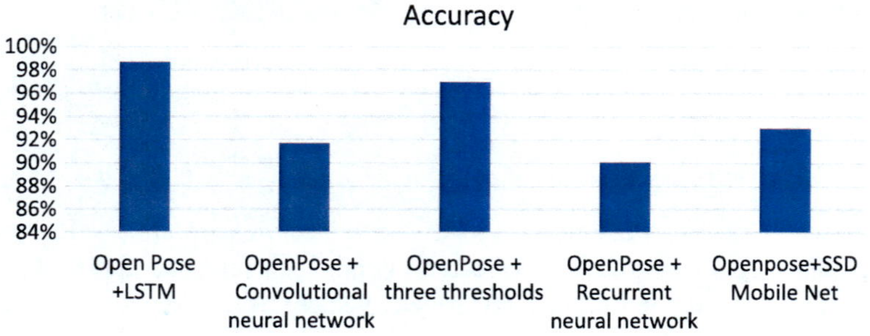

Fig. 5 Performance comparison of ML algorithm

4 Pose Estimation and Fall Detection

To determine and warn of the occurrence of a fall, we use the power and capability of posture estimation algorithms. The subject's stance is constantly monitored by the camera-based sensors. This gives us a set of 17 essential keypoints of a human body. These keypoints are used for determining the angle of the human body with respect to the surroundings, in particular the ground.

We estimate the angle in all conditions. We also keep a track of the timings between any pose changes. If the position of a person changes from a position like, and not limited to, Fig. 6a to a position like Fig. 6b, we predict it to be a fall. The angle formed by the subject in Fig. 6a is close to 90° and in a lying position Fig. 6b, it is closer to either 0° or 180°. Our approach classifies a change of body angle to 0° or 180° as a fall. It is possible that the subject has voluntarily lied down. To rule out any such misclassifications, the approach takes into account the time taken for the pose change. Since falls tend to last for a smaller duration as compared to voluntary movements, if the pose changes to Fig. 6b in a very small duration of time, our system classifies the action as a fall and alerts the attendee. We set the threshold of fall to a duration less than what it takes for an average person to lie down voluntarily. When the subject's pose change time is less than a predefined threshold, the system detects it as a fall.

In addition to the angle of the body, our system also assumes a sphere of presence for certain body parts on the subject's body. Some of these parts include the nose,

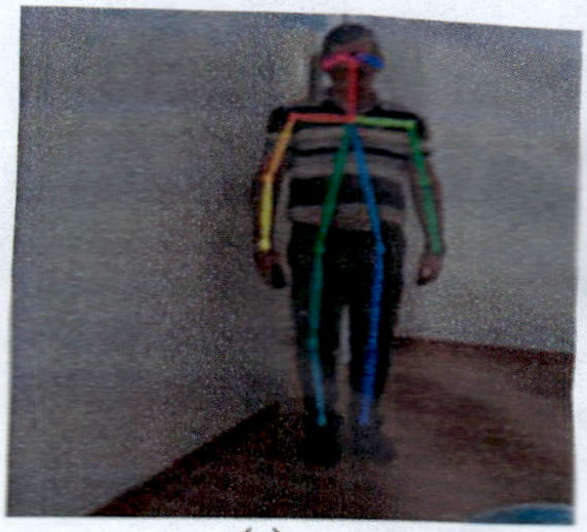

(a)

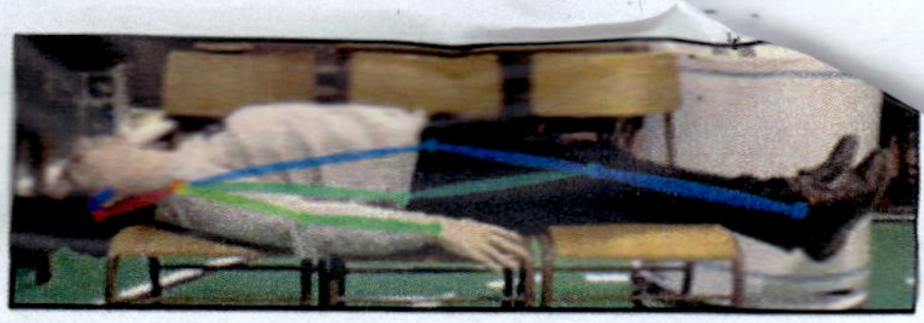

(b)

Fig. 6 **a** Standing person pose. **b** Lying person pose

shoulders, eyes etc. Since our pose estimation algorithm already estimates these keypoints, we can keep a track of them to estimate a fall. If any of these parts moves out of the sphere of presence in a time less than the predefined threshold, we can classify it as a fall.

5 Privacy Concerns Related to Camera-Based Sensors

In most cases, video surveillance systems are implemented to improve the safety and security of people or property in the monitored regions. Robbery, vandalism, stealing, and terrorism are all common threats. Other application scenarios, such as home monitoring or assisted living, are more intimate and private. However, due to modern embedded systems' onboard processing capabilities, it is now possible to compensate for this privacy loss by making security and privacy protection built-in to video surveillance cameras.

Traditionally, privacy-preserving computer vision systems guarantee privacy by applying existing software-based privacy algorithms after capture. Single and multi-aperture sensor designs have come into consideration which have been presented in the domain of fixed privacy optics [35]. The advantages of fixed privacy optics are twofold. Firstly, they do not contain any of the cyber threats related to software-based sensors, since they are devoid of electronics of any kind. Next, due to the fact that the optics filter any incident light-field directly, the number of onboard processes reduces significantly and the onboard power requirements are reduced to zero. A programmable privacy optics system consists of an alignment sensor having the ability to pre-capture privacy alignment and an output sensor (roughly equivalent to an ideal pinhole camera), the seeing path of which is split between the scene and an active optical mask, such as a projector or electronic display.

Masking sensitive targets involves five steps with this setup:

(1) [illegible] alignment sensor, capture an alignment image.
(2) [illegible]e the alignment image for target segmentation.
(3) Create privacy masks using target segmentations.
(4) Using the output sensor, capture a private image.

The system can also use thermal cameras to determine poses instead of RGB cameras. Thermal cameras use limited private information and are comparatively much safer to normal cameras, when considering privacy. A Stanford-based research [36] showed promising results in the domain of thermal imaging-based pose estimation. The results improved manifold when a depth camera was used in conjunction with the thermal camera.

6 Conclusion

Injuries due to abrupt fall of elderly people is a matter of serious concern in nursing homes and hospitals. We have reviewed many of the previous works in this area and also compared the efficiency and accuracy of the systems. Through this, we were able to determine that most of the research presented is based on hardware-based approaches which are comparatively expensive and difficult to setup. The domain of camera-based sensors has seen less research as compared to its counterparts. We have also discussed the most used pose estimation methods to determine whether a person has fallen. It keeps track of a person's body angle to classify movements as a fall and tracks time to rule out any false positives. We provide an overview of some of the leading pose estimation algorithms and also determined that OpenPose works best with LSTM with 98.7% accuracy, and PoseNet works best with MLP with 99.84% accuracy. Lastly, we also explained the privacy concerns which arise due to the use of camera-based sensors. We provide some alternatives to existing systems to set aside any such doubts.

References

1. Rastogi, S., & Singh, J. (2021). A systematic review on machine learning for fall detection system. *Computational Intelligence*.
2. Núñez-Marcos, A., Azkune, G., & Arganda-Carreras, I. (2017). Vision-based fall detection with convolutional neural networks. In *Wireless communications and mobile computing*.
3. Ambrose, A. F., Paul, G., & Hausdorff, J. M. (2013). Risk factors for falls among older adults: A review of the literature. *Maturitas, 75*(1), 51–61.
4. Gates, S., Fisher, J. D., Cooke, M. W., Carter, Y. H., & Lamb, S. E. (2008). Multifactorial assessment and targeted intervention for preventing falls and injuries among older people in community and emergency care settings: Systematic review and meta-analysis. *BMJ, 336*(7636), 130–133.

5. Vallabh, P., & Malekian, R. (2018). Fall detection monitoring systems: A comprehensive review. *Journal of Ambient Intelligence and Humanized Computing, 9*(6), 1809–1833.
6. Miaou, S. G., Sung, P. H., & Huang, C. Y. (2006, April). A customized human fall detection system using omni-camera images and personal information. In *1st Transdisciplinary Conference on Distributed Diagnosis and Home Healthcare, 2006. D2H2* (pp. 39–42). IEEE.
7. Dai, J., Bai, X., Yang, Z., Shen, Z., & Xuan, D. (2010). Mobile phone-based pervasive fall detection. *Personal and Ubiquitous Computing, 14*(7), 633–643.
8. Tao, S., Kudo, M., & Nonaka, H. (2012). Privacy-preserved behavior analysis and fall detection by an infrared ceiling sensor network. *Sensors, 12*(12), 16920–16936.
9. Rimminen, H., Lindström, J., Linnavuo, M., & Sepponen, R. (2010). Detection of falls among the elderly by a floor sensor using the electric near field. *IEEE Transactions on Information Technology in Biomedicine, 14*(6), 1475–1476.
10. Putra, I. P. E. S., Brusey, J., Gaura, E., & Vesilo, R. (2018). An event-triggered machine learning approach for accelerometer-based fall detection. *Sensors, 18*(1), 20.
11. Droghini, D., Ferretti, D., Principi, E., Squartini, S., & Piazza, F. (2017). A combined one-class SVM and template-matching approach for user-aided human fall detection by means of floor acoustic features. In *Computational intelligence and neuroscience.*
12. Sannino, G., De Falco, I., & De Pietro, G. (2015). A supervised approach to automatically extract a set of rules to support fall detection in an mHealth system. *Applied Soft Computing, 34*, 205–216.
13. Greene, S., Thapliyal, H., & Carpenter, D. (2016, December). IoT-based fall detection for smart home environments. In *2016 IEEE International Symposium on Nanoelectronic and Information Systems (iNIS)* (pp. 23–28). IEEE.
14. Bosch-Jorge, M., Sánchez-Salmerón, A. J., Valera, Á., & Ricolfe-Viala, C. (2014). Fall detection based on the gravity vector using a wide-angle camera. *Expert Systems with Applications, 41*(17), 7980–7986.
15. Cao, H., Wu, S., Zhou, Z., Lin, C. C., Yang, C. Y., Lee, S. T., & Wu, C. T. (2016, August). A fall detection method based on acceleration data and hidden Markov model. In *2016 IEEE International Conference on Signal and Image Processing (ICSIP)* (pp. 684–689). IEEE.
16. Chen, W., Jiang, Z., Guo, H., & Ni, X. (2020). Fall detection based on key points of human-skeleton using openpose. *Symmetry, 12*(5), 744.
17. Jeong, S., Kang, S., & Chun, I. (2019, June). Human-skeleton based fall-detection method using LSTM for manufacturing industries. In *2019 34th International Technical Conference on Circuits/Systems, Computers and Communications (ITC-CSCC)* (pp. 1–4). IEEE.
18. Kang, Y. K., Kang, H. Y., & Weon, D. S. (2020). Fall detection based on human skeleton keypoints using GRU. *International Journal of Internet, Broadcasting and Communication, 12*(4), 83–92.
19. Gibello Foglio, D. (2021). Pose classification for assistive unmanned vehicles with deep learning at the edge [Doctoral dissertation, Politecnico di Torino].
20. Lin, C. B., Dong, Z., Kuan, W. K., & Huang, Y. F. (2021). A framework for fall detection based on OpenPose Skeleton and LSTM/GRU models. *Applied Sciences, 11*(1), 329.
21. Zhang, C., Lai, C. F., Lai, Y. H., Wu, Z. W., & Chao, H. C. (2017). An inferential real-time falling posture reconstruction for Internet of healthcare things. *Journal of Network and Computer Applications, 89*, 86–95.
22. Foroughi, H., Rezvanian, A., & Paziraee, A. (2008, December). Robust fall detection using human shape and multi-class support vector machine. In *2008 Sixth Indian Conference on Computer Vision, Graphics and Image Processing* (pp. 413–420). IEEE.
23. Rougier, C., Meunier, J., St-Arnaud, A., & Rousseau, J. (2007, May). Fall detection from human shape and motion history using video surveillance. In *21st International Conference on Advanced Information Networking and Applications Workshops (AINAW'07)* (Vol. 2, pp. 875–880). IEEE.
24. Rougier, C., Meunier, J., St-Arnaud, A., & Rousseau, J. (2006, August). Monocular 3D head tracking to detect falls of elderly people. In *2006 International Conference of the IEEE Engineering in Medicine and Biology Society* (pp. 6384–6387). IEEE.

25. Chen, O. T. C., & Kuo, C. J. (2014, August). Self-adaptive fall-detection apparatus embedded in glasses. In *2014 36th Annual International Conference of the IEEE Engineering in Medicine and Biology Society* (pp. 4623–4626). IEEE.
26. Kwolek, B., & Kepski, M. (2014). Human fall detection on embedded platform using depth maps and wireless accelerometer. *Computer Methods and Programs in Biomedicine, 117*(3), 489–501.
27. Haralick, R. M., Joo, H., Lee, C. N., Zhuang, X., Vaidya, V. G., & Kim, M. B. (1989). Pose estimation from corresponding point data. *IEEE Transactions on Systems, Man, and Cybernetics, 19*(6), 1426–1446.
28. Toshev, A., & Szegedy, C. (2014). Deeppose: Human pose estimation via deep neural networks. In *Proceedings of the IEEE Conference on Computer Vision and Pattern Recognition* (pp. 1653–1660).
29. Divya, R., SCET, T., Riya, T. B., Johns, R., Sreelakshmi, T. J., & Davies, T. (2021). *Fall detection using OpenPose.*
30. Papandreou, G., Zhu, T., Chen, L. C., Gidaris, S., Tompson, J., & Murphy, K. (2018). Personlab: Person pose estimation and instance segmentation with a bottom-up, part-based, geometric embedding model. In *Proceedings of the European Conference on Computer Vision (ECCV)* (pp. 269–286).
31. Youssfi Alaoui, A., Tabii, Y., Oulad Haj Thami, R., Daoudi, M., Berretti, S., & Pala, P. (2021). Fall detection of elderly people using the manifold of positive semidefinite matrices. *Journal of Imaging, 7*(7), 109.
32. Cao, Z., Hidalgo, G., Simon, T., Wei, S. E., & Sheikh, Y. (2019). OpenPose: Realtime multi-person 2D pose estimation using part affinity fields. *IEEE Transactions on Pattern Analysis and Machine Intelligence, 43*(1), 172–186.
33. Xu, Q., Huang, G., Yu, M., & Guo, Y. (2020). Fall prediction based on key points of human bones. *Physica A: Statistical Mechanics and its Applications, 540*, 123205.
34. Sun, G., & Wang, Z. (2020, April). Fall detection algorithm for the elderly based on human posture estimation. In *2020 Asia-Pacific Conference on Image Processing, Electronics and Computers (IPEC)* (pp. 172–176). IEEE.
35. Pittaluga, F., & Koppal, S. J. (2016). Pre-capture privacy for small vision sensors. *IEEE Transactions on Pattern Analysis and Machine Intelligence, 39*(11), 2215–2226.
36. Mehra, R., Chetty, M., & Kamalu, J. (2017). *Multiperson pose estimation using thermal and depth modalities* [Technical Report 1]. Department of Computer Science, Stanford University, Stanford, CA, USA.

Study and Develop a Convolutional Neural Network for MNIST Handwritten Digit Classification

Disha Jayswal, Brijeshkumar Y. Panchal, Bansari Patel, Nidhi Acharya, Rikin Nayak, and Parth Goel

Abstract The goal of this analysis has been on the development of handwritten digit recognition with the use of the MNIST dataset. In the latest days, the identification of handwritten digits has become a challenging research topic in machine learning. Due to physically formed digits having varying lengths, widths, orientations, and positions. It may be utilized in several ways, such as the amount and signature on bank checks, the location of postal and tax papers, and so on. This research used CNN for recognition. Total four steps followed by pre-processing, feature extraction, training CNN, classification, and recognition. Along with its great higher accuracy, CNN outperforms other methods in detecting essential characteristics without the need for human intervention. On top of that, it incorporates unique levels of convolution and pooling processes. Through CNN, 97.78% accuracy was obtained.

Keywords Convolutional neural network (CNN) · Handwritten digit recognition · MNIST dataset · Neural network

D. Jayswal · B. Y. Panchal (✉) · B. Patel · P. Goel
Department of Computer Science and Engineering, Faculty of Technology and Engineering (FTE), Devang Patel Institute of Advance Technology and Research (DEPSTAR), Charotar University of Science and Technology (CHARUSAT), Changa, India
e-mail: panchalbrijesh02@gmail.com

N. Acharya
Department of Computer Engineering, Faculty of Technology and Engineering (FTE), Devang Patel Institute of Advance Technology and Research (DEPSTAR), Charotar University of Science and Technology (CHARUSAT), Changa, India

R. Nayak
V. T. Patel Department of Electronics and Communication Engineering, CHARUSAT Space Research and Technology Center, Charotar University of Science and Technology, (CHARUSAT), CHARUSAT Campus, Changa 388421, India

P. K. Singh et al. (eds.), *Proceedings of Third International Conference on Computing, Communications, and Cyber-Security*, Lecture Notes in Networks and Systems 421,
https://doi.org/10.1007/978-981-19-1142-2_32

1 Introduction

Handwritten digit identification is a difficult task. The aspect that complicates the situation is the natural diversity in syntaxes at various times. As a result, developing a general recognition system process of tracking numbers and compositions produced by a variety of authors is also not possible. Yet, one of the really difficult issues for this assignment is identifying the most useful characteristics with the strong discriminating capacity to increase accuracy rate while minimizing potential. This is an important task for which conventional databases exist, allowing alternative techniques to be tested and verified.

CNN has recently emerged as one of the useful approaches, gambling a key function in more than a few of new fulfillment and traumatic ML knowledge of packages consisting of mission ImageNet item identification, photo segmentation, and face recognition. As a result, here researcher selected CNN for handwritten digit recognition.

Recognition is recognizing or differentiating an object or a person from previous experiences or learning. So, by this, it can be made out easily that handwritten recognition is recognizing or identifying the digits of any document [1]. The MNIST handwritten digit classification problem is a well-known dataset utilized in computer vision and deep learning.

The MNIST dataset is utilized as a database of different handwritten digits. MNIST is a vast database of handwritten numeric or digits that are utilized to train and test machine learning algorithms. The training and testing images in this dataset total 60,000 and 10,000 photos, respectively.

The following are how the paper is structured: Sect. 2 has CNN modeling for the classification of the handwritten digit. Section 3 has MNIST dataset, Sect. 4 has literature survey, and Sect. 5 has experimental results and discussion, then conclusion and future work.

2 CNN Modeling for Classification of Handwritten Digits

In artificial intelligence, CNNs are a type of feed-forward neural network frequently employed for image recognition. CNN takes data in the form of multidimensional arrays as input. It performs admirably when dealing with enormous amounts of tagged data. The receptive field is what CNN uses to extract every piece of the input image. It assigns weights to each neuron based on the significance of the receptive field. As a result, each neuron can tell itself apart from the others. The architecture of layers in CNN.

In Fig. 1, we can see a simple CNN model 1. The input layer is the initial layer, with a 28-by-28-pixel input image. Then there's the convolution layer, which may combine with the input image to produce four feature maps. The pooling layer is the third layer. It calculates the input feature maps' local average or maximum. The next

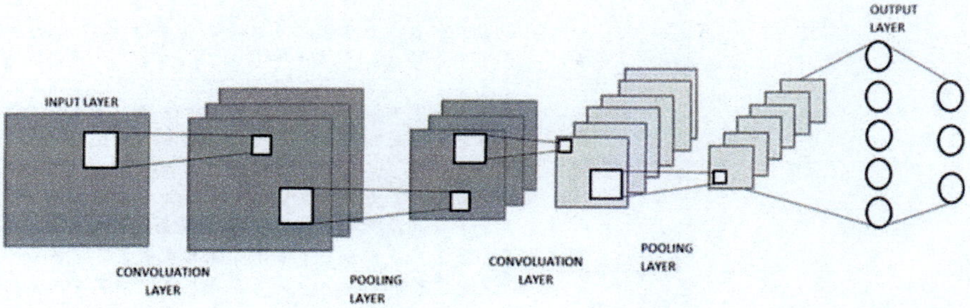

Fig. 1 Convolutional neural network

repeated except for the number and size of convolution kernels, the convolution layer, and the pooling layer work similarly to the preceding ones. Eventually, the output layer is a fully linked layer where the classifier's outcome is the output neurons' largest value [2].

2.1 Advantages of CNN

CNN is widely used replacing various other algorithms these days. As CNN works way better than various other algorithms because of its high computational efficiency, detects [3] the important features without any human supervision. Upon all this, it uses unique layers of convolution, and pooling operations have been inculcated.

3 Dataset MNIST

The MNIST dataset, which was published by Y. LeCun of New York University's Courant Institute, stands for Modified National Institute of Standards and Technology. It's made up of 60,000 squares of 2828-pixel grayscale [4] handwritten numbers ranging from 0 to 9. This dataset contains 60,000 training images and 10,000 testing images. For the test set, more than 250 different writers were picked for handwritten [5] data samples (Fig. 2).

4 Literature Survey

See Table 1.

Fig. 2 MNIST dataset

5 Experimental Result and Discussion

The basic idea of this paper is to get the best accuracy possible with the CNN algorithm. Hence, there is a long procedure [14] that acts behind it consisting of various steps as shown in Fig. 3.

The above figure illustrates the architecture diagram of the proposed system. It contains four stages starting from taking the input dataset and ending with giving the output of the recognized digit. The four stages are as follows: [1]

A. Pre-processing
B. Feature extraction
C. Training CNN
D. Classification and recognition.

A. Pre-processing

Various tasks on the input image must be completed during this pre-processing step. It is defined in such a way that binarization converts a grayscale image to a binary image [5].

Essentially, the training set photos will be thresholded into a binary image to reduce the amount of data [15].

B. Feature Extraction

Following the conclusion of the pre-processed images are now represented in matrix form, which includes pixels from extremely large images. It aids in obtaining the necessary digit information from photos. Feature extraction is the term for this activity. The data redundancy is removed at this stage [1].

C. Training CNN

Training starts from the very first layer which is the input layer where the MNIST dataset is a monochromic picture with the 28 × 28 size is taken. Then there's the convolution layer, which may combine with the input image to produce four feature maps. Next is the pooling layer, in which the pooling computation [16] will reduce the extension of the data.

Table 1 Literature survey

S. No.	Title	Year of publication	Method or algorithm or techniques	Accuracy (%)
1	An Efficient CNN Model for Automated Digital Handwritten Digit Classification [6]	April 2021	CNN architectures (training and validation), MNIST dataset	99.93
2	Comparative Analysis of Algorithms Used in Handwritten Digit [7]	June 2021	Decision tree, logistic regression, *k*-nearest neighbors (KNN), and deep learning algorithm CNN	86.6, 92.6, 96.89, 99
3	Handwritten Digit Recognizer using Deep Neural Network [7]	April 2021	Deep neural network	99.19
4	Convolutional neural network-based ensemble methods to recognize Bangla handwritten character [8]	June 2021	CNN	98.68
5	Evaluating Machine Learning Models for Handwriting Recognition-based Systems under Local Differential Privacy [9]	2021	Machine learning models	97
6	Hybrid CNN-SVM Classifier for Handwritten Digit Recognition [10]	2020	Convolutional neural networks (CNN) and support vector machine (SVM)	99.28
7	Handwritten Digit Recognition of MNIST dataset using Deep Learning state-of-the-art Artificial Neural Network and CNN (CNN) [11]	2021	Deep learning state-of-the-art artificial neural network (ANN) and convolutional neural network (CNN)	80

(continued)

Table 1 (continued)

S. No.	Title	Year of publication	Method or algorithm or techniques	Accuracy (%)
8	Handwritten Digit Recognition with Feed-Forward Multi-Layer Perceptron and Convolutional Neural Network Architectures [12]	2020	Feed-forward multi-layer perceptron and convolutional neural network architectures	97.44, 98.76
9	Handwritten Digit Recognition by Deep Learning for Automatic Entering of Academic Transcripts [13]	2020	Deep learning for automatic entering of academic transcripts	98.01
10	Implementation of CNN for Handwritten Digit Recognition [14]	2020	FPGA implementation of CNN	97.57

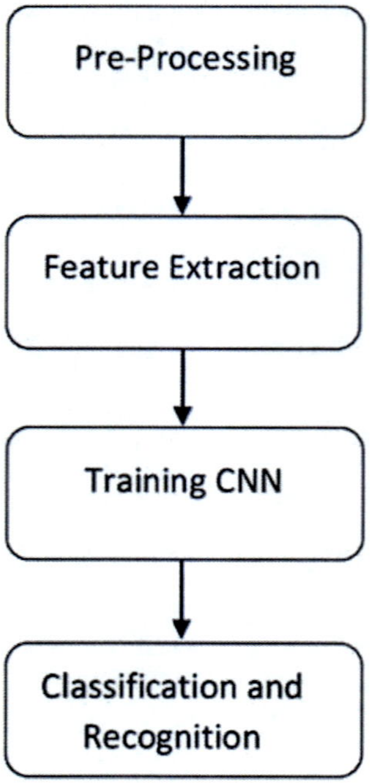

Fig. 3 Block diagram of proposed work

```
Epoch 1/10
118/118 [==============================] - 2s 13ms/step - loss: 1.0223 - accuracy: 0.6974 - val_loss: 0.2267 - val_accuracy: 0.9325
Epoch 2/10
118/118 [==============================] - 1s 11ms/step - loss: 0.2398 - accuracy: 0.9308 - val_loss: 0.1549 - val_accuracy: 0.9528
Epoch 3/10
118/118 [==============================] - 1s 11ms/step - loss: 0.1704 - accuracy: 0.9510 - val_loss: 0.1232 - val_accuracy: 0.9624
Epoch 4/10
118/118 [==============================] - 1s 11ms/step - loss: 0.1301 - accuracy: 0.9617 - val_loss: 0.1037 - val_accuracy: 0.9678
Epoch 5/10
118/118 [==============================] - 1s 11ms/step - loss: 0.1054 - accuracy: 0.9694 - val_loss: 0.0949 - val_accuracy: 0.9702
Epoch 6/10
118/118 [==============================] - 1s 11ms/step - loss: 0.0859 - accuracy: 0.9738 - val_loss: 0.0832 - val_accuracy: 0.9737
Epoch 7/10
118/118 [==============================] - 1s 10ms/step - loss: 0.0736 - accuracy: 0.9778 - val_loss: 0.0828 - val_accuracy: 0.9735
Epoch 8/10
118/118 [==============================] - 1s 10ms/step - loss: 0.0619 - accuracy: 0.9819 - val_loss: 0.0773 - val_accuracy: 0.9760
Epoch 9/10
118/118 [==============================] - 1s 11ms/step - loss: 0.0580 - accuracy: 0.9823 - val_loss: 0.0759 - val_accuracy: 0.9752
Epoch 10/10
118/118 [==============================] - 1s 11ms/step - loss: 0.0446 - accuracy: 0.9865 - val_loss: 0.0726 - val_accuracy: 0.9769
```

Fig. 4 Training CNN and the improved accuracies during each epoch

Again comes another set of convolution layer and pooling [17] layers which have similar operation patterns, except for the fact that the numeral amount and size of convolution kernels. Our final layer that is the output layer which is fully connected layer as the name suggests it combines all the neuron to produce and output where the result of the classifier is the maximum value of output neurons [2].

One model has been constructed; it is required to build and apply it. During the fitting phase, the algorithm will go over the dataset and grasp the relationships. This will educate many more times as specified along with the procedure. We've established ten epochs in our example. Throughout the process, the CNN model will learn as well as making errors. There is indeed a cost for [18] each error made by the model, which is reflected in the lower number for each epoch. In summary, by the conclusion of the last epoch, the model should provide the least amount of losses or as much precision as feasible (Fig. 4).

D. Classification and Recognition

Figure 5 is the tabular representation of the confusion matrix which shows digits 0–9 as class 1–10 respectively that mean class 1 corresponds to 0, class 2 corresponds to 1, class 3 corresponds to 2, and goes on. Vertical rows represent classifier results, and horizontal columns represent our true data. Here classifier results mean the value of digit which is recognized by the classifier and truth data is the actual value of the digit. The diagonally mentioned figure shows the number of correct predictions of the classifier corresponding to true values. Let's understand it better by taking cell of (class 1(0), class1(0)) which gives 970, which means that there are 970 right predictions of zeroes. Now for considering values other than diagonal shows that how many values are predicted as wrong and what digit they have been predicted. Let's understand it better by taking cell of (class 9(8), class 7(6)) which gives 6, it means that true value 8 has been wrongly predicted as 6 for 6 times (Fig. 5).

Figure 6 is the graphical representation of our model accuracy corresponding to the number of epochs taken. The numeral epoch is a hyperparameter that interprets the number of times that the learning algorithm will work through the whole dataset of training. And here we can see a general pattern that as the number of epochs for training and testing dataset increases corresponding to that accuracy also increases.

	Truth data											
		Class 1	Class 2	Class 3	Class 4	Class 5	Class 6	Class 7	Class 8	Class 9	Class 10	Classification overall	[illegible]
Classifier results	Class 1	970	0	8	0	2	2	4	2	3	3	994	97.586%
	Class 2	0	1122	2	1	0	0	2	5	1	5	1138	98.594%
	Class 3	1	2	1005	2	0	0	0	12	1	1	1024	98.145%
	Class 4	0	2	5	990	0	7	1	4	3	3	1015	97.537%
	Class 5	0	0	1	0	957	2	2	1	3	8	974	98.255%
	Class 6	2	0	1	3	0	868	2	0	1	2	879	98.749%
	Class 7	4	3	3	0	11	8	947	0	6	1	983	96.338%
	Class 8	1	0	3	5	2	1	0	994	1	5	1012	98.221%
	Class 9	1	6	4	6	1	3	0	3	951	7	982	96.843%
	Class 10	1	0	0	3	9	1	0	7	4	974	999	97.497%
	Truth overall	980	1135	1032	1010	982	892	958	1028	974	1009	10000	
		98.98%	98.855%	97.384%	98.02%	97.454%	97.309%	98.852%	96.693%	97.639%	96.531%		

Overall accuracy (OA): 97.78%

Kappa[1]: 0.975

Fig. 5 Confutation matrix of CNN classifier

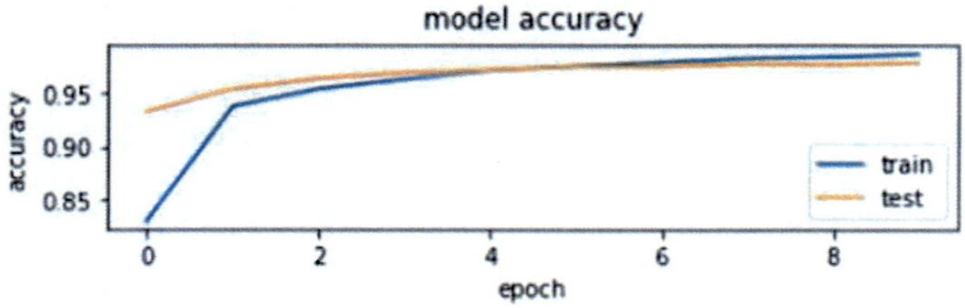

Fig. 6 Model accuracy of proposed work

Figure 7 is the graphical representation of our model loss corresponding to the number of epochs taken. The numeral epochs are a hyperparameter that interprets the number of times that the learning algorithm will work through the whole dataset of training. And here we can see a general pattern that as the number of epochs for training and testing dataset increases corresponding to that loss decreases.

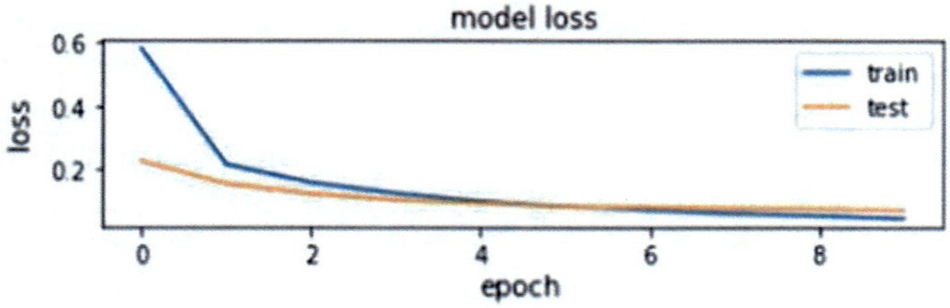

Fig. 7 Model loss of proposed work

6 Conclusion

It was to be found that the algorithm CNN which has given the accuracy of 97.78% works better than various other algorithms. CNN provides tremendous computing efficiency while also detecting significant traits without the need for human intervention. CNN being a special architecture to detect complex features in data gives us the convenience to perform recognition. Upon all this, it also has one of the unique features of combining various convolution and pooling layers which help us to improve accuracy. CNN models can now run on any device, making them globally appealing by combining two layers of convolutions and pooling each.

7 Future Work

Future efforts can take a look at the effectiveness in gaining in-depth knowledge and put in it to greater complicated problems under image recognition. Such that an easy-to-use application can be created for mobile phones or pc which can take in input and recognize it and gives us the identity of the digit input.

The obtained following outcomes can be made way more detailed and accurate by using numerous amounts of convolution layers and a huge number of hidden neurons. And also, accuracy can be increased by using some hybrid model which consists of more than one algorithm combinedly. In the future, this project can be inculcated with real-time data using real-time handwritings of humans.

References

1. Rudraswamimath, V.R., & Bhavanishankar, K. (2019). Handwritten digit recognition using CNN. *International Journal of Innovative Science and Research Technology, 4*, 182–187.
2. El Kessab, B., Daoui, C., Bouikhalene, B., Fakir, M., & Moro, K. (2013). Extraction method of handwritten digit recognition tested on the MNIST database. *International Journal of Advanced Science and Technology, 50*, 99–110.
3. LeCun, Y., Jackel, L., Bottou, L., Brunot, A., Cortes, C., Denker, J., Drucker, H., Guyon, I., Muller, U., Sackinger, E., Simard, P. & Vapnik, V. (1995). Comparison of learning algorithms for handwritten digit recognition. In *International Conference on Artificial Neural Networks.*
4. Abu Ghosh, M. M., & Maghari, A. Y. (2017). A comparative study on handwriting digit recognition using neural networks. In *International Conference on Promising Electronic Technologies (ICPET)* (pp. 77–81).
5. Bohara, M., Patel, K., Patel, B., & Desai, J. (2021, September). An AI based web portal for cotton price analysis and prediction. In *3rd International Conference on Integrated Intelligent Computing Communication and Security (ICIIC 2021)* (pp. 33–39). Atlantis Press.
6. Athila, V. A., & Chandran, A. S. (2021). *Comparative analysis of algorithms used in handwritten digit recognition.*
7. Nikesh, G. S., Amruth, T., Reddy, B. A., Rajashekhar, K., & Surya, N. J. (2021). *Handwritten digit recognizer using deep neural network.*

8. Shahid, A. R., & Talukder, S. *Evaluating machine learning models for handwriting recognition-based systems under local differential privacy.*
9. Ahlawat, S., & Choudhary, A. (2020). Hybrid CNN-SVM classifier for handwritten digit recognition. *Procedia Computer Science, 167*, 2554–2560.
10. Beohar, D., & Rasool, A. (2021, March). Handwritten digit recognition of MNIST dataset using deep learning state-of-the-art artificial neural network (ANN) and convolutional neural network (CNN). In *2021 International Conference on Emerging Smart Computing and Informatics (ESCI)* (pp. 542–548). IEEE.
11. Harikrishnan, A., Sethi, S., & Pandey, R. (2020, March). Handwritten digit recognition with feed-forward multi-layer perceptron and convolutional neural network architectures. In *2020 2nd International Conference on Innovative Mechanisms for Industry Applications (ICIMIA)* (pp. 398–402). IEEE.
12. Nouri, H. E. (2020, October). Handwritten digit recognition by deep learning for automatic entering of academic transcripts. In *Proceedings of the Computational Methods in Systems and Software* (pp. 575–584). Springer.
13. Xiao, R., Shi, J., & Zhang, C. (2020, June). FPGA implementation of CNN for handwritten digit recognition. In *2020 IEEE 4th Information Technology, Networking, Electronic and Automation Control Conference (ITNEC)* (Vol. 1, pp. 1128–1133). IEEE.
14. Hossainm M. A., Ali, M. M. (2019). Recognition of handwritten digit using convolutional neural network (CNN). *Global Journal of Computer Science and Technology: D Neural and Artificial Intelligence, 19*, 27–33
15. Vinjit, B. M., Bhojak, M. K., Kumar, S., & Nikam, G. (2021). Implementation of handwritten digit recognizer using CNN. In *Workshop on Advances in Computational Intelligence at ISIC.*
16. Singh, M., & Rahul (2020). Handwritten digit recognition using machine learning. *International Research Journal of Engineering and Technology (IRJET), 07*, 921–925
17. Biswas, A., & Islam, M. S. (2021). An efficient CNN model for automated digital handwritten digit classification. *Journal of Information Systems Engineering and Business Intelligence, 7*(1), 42–55.
18. Gope, B., Pande, S., Karale, N., Dharmale, S., & Umekar, P. (2021). Handwritten digits identification using mnist database via machine learning models. *IOP Conference Series: Materials Science and Engineering, 1022*, 1–12.

Unravel the Outlier Detection for Indian Ayurvedic Plant Organ Image Dataset

Meera Kansara and Ajay Parikh

Abstract Image-based outlier detection has been a fundamental research problem for machine learning and computer vision researchers. This paper unravels the outlier detection process for the data preparation framework of the Indian Ayurvedic plant organ image dataset. While creating dataset the outlier images might get introduce due to human or device errors. Identification and rectification of such outlier images are crucial part for creating clean dataset. This paper evaluated and compared four well-known and state-of-the-art outlier detection algorithms, namely Isolation Forest, Local Outlier Factor, Histogram-Based Outlier Score, and One-Class Support Vector Machine for detecting the outliers from the dataset of Indian Ayurvedic plant organ images. For this experiment dataset containing 690 images of "*Centella asiatica*" was used and augmented to generate more image samples. In total, 21 morphological, geometric, color, and texture features have been extracted from each plant organ image. The experiment shows the isolation forest giving superior results with 91% accuracy, at the same time Histogram-Based Outlier Score proves to be the fastest in execution time.

Keywords Outlier detection · Indian Ayurvedic plant identification · Image dataset · Image processing · Machine learning

1 Introduction

Outlier detection plays a vital role in the data preparation process for machine learning and deep learning-based systems. Outlier detection refers to the problem of discovering patterns or data points in dataset that do not confirm the normal pattern. This experiment is part of the process of creating image dataset for Indian Ayurvedic plant organs. Outliers in the context of this research are the medicinal plant organ images that do not belongs to the specific class or specific organ. While capturing images for creating dataset, due to device or human error outliers gets introduces,

M. Kansara (✉) · A. Parikh
Gujarat Vidyapith, Ahmedabad, India
e-mail: meeraj.kansara@gmail.com

P. K. Singh et al. (eds.), *Proceedings of Third International Conference on Computing, Communications, and Cyber-Security*, Lecture Notes in Networks and Systems 421,
https://doi.org/10.1007/978-981-19-1142-2_33

these outliers can have a huge and adverse impact on the results and final prediction. Hence, identification and elimination of this outliers are necessary part of the dataset preparation process.

In recent time, many outlier detection approaches have been proposed in order to identify outliers efficiently [1]. These approaches are broadly categorized in to (1) statistics-based approaches [2, 3] where a statistical technique had been used to detect outliers. (2) Density-based approaches [4, 5] in which outliers are detected by evaluating the distances to their nearest neighbors. (3) Clustering-based approaches which groups similar data points to form individual clusters and considers clusters of small size as outliers [6]. (4) Model-based approach, where a model is learned from a set of training data instances and nonconformity from the model is considered as outliers [7]. In this paper, four state-of-the-art outlier detection approaches, namely Isolation Forest (IF), Local Outlier Factor (LOF), Histogram-Based Outlier Detection (HBOS), and One-Class SVM (OCS), have been evaluated on the dataset of Indian Ayurvedic plant organ images. The dataset for this experiment was created using mobile-based image capturing tool developed by [8].

This paper is organized in the following sections, Sect. 2 enlists related work in the area of outlier detection. Section 3 describes proposed work, and experiments we have carried out for the outlier detection. Section 4 discuss experimental results and outcome. Section 5 is the conclusion and directives for the future work.

2 Related Work

In recent time, lot of attention is being paid on data preparation techniques. Outlier detection and handling is the important part of data preparation process for computer vision and machine learning task. Many efficient and diverse algorithms have been proposed for the outlier or anomaly detection. The algorithms vary in techniques and complexities. Li et al. [9] proposed hyperspectral anomaly detection method using kernel Isolation Forest, the research mapped hyperspectral data into kernel space and extracted first *k* components using principal component analysis (PCA) and then applied Isolation Forest to isolate outliers from normal observations. Cheng et al. [10] have created ensembled and progressive model using the Isolation Forest and Local Outlier Factor for anomaly detection. Kriegel et al. [11] proposed and outlined the Local Outlier Probability (LoOP) outlier detection model that takes advantage of the idea of local, density-based outlier scoring and a probabilistic, statistically oriented approach. Another interesting approach defined by [12], the novel Local Distance-based Outlier Factor (LDOF) which measure the score or outlier-ness of samples in primarily scattered datasets. LDOF mainly considered the relative location of a sample to its neighbors and determine the degree to which the sample deviates from its neighborhood samples. In recent time, [13] proposed the training of anomaly detectors against an auxiliary dataset of outliers, they named the approach as OE—Outlier Exposure and used anomaly detectors to detect unseen outliers. Goldstein et al. [14] carried out comparative analysis of different unsupervised anomaly detection algorithms on different datasets belonging to various domains. Sehwag et al. [15] developed Self-Supervised Outlier Detector (SSD) based on unlabeled in-distribution data with self-supervised representation learning followed by a Mahalanobis distance-based detection in the feature space. Elmogy et al. [16] proposed the first clustering-based outlier detection framework On the Fly Clustering-Based Outlier Detection (OFCOD). This framework effectually finds out outliers within huge datasets. Chen et al. [17] proposed novel image-based outlier detection method by combining autoencoder with Adaboost (ADAE), here many weak autoencoders were ensembled in order to capture the statistical correlations among the features of normal data. Shahid et al. [18] had carried out detailed experimental analysis of One-Class support vector machine formulations like, hyper-plane, hyper-sphere, quarter-sphere and hyper-ellipsoidal and used these formulations to separate outlier observations from normal data points. Cao et al. [19] proposed deep neural forest-based approach that synchronized the image classification and outlier detection from image data, the proposed IOD approach was able to capture more than 90% of outliers from datasets.

In this paper, we carried out experimental analysis of four states of the art and well-known algorithms, namely Isolation Forest [20], Local Outlier Factor [21], Histogram-Based Outlier Score [22], and One-Class SVM [23] on the sample leaf dataset of "*Centella asiatica*", we extracted various morphological and geometric features along with texture and color features and carried out experimental analysis.

3 Proposed Work

This section is mainly divided into three subsections. Section 3.1 explains and outlines dataset used for this experiment as well as preprocessing and augmentation. Section 3.2 is feature extraction section, where we have extracted various morphological and geometric features along with texture and color features from each image of dataset. As our work is for creation of Indian Ayurvedic plant image dataset, the texture and color of the plant images also play vital role in detecting outliers. Section 3.3 is an experiment section where state-of-the-art techniques have been evaluated and compared for performance.

3.1 Dataset

For this experiment, we have captured images of "*Centella asiatica*" using mobile-based image capturing tool. The mobile-based tool is convenient, easy to use, and also provides visual guidelines for capturing images of plant organs. We captured 690 images of "*Centella asiatica*" from different locations of Gujarat. The image capturing was carried out with the natural background and in non-destructive way, where plant organ has not been plucked or destroyed. To create a dataset for outlier

Fig. 1 Sample images from the dataset of "*Centella asiatica*" with outliers

detection experiment, we have also taken some other plant organ images with same image capturing tool (Fig. 1 shows a sample images from the dataset), these images are with contaminants and algorithm should be able to identify all such outliers efficiently.

We performed basic image preprocessing for each plant organ image. The main purpose of preprocessing the images is to enhance the visual appearance of the image and to improve feature extraction. This is done by removing unwanted noise, image smoothing, enhancing quality of image. The first step in preprocessing is by sharpening the RGB image. The sharpening of the image improves the image appearance and also, the edge points of an image are enhanced. After preprocessing, the images had been augmented. Image augmentation is the process of generating new data points based on the existing data points. The images in the dataset have been augmented to generate more samples. The augmented images were generated and added to the dataset of outlier detection. Rotation, transformation, scaling, flipping, and other augmentation techniques had been applied to each image. After augmentation, the dataset contains 6907 images including outlier images.

3.2 Feature Extraction

The feature extraction is the crucial part of the experiment. In total, 21 features were extracted from each plant organ image. Some of the features requires to convert organ image into gray scale and subsequently into binary image. Some features like color feature extraction requires colored images. The experiment used various image processing techniques to extract a set of features. Based on the work carried out by [24], Table 1 enlists the major features extracted from each organ image.

Along with the shape features, various color features like mean and standard deviation per color channel have been extracted. Figure 2 shows example of channel separation of RGB image to calculate mean value per color channel. Texture features were extracted using gray-level co-occurrence matrix (GLCM).

Table 1 Major features extracted from each image

Feature	Description
Area	Area is the number of pixels in the region of the plant organ
Perimeter	Perimeter is the sum of the distances between each adjoining pair of pixels around the border of the plant organ
Major axis length	Major axis length is the length of the line segment joining the base and the tip of the plant organ
Minor axis length	Minor axis length is the maximum width that is vertical to the major axis length
Aspect ratio	Aspect ratio also known as slimness is the ratio of major axis length to minor axis length that denotes narrow or wide leaf or flower characteristics
Solidity	Solidity is the ratio of the area of the convex hull and area of the binary image of the plant organ
Eccentricity	Eccentricity is the ratio of the distance between the foci of the ellipse and its major axis length
Isoperimetric quotient	Isoperimetric quotient is the ratio of the area of the organ to the area of a circle having the same perimeter
Convex hull	Convex hull or convex area is the smallest region that fulfill two conditions: (a) it is convex and (b) it contains region of the organ

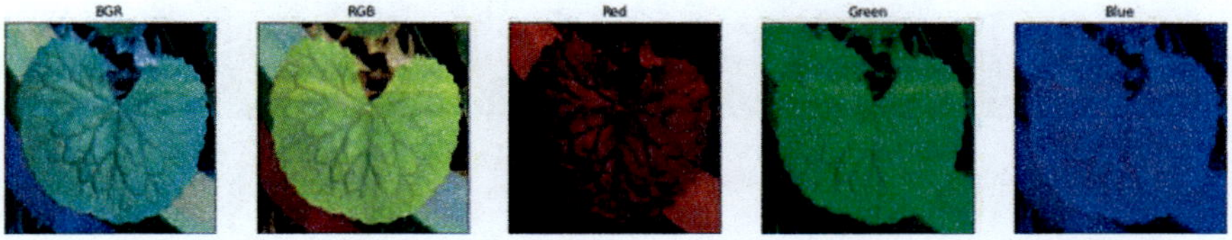

Fig. 2 Example of channel separation for "*Centella asiatica*" leaf image

3.3 Experiments

For this research, we have experimented with following state-of-the-art outlier detection algorithms.

Isolation Forest (IF)
Isolation Forest is an unsupervised tree-based algorithm which is widely used for outlier detection in the field of machine learning. The IF "isolates" data points by randomly choosing a feature out of feature set. Algorithm also randomly select a split value from the maximum and minimum value for the selected feature. The random partitioning of features produces smaller paths in trees for the outlier data values and separate them from the normal data.

The algorithm has been tested on the features extracted from dataset mentioned in Sect. 3.1. With the contamination parameter set to 0.1, the Isolation Forest is giving 91% accuracy. The result of the Isolation Forest is displayed in Fig. 3.

Local Outlier Factor

The Local Outlier Factor algorithm basically calculates the local density deviation of a sample with respect to its surrounding or neighboring samples. LOF considers the sample "outlier" which has lower density as compared to its neighbors. LOF is giving 85.55% accuracy. The results of LOF have been displayed in Fig. 4.

Fig. 3 Outliers isolated by Isolation Forest algorithm

Fig. 4 Outliers isolated by Local Outlier Factor

Fig. 5 Outlier isolated by Histogram-Based Outlier Score

Histogram-Based Outlier Score (HBOS)
In histogram-Based Outlier Score algorithm, for individual feature or dimension a univariate histogram is computed, where the height of individual bin represents a density estimate. Then comes the normalization step where histograms are normalized to make maximum height of 1.0. This is done to give each feature an equal weight for the outlier score. The HBOS is giving 85.55% accuracy. The results of HBOS have been displayed in Fig. 5.

One-Class SVM (OCS)
One-Class SVM algorithm developed for binary classification can also be used for the outlier or anomaly detection. OCS algorithm considers the density of the majority of samples and samples falling on the extreme of the density will be classified as an outliers or anomalies. One-Class SVM in this experiment gives 71.42% of accuracy. The top 6 outlier images isolated by OCS are displayed in Fig. 6.

4 Results and Discussion

The experiment of outlier detection was carried out for the purpose of creating dataset of Indian Ayurvedic plant organ images. Outlier detection process is an important part of data preparation framework for Indian Ayurvedic plant organ image dataset. The accurate and precise algorithm not only eliminates contamination but also improves results and accuracy of final identification process. As outlined in Sect. 3, experiments were carried out using the sample dataset of "*Centella asiatica*" with contaminations in form of other plant organ images. For this experiment, we carried out threefold cross-validation, we randomly shuffle the dataset into three sets *ce0, ce1, ce2* such

Fig. 6 Outlier isolated by One-Class SVM

Table 2 Accuracy of IF, LOF, OCS, and HBOS on sample dataset

Algorithm	Accuracy in %
Isolation Forest	91.00
Local Outlier Factor	85.55
One-Class SVM	71.42
Histogram-Based Outlier Score	85.55

that *ce0, ce1,* and *ce2* are of equal size. The result for the outlier detection algorithms was averaged over three trials. The result of experiment is listed in Table 2.

As listed in Table 2, One-Class SVM obtained accuracy score of 71.42%, whereas both LOF and HBOS give accuracy of 85.55%. In spite of giving equal accuracy the HBOS is much faster than LOF in execution time, because of high time complexity, LOF is less appropriate for large-scale high-dimensional datasets like plant organ image datasets. HBOS performed low in detecting local outliers. The highest accuracy of 91% is achieved with Isolation Forest. IF is an ensemble-based unsupervised outlier detection algorithm with high precision and linear time complexity. The dataset was high dimensional, noisy with intra-class variability, as contains field images of plant organ which poses major challenge for density-based algorithms, as IF is tree based and work on random partitioning of features, IF is more suitable for this dataset and feature set. IF can be scaled up to handle high-dimensional as well as large datasets. Overall low accuracy of the algorithms could be due to the fundamental nature of the problem domain, for example, varied appearance of plant organ images (as shown in Fig. 1) and complex structure of plant organs as well as high intra-class variability and low inter-class variances.

5 Conclusion

This paper carried out experiments for outlier detection, which is important and integral part of data preparation framework for creating dataset of Indian Ayurvedic Plant Organ Image dataset. For this purpose, sample dataset of 690 images of "*Centella asiatica*" with outliers have been used. In total, 21 morphological, geometric, color, and texture features have been extracted. Four well-known algorithms, namely Isolation Forest, Local Outlier Factor, One-Class SVM and Histogram-Based Outlier Score, have been evaluated. Isolation Forest found to give best accuracy of 91%. The low quality of data and existence of noise makes outlier detection process challenging. The limitation of this experiment is the result is highly dependent upon the extracted features. More robust and distinctive features may improve the accuracy of the outlier detection task.

For the future work, more experiments could be carried out with different outlier algorithms and more efficient features could be extracted to elevate accuracy. The clustering-based outlier detection techniques could be explored and neural network could also be used instead of feature extractor to improve accuracy and efficiency. The use of autoencoders could be experimented. The dataset could be extended and experiments could be carried out for multiple plant organs.

Acknowledgements Authors extend their sincere thanks to Dr. Minubhai Purabia, retired Professor Department of Botany, South Gujarat University, for his continuous support and providing domain knowledge. We are thankful to Late Dr. Haresh L. Dhaduk from Anand Agriculture University for facilitating researchers with sample collection and support.

References

1. Boukerche, A., Zheng, L., & Alfandi, O. (2020). Outlier detection: Methods, models, and classification. *ACM Computing Surveys, 53* (3) (2020). https://doi.org/10.1145/3381028
2. Navarro-Esteban, P., & Cuesta-Albertos, J. A. (2021). High-dimensional outlier detection using random projections. *TEST*. https://doi.org/10.1007/s11749-020-00750-y
3. Eskin, E. (2008). Anomaly detection over noisy data using learned probability distributions. In *Proceedings of the Seventeenth International Conference on Machine Learning* (pp. 255–262).
4. Tang, B., & He, H. (2017). A local density-based approach for outlier detection. *Neurocomputing, 241*, 171–180. https://doi.org/10.1016/j.neucom.2017.02.039
5. Ma, M. X., Ngan, H. Y., & Liu, W. (2016). Density-based outlier detection by local outlier factor on largescale traffic data. *Electronic Imaging, 14*, 1–4. https://doi.org/10.2352/issn.2470-1173.2016.14.ipmva-385
6. Christy, A., Gandhi, G. M., & Vaithyasubramanian, S. (2015). Cluster based outlier detection algorithm for healthcare data. *Procedia Computer Science, 50*, 209–215. https://doi.org/10.1016/j.procs.2015.04.058
7. Liu, B., Xiao, Y., Cao, L., Hao, Z., & Deng, F. (2012). SVDD-based outlier detection on uncertain data. *Knowledge and Information Systems, 34*(3), 597–618. https://doi.org/10.1007/s10115-012-0484-y

8. Kansara, M., & Parikh, A. (2020). Indian Ayurvedic plant identification using multi-organ image analytics: Creation of image dataset of Indian medicinal plant organs. *SSRN Electronic Journal*. https://doi.org/10.2139/ssrn.3563074
9. Li, S., Zhang, K., Duan, P., & Kang, X. (2020). Hyperspectral Anomaly detection with kernel isolation forest. *IEEE Transactions on Geoscience and Remote Sensing, 58*(1), 319–329. https://doi.org/10.1109/tgrs.2019.2936308
10. Cheng, Z., Zou, C., & Dong, J. (2019). Outlier detection using isolation forest and local outlier factor. In *Proceedings of the Conference on Research in Adaptive and Convergent Systems* (2019). https://doi.org/10.1145/3338840.3355641
11. Kriegel, H. P., Kröger, P., Schubert, E., & Zimek, A. (2009). LoOP. In *Proceeding of the 18th ACM Conference on Information and Knowledge Management—CIKM'09*. https://doi.org/10.1145/1645953.1646195
12. Zhang, K., Hutter, M., & Jin, H. (2009). A new local distance-based outlier detection approach for scattered real-world data. In *Advances in knowledge discovery and data mining* (pp. 813–822). https://doi.org/10.1007/978-3-642-01307-2_84
13. Hendrycks, D., Mazeika, M., & Dietterich, T. G. (2019). *Deep anomaly detection with outlier exposure*. Opgehaal van. CoRR, abs/1812.04606. http://arxiv.org/abs/1812.04606
14. Goldstein, M., & Uchida, S. (2016). A comparative evaluation of unsupervised anomaly detection algorithms for multivariate data. *PLoS ONE, 11*(4), e0152173. https://doi.org/10.1371/journal.pone.0152173
15. Sehwag, V., Chiang, M., & Mittal, P. (2021). *SSD: A unified framework for self-supervised outlier detection*. Opgehaal van. CoRR, abs/2103.12051. https://arxiv.org/abs/2103.12051
16. Elmogy, A., Rizk, H., & Sarhan, A. M. (2020). OFCOD: On the fly clustering based outlier detection framework. *Data, 6*(1), 1. https://doi.org/10.3390/data6010001
17. Chen, Z., Yeo, C. K., Lee, B. S., Lau, C. T., & Jin, Y. (2018). Evolutionary multi-objective optimization based ensemble autoencoders for image outlier detection. *Neurocomputing, 309*, 192–200. https://doi.org/10.1016/j.neucom.2018.05.012
18. Shahid, N., Naqvi, I. H., & Qaisar, S. B. (2013). One-class support vector machines: Analysis of outlier detection for wireless sensor networks in harsh environments. *Artificial Intelligence Review, 43*(4), 515–563 (2013). https://doi.org/10.1007/s10462-013-9395-x
19. Cao, L., Yan, Y., Madden, S., & Rundensteiner, E. (2019). *Outlier detection from image data*. Opgehaal van. https://openreview.net/forum?id=HygTE309t7
20. Liu, F. T., Ting, K. M., & Zhou, Z. H. (2008). Isolation forest. In *2008 Eighth IEEE International Conference on Data Mining*. https://doi.org/10.1109/icdm.2008.17
21. Breunig, M. M., Kriegel, H. P., Ng, R. T., & Sander, J. (2000). LOF. In *Proceedings of the 2000 ACM SIGMOD International Conference on Management of Data—SIGMOD*. https://doi.org/10.1145/342009.335388
22. Goldstein, M., & Dengel, A. (2012). Histogram-based outlier score (HBOS): A fast unsupervised anomaly detection algorithm. In *Poster and Demo Track of the 35th German Conference on Artificial Intelligence (KI-2012)* (pp. 59–63).
23. Schölkopf, B., Platt, J. C., Shawe-Taylor, J., Smola, A. J., & Williamson, R. C. (2001). Estimating the support of a high-dimensional distribution. *Neural Computation, 13*(7), 1443–1471. https://doi.org/10.1162/089976601750264965
24. Wäldchen, J., & Mäder, P. (2018). Plant species identification using computer vision techniques: A systematic literature review. *Archives of Computational Methods in Engineering. State of the Art Reviews, 25*(2), 507–543.

A Review on Service Delivery in Tourism and Hospitality Industry Through Artificial Intelligence

Yashwant Singh Rawal, Harvinder Soni, Rakesh Dani, and Purnendu Bagchi

Abstract AI in service industry like tourism and hospitality is changing at an impressive pace and has uncovered new research opportunities. It has been progressively reshaping the service industry and has led to significant innovations in this sector. This study focuses on the systematic review of artificial intelligence in delivery of service in the field of tourism and hospitality. The purpose of the paper is to explore and signify the relevance of artificial intelligence in tourism and hospitality industry in the contemporary times to meet the challenges posed by the pandemic and to ensure speed and accuracy in service delivery for enriching guest experience and sustaining competition. Paper mainly discusses about the optimum use of artificial intelligence through the adoption of AI technology in service delivery in tourism and hospitality industry. The use of AI-enabled tools like chatbots, smart rooms with voice control system, facial recognition technology, robots, operational analysis, and virtual reality in hospitality industry has been analyzed. This study also explores the acceptance of AI by the customer in the tourism and hospitality industry.

Keywords Artificial intelligence · Hospitality and tourism industry · Service delivery

Y. S. Rawal (✉)
Amity University Rajasthan, Jaipur, India
e-mail: yashwantr84@gmail.com; ysrawal@jpr.amity.edu

H. Soni
Taxila Business School, Jaipur, India

R. Dani
Graphic Era Deemed to be University, Dehradun, India

P. Bagchi
Amity University, Kolkata, India

P. K. Singh et al. (eds.), *Proceedings of Third International Conference on Computing, Communications, and Cyber-Security*, Lecture Notes in Networks and Systems 421,
https://doi.org/10.1007/978-981-19-1142-2_34

1 Introduction

AI in tourism and hospitality is playing the role of game changer [1]. AI has led to innovative solutions in the service industry which have helped to reinvent the industry in terms of enhanced quality of service and improved organizational performance. The AI-based advances can be placed being used in human services to help organizations lighten impressive authoritative weight and average time for more basic duties by improving dynamic and making less expensive and quicker delivery services [2].

Today, many industries have embarked on the path of digitization and are embracing distinctive information technology solutions for adding greater value for the consumer, to fuel development and drive income. Due to the increasing demand and expectations of the guests, the hospitality business needs novel solutions not only to enhance guest satisfaction but also to maintain a competitive edge in the market. Apart from chatbots, the use of AI in tourism and hospitality industry is still at the inception stage, and as compared to other sectors such as banking, health care, and ecommerce, this sector does not have a very robust perspective for AI-enabled solutions [3]. Although hospitality industry does not often encourage the adoption of artificial intelligence applications, yet, it cannot be denied that these technologies have become a part of our lives and have become necessary for survival and growth of business in the present age of technology [4].

By the use of AI technology, the hotel or travel agencies can improve the customer experience as well as reduce the cost and help in providing more accurate data for taking the right decision [5].

Moreover, the global pandemic has adversely affected hotel industry, and it has become highly unsafe for the guests to travel, stay, and dine in hotels as their health is at risk due to the contagious corona virus. The purpose of this paper is to signify the relevance of artificial intelligence in effectivizing the service delivery in tourism and hospitality industry in the contemporary times; where on one hand, the world is confronting the COVID-19 crisis, and physical distancing is the need of the hour, and on the other hand, speed and accuracy through the use of information technology have become imperative for enriching guest experience and sustaining competition. AI techniques would enhance guest satisfaction by providing hygienic, safe, and healthy stay.

2 Leveraging AI-Enabled Tools in Hospitality Industry

This section presents the way in which AI will effectivize service delivery in hospitality industry. For effectivization of service delivery in tourism and hospitality industry, the following AI-enabled tools can be leveraged.

2.1 Chatbots

Chabot is a type of computer software which is used to answer the questions asked by the customer. It is computer software which is text based or can use speech recognition to answer the customer's questions [6]. With the help of chatbots, booking.com is providing 24 * 7 services in more than 42 languages which increased the sale of the company [7]. Chatbots are most regularly connected with on the online or telephonic exchanges. It is conspicuous that chatbots in accommodation are changing the manner in which booking requests are prepared. Conversational bots could supplant the front work area staff that handles booking-related questions. These insightful chatbots are modified to make reenacted discussion (text/voice) in local language, empowering controlled, brief, and proficient cooperation's among people and machines. Artificial intelligence chatbots have been used via social media also, permitting customer to ask inquiries and acquire practically momentary reactions. This service is provided 24 * 7. It is very useful in hotel industry. It gives the kind of reaction in times in which it is practically difficult to connect or put with human-to-human interaction [8].

Chatbots provide an added advantage over human assistance specialists in that they can handle basically a limitless magnitude of transactions virtually and at the same time store huge pool of the information with minimum threat of errors, impact of variations in emotions or fatigue [9]. It is indicated by TMC trends research by MTT, 43% of travel companies mean to put resources into chabot advances. Chabot-empowered individual travel assistance is unquestionably a focal point for overseas business travel advancement [8, 9].

2.2 Smart Room with Voice Control System

Now-a-days, there is a trend of automation and providing hyper-personalization service. Therefore, the hotels make the provision for guests to enable them to choose room amenities as per their preferences. The Aloft Santa Clara hotel in San Jose implemented the booking of voice-activated rooms with an IPad installed with custom Aloft app [10]. With the voice command guest can turn on the lightings, play music, play video on YouTube, play movie. A personalized temperature control provided by smart thermostat gives unique experience to the guest interacting with their room [9, 10].

2.3 Facial Recognition

Facial recognition is the technology which is used to recognize a customer in a digital image or video as in the check-in process it could be used to automatically recognize the guest. However, it can be used for recognizing a particular person and also to keep a track on the number of persons in a specific area. It can also recognize emotions of individuals passing by a specific point (e.g., contentment of guests who are exiting after dinner from the restaurant) [11, 12].

Facial expression, body gesture, emotional appeal help to understand the customer's preference and needs.

Facial recognition technology is also used in airline industry to provide smooth check-ins at airport without any document verification by the immigration department, custom office, and other stations at airport [11, 12]. This technology allows guests to complete registration for without standing in lines at the front desk. Two hotels of Marriott International started facial recognition check-in machines on trial basis [12].

2.4 *Operational Analysis*

AI can be used within the tourism and hospitality industry away from pure customer service in data analysis. With the help of this, large amount of data can be quickly sorted and can help in drawing important conclusions about the customer or potential customer [13]. AI can help to extract the important conclusion from the large amount of data about the customer or potential customers. AI saves the hotel staff endless hours of studying customer surveys and feedback [14]. By implementing AI technologies, hotels can examine guest date to deliver a more customized service experience through which they can improve the credibility of their hotel and increase the guest satisfaction. With the help of AI-based apps, one can predict about the guests, as these apps record the detailed data about hotel's guests which include their habits, behavior, preferences, consuming pattern, and many more. With the help of AI technology, hotel can predict any problem, issue, needs, and wants proactively. Therefore, hoteliers can customize the offer on the basis of forecasting which will lead to improvement in the guest satisfaction level, loyalty of customer and brand [15].

2.5 *Virtual Reality*

In hotel industry, there is a major gap between customer and product; customer is far away from the product, not aware about the hotel and other facilities, product, ambience of the hotel [15, 16].

In the present scenario, customers prefer to experience and explore the tourist spots before they visit those spots through searching-related information on the Internet like customers reviews [16]. On Web site, few pictures and videos are available which may not enough be to attract the customer to visit a specific spot. Very often, it leads to dubiety in the customers. This issue can be resolved through virtual technology

Fig. 1 Virtual reality in hospitality [18]

which can be used in the tourism and hospitality industry for demonstrating tourist spots and hotel location and view by using 3D videos [17] (Fig. 1).

Virtual reality (VR) technology commonly makes use of the VR head set to create simulated surroundings. This simulated environment offers an experience of virtual reality. Consumers are using the virtual technology experience in the digital world [19]. Tourism and hotel industry uses various virtual reality applications like virtual hotel tours, virtual travel experiences, and virtual booking interface to demonstrate the products. It gives a real-time experience to the customers regarding the major amenities provided by the hotel [19]. With the appearance of virtual reality advancements, there can be a drastic difference in the accessibility of resources and data for the customer based on which they can take informed decision about their movement in advance [20].

2.6 Robots

Robots in hospitality and tourism industry are making the strong use of AI technology. It is one of the most transformative technical implications in hospitality industry [20]. ALO a robot service assistant at Aloft hotel uses to deliver the room amenities like dental kit, shaving kit, bath linen, bed sheets, and shampoo and collect dirty laundry from guest's room [20]. Robot receptionists have become something moving in the travel industry, which has immediate ramifications on the nature of interaction and experience of the customer. They pay added attention to the provisions of room service so that the guest would not confront any problem during checking-in to the hotel room [21], while robotic butlers and in-room voice have some very obvious advantages at the front end of hospitality [22]. In some of the Marriott group's hotels, Alexa robot has been getting a lot of attention.

In air lines also, the use and impact of robots are enhancing, and they are being used for assisting and guiding their customer. Few of the main benefits of robots in the travel industry incorporate upgraded client experience, simplification of the work process, and improved productivity of the travel industry business [22].

2.7 *Attitude of Customer or Acceptance*

Despite the fact that few examinations have effectively searched distinctive AI-related themes, there is as yet not a far-reaching hypothetical and reasonable structure which can be utilized to describe the process of generation of customer attitude toward the utilization of AI gadgets and the most basic determinates of AI acceptance. In past studies related to AI topics, it is noticed that still there is a lack of extensive theoretical and conceptual base that can be used to understand how the customers develop their attitude toward the use of AI devices and the pivotal determinates leading to a positive affirmation toward AI. Over the last 50 years, limited acceptance of AI and automation has been the key challenge, but the limitations have historically centered on utility and capacity in the service context [23].

Use of AI devices to deliver service which used to be done by the human being may challenge customer's perceptions regarding service by the reason that customers may have a feeling of being devalued as they have to communicate with an AI device which lacks human touch [24]. Technology-based service can impact both negative and positive. It depends on the customer's level of technology readiness, comfort, use optimism and insecurity [25]. The issue of ethicality arises when the highly advanced chatbots like human lead people to believe that they are communicating with the human being, but in reality, they interact with machine. Hence, the customers may feel that they are undervalued by the company [26, 27].

2.8 *AI in Tourism and Hospitality*

See Fig. 2.

3 Findings and Discussion

Artificial intelligence has opened up novel avenues in tourism and hospitality industry and has resulted in enhancement of the quality of services offered. AI-enabled technology not only improves guest experience but also results in cost-effectiveness in service industry.

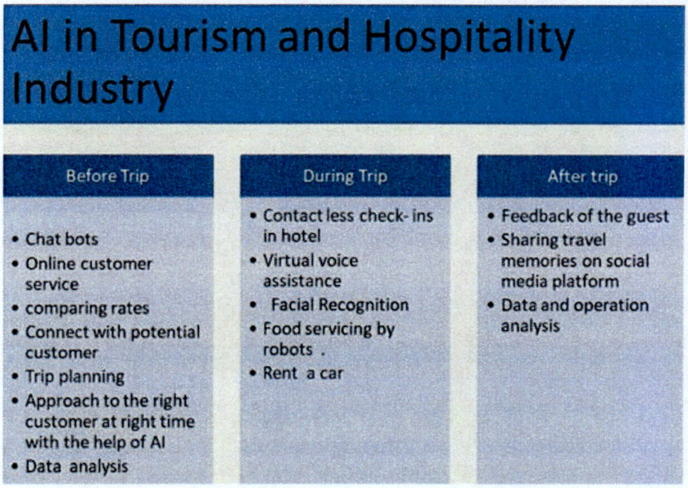

Fig. 2 AI in tourism and hospitality industry. *Source* Author's own

Chatbots have enabled the communication of customer's queries and booking of accommodation by guests in a speedy and efficient manner. Smart room with voice control system has enriched the room experience of guests, and facial recognition technology has enabled recognition and counting of persons together with the understanding of their emotions. Artificial intelligence helps in operational analysis of data which minimizes the time spent on customer survey and feedback. With the help of AI technology, hotel can forecast any problem, issue, needs, and wants even before they arise and provide personalized services. Various virtual reality applications like virtual hotel tours, virtual travel experiences, and virtual booking interface are used in tourism and hotel industry to demonstrate the products. This provides a real-time experience to the customer about the facilities in the hotel and takes a better decision. If the current situation is compared with the earlier one, then it is found that service delivery, booking, virtual visiting, amenities, and contactless secure services are available with the help of technological trends. Artificial intelligence has enabled the provision of all types of facilities at customer's finger tips.

4 Future Scope of Research

Though a number of AI techniques are available to hospitality industry, but, the main concern arises with respect to the security and efficient outcomes of the used techniques. For this, the use of artificial neural network, random forest, decision tree, K-nearest neighbor, etc., to provide efficient and specific outputs is suggested. In future scope, we can incorporate any of the abovementioned algorithms for best outcomes. These algorithms will definitely predict the outcomes and visitors' availability depending on the current amenities provided to customers.

5 Conclusion

AI can play a significant role in servicing and retention of customers by providing round the clock service [28]. In the pre-stay, customer segmentation, customer service, data analytics, optimization of campaign, and many other marketing tasks can be accomplished successfully and effectively with the AI tools.

During the pandemic, hotels need to update the internal and external communications used by technologies as the guests are highly concerned with cleanliness and hygiene in and around the hotel where they plan to stay. "Clean and hygiene-COVID free" marketing campaign with the application of artificial intelligence can be launched by [29, 30] smart mobile searches including cache of keywords for managing restaurant reservations through search engine assistant can further effectivize service delivery and performance. During the hotel stay, safety of guests can be ensured as cleanliness, hygiene, and safety through the use of AI, robots, and automation which play a major role in this respect. During the pandemic for cleaning and sanitization tasks and assisting the hotel staff in accomplishing their duties from check-in to check-out process, mechanical AI is of great utility [31].

In the era of the technology, the increase in AI impression in travel is a positive indication since it demonstrates that the industry can utilize the cutting-edge technology to increase the productivity which can result in the improvement of the level of customer satisfaction. Through AI technology, the hotel or travel agencies can improve the customer experience as well as reduce the cost and help in providing more accurate data for taking the right decision at right time. Simultaneously, the travel industry business endeavors can have better control on the operation.

References

1. Pillai, R., & Sivathanu, B. (2020). Adoption of AI-based chatbots for hospitality and tourism. *International Journal of Contemporary Hospitality Management, 32*(10), 3199–3226. https://doi.org/10.1108/IJCHM-04-2020-0259
2. Reis, J., Amorim, M., Cohen, Y., & Rodrigues, M. (2020). Artificial intelligence in service delivery systems: A systematic literature review. In Á. Rocha, H. Adeli, L. P. Reis, S. Costanzo, I. Orovic, & F. Moreira (Eds.), *Trends and innovations in information systems and technologies* (pp. 222–233). Springer International Publishing.
3. Saleiro, P., Rodolfa, K. T., & Ghani, R. (2020). Dealing with bias and fairness in data science systems. In *Proceedings of the 26th ACMSIGKDD International Conference on Knowledge Discovery and Data Mining.* https://doi.org/10.1145/3394486.3406708
4. Badicio, M. (2019, November). *AI in the travel and tourism industry—Current applications.* Retrieved 2021, from Emerj.com. https://emerj.com/ai-sector-overviews/ai-travel-tourism-industry-currentapplications/

5. Artificial Intelligence Insight. (2020). *Improve customer experience in your hospitality business*. Retrieved March 2021, from https://insights.ehotelier.com/. https://insights.ehotelier.com/insights/2020/05/11/artificial-intelligence-improve-customer-experience-in-your-hospitality-business/
6. Li, M., Yin, D., Qiu, H., & Bai, B. (2021). A systematic review of AI technology-based service encounters: Implications for hospitality and tourism operations. *International Journal of Hospitality Management, 95*, 102930. https://doi.org/10.1016/j.ijhm.2021.102930
7. Varsha, P. S., & Akter, S. (2021). The impact of artificial intelligence on branding: A bibliometric analysis (1982–2019). *Journal of Global Information Management, 29* (4), 26.
8. State of Artificial Intelligence in Travel. (2020). Retrieved March 2021, from www.hospitalitynet.org. https://www.hospitalitynet.org/news/4097065.html
9. Chi, O. H., Denton, G., & Gursoy, D. (2020). Artificially intelligent device use in service delivery: A systematic review, synthesis, and research agenda. *Journal of Hospitality Marketing and Management, 29*(7), 757–786. https://doi.org/10.1080/19368623.2020.1721394
10. How the Hospitality Industry Uses Performance-enhancing Artificial Intelligence and Data Science. (2018). Retrieved April 2021. https://www.altexsoft.com/. https://www.altexsoft.com/blog/datascience/how-the-hospitality-industry-uses-performance-enhancing-artificial-intelligence-and-data-science/
11. Chang, H.-L., & Yang, C.H. (2008). Do airline self-service check-in kiosks meet the needs of passengers? *Tourism Management, 29*(5), 980–993.
12. Patel, V. (2018). *Airport passenger processing technology: A biometric airport journey*. Available at: https://commons.erau.edu/edt/385/. Accessed September 5, 2019.
13. Sharma, S., & Rawal, Y. S. (2021). The possibilities of artificial intelligence in the hotel industry. In X. Z. Gao, R. Kumar, S. Srivastava, & B.P. Soni (Eds.), *Applications of artificial intelligence in engineering. Algorithms for intelligent systems*. Springer. https://doi.org/10.1007/978-981-33-4604-8_53
14. Guttentag, D. A. (2010). Virtual reality: Applications and implications for tourism. *Tourism Management, 31*(5), 637–651. https://doi.org/10.1016/j.tourman.2009.07.003
15. Barnes, S. (2016). *Understanding virtual reality in marketing: Nature, implications and potential*. Available at SSRN: https://doi.org/10.2139/ssrn.2909100
16. Kim, J., & Hardin, A. (2010). The impact of virtual worlds on word-of-mouth: Improving social networking and service scape in the hospitality industry. *Journal of Hospitality Marketing and Management, 19*(7), 735–753. https://doi.org/10.1080/19368623.2010.508005
17. Samala, N., Katkam, B. S., Bellamkonda, R.S., & Rodriguez, R.V. (2020). Impact of AI and robotics in the tourism sector: A critical insight. *Journal of Tourism Futures*. https://doi.org/10.1108/JTF-07-2019-0065
18. Travel and Tourism with Augmented Reality. (2021). Retrieved from Zumoko.com: https://www.zumoko.com/travel-and-tourism-withaugmented-reality/
19. Tuomi, A., Iis, P. T., & Stienmetz, J. (2021). Applications and implications of service robots in hospitality. *The Cornell Hospitality Quarterly, 62* (2), 232–247.
20. Seo, K. H., & Lee, J. H. (2021). The emergence of service robots at restaurants: Integrating trust, perceived risk, and satisfaction. *Sustainability, 13* (8), 1–16.
21. Lin, H., Chi, O. H., & Gursoy, D. (2019). Antecedents of customers' acceptance of artificially intelligent robotic device use in hospitality services. *Journal of Hospitality Marketing and Management*. https://doi.org/10.1080/19368623.2020.1685053
22. Granger, B. (2017). *How artificial intelligence is revolutionizing hospitality*. Retrieved May 2021, from https://www.hospitalityupgrade.com/techTalk/September-2017/How-ArtificialIntellgenceis-Revolutionizing-Hos/
23. Ackerman, E. (2016). Study: Nobody wants social robots that look like humans because they threaten our identity. *IEEE Spectrum*, 1–5.
24. Belanche, D., Casaló, L. V., Flavián, C., & Schepers, J. (2020). Service robot implementation: A theoretical framework and research agenda. *The Service Industries Journal, 40*(3–4), 203–225. https://doi.org/10.1080/02642069.2019.1672666

25. Sharma, S., Rawal, Y. S., Pal, S., & Dani, R. (2021). Fairness, accountability, sustainability, transparency (FAST) of artificial intelligence in terms of hospitality industry. In: ICT systems and sustainability. Lecture notes in network system. Springer (in press)
26. Handoff, T. (2020) The ethics of AI ethics. An evaluation of guidelines. In *Minds and machines* (pp. 1–22). [Online]. Available at: https://arxiv.org/pdf/1903.03425
27. Hamon, R., Junklewitz, H., & Sanchez, I. (2020). *Robustness and explainability of artificial intelligence—From technical to policy solutions*. EUR 30040, Publications Office of the European Union, Luxembourg. ISBN 978-92-79-14660-5 (online), https://doi.org/10.2760/57493 (online). JRC119336.
28. Kılıçhan, R., & Yılmaz, M. (2020). Artificial intelligence and robotic technologies in tourism and hospitality industry. *Erciyes Üniversitesi Sosyal Bilimler Enstitüsü Dergisi L, 2020*(3), 353–380.
29. Davenport, T. H., & Ronanki, R. (2018). Artificial intelligence for the real world: Don't start with moon shots. *Harvard Business Review, 96*, 108–116.
30. Vishnoi, S. K., Bagga, T., Sharma, A., & Wani, S. N. (2018). Artificial intelligence enabled marketing solutions: A review. *Indian Journal of Economics and Business*, 167–177.
31. Ivanov, S. H., Webster, C., Stoilova, E., & Slobodskoy, D. (2020). Biosecurity, crisis management, automation technologies and economic performance of travel, tourism and hospitality companies—A conceptual framework. *Tourism Economics*. https://doi.org/10.1177/1354816620946541

MegaMart Sales Prediction Using Machine Learning Techniques

Gopal Gupta, Kanchan Lata Gupta, and Gaurav Kansal

Abstract These days online shopping and MegaMarts record their sales and purchase data of each and every item. As the competition between various stores is increasing rapidly, it is necessary to predict future demand of each product at various stores for the customers. This data contain various attributes related to product like its ID, store ID, weight of product, visibility percentage of product, its fat content, its type, location of store, etc. This data are then analyzed to detect the further, anomalies and frequent patterns in the data. After analyzing data, it is processed so as to give us exact report for sales of each product. Then, final data can be used for predicting future sales using different machine learning techniques. We apply different machine learning models like 'linear regression', 'decision tree', 'random forest', 'ridge regression', and 'XGBoost model' to predict outlet sales. We found out that XGBoost gives us the best accuracy. With this predicted sales, MegaMart can observe the various patterns that should be changed to ensure its success in business.

Keywords Machine learning · Data visualization · Forecasting · Sales

1 Introduction

Nowadays, we surrounded by shopping centres such as food-mart, big malls, and MegaMarts, and all shopping centres are in race to increase the sales. They all are advanced nowadays to record all their transaction, sales, and purchase electronically. To increase the sales, it is necessary them to predict the sales item [1]. The data dataset has various features, and some of them are dependent and some independent. The

G. Gupta (✉) · G. Kansal
ABES Engineering College, Ghaziabad, Uttar Pradesh, India
e-mail: gopalgupta@abes.ac.in

G. Kansal
e-mail: gkansal@abes.ac.in

K. L. Gupta
Institute of Engineering and Technology, Lucknow, India

P. K. Singh et al. (eds.), *Proceedings of Third International Conference on Computing, Communications, and Cyber-Security*, Lecture Notes in Networks and Systems 421,
https://doi.org/10.1007/978-981-19-1142-2_35

data can be used for forecasting future sales by applying machine learning approaches [2]. For this, we have first fit our model with training data and then predicted the item outlet sales.

In [3], author applied the multi-objective evolutionary algorithm for feature selection of sales prediction in online advertising.

Warnakulasooriya et al. [4] proposed a system to predict the future price and demand of vegetable in retail. They applied and compare different machine learning algorithm and find XGBoost gives best result in compare to LSTM, SARIMA, and ARIMA.

Liu et al. [5] apply different machine learning algorithm on food sales prediction (Japanese Chain Supermarket) and find we can apply and forecast sales in supermarket.

In this report, we are going to perform analysis of sales data using Python library. We plan to forecast the sales item. Data exploration, data transformation, and feature engineering play a vital role in predicting accurate results [6]. Relationship between sales and retail stocks at the aggregate level depends to a large extent on the accurate forecasting of retail sales [7]. In section two, we discussed methodology of this work.

The work is divided into four section; in Sect. 1, we talk about different paper applied machine learning algorithm on different type of dataset and areas. In Sect. 2, we apply data analytics steps to find relation and describe the data, i.e., what the data are. In Sect. 3, we apply different machine learning algorithm and compare it. In Sect. 4, a conclusion.

2 Methodology

The following are the stepwise methodology to how to complete this work.

2.1 Dataset Description

Dataset has sales data from the year 2013, for 1559 products across different stores in different cities. Test dataset contains 5681 rows and 11 attributes, whilst trail dataset has 8523 rows and 12 attributes [8]. Train dataset has an extra Item_Outlet_Sales column. Description of all the attributes used in our dataset is given in Table 1.

Both train and test dataset have some null value, so first, we combine both dataset and perform the pre-processing task to clean and the data.

Table 1 Train dataset

#	Column	Count	Dtype	Description
1	Item_Identifier	8523	Object	Product-ID
2	Item_Weight	7060	Float64	Product weight
3	Item_Fat_Content	8523	Object	It contains fat level (low fat or regular)
4	Item_Visibility	8523	Float64	% of total display area of all products
5	Item_Type	8523	Object	Category of item
6	Item_MRP	8523	Float64	MRP of item
7	Outlet_Identifier	8523	Object	Unique ID of store
8	Outlet_Establishment_Year	8523	Object	Date at which store established
9	Outlet_Size	6113	Object	Area of store
10	Outlet_Location_Type	8523	Object	Type of city
11	Outlet_Type	8523	Object	Grocery store or supermarket
12	Item_Outlet_Sales	8523	Float64	Sales of the product in the particulate store

2.2 Data Pre-processing

Data pre-processing is a techniques of data mining which includes preparation and transformation of data in appropriate form before applying analysis and prediction [9]. In data pre-processing, we perform following tasks

(i) Data cleaning—handling missing value, ignore the tuple
(ii) Data transformation—normalization, attribute selection
(iii) Data reduction—aggregation, attribute subset selection, dimensionality reduction.

Item_Weight attributes have total 30% null data, so we can fill these missing values by mean value of the 'Item_Weight' column to clean the data. Outlet_size attribute is important feature and 35% null data hence we use median to fill the missing values because Outlet_size attribute has categorical values.

2.3 Univariate Analysis

It is used to describe the data. It takes data, summarizes that data, and finds patterns in the data. From this, we can get an idea of the distribution of numerical variables the outliers of our dataset [10].

Distribution of columns with numerical values—Here, we have plotted boxplot for various attributes so as to observe their distribution.

From Fig. 1, we observe that data are not normally distributed and here are many outliers which we need to remove.

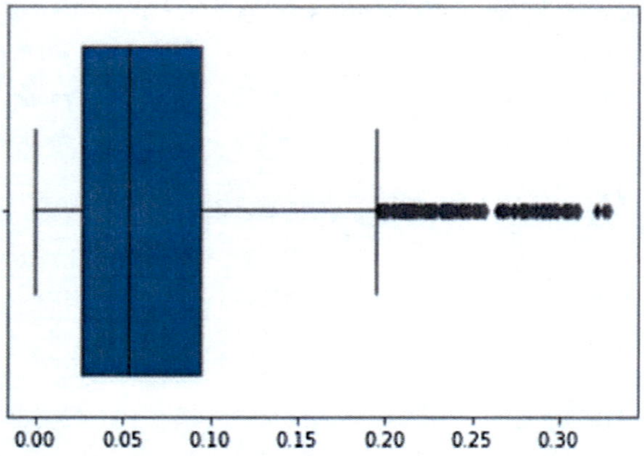

Fig. 1 Distribution of Item_Visibiltiy

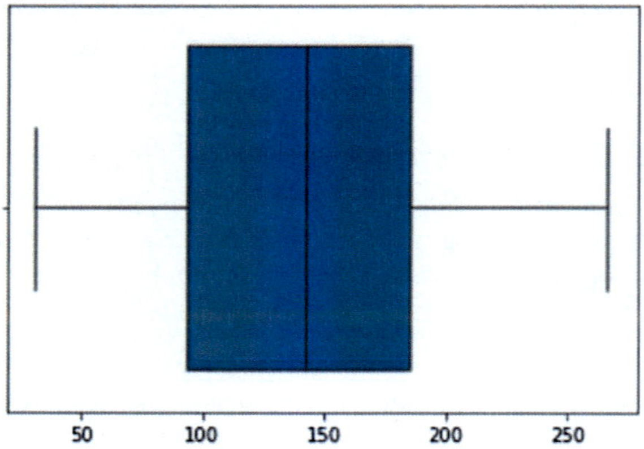

Fig. 2 Distribution Item_MRP

Item_MRP attribute is normally distributed with no outiers in it as shown in Fig. 2.

Outlet_Establishment_Year attribute is also normally distributed with no outiers in it as shown in Fig. 3. As shown in figure, outlet establishment year data ranges in between 1985 and 2010, and mostly, data are less than 1999.

From Fig. 4, we can easily observe that distribution of outlet sales has some outlier. Before apply any machine learning algorithm, we should first detect and remove the outlier.

2.4 *Bivariate Analysis*

Distribution of Item_Weight with Item_Outlet_Sales

It shows the distribution of Item_Weight with respect to Item_Outlet_Sales. From Fig. 5, we can analyze which type of product have maximum effect on sales of BigMart stores.

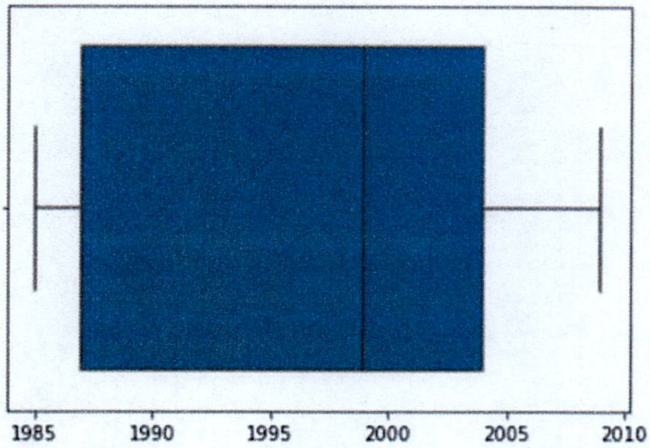

Fig. 3 Distribution Outlet_Establishement_Year

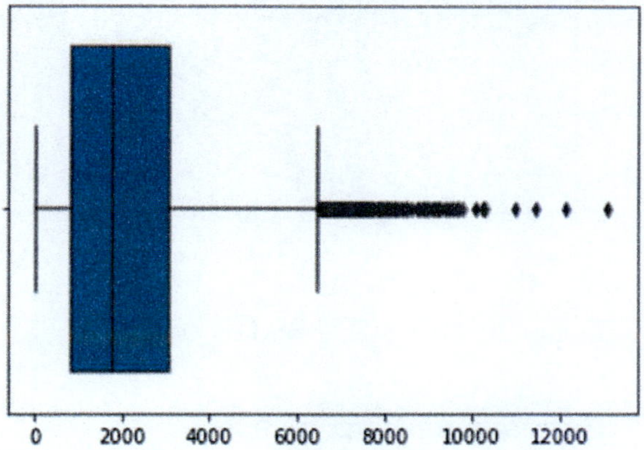

Fig. 4 Distribution Item_Outlet_Sales

Distribution of Item_Visibility with Item_Outlet_Sales

The sales will be impact by location of product. The items which are at front entrance will first catch the eye of customer than the ones in back. This was the assumption but according to graph in Fig. 6, the products which are more visible have less sales. This might be due a large number of daily use products. Which do not need eye catchy visibility. Furthermore, we observe that some items have zero Item_Visibility which is not possible as zero means item is not there in the outlet but it is there.

Distribution of Item_Fat_Content with Item_Outlet_Sales

Daily use products could have a higher probability to sell in compared to the specific use products. In Fig. 7, 'low fat' products have higher sales values than 'regular' products. And moreover, we need to treat them as it has only two categories which are miscoded as 'LF' and 'low fat' instead of 'low fat' and 'regular' are miscoded as 'reg'.

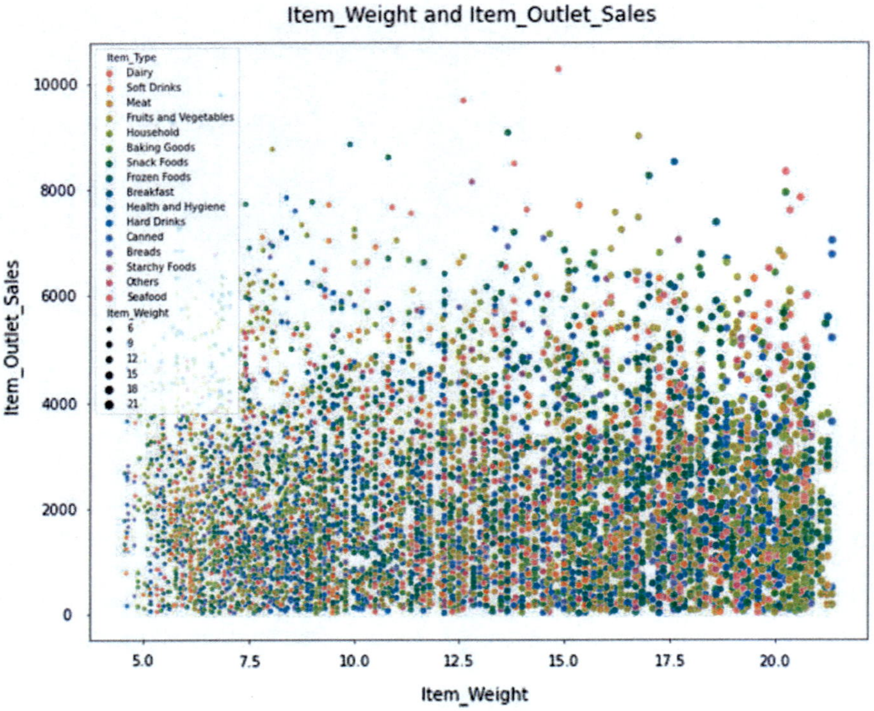

Fig. 5 Distribution of Item_Weight with Item_Outlet_Sales

2.5 Correlation Matrix

Correlation matrix gives clear picture of relation between variable as shown in Fig. 8. Item_outlet_sales variable is our dependent variable. We can observe that Item_MRP has strong relationship with Item_outlet_sales variable, and the relationship is positive. The contribution of Item_MRP will be high in machine learning model. Item_Visibility features had the lowest correlation with our target variable.

Hence, the less visible the product is in the store the higher the price will be. Outlet_establishment_Year has negative and very less correlation with our target variable. That shows weak involvement in prediction.

3 Implementation and Result

The prediction of work has been tested using various machine learning algorithms [11–13], and we can see that different algorithm performs differently on this dataset.

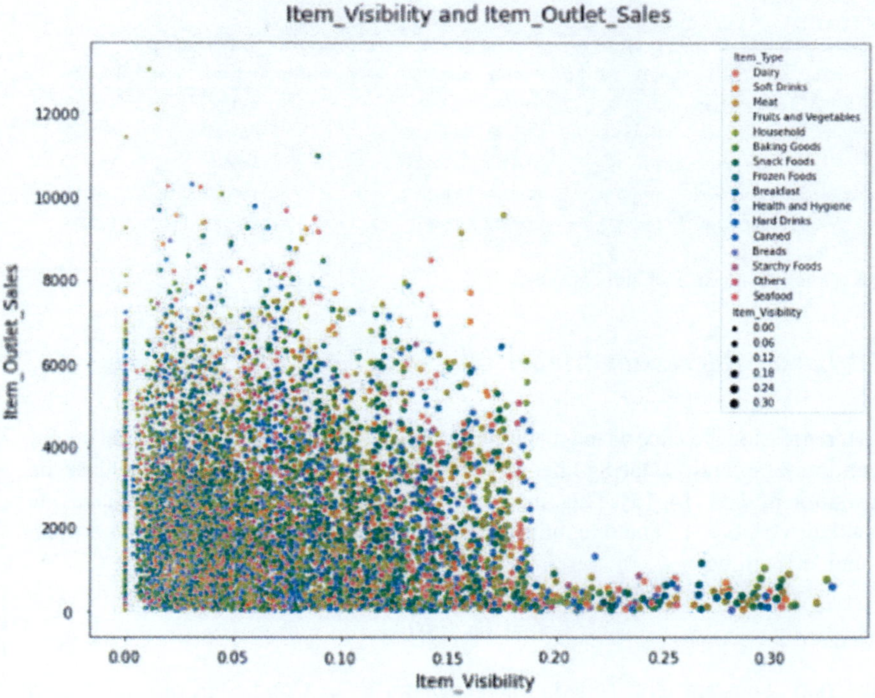

Fig. 6 Distribution of Item_Visibility with Item_Outlet_Sales

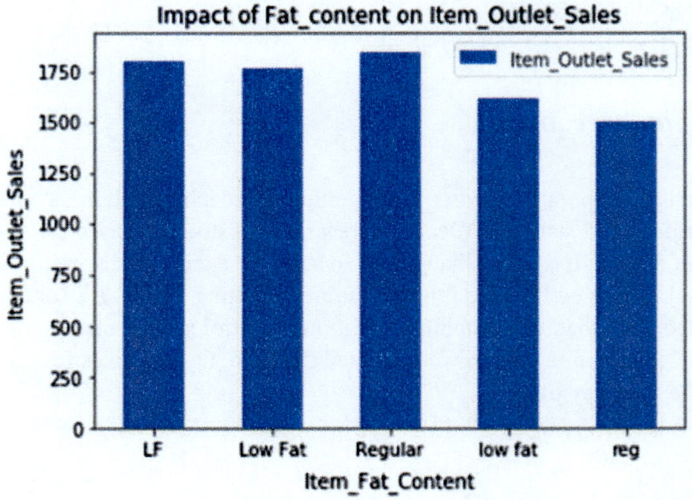

Fig. 7 Distribution of Item_Fat_Content with Item_Outlet_Sales

```
Train.corr()
```

	Item_Weight	Item_Visibility	Item_MRP	Outlet_Establishment_Year	Item_Outlet_Sales
Item_Weight	1.000000	-0.014048	0.027141	-0.011588	0.014123
Item_Visibility	-0.014048	1.000000	-0.001315	-0.074834	-0.128625
Item_MRP	0.027141	-0.001315	1.000000	0.005020	0.567574
Outlet_Establishment_Year	-0.011588	-0.074834	0.005020	1.000000	-0.049135
Item_Outlet_Sales	0.014123	-0.128625	0.567574	-0.049135	1.000000

Fig. 8 Correlation matrix of numerical data

3.1 Linear Regression Model

Linear regression is a supervised machine learning algorithm used to predicted output as continuous quantity. During linear regression, our objective is to fit a line over the distribution of data [14, 15]. This line is nearest to most of the points. If we have one dependent variable 'Y' and one independent variable 'X', then relationship between 'X' and 'Y' will be

$$Y = B_0 + B_1 X$$

where B_0 is constant term (y axis intercept) and B_1 = Coefficient of relationship between 'X' and 'Y' (i.e., slope).

From this linear regression model, we predicted the Item_Outlet_Sales for test data, and we get the accuracy of 51%.

3.2 Decision Tree Model

Decision tree is most popular predictive modelling approaches used in data mining, machine learning, and statistics. Decision trees are constructed based on different condition over dataset. It is a tree-like graph. To build the decision tree, we use CART algorithm [11]. It is used because it mimic human thinking ability. It is one way to display an algorithm that only contains conditional control statements.

From this decision tree model, we predicted the Item_Outlet_Sales for test data, and we get the accuracy of 61%.

3.3 Random Forest Model

Random forest is a bagging technique and not a boosting technique. The trees in random forests are run in parallel. There is no interaction between these trees whilst

Table 2 Comparison of machine learning model

#	Model	Accuracy (%)
1	Linear regression	51
2	Decision tree	61
3	Random forest	61
4	XGBoost	87

building the trees. Boosting means teamwork. Random forest has ability to give excellent performance when the number of variable is much higher than the number of observation [12].

From this random forest model, we predicted the Item_Outlet_Sales for test data, and we get the accuracy of 61%.

3.4 XGBoost Model

XGBoost stands for 'extreme gradient boosting', where the term 'gradient boosting' originates from the paper greedy function approximation: A gradient boosting machine, by Friedman [13]. We think this explanation is cleaner, more formal, and motivates the model formulation used in XGBoost.

From this, we get the accuracy 87.0 which is the best accuracy in all the models which we use (Table 2).

4 Conclusion

In this digitally connected world, every MegaMart keen to know the customer demands beforehand to avoid the shortfall of sale items in all the seasons. And as the trend of online shopping is increasing, the companies or the MegaMart are predicting more accurately the demand of product sales or user demands to ensure success in their businesses. At enterprise level, extensive research is happening for accurate sales prediction.

In this research work, we try supervised predictive model like linear regression, decision tree, random forest, and XGBoost model shown excellent accuracy. Whilst doing any predictive models, the data wrangling is most important task. We can easily see in correlation matrix that only one attribute has strong relation with our independent variable. If we want to make good predictive model, then we need more dependent data and attribute for good accuracy.

References

1. Meghana, N., Chatradi, P., Avinash Chakravarthy, V., Kalavala, S. M., & Neetha, K. S. (2020). Improvizing big market sales. *Journal of Xi'an University of Architecture and Technology, XII* (IV), 4307.
2. Shukla, R., & Yadav, V. (2020). Input data characterization using machine learning and deep. In *1st International Conference on Computational Research and Data Analytics (ICCRDA-2020)*, October 24, 2020.
3. Jiménez, F., Sánchez, G., García, J. M., Sciavicco, G., & Miralles, L. (2017). Multi-objective evolutionary feature selection for online sales forecasting. *Neurocomputing, 234*, 75–92.
4. Warnakulasooriya, H., Senarathna, J., Peiris, P., Fernando, S., & Kasthurirathna, D. (2020). Supermarket retail-based demand and price prediction of vegetables. In *2020 20th International Conference on Advances in ICT for Emerging Regions (ICTer)* (pp. 308–309). IEEE.
5. Liu, X., & Ichise, R. (2017). Food sales prediction with meteorological data—A case study of a Japanese chain supermarket. In Y. Tan, H. Takagi, & Y. Shi (Eds.), *Data mining and big data. DMBD 2017. Lecture notes in computer science* (Vol. 10387). Springer.
6. Kumari, P., Pamula, R., & Jain, P. K. (2018). A two-level statistical model for big mart sales prediction. In *2018 International Conference on Computing, Power and Communication Technologies (GUCON)* Greater Noida.
7. Barksdale, H., & Hilliard, J. (1975). A cross-spectral analysis of retail inventories and sales. *Journal of Business, 3*(48), 365–382.
8. BigMart Sales Data, Kaggle. (2019). [Online]. Available: https://www.kaggle.com/brijbhushannanda1979/bigmart-sales-data
9. Suad, A. A., & Wesam, S. B. (2017). Review of data preprocessing techniques in data mining. *Journal of Engineering and Applied Science, 16*(12), 4102–4107.
10. Ho, R. (2006). *Handbook of univariate and multivariate data analysis and interpretation*. CRC Press.
11. Myles, A. J., Feudale, R. N., Liu, Y., Woody, N. A., & Brown, S. D. (2004). An introduction to decision tree modeling. *Journal of Chemometrics, 18* (6), 275–285.
12. Janitza, S., Kruppa, J., König, I.R., & Boulesteix, A.-L. (2012). Overview of random forest methodology and practical guidance with emphasis on computational biology and bioinformatics. *WIREs Data Mining and Knowledge Discovery, 2* (6), 493–507.
13. Yadav, V., & Rahul, M. (2021). A new efficient method for the detection of intrusion in 5G and beyond networks using machine learning. *Journal of Scientific and Industrial Research, 80*(1), 60–65.
14. Peck, E. A., Vining, G. G. & Montgomery, D. C. (2021). *Introduction to linear regression analysis*. Wiley.
15. Yadav, V., & Shukla, R. (2019). Human behavioral analyzer using machine learning technique. *International Journal of Engineering and Advanced Technology (IJEAT), 8*(6), 5150–5154.

Collaborative Filtering-Based Music Recommendation in View of Negative Feedback System

Jai Prakash Verma, Pronaya Bhattacharya, Aarav Singh Rathor, Jaymin Shah, and Sudeep Tanwar

Abstract Recommender systems (RS) are information filtering algorithms that suggest users items that they might be interested in. In this paper, the authors have proposed a content-based approach that maintains fresh recommendations in a music recommendation ecosystem that improves by suggesting new recommendations. A collaborative filtering system has been proposed alongside a negative feedback system (NFS). This results in a much newer array of song recommendations based only on the songs which the user likes, and due to NFS, it can be easily recognized by the user with the precision of 16.78%. Analysis of the results reveals that the song recommendations made by the newly proposed system have a significantly lower intersection with songs that users play from general playlists and available music datasets. Thus, the proposed system allows users to discover new recommendations every time they use the NFS recommendation algorithm and thus performs better compared to the old content-based algorithms, such as popularity-based filtering mechanisms.

Keywords Big data analytics · Data mining · Deep learning · Machine learning · Recommendation system · Text data analytics

J. P. Verma · P. Bhattacharya (✉) · A. S. Rathor · J. Shah · S. Tanwar
Department of Computer Science and Engineering, Institute of Technology, Nirma University, Ahmedabad, Gujarat, India
e-mail: pronoya.bhattacharya@nirmauni.ac.in

J. P. Verma
e-mail: jaiprakash.verma@nirmauni.ac.in

A. S. Rathor
e-mail: 16BCE131@nirmauni.ac.in

J. Shah
e-mail: 16BCE075@nirmauni.ac.in

S. Tanwar
e-mail: sudeep.tanwar@nirmauni.ac.in

P. K. Singh et al. (eds.), *Proceedings of Third International Conference on Computing, Communications, and Cyber-Security*, Lecture Notes in Networks and Systems 421,
https://doi.org/10.1007/978-981-19-1142-2_36

1 Introduction

Recommender systems (RSs), in basic terms, gather user information, study it and finally use the information to predict what items the user would also like to buy, watch, read or listen to based on the application domain [2]. RS are widely adopted in a range of diverse domains, from e-commerce sites, movies on video streaming services, articles to read on print-media and electronic websites, and many more [17]. Thus, RSs prove to be vital for the user, as it leverages users to find meaningful content from a large pool, according to the interest. Thus, it allows a narrowing of content selection based on filtering mechanism and picks only the relevant content. In music recommender ecosystems (MRS), the challenge is to provide the user with selective playlists based on previous playlists and listen to behaviour, based on genre, demographics, and language. Researchers globally have proposed content filters for MRS to address the inherent shortcomings of higher accuracy in the system, at fewer iterations.

Secondly, with MRS, there is an abundance of digital content available, and it becomes difficult to mine selective content as items of interest for the user [2]. For the same, MRS employs prediction algorithms that tell whether a user likes the item (songs/video content), or not, based on previous selection history. If accurate and efficient predictions are provided to the user, they can prove to be very beneficial for both the service provider and the user [15]. To establish the importance of MRS in today's digital content wave, we present some motivating examples.

1. In YouTube streaming apps, based on the creator profile, and searched content historical tags, RS selects the videos of interest. Statistically, it is found out that RS accounts for 70% of the overall hits and watch time that users spend on watching videos on YouTube [1].
2. According to McKinsey, Amazon has a very strong RS, and $\approx$35% of its sales is attributed to the item selection thrown by RS [7].
3. According to sigmoidal [12], on over-the-top (OTT) platforms, like Netflix, 75% of the streamed content comes from RS.

1.1 MRS in Indian Context: A Statistical Mapping

India is one of the fastest countries that have transitioned towards digital connectivity. Currently, 12% of worldwide global users, out of 3.8 billion Internet users. Currently, China leads the Internet user base with 21% Internet users. Figure 1a presents the details of expected number of music listeners in India. The shown values indicate a linear trend, and it is expected that the number of Internet users in India is expected to reach a figure of 100.3 million by 2023. Thus, the content-aware digital wave is exponentially on the rise, and thus industries have shifted towards the builds of resilient and accurate MRS ecosystems.

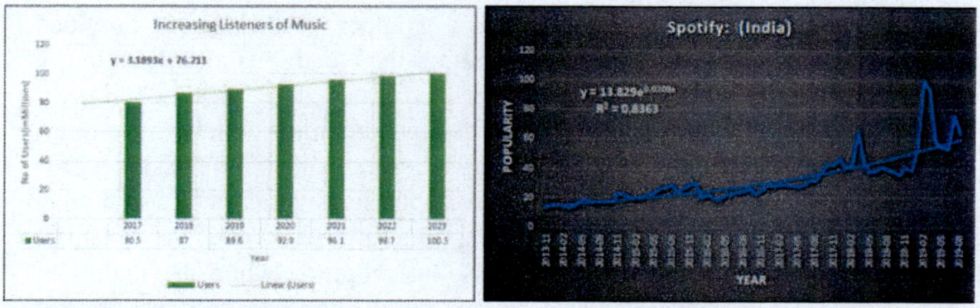

(a) Expected linear growth of users in the music industry

(b) Growing Popularity of MRS to selectively filter personalised content

Fig. 1 *Spotify* case: the music content wave and expected growth of music industry by 2023

Statistically, the Indian music industry is estimated at 14.2 billion rupees in 2018, at a CAGR of 10.8%. Currently, there is a base of 150 million music listeners on different music apps like *Spotify*. It is found out that an average user spent 21.5 h on watching digital content and listen to their favourite playlists every week.

Spotify, which is considered to be the most popular music streaming application worldwide, has achieved 2 million subscribers in the Indian market. *Spotify* uses MRS as a recommender engine and selectively presents personalized content to users. Figure 1b presents the details. It is evident from the graph that we observe a spike in popularity growth since its inception. There is an exponential increase in the growth of popularity. The R2 value calculated indicates that the data closely fits the exponential model. Thus, there is a huge demand for MRS in the digital industry. The inherent benefits of their usage are as follows:

1. RS builds up the user base based on predictive and usage behaviour. As they are based on past usage patterns, most of the times they hit the correct result and increase the user experience.
2. RS through collaborative filtering approach builds up new recommendations on the basis of similarity of users and content.
3. The RS is dynamic and changed based on the change in user behaviour and search traffics. Thus, RS always reflects the most updated content to the user.

2 State of the Art

The section presents the existing state-of-the-art schemes pertaining to RS for digital industry. For example, Jorro-Aragoneses et al. [5] proposed a framework named as *RecoLibry-core*, built-in Java to create MRS through components. The framework is presented as a tool and is included in the *RecoLibry* Suite. The framework addressed two key challenges in existing MRS frameworks. Firstly, it is oriented to a single type of recommendation method, and secondly, it consists of users that have previous knowledge of the *RecoLibry-core* and acts as a wrapper of components provided by third-party frameworks. It helps by integrating the existing frameworks into a homogeneous set of components.

Yoshizaki et al. [19] proposed an MRS method by combining collaborative filtering and MRS process based on impression words. Several pairs of impression words are given to the user, and the user scores the impression words on a seven-level scale. These seven-level scores are then converted to a three-level score. Yoshii et al. [18] have proposed a hybrid MRS that ranks musical pieces while also maintaining collaborative and content-based data. Collaborative filtering cannot recommend non-rated pieces, and content-based filtering does not have a good accuracy as it is based on the heuristic that the users favourite pieces would have similar music content even though there may be exceptions. A probabilistic generative model is proposed to attain higher recommendation accuracy and reach a wider variety of songs. Horsburgh et al. [3] presented a method that deals with defining pseudo-tag representations from content and then is used into a hybrid RS. They defined non-musical properties of tracks based on similar tags and constructed the pseudo-tag representation. Valcarce et al. [14] proposed *Prefs2vec*, which is a word embedding technique for representing users and items for memory-based RS. *Prefs2vec* model allows a quick update of embeddings through using memory-based algorithms and provides incremental recommendations. A variant of dropout for regularization is also used to reduce overfitting, which improves the model performance.

Schedl et al. [11] presented the recent challenges in MRS and proposed techniques to solve them. The major aspects proposed are author profiles, items, duration of each item, sequential consumption of songs, previous recommendation histories, emotions and content tags. To address the issues of cold start and sparsity problems, the paper proposed strategies like hybridization, cross-domain recommendation and active learning mechanisms. Authors in [8] presented an MRS based on automated playlists in which songs of a similar kind are played without any interruption, and no finite length is set on the generated playlist. Haoting et al. [6] presented a triplet network that considers both the negative and a positive response to learning the representation between users and items. Distance between user preference and positive feedback items is considered more close as compared to negative feedback items and users.

Hu and Ogihara [4] have introduced a new technique through which the system recommends suitable tracks from a collection of songs to the user based on certain parameters. It focuses on the problem to manage a large number of tracks in the playlist. The main aim is to minimize the user's effort and right song selection. The system evaluates the user attitude towards the song by portioning the playing time and the listen time. In the case of skips, it is not added to recommendations. The parameters include genre, year, freshness and pattern. Techniques such as the forgetting curve and Gaussian mixture model are used. Singhal et al. [13] have suggested deep learning methods to support the lack of information at the start of the recommendation system or due to the spare user-item rating matrix. The scheme used a deep convolutional neural network (DeepCNN) to generate the latent factors for the songs from audio and fill when no data is available. The model combines deep belief

Table 1 Comparative analysis of proposed scheme with existing state-of-the-art approaches

Author	Year	1	2	3	4	Technique	Model	Cons
Yoshii et al. [18]	2008	Y	Y	Y	N	Piece-wise rating score	Probabilistic generative	Audio features like music tempi, pitches and rhythimic patterns are not considered in MRS
Hu and Ogihara [4]	2011	N	Y	Y	N	Autoagressive integrated moving average (ARIMA)	Forgetting curve	Mixing recommendatons for multiple users are not considered in ARIMA evaluation
Yoshizaki et al. [19]	2013	Y	N	Y	Y	Collaborative filtering	Fitting curve	Content-based approach is not discussed
Horsburgh et al. [3]	2015	Y	N	N	Y	Hybrid recommender that augments sparse tags	Pesudo-tag sparse representation	Cold-start problems are not addressed
Singhal et al. [13]	2017	Y	N	N	Y	Combination of collaborative and content filters	Long short-term memory	Real-time user interactions in RS not considered
Schedl et al. [11]	2018	Y	N	Y	Y	A tutorial approach to different recommender systems	Situation and context aware	Sparsity of user data and emotion tagging not discussed
Jorro-Aragoneses et al. [5]	2019	Y	Y	N	N	RecoLibry Suite with third-party Java integrations	Collaborative filtering	Platform independence is not considered
Valcarce et al. [14]	2019	N	Y	N	N	Word vector embedding	Bag of words with regularization dropout	Finite length playlists matching with word vectors are not considered, rather an infinite playlist is assumed
Haoting et al. [6]	2019	N	Y	Y	N	Triplet is considered: user preference, positive item and negative item sets	Piece-wise element difference and latent common space	Subnetwork distance calculation among items not considered
Wang et al. [17]	2020	Y	Y	Y	N	Heterogeneous information network-based MRS (HIN-MRS)	Collaborative item-based filtering	Complex feature of user preferences is not considered in HIN
Wang [16]	2020	Y	N	Y	N	Hybrid RS based on weighted combination and filtering approaches	Gaussian mixture model	Content-based approach is not discussed

(continued)

Table 1 (continued)

Author	Year	1	2	3	4	Technique	Model	Cons
Melchiorre et al. [8]	2021	N	Y	Y	Y	Demographic-based RS with notion of fairness	Equal opportunity metric	Consider gender as a binary construct, thus the model is over-simplified, and bias is induced. Moreover, the inherent complexity and gender fairness in datasets are not balanced
Proposed	2021	Y	Y	Y	Y	Negative recommendation system to update recommendations at real time and ensure freshness in content	Hybrid approach, content and collaborative	Twofold cross-validation and information symmetry are not considered

1. Collaborative filtering. 2. Public datasets 3. MRS 4. Content filtering, Y-parameter considered, N-not considered

networks and a probabilistic graphical model to simultaneously learn from the audio content and generates personalized recommendations. The scheme is validated over the echo nest taste profile dataset. Wang et al. [17] proposed a scheme named heterogeneous information network-based MRS (HIN-MRS) that considers contextual factors, user personalized preferences and topic recommender to build a user satisfaction model. The model considers 10,000 music playlists and uses collaborative filtering through an item-based filter algorithm. Wang [16] proposed a collaborative filtering approach and the wonton recommendation algorithm on different music genres and proposed a hybrid RS based on the weighted combination and filtering approaches. The authors considered a Gaussian mixture model where the audio signals are transformed through fast Fourier transformation with triangle windowing property. The proposed results yield better results than simple collaborative MRS. Table 1 presents the comparative analysis of the proposed scheme with other similar schemes.

3 The Proposed MRS Architecture

In this section, we present the proposed MRS architecture. The architecture consists of available datasets and user profiling from where it is classified whether the user is new, or an existing user. Based on the user data, we perform the content, popularity and collaborative filtering mechanisms. A negative feedback system (NFS) is

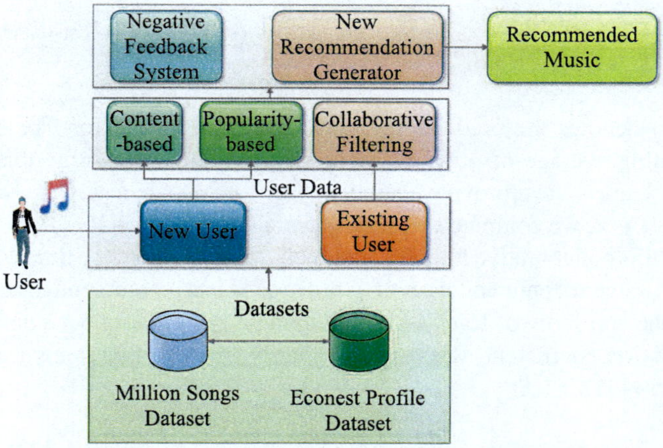

Fig. 2 Proposed MRS architecture

introduced that keeps the freshness in recommendations. Figure 2 presents the details as follows. In the proposed system, we consider target users $U = \{U_1, U_2, \ldots, U_n\}$, who wish to listen to a particular musical item **M**. We consider a streaming server S_M that sends q audio packets $\{A_1, A_2, \ldots, A_q\}$ of **M**. The media player at U_n arranges the packets based on sequence numbers. Any MRS object is to rank the musical packets, as a piece-wise sequence that is not rated by U_n. We consider the sequences as $\{S_{A_1}, S_{A_2}, \ldots, S_{A_q}\}$, with the trivial conditions $A_q \in \mathbf{M}$, and $U_n \in U$.

We observe the user rating data $\text{Rat}(\mathbf{M})$ in the ecosystem and consider the score-set as a scale unit from 0 to 5, where 0 means song is disliked and 5 presents song is liked. We form a rating matrix, denoted as $M[U_n, \mathbf{M}]$. In case U_n has not rated **M**, the rating is set to ϕ, which denotes an empty rating. Based on $M[U_n, \mathbf{M}]$, we assume that piece-wise contents are presented as single vector units $V(\mathbf{M})$. For every n user, the vectors are extracted and then collected. Recommendations are then made based on the memory-driven method on $M[U_n, \mathbf{M}]$, due to its lower computational cost than the model-based approach. We next present the collaborative and content-based filtering approach.

3.1 Memory-Based Collaborative Filtering

In this, we predict the unknown $\text{Rat}(\mathbf{M})$ score, i.e. which is not rated and has value ϕ. For the same, we consider the score of other $n-1$ users U_n and apply heuristics to predict the rating score of any nth user. For any U_n, the predicted score is given as follows:

$$\mathrm{Rat}(U_n) = \mathrm{Avg}(U_n) + k \sum_{U_q \neq U_w, U_q, U_w \in U} w_{U_q,U_w}(\mathrm{Rat}_{U_k}(\mathbf{M}) - \mathrm{Rat}_{U_q}(\mathbf{M})) \quad (1)$$

where $\mathrm{Rat}(U_n)$ denotes the predicted collaborative rating score for U_n, $\mathrm{Avg}(U_n)$ denotes the rating average of previous pieces by U_n, and we assume that out of $(n-1)$ users, k users have provided rating scores for $\mathbf{M}$, where $k \subseteq (n-1)$. For any users $\{q, w\} \in k$, we compute the overall average rating.

To improve this collaborative filtering approach, we consider every time U_n makes a rating $\mathrm{Rat}(U_n)$, we recommend a set of p other songs as recommendations to U_n, computed by the similarity of $\mathrm{Rat}(U_n)$. To compute this similarity, we consider the Pearson correlation coefficient, where the similarity between two items i and i' is defined as follows [18]:

$$w_{i,i'} = \frac{\sum_m (\mathrm{Rat}(i, \mathbf{M}) - \overline{\mathrm{Rat}(i, \mathbf{M})}) \sum_m (\mathrm{Rat}(i', \mathbf{M}) - \overline{\mathrm{Rat}(i', \mathbf{M})})}{\sqrt{\sum_m (\mathrm{Rat}(i, \mathbf{M}) - \overline{\mathrm{Rat}(i, \mathbf{M})})^2 \sum_m (\mathrm{Rat}(i', \mathbf{M}) - \overline{\mathrm{Rat}(i', \mathbf{M})})^2}} \quad (2)$$

Thus, by keeping the similarity computation, we recommend user U_n p similar songs by taking the same computation for p iterative loops. Whenever U_n recommends any similarity item i, the recommendation is removed from the similarity measure while computing the Pearson correlation. Thus, by updating the value of the correlation coefficient, we assure that U_n is not recommended any $\mathbf{M}$ that is already rated by U_n itself. Once all the p recommendations are exhausted, we set back the similarity score to 0 and repeat the overall process. Next, we discuss the content-based recommender model.

3.2 Memory-Based Content RS

In memory-based content RS, we focus on the similarity of musical content, denoted as $C(\mathbf{M})$. On the basis of $C(\mathbf{M})$, we present the preferences in content space V. Next, we denote the content piece length of size $|q|$ units, where $\alpha_m = \{\alpha_{(m,1)}, \alpha_{(m,2)}, \ldots, \alpha_{(m,q)}\}$. We define two categories of rating scores, positive and negative, represented by $\mathrm{Pos}(\mathrm{Rat}(\mathbf{M}))$, and $\mathrm{Neg}(\mathrm{Rat}(\mathbf{M}))$, respectively. We consider $\mathrm{Pos}(\mathrm{Rat}(\mathbf{M})) = \{4, 5\}, \forall V$. A score of 4 denotes good recommendation for user, and 5 denotes excellent recommendation for U_n. Similarly, $\mathrm{Neg}(\mathrm{Rat}(\mathbf{M})) = \{0, 1, 2, 3\}$ is defined as set of negative scores, where 0 denotes not at all the user preference and 3 denotes slight user preference. Based on this target formulation, we define the content-based rules as follows:

- If $\mathrm{Pos}(\mathrm{Rat}(\mathbf{M})) = \phi$, we assign the set of content vectors $V_u^+ : \{C(\mathrm{Pos}(\mathrm{Rat}(\mathbf{M})))$ to present the musical preference of U_n, and $V_u^- : \{C(\mathrm{Neg}(\mathrm{Rat}(\mathbf{M})))$ to represent the dislike of U_n.

- Based on V_u^+ and V_u^-, the similarity scores S_u^+ are computed for V_u^+, and S_u^-, for V_u^-, respectively. The final similarity score based on content C is then denoted as $S_{(U_n)}$.
- The musical pieces that are not been rated for U_n are then rated based on pair (C, V_U^+, V_U^-), where we measure the similarity based on the cosine measurements. Based on the same, we present the binary categorization models as follows:

$$B_u^+ = \prod_{\beta} u_+(\beta|u)^{(C,V_U^+)}$$
$$B_u^- = \prod_{\beta} u_-(\beta|u)^{(C,V_U^-)} \quad (3)$$

3.3 The Proposed Negative Feedback System

In this scenario, we consider for U_n $C(\mathrm{Neg}(\mathrm{Rat}(\mathbf{M})))$ and V_u^- and look at music titles m_t, where $m_t > 1$, we consider them as liked song and assign $\mathrm{Pos}(\mathrm{Rat}(\mathbf{M}))$, and $\mathrm{Neg}(\mathrm{Rat}(\mathbf{M}))$ for disliked songs. Based on the same, we perform the piece-wise classification and apply item–item filtering sets. The final set is then performed by constructing the set difference as follows:

$$m_t = m_{t^+} - m_{t^-} \quad (4)$$

4 Implementation and Execution

In this section, we present the experimental analysis of the proposed model.

4.1 Environment Set-up

For experimental analysis, the proposed model is implemented with the PC hardware set-up as following configurations: CPU: Intel Core *i7* fifth-generation processor, RAM: 8 GB, Operating System: Microsoft Windows 10. The dataset analysis is done on Jupyter Notebook *v6.0.2*.

4.2 Public Datasets

The dataset which has been used is made from the inner join of two datasets. The first is a subset of the million song dataset [10], which contains the audio features and metadata for 10,000 songs. This dataset has been merged along with the taste profile subset [9], which is the official user dataset of the million song dataset. Both of them are converted into comma-separated values (CSV) files. This dataset contains actual user-play counts, and the songs have already been matched with the million song dataset. A final CSV file has been constructed using inner join on *song_ID* and then dropping the duplicate rows.

4.3 Characteristics of Dataset

On plotting different histograms based on the different attributes of the data, analysis of the important features of the data can be done, i.e. year, tempo, duration, and key signature. Figure 3 presents the details of the dataset.

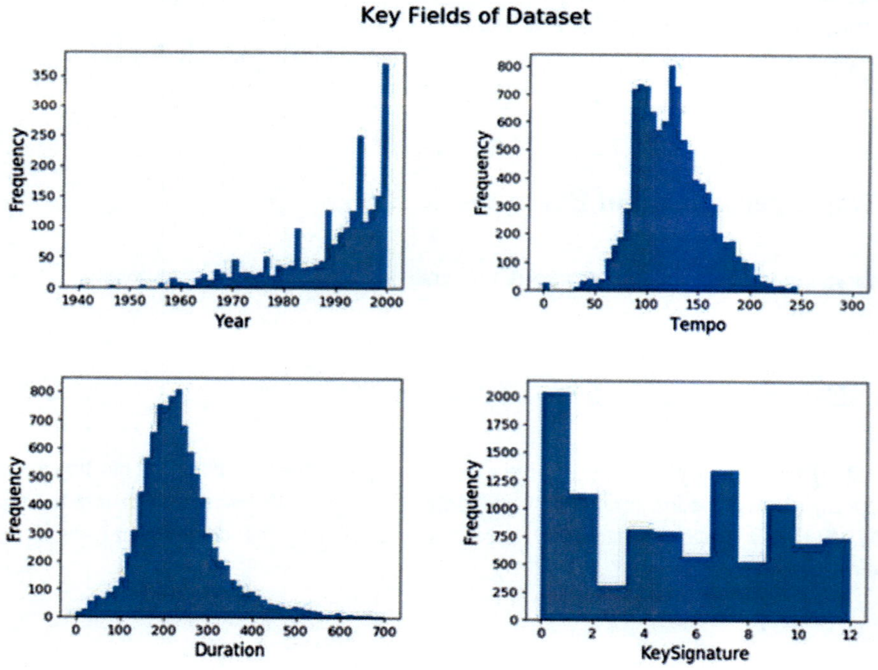

Fig. 3 Characteristics of dataset

With the help of the group by and sum function on the title, listen count of each song has been found which is in the dataset. A histogram has been made based on different listening counts for each song.

4.4 Results and Discussion

For popularity-based recommendations based on the listening to count, we have filtered the top 50 songs from the dataset which are trending songs. Similarly, for the content-based recommendation, we have recommended the songs to the user based on different attributes of the dataset. In our instance, we have recommended songs to the user based on the artist.

For the collaborative filtering approach, for each song, listened to by the user we will apply item–item filtering and get the needed recommendations for each song. For applying item–item filtering, we have constructed a co-occurrence matrix, where the rows are users and columns correspond to the song. Each cell value represents the listen count in the co-occurrence matrix. Now, after getting the recommendation for each song, we will finally merge the recommended songs and output the top ten results from them, which will then be recommended to the user. Figure 4a presents the details of the plot of users U_n that listen to recommended songs and the frequency of listening to a particular song.

Fresh recommendation works as an add-on for each recommendation technique where we can recommend different songs to the user each time. For recommending different songs, we can simply increment the count by n or we can select some random songs from the recommended songs list so that the user does not get the same result each time.

NFS is an update over the collaborative approach. For negative feedback, we will apply collaborative filtering twice: one for the liked songs (listen to count greater than 1) and disliked songs (listen to count equal to 1). Figure 4b presents the details

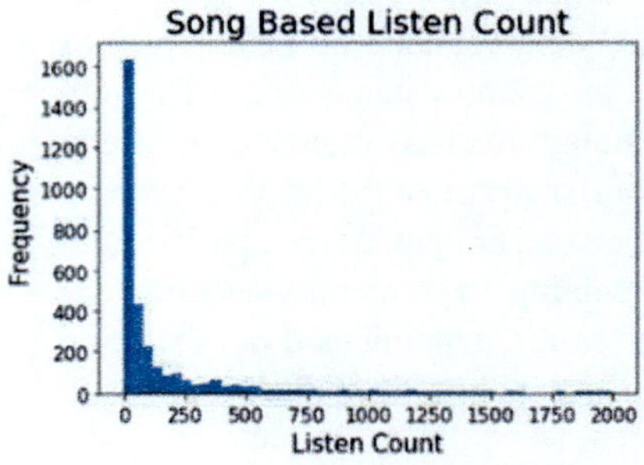

(a) Song Count Evaluation

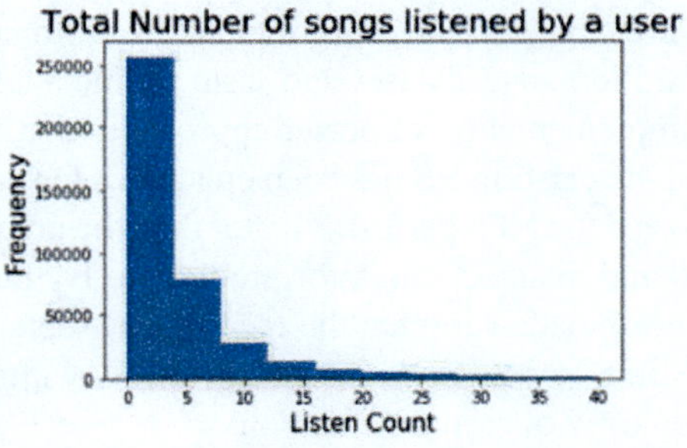

(b) Listen Count of recommended songs

Fig. 4 Performance evaluation

Table 2 Precision and recall @ 10 (in percentage)

Technique	Precision	Recall
Popularity	1.6906	15.0479
Collaborative filtering	17.3261	2.0143
Negative feedback system	16.7865	1.9424

of the total number of listened songs. Now, after getting the results on both the parts, we will apply the set difference on liked songs recommendation and disliked song recommendation to get the final result. Thus, the user will be recommended songs based on the songs liked by the user while abandoning the disliked songs.

On observation of these results, it can be observed that recall while considering negative feedback along with collaborative techniques is lower than when we use the content-based popularity approach. This means that the intersection was found to be considerably lower than when making popularity-based recommendations. Hence, it can be concluded that the newly proposed recommendation system is making much newer recommendations compared to the old content-based popularity approach which recommends the same songs each time resulting in a high intersection value and hence recalls. The details of the precision and recall values are presented in Table 2.

5 Conclusion

Collaborative systems present much better results than content-based systems and thus are applied in a wide range of scenarios. They utilize user-item interactions to predict items of interest. In the collaborative approach, grouping between similar user profiles takes place to share the information in the profiles. This helps in making recommendations among users of the same group and results in a high probability of surprising data items coming forward. In the article, both content and collaborative techniques have been applied on a merged dataset made up of a subset of the million song dataset and taste profile subset to recommended songs. Through the implementation of newer approaches that maintain fresh recommendations and NFS, the existing RS has been enhanced. On careful analysis of the results, it can be observed the NFS performs better than the predefined techniques as it considers both liked and disliked songs for prediction. By maintaining fresh recommendations, it has been made sure that the recommendations given are dynamic and not the same each time recommendations are given allowing the user to be able to discover a much wider array of recommendations.

References

1. Alvarado, O., Heuer, H., Vanden Abeele, V., Breiter, A., & Verbert, K. (2020). Middle-aged video consumers' beliefs about algorithmic recommendations on youtube. *Proceedings of the ACM on Human-Computer Interaction, 4*(CSCW2). https://doi.org/10.1145/3415192
2. Ben Sassi, I., Ben Yahia, S., & Liiv, I. (2021). MORec: At the crossroads of context-aware and multi-criteria decision making for online music recommendation. *Expert Systems with Applications, 183*, 115375.
3. Horsburgh, B., Craw, S., & Massie, S. (2015). Learning pseudo-tags to augment sparse tagging in hybrid music recommender systems. *Artificial Intelligence, 219*, 25–39.
4. Hu, Y., & Ogihara, M. (2011). Nextone player: A music recommendation system based on user behavior. In *12th International Society for Music Information Retrieval Conference (ISMIR 2011)* (pp. 103–118)
5. Jorro-Aragoneses, J. L., Recio-García, J. A., Díaz-Agudo, B., & Jimenez-Díaz, G. (2019). Recolibry-core: A component-based framework for building recommender systems. *Knowledge-Based Systems, 182*, 104854.
6. Liang, H., Zeng, D., Yu, Y., & Oyama, K. (2019). Personalized music recommendation with triplet network. eprint arXiv:1908.03738
7. McKinsey, P. (2021). Our new research shows that more than 70% of shoppers plan to participate in amazon prime day 2019. https://www.mckinsey.com/business-functions/marketing-and-sales/solutions/periscope/news/press-releases/more-than-70-percent-of-shoppers-planning-to-participate-in-amazon-prime-day-2019-new-research-from-periscope-by-mckinsey-reveals
8. Melchiorre, A. B., Rekabsaz, N., Parada-Cabaleiro, E., Brandl, S., Lesota, O., & Schedl, M. (2021). Investigating gender fairness of recommendation algorithms in the music domain. *Information Processing & Management, 58*(5), 102666.
9. Nest, T. E. (2021). *The echo nest taste profile subset*. http://millionsongdataset.com/tasteprofile/
10. Nest, T.E. (2021). *Million songs dataset*. http://millionsongdataset.com/
11. Schedl, M., Zamani, H., Chen, C. W., Deldjoo, Y., & Elahi, M. (2018). Current challenges and visions in music recommender systems research. *International Journal of Multimedia Information Retrieval, 7*(2), 95–116. https://doi.org/10.1007/s13735-018-0154-2
12. *SigMoidal: Recommendation systems—How companies are making money*. (2021). https://sigmoidal.io/recommender-systems-recommendation-engine/
13. Singhal, A., Sinha, P., & Pant, R. (2017). Use of deep learning in modern recommendation system: A summary of recent works. *International Journal of Computer Applications, 180*(7), 17–22.
14. Valcarce, D., Landin, A., & Parapar, J. (2019). Álvaro Barreiro: Collaborative filtering embeddings for memory-based recommender systems. *Engineering Applications of Artificial Intelligence, 85*, 347–356.
15. Varez, P., Zarazaga-Soria, F., & Baldassarri, S. (2020). Mobile music recommendations for runners based on location and emotions: The DJ-running system. *Pervasive and Mobile Computing, 67*, 101242.
16. Wang, L. (2020). Design and implementation of hybrid music recommendation system based on music gene. In *Proceedings of the 2020 International Conference on Computers, Information Processing and Advanced Education. CIPAE 2020* (pp. 121–124). Association for Computing Machinery. https://doi.org/10.1145/3419635.3419669.
17. Wang, R., Ma, X., Jiang, C., Ye, Y., & Zhang, Y. (2020). Heterogeneous information network-based music recommendation system in mobile networks. *Computer Communications, 150*, 429–437.
18. Yoshii, K., Goto, M., Komatani, K., Ogata, T., & Okuno, H. G. (2008). An efficient hybrid music recommender system using an incrementally trainable probabilistic generative model. *IEEE Transactions on Audio, Speech, and Language Processing, 16*(2), 435–447. https://doi.org/10.1109/TASL.2007.911503

19. Yoshizaki, S., Yoshitomi, Y., Koro, C., & Asada, T. (2013). Music recommendation hybrid system for improving recognition ability using collaborative filtering and impression words. *Artificial Life and Robotics, 18*(1), 109–116. https://doi.org/10.1007/s10015-013-0107-z.

Internet of Things-Based e-Health Care: Key Challenges and Recommended Solutions for Future

Gadiparthy Harika Sai, Khushboo Tripathi, and Amit Kumar Tyagi

Abstract Internet of Things (IoT) has changed the way of living today. Today, Internet connected things (ICT) are increasing at a rapid rate and connecting with devices to reduce load from human being. Irrespective of the sector, the IoT devices are everywhere taking care of everything from the agriculture sector to the sector of manufacturing. But, due to the global COVID 19 pandemic, the sector of health care demands the major use of IoT today. Due to the prevailing pandemic, healthcare professionals also choose to treat the patients virtually rather than treating them physically. IoT plays a major role here. But, most of the application providers or service providers or any other system involving IoT devices for generating and storing data may become a way of leak of information or stolen by a third party for black mailing or financial gain thus leading to privacy and security leak of the user. This work includes all such views with various issues and recommended solutions for the same. Also, other security and privacy requirements and corresponding solutions are also included to provide future researchers a solid base and a clear depth in knowledge regarding the security and privacy issues and solutions required.

Keywords Smart health care · Challenges · Internet of Things (IoT)-cloud-based health care · Wearable devices · Medical Internet of Things (MIoT)

G. H. Sai · A. K. Tyagi (✉)
School of Computer Science and Engineering, Vellore Institute of Technology, Chennai, Tamil Nadu 600127, India
e-mail: amitkrtyagi025@gmail.com

K. Tripathi
Department of Computer Science and Engineering, Amity University Haryana, Gurgaon, India

A. K. Tyagi
Centre for Advanced Data Science, Vellore Institute of Technology, Chennai, Tamil Nadu 600127, India

P. K. Singh et al. (eds.), *Proceedings of Third International Conference on Computing, Communications, and Cyber-Security*, Lecture Notes in Networks and Systems 421,
https://doi.org/10.1007/978-981-19-1142-2_37

1 Internet of Things—Introduction

IoT has become the new environment of computation with all the software services, sensors, and equipment involved. The growth of IoT in near future seems to be very high. IoT can expand its services to almost every existing sector. Though the growth has been very useful to the people, the problems of privacy make it distant to many other people. For example, when a customer visits a store, his/her picture is captured and the face detection services identify the name and the corresponding RFID tags aid in locating. Not only the person is tracked with the help of these but also his location privacy is leaked. In spite of many existing privacy protecting strategies like anonymization, utility trade-offs, and also imposing legal restrictions on data extraction, privacy still stands to be on top of the challenges in IoT.

The two types of privacy protection strategies in IoT include: rule-based approaches and architectural-based approaches [1]. The models involving rule-based approach are mainly for those environments that are closed. These models mainly involve applying the rules over shared information and protecting the privacy. But, since the IoT is considered to be an open environment, these models are not suitable for IoT. Architectural-based approaches include anonymization, utility trade-off techniques, and proxy based approaches. However, the anonymization techniques are actually limited to only like information collection or attempt to steal the information from the collector. The main assumption here is that the information collector is a trusted party. These techniques involve in protecting the information like participation of the devices or the so called "things" in information collection, but they do not matter about the disclosure of the already collected information by the "things."

Discussing about the e-health care, health sector is one of the most prominent sectors. E-health care involves providing health services remotely through servers without actually doctor visiting the patient physically. A quick dive into the fundamental concepts in innovative e-health systems are as follows:

- Wearable Devices: Devices like fitness bands, blood pressure monitoring, heart rate monitoring devices, and pulse monitor are very helpful for patients in this e-health sector. IoT has also helped people like elders with a tracking device which they can wear, and their people can track them in case of any emergency. Also, with the help of these wearable devices, the doctors can keep a track of their patients and their records. They can reach them in case of any emergency or sudden medical attention. Additionally, the data that are produced by these IoT devices aid the doctors in maintaining records of their patients. Not only tracking of patients but also tracking of some sensor-enabled medical equipment like wheel chairs, oxygen pumps, and nebulizers is possible though IoT-based applications.
- Ambient Assisted Living (AAL): The placing of smart devices (or IoTs) for senior citizens in an environment would be more helpful in assisting them and caring for them (in case of any health emergency). These devices or applications can also be helpful in monitoring the patient's activity and some vivid parameters like blood pressure, oxygen level, and temperature, and in case of any emergency, the nearest hospital or a clinic is given the alert.

- Internet of Health Things (IoHT): Smart devices that are integrated with cloud computing, used in the health sector for analyzing the patient's data for providing better healthcare solutions to the patients, and monitor the patient in real time. The patient data that are collected can be analyzed and diagnosed immediately thus making the treatment to happen faster. Apart from these advantages of IoHT, the main disadvantage is that the data are still prone to privacy and security attacks [2].

Organization of the work: This paper proposes complete information about Medical Internet of Things. The rest of the paper is structured as follows. Further, Sect. 2 presents security and privacy requirements toward IoT. Section 3 discusses existing solutions in the field of MIoT. Section 4 discusses Internet of Things healthcare security. Then, Sect. 5 presents the Internet of Things-based healthcare technologies. Section 6 presents the future challenges involved. Finally, the work is concluded in Sect. 7 with explanation of future work in Sect. 8.

2 Security and Privacy Requirement Toward Internet of Things-Based Applications

In the smart era that is prevailing today, IoT has a lot of applications in many sectors like agriculture, medical, manufacturing, and logistics. The IoT devices that are involved here deal with a lot of data belonging to a user. The devices may share the sensitive information related to user to any service provider or a perpetuator leading to privacy and security threats [3]. In this section, several security and privacy requirements are discussed regarding the IoT-based smart healthcare applications.

2.1 Security Requirements for IoT-Based Health Care/Medical Internet of Things (MIoT)

Security and privacy play a major role in MIoT. MIoT devices produce, transmit, and store a lot of data related to user which is highly confidential and sensitive as well. Any security attack on the network of the medical system can lead to very harmful consequences. Also, the patient's private information is present everywhere, i.e., in all the levels of data collection, transmission, and storage. The following four requirements should be considered in developing the required privacy and security models for MIoT:

- Data Integrity: Data integrity refers to the accuracy, reliability, and trustworthiness of a particular data throughout its lifecycle. Data integrity can be referred in terms of both state and a process. In terms of a state, data integrity defines a dataset which is valid and also accurate. In terms of a process, data integrity refers to measures that are used to secure the validity or correctness of the data that is being worked upon. Data integrity is mainly of four types:

i. Domain Integrity: Domain integrity refers to a specific range of values that are going to be accepted and stored in a specific column within the database being worked on.
ii. Entity Integrity: Entity integrity involves the way that makes sure that every row in the table of the database has a unique and a non-null primary key value.
iii. Referential Integrity: Referential integrity is concerned about the relationship between tables in the database.
iv. User-defined Integrity: It involves the rules that are created by the user to fit her/his needed requirements.

- Data Usability: Data usability ensures that authorized users or systems can make use of the data or the data systems involved. Any access of data by an unauthorized users or system can further lead to the destruction of data usability.
- Data Auditing: Data auditing plays a major role in the security of Medical Internet of Things. Audit of the medical data regularly in an efficient way is a coherent means to check on the use of resources and also track any abnormal or mysterious events that are occurring or about to occur. Adding to this, the cloud service providers turn out to be untrusted after a period of time, and this will definitely require refined auditing methods to take care of.
- Patient Information Privacy: The information related to the patient can actually be categorized into two: general and the sensitive information. General data include the basic details of the patients such as their name, age, and address, whereas the sensitive information may include the details of fertility status, sexual functioning, genetic information, drug addiction, and other personal details of the patient. So, it is very clear that this sensitive information has to be kept private and should never be leaked to unauthorized systems or users which may result in a huge loss to the patient [4].

2.2 *Security Requirements for IoT-Cloud-Based e-Health Systems*

Internet of Things and cloud computing are two different technologies but are mutually related to each other in several applications [5]. The connection of the IoT devices among themselves leads to generation of a huge amount of big data, and this data are stored on the cloud for further requirements. These cloud-based services are being provided by many companies like Amazon, Google, and IBM. It is problem-free when the company follows all the security and privacy standards. But, on the other side, there are chances of information leaking, security breaches when the security standards are not met. So, defining the security requirements for IoT-cloud-based e-Health systems is a must. A summary on security requirements for the same is as follows:

- Data gathering, processing, and usage have to be done only in accordance with the law and not by any illegal means.
- Minimum security and privacy protection for the data are a must.
- All the IoT devices that are connected to a particular network should be able to transmit the data and receive the same without destroying the data accuracy and integrity.
- All the protocols involving the collection of data, transmission, and usage must be defined clearly according to the prevailing standards. This will actually improve the trust of the patients/users toward the system.

Key elements that have to be kept in mind while securing the IoT-cloud-based e-health systems are as follows:

i. Confidentiality: Confidentiality is a way of ensuring that any kind of data or other exchanges between the sender and receiver are protected against any kind of malicious or suspicious usage. Confidentiality should be guaranteed at different levels of the communication network, i.e., the data have to confidential when it is being exchanged between any two IoT devices in the network, IoT device, and cloud computing to e-health systems or even between the system and the end user.
ii. Data Integrity: Data integrity refers to the trustworthiness, accuracy of the data throughout its lifecycle. In this particular IoT-cloud-based e-health system, data integrity check can be done at each node involving transmission of data between a sender and a receiver.
iii. Availability: It ensures that the data are available to the authorized users but not to any other suspicious or unauthorized users at any stage of the data lifecycle, i.e., generation, processing, transmission.
iv. Access Control: This refers to the controlling of access and authorization to protected data by actually evaluating or enforcing the access required.
v. Anonymization: The use of anonymous access helps in protecting the user's security and privacy without letting the details passed on to the perpetrators.
vi. Authentication: The very important security element in any system. Verification and validation of users details before letting them access the data or the system help in majorly reducing the identity thefts or data breaches.
vii. Resistance Attraction: Resistance attraction ensures that any attacks from unauthorized users or systems are prevented or avoided.

2.3 Privacy Requirements for IoT-Cloud-Based e-Health Systems

Privacy of the patient's information has to be maintained throughout its life cycle. The internal privacy policies judge who can access, use, or view the sensitive and confidential data belonging to the patient. The most important is the protection of patient's sensitive data from leakage or unauthorized access or use. There are several methods that can safeguard the data like cloud computing, anonymization of data, and tracking the data exchange. These methods can be useful to some extent to identify or track the suspicious or malicious actions happening. There are a plethora of privacy protection measures coming up these days, but these have to be tailored separately to every need of privacy protection for better results [6]. For example, the e-health system offers several applications like patient tracking, remote monitoring, and artificial intelligence-based diagnosis. All of these services should be provided with respective potential privacy protection measures. The users have become more cautious about their data on any kind of system, especially when it comes to their medical data since this carries a lot of sensitive information about the users. If there is privacy leak from the system, this makes the system less trustworthy to the users. Strengthening of the privacy of IoT-cloud-based e-health system can be done by including privacy by design (PbD) [7] along with the following measures:

- Location Privacy: Applications involving big data networks make it mandatory to seek location information for the data being used. To prevent the loss or leakage of the location information, related effective privacy measures and strategies are used [8–11].
- Data Lifecycle Protection: This ensures that the necessary security measures needed for privacy are taken in all the different stages of the data life cycle right from retaining the data security till it is destroyed safely after the processing and usage.
- Default Privacy: Privacy as a default setting means that the privacy of a user is preserved in all situations even without his/her intervention [12]. This means that no action or work is needed from the user's end to protect his/her privacy since the privacy comes built in with the system, whereas the traditional systems require the user to take the basic steps toward privacy protection.
- Embedded Privacy: Embedding privacy into the design of the system or the architecture of the IoT systems makes it a core function of the system.
- Robustness: This ensures that all the security requirements are met at all stages of the data lifecycle in order to protect the privacy for the IT systems.
- Visibility and Transparency: Privacy just does not mean protecting the data but also maintain the trust of the users. This is where the factors visibility and transparency come into play. The operations that are being performed on the data should remain visible to the users. These factors make sure that the actions including collection, analyzing, processing, and transmission of personal data are maintained a record of and available to the users for their purposes of accountability.

Both the security and privacy measures should accord with the international standards of risk management techniques and methods in order to provide a hassle-free environment and get adopted. Data privacy is also a fundamental strategy that is concerned about protecting user data. One of the major requirements in this context is the data protection through design known as privacy by design (PbD) which is playing a key role in securing and safeguarding privacy in many of the major technological systems.

2.4 *System Requirements of IoT-Cloud-Based e-Health Systems*

Other system requirements concerning with the IoT-cloud-based e-health systems are as follows:

- Secure Protocols: These protocols aid in establishing secure and safe computer network connections and improve the security of the entire network. The application of such secure protocols not only ensures the security of the network but also enhance the security of information.
- Secure communication: This is guaranteed by specific cryptosystems and also aids in making sure of the confidentiality of the data. A secure communication is meant to protect the data transmissions from any kind of malicious exploitation by the perpetrators.
- Secure Transmission: This ensures that the data transmission is happening without any malicious or suspicious users causing harm to the system both internally and externally. The security of the data transmitted can be achieved through necessary cryptographic mechanisms.
- Data Encryption: This ensures that the data are protected throughout its lifecycle by encoding the entire data that are being worked with. This also aids in avoiding security breaches of the raw data available through encoding.

3 Existing Solutions for Internet of Things-Based Health Care/Medical Internet of Things (MIoT)

We have discussed several security and privacy requirements of IoT-cloud-based e-health systems and IoT-based health care as well. In both the healthcare systems, there is a chance of privacy leakage or security breaches possible. Also, there are chances for stealing of information from the system by perpetrators. Let's take a quick dive into some of the existing solutions for the above mentioned issues:

i. Data Anonymization: Data anonymization [13] refers to the process of protecting sensitive information by either erasing or encrypting the identifiers that play a role of connectors between the individual and the data. A few data anonymization techniques that are used are [14]:

    - Data masking: Masking the data with altered values.
    - Pseudonymization: Replacing the identifiers that are private with fake identifiers.

- Generalization: Removing some of the data intentionally in order to make it less identifiable.
- Data Swapping: Rearranging the attribute values of the dataset or the database so that they do not exactly correlate with the original records.

ii. Data Encryption: Cryptography is the main and basic technology that is being used in the data encryption process where the data are encoded and then used in its life cycle of processing, analysis, transmission, etc.

iii. Access Control: Access control is a technique that is used to regulate who can access the data in a particular computing environment or the system. This is a fundamental yet efficient security concept that aids in minimizing the risk to the related business or the organization. The two major types of access control are as follows:

  - Physical access control: This ensures the access control to campuses, rooms, buildings, or any physical assets or devices.
  - Logical access control: This ensures the access control over the computer network, files, systems, and data.

iv. Trusted Third-party Auditing: The cloud servers that are being used cannot be fully trusted. There may be a possibility of loss of data integrity and consistency in case of any data corruption or any deletion without the notice of users. Here, the trusted third party comes into play. This trusted third party [15] with a good reputation provides proper and accurate auditing results which results in accountability of the cloud service providers.

v. Data Search: In terms of protection of data privacy over the system or cloud, data should be initially encrypted. This overcomes the existing traditional plaintext keyword searches. So, enabling an accurate encrypted cloud data search will be of a great importance toward the protection of privacy.

vii. Blockchain: Blockchain is a system in which a record of actions is actually maintained across several linked computers in a network. Use of blockchain technology in the IoT-based healthcare systems aids in increasing the transparency between the doctors and patients, also ensures efficient collaboration between different health organizations and also smart contracts. Also, this helps in resisting failure and data fragmentation. But, at the same time, blockchain technologies are prone to attacks because of their transparency.

Few interesting enhancement and solutions toward IoT-based health care have been discussed in [16, 17].

Security of Electronic Healthcare Records (HERs) Systems

Electronic healthcare records (EHRs) consist of mainly the medical history of the patient, his/her statistical laboratory test results, etc. Security and privacy of these data have to be ensured properly and is crucial to save these from any kind of malicious security or privacy attacks. Adding to these, there are a number of challenges in building and deploying the healthcare systems. Because such models are vulnerable to several kinds of cyber- attacks and the users are much concerned about these cyberattacks and required efficient and effective solutions for such cyber-attacks. Similarly, the EHRs are prone to several kinds of security attacks [18]. Therefore, the following requirements are to be met based on the relevant standards when implementing the secure electronic healthcare records in the future:

- Accuracy and data integrity
- Privacy and security of the data dealing with
- An efficient data sharing mechanism
- Accurate and proper auditing and accountability of data
- Ability that the patients can control their own EHRs, i.e., monitoring them, checking records frequently, etc.

Hence, security of EHR records can be found in detail in [16].

4 Internet of Things Healthcare Security

With the rapid development of the IoT and its applications over the recent years, the healthcare sector is also expected to witness the applications of IoT majorly in the coming future. All the devices involved in the healthcare or the medical sector are also expected to deal with the integration of IoT. Though this leads to many kind of applications and makes it easier for both doctors and patients in the sector, it has its drawbacks of security and privacy challenges as discussed earlier in the paper. To completely facilitate the adoption of IoT into health sector, it is also important to know about the threat models, attack taxonomy, and possible countermeasures related which are discussed further below:

Threat Model

Both the IoT health devices and the network being used by them is prone to different kinds of security attacks. One case can be the expansion of the current network, cloud networks, and services. Second case could be the increase in the communication between the IoT devices over the network, cloud services, and applications. Another scenario would be in the in-device hardware and software limitations. Threats can be raised from both within the network or outside the network. If an attack or a threat arises from a health device in a proximal network, then the risk related would be more severe. Also, determining the malicious or suspicious device causing this would be very difficult within a proximal network.

An Attack Taxonomy

With the increasing advancements in the technology field, not only they are becoming advantageous to people but also to the perpetrators increasing their ability to introduce several types of security attacks and threats into the networks [19] or the system devices. Some of the threats are predictable and tangible, whereas it is even harder to predict many of the other threats. The major types include: attacks based on network properties, attacks based on host properties, and attacks based on information

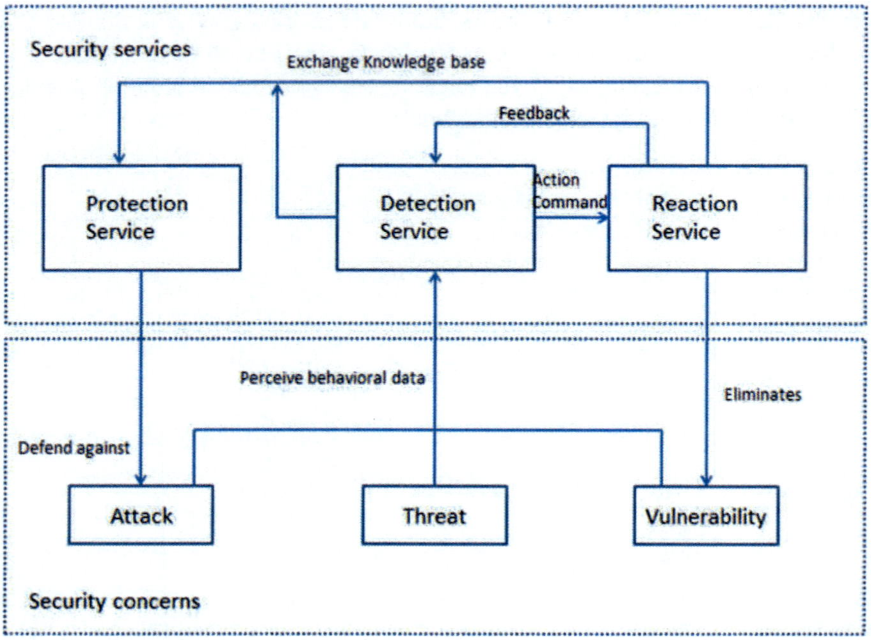

Fig. 1 Collaborative scheme for providing security services

disruptions. A security model for IoT-based health care is discussed in Fig. 1, or Fig. 1 represents a security collaboration scheme for the following security services [20]:

- Protection services: designed to reduce attacks.
- Detection services: These services will be receiving data from the applications involving health care periodically and analyze the captured data, detecting if there is any anomaly being involved.
- Reaction services: This specific type of services help the health entities in surviving all the attacks with the help of defense mechanisms.

5 Internet of Things-Based Healthcare Technologies

There are actually many prevailing technologies strengthening the Internet of Things-based health care. So, preparing an explicit list on this is definitely a tougher job. So, a brief idea on the core technologies available is given below:

i. Cloud Computing: Integration of IoT-based health care with cloud computing has enormous advantages. Some of them include access to shared resources, ability to increase the storage capacity for the data being worked with, and another important added advantage would be providing services upon request over the network. All these advantages together make it easier for the operations to get executed and thus also resulting in good efficiency.

ii. Big Data: Big data aid in accommodating huge amounts of data be it generated by the medical sensors or the IoT devices involved in the IoT-based health care or also the data that are being transmitted over the network among the IoT devices. In addition to these advantages, big data also provide a number of tools to actually improve the necessary health diagnosis in terms of efficiency.

iii. Grid Computing: In general terms, grid computing involves working of a network of computers under a single working protocol acting almost like a super virtual computer to perform a specified task which may be difficult for a single machine or computer to execute or achieve [21]. With this particular application of grid computing, it can be used in the field of IoT-based health care addressing the insufficient computational capability of the medical sensors or the devices that are aiding in the system. Grid computing can also be viewed as a backbone for the cloud computing.

iv. Augmented Reality: Augmented reality (AR) which is a part of IoT can play a major role in the IoT-based health care. Since the IoT-based healthcare systems mostly involve remote monitoring and diagnosing, AR comes into play here thus aiding the doctors and the system in performing remote monitoring and the needful.

v. Wearables: Wearables also play a major role in the IoT-based healthcare systems. The main advantages of these would actually be patient engagement, ability to track the patients in case of any emergency. This is also a way helpful to the senior citizens who can be tracked in case of any medical emergency, and the needful treatment can be provided, or a nearby hospital or clinic can be informed of this situation.

vi. Networks: Networks are the basic necessity of any IoT-based healthcare system. All sorts of networks ranging from short-range communications to long-range communication networks aid in the system being a part of the infrastructure of the IoT-based healthcare network. All the data transmissions or any tracking information could be shared only through these connected networks. In addition to these, the introduction of ultra-wide bands or RFID tags into the network can actually help in designing the low-power-based medical sensors and also aid in communication protocols.

vii. Ambient Intelligence: Ambient intelligence basically involves the electronic environments that are able to respond to the presence of people and are sensitive as well [22]. Since the end users of the IoT-based healthcare systems are actually humans, ambient intelligence can play an effective role over here.

6 Future Challenges Involved Internet of Things-Based Health Care

Anyone involving in the development of the security and privacy of the Medical Internet of Things (MIoT) should take the following into account:

i. Network Insecurity: Keeping in mind of various parameters like the low cost or convenience, many devices and services depend on the wireless networks like Wi-Fi which are actually prone to any unauthorized access or several intrusions taking place. They may also be vulnerable to security attacks like man-in-the-middle attack and denial of Service arracks easily. Adding to these, the free wireless networks available publicly mostly do not adhere to the standards of security and thus resulting in more chances of any kind of security attacks
ii. Lightweight protocols: Any low-cost devices or any kind of software applications should follow specific set of policies and rules in order to provide their services. Failing to do so would result in a huge loss causing security attacks over the network. In present days, the security and the cost are directly proportional, i.e., if we want to provide a high-level security for a network or a device, then the cost requirements would also be extremely high. This is not always possible in MIoT. So, developing lightweight protocols [23] for security at different levels is one good option in the future.
iii. Data Sharing: Though there is a day-to-day development in the fields of medical information technology, the problems of security and privacy are still being revolved around. The issue of information leakage seems to be in an active state even now. The information would have to be shared between different systems of MIoT in the future. Since the data are collected from different sources, it is not really possible to completely unify the data management. Any kind of disclosure or unauthorized sharing of the patient data would cause a serious loss to the patient and remains as a security issue in MIoT system.

7 Conclusion

With the recent advancements in technology and introduction of several medical devices and related software applications, large amounts of data are being generated and stored. In present days, the importance of data is on the high. With huge amounts of data over the networks, the problems of security and privacy attacks are also on rise. The ways of protecting data security and privacy at all the different stages of the data lifecycle would be of a great importance in the future research. Starting off with the IoT security and privacy requirements of the IoT-based health care and IoT-cloud-based e-health systems, this paper discusses many requirements including the security requirements, privacy requirements, system requirements, and the problems being faced and the possible solutions for the same. Medical Internet of Things (MIoT) is given a good importance in the paper. The recent advancements and innovations being made in the field of IoT have changed the lives and networks a lot connecting a plethora of devices. The IoT-based healthcare systems, IoT-cloud-based e-health systems, Medical Internet of things (MIoT) all come under the applications of the same providing remote medical facilities from anywhere in the world thus reducing the cost and getting better patient outcomes when compared to the traditional modes. However, security and privacy of these systems are still vulnerable. So, the researchers should focus on these aspects and provide possible other different solutions to these issues in the future.

8 Future Work

Due to the prevailing pandemic everywhere in the world, many sectors are not able to provide services to the people properly. Healthcare sector also comes into this list. Afraid of the conditions outside and the increase in number of Covid-19 cases everywhere, people are not willing to go to the hospitals or clinics especially. So, this is resulting in a lot of unaddressed medical cases everywhere, especially in rural areas where the lack of basic transportation facilities is adding up to this. In such situations, IoT-based healthcare systems can play a major role. Bringing the IoT-based healthcare systems to rural areas can majorly aid in addressing the medical cases of the people over there remotely without actually needing them to visit the hospital or clinic physically unless it is a serious medical issue that cannot be addressed remotely. Also, the applications of this IoT-based healthcare system like the wearables can make it easier for the doctors to track their patients' status and records easily in a remote manner.

References

1. Samani, A., Ghenniwa, H. H., & Wahaishi, A. (2015). Privacy in internet of things: A model and protection framework. *Procedia Computer Science, 52*, 606–613. https://doi.org/10.1016/j.procs.2015.05.046
2. Fernandez, F., & Pallis, G. C. (2014). Opportunities and challenges of the internet of things for healthcare: Systems engineering perspective. In *2014 4th International Conference on Wireless Mobile Communication and Healthcare—Transforming Healthcare Through Innovations in Mobile and Wireless Technologies (MOBIHEALTH)* (pp. 263–266). https://doi.org/10.1109/MOBIHEALTH.2014.7015961
3. Oh, S., & Kim, Y. (2017). Security requirements analysis for the IoT. In *2017 International Conference on Platform Technology and Service (PlatCon)* (pp. 1–6). https://doi.org/10.1109/PlatCon.2017.7883727
4. Hodge, Jr., J. G., Gostin, L. O., & Jacobson, P. D. (1999). Legal issues concerning electronic health information: Privacy, quality, and liability. *JAMA, 282,* (15), 1466–1471.
5. Malik, A., & Om, H. (2018). Cloud computing and internet of things integration: Architecture, applications, issues, and challenges. In *Sustainable cloud and energy services* (pp. 1–24). Springer.

6. Sahmim, S., & Gharsellaoui, H. (2017). Privacy and security in internet-based computing: Cloud computing, internet of things, cloud of things: A review. *Procedia Computer Science, 112*, 1516–1522.
7. Cavoukian, A. (2009). *Privacy by design.*
8. Wang, L., & Meng, X.-F. (2014). Location privacy preservation in big data era: A survey. *Journal of Software, 25*(4), 693–712.
9. Nair, M. M., & Tyagi, A. K. (2021). Privacy: History, statistics, policy, laws, preservation and threat analysis. *Journal of Information Assurance and Security, 16* (1), 24–34, 11p.
10. Tyagi, A. K., & Shamila, M. (2019). Spy in the crowd: How user's privacy is getting affected with the integration of internet of thing's devices (March 20, 2019). In *Proceedings of International Conference on Sustainable Computing in Science, Technology and Management (SUSCOM).* Amity University Rajasthan, February 26–28, 2019.
11. Tyagi, A. K., Rekha, G., & Sreenath, N. (2020) Beyond the hype: Internet of things concepts, security and privacy concerns. In Satapathy, S., Raju, K., Shyamala, K., Krishna, D., & M. Favorskaya (Eds.), *Advances in decision sciences, image processing, security and computer vision. ICETE 2019. Learning and analytics in intelligent systems* (Vol. 3). Springer. https://doi.org/10.1007/978-3-030-24322-7_50
12. Willis, L. E. (2014). Why not privacy by default. *Berkeley Technology Law Journal, 29*, 61.
13. Bayardo, R. J., & Agrawal, R. (2005). Data privacy through optimal *k*-anonymization. In *21st International Conference on Data Engineering (ICDE'05).* IEEE.
14. https://www.imperva.com/learn/data-security/anonymization/
15. Nigel, J., Mitchell, C., & Walker, M. (1995). A proposed architecture for trusted third party services. In *International conference on cryptography: Policy and algorithms.* Springer.
16. Tyagi, A. K., Gupta, M., Aswathy, S. U., & Ved, C. (2021). Healthcare solutions for smart era: An useful explanation from user's perspective. In *Recent trends in blockchain for information systems security and privacy.* CRC Press.
17. Nair, M. M., Tyagi, A. K., & Sreenath, N. (2021). The future with industry 4.0 at the core of society 5.0: Open issues, future opportunities and challenges. In *2021 International Conference on Computer Communication and Informatics (ICCCI)* (pp. 1–7). https://doi.org/10.1109/ICCCI50826.2021.9402498
18. Fernández-Alemán, J. L., et al. (2013). Security and privacy in electronic health records: A systematic literature review. *Journal of Biomedical Informatics, 46* (3), 541–562.
19. Mayzaud, A., Badonnel, R., & Chrisment, I. (2016). A Taxonomy of attacks in RPL-based internet of things. *International Journal of Network Security, 18*(3), 459–473.
20. Singh, S., Ra, I.-H., Meng, W., Kaur, M., & Cho, G. H. (2019). SH-BlockCC: A secure and efficient Internet of things smart home architecture based on cloud computing and blockchain technology. *International Journal of Distributed Sensor Networks.*
21. Joseph, J. (2004). *Grid computing.* Pearson Education India.
22. Aarts, E., & Wichert, R. (2009). Ambient intelligence. In *Technology guide* (pp. 244–249). Springer.
23. Khan, M. N., Rao, A., & Camtepe, S. (2020). Lightweight cryptographic protocols for IoT-constrained devices: A survey. *IEEE Internet of Things Journal, 8* (6), 4132–4156.

Deep Learning and Machine Intelligence for Operational Management of Strategic Planning

Anupam Kumar Sharma, Prashant Singh, Prashant Vats, and Dhyanendra Jain

Abstract Currently, industries and businesses are adopting the concept of AI in the technological frontier, while some are opposing the progress. Many financiers are putting more money into AI businesses intending to just see AI adoption in marketing grow at a rapid pace since they are ready to pay for AI equipment, applications, and interfaces. Facebook, Google, and other Internet behemoths are developing tools to kick-start targeted advertising and improved searching. Nevertheless, gaining an understanding of how conventional businesses in the retailing, medical, and telecoms industries spend their own money on AI initiatives is important. Concerning machine learning, the next digitalization frontier is intended to be unleashed using AI. As a result, businesses should be prepared for this type of development since it provides a real-world edge to the corporate sector as a result of the forthcoming digital changes. This article focuses on five AI technical innovations: self-driving cars, computerized visions, robotic systems, deep learning, and virtual assistants, which cover a wide range of current AI breakthroughs and acquiring knowledge. AI development is rising all the time, with Baidu and Google now leading the market. Globally, we estimate that the technical behemoths spent around $2 trillion on AI alone and in 2021. Approximately, 90% of the total funds has been committed to R&D, with the remaining 10% going into AI acquisition. Grants from PE and VC firms, as well as financing and startup funds, have grown considerably. Deep learning continues to be a growing technology investment with a substantial presence in both the internal and external corporate worlds.

Keywords Machine learning · Artificial intelligence · Data intelligence · Organizational optimization · Automated manufacturing processes

A. K. Sharma (✉) · P. Singh · P. Vats · D. Jain
Dr. Akhilesh Das Gupta Institute of Technology and Management, Guru Gobind Singh Indraprastha University, New Delhi, India
e-mail: anupam.sharma@adgitmdelhi.ac.in

P. K. Singh et al. (eds.), *Proceedings of Third International Conference on Computing, Communications, and Cyber-Security*, Lecture Notes in Networks and Systems 421,
https://doi.org/10.1007/978-981-19-1142-2_38

1 Introduction

Machine intelligence application in the technology sector, depending on digitalized boundaries, is often in the prototype phase. Fewer firms have incorporated the concept of AI into their business processes. Around thirty thousand intelligence C-level executives from 10 countries are now adopting the AI mindset on a bigger scale because they want to see it as a critical component of their organization. Many businesses claim that there have been ambiguities in their company operations that should be addressed to increase their financial return. In this context, a comprehensive examination involving large corporations was performed to demonstrate the commercial value of contemporary technology in their corporate operations. The study found a substantial difference between firms that employed AI previously and those that do today.

The ultimate users of artificial intelligence are business categories at the top of the MGI's company digital index, such as telecommunications and advanced tech or banking sectors. These segments are particularly important AI development targets since they use technology across different functions in their business activity. Industries employ machine learning to build self-driving automobiles, which are intended to improve transportation; for instance, financial companies want to use customer knowledge to enhance associated service offerings based on technology. The abovementioned technological developments in artificial intelligence may provide genuine benefits to subscribers, which may be a key element in technology changes. According to this study, AI innovators in the preceding digital landscape that combine solid suggested by higher with preventive approaches have a profitable revenue sector and projected performances element with a lot of companies that are developing to a potential. Electricity companies, retail outlets, health facilities, educational establishments, and industrial firms use the idea of AI to improve their commercial operations such as procurement, predicting, automating, and streamlining. These steps are critical in the processing of creating promotional strategies, and how to enhance consumer involvement (Fig. 1).

The dependency of machine learning on technical boundaries, and the notion that the idea should be based on separate datasets, suggests that organizations cannot merely use the shortcut in their commercial ventures. Companies cannot afford to put off the process of advancing toward digital possibilities, which includes the use of AI. Entrants of technology throughout the last few generations are now focusing on building a competitive edge by using AI. Businesses must focus on particular core components of explanatory and virtual execution that are essential in identifying and determinants affecting instances, assessing sets of data in the ecological system, acquiring AI machinery, trying to engage findings resulting, and adopting diverse cultures as part of a significant policy developed by these early adopters. This study specifically recognizes the reality that senior brass investigates and involves management and technical specialists in the assessment of frictionless availability of information, which would be a major facilitator of industrial innovation. The concept of AI guarantees businesses of substantial benefits, but it also contains certain

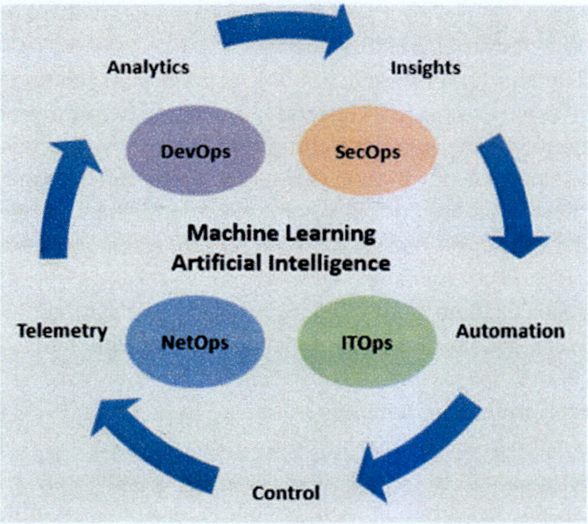

Fig. 1 To demonstrate the use of machine learning and AI in operations management

potential problems that are seen in many corporate operations, among programmers, employees, and the authorities.

2 Machine Learning Preparedness for Organizations

The debate over the risks and advantages of machine learning is heating up because it has the potential to enable machines to imitate human attributes and consciousness while operating automobiles, strengthen corporate espionage, enhance employee productivity, and infringe on customer privacy. According to studies, the population that interacts with risky and monotonous tasks often deprives itself of its living. Machine intelligence is now being used at a far higher rate than it was in previous decades. Because its extravagant hopes and disillusionment are obvious, the background of AI is progressively advancing, which is dissimilar from what is currently observed.

On either side, AI applications are increasingly concentrating on delivering real-world advantages in the commercial sphere [1]. Several factors that begin this development have indeed been addressed in the design's execution. The computational power is rapidly increasing, and programs are getting increasingly complicated, but still important when utilized in the commercial sector. As a consequence, this innovation enables businesses to generate a considerable volume of simulated intelligence data for use in different business processes. On a given day, several terabytes are used. As a consequence, enterprises at the digital boundaries, such as internet giants like Google and Amazon, are ingesting a substantial quantity of information based on machine intelligence. We anticipate that approximately thirty billion dollars will be spent in 2018 alone, which includes a substantial portion of merger and acquisition.

Private company entrepreneurs are becoming involved in the progress that examines the use of artificial intelligence.

The startup and grant funding generates an incremental one billion US dollars in invested capital [2]. Currently, the majority of progress in the implementation of human awareness is focused on AI technology. In their experimental procedure, a wide range of novel applications is possible. There are a few items in the market segments that are critical when used as an instant accelerator of technical development in the corporate sector. As a reason, researchers are separated according to the possibilities of machine learning.

A lot of them may have reached a broad agreement on the promise of AI, but they are becoming increasingly wary about the actual financial valuation. This is a type of disagreement that is reflected in a considerable difference in the current market projection, which is estimated to be around 600 trillion dollars by 2030 [3]. Given the value of the portfolio to be realized out from the element of machine learning, the most conservative assessment will indicate that enterprises are entering the second essential factor in the boom period. The commercial expertise depending on the deployment of AI suggests that boom instances are essentially improbable. To give a potentially significant viewpoint, we had chosen to investigate how users have used the concept of intelligence in their company's activities. This article presents a description of the current state of the continuously changing pace of AI in the corporate world, seen through the eyes of users and providers who wish to get a comprehensive understanding of AI's capabilities.

To begin, we assessed the financial panorama, including comprising the organization's corporate expenditures in R&D operations, mobilization of mergers and acquisitions, and money received through venture funding (VC), which comprises investment management (PE) firms [4]. Following that we looked just at the part that focuses on customer desire, as well as case evaluations and an examination of the significance of firms that use artificial intelligence in their company operations. Acceptance of AI is focused on the use of strategies to reduce obstacles in a company's activities, finances, customer groups, and the advantages that arise with AI application.

In essence, AI is interested in computers' potential to objectively display knowledge similar to that of humans. This type of assessment involves the capacity to address prospective difficulties without relying on hand-coded computer languages with numerous details. In essence, AI is preoccupied with computers' potential to objectively display knowledge similar to that of humans. This type of knowledge involves the capacity to address prospective difficulties without relying on hand-coded computer languages with numerous details. There are numerous methods for categorizing AI and machine learning, but it is kind of challenging to organize the summary, which is profoundly collectively and solely exhausting because users typically contest and mix several different technologies to create alternatives that concentrate on trying to mitigate genuine problems. The development of innovations is thought to be autonomous and can be classified depending on some other innovations and intelligent systems. Several methods categorize AI and machine learning based on fundamental capabilities such as voice, image, and text categorization, whereas others categorize them related to organizational implications such as surveillance and economics.

Concentrating on how AI could be accurately reconstructed is feasible for a variety of reasons. One among them is that it encompasses a variety of applications and technology; some of which are expansions of approaches used in previous decades. Furthermore, there are still no universally accepted conceptions of cognition, which implies that conceptions of artificial intelligent change as users become increasingly reliant on previous advancements. Tesler's hypothesis, which applies to intelligence analyzers Larry Tesler, admits that machine learning represents certain breakthroughs that have yet to be digested.

The AI-based technologies praised in this piece are classified are called narrow AI, which does the most specific jobs, as opposed to natural AI, which attempts to perform cognitive tasks that humans can also perform [5]. This article values machine learning and artificial intelligence that are restricted owing toward the relatively close of the company that is effectively dependent on the advent of such AGI. As a result, there is indeed a variety of AI technical structures that can minimize future difficulties in the corporate sector.

There are five distinct kinds of automated processes, each of which is a critical component of machine intelligence innovation. Such key classifiers involve independent and robotic systems vehicles, dialect, data processing imaginings, deep learning, as well as virtual assistants, which take into account the use of methodologies in the stages of studying data without any need for rule-based software applications to create instructions for performing certain tasks. Certain techniques, such as programing languages and visualizations, are connected to the implementation of information from other realms, such as text processing, ontology, language understanding, and voice recognition procedures.

Certain innovations are associated with the work of analyzing data computational intelligence, while many are associated with data action necessary for virtual personal assistants, driverless cars, and robots. Such technology makes use of computer coding to connect with humans. The main parts of the growth of AI technologies that were used in recent times have indeed been the study of machine learning, and the subcategories are known as deep learning. These innovations have piqued the interest of consumers, resulting in a substantial proportion of AI investments of around 60% as of 2016 [6].

3 As in Present Era, with Idea of AI Technology

The AI concept was embraced almost soon when humans began to build electronic digitalized computers. Unlike digitalized technology, AI has considerably surfed the waves of doom and hype, with one oddity: AI hasn't experienced widespread commercialization. When AI is carefully examined, it is progressively evolving. AI-powered robots can now execute a wide range of activities, including beginning data synthesis, identifying a wide range of complicated structures, and making major predictions and judgments that were previously performed solely by individuals. Furthermore, AI capabilities have grown substantially, implying that AI is now used in a wide range of commercial areas. It is also crucial to realize that machine learning has significant limits.

For example, when computers are given training based on specific information [7], they are likely to be biased. As a result, users must train the computers with a wide range of datasets to see the required development. Since 2000, these technical breakthroughs have enabled machine learning to be applied in numerous business areas to begin profound machine learning techniques, among many other essential technological improvements [8]. Such techniques have created use of the capacity of complex and diverse sources of numbers, improved algorithms which presumably positioned patterns in the data, enhanced research and innovation budgeting, and essential 3D graphics stream processors that could be used in the incorporation of advanced structures of arithmetical computational power. This same graphic system processor GSP that is schemed for integrated chips was developed for computer games, with the capacity to construct envisions thirty to ninety 30% faster than the quick version that was used in the research in [8].

The technical advancements in the performance of GSP have substantially increased that allows the students to learn relevant in the machine learning field to increase five to six-fold every twenty months of implementing new technologies. Increased data, the world generates roughly two billion GB each day, which would be correctly instantiated owing to the application of techniques that properly recognize responses. ML architectures enabled by BitTorrent data have significantly reduced the number of errors in software applications. The AI concept was embraced almost soon when humans began to build computerized digitally enhanced electronics. Despite digitalized technology, AI has considerably surfed the waves of doom and hype, with one oddity: AI has not experienced widespread commercialization. When AI is carefully examined, it is progressively evolving. AI-powered robots can now execute a wide range of activities, including beginning to deliver the information, identifying a wide range of complicated structures, and making major predictions and judgments that have been previously performed solely by individuals.

The IT sector is affected by major corporations' domestic deployment of innovations. Significantly, it has been estimated that this progress amounted to around seventeen billion us dollars alone in 2016, whereas external technical development from PE companies, venture capital funds, and startup investment was estimated to have cost approximately nine billion US dollars in that year. According to current spending data, the concept of machine learning and its implementation has been restricted to a specific aspect. This is crucial because the benefit realized from the expenditure, which links research and development to an improvement of performance in modern-day businesses. Nevertheless, because of the enormous slowness of analytic and digitalized data concerning the economy, there is indeed a large need for machine learning and artificial intelligence in companies. And over three hundred organizations from around the world discovered that company executives encourage the use of machine learning when running their apps.

So, according to current spending data, the concept of machine learning and its implementation has been restricted to a specific aspect. This is crucial because the benefit realized from the expenditure, which links research and development to betterment in modern-day businesses. Nevertheless, because of the enormous slowness of insights and digitalized information affecting the economy, there is indeed a large need for machine learning and artificial intelligence in companies. And over three hundred organizations from around the world discovered that company executives encourage the use of machine learning when running their apps.

Nevertheless, some businesses find it extremely difficult to incorporate such technologies into their operations of the company since calculating the financial return from the deployment of these innovations is challenging. Many expenditures aimed at adopting AI have included internal spending, such as the mobilization of research and development activities.

Furthermore, financially constrained companies such as Google, Facebook, and Microsoft are concentrating on how they will better integrate the capabilities inside while continuing working on AI. Businesses like Amazon are focusing on voice commands and robotic systems when launching salesforces based on agents and intelligent machines. Nissan, Tesla, BMW, and Toyota are all involved. BMW, Tesla, and Toyota are just a few of the companies that are using machine intelligence and robotics to create self-driving cars. Toyota, for example, has put aside around one billion dollars to build an institute focused only on machine learning for automated cars and robots. Firms like ABB, GE, SIEMENS, and BOSCH have also benefited from creating software of robots and deep learning, which focuses on creating particular shows the relationship to its core competency.

On either hand, IBM has concentrated on spending around two billion US dollars in developing Watson's intelligent computing capabilities that mimic IOT technologies (IoT) [9]. More than two eras, Baidu also has spent around one billion US dollars in machine intelligence development. During the last decade, the country's biggest high tech firms and worldwide sophisticated industries have completed over a hundred mergers and acquisitions. For example, in the age of technological release, Google completed 24 tasks, which comprised approximately eight computing images and approximately seven programming language understanding.

The next dominating company is applying, which had accomplished around 9 jobs like deep learning, machine translation, and data analysis. Firms are continuously refining their emphasis to better use technology in whatever operations they are involved in. A corporation like Facebook has built a machine learning lab in France, which will augment similar arrangements in San Francisco and New York City. This one will make it very difficult for companies to successfully attract talented academics in Europe. Google also has done spend around $4 million in machine learning at Québec College, which would be a research lab within the Montréal school. Intel has also made a one-billion-dollar donation to Georgia Technical University to establish deep learning and computer security research centers [10]. NVIDIA is also working with Taipei University to establish machine intelligence facilities in Taipei. Machine intelligence progress is now regarded to be in its early phases in attempt to meet up with the new technological age. AI, for example, has garnered around 3% of all venture capital by financial value by 2016, while an information filtering system is estimated to account for approximately 60%.

4 Adopting Machine Learning on the Cyber-Horizon

Currently, industries and businesses are adopting the concept of AI in the technological frontier, while some are opposing the progress. Many financiers are putting more money into AI businesses intending to just see AI adoption in marketing grow at a rapid pace since they are ready to pay for AI equipment, applications, and interfaces. Facebook, Google, and other Internet behemoths are developing tools to kick-start targeted advertising and improved searching. Nevertheless, gaining an understanding of how conventional businesses in the retailing, medical, and telecoms industries spend their own money on AI initiatives is important. To that end, it is necessary to perform an analysis to have a better understanding of the situation.

On average, only, a few companies have incorporated AI on a large scale in their value chains. The bulk of significant technology players is familiar with AI technologies that are still in the pioneering and experimentation stages. According to a 2018 report of around three hundred participants, 20% of them stated that they have embraced at least one AI-centered product on a substantial scale. 10% of participants said they had implemented and over three AI and machine learning, while 9% said they had engaged in computer vision.

Given the current new advancement, it is predicted that businesses would adopt the trends of technology change that are adopted both by early to mid-tech buyers. The previous AI trend of ICT adoption was thought to be broad in line for firms wanting to use digitalization technology. These major actors are regarded as the forerunners of the previous electronic wave of machine learning. Nevertheless, such individuals are regarded as the forerunners of the first generation of the digital world of machine learning. The very first distinguishing feature would be that the new buyers are centered on segments that spend a substantial portion of the linked technology such as cloud and data solutions. These categories are at a technological tipping point in terms of digital consumption and resources. This general conclusion, because it implies, restricted discoveries of businesses and industries are now getting to grips with kinds of improvements visible in corporations to develop more tech-oriented items.

The second feature is segment independent, with large organizations focusing on investment in machine learning faster than small businesses. This is a basic type of digital expenditure in which small and quasi firms continue to underperform in respect of tech considerations they must make while focusing on innovations. Third, new buyers of innovation are not concentrating on a single type of system. Nevertheless, they had expanded to the point where they use a variety of AI technologies in different enterprise applications. The fourth feature is derived based on expenditure that absorbs fundamental company operations. First, early computer consumers have adopted the size to be encouraged in terms of AI development potential since they possibly reduce capital investment. Machine learning is associated with more than just automated processes. Businesses, on the other hand, use technological development as a kind of product and service development. As a consequence, this seems to be the situation for early investors of digitalization technology, indicating that machine intelligence advances would be regarded as alternative sources of corporate productivity aimed at closing the income disparity between high-performing firms and low-performing ones.

5 Discussion

Operational management in the distribution chain is critical for transporting items across greater distances and for connectivity among many parties such as raw material procurement, manufacturers, merchants, logistics firms, and customers. As a result, an efficient and effective operational distribution chain (ODC) ensures that these interconnections are established properly, promptly, and also at the lowest possible cost. Exchange of information, system integration, and cooperation are important success elements for ODC. As a result, ODC has to be digitalized and progressively reliant on technologies in the context of IoT and monitors across the supply chain, allowing users to gather information in a real-timed environment. Our research is motivated by the current increase in artificial intelligence and machine learning. Many studies show that AI has indeed been widely implemented in supply chain management and has produced the highest value in manufacturing enterprises. The outcomes of widespread AI use have played an important role in enhancing logistics operations. In general, machine learning and AI implementations were some of the most intriguing and important contemporary study topics. AI is used not just in everyday living but also in logistics and operational management. AI-based operational chain is a completely integrated technology and production system that uses knowledge and smart technologies to realize understanding, networking, harmony, unification, and automating. Strategic sourcing is now becoming an autonomous operational supply chain via the help of artificial intelligence, with both the qualities of someone being self-governing, self-optimizing, self-awarded, self-determining. Our research fills an essential void in the existing literature by examining how artificial intelligence may be used in operational management and how it might assist to enhance organizational efficiency.

The notion of distribution chain originated long before the commodities themselves, and it is as venerable as the commodities themselves. The distribution network is a complicated and interconnected notion that encompasses the whole production and assembly routes from suppliers of raw materials to wholesalers and, finally, to the final consumer. Usually, the supplier chain aims to meet customer needs, enhance reactivity, and establish connections among various participants. Following business organization, business responsibilities, and customers, the entire supply chain is now becoming increasingly dispersed, varied, and accessible. For several enterprises, the main issue is that the transparency of the whole distribution network and the availability of detail accessible inside the firm are not optimum [11]. As a result, the main purpose of the operational system is to digitalize the business operations, connect many participants and resources to guarantee that the goods are in sync with the requirements of the consumers, and accomplish whole systems' comparative advantage objectives [12]. Many conventional IT solutions, such as enterprise resource planning (ERP), production planning and control (PPC), manufacturing execution system (MES), supervisory control and data acquisition (SCADA), and others, are devoted to assisting different business activities in logistics operations [13]. To regulate production throughout whole distribution networks, sophisticated technology has been powered by artificial intelligence in nearly every organizational process [14].

Notwithstanding, due to the evolving of the distribution chain, customer's changing trends, unorganized decisional problems, and the continuously shifting abilities of the company procedures, these fragmentary remedies are not 'smart' sufficiently and are not capable of acting rational way depends on the nature, also are not very appropriate for the current operations management. It is critical to function with the peak effectiveness in all main operations and strategic activities in the supply chain network to build an intelligent, quick, and excellent business signaling pathway. As a result, increasingly complex IT solutions are essential to cope with nonlinear and non, high variability corporate operations challenges in digitalization [15].

Our work is novel in that it synthesizes the most current findings while also investigating four real-world operational management cases using AI and machine learning: customer relationship management, production scheduling, quality assurance, and management services. The conclusions offer a comprehensive and relevant knowledge of the use of AI for operations management in dynamic contexts. Professionals in machine learning are developing innovative intelligence solutions that are intended to give possible explanations to real-world issues. The use of machine learning in corporate operations is associated with a significant kind of senior corporate management. Companies that have used AI technology efforts have seen a significant increase in the quality of their services. There at the meantime, industry AI deployment is substantially unequal, reflecting several features of AI technological investments.

6 Conclusions

This study contributes to both theoretical and management approaches that will be important in the future. Because this research was done through experimental case investigations, it has the potential to establish the groundwork for the growth and beginnings of AI in quality management, as well as affect the operational management's commercial effectiveness. This study may potentially open up new avenues for future investigation. A future study might also look at the strategic and managerial elements that influence the acceptance of an AI operational approach in production and supply chain management. While AI offers great promise in production and supply chain management, this has a long road ahead to go before its true worth is realized. This study admits that the tech behemoths that have embraced these digital tools are continuing to welcome the impending storms of digitalization. This type of engagement is now regarded as a novel by other small businesses that are continuously implementing corporate social technology efforts. However, this demonstrates that machine learning will speed to the category of digitally enhanced breakthroughs in the long term, increasing the range among tech pioneers and slackers in sectors, regions of the world, and businesses. Financial institutions, telecommunications businesses, and slightly elevated businesses have a long history of digitally enhanced AI development and will be continuing to do so.

References

1. Võhandu, L. (1994). Artificial intelligence frontiers in statistics. *Engineering Applications of Artificial Intelligence, 7*(1), 87.
2. Kharin, Y. (1994). Artificial intelligence frontiers in statistics AI and statistics III. *Knowledge-Based Systems, 7*(1), 57–58.
3. Rao, M. (1992). Frontiers and challenges of intelligent process control. *Engineering Applications of Artificial Intelligence, 5*(6), 475–481.
4. Anandakumar, H., & Umamaheswari, K. (2017). Supervised machine learning techniques in cognitive radio networks during cooperative spectrum handovers. *Cluster Computing, 20*(2), 1505–1515.
5. Anandakumar, H., & Umamaheswari, K. (2018). A bio-inspired swarm intelligence technique for social aware cognitive radio handovers. *Computers and Electrical Engineering, 71*, 925–937.
6. P. Giudici, Fintech Risk Management: A Research Challenge for Artificial Intelligence in Finance. *Frontiers in Artificial Intelligence, 1.*
7. S. O'Halloran, N. Nowaczyk, An artificial intelligence approach to regulating systemic risk. *Frontiers in Artificial Intelligence, 2.*
8. S. Bredt, Artificial intelligence (AI) in the financial sector—Potential and public strategies. *Frontiers in Artificial Intelligence, 2.*
9. Dubois, D., & Prade, H. (2003). Fuzzy set and possibility theory-based methods in artificial intelligence. *Artificial Intelligence, 148*(1–2), 1–9.
10. Chittaro, L., & Ranon, R. (2004). Hierarchical model-based diagnosis based on structural abstraction. *Artificial Intelligence, 155*(1–2), 147–182.
11. Singh, S. K., Rathore, S., & Park, J. H. (2020). Block IoT intelligence: A blockchain-enabled intelligent IoT architecture with artificial intelligence. *Future Generation Computer Systems, 110*, 721–743.
12. Tammela, I., Canen, A. G., & Helo, P. (2008). Time-based competition and multiculturalism: A comparative approach to the Brazilian, Danish and Finnish Furniture Industries. *Management Decision, 46*(3), 349–364.
13. Haas, A. (2020). Logistics and supply chain intelligence. In A. Kolinski, D. Dujak, & P. Golinska-Dawson (Eds.), *Integration of information flow for greening supply chain management* (pp. 111–129). Springer.
14. Schiavone, F., & Sprenger, S. (2017). Operations management and digital technologies. *Production Planning and Control, 28*(16), 1281–1283.
15. Seyedghorban, Z., Tahernejad, H., Meriton, R., & Graham, G. (2020). Supply chain digitalization: Past, present and future. *Production Planning and Control, 31*(2–3), 96–114.

Machine Learning-Enabled Estimation System Using Fuzzy Cognitive Mapping: A Review

Ashutosh Sharma and Alexey Tselykh

Abstract With a growing interest in Explainable Artificial Intelligence, the fuzzy cognitive maps (FCMs) have proved to be a simple yet powerful tool for causal reasoning and decision making. It is a hybrid methodology that combines the aspects of recurrent neural network and fuzzy logic. In this paper, we elaborate a FCM technique for Web effort estimation which is a critical challenge in software engineering. Web applications market size is giant and ever-growing. Today, Web applications are becoming more refined as it is not only for uploading and fetching the data but also gathering data from various sources and analyzing by the ML techniques. The mean square error (MSE) is measured and analyzed to show the superiority of FCM estimation technique. It is analyzed that the project characteristics presence should not be ignored by the effort estimation technique selection. On software estimation technique recommendation, there is 70% success probability by the FCM approach.

Keywords Project effort estimation · Fuzzy cognitive mapping · Web applications · Machine learning techniques

1 Introduction

With the increase in advanced communication and information, there is rapid growth of the fuzzy cognitive maps (FCMs) and it became more popular. It is one of the kinds of recurrent neural network which carries the fuzzy logic aspects. The system mimicking is allowed by the FCM and the phenomenon with the utilization of the causal relationships [1]. In complex systems, decision making and the modeling of system are done effectively by the FCM systems. For the various application domains, the FCM systems are utilized, for example, pattern recognition, decision support tool,

A. Sharma (✉) · A. Tselykh
Institute of Computer Technology and Information Security, Southern Federal University, Taganrog, Russia
e-mail: ashutosh@sfedu.ru

A. Tselykh
e-mail: tselykh@sfedu.ru

P. K. Singh et al. (eds.), *Proceedings of Third International Conference on Computing, Communications, and Cyber-Security*, Lecture Notes in Networks and Systems 421,
https://doi.org/10.1007/978-981-19-1142-2_39

and the socio-economic development planning. The FCM is an efficient technique for numerical and the forecasting of time series. The FCM system has mainly two types: manual FCM and the automated FCM [2]. The way for the FCM formation is the only difference between these two types of FCMs. The experts manually produce the manual FCMs, and the information sources produced the automated FCMs numerically. Manual production of FCM becomes difficult sometimes when the interference of the experts could not be enough for the problem solution. Because of the difficulties in manual FCM generation, the computational methods development for FCM learning is needed for the FCMs automation [3, 4]. In every field, the prediction is the dynamic issue, and it is considered as very complex which exhibits high uncertainty. This is a challenging problem which leads to large number of approaches and produced accurate results. In the research community, substantial efforts are recorded which focused on two aspects: prediction's important aspect is accuracy and the timelines for delivering the accurate prediction. The framework general application is demonstrated by utilizing the FCM and the framework validation involves the framework's concrete applications [5–7]. In the industry environment, different organization studies are required for the framework concrete applications. The FCM main merits are the capabilities of flexibility and adaptability. The FCM interest is maximum in the industries, on the part of researchers, robotics, medicines, and information technologies.

The Web applications presence is universal and Web-centric applications are predictable to come due to perpetual need of business [8–10]. Today, Web applications becoming more refined as it is not only for uploading and fetching the data but also gather data from various resources and analyzing by the artificial intelligence. The ML-based software estimation techniques are very popular for the parametric and the nonparametric environment for accuracy in prediction. The historical project data availability is the ML techniques essential necessity [11, 12]. The project characteristics spent on the project building like lines of code utilized in project, class, and the actual effort are usually contained in the data. The effort estimation is directly provided by the software cost estimation by the project's characteristics and for the context, ML-based techniques are required to be trained. When trained with the same dataset, estimation of all the ML techniques is not equal, some of them are superior comparatively [13, 14]. Verification is done by the ML technique's performance parameter examination. However, for individual project parameter configurations, superior most ML technique performance is also lagging sometimes. In the given area, extraction of knowledge from an expert is denoted by the FCM. In the FCM drawing, the better consistency is possible in the system modeling if more than one expert is utilized. The final model reliability is improved allowed by the different expert's aggregation process. The process of two FCM aggregations is presented in Fig. 1 where investigated systems are describing by different concepts.

The final map made by this process is less susceptible to the erroneous beliefs. It is advantageous if there is possibility of FCM aggregation into the knowledge structure. In FCM, there is no limitation of experts, more the experts, higher the reliability. The average of sample size will congregate with underlying matrix if the sample size increases. The simplest procedure is based on the causal weight

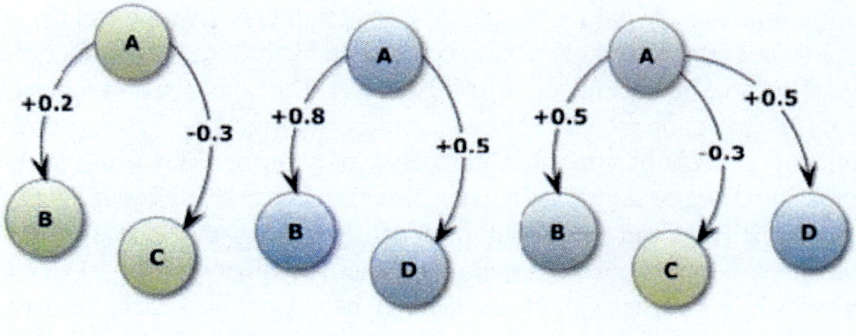

Fig. 1 Typical FCM process of aggregation

matrixes mathematical transformation. The same concepts characterized the maps then the entire system represented FCM aggregation can be calculated easily as their causal matrixes weighted average or median. But regarding the concept, the different experts have different opinions in the modeling. So, each causal matrix is augmented by new rows and columns, and the operations are performed over them.

The complex relationships within an environment method are provided by the FCM graph structures for the environment improvement. The problem with no data is modeled by utilizing the FCMs. The what-if analysis is done by utilizing the FCM where alternative scenarios are considered. The environment behavior is represented by the concept nodes within FCM. The arcs are utilized for the concept connection which shows the concept relations. The FCM development is based on expert's knowledge utilization forming environment framework. The concepts are identified by utilization of expert knowledge. The framework is analyzed by utilizing the FCM for demonstration. The framework validation in contrast involves the concrete applications framework. It requires different organizations case study in an industry environment. This paper provides the overview of the technique having organized way for the Web application estimation technique selection from the predefined set. The performance parameter, i.e., MSE is measured and analyzed for technique estimation indication for well performance. The huge resources are put for the Web application projects development. Thus, the optimum cost estimation technique selection is done by the well supervisory method. For the other organization stakeholders, it can prove to be a savior.

The paper is organized as follows. Section 2 offers an overview of the FCM-based method including exhaustive literature survey and the methodology. The FCM-based methodology given in Sects. 3 and 4 concludes the paper.

2 Related Work

For the nonlinear and the complex systems, the FCMs are very powerful tool as they are directed graphs with feedbacks. The experts create the FCM models, and the building of these models is hard with the increasing number of variables [15]. The fuzzy cognitive map automated generation is provided by the various methods in the last decades. The large number of variables handling is difficult by the presented technique, and it is the main drawback. The new strategy is brought out by the presented method called as concept by concepts approach (CbC) for the FCM learning. With the high precision, large-sized FCM model generation is enabled by utilizing the historical data. The FCM and Agent-Based Modeling (ABMs) methods are complementary which represent interacting agents over time. The agent subjective behaviors encapsulate which is not specifying by the method, but it lacks the scaling ability to a population [16]. The emerging practice review is presented which combines the FCM and the ABM. Three different high-level architectures are revealed by the analysis for the combined utilization of these two methods. The authors in this paper presented the prediction model which combines the FCM and Support Vector Machines (SVM) for increment of the accuracy [17]. The FCM part utilizes by the presented technique for the correlation pattern discovery which exists between the data variables and latent variables are formed. The prediction capabilities are also improved when the variables are fed to the SVM part. The demonstration of the hybrid model efficacy is done on the different domains. The efficiency of the presented model is better as compared to the existing models.

A new "Structure Optimization Genetic Algorithm (SOGA)" is presented for decision support system modeling [18]. The FCM model is constructed and optimized automatically by the presented approach. The error function is defined by the SOGA for the FCM complexity as there are number of connections between them. For fuzzy cognitive maps learning, SOGA is utilized for the analysis purpose. SOGA is compared with the FCM learning algorithms and simulations were done. The FCM model structure is reduced significantly by the utilization of the SOGA as obtained by the results. The model complexity construction is difficult task; hence, the effective techniques are required [19]. Based on a fuzzy cognitive map (FCM), the nested structure is created at the higher map which is decomposed into other FCM. The whole nested structure is restructured through the dynamic optimization process to derive relationships between map concepts. The hidden relationship discovering is allowed by this process among map concepts. The suggested nested approach application is presented by the paper for the forecasting of the time series and the decision making. A framework is detailed in this paper for the test manager assistant for the AI technique evaluation in testing the software [20]. The framework evaluation is employed by the fuzzy cognitive maps (FCMs) and the decision analysis is made easier. The presented technique effectiveness is shown by the simulations analysis.

FCM has emerged gradually as a powerful tool and the applicability of the simulation mechanism for different applications. The system states are investigated by the different methods and FCMs are constructing for the complex systems [21]. The fuzzy neural network is presented in this paper for the FCMs learning ability. The conventional FCMs are incorporated by the presented approach with the membership functions determination and the causalities quantification. The causalities are described by the mutual subset-hood in the fuzzy neural network. The effectiveness of the presented approach is confirmed by the simulation. Authors in this paper describe the FCMs for the autonomous entities modeling in the dynamic environment [22]. The FCM general design is offered in this paper for the autonomous agent's decision making, and this concept is categorized into three classes like requirement, activities, and the states. The feature supporting decision making is enabled by the classification such as sensors input processing. The presented method is utilized as a decision making for the simulation of human activities in the ambient model, and the scenario-oriented mechanism is combined for proving the modularity.

In the economic modeling, the FCM is a significant tool, and its aim is the investigation of the genetic algorithm-based FCM utilization [23]. The FCM models predict the complex financial system clearly in this study. The benefits of FCM applications are confirmed in this study for the researchers and policy makers. This paper presents the Thayer's emotion model and FCM-based proposal for the forecasting artificial emotions [24]. An innovative method is provided by the author for artificial emotions forecasting and for the affective decision system design. An experiment included in this work with different artificial scenarios for the proposal testing. Different emotions are generated by each scenario to artificial model accordingly. Authors in this paper detail the cheap sensors development and the transfer possibility of data [25]. The sensor movement possibility from the robots with IoT is presented, and the IoT concept is modified in intelligent space. The distributed networked sensor potential is clarified with the route tracking example, and by utilizing the FCM, data is processed for the robotic navigation. The adaptation approach modifications are presented, namely particle swarm optimization and migration algorithm.

Fuzzy cognitive map (FCM) penetrates to areas as control systems which include the robotics characterized by its distributiveness [26]. The needs for a robot control system are specified by the authors, and the defined tasks are divided into different decision levels. Author details the various machine learning techniques utilized for the project efforts based on the historical project-related dataset. The project characteristic array is consumed by these techniques for the project cost estimation. The project cost is determined by the right technique selection, and it is very significant [27]. The FCM approach is presented for the machine learning-based software estimation technique recommendation. The different estimation techniques are utilized by the current analysis, and the supremacy of one estimation technique is declared by the mean square error. The project characteristics' presence should not be ignored by the effort estimation technique as shown by the experimental results.

For modeling and simulations, FCM has become an important soft computing method in recent years [28]. The concepts are involved in it as they are recurrent structures which describe the causal connections. The swarm intelligence-based two abstract models are described in this paper by the authors for FCM characterization. The accurate maps are obtained at the end that allows the system simulation and relevant knowledge extraction. Authors presented the FCM approach for ecological creation with expert's knowledge [29]. These maps are system's qualitative models which contain variables and their relationships.

In environmental management applications, utilization of cognitive mapping research is described. Different stakeholder perception is examined to facilitate the environmental management plans development. Many advantages are offered by the FCM for the modeling which includes the model aggregate variables. The complex relationships are modeled by the process and different knowledge sources are combined. With the IoT of green supply chain perspective, IoT actualization in difficult system for "green inventory management link" with the world [30]. The complex system modeling utilization is hard. The lack of autonomous mechanism and the stable integrated architecture are the two problems of IoT-enabled system. The IoT system is presented in green system utilizing the FCM method. The complex system simulation is the aim by linking digital objects, while the performance efficiency is enhanced for the management. The FCM approach evaluation is presented for improvement of work. The total effect performance is compared by the experiments as assessed by the methods utilizing the IoT-enabled FCM. The fuzzy factor comparison method is developed in this paper for the feasible and the sustainable material selection [31]. The alternative building materials are evaluated carried out through the building information modeling, and the 73% of average result is observed achieved by the alternative methods. The Building Information Modeling-enabled energy management system is presented in this paper by IoT for metro rail station. The operational cooling load of 25% is reduced as observed by the results by the presented system. Many researchers have worked on the FCM-based methods for different applications as tabulated in Table 1.

3 Machine Learning-Based Web Effort Estimation

3.1 Existing Framework

Between the conventional software and Web applications, there are significant differences as demonstrated by the Web software practitioner experience. A Web-based application development is separated from the traditional software development projects from the various elements. The Web applications have many characteristics like network intensiveness, concurrency, content sensitivity, availability, etc.

The HTTP utilization, Web multifaceted nature to the client experience and typical Web application protocols are components portion and development unpredictability is decided by it. The Web-based applications rising significance in business domains leads many researchers Web effort estimation distinguish as a predominantly valuable area [40, 41]. The Web effort estimation has four divergent approaches like "case-based reasoning (CBR), stepwise regression (FSR), classification and regression trees (CART), and Bayesian networks (BNs)". The Tabu search meta-heuristic methodology utilization is examined by the researchers in conjunction for the parameters precise selection. It is used for the Web application effort estimation [42].

Many researchers are led by the applications based on the Web in different domains, and it is rising significantly. It distinguishes the Web applications effort estimation as a current research valuable area. Four divergent approaches are utilized by the researchers for the estimation of Web efforts. The case-based reasoning (CBR),

Table 1 Different existing FCM-based methods with their merits and demerits

References	Methodology	Advantages	Disadvantages
[32]	Agile MOW approach for software cost estimation process modeling	Identifies the difference between conventional and Web projects It enhances the visibility level in the planning stages Speedy and reliable effort estimations of Agile-based Web development projects	Time consuming Computationally complex
[33]	FCM-based classifier with a fully connected map structure is proposed	Executed few FCM iterations Pipelines built performance from FCM-based data transformer FCM-based classifier performance and its capability to improve data is confirmed	Size of datasets is limited that could be processed in acceptable time
[34]	Algorithms that support learning of FCMs from data are developed	Efficient and accurate semi-automated algorithms	Model formation from data is a complex task
[35]	Fuzzy cognitive mapping (FCM) is presented as a participatory method for understanding social-ecological systems (SESs)	FCM requires understanding a terms and following simply logical heuristics Facilitate the discourse with governing agencies	High computational complexity
[36]	For training FCMs, the utilization of the un-supervised Hebbian algorithm is presented to nonlinear units	Modifies its fuzzy causal Web The deficiencies that appear in operation of FCMs	Not effective more complex problems Time consuming
[37]	Brand-new approach for student performance prediction is presented utilizing the learning fuzzy cognitive map (LFCM) approach	Work well with large datasets	They face challenges dealing with small sample sizes
[38]	The impact of pixel-based ML techniques, i.e., fuzzy-*c*-means clustering method (FCM) and the artificial neural network (ANN) and support vector machine (SVM) are investigated	The FCM-based framework achieved the highest performance Highest accuracy and precision	Image blurring effect causes ambiguous outlines of tumors Affect the segmentation accuracy of the automated frameworks

(continued)

Table 1 (continued)

References	Methodology	Advantages	Disadvantages
[39]	New FCM model is presented based on deep learning neural networks	High prediction accuracy High learning efficiency	Cost ineffective

forward stepwise regression (FSR), classification and regression trees (CART), and Bayesian networks (BNs) are the different approaches. The Fuzzy ID3 decision effectiveness is also checked in the effort estimation. The standard fuzzy set concepts are incorporated by the standard research in ID3 decision tree. An unmarked method is also demonstrated for the Web-based project effort determination utilizing a content management framework (CMF) [43, 44]. Several researchers are influenced by the mobile applications for the various issues identifications and for the efficient mobile applications development, their solutions are utilized. To implement the presented technique, there are different steps for processing, and which are described in detail. The entire methodology is represented pictorially in Fig. 2.

Step 1: First, dataset is viewed carefully, and its suitability is verified with the ML-based assessment approaches. The dataset is prepared from the projects of the student in the university. There are 125 projects contained in the data having different attributes. Number of Function Points (NFP), Number of Multimedia Files (NMM), Number of Building Blocks (NBB), and Number of XML, HTML, and

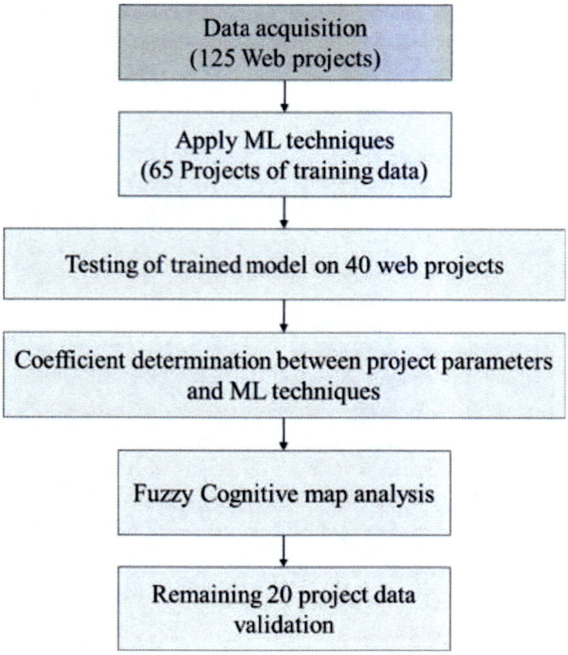

Fig. 2 Schematic diagram of the presented technique [27]

query Language links (NHL). The segregation of the project details in the functional attributes is preferred generally for the analysis based on ML. The attributed projects are specific to the applications of the Web, and utilization is repeatedly effort estimation of Web application.

Step 2: For training estimation models, five generally utilized ML techniques are recognized for the estimation of the training models on the database. The "Case-Based Reasoning (CBR), Random Forest (RF), Artificial Neural Network (ANN), Support Vector Machine (SVM), and Multiple Linear Regression (MLR)" are the different techniques which are utilized.

Step 3: The over-fitting irregularity is overcome by the ML technique which are trained on data of 65 projects and then tested on the 40 projects. For the validation purpose, rests of 20 projects are utilized. The ML techniques are compared by utilizing the mean square error (MSE) as performance metrics. The actual effort and the estimated effort difference were calculated for each of 40 testing data tuples. The MSE values were calculated after the error terms square averaging.

Step 4: Between the different employed techniques and different project characteristics, correlation values between the MSEs are calculated which are established in this step of processing. It is observed that the error is strongly influenced by the NFP strongly and positively in ANN. It is inferred that the higher the applications functional points, lesser the ANN importance. Similarly, as an estimation technique choice, number of multimedia files presence attracts the random forest.

Step 5: The FCMs are built by utilizing the correlation results for the further analysis. The analysis of the FCM is required because a project characteristic is not present in the project configuration. In every project configuration, amount of each driving factor is present. Thus, assessment of the overall system is needed to take the decision that which technique is preferred for configuration of the projects.

Step 6: In the last step, validation is done after the analysis of FCM on the 20 projects' data.

3.2 Recommendation System Based on FCM

Relationship between the variables is represented by the directed graphs which is called FCM. The FCM is utilized for the determination of the causal relationships between the characteristics of project and the ML technique performance [45]. Figure 3 represents the FCM recommendation system.

The existing relationships insight is got by the project variables and ML technique performance's correlation coefficient. The driving factors and the receiving factors are two classes of factors utilized as project parameters and ML techniques. The positive and negative relationship between the two factors and the related degree is determined by the correlation coefficient.

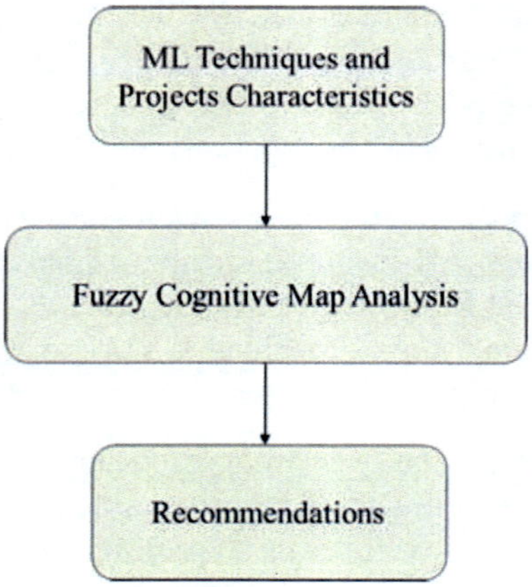

Fig. 3 Recommendation system based on FCM

The one factor has the degree of influence [46] that is represented by utilizing the fuzzy triangular numbers on another factor. For example, the dominance of NMM on RF is strong for the NMM, and RF factors and the $(5-\delta, 5, 5+\delta)$ are the assigned fuzzy numbers. From the driving factor, the directed graphs are drawn toward one or more receiving factors [47, 48]. The degree of influence is showed by the fuzzy number that the respective receiving factors are exercised by the driving factor.

After the justified relationships labeling and the establishment, the corresponding FCM is presented in Fig. 4 and Table 2 shows the fuzzy number of each cell entry.

The influence of the NMM driving factor is strong over the MLR, SVM, RF, and CBR, while in case of SVM and CBR, impact is positive. For MLR and RF, it is negative. The NHL links influence the error in SVM while on the ANN, RF, and CBR, it has negative influence. The SVM technique is better for estimation if the

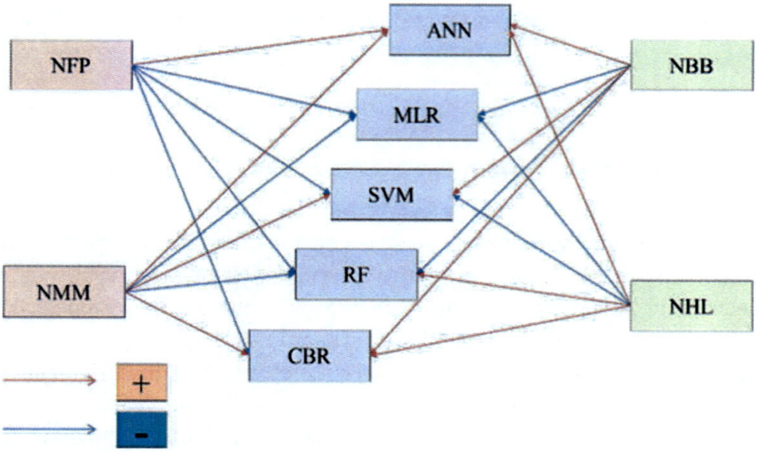

Fig. 4 Relationships among project factors shown by FCM [27]

Table 2 Corresponding to Fig. 4 adjacency matrix [27]

	ANN	MLR	SVM	RF	CBR
NFP	5	−3	−3	−3	−5
NBB	−3	5	−5	3	−5
NMM	3	−5	5	−5	5
NHL	−5	3	5	−5	−5

hyperlinks presence is more in a Web application. Afterward the creation of cognitive maps and the matrix adjacency, steady state can be shown by the simulations [27].

It is observed from the table that the recommended ML techniques didn't consider as a winner. If the half marks are given for being laggard marginally, the FCM approach success probability in ML technique is close to 70% for the further research motivation. The FCM approach modeling other than social sciences is done. The presented technique importance is understood by the fact that it is not Web estimation technique itself, but the systematic way is provided for the best estimation technique selection from the predefined set. The Web application project correct estimation is the critical requirement of the business. The performance metrics like mean square error (MSE) indicates that the performance of the estimated technique is well in various cases. For individual project configuration, the best estimation technique is not advised. The presented technique merit is that the overall system equilibrium is considered for the underpinning factors mutual influence derivation. It is also advantageous that the FCM techniques are augmented by the presented technique for the abstract relationships between factors or variables representation, and it counters the human-borne ambiguities effect.

3.3 Research Gaps

Various research gaps are found in the state-of-the-art techniques.

- The complex relationships within an environment method are provided by the FCM graph structures for the environment improvement.
- Existing techniques are time consuming and computationally complex.
- Some techniques are limited to the size of dataset as they face challenges dealing with small sample sizes.

4 Conclusion and Future Scope

The abstract relationships' transformation into the quantitative measures by the FCM is utilized for the software assessment approaches recommendation. The correlation values are mapped into the fuzzy numbers by the linguistic values. The state vector and adjacency matrix are then represented by utilizing the fuzzy numbers. The success probability is obtained after software estimation technique recommendation which is close to 70%. The FCM is the potential technique to count on when the scientific knowledge is very complex. Only effects are shown by the analysis which is the main FCM's demerit. The co-occurrence of multiple causes can't be deal with the FCM, and the causal factor combined impact is important. The main error of FCM is that the if–then statements can code in it. However, only synergistic interaction can be produced by FCM among the factors. The FCM can be extended in the future for finding the complex abstract relationships between the issues of the mobile application. The success and the failure of the app are governed by the factors of the mobile apps.

Acknowledgements The research is supported by postdoc fellowship granted by the Institute of Computer Technologies and Information Security, Southern Federal University, project No. PD/20-03-KT.

References

1. Meliadou, A., Santoro, F., Nader, M. R., Abou Dagher, M., Al Indary, S., & Abi Salloum, B. (2012). Prioritising coastal zone management issues through fuzzy cognitive mapping approach. *Journal of Environmental Management, 97*, 56–68.
2. Reckien, D. (2014). Weather extremes and street life in India—Implications of fuzzy cognitive mapping as a new tool for semi-quantitative impact assessment and ranking of adaptation measures. *Global Environmental Change, 26*, 1–13.
3. Gray, S. A., Gray, S., De Kok, J. L., Helfgott, A. E., O'Dwyer, B., Jordan, R., & Nyaki, A. (2015). Using fuzzy cognitive mapping as a participatory approach to analyze change, preferred states, and perceived resilience of social-ecological systems. *Ecology and Society, 20*(2).
4. Lousada, A. L., Ferreira, F. A., Meidutė-Kavaliauskienė, I., Spahr, R. W., Sunderman, M. A., & Pereira, L. F. (2021). A sociotechnical approach to causes of urban blight using fuzzy cognitive mapping and system dynamics. *Cities, 108*, 102963.
5. Jetter, A., & Schweinfort, W. (2011). Building scenarios with fuzzy cognitive maps: An exploratory study of solar energy. *Futures, 43*(1), 52–66.
6. Christen, B., Kjeldsen, C., Dalgaard, T., & Martin-Ortega, J. (2015). Can fuzzy cognitive mapping help in agricultural policy design and communication? *Land Use Policy, 45*, 64–75.
7. Assunção, E. R. G. T. R., Ferreira, F. A. F., Meidutė-Kavaliauskienė, I., Zopounidis, C., Pereira, L. F., & Correia, R. J. C. (2020). Rethinking urban sustainability using fuzzy cognitive mapping and system dynamics. *International Journal of Sustainable Development and World Ecology, 27*(3), 261–275.
8. Goswami, R., Roy, K., Dutta, S., Ray, K., Sarkar, S., Brahmachari, K., Nanda, M. K., Mainuddin, M., Banerjee, H., Timsina, J., & Majumdar, K. (2021). Multi-faceted impact and outcome of COVID-19 on smallholder agricultural systems: Integrating qualitative research and fuzzy cognitive mapping to explore resilient strategies. *Agricultural Systems, 189*, 103051.
9. Ziv, G., Watson, E., Young, D., Howard, D. C., Larcom, S. T., & Tanentzap, A. J. (2018). The potential impact of Brexit on the energy, water and food nexus in the UK: A fuzzy cognitive mapping approach. *Applied Energy, 210*, 487–498.
10. Morone, P., Falcone, P. M., & Lopolito, A. (2019). How to promote a new and sustainable food consumption model: A fuzzy cognitive map study. *Journal of Cleaner Production, 208*, 563–574.

11. Pluchinotta, I., Esposito, D., & Camarda, D. (2019). Fuzzy cognitive mapping to support multi-agent decisions in development of urban policymaking. *Sustainable Cities and Society, 46*, 101402.
12. Martinez, P., Blanco, M., & Castro-Campos, B. (2018). The water–energy–food nexus: A fuzzy-cognitive mapping approach to support nexus-compliant policies in Andalusia (Spain). *Water, 10*(5), 664.
13. Pereira, I. P., Ferreira, F. A., Pereira, L. F., Govindan, K., Meidutė-Kavaliauskienė, I., & Correia, R. J. (2020). A fuzzy cognitive mapping-system dynamics approach to energy-change impacts on the sustainability of small and medium-sized enterprises. *Journal of Cleaner Production, 256*, 120154.
14. van der Sluis, T., Arts, B., Kok, K., Bogers, M., Busck, A. G., Sepp, K., Ramos, I. L., Pavlis, E., Geamana, N., & Crouzat, E. (2019). Drivers of European landscape change: Stakeholders' perspectives through fuzzy cognitive mapping. *Landscape Research, 44*(4), 458–476.
15. Dodurka, M. F., Yesil, E., Ozturk, C., Sakalli, A., & Guzay, C. (2013, September). Concept by concept learning of fuzzy cognitive maps. In *IFIP International Conference on Artificial Intelligence Applications and Innovations* (pp. 577–586). Springer.
16. Davis, C. W., Giabbanelli, P. J., & Jetter, A. J. (2019, December). The intersection of agent based models and fuzzy cognitive maps: A review of an emerging hybrid modeling practice. In *2019 Winter Simulation Conference (WSC)* (pp. 1292–1303). IEEE.
17. Christodoulou, P., Christoforou, A., & Andreou, A. S. (2017, April). A hybrid prediction model integrating fuzzy cognitive maps with support vector machines. In *International Conference on Enterprise Information Systems* (Vol. 2, pp. 554–564). SciTePress.
18. Poczęta, K., Yastrebov, A., & Papageorgiou, E. I. (2015, September). Learning fuzzy cognitive maps using structure optimization genetic algorithm. In *2015 Federated Conference on Computer Science and Information Systems (FedCSIS)* (pp. 547–554). IEEE.
19. Poczeta, K., Papageorgiou, E. I., & Gerogiannis, V. C. (2020). Fuzzy cognitive maps optimization for decision making and prediction. *Mathematics, 8*(11), 2059.
20. Larkman, D., Mohammadian, M., Balachandran, B., & Jentzsch, R. (2010, October). Fuzzy cognitive map for software testing using artificial intelligence techniques. In *IFIP International Conference on Artificial Intelligence Applications and Innovations* (pp. 328–335). Springer.
21. Song, H., Miao, C., Roel, W., Shen, Z., & Catthoor, F. (2009). Implementation of fuzzy cognitive maps based on fuzzy neural network and application in prediction of time series. *IEEE Transactions on Fuzzy Systems, 18*(2), 233–250.
22. Nachazel, T. (2021). Fuzzy cognitive maps for decision-making in dynamic environments. *Genetic Programming and Evolvable Machines, 22*, 101–135.
23. Sammour, G., Alghzawi, A., & Vanhoof, K. (2020). A fuzzy cognitive map approach to investigate the sustainability of the social security system in Jordan. In *ICEIS* (Vol. 1, pp. 481–489).
24. Salmeron, J. L. (2012). Fuzzy cognitive maps for artificial emotions forecasting. *Applied Soft Computing, 12*(12), 3704–3710.
25. Vaščák, J., Pomšár, L., Papcun, P., Kajáti, E., & Zolotová, I. (2021). Means of IoT and fuzzy cognitive maps in reactive navigation of ubiquitous robots. *Electronics, 10*(7), 809.
26. Vaščák, J., & Reyes, N. H. (2014). Use and perspectives of fuzzy cognitive maps in robotics. In *Fuzzy cognitive maps for applied sciences and engineering* (pp. 253–266). Springer.
27. Pandey, P., & Litoriya, R. (2020). Fuzzy cognitive mapping analysis to recommend machine learning-based effort estimation technique for Web applications. *International Journal of Fuzzy Systems*, 1–12.
28. Napoles, G., Grau, I., Pérez-García, R., & Bello, R. (2013). Learning of fuzzy cognitive maps for simulation and knowledge discovery. *Studies on knowledge discovery, knowledge management and decision making* (pp. 27–36). Atlantis Press.
29. Özesmi, U., & Özesmi, S. L. (2004). Ecological models based on people's knowledge: A multi-step fuzzy cognitive mapping approach. *Ecological Modelling, 176*(1–2), 43–64.
30. Chen, R. Y. (2015). Intelligent IoT-enabled system in green supply chain using integrated FCM method. *International Journal of Business Analytics (IJBAN), 2*(3), 47–66.

31. Bapat, H., Sarkar, D., & Gujar, R. (2021). Application of integrated fuzzy FCM-BIM-IoT for sustainable material selection and energy management of metro rail station box project in western India. *Innovative Infrastructure Solutions, 6*(2), 1–18.
32. Litoriya, R., & Kothari, A. (2013). An efficient approach for agile Web based project estimation: AgileMOW.
33. Szwed, P. (2021). Classification and feature transformation with fuzzy cognitive maps. *Applied Soft Computing, 105*, 107271.
34. Stach, W., Kurgan, L., & Pedrycz, W. (2010). Expert-based and computational methods for developing fuzzy cognitive maps. In *Fuzzy cognitive maps* (pp. 23–41). Springer.
35. Gray, S., & Scyphers, S. (2015). Using fuzzy cognitive mapping as a participatory approach to measure change, preferred states and perceived resilience of social-ecological systems. *Ecology and Society, 20*(2).
36. Papageorgiou, E., Stylios, C., & Groumpos, P. (2003, December). Fuzzy cognitive map learning based on nonlinear Hebbian rule. In *Australasian joint conference on artificial intelligence* (pp. 256–268). Springer.
37. Mansouri, T., ZareRavasan, A., & Ashrafi, A. (2021). A learning fuzzy cognitive map (LFCM) approach to predict student performance. *Journal of Information Technology Education: Research, 20*, 221–243.
38. Kawata, Y., Arimura, H., Ikushima, K., Jin, Z., Morita, K., Tokunaga, C., Yabu-Uchi, H., Shioyama, Y., Sasaki, T., Honda, H., & Sasaki, M. (2017). Impact of pixel-based machine-learning techniques on automated frameworks for delineation of gross tumor volume regions for stereotactic body radiation therapy. *Physica Medica, 42*, 141–149.
39. Liu, L., Ma, M., & Cui, J. (2017). A novel model-based on FCM–LM algorithm for prediction of protein folding rate. *Journal of Bioinformatics and Computational Biology, 15*(04), 1750012.
40. Aravindakshan, S., Krupnik, T. J., Shahrin, S., Tittonell, P., Siddique, K. H., Ditzler, L., & Groot, J. C. (2021). Socio-cognitive constraints and opportunities for sustainable intensification in South Asia: Insights from fuzzy cognitive mapping in coastal Bangladesh. *Environment, Development and Sustainability*, 1–29.
41. Pacilly, F. C., Groot, J. C., Hofstede, G. J., Schaap, B. F., & van Bueren, E. T. L. (2016). Analysing potato late blight control as a social-ecological system using fuzzy cognitive mapping. *Agronomy for Sustainable Development, 36*(2), 35.
42. Dias, S. B., Hadjileontiadou, S. J., Hadjileontiadis, L. J., & Diniz, J. A. (2015). Fuzzy cognitive mapping of LMS users' quality of interaction within higher education blended-learning environment. *Expert Systems with Applications, 42*(21), 7399–7423.
43. O'Garra, T., Reckien, D., Pfirman, S., Bachrach Simon, E., Bachman, G., Brunacini, J., & Lee, J. (2021). Impact of gameplay vs. reading on mental models of social-ecological systems: A fuzzy cognitive mapping approach. *Ecology and Society, 26*(2).
44. Abrantes, J. A., Ferreira, F. A., Zopounidis, C., Pereira, L. F., & Meidutė-Kavaliauskienė, I. (2020). Analyzing ethical practices in the public healthcare sector using fuzzy cognitive mapping. *Journal of Multi-Criteria Decision Analysis.*
45. Ribeiro, M. I., Ferreira, F. A., Jalali, M. S., & Meidutė-Kavaliauskienė, I. (2017). A fuzzy knowledge-based framework for risk assessment of residential real estate investments. *Technological and Economic Development of Economy, 23*(1), 140–156.
46. Tselykh, A., Vasilev, V., & Tselykh, L. (2020). Assessment of influence productivity in cognitive models. *Artificial Intelligence Review, 53*, 5383–5409.
47. Sarmiento, I., Paredes-Solís, S., Loutfi, D., Dion, A., Cockcroft, A., & Andersson, N. (2020). Fuzzy cognitive mapping and soft models of Indigenous knowledge on maternal health in Guerrero, Mexico. *BMC Medical Research Methodology, 20*, 1–16.
48. Yamagishi, K., Ocampo, L., Abellana, D. P., Tanaid, R. A., Tiu, A. M., Medalla, M. E., Selerio Jr., E., Go, C., Olorvida, R. C., Maupo, A., Maskariño, D., & Tantoo, E. (2021). The impact of social media marketing strategies on promoting sustainability of tourism with fuzzy cognitive mapping: A case of Kalanggaman Island (Philippines). *Environment, Development and Sustainability*, 1–33.

Latest Electrical and Electronics Trends

Energy Efficiency in IoT-Based Smart Healthcare

Pallavi Sangra, Bharti Rana, and Yashwant Singh

Abstract IoT-based smart healthcare is a new technological shift toward efficient, convenient, cheaper, and faster medical services using artificial learning, big data analytics, sensor technologies, and cloud computing. Smart healthcare reduces the time and cost to avail the services region-wise rather than to get the clinical services at distant locations. Along with this, smart healthcare poses many challenging issues. One of the greatest challenges in smart healthcare is to accomplish the energy-efficient services as smart nodes are energy constrained. Therefore, this study addresses the energy consumption challenges in smart healthcare sector. Then, we focus on energy preserving mechanisms to reduce the energy consumption. The various energy-conserving mechanisms such as intelligent techniques, duty cycling techniques, collision resolution techniques, and edge techniques in context to smart healthcare have been discussed for reducing energy consumption.

Keywords Energy efficiency · Energy consumption · Internet of things · Smart healthcare

1 Introduction

Smart healthcare is not only a technological advancement but also a global revolution. The revolution from traditional healthcare to smart healthcare shifts to person-centric care from disease-centric, to regional services from clinical services, and to personalized care from general care. IoT-based smart healthcare aims to meet the on-demand medical needs of people. Smart healthcare greatly improves the efficiency of medical services and regional health services [1]. Smart healthcare exemplifies the intelligent infrastructure including smart devices to take informed decisions and better resource management.

P. Sangra · B. Rana · Y. Singh (✉)
Department of Computer Science and Information Technology, Central University of Jammu, Samba, Jammu and Kashmir 181143, India
e-mail: yastwant.csit@cujammu.ac.in

P. K. Singh et al. (eds.), *Proceedings of Third International Conference on Computing, Communications, and Cyber-Security*, Lecture Notes in Networks and Systems 421,
https://doi.org/10.1007/978-981-19-1142-2_40

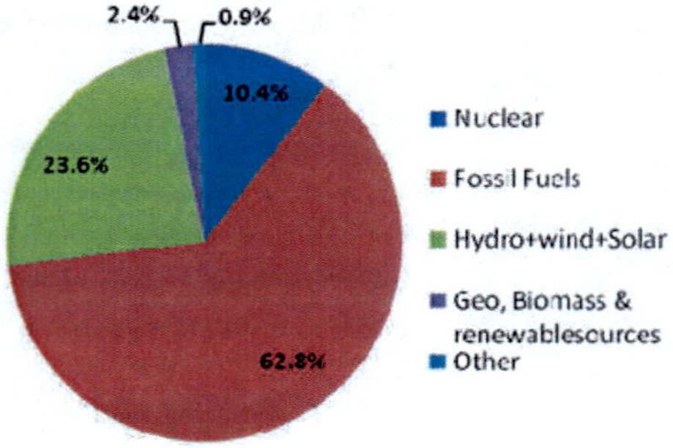

Fig. 1 World power consumption statistics (2019) [3]

Smart healthcare is a concept originated from the "smart planet" introduced by IBM in 2009 [2]. Smart healthcare forms a body area network to transmit and perceive the information from body sensors through Internet of things technology. The transmitted information is sent to the cloud which is processed by using super computers. Smart healthcare utilizes wearable devices, mobile Internet, and IoT to access the information dynamically. Therefore, smart healthcare links various institutions, people, and medical services to perform the functionalities smoothly and intelligently. Shortly, smart healthcare is the higher level of digitization in the medical sector.

Though the smart healthcare eases the early prevention and detection of diseases, it is also prone to severe issues. One of the critical issues is energy consumption. The energy consumption is continuously increasing because of the sensing and transmitting mechanism by using body sensors, power constraints of smart devices, cloud analysis of data, collisions, and battery drainage attacks. Other challenges in smart healthcare are heterogeneity, security and privacy of data, resource management, and identification and allocation. In this study, we would address the existing smart healthcare challenges and the energy preserving mechanisms in smart healthcare. As per the statistical analysis performed by the world power consumption [3] in year 2019, it has been estimated that the power consumption from fossil fuels itself contributed to 63%. And, the power consumption from other sources is 37% (Fig. 1).

The rest of the study is structured as follows: Sect. 2 distinguishes the WSN and the IoT. Section 3 discusses the related work. Section 4 presents the foreseeable energy consumption challenges in smart healthcare. Section 5 provides the energy conserving mechanisms in IoT-based smart systems. Section 6 concludes the article with future scope.

2 IoT Versus WSN

IoT is a higher level technology than WSN. In other words, WSN is frequently utilized as part of an IoT system. In an IoT system, a large number of sensors in a mesh network can be utilized to collect the data and communicate it to the Internet

Table 1 Distinguishing features of IoT and WSN

Features	Internet of things (IoT)	Wireless sensor network (WSN)
Connectivity	Sensors have Internet access and broadcast the data immediately to the Internet [4]	Nodes are not directly connected to the Internet; instead, they route traffic to sink nodes. WSN nodes are not required to be connected to the Internet [5]
Devices	Sensors, humans, cameras, computers, and phones are Internet of things. These gadgets may post their data to the Internet, making it available to other users. Sensors are a type of gadget that collects data on the WSN	Sensors are a type of gadgets that collects data on the WSN
Things identification	Things can be identified because of the unique IP address	Things cannot be identified
Device support	Heterogeneous devices participate in the network	Homogeneous devices participate in the network
Range	Universal network	Subset of IoT
mobility	High mobility of nodes are supported	Mobility of nodes are not so high

via a routers. The term "wireless sensor network" is not quite as broad as "internet of things. A wireless sensor network (WSN) is a network comprises of exclusively wireless sensors. The network could no longer be called a "wireless sensor network" if it includes a wired sensor. This is not the case with the Internet of things. An IoT device is essentially any electronic and digital gadget that can connect to the Internet. The distinguishing features of IoT and WSN are depicted in Table 1.

3 Related Work

Atzori et al. [6] explored wireless and wired tracking technologies for identification, and developed communication protocols to give distributed intelligence for smart things. Many visions of the Internet of things paradigm are presented, as well as enabling technologies. Various aspects of IoT are considered, and the main research challenges for future countermeasures are also identified. Without regard for energy efficiency, the author considers industrial automation in IoT.

Mukherjee et al. [7] discussed the healthcare technology and network platforms that were already in place. The author offers alternatives to current security and privacy concerns. The proposed technique was put to the test in two separate scenarios: at home and in the hospital. The proposed solution was tested with a prototype, and the results were compared to existing approaches. Furthermore, the author addresses the role of contemporary technologies in the healthcare domain, such as ambient intelligence, big data, and wearable. The authors provide an overview of various IoT and e-healthcare rules and policies from around the world, as well as some potential research venues for IoT-based healthcare (Table 2).

Moosavi et al. [14] offered a smart gateway-based energy-efficient and secure authentication architecture for IoT healthcare systems. IoT sensor nodes do not require authentication for distant healthcare providers due to the availability of smart gateways. The addition of a smart gateway lightens the sensor node's load. As a result, security and efficiency are enhanced. The article includes a review of related research as well as a case study of a Malaysian government hospital. Based on the findings, a model is proposed that is said to serve as a fundamental premise for protecting IoT-based healthcare systems from existing security threats.

Feng et al. [9] presented a fog-based IoT healthcare platform to reduce fog node energy usage. Fog computing is one of the potential options for lowering the delay of multi-hop data connection, managing resources, and enhancing service flexibility in the healthcare area. The authors analyzed and suggested critical big data infrastructure services that should be available in fog devices for healthcare big data analytics. The experiment's parameters were network delay and energy consumption. Because the same communication link is shared in the cloud by several healthcare applications, the average network delay increases in cloud-based healthcare.

Mukherjee et al. [7] employ wireless body sensor nodes (WBSN) to monitor patients' health in real time outside of the hospital setting. The WBS uses sensors to collect signals from a patient's body and wireless transmitters to broadcast the signals in real time to a server in the private/public cloud. Energy-constrained processes consume a significant amount of energy, limiting their operating lifetime. For sparse encoding of bio-signals, the author proposed a real-time encoding technique that conducts iterative thresholding and approximation of wavelet coefficients (ECG signals). As a result, energy and bandwidth use are reduced.

4 Smart Healthcare: Energy Consumption Challenges

Due to the miniaturization and power constraints of smart nodes, IoT systems are prone to several energy consumption issues as shown in Fig. 2.

(a) *Big Data Management:* According to Gartner, the number of connected devices will reach 50 billion by 2020 and will more than double in the next years. As the number of smart devices grows tremendously, so will the large amount of data generated. The amount of data created by these devices is likewise fast increasing, which is referred to as "Big Data" [16]. The velocity, volume, and variety of big data are its distinguishing characteristics. Furthermore, the information is unstructured, structured, and semi-structured. In healthcare, time and expense of processing heterogeneous data increase overall energy usage.

(b) *Security and Privacy:* For businesses, industries, and large organizations, IoT data has become a monetary value. Because of security and privacy flaws,

Table 2 Comparative analysis of energy efficiency consideration in the state-of-the-art smart healthcare

Author	Year	Contribution	Energy-efficiency	Research method	Merits	Demerits
Mukherjee et al. [7]	2020	Designed energy-efficient scheme for smart healthcare monitoring system	Considered	Arduino UNO R3	Patient monitoring in real-time. Reduces the amount of time to access the patient's information	There is no implementation on real-time hardware
Humayun et al. [8]	2020	A detailed framework for energy efficiency and security	Considered	Cooja Contiki simulator	Avoid replay attacks and impersonation	There is no validation
Feng et al. [9]	2020	Energy-efficient framework for IoT healthcare solutions	Considered	IFogSim	Security management of keys, end-to-end security. Reduced cost of communication	Validation and implantation is missing
Majumdar et al. [10]	2020	A detailed framework for energy-efficient smart healthcare by using genetic algorithm	Considered	Cluster model	Data privacy and data protection in case of security breach	The presented system should be put to the test on a real-time IoT-based healthcare app
Almulhim et al. [11]	2019	Provided a secure and energy-efficient scheme	Considered	Contiki Cooja simulator	Secure transmission of healthcare data. Communication distance is reduced, thus enhances energy efficiency	Any malfunctioning of nodes result in serious network outage and security breaches
Zoukz et al. [12]	2019	Developed the energy-efficient for smart healthcare system	Not considered	MikroC Pro	Patient monitoring in real time. Reduce the time to generate keys	There is a lack of real-time hardware implementation

(continued)

Table 2 (continued)

Author	Year	Contribution	Energy-efficiency	Research method	Merits	Demerits
Zakaria et al. [13]	2019	Developed the secure model for IoT healthcare environment	Not considered	Control objectives for information technology (COBIT)	Identification of privacy and security risk factors	Validation of energy-efficient reviewed is not presented
Moosavi et al. [14]	2018	presented the end-to-end security for smart healthcare system	Considered	Contiki, Cooja simulator	Important security requirements have been identified. Prototype creation is based on a performance analysis of existing end-to-end security solutions	Only cloud-based theoretical model is presented
Luo et al. [15]	2018	To offer various privacy protector for smart healthcare systems	Not considered	MATLAB	Servers experience a security compromise	The presented system should be put to the test on a real-time IoT-based healthcare app

Fig. 2 Energy consumption challenges in smart healthcare system

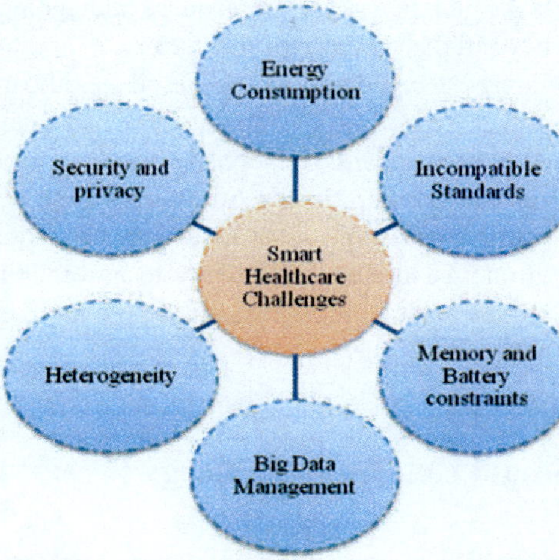

healthcare data can readily be targeted by attackers. Hacking data, for example, can reveal information about a person's buying habits, regular activities, and financial transactions. DoS attacks, man-in-the-middle assaults, Sybil attacks, spoofing, and data forging are all threats that IoT nodes in healthcare are vulnerable to. As a result, finding a security solution that uses the least amount of resources while still providing enough protection is a difficult issue. Because of the security dangers, the majority of the energy is used to execute compute operations that are not required to keep the system running [17].

(c) *Heterogeneity:* Interoperability defines the compatibility among subsystems to interact with each other. The major challenge in IoT is to enable communication among each other. IoT devices are being developed by various manufacturers which induces heterogeneity among IoT products. Heterogeneity arises from diverse technologies, diverse standards, diverse architectures, and diverse formats of language [18].

(d) *Incompatible Standards*: There are no open standards for the manufacturers to develop universal hardware products of IoT. Some efforts have been made by the IEEE P2413 Standard for an Architecture infrastructure for the Internet of things [19]. Several standardized bodies have presented their architectures such as IoT world Forum, and ETSI. In addition, diverse IoT architectures such as machine-to-machine, three-layer, five-layer, and seven-layer reference architecture. But there is no standardized architecture of IoT till now. For the proper realization of IoT, there must be compatible standards, protocols, and technologies.

(e) *Memory and Battery Constraints*: Low-speed CPUs are built into IoT devices. The speed of devices are very low. These devices are not built to execute computations quickly [20]. IoT devices have limited memory. Smart devices are also enabled via system software or an embedded operating system. Complicated tasks like data analysis, resource management, and process scheduling may be beyond the memory of the user.

(f) *Energy Consumption*: IoT-based healthcare devices have minimum battery power. (e.g., body temperature sensors and blood pressure sensors). Smart devices in healthcare conserve energy by switching on/off the power. The sensors are kept in the off state in case of no data sensing. In addition, the energy harvesting systems are not so efficient to convert renewable sources of energy to electrical energy to power IoT nodes [21]. The intelligent techniques must be incorporated to save more energy in resource-constrained IoT networks.

5 Smart Healthcare: Energy Preserving Mechanisms

Smart healthcare employs many energy-preserving mechanisms including intelligent techniques, edge/fog computing, energy harvesting, duty-cycling, collision resolution, and compression techniques as shown in Fig. 3.

(a) *Intelligent Techniques*: Intelligent techniques like artificial intelligence, machine learning, deep learning, and reinforcement learning have a great potential to predict the critical diseases and abnormal conditions of patient's in smart healthcare beforehand. Diseases are predicted based on the supervised and unsupervised learning by forming classification and clustering of data. Currently, intelligent techniques with data science are used in medical image processing, epidemic outbreak prediction, personalized healthcare, cost management, power consumption estimation, and disease prediction [22]. Recently, a joint AI-based recommendation system is leveraged by

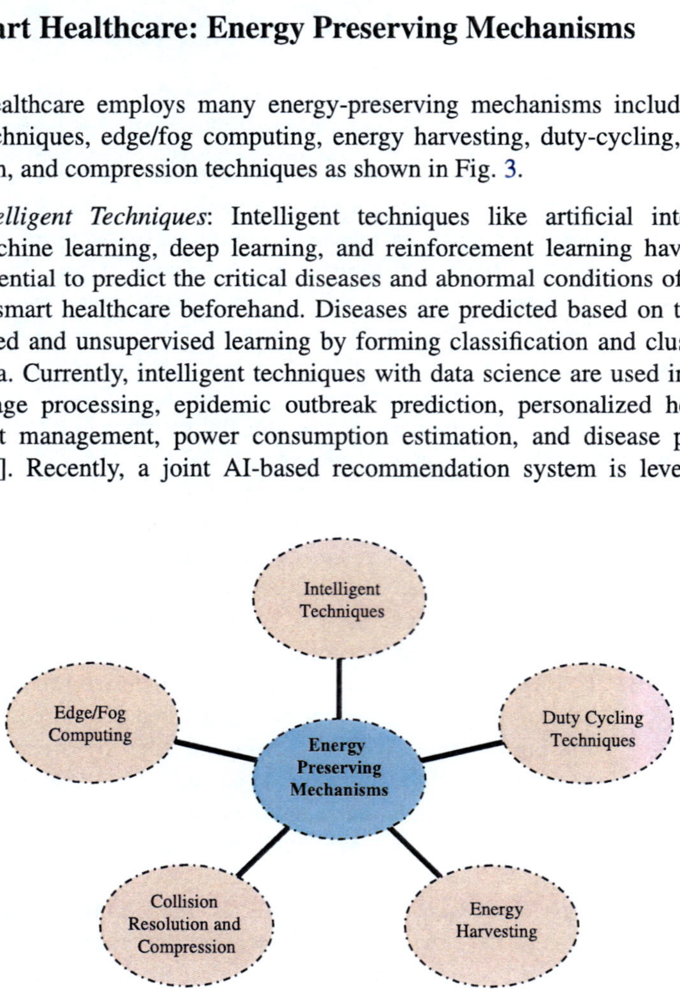

Fig. 3 Energy-preserving mechanisms in smart healthcare system

Google and DeepMind for energy consumption improvement in data centers. Google's DeepMind recommendation system with IoT aims to reduce energy consumption by 30% and control the cooling system independently.

(b) *Edge/Fog Computing:* Fog computing is a distributed concept in which some computations are done on edge or fog nodes and some data is offloaded for remote computation. Fog computing performs data processing in smart device itself using smart routers and gateways. Fog computing [23] increases the energy efficiency, throughput, privacy, and security because of the near processing of data from where the data is generated IoT based smart healthcare. Without taking into account the fog computing as happens traditionally, the whole data processing and analytics are performed on cloud. The processing on cloud incurs extra latency, low throughput, high bandwidth usage, and more energy consumption. IoT verticals, orchestration layer, and the abstraction layer forms are the basic entities in the fog environment. Therefore, in context to smart healthcare, fog and edge analytics are more relevant for real-time analysis of data.

(c) *Energy Harvesting*: Energy harvesting involves the conversion of ambient energy to power smart nodes from external sources. Energy harvesting is done in IoT systems for uninterrupted supply of power and prevents the node outage. Harvesting of energy can be performed in several ways: solar harvesting, piezoelectric, hydro, wind, and radiofrequency harvesting. Conversion of energy from its basic form to electrical energy requires efficient management systems for maximum energy conversion. Therefore, the energy harvesting depends upon the capacity of the battery, recharging time, energy, efficiency of conversion systems, and availability of the external renewable source to generate electricity [24].

(d) *Duty Cycling Techniques*: Duty cycling techniques are the most used techniques to conserve energy in IoT systems. Duty cycling is the mechanism to keep the node in either on state or off state. Nodes are kept in on state during data transmission and sensing the channel. When a node does not have any data to send, nodes are generally put in off state. Duty cycling is performed on media access control (MAC) layer. An IoT network uses the IEEE 802.15.4 communication standard on MAC layer. Therefore, duty cycling management and coordination requires controlling the Superframe order, beacon order, and personal area network (PAN) coordinator [25]. The MAC-based duty cycling techniques include T-MAC, Z-MAC, ZISENSE, etc. Currently, researchers are working to combine the reinforcement learning for MAC-based adaptive duty cycling in IoT systems.

(e) *Collision resolution and Compression*: The huge amount of data generated from wearable sensors and healthcare devices results in collision of data that consumes more energy. So, collision resolution and compression techniques are used for energy minimization by reducing the collisions and compressing large amount of data. The light weight SZ compression technique deals with the gigantic amount of data. The lightweight SZ compression technique only considers the floating data types by discarding the other data types, especially

Table 3 Energy-conserving mechanisms in smart healthcare system

Techniques	Functionality	Algorithms
Intelligent techniques	Modern technologies such as the Internet of things (IoT), machine learning, deep learning, and reinforcement learning are used to create a smart healthcare system	• Fuzzy neural network [26] • Elfes probability sensing • Routing algorithm [16]
Energy harvesting	Energy harvesting is critical for improving the efficiency and longevity of IoT devices. Mechanical, aeroelastic, wind, solar, radiofrequency, and pyroelectric energy harvesting mechanisms	• Energy-harvesting-aware routing algorithm (EHARA) [27] • Optimal energy allocation (OEA) [28]
Duty cycling	To save energy, a sensor node's duty cycle is a smart idea. The duty cycle is a system that repeats sleep and wake intervals on a regular basis	• Improved duty cycling (IDC) [29] • Delay constrained duty cycle scheduling (DDS) [30]
Collision resolution	For an IoT system, a collision resolution mechanism is needed to maintain enormous random access (RA)	• Decoding of three or more LoRa signals that are slightly desynchronized [31] • Preamble generation scheme [32]
Compression	The technique of lowering the amount of data necessary to express a given amount of information is known as data compression	• Elias [33] • Minimalist, adaptive, and streaming (MAS) [34] • Tiny anomaly compressor

for tiny IoT devices and for easier compilation by making the code small. Similarly, a collision resolution technique is presented for periodic communication through access points in an IoT network. Each device is allowed to transmit the data in one time slot to reduce collisions. Sequential lossless entropy compression scheme was used to reduce the amount of transmitted data, thereby reducing the transmitted power also (Table 3).

6 Conclusion

Smart healthcare is rapidly growing worldwide to deliver the on-demand services. But, the limited use of energy in an efficient manner is somehow lacking in smart healthcare. In this study, we focused on reducing the energy consumption in IoT-based smart healthcare. Intelligent techniques and edge analysis must be taken into account for real-time responses and energy consumption minimization. Shifting the data analysis on edge further reduces the latency, bandwidth usage, and security attacks. Also, incorporating adaptive duty cycling techniques based on reinforcement learning must be focused in the future to enhance energy conservation. In case of energy harvesting mechanisms, efficient conversion management systems must be developed for maximum energy conversion to power IoT nodes. Intelligent models for early prediction of energy must be designed for smart healthcare in future.

References

1. Alabdulatif, A., Khalil, I., Yi, X., & Guizani, M. (2019). Secure edge of things for smart healthcare surveillance framework. *IEEE Access, PP*(c), 1. https://doi.org/10.1109/ACCESS.2019.2899323
2. Tian, S. et al. (2019). Smart healthcare: making medical care more intelligent. *Global Health Journal*, 0–3. https://doi.org/10.1016/j.glohj.2019.07.001
3. World total final consumption—World energy data.
4. Jeong, S., Shen, J., & Ahn, B. (2021). *Research article a study on smart healthcare monitoring using IoT based on Blockchain* (Vol. 2021).
5. Xu, H., Chen, X., Zhu, F., & Li, P. (2021). *Research article a novel security authentication protocol based on physical unclonable function for RFID healthcare systems* (Vol. 2021).
6. Atzori, M. (2018). Blockchain-based architectures for the internet of things: A survey. *SSRN Electronic Journal.* https://doi.org/10.2139/SSRN.2846810
7. Almulhim, M., Islam, N., & Zaman, N. (2019). A lightweight and secure authentication scheme for IoT based E-health applications.
8. Ghosh, A., Raha, A., & Mukherjee, A. (2020). Energy-efficient IoT-health monitoring system using approximate computing. *Internet of Things, 9*, 100166. https://doi.org/10.1016/j.iot.2020.100166
9. "Humayun, M., Jhanjhi, N. Z., & Alamri, M. Z. (2020). Smart secure and energy efficient scheme for E-health applications using IoT: A review. *IJCSNS International Journal of Computer Science and Network Security, 6*, 20.
10. Feng, C., Adnan, M., Ahmad, A., Ullah, A., Khan, H. U. (2020). Towards energy-efficient framework for IoT big data healthcare solutions. https://doi.org/10.1155/2020/7063681
11. Majumdar, A., Debnath, T., Biswas, A., Sood, S. K., & Baishnab, K. L. (2020). An energy efficient e-healthcare framework supported by novel EO-μGA (extremal optimization tuned micro-genetic algorithm). *Information Systems Frontiers.* https://doi.org/10.1007/S10796-020-10016-5
12. El Zouka, H. A., & Hosni, M. M. (2019). Secure IoT communications for smart healthcare monitoring system. *Internet of Things, 13*, 100036. https://doi.org/10.1016/j.iot.2019.01.003
13. Zakaria, H., Abu Bakar, N. A., Hassan, N. H., Yaacob, S. (2019). IoT security risk management model for secured practice in healthcare environment. *Procedia Computer Science, 161*, 1241–1248. https://doi.org/10.1016/j.procs.2019.11.238
14. Moosavi, S. R., Nigussie, E., Levorato, M., Virtanen, S., & Isoaho, J. (2018). Performance analysis of end-to-end security schemes in healthcare IoT. *Procedia Computer Science, 130*, 432–439. https://doi.org/10.1016/J.PROCS.2018.04.064
15. Luo, E., Bhuiyan, M. Z. A., Wang, G., Rahman, M. A., Wu, J., & Atiquzzaman, M. (2018). PrivacyProtector: Privacy-protected patient data collection in IoT-based healthcare systems. *IEEE Communications Magazine, 56*(2), 163–168. https://doi.org/10.1109/MCOM.2018.1700364
16. Pullmann, J., & Macko, D. (2019). Increasing energy efficiency by minimizing collisions in long-range IoT networks. In *2019 42nd International Conference on Telecommunications and Signal Processing* (pp. 178–181).
17. Ray, P. P. (2018). A survey on internet of things architectures. *Journal of King Saud University—Computer and Information Sciences, 30*(3), 291–319. https://doi.org/10.1016/j.jksuci.2016.10.003

18. Yin, Y., Zeng, Y., Chen, X., & Fan, Y. (2016). The internet of things in healthcare: An overview. *Journal of Industrial Information Integration, 1*, 3–13. https://doi.org/10.1016/j.jii.2016.03.004
19. Baker, S. B., Xiang, W., & Atkinson, I. (2017). Internet of things for smart healthcare: Technologies, challenges, and opportunities. *IEEE Access, 5*, 26521–26544. https://doi.org/10.1109/ACCESS.2017.2775180
20. Rana, B., Singh, Y., & Singh, P. K. (2020). A systematic survey on internet of things: Energy efficiency and interoperability perspective. *Transactions on Emerging Telecommunications Technologies*. https://doi.org/10.1002/ETT.4166
21. Poongodi, T., Balamurugan, B., Sanjeevikumar, P., & Holm-Nielsen, J. B. (2019). *Internet of things (IoT) and E-healthcare system—A short review on challenges call for book chapters: Deregulated electricity market: A smart grid perspective view project electric vehicle involvement and its adaptation in smart grid view project* [Online]. Available: https://www.researchgate.net/publication/331876656
22. Comput, J. P. D., García-martín, E., Faviola, C., Riley, G., & Grahn, H. (2019). Estimation of energy consumption in machine learning. *Journal of Parallel and Distributed Computing, 134*, 75–88. https://doi.org/10.1016/j.jpdc.2019.07.007
23. Xhafa, F., et al. (2020). *Editorial board editor-in-chief* [Online]. Available: www.elsevier.com/locate/iot
24. Haimour, J., & Abu-Sharkh, O. (2019). Energy efficient sleep/wake-up techniques for IOT: A survey. In *2019 IEEE Jordan International Joint Conference on Electrical Engineering and Information Technology. JEEIT 2019—Proceedings* (pp. 459–464), May 2019. https://doi.org/10.1109/JEEIT.2019.8717372
25. Hameed, K., Bajwa, I., Ramzan, S., Anwar, W., & Khan, A. (2020). An intelligent IoT based healthcare system using fuzzy neural networks. In *Scientific programming* (Vol. 2020, pp. 1–15). Available: https://doi.org/10.1155/2020/8836927
26. Malchi, S., Kallam, S., Al-Turjman, F., & Patan, R. (2021). A trust-based fuzzy neural network for smart data fusion in internet of things. *Computers and Electrical Engineering, 89*, 106901. Available: https://doi.org/10.1016/j.compeleceng.2020.106901
27. Steffi, R. (2021). IoT enabled cluster based energy aware routing protocol in WSN. *Innovations in Information and Communication Technology Series*, 93–102. Available: https://doi.org/10.46532/978-81-950008-7-6_009
28. Kosunalp, S. (2016). A new energy prediction algorithm for energy-harvesting wireless sensor networks with Q-learning. *IEEE Access, 4*, 5755–5763. Available: https://doi.org/10.1109/access.2016.2606541
29. Dhall, R., Agrawal, H. (2018). An improved energy efficient duty cycling algorithm for IoT based precision agriculture. *Procedia Computer Science, 141*, 135–142. Available: https://doi.org/10.1016/j.procs.2018.10.159
30. Vu, D., Dao, T., Yoon, S.: DDS: A delay-constrained duty-cycle scheduling algorithm in wireless sensor networks. *Electronics, 7*(11), 306. Available: https://doi.org/10.3390/electronics7110306
31. Onishi, T., Li, A., Kim, S., & Hasegawa, M. (2021). A reinforcement learning based collision avoidance mechanism to superposed LoRa signals in distributed massive IoT systems. *IEICE Communications Express, 10*(5), 289–294. Available: https://doi.org/10.1587/comex.2021xbl0033
32. Zhong, A., Li, Z., Wang, R., Li, X., & Guo, B. (2021). Preamble design and collision resolution in a massive access IoT system. *Sensors, 21*(1), 250. Available: https://doi.org/10.3390/s21010250
33. Guberovic, E., Kristo, F., Krivic, P., & Cavrak, I. (2019). Assessing compression algorithms on IoT sensor nodes. In *2019 42nd International Convention on Information and Communication Technology, Electronics and Microelectronics (MIPRO)*. Available: https://doi.org/10.23919/mipro.2019.8756995. Accessed August 22, 2021.

34. Abuda, C., Caya, M., Cruz, F., & Uy, F. (2018). Compression of wireless sensor node data for transmission based on minimalist, adaptive, and streaming compression algorithm. In *2018 IEEE 10th International Conference on Humanoid, Nanotechnology, Information Technology, Communication and Control, Environment and Management (HNICEM), 2018*. Available: https://doi.org/10.1109/hnicem.2018.8666320. Accessed August 22, 2021.

T-Shaped MIMO Microstrip Patch Antenna for C-Band Applications

Piyush Kumar

Abstract The demand of high capacity in the wireless communication systems is increasing as the users and applications are continuously increasing. The capacity of the wireless systems can be enhanced using multiple input multiple output (MIMO) technology. This article presents the design of a T-shaped MIMO microstrip patch antenna (TSMMPA) for C-band applications. The T-shaped structure provides the high gain (GN) and directivity (DY), and the two orthogonal T-shaped structures make the antenna (ANA) suitable for 2 × 2 MIMO communication systems. The ANA structure is simulated and optimized using CST microwave studio software (CSTMSS). The proposed ANA provides the wide bandwidth (BH), wide 3-dB beamwidth, high efficiency (EFY), and easy to fabricate. The presented TSMMPA operates at the C-band frequencies.

Keywords MIMO · T-shaped ANA · Reflection coefficient · Isolation · C-band

1 Introduction

In recent years, the number of users and applications in wireless communications systems are increasing rapidly. To fulfill this demand of the wireless systems, the wireless systems with high capacity and BH are required. MIMO wireless systems provide the high capacity as different ports (PTs) are used to transmit/receive signals within the same BH [1]. Microstrip ANAs are suitable for many applications due to their several advantages such as light weight, compact size, and easy to fabricate. [2, 3].

These ANAs suffer with the low GN and small BH [4]. By using metamaterials [5, 6], gap-coupling [7, 8], frequency selective surface (FSS) [9, 10], defected ground structure (DGS) [11–13], DGS with reflector surface [14], partial ground [15–18],

P. Kumar (✉)
Discipline of Electrical, Electronic and Computer Engineering, School of Engineering, Howard College Campus, University of KwaZulu-Natal, King George V Avenue, Durban 4041, South Africa
e-mail: pkumar_123@yahoo.com

P. K. Singh et al. (eds.), *Proceedings of Third International Conference on Computing, Communications, and Cyber-Security*, Lecture Notes in Networks and Systems 421,
https://doi.org/10.1007/978-981-19-1142-2_41

the ANA parameters can be improved. In [5], Zhang et al. presented the review of metamaterial-based wearable ANAs for wireless body area networks applications. In [6], Pattar et al. proposed a complimentary split ring resonator metamaterial-based ANAs for RADAR applications. Using metamaterial, the mutual coupling was reduced, and the GN was enhanced. A reduction of 6 dB in mutual coupling was achieved. In [7], Kumar proposed the numerical model for calculating the resonant frequencies of the gap-coupled ring microstrip ANAs. It was shown that how gap-coupled ring ANA produces dual frequencies. In [8], Kumar and Singh proposed the numerical computation of gap-coupled circular patch ANAs. It is shown that how the gap-coupled circular patch ANAs generate dual resonances. In [9], Adeline Mellita et al. proposed a dual-band FSS-based. The GN of the dual-band ANA was enhanced by 50%. In [10], Evangelista et al. proposed the FSS was used as a superstrate for enhancing the GN of the ANA. In [11], Mabaso and Kumar proposed the dual-band patch ANA with DGS for Bluetooth and wireless local area networks applications. A maximum GN of 7.398 dB was achieved. In [12], Nigam et al. proposed a compact microstrip patch ANA for ultra-wideband applications. The ANA is suitable for 9.4 GHz RADAR applications. In [13], Gaid et al. proposed a DGS-based microstrip ANA for multi-band applications. In [14], Ngobese and Kumar proposed a four element rectangular patch array for 5 GHz wireless local area networks application. The GN of the array was enhanced using DGS and a reflector surface. In [15], Kumar and Masa-campos designed a partial ground-based ultra-wideband ANA. The two ANAs were designed for upper and lower frequency ranges of the ultra-wideband. The combined ANA structure operated for the entire ultra-wideband. In [16], Mahmud et al. proposed a parasitic element and partial ground plane-based ANA for ultra-wideband applications. In [17, 18], the partial ground plane was utilized for designing the ANA for ultra-wideband applications. The T-shaped ANAs provide the better GN as compared to the dipole ANAs [19]. With this motivation, this paper is focused on the design of the TSMMPA.

The T-shaped microstrip ANA for MIMO wireless systems is presented in this paper. The T-shaped microstrip ANA for single input single output was designed in [20]. The two T-shaped microstrip ANAs are placed orthogonally to each other to minimize the isolation between the PTs. The TSMMPA provides the wide BH, wide 3 dB beamwidth, high EFY, reasonable GN, and DY and is suitable for C-band applications.

The structure of the paper is as follows. The geometrical configuration of the TSMMPA is presented in Sect. 2. The simulated parameters of the TSMMPA are discussed in Sect. 3. The conclusion of the work is given in Sect. 4.

2 Geometry of the TSMMPA

The geometrical configuration of the TSMMPA is shown in Fig. 1. The two T-shaped ANAs are placed orthogonal to each other, as shown in Fig. 1, to achieve the dual polarization and high isolation between the PTs. The top view of the TSMMPA and

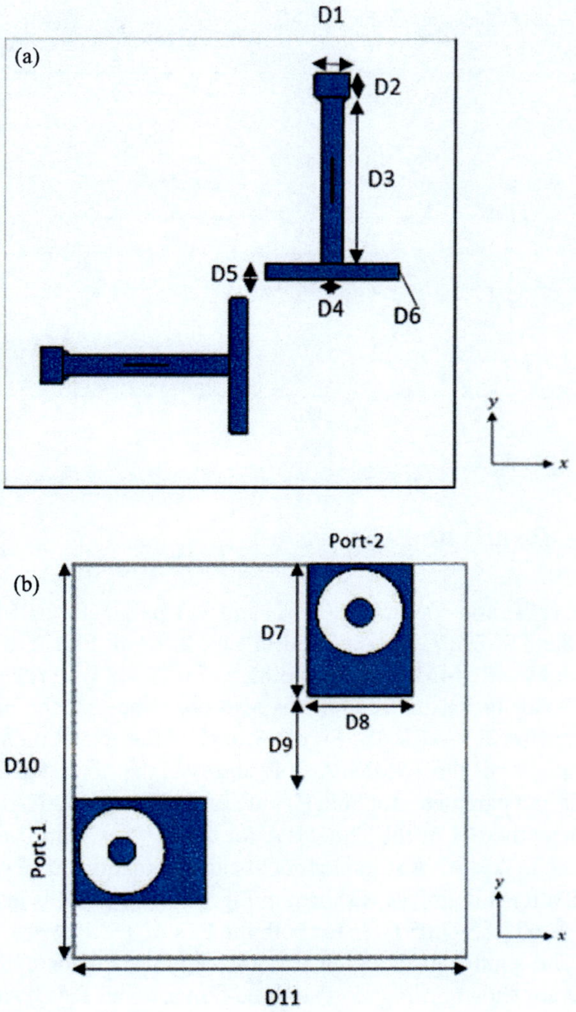

Fig. 1 Geometry of the TSMMPA, **a** top view, **b** bottom view

bottom view of the TSMMPA are shown in Fig. 1a, b, respectively. The T-shaped structure provides the high GN and the partial ground plane in the structure helps to achieve the wideband operation. The ANA is designed on the FR4 substrate. The thickness of the substrate is 1.58 mm with dielectric constant of 4.4. The proposed TSMMPA is designed and optimized using the CSTMSS. The optimized dimensional parameters of the TSMMPA in terms of center wavelength (λ_0) are given in Table 1.

Table 1 Dimensions of the TSMMPA

S. No.	Parameter	Value
1	D1	0.053 λ_0
2	D2	0.034 λ_0
3	D3	0.253 λ_0
4	D4	0.032 λ_0
5	D5	0.052 λ_0
6	D6	0.025 λ_0
7	D7	0.229 λ_0
8	D8	0.178 λ_0
9	D9	0.177 λ_0
10	D10	0.678 λ_0
11	D11	0.678 λ_0

3 Results and Discussion

The reflection coefficient (PT-1) and S21 parameter with the variation of frequency of the TSMMPA are depicted in Fig. 2. From Fig. 2, it can be observed that the BH of the TSMMPA is from 3.85 to 7.05 GHz with reflection coefficient less than −10 dB. From this figure, it is also observed that the minimum value of the S21 parameter is −51.8 dB. Figure 3 depicts the reflection coefficient (PT-2) and S12 parameter of the TSMMPA. It is observed that the reflection coefficient and mutual coupling parameter for both PTs are same as the two ANAs are placed symmetrically. Hence, the BH of the TSMMPA for both PTs is same, i.e., from 3.85 to 7.05 GHz, which makes ANA suitable for C-band applications. The voltage standing wave ratio (VSWR) at both PTs is shown in Fig. 4. The VSWR is less than 2 for the frequency range of 3.85–7.05 GHz for both the PTs of TSMMPA.

The simulated radiation patterns (RADPATs) for both PTs at various frequencies are shown in Fig. 5. The three-dimensional RADPAT at 5.5 GHz, the normalized RADPAT at 4.5 GHz, the normalized RADPAT 5.5 GHz, and the normalized

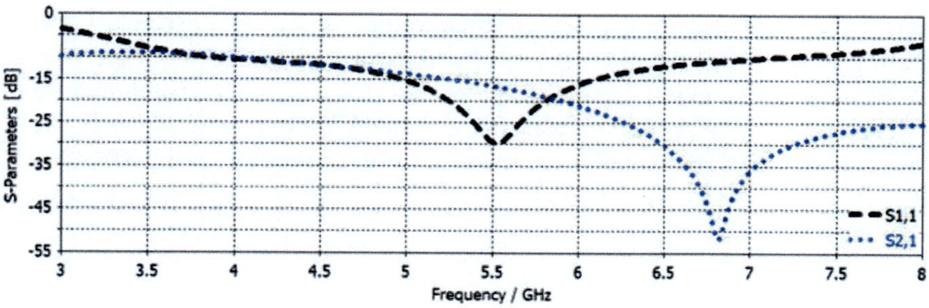

Fig. 2 S11 and S21 parameters of the TSMMPA

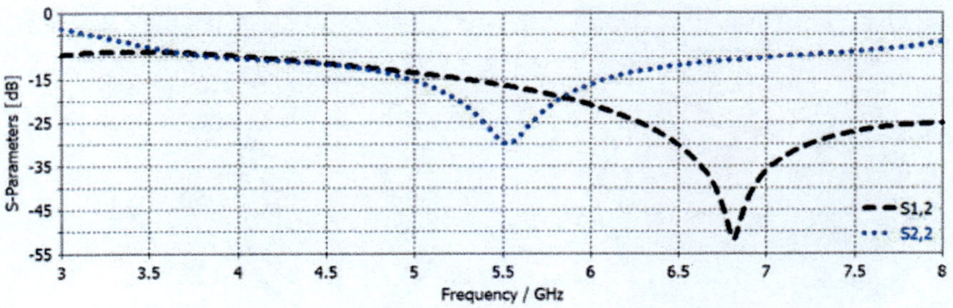

Fig. 3 S22 and S12 parameters of the TSMMPA

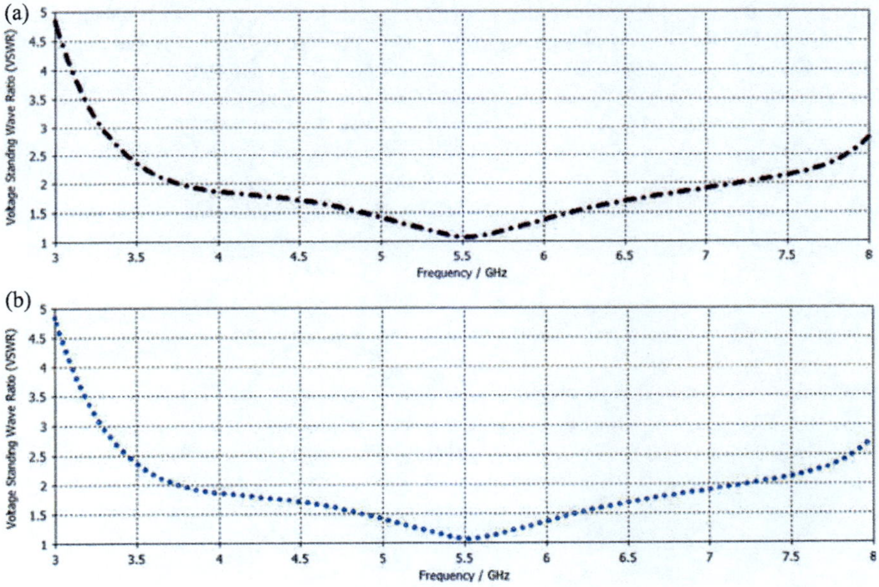

Fig. 4 VSWR of the TSMMPA at **a** PT-1, **b** PT-2

RADPAT at 6.5 GHz for PT-1 are shown in Fig. 5a–d, respectively. The three-dimensional RADPAT at 5.5 GHz, the normalized RADPAT at 4.5 GHz, the normalized radiation RADPAT 5.5 GHz, and the normalized RADPAT at 6.5 GHz for PT-2 are shown in Fig. 5e–h, respectively. From these RADPATs, it can be seen that the RADPATs are of the shape of figure of eight and with wide beamwidth. Various parameters of the TSMMPA are summarized in Table 2. The maximum GN, maximum DY, radiation EFY, and total EFY for both PTs at 5.5 GHz of the TSMMPA are 3.276 dB, 3.689 dBi, −0.4129 dB, and −0.5168 dB, respectively. The RADPAT parameters of the designed TSMMPA are given in Table 3. The proposed TSMMPA gives a broad 3 dB beamwidth of 311.8° and 89.9° in phi = 90° plane and phi = 0° plane for PT-1 at 5.5 GHz, respectively. The 3-dB beamwidth in phi = 90°

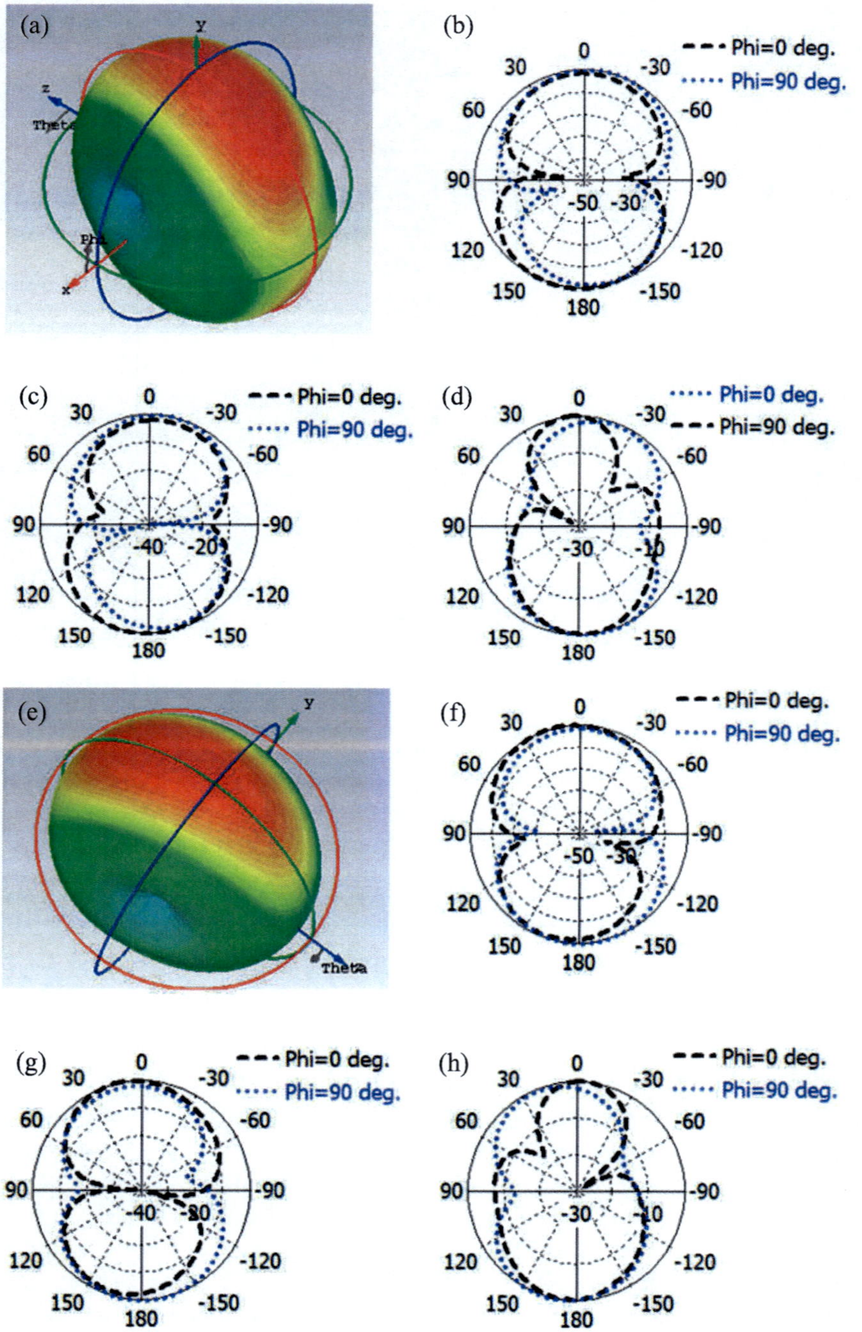

Fig. 5 RADPATs of the TSMMPA, **a** 3D at 5.5 GHz (PT-1), **b** at 4.5 GHz (PT-1), **c** at 5.5 GHz (PT-1), **d** at 6.5 GHz (PT-1), **e** 3D at 5.5 GHz (PT-2), **f** at 4.5 GHz (PT-2), **g** at 5.5 GHz (PT-2), **h** at 6.5 GHz (PT-2)

Table 2 Simulated parameters of the TSMMPA

S. No.	Parameter	Value
1	BH (PT-1)	3.85–7.05 GHz
2	Center frequency (PT-1)	5.45 GHz
3	Fractional BH (PT-1)	58.716%
4	BH (PT-2)	3.85–7.05 GHz
5	Center frequency (PT-2)	5.45 GHz
6	Fractional BH (PT-2)	58.716%
7	Minimum S21	−51.8 dB
8	GN (PT-1) at 5.5 GHz	3.276 dB
9	DY (PT-1) at 5.5 GHz	3.689 dBi
10	Radiation EFY (PT-1) at 5.5 GHz	−0.4129 dB
11	Total EFY (PT-1) at 5.5 GHz	−0.5168 dB
12	GN (PT-2) at 5.5 GHz	3.276 dB
13	DY (PT-2) at 5.5 GHz	3.689 dBi
14	Radiation EFY (PT-2) at 5.5 GHz	−0.4129 dB
15	Total EFY (PT-2) at 5.5 GHz	−0.5168 dB

Table 3 Simulated RADPAT parameters of the TSMMPA at 5.5 GHz

S. No.	Parameter	Plane (°)	Value (°)
1	Main lobe direction (PT-1)	Phi = 0	178
2	Main lobe direction (PT-1)	Phi = 90	113
3	Main lobe direction (PT-2)	Phi = 0	113
4	Main lobe direction (PT-2)	Phi = 90	178
5	3-dB beam width (PT-1)	Phi = 0	89.9
6	3-dB beam width (PT-1)	Phi = 90	311.8
7	3-dB beam width (PT-2)	Phi = 0	311.8
8	3-dB beam width (PT-2)	Phi = 90	89.9

plane and phi = 0° plane for PT-2 is 89.9° and 311.8° at 5.5 GHz, respectively. The variation of simulated maximum GN, maximum DY, radiation EFY, and total EFY with frequency of the TSMMPA is shown in Fig. 6. The maximum GN, maximum DY, radiation EFY, and total EFY of the TSMMPA at 6.5 GHz are 3.604 dB, 4.179 dBi, -0.5755 dB, and −0.87 dB, respectively. From various parameters of the ANA, it is observed that the proposed TSMMPA is suitable for C-band MIMO systems.

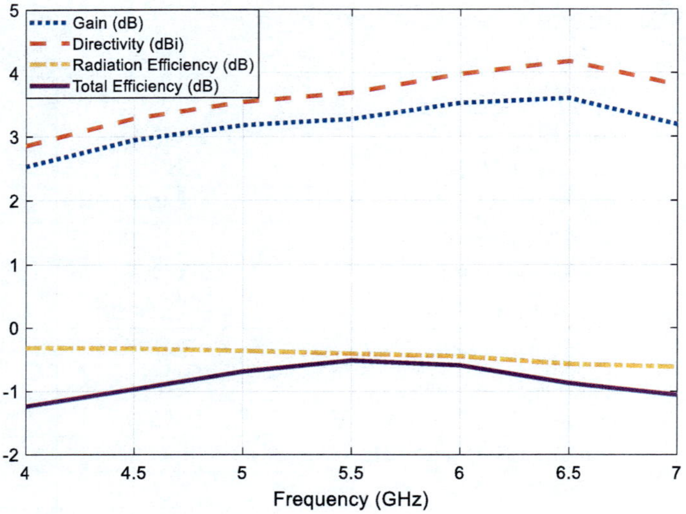

Fig. 6 GN, DY, radiation EFY, and total EFY of the TSMMPA

4 Conclusion

The design of the TSMMPA has been presented in this paper. The proposed TSMMPA utilizes two orthogonally placed T-shaped structures in order to make ANA suitable for MIMO systems. The partial ground plane is used to obtain the wideband operation. The proposed TSMMPA is a wideband ANA with the BH of 3.85–7.05 GHz. The minimum coupling coefficient achieved by the TSMMPA is −51.8 dB. The ANA provides the maximum GN of 3.604 dB and the maximum DY of 4.179 dBi. The ANA parameters confirm the suitability of the TSMMPA in wideband MIMO wireless systems.

References

1. What is MIMO Wireless Technology. https://www.electronics-notes.com/articles/antennas-propagation/mimo/what-is-mimo-multiple-input-multiple-output-wireless-technology.php. Last accessed February 03, 2021.
2. Balanis, C. A. (2005). *Antenna theory-analysis and design*. Wiley.
3. Garg, R., Bhartia, P., Bahl, I., & Ittipiboon, A. (2001). *Microstrip antenna design handbook*. Artech House Publishers.
4. Bankey, V., & Kumar, N. A. (2015). Design and performance issues of microstrip antennas. *International Journal of Scientific and Engineering Research, 6*(3).
5. Zhang, K., Soh, P. J., & Yan, S. (2021). Meta-wearable antennas—A review of metamaeterail based antennas in wireless body area networks. *Materials, 14*, 149.
6. Pattar, D., Dongaokar, P., Nisha, S. L., & Amith, S. (2020). Design and implementation of metamaterial based Patch antenna. In *2020 IEEE International Conference for Innovation in Technology (INOCON)* (pp. 1–4).

7. Kumar, P. (2014). Computation of resonant frequency of gap-coupled ring microstrip antennas. *International Journal of Automation and Computing, 11*(6), 671–675.
8. Kumar, P., & Singh, G. (2009). Theoretical investigation of the input impedance of gap-coupled circular microstrip patch antennas. *Journal of Infrared, Millimeter, and Terahertz Waves, 30*(11), 1148–1160.
9. Adeline Mellita, R., Chandu, D. S., & Karthikeyan, S. S. (2018). A compact dual-band frequency selective surface for gain enhancement of a dual-band antenna. *Lecture Notes in Electrical Engineering, 472*, 607–614.
10. Evangelista, T. S., Neto, A. G., & Serres, A. J. R. (2021). Improved microstrip antenna with FSS superstrate for 5G NR applications. In *2021 15th European Conference on Antennas and Propagation (EuCAP).*
11. Mabaso, M., & Kumar, P. (2018). A dual band patch antenna for Bluetooth and wireless local area networks applications. *International Journal of Microwave and Optical Technology, 13*(5), 393–400.
12. Nigam, H., Mathur, M., & Arora, M.: Design of a compact MIMO antenna for RADAR applications using DGS technology. *Algorithms for Intelligent Systems*, 199–205, (2019)
13. Gaid, A. S. A., Sallam, A. A., Qasem, M. H. M., Abbas, M. S. G., & Aoun, A. M. H. (2021). A circular multiband microstrip patch antenna with DGS for WLAN/WiMAX/Bluetooth/UMTS/LTE. *Lecture Notes on Data Engineering and Communications Technologies, 72*, 647–658.
14. Ngobese, B. W., & Kumar, P. (2018). A high gain microstrip patch array for 5 GHz WLAN applications. *Advanced Electromagnetics, 7*(3), 93–98.
15. Kumar, P., & Masa-Campos, J. L. (2014). Dual polarized microstrip patch antennas for ultra wideband applications. *Microwave and Optical Technology Letters, 56*(9), 2174–2179.
16. Mahmud, M. Z., et al. (2019). A parasitic resonator-based diamond-shaped microstrip antenna for microwave imaging applications. *Electronics, 8*(4), 434.
17. Kumar, P. (2017). Design of low cross-polarized patch antenna for ultra-wideband applications. *International Journal on Communication Antenna and Propagation, 7*(4), 265–270.
18. Kumar, P., & Masa-Campos, J. L. (2016). Dual polarized monopole patch antennas for UWB applications with elimination of WLAN signals. *Advanced Electromagnetics, 5*(1), 46–52.
19. Singh, D., Gangwar, S. P., & Verma, A. K. (2016). 'T' shaped bended type antenna for wireless communication. In *2016 International Conference on Emerging Trends in Electrical Electronics and Sustainable Energy Systems (ICETEESES).*
20. Kumar, P. (2017). A T shaped microstrip antenna for wireless local area network (WLAN) applications. In *Proceedings of 2017 International Conference on Multimedia, Signal Processing and Communication Technologies*, 147–150

Eye Disease Detection Using Transfer Learning on VGG16

Aditi Arora, Shivam Gupta, Shivani Singh, and Jaya Dubey

Abstract Deep learning has emerged as a breakthrough technology in varied fields like health care, computer vision, natural language processing and many more. Ocular infections like diabetic macular edema (DME), choroidal neovascularization (CNV) and DRUSEN are commonly found eye diseases in humans and can lead to temporary or permanent loss of eyesight. The optical coherence tomography (OCT) technique is often used for the preliminary screening of mentioned ocular ailments and provides high resolution cross-sectional imaging. In this work, we have focused on classification of normal and abnormal optical coherence tomography by making use of visual geometry group (VGG16) convolution neural network (CNN) model for prompt diagnosis and timely proper medical treatment of the eye diseases mentioned. OCT image is high resolution imaging technique capable of capturing microstructures within human eye. Here, we endeavored to develop a CNN model for classifying OCT images into normal and abnormal category. Our model achieves an accuracy of 99% and precision of 98.8% which is quite improved results in comparison with other state-of-the-art works that we reviewed.

Keywords Optical coherence tomography · Choroidal neovascularization · Diabetic retinopathy · Diabetic macular edema · DRUSEN · CNN

A. Arora (✉) · S. Gupta · S. Singh · J. Dubey
ABES Engineering College, Ghaziabad, Uttar Pradesh, India
e-mail: aditi.arora@abes.ac.in

S. Gupta
e-mail: shivam.17bcs1065@abes.ac.in

S. Singh
e-mail: shivani.17bcs1180@abes.ac.in

J. Dubey
e-mail: jaya.18bcs3010@abes.ac.in

P. K. Singh et al. (eds.), *Proceedings of Third International Conference on Computing, Communications, and Cyber-Security*, Lecture Notes in Networks and Systems 421,
https://doi.org/10.1007/978-981-19-1142-2_42

1 Introduction

Artificial intelligence (AI) and deep learning (DL) techniques have increasingly ubiquitous applications in various fields like medical diagnostics and tests. In ophthalmology, one can precisely recognize diseases as a professional with the help of AI services [1]. Medical imaging provides essential traces for detecting many diseases and thus helps doctors and clinicians. Retinal diseases are known causes of visual impairment and blindness, hence raises a lot of concerns for researchers in medical domain. In most parts of the developing world, emphasis is laid on avoidable causes of blindness which are more common. Efforts have been made for early treatments for such eye diseases causing blindness. There is a need for such a system which can provide easier, more cost-effective and achieve higher accuracy with advance technology available in the current time. Healthcare managers have invested more in preventing or treating blindness from eye disorders that are easily handled than procuring advance technology toward the treatment of diseases that were considered less common and more expensive to handle. The occurrence of retinal degeneration diseases related to age factor which is also a major reason of blindness is further growing [2]. OCT images are very helpful in directing the professionals for detection of eye diseases. OCT is considered as a noninvasive technique of imaging which plays prominent role in conducting the administration of anti-VGF by accommodating full cross-section scan of retina [3]. Automated image detection can provide the constant pre-diagnosis on inspecting the OCT image of retina. Thus, we explored the methods of deep learning to do the task systematically and efficiently. In the following work, we will present a retinal disease detection system capable of classifying images. In this work, we have analyzed optical coherence tomography (OCT) images with the help of deep learning model for identifying patients who are suffering from diseases like DRUSEN, CNV and DME.

In our proposed solution, we have used transfer learning on VGG16 model. In transfer learning, multiple layers from a pre-trained model is used in a new model for a related problem. VGG16 stands for very deep convolutional neural network. VGG16 consists of 13 convolution layers, five max pooling layers and at last three fully connected layers. The layers which are having tunable parameter are 13 convolution layers and three fully connected layers which makes it 16, hence referred to as VGG16. The size of max pooling window is 2×2. There are 4096 channels present in first two connected layers, and last fully connected layer consists of 1000 channels. The last layer is a softmax layer with 1000 channels, one for each category in ImageNet database. Hidden layer has ReLU as an activation function.

The color images of dimensions $224 \times 224 \times 3$ are provided as an input to the architecture, where the dimensions are represented as height, width and channel. For constructing layer 1, input layer was convolved with 64 $3 \times 3 \times 3$ kernels and then convolution performed in layer 1 again with 64 $3 \times 3 \times 64$ kernels. After that layer 3 was generated with the help of 2×2 maxpool operation. The same operation was performed until layer 18, where a $7 \times 7 \times 512$ number of neurons were produced. From layer 19 to 21, the fully connected layers have been arranged. The neurons which are presented in layer 18 are thoroughly attached with 4096 neurons in layer number 19. Thereafter, layer 19 with 4096 neurons is entirely attached with layer 20. At the end, layer 20 is coupled with layer 21(the output layer) [4, 5].

The model is pre-trained over 10 million of images with ImageNet dataset and learnt how to detect generic features such as edge and roundness from images. This model can be downloaded as a feature extractor and combined with customer dense and output layer to predict the output.

2 Related Work

Maheshwari et al. [6] in their work showed the automated diagnosis of glaucoma. Classification was done on the basis of features with a high accuracy. Various classification kernels were tested, and it was discovered that RBF and Morlet wavelet kernels had the best results. The same concept can be applied to the diagnosis of other disorders such as diabetic retinopathy, fatty liver disease and thyroid disease. The author achieved an accuracy of 98.33.

Abbas [7] in his work proposed a model named glaucoma deep system for detecting glaucoma eye disease. The model was illustrated with three steps. First, the extraction of features through multilayer from raw pixel intensities on 1200 retinal images taken from public and private datasets was done through CNN unsupervised architecture. Secondly, the selection of deep discriminative features based on the annotated training dataset has been done with the deep belief network mode, and at last, the differentiation of glaucoma and non-glaucoma images was done with the help of softmax linear classifier. The results showed that his approach achieved an accuracy of 99%, precision of 84% and sensitivity of 84.50%. By comparing their own model with state-of-the-art system found that nodular deep system generated highly substantial results and was able to recognize the glaucoma eye disease.

Gargeya et al. [8] made an attempt for auto-detection of diabetic retinopathy (DR) by developing and evaluating a deep learning algorithm. They designed the algorithm in such a way so that it processed the fundus images (colored) and then classified them into no retinopathy (healthy) or having (DR). The information learned was visualized through an auto-generated heat map and underlining subregions. In their work, they have used the region covering the receiver characteristic curve as a parameter to calculate the precision, reporting associative sensitivity and specificity metric. For external validation, they tested their model through the MESSIDOR 2 and E-OPTHA (databases) which are available publically. The authors achieved a result of 0.97 AUC with 94% sensitivity and 98% specificity.

Awais et al. [9] in their work focused on detection of diabetic macular edema (DME).They performed a classification of optical tomography images (OCT) volumes whether they are normal or abnormal by making use of a pre-trained convolution neural network (CNN). For improving image classification with high performance, they adopted deep learning. OCT provides high resolution images for DME detection. By using VGG16, they performed the extraction of features at different layer of network. Different classifiers have been used for classification on the basis of the feature. The authors achieved results with an accuracy value of 87.5%, sensitivity equal to 93.55% and specificity value 81%.

Al-Bander et al. [10] in their work proposed a methodology for detecting the glaucoma diseases causing blindness. They addressed in their work that deep learning techniques can be used to develop an automated feature learning technique to detect glaucoma disease in retinal fundus images (colored). For taking the decision in distinction of normal and glaucomatous pattern, they have developed a fully automated system. Through convolution neural network (CNN), the extraction of features done automatically from raw images and then through SVM classifier the images was classified into normal or abnormal. By comparing their methods with existing equivalent methods, they have found that their method has lower computational cost and gives an accuracy value: 88.2%, specificity: 90.8% and sensitivity: 85%.

Malik et al. [11] made an attempt in developing a framework for capturing diagnostic information in some universal format so that by using machine learning algorithm, diseases can be predicted by analyzing symptoms. To study and analyze patient's data on the basis of features like age, disorder history and clinical observations, multiple machine learning algorithms have been used like Naive Bayes, neural network and random forest. They have designed system such that the evolution takes place through self-learning in it by including some new classifications for symptoms and further diagnosis. The authors achieved more than 90% predication rate with the help of random forest and decision tree algorithm in comparison with more intricate methods like neural networks and Naïve Bayes algorithm.

Bajwa et al. [12] defined a two-stage framework for the detection of optic disk which is present in retina to be in a healthy state or suffered from glaucomatous. In their work, they laid emphasis on localization and classification of glaucoma with AUC. They have achieved localization with the help of region-based convolutional neural networks (R-CNNs), while the deep CNN has been used for the classification of computed disk into glaucomatous/healthy. This method is complex in computation because it is a two-stage process for localizing and classifying the glaucoma. The have observed that if they increase network hierarchy, then it adversely affects the performance, and as a result, it shows the loss of the discriminative set of features. Their approach achieved an accuracy level of 79.39% ± 3.42% and a precision of 77.97% ± 3.78%.

Nazir et al. [13] in their work proposed a novel methodology for detection of diabetes-based eye disease, i.e., glaucoma, diabetic retinopathy and diabetic macular edema. These types of diseases gradually harm the eye retina. For detecting such diseases, the authors in their work proposed a technique comprising two main phases: First is disease detection and localization, and the second is segmenting localized regions. With the help of fast region-based convolutional neural network (FRCNN) algorithm along with fuzzy K-means clustering (FKM), automated diseases localization and segmentation approaches are being presented. For localization, the FRCNN is trained over the annotated images and then segmented out by FKM clustering. The performance was evaluated on datasets like Diaretdb1, MESSIDOR, DR-HAGIS

Table 1 Summary of related work showing the shortcomings of approaches used

References	Year	Purpose	Approach used	Finding	Shortcoming
Shanthi and Sabeenian [14]	2019	Diabetic retinopathy (DR) fundus images classification according to severity	Deep learning (AlexNet CNN model)	The described method has accomplished an accuracy value of 96.6%, precision value of 86% and sensitivity of 96%	Experiments were performed on small dataset
Malik et al. [7]	2019	Eye diseases classification using machine learning based on different parameter like age, illness history and clinical observation	Machine learning algorithms (random forest, decision tree, Naive Bayes and neural networks)	Achieved more than 90% prediction rate with random forest and decision tree as compared to Naïve Bayes and neural networks	Accuracy (81.3%) and precision (8 1.6%) in Naïve Bayes case need further improvement
Nazir et al. [6]	2020	Analyze retinal images for detection of diabetes-based eye diseases	Used two-phase technique for localization of diseases and segmentation based on FRCNN with fuzzy *K*-means (FKM) clustering	The proposed method ensembles the result with means IoU of 0.95 and the map value higher than 0.94 and an accuracy of 95.2%	The proposed method is computationally complicated

and HRF and ensembled the results with map value greater than 0.94 and mean IoU value equal to 0.95.

We have summarized the related work done highlighting the findings and shortcomings in Table 1.

3 Methodology

The methodology used in our work for classification OCT images into CNV, DME, DRUSEN and normal is shown in Fig. 1. The dataset used in the proposed method is a public dataset named retinal optical coherence tomography (OCT) images taken from Kaggle platform [15]. The published dataset contains 84,484 OCT images. The dataset is categorized into two folders: one for training set and another for test set. Both folders contain subfolder for each image category (normal, CNV, DME

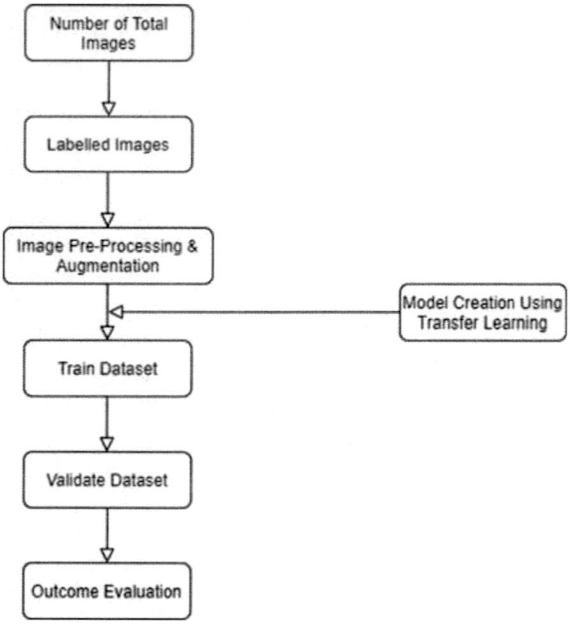

Fig. 1 Flow diagram of our proposed method for classification of OCT images into CNV, DME, DRUSEN and normal

and DRUSEN). A total of 83,484 images (JPEG) are used for training the model in which 37,205 are of CNV,11,348 are of DME, 8616 of DRUSEN and 26,315 are normal images, respectivelyn and the test dataset contains 1000 images (250 DME, 250 CNV, 250 DRUSEN and 250 normal).

To bring uniformity and removing noise from the images, various image preprocessing techniques are available. So before training, proposed solution requires the pre-treatment of image. Assuming that during the collection, images of retina got tilted, distorted or misshaped, thus required to get a uniform image through diminishing and cutting the images. As VGG16 architecture takes input image of size 224 × 224 RGB image, thus all images must be converted to this size.

Deep learning model performs better when it has large dataset to be trained. Image augmentation is a technique for artificially expanding the size of dataset by modifying the images in dataset. The image transformation techniques used in this paper include image rotation, image shifting, horizontal flipping, zooming and adjusting the brightness of images.

4 Experimental Work and Statistical Analysis

In this paper, we have used VGG16 model based on transfer learning. Transfer learning is a task where we take leverage of learning obtained from previous problem.

Transfer learning is used where the dataset is not sufficient for training the model from initial/base.

Initially, VGG16 model was loaded with last three layers with random weights, this baseline model was trained by training only the last three layers, and rest layers were initialized with pre-trained weight from ImageNet dataset.

Here, we used the softmax activation function, and trainable parameter was set to false. We also used the early stop algorithm to stop model from over training. This is one of the major challenges during the training of deep learning model to decide how long should a model be trained. Too long training will over fit the model on training dataset while less training will under fit the model. In both cases, it leads to poor performance.

During the training the model, there is a point where model reaches a point where it stops generalizing and actually starts learning the statistical noises present in the training dataset. Early stopping is like a trigger will be fired and stop the training. It monitors the performance of model after a specified number of epochs, and when performance decreases, it stops the training and saves the model. This step is known as feature extraction.

Final step was the fine-tuning where all the layers of model are unfroze and retrains the model as a whole with complete dataset with early stop algorithm. Here, we set trainable parameter to be true. Learning rate of model could be low. Using fine-tuning approach, there could be significant increase in output parameter.

As the application of this paper lies in medical field, there can be serious consequences in case of misdiagnosis. Thus, model should be evaluated on certain grounds. For evaluation of model, we calculated the classification accuracy, classification sensitivity and classification specificity of validation dataset using Scikit and NumPy module of Python.

Area under the receiver operating characteristic (AUC ROC) curve was also drawn between the true positive rate (TPR) and false positive rate (FPR).

$$\text{TPR/Recall/Sensitivity} = \frac{\text{TP}}{\text{TP} + \text{FN}} \tag{1}$$

$$\text{FPR} = 1 - \text{Specificity} = \frac{\text{FP}}{\text{TN} + \text{FP}} \tag{2}$$

$$\text{Specificity} = \frac{\text{TN}}{\text{TN} + \text{FP}} \tag{3}$$

$$\text{Accuracy} = \frac{\text{TN} + \text{TP}}{\text{TN} + \text{FP} + \text{TP} + \text{FN}} \tag{4}$$

Table 2 Values of different parameters for the types of diseases taken into consideration

	Precision	Recall	F1-score	Support
CNV	0.96	1.00	0.98	242
DME	1.00	1.00	1.00	242
DRUSEN	1.00	0.96	0.98	242
NORMAL accuracy			0.99	968
Macro average	0.99	0.99	0.99	968
Weighted average	0.99	0.99	0.99	968

Fig. 2 Confusion matrix

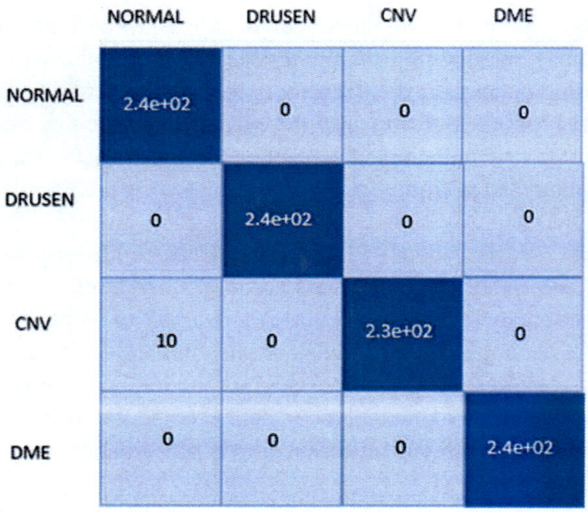

5 Results

Our proposed solution used transfer learning method of deep learning and VGG16 CNN model for classifying OCT images of eyes into CNV, DRUSEN, DME and normal. And further we have used early stop algorithm for finding the stagnant accuracy, and at last, we have done fine-tuning. This presented approach shows high performance with an accuracy of 99% and precision of 98.8% as shown in Table 2 (Fig. 2).

6 Conclusion and Future Scope

In today's scenario, deep learning comes out as an emerging technology in computer vision. Eye diseases like CNV, DME and DRUSEN are commonly found diseases in humans and can be detected with the help of computer vision. In this paper, we have proposed a method using deep learning which is based on VGG16 CNN model which we have used for classification of OCT images into four classes, namely CNV, DRUSEN, DME and normal. Results show significant effectiveness in classification. We have achieved a result with accuracy 99% and precision 98.8%. In the future, this model can be incorporated into software that can be used by clinical experts for screening purposes. Some other important questions that can be addressed in future by medical researchers are that in what other clinical situations OCT technique can be used. Moreover, the exceptional capabilities of OCT imaging envisage the potential to have noteworthy impact on the diagnosis and clinical treatment of many other diseases.

References

1. Kermany, D. S., Goldbaum, M., Cai, W., Valentim, C. C., Liang, H., Baxter, S. L., et al. (2018). Identifying medical diagnoses and treatable diseases by image-based deep learning. *Cell, 172*(5), 1122–1131.
2. Kocur, I., & Resnikoff, S. (2002). Visual impairment and blindness in Europe and their prevention. *British Journal of Ophthalmology, 86*(7), 716–722.
3. Fujimoto, J. G., Pitris, C., Boppart, S. A., & Brezinski, M. E. (2000). Optical coherence tomography: An emerging technology for biomedical imaging and optical biopsy. *Neoplasia, 2*(1–2), 9–25.
4. Li, F., Chen, H., Liu, Z., Zhang, X., & Wu, Z. (2019). Fully automated detection of retinal disorders by image-based deep learning. *Graefe's Archive for Clinical and Experimental Ophthalmology, 257*(3), 495–505.
5. Ferguson, M., Ak, R., Lee, Y. T. T., & Law, K. H. (2017). Automatic localization of casting defects with convolutional neural networks. In *IEEE International Conference on Big Data* (pp. 1726–1735). IEEE.
6. Maheshwari, S., Pachori, R. B., & Acharya, U. R. (2016). Automated diagnosis of glaucoma using empirical wavelet transform and correntropy features extracted from fundus images. *IEEE Journal of Biomedical and Health Informatics, 21*(3), 803–813.
7. Abbas, Q. (2017). Glaucoma-deep: Detection of glaucoma eye disease on retinal fundus images using deep learning. *International Journal of Advanced Computer Science and Applications, 8*(6), 41–45.
8. Gargeya, R., & Leng, T. (2017). Automated identification of diabetic retinopathy using deep learning. *Ophthalmology, 124*(7), 962–969.
9. Awais, M., Müller, H., Tang, T. B., & Meriaudeau, F. (2017). Classification of sd-oct images using a deep learning approach. In *IEEE International Conference on Signal and Image Processing Applications (ICSIPA)* (pp. 489–492). IEEE.
10. Al-Bander, B., Al-Nuaimy, W., Al-Taee, M. A., & Zheng, Y. (2017). Automated glaucoma diagnosis using deep learning approach. In *14th International Multi-Conference on Systems, Signals and Devices (SSD)* (pp. 207–210). IEEE.
11. Malik, S., Kanwal, N., Asghar, M. N., Sadiq, M. A. A., Karamat, I., & Fleury, M.: Data driven approach for eye disease classification with machine learning. *Applied Sciences, 9*(14), 2789.
12. Bajwa, M. N., Malik, M. I., Siddiqui, S. A., Dengel, A., Shafait, F., Neumeier, W., & Ahmed, S. (2019). Two-stage framework for optic disc localization and glaucoma classification in retinal fundus images using deep learning. *BMC Medical Informatics and Decision Making, 19*(1), 1–16.
13. Nazir, T., Irtaza, A., Javed, A., Malik, H., Hussain, D., & Naqvi, R. A.: Retinal image analysis for diabetes-based eye disease detection using deep learning. *Applied Sciences, 10*(18), 6185 (2020)

14. Shanthi, T., & Sabeenian, R. S. (2019). Modified Alexnet architecture for classification of diabetic retinopathy images. *Computers and Electrical Engineering, 76*, 56–64.
15. https://www.kaggle.com/paultimothymooney/kermany2018. Last accessed October 26, 2021.

Text-Based Automatic Personality Recognition: Recent Developments

Sumiya Mushtaq and Neerendra Kumar

Abstract The use of computation in personality recognition has been explored for several decades now. As such, it is possible to derive personality from the data available on social media, telecommunication signals, and every signal obtained from human–machine interaction. Personality computation has been explored in two major domains: social signal processing and human–computer interaction. Automatic personality trait recognition from textual context is an emerging research topic that has gotten considerable attention in the area of natural language processing (NLP). In this survey, we reviewed the existing works in the field of automatic personality detection from texts and provided a comparative analysis. We identified some open research gaps and discussed major issues presented in existing literature, including issues with current datasets, techniques, personality features, and personality models employed, as well as how they can be bettered in the future.

Keywords Personality recognition · Natural language processing · Computational linguistics · Machine learning · Deep learning

1 Introduction

Every facet of a person's life is reflected in his or her personality. Personality is a psychological construct that uses individual qualities to explain a wide range of human behaviors [1]. Automatic personality recognition is an emerging research field. It deals with the identification of a target individual's personality type from

S. Mushtaq (✉) · N. Kumar
Department of Computer Science and Information Technology, Central University of Jammu, Samba, Jammu and Kashmir 181143, India
e-mail: sumiya7.mushtaq@gmail.com

N. Kumar
e-mail: neerendra.csit@cujammu.ac.in

P. K. Singh et al. (eds.), *Proceedings of Third International Conference on Computing, Communications, and Cyber-Security*, Lecture Notes in Networks and Systems 421,
https://doi.org/10.1007/978-981-19-1142-2_43

a variety of sources, including text, audio, video, and social media, using computational approaches. Language psychology demonstrates that psychological variables including emotions, prestige, interpersonal attitudes, reputation, and personality factors influence word choice in addition to meaning [2]. So it makes it possible to predict personality from texts which have recently attracted a lot of interest in the computational linguistics and natural language processing sectors, for example, in Fig. 1, a machine detects a person's personality based on their textual qualities and then assigns personality attributes to them. Advanced machine learning algorithms provide the tools required to assess massive volumes of data from various sources [3–5] and can be utilized to forecast outcomes of personality assessments, and these techniques assess personality in an unobtrusive, accurate, and consistent way. Personality recognition can be used in real-life scenarios [6], including recommendation systems, personalization of production services, employment screenings, social network analysis, and sentiment analysis. Personality detection from texts has risen in popularity in recent years, but more research is needed to maximize the benefit of automated personality recognition, which is still in its infancy. This paper attempts to examine all recent topics for measuring personality using text, as well as their performance evaluation, commonly used datasets, and open research challenges and gaps.

Motivation and Contribution

The significant advancements in the field of text-based personality classification inspired this review, so the authors identified, analyzed, and assessed key works in this subject for this study. We provide an overview of the most often utilized personality models and datasets. We presented an up-to-date review of available text-based personality recognition approaches with their comparative analysis. We also identified the major flaws in present approaches and provided potential fixes.

The remainder of the paper is organized as follows: The personality models are discussed in Sect. 2, the research methodology is provided in Sect. 3, the literature review and comparative analysis are presented in Sect. 4, the open research challenges

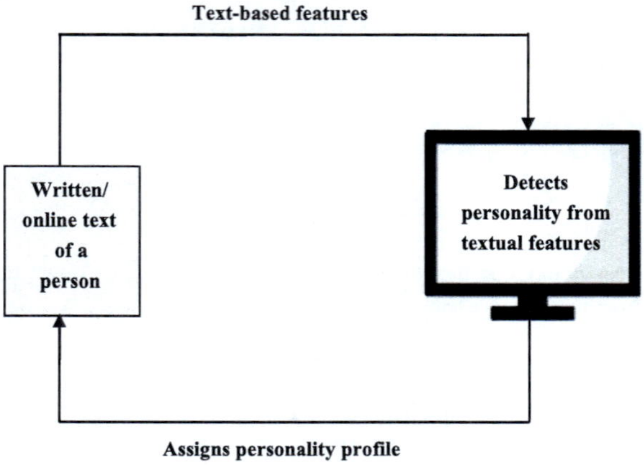

Fig. 1 Automatic personality detection from text

and gaps are discussed in Sect. 5, future trends are covered in discussion section in Sect. 6, and the conclusion is drawn in Sect. 7.

2 Personality Models

The Big-Five model [7] and the Myers–Briggs Type Indicator (MBTI model) [8] are the most widely utilized personality models. According to the Big-Five theory, five major dimensions can be utilized to assess a person's personality. Openness, conscientiousness, extraversion, agreeableness, and neuroticism are the Big-Five dimensions (OCEAN) (Table 1). MBTI creates a topology of the individual based on the combination of four unique functions, such as introversion, intuition, feeling, and perception (INFP) or extraversion, sensing, thinking, and judgment (ESTJ) (Table 2).

Table 1 Dimensions of Big-Five model

Traits	Description
Openness	This group of people is imaginative and innovative
Conscientiousness	This group of people is honest, well organized, and professional
Extraversion	This group of people is active and enjoys social interactions
Agreeableness	This group of people is friendly and polite
Neuroticism	This group of people is frequently anxious and gloomy

Table 2 Dimensions of MBTI model

Personality dimension	Description
Extraversion/introversion	Discusses how people direct their energy in response to their encounters with the world
Sensing/intuition	Discusses how a person collects and processes information from the environment
Thinking/feeling	Discusses the process by which people make decisions depending on the knowledge they have acquired
Judgment/perception	Discusses how people respond to the world, whether it is by managing it or keeping your eyes peeled for incoming knowledge

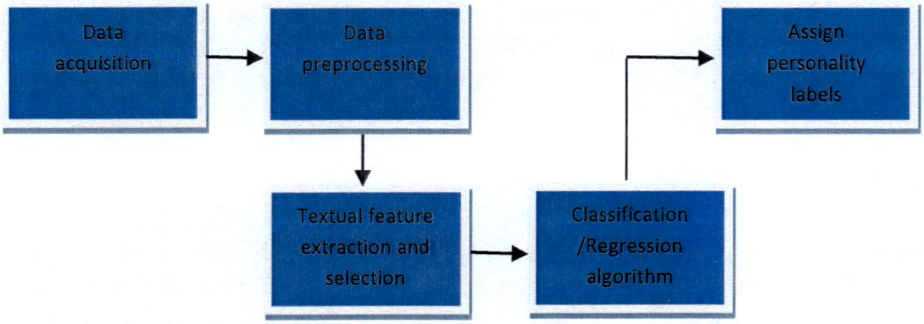

Fig. 2 Methodology for detecting personality from the text

3 Research Methodology

To get at the desired solution to any problem, a methodical strategy must be used. Figure 2 depicts the research methodology used to discern personality from the text. For conducting experiments, the initial step is to collect data or choose a dataset. The preprocessing step is used after the data collection to deal with any redundant or undesired values in the dataset. The next step is to extract text-based features and then choose the relevant ones. Then these selected features are mapped into personality traits using machine/deep learning methods. Finally, a personality class/label is assigned based on the personality model used to the subject present in the dataset.

4 Literature Review

Various deep/machine learning approaches can be used to determine personality in text. This section provides a review of the strategies used to determine personality from the text as well as their comparative analysis.

In 2018, Bharadwaj et al. [9] attempted to generate the personality profile for a person based on their social media text/tweets. Firstly, they perform text preprocessing, and then for classification, they utilized various machine learning approaches (SVM, Naive Bayes (NB), neural net). The results revealed that SVM outperformed the other two methods. In the same year, Chaudhary et al. [10] used the Kaggle dataset to test the performance of various classifiers [NB, SVM, logistic regression (LR), random forest (RF)] in predicting user personality. In terms of accuracy and F-measure, the results revealed that the logistic regression approach is the most effective classifier. In the same year, Xue et al. [11] presented a textual context personality recognition method based on deep learning. They suggested a hierarchical structure based on AttRCNN for this purpose, which can learn deep semantic properties using user posts. The results are promising that demonstrate that the suggested deep semantic attributes surpass the baseline characteristics. In 2019, Yamada et al. [12]

conducted research on Twitter user postings in the Japanese language and found that textual factors in MBTI modeling are more predictive than behavioral features. They also demonstrate the importance of considering user actions when determining users' personalities who do not post on a regular basis. In the same year, Kunte et al. [13] performed experiments to estimate the users' personalities from textual data using a real-time dataset. After preprocessing step, different classifiers such as AdaBoost, multinomial Naive Bayes, and linear discriminate analysis (LDA) algorithms are employed to classify personalities, and the results revealed multinomial Naive Bayes surpasses LDA and AdaBoost. In the same year, Ergu et al. [14] conducted studies using Turkish tweets to analyze personality. They used a variety of machine learning models (KNN, decision tree (DT), RF, AdaBoost, stochastic gradient descent (SGD), gradient boosting (GB), and SVM) to achieve this goal, and when models were trained on users' most recent 50 tweets, they attained accuracies ranging from 0.76 to 0.97. In 2020, Rohit et al. [15] conducted research to determine a user's personality based on their social media profile status information. Then they further categorized the users' personalities into one of the OCEAN model's categories based on the analysis results, and their model achieved good accuracy. In the same year, Mehta et al. [16] developed a unique deep learning-based approach to assess personality from the essays dataset for OCEAN traits and the Kaggle dataset for MBTI using language modeling features in conjunction with conventional psycholinguistic features. For both datasets, the results demonstrated that language model embeddings frequently outperformed traditional psycholinguistic features, and suggested models outperformed state of the art. In the same year, Kazameini et al. [17] created an efficient and robust text-based deep learning model for predicting personality. They used BERT to extract contextualized embeddings and fed these contextualized embeddings, together with psycholinguistic features, to a Bagged SVM classifier, outperforming the state of the art. In the same year, Jayaratne et al. [18] used open-vocabulary natural language processing techniques combined with machine learning to detect a person's personality based on their use of words during a job interview. For this purpose, they employed five alternative text representation approaches to generate regression models for each HEXACO trait. The accuracy of text representation based on terms and topics was the best. They next tested their model on a sample of 117 volunteers, who scored the individual trait descriptors created as a result of the model's outputs using the yes/no/maybe agreement scale, and each of the six personality traits received an average of 87.83% agreement from the participants. In 2021, Wang et al. [19] used text mining techniques to classify people's proactive personalities. Short-answer questions and Weibo text datasets were employed to accomplish this task, and for classification, various machine learning algorithms were used out of which SVM performed better. In the same year, Asghar et al. [20] provided a method for separating psychopath and non-psychopath characteristics in the input text. For this purpose, they utilized BILSTM with a larger dataset to efficiently classify the input into a psychopath and non-psychopath categories. In the same year, Xue et al. [21], created a unique semantic-enhanced personality recognition neural network (SEPRNN) that can identify numerous personality features. They employed context learning-based word-level semantic representation followed by a fully connected layer to acquire text's higher-level semantics, and when compared to different baselines, their suggested strategy improves accuracy significantly. In the same year, Demerdash et al. [22] presented a deep learning approach for personality evaluation using fusion approaches pre-trained language models for transfer learning, and their proposed model outperforms the baseline results. In the same year, Christian et al. [23] proposed a new feature extraction strategy for various social networking data sources based on a multi-model deep learning architecture paired with numerous pre-trained language models such as BERT, RoBERTa, and XLNet, to develop personality prediction systems, and their model achieved good accuracy.

Comparative Analysis
Table 3 provides a comparison of the existing techniques for detecting personality from text, and Table 4 lists some of the most often used datasets for detecting personality from the text.

5 Open Research Challenges

Our review and analysis of many studies have revealed research concerns in text-based personality recognition that are accessible to further exploration by the research community. The following are the various research challenges:

i. **Shared datasets**: There is a scarcity of open-source datasets for the task of automatic personality detection, and this limitation can be overcome by developing new and large shared datasets for personality detection.
ii. **Data integrity**: The most serious issue is the integrity of sample data that is presented as input. If the data's integrity is in question, the outputs cannot be trusted. So, input data should be reliable and derived from natural circumstances.
iii. **Personality models**: Current personality models are built on a fundamental set of personality characteristics; they only cover a limited number of personality traits when assessing personality from social media text. So, increasing the dimensions of personality models can help to solve this problem.
iv. **Human behavior is situational**: Humans behave in different ways depending on the scenario. As a result, while predicting social media behavior, we must include a plethora of additional psychological elements before declaring the prediction results to be reliable.
v. **Ethical issues**: When it comes to determining a person's personality based on their social media context, there are several ethical considerations because the data comes from social media platforms that are deemed extremely private. The most serious problem is the unauthorized use of social media data. Data protection laws can help to handle this issue.
vi. **Fairness issues**: Another important concern is whether automatic personality detection generates fair outcomes and if it works effectively for people

Table 3 Analysis of various text-based personality detection approaches

References	Aim	Dataset used	Techniques used	Results	Limitation/future work
[9]	Creating personality profiles of users based on their social media posts/tweets	MBTI dataset	SVM, NB, Neural net	Accuracy (Acc) up to 88% achieved	Results can be improved by using more cutting-edge methods
[10]	Evaluated performance of various classifiers in predicting Kaggle users personalities	MBTI dataset	SVM, RF, NB, LR	Acc = 65%	Used only classic machine learning methods
[11]	Identify personality traits in user posts using the Big-Five model	MyPersonality dataset	SVR, RF, Multi-layer perceptron (MLP), GB, AttRCNN	Acc = 89%	To incorporate deep semantic features in more frameworks
[12]	Analyze the impact of text-based features on users behavior for predicting personality	Japanese tweets	Linear SVM	AUC_{avg} up to 0.6903	To analyze more textual features for personality detection
[13]	Machine learning-based personality prediction from Twitter	Real-time Twitter dataset	LDA, AdaBoost, multinomial NB	Acc up to 73.43%	To improve the model's accuracy
[14]	Machine learning-based Turkish personality analysis	Turkish Twitter data (51 participants)	SGD, SVM, GB, RF, KNN, DT, AdaBoost	Acc up to 0.978	Includes less number of features and users
[15]	Facebook status text-based personality analysis	MyPersonality, MongoDB (for storing status)	RF classifier and regressor	Acc = 64.25%	Using more reliable data acquisition can enhance performance

(continued)

Table 3 (continued)

References	Aim	Dataset used	Techniques used	Results	Limitation/future work
[16]	Personality trait prediction based on language	Kaggle dataset, Essays dataset	SVM, MLP, LR	Acc_{avg} up to 77.1%	To focus on more enhanced deep network settings
[17]	Deep learning-based personality prediction from text	Essays dataset	BERT, bagged SVM	$Acc_{avg} = 59.03$	Advanced deep learning models can enhance accuracy
[18]	Inferring personality using text-based answers to interview	PredictiveHire[1] FirstInterview (™)	RF regressor	Acc up to 87.83%	The quality of output needs to be improved
[19]	Predict persons proactive personality	Weibo text and short-answer dataset	SVM, XGBoost, KNN, Naive Bayes, LR	Acc up to 0.896	Limited sample size, only focus on proactive personality
[20]	Identifying psychopaths using user-generated textual contexts	Social media data (601 samples)	BILSTM	Acc = 0.85%	Limited dataset size, focus only on English content

(continued)

Table 3 (continued)

References	Aim	Dataset used	Techniques used	Results	Limitation/future work
[21]	To estimate personality traits with multiple labels	YouTube dataset, essays dataset	SEPRNN	Acc up to 78%	To explore more personality models
[22]	Transfer learning-based personality detection	MyPersonality dataset, Essays dataset	BERT, Elmo, ULMFiT	Acc_{avg} up to 73.91%	To include other emotive notions and subjective factors
[23]	Personality detection using diverse social media data sources	MyPersonality dataset, Twitter dataset	Feed-forward neural network, BERT, XLNet, RoBERTa	Acc up to 88.49%	Using larger datasets and different pre-trained models can improve accuracy

Table 4 Commonly used datasets for textual-based personality detection

Dataset	Description
Stream-of-consciousness essay dataset [24]	Comprises 2468 student essays labeled with the writers' Big-Five personality traits
Kaggle MBTI dataset [25]	There are 8675 rows in all, each representing a different user, containing users' latest 50 posts as well as their MBTI personality type. So total records: 422,845
MyPersonality dataset [26]	250 Facebook users, 9917 statuses labeled with Big-Five traits

of different races, genders, and cultures. Due to systematic disparities in access to personalized services or targeted manipulation, any algorithm-based discrimination could amplify socioeconomic inequities.

vii. **Methodological issues**: According to a survey of the current literature, the majority of previous studies attempted to detect personality traits using machine learning approaches with manual feature extraction, which is a time-consuming task. Deep learning is a promising alternative because it automatically extracts features. Deep learning approaches outperform machine learning techniques by successfully capturing hidden representations; therefore, researchers should focus on them in the future.

6 Discussion

This section delves into the specific issues raised in Sect. 4, as well as their future developments.

Detecting personality from text is one of the research areas that demands a lot of attention. Although a significant amount of research has already been accomplished, more work is still needed to improve prediction performance. Thus, different specifications, such as data sources, feature extraction, and methodologies, must be improved. Increasing the dataset size and improving feature extraction can help enhance performance [11]. But when the dataset increases in size, labeling the data becomes expensive and impractical, so utilizing explicit resources and clustering the text, unsupervised learning [27] can be utilized to make personality predictions. Similarly, introducing new features and improved feature extraction techniques can help boost performance [11, 12, 14]. The performance may also be improved by incorporating new and improved approaches like in [10], and the XGBoost method, which has won most Kaggle and other dataset challenges, could help enhance the results even more. More cutting-edge approaches can be incorporated to improve the situation even more [9]. Another unanswered question is which personality model is the most effective. Even though personality models have recently been researched, their significance in text processing has received less attention. Also, existing personality models must be refined to better fit social media users' vocabulary, which includes emojis, slang terminology, and informal language constructs such as brief lyrical verses. Furthermore, personality evolves with time; thus by examining the sample data over a longer period, consistency may be considered. So to enhance the accuracy of the forecast based on the user's data, it is necessary to obtain the user's data over a longer period.

We find out the following issues: (a) Traditionally, personality is assessed by responding to a series of questions in a questionnaire. It takes a long time to manually annotate data. Also, the survey's accuracy in correctly identifying a person's personality remains under doubt. (b) In terms of the personality recognition model, it is critical to increase performance [10]. (c) Existing datasets limit performance in personality trait classification [14, 19]. (d) When the data size and the number of features increase in size, the accuracy of the machine learning algorithm drops [10]. (e) Personality is quantified via continuous scores in psychological personality tests, but current datasets only provide personality scores in manually binned format [16]. (f) The method of evaluating model performance by testing alternative model settings on the entire dataset might cause a model's performance to be overestimated [16]. (g) There are not enough manual fact-checking profiles to train a deep neural network with [11]. For each of these issues, the following solutions have been proposed: (a) It is important to explore automatic labeling and more precise and effective ways of annotating personality traits. (b) More classification techniques, as well as generating unique and more durable features and using improved feature extraction approaches, should be evaluated in order to attain better results. (c) To aid advances in personality identification, researchers should work together to create a standardized dataset. (d) Deep learning could be utilized to maintain the algorithm's accuracy because it can extract latent information automatically with minimum manual engineering. (e) Further studies should seek to employ datasets with continuous personality trait scores. (f) Future research should use a stacked cross-validation strategy to examine various model configurations. (g) The digital application is required to perform a real-time personality check.

7 Conclusion

In perceptual sentiment analysis, detecting user behavior and personality is a difficult and time-consuming task. Researchers are increasingly turning to textual-based personality prediction. Numerous researches on predicting personality from input text have already been undertaken. This paper examines recent advances in recognizing personality from user-generated textual contexts. We presented an up-to-date overview of existing text-based personality recognition methods. We also discussed important concerns with current datasets, personality models, features, and approaches, as well as potential solutions. According to the study, personality from texts can be accurately determined by utilizing a machine or deep learning approach. However, after examining many research papers, the study suggests that deep learning approaches outperform machine learning techniques in terms of performance. As a result, more deep architectures are needed to improve the efficiency of existing systems. Future research should continue to focus on developing deep learning-based models and novel frameworks that can map exceedingly complicated functions and improve the performance of existing systems.

References

1. Corr, P. J., & Matthews, G. (Eds.). (2009). *The Cambridge handbook of personality psychology*. Cambridge University Press.
2. Tausczik, Y. R., & Pennebaker, J. W. (2010). The psychological meaning of words: LIWC and computerized text analysis methods. *Journal of Language and Social Psychology, 29*(1), 24–54.
3. Vinciarelli, A., & Mohammadi, G. (2014). A survey of personality computing. *IEEE Transactions on Affective Computing, 5*(3), 273–291.
4. Stachl, C., Pargent, F., Hilbert, S., Harari, G., Schoedel, R., Vaid, S., Gosling, S., & Bühner, M. (2019). Personality research and assessment in the era of machine learning. *European Journal of Personality, 34*(5), 613–631.
5. Stachl, C., Boyd, R. L., Horstmann, K. T., Khambatta, P., Matz, S. C., & Harari, G. M. (2021). Computational personality assessment. *Personality Science, 2*, 1–22.
6. Mehta, Y., Majumder, N., Gelbukh, A., & Cambria, E. (2020). Recent trends in deep learning based personality detection. *Artificial Intelligence Review, 53*(4), 2313–2339.
7. Mccrae, R. R., & John, O. P. (1992). An introduction to the five-factor model and its applications. *Journal of Personality, 60*(2), 175–532.
8. Myers, I. B., & Myers, P. B. (1995). *Gifts differing: Understanding personality type*. Davies-Black Publishing.
9. Bharadwaj, S., Sridhar, S., Choudhary, R., & Srinath, R. (2018). Persona traits identification based on Myers-Briggs type indicator (MBTI)—A text classification approach. In *2018 International Conference on Advances in Computing, Communications, and Informatics (ICACCI)* (pp. 1076–1082). IEEE.
10. Chaudhary, S., Singh, R., Hasan, S. T., & Kaur, M. I. (2018). Comparative study of different classifiers for Myers-Brigg personality prediction model. *International Research Journal of Engineering and Technology, 5*(5), 1410–1413.
11. Xue, D., Wu, L., Hong, Z., Guo, S., Gao, L., Wu, Z., Zhong, X., & Sun, J. (2018). Deep learning-based personality recognition from text posts of online social networks. *Applied Intelligence, 48*(11), 4232–4246.
12. Yamada, K., Sasano, R., & Takeda, K. (2019). Incorporating textual information on user behavior for personality prediction. In *Proceedings of the 57th Annual. Meeting of the Association for. Computational Linguistics: Student Research Workshop* (pp. 177–182).
13. Kunte, A. V., & Panicker, S. (2019). Using textual data for personality prediction: A machine learning approach. In *2019 4th International Conference on Information Systems and Computer Networks (ISCON)* (pp. 529–533). IEEE.
14. Ergu, İ. (2019). *Twitter Verisi ve Makine Ö ğ renmesi Modelleriyle Ki ş ilik Tahminleme* [Predicting personality with Twitter data and machine learning models]. In *2019 Innovations in Intelligent Systems and Applications Conference (ASYU)* (pp. 1–5). IEEE.
15. Rohit, G. V., Bharadwaj, K.R., Hemanth, R., Pruthvi, B., & Manoj Kumar, M. V. (2020). Machine intelligence-based personality prediction using social profile data. In *2020 3rd International Conference on Smart Systems and Inventive Technology (ICSSIT)* (pp. 1003–1008). IEEE.

16. Mehta, Y., Fatehi, S., Kazameini, A., Stachl, C., Cambria, E., & Eetemadi, S. (2020). Bottom-up and top-down: Predicting personality with psycholinguistic and language model features. In *2020 IEEE International Conference on Data Mining (ICDM)* (pp. 1184–1189). IEEE.
17. Kazameini, A., Fatehi, S., Mehta, Y., Eetemadi, S., & Cambria, E. (2020). *Personality trait detection using bagged SVM over BERT word embedding ensembles*. arXiv preprint arXiv: 2010.01309
18. Jayaratne, M., & Jayatilleke, B. (2020). Predicting personality using answers to open-ended interview questions. *IEEE Access, 8*, 115345–115355. https://doi.org/10.1109/ACCESS.2020.3004002
19. Wang, P., Yan, M., Zhan, X., Tian, M., Si, Y., Sun, Y., Jiao, L., & Wu, X. (2021). Predicting self-reported proactive personality classification with Weibo text and short answer text. *IEEE Access, 9*, 77203–77211.
20. Asghar, J., Akbar, S., Asghar, M. Z., Ahmad, B., Al-Rakhami, M. S., & Gumaei, A. (2021). Detection and classification of psychopathic personality trait from social media text using deep learning model. *Computational and Mathematical. Methods in Medicine, 2021*, 1–10.
21. Xue, X., Feng, J., & Sun, X. (2021). Semantic-enhanced sequential modeling for personality trait recognition from texts. *Applied Intelligence, 51*, 7705–7717.
22. El-Demerdash, K., El-Khoribi, R.A., Ismail Shoman, M.A., & Abdou, S. (2021). Deep learning based fusion strategies for personality prediction. *Egyptian Informatics Journal.*
23. Christian, H., Suhartono, D., Chowanda, A., & Zamli, K. Z. (2021). Text-based personality prediction from multiple social media data sources using pre-trained language model and model averaging. *Journal of Big Data, 8*(1), 1–20.
24. Pennebaker, J. W., & King, L. A. (1999). Linguistic styles: Language use as an individual difference. *Journal of Personality and Social Psychology, 77*(6), 1296–1312.
25. Kaggle MBTI Dataset. https://www.kaggle.com/datasets. Last accessed October 21, 2021.
26. MyPersonality Project. https://sites.google.com/michalkosinski.com/mypersonality. Last accessed October 21, 2021.
27. Yu, J., & Markov, K. (2017). Deep learning-based personality recognition from facebook status updates. In *IEEE 8th International Conference on Awareness Science and Technology (iCAST)*, pp. 68–74. IEEE.

Use of a Precious Commodity—'Time' for Building Skills by Teachers for Online Teaching During Pandemic by Using Decision Tree and SVM Algorithm of Machine Learning

Bharti Khemani, Jewel Sabhani, and Mala Goplani

Abstract The competency to perform a particular task effectively and efficiently is what we call a developed skill. Skills could be of any type: communication, leadership, interpersonal, problem solving, decision making, etc. This crucial period of the pandemic has brought along the threats and challenges and several opportunities with it. A chance to learn something new, think out of the box, be creative, convert our idle time into a quality one, etc. All this has given rise to using our time for some productive purpose. For months, we have been facing this pandemic, and 'Work from Home' is the policy adopted by almost every company, firm, and educational institution. And this has given all the employees working from home an opportunity to put their saved time into something innovative and productive. So, this study has emphasized the usage of time for skill development by teachers of the educational institutions of Mumbai for online teaching during the period of the COVID-19 pandemic through different training programs. This study is based on the primary data that has been collected from the teachers aged from 30 to 60 and above. Also, its results state that the skills which are required by the teachers for their effective teaching–learning process are developed successfully, and the majority of the faculties have improved their technical skills as well, which in turn have enabled them to adopt new and innovative teaching techniques.

Keywords Time · COVID-19 pandemic · Skill development · Online teaching · Decision tree · Machine learning · SVM · Classification

B. Khemani (✉)
A. P. Shah Institute of Technology, Thane, Mumbai, India
e-mail: khemanibharti23@gmail.com

J. Sabhani · M. Goplani
HVPS Ramniranjan Jhunjhunwala College Arts, Science, and Commerce, Mumbai, India

P. K. Singh et al. (eds.), *Proceedings of Third International Conference on Computing, Communications, and Cyber-Security*, Lecture Notes in Networks and Systems 421,
https://doi.org/10.1007/978-981-19-1142-2_44

1 Introduction

The COVID-19 pandemic has transformed traditional living into the online mode in every field, including teaching. It has made us adapt to all the changes and challenges that come across the path of success. This pandemic period has given the teachers of educational institutions several opportunities like learning to work online, engagement of students in virtual classes, making students learn to use different online learning platforms or LMS platforms to have easy access to virtual lectures and study material, students' virtual participation in various activities for their growth, etc. All this requires a teacher to develop new and innovative skills to perform better in the workplace. Also, there are several problems faced by teachers during the online teaching process, such as non-availability of a good Internet connection or electronic gadgets (laptop and computer), difficulty in operating the digital teaching applications (Zoom, Google Meet, etc.), providing online ready to use content to students, etc. These obstacles can be removed by developing required skills by teachers to conduct smooth online classes through faculty development programs (FDPs), refresher and training programs, short-term courses, webinars, workshops, etc. The study has elaborated the use of 'Time' (lockdown period) by teachers to develop their skills required for the online teaching process by attending training sessions or programs conducted by the institutes and different organizations. The study mentions the positive effects that a faculty gains after investing one's time in such training programs. A few of them are video streaming, downloading, video uploading, effective use of different LMS platforms or digital teaching platforms, easy and optimum use of excel, google drive, and Google Docs. It also highlights the interconnection between the skills that a faculty develops in oneself and their applicability into one's profession.

2 Review of Literature

The paper provides suggestions on teaching online courses that would result in more students' engagement and learning [1]. Practical aspects are considered for changing teaching strategies for a better online environment, including pre-preparing students, promoting learning through discussion boards, managing communication, incorporating multimedia, and evaluating the course. The paper highlights students' performance as measured by grade, which is independent of the instruction mode [2]. What is more challenging in research method classes compared to other public administration classes is 'persistence'? In addition to this, participation may be less pressurizing, and perhaps, quality and quantity of interaction be enhanced in online classes. The compelling circumstances of the overall COVID-19 pandemic and the subsequent lockdown made the institutions and the teacher–trainees dependent and involved them in the innovative teaching–learning methods [3]. This paper presents the perspectives of teacher–trainees for the current transition to online teaching–learning methods and the influence or impact of the home environment on it during the lockdown imposed due to the COVID-19 pandemic.

The paper emphasized the impact of the COVID-19 pandemic on various sectors like business organizations, religious and spiritual bodies, educational institutions, functioning of households, etc., which made us decide about the changes we need to bring in our activities to 'survive'; therefore, we often try to find the solution for the problems that are associated with current and future time with the help of 'innovations' [4]. The same has been the case in our formal and informal learning mechanisms during this period of 'lockdown'. This paper analyzes the same through the observational study that involves how our education system has changed and its future success and failures.

3 Objectives

(1) To identify the different types of skills developed by teachers during the pandemic.
(2) To evaluate the no. of hours devoted to developing skills by attending different webinars, FDPs, orientation programs, refresher programs, etc.
(3) To analyze an interconnection between skills developed and teachers' profession (whether those skills can be applied to one's profession).
(4) To examine the positive effects of developing those skills in one's profession regarding promotion, salary hike, enhanced technical knowledge, etc.
(5) To analyze the degree of involvement in learning new skills to cope up with technological advancements.

Hypothesis

i. There is no association between the number of hours spent in a day and the technical glitches faced by the respondents.
ii. There is no relationship between age and the time spent in skills development.

4 Data Analysis

The number of faculties from colleges among women is 66 (66.67%) and that of males is 33 (33.33%), and the total is 99. Likewise, the number of faculties from schools, universities, and other categories like coaching classes among women is 4(75%), 3 (75%), and 2 (40%), respectively, and among males are 1(25%), 1(25%), and 3(60%), respectively. One hundred and thirteen responses were received in total, out of which 38 were males and 75 were females (Table 1).

From Fig. 1, it is clear 64.61% (73) of teachers aged between 30 and 40 years are found to spend the most significant number of hours enhancing their skills. And

Table 1 Number of responses in terms of gender and respondents profile

Gender/respondents profile	Male	Female	Total
College	33 (33.33%)	66 (66.67%)	99 (87.61%)
University	1 (25%)	3 (75%)	4 (3.55%)
School	1 (25%)	4 (75%)	5 (4.42%)
Other (coaching institute/private tuitions)	3 (60%)	2 (40%)	5 (4.42%)
Total	38 (33.62%)	75 (66.38%)	113

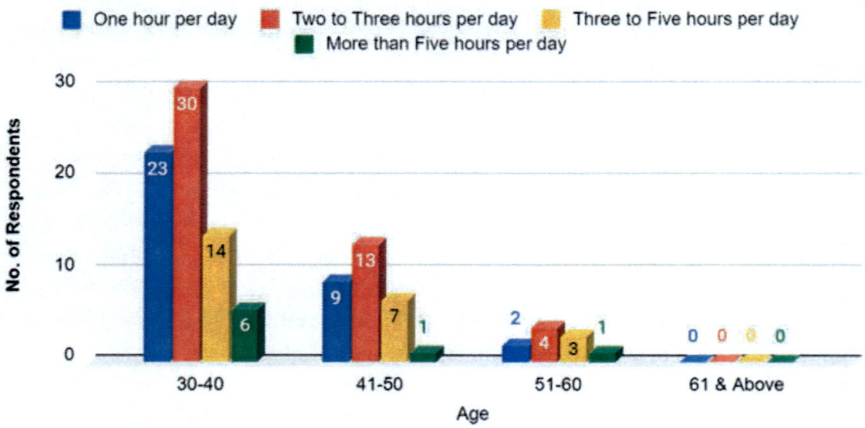

Fig. 1 Number of hours devoted by the respondents in terms of age profile

only 7.96% (10) of teachers aged between 51 and 60 responded that they spend very little time on skill development. From the age group of 30 to 40 years—30 people, from 41 to 50 years—13 people and from the age group of 51 to 60—4 people found to be spending 2–3 h per day which is maximum in all the categories. No one fell in the 61 and above age group criteria, which means that faculties belonging to 61 and above age group do not spend any no. of the hour for enhancing their skills.

From Fig. 2, it is clear that 103 teachers enhanced their skills of being able to engage their students virtually effectively and enhanced their presentation skills in the online classroom as well; 111 teachers developed the skills of using the different digital teaching applications like Zoom, Google meet, etc.; 99 facilitators developed their computer skills. Ninety-three of them developed listening skills, and 85 of them set their communication skills. Therefore, it was noticed that the teachers were found to have successfully enhanced their skills by attending these different training programs.

It is clear from Fig. 3 that the maximum number of hours devoted by the faculties is on webinars. It also observed that the maximum number of faculties used to spend 2–3 h on different programs, viz 40 in FDPs, 47 in webinar, 30 in orientation programs, 19 in refresher programs, 31 in short-term courses, and 34 got in-house

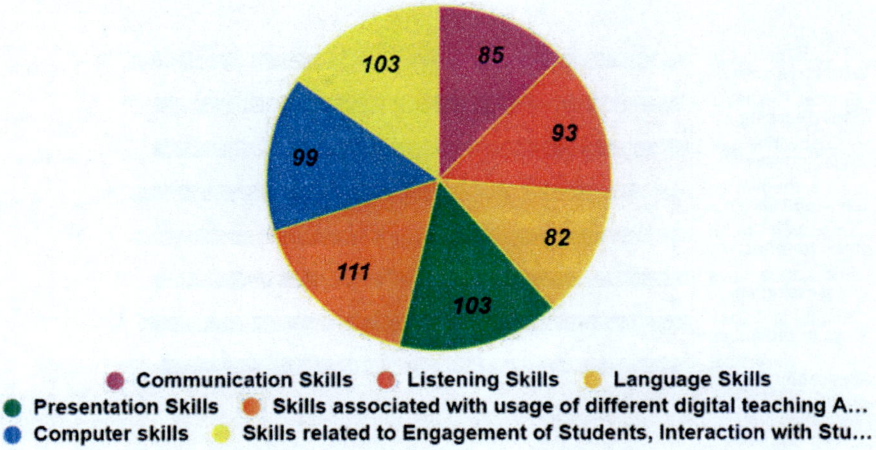

Fig. 2 Skills developed after undergoing refreshers/orientation programs, FDPs, webinars, etc.

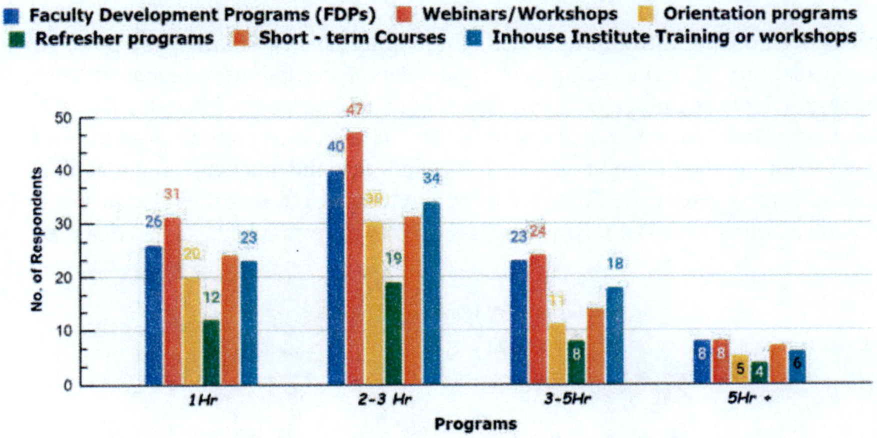

Fig. 3 Number of hours devoted by the respondents to attending different programs

training. However, very few teachers are found to be spending 5 h and above in various training programs.

From Fig. 4, it is clear that 111 faculties, i.e., 98.23%, have improved their technical knowledge, 105 teachers achieved greater flexibility in the teaching–learning process and also learned the techniques of an innovative and engaging way of teaching and assessment; 109 (96.46%) faculties have gained insights for using their innovative teaching–learning tools and materials, while 110 teachers learned about different digital tools for online learning. One hundred and eight faculties improved their skills related to online content development.

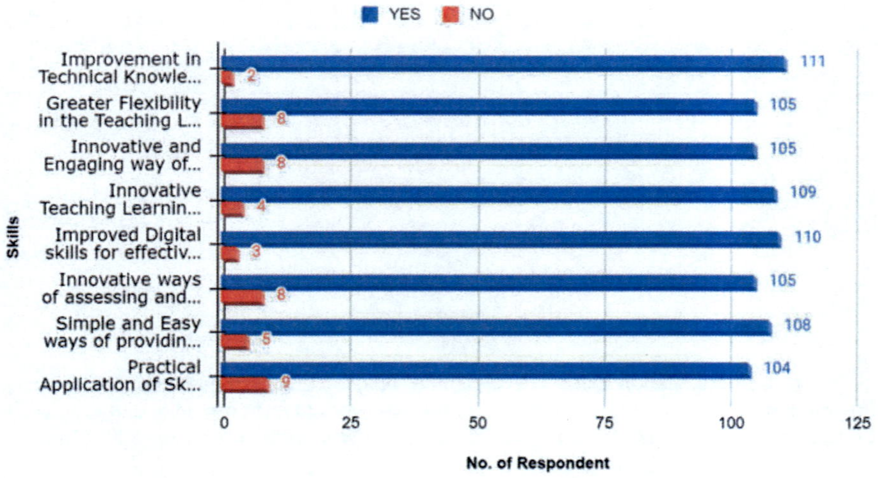

Fig. 4 Positive effects of developing the skills

It has been observed from Fig. 5 that 55 faculties have been facing the problem of the availability of digital equipment (phone/laptop/computer), whereas 38.26% of the respondents do not face any problem related to equipment. Likewise, 76.16% of the respondents have faced the issue of less face-to-face communications and interaction during the online programs. Fifty-eight respondents expressed that their mental health has deteriorated because of online training because they had to spend more time learning such skills. Sixty-eight (60.18%) of respondents agreed that they

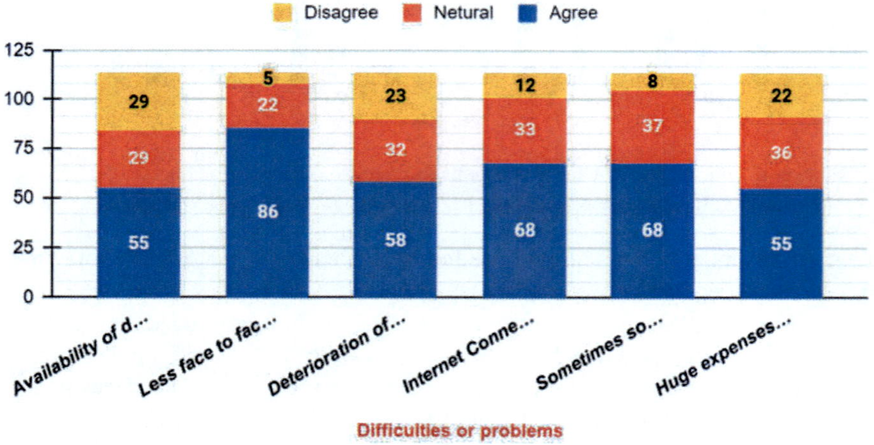

Fig. 5 Difficulties or problems faced by the respondents for developing skills through FDPs, webinars, refresher programs, etc., during COVID-19 pandemic

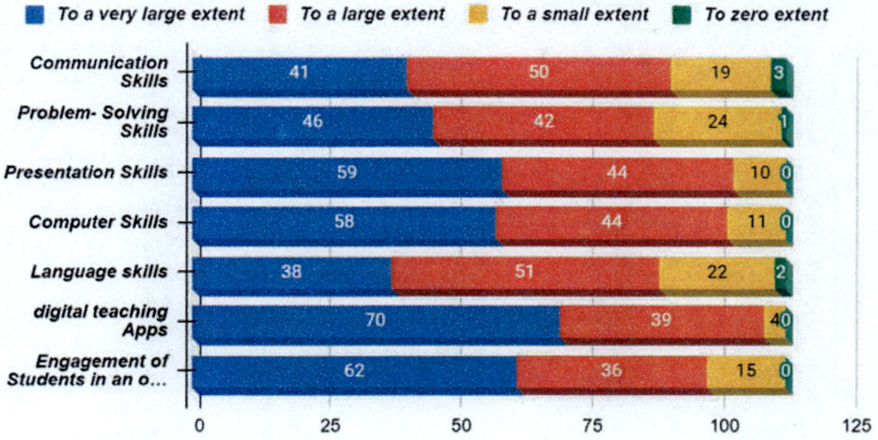

Fig. 6 Extent of applicability of these developed skills in the teaching–learning process

had faced Internet connectivity issues, and 55 agreed that they had spent more on getting a good Internet connection.

It has been noticed from Fig. 6 that a maximum of 70 respondents (61.94%) have agreed to a very large extent that they have developed their skills in digital teaching applications. In contrast, very few, i.e., only 38, agreed to a very large extent that they have developed language skills through these programs. The majority of the respondents, i.e., 50, 42, 44, 44, 51, 39, and 36, believed to a large extent that they developed their communication skills, problem-solving skills, presentation skills, computer skills, language skills, digital teaching applications, and engagement of students in an online class, respectively. However, very few fall under the zero category.

Figure 7 shows that because of the development of required skills during the pandemic, there is a vast and positive impact on different activities conducted during the online teaching process. 93.80% of faculties have started taking quizzes. Seventy-two teachers are engaging their online lectures with the help of gamification methods. Again 95.58% of faculties are taking their online classes with the help of PPTs. 92.03% of teachers are sharing the videos for a better understanding of the topics. Only nine seemed to disagree with this. One hundred and five faculties involve their students in different virtual activities.

It has been seen from Fig. 8 that the majority of the respondents have got positive outcomes by attending the different workshops and training programs for their skill development. One hundred and five respondents agreed that they had learned the techniques of smooth online lecture delivery (Microsoft Teams,

Google Meets, Zoom, etc.) followed by 103 who learned the techniques of innovative and easy assessment and evaluation. For the video editing and compression and video hosting, streaming, and downloading, 78 respondents got a positive outcome. Ninety-nine respondents (87.61%) have learned the techniques of optimum utilization of Google Drive (For Reports, Content, etc.)

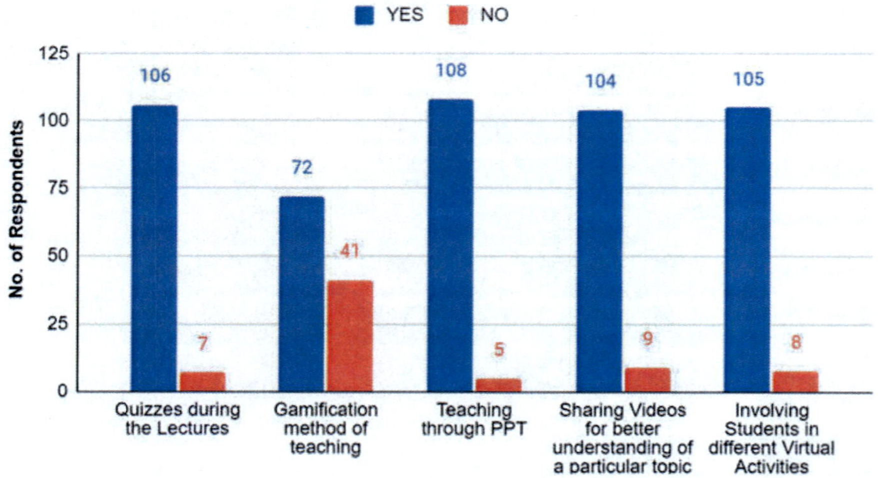

Fig. 7 Activities carried out by faculties for students in the online teaching process

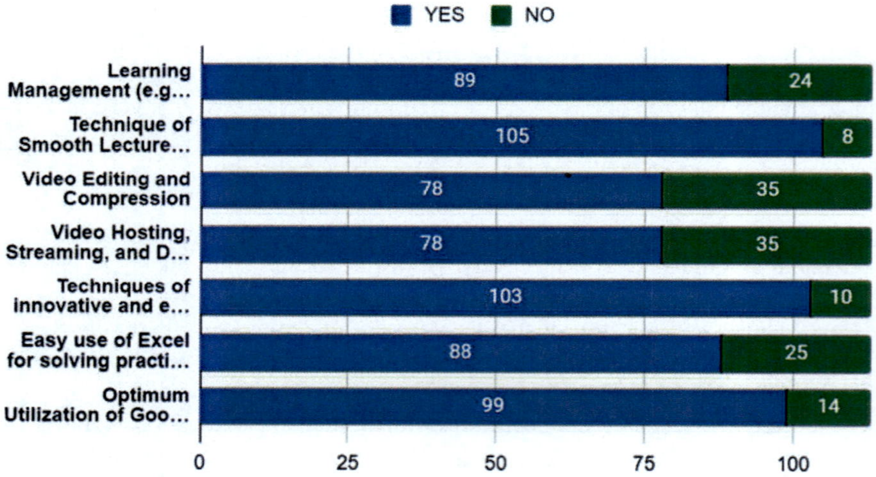

Fig. 8 Highlights of the outcomes from FDPs, webinars, workshops, professional courses, etc.

From Fig. 9, out of 113 respondents, majority, i.e., 72.6% respondents, selected the blended teaching–learning method, which should be adopted to keep a balance between online and offline teaching methods, followed by only 18.6% who agreed to continue with online teaching as it is easy and convenient after learning all new required skills and very few (8.8%) selected the traditional chuck duster method of teaching as they believe it is far better than the online method. It can be predicted easily that currently, the faculties prefer more blended teaching for their teaching–learning process.

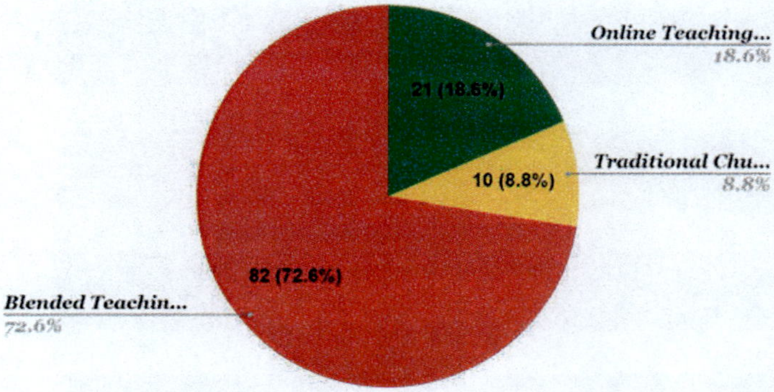

Fig. 9 Most effective and chosen way of the teaching–learning process

5 Classification Algorithm of Machine Learning

This paper uses decision tree (DT) and support vector machine (SVM) algorithms. From Fig. 10, we can see we divided the whole data into two sub-parts, i.e., training and testing, with the help of algorithm, we calculated the accuracy, and DT and SVM have supervised learning algorithms. Both learning methods are used for classification and regression tasks, but we have used both algorithms as classification

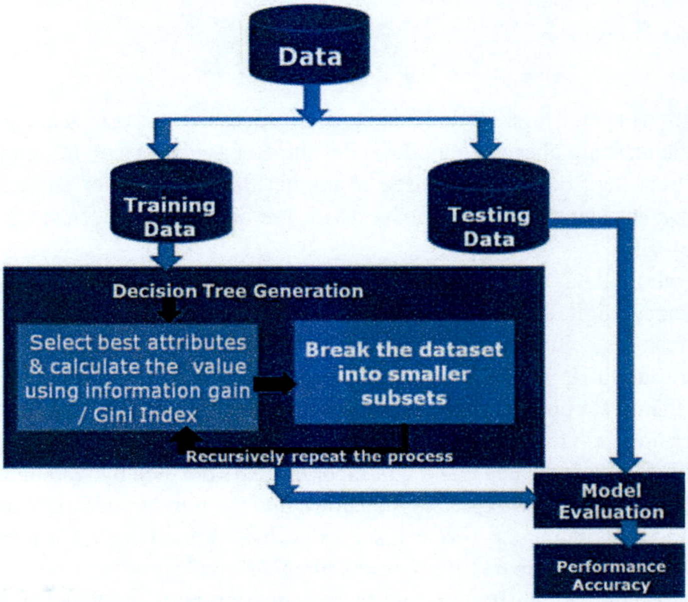

Fig. 10 Flow of classification algorithms

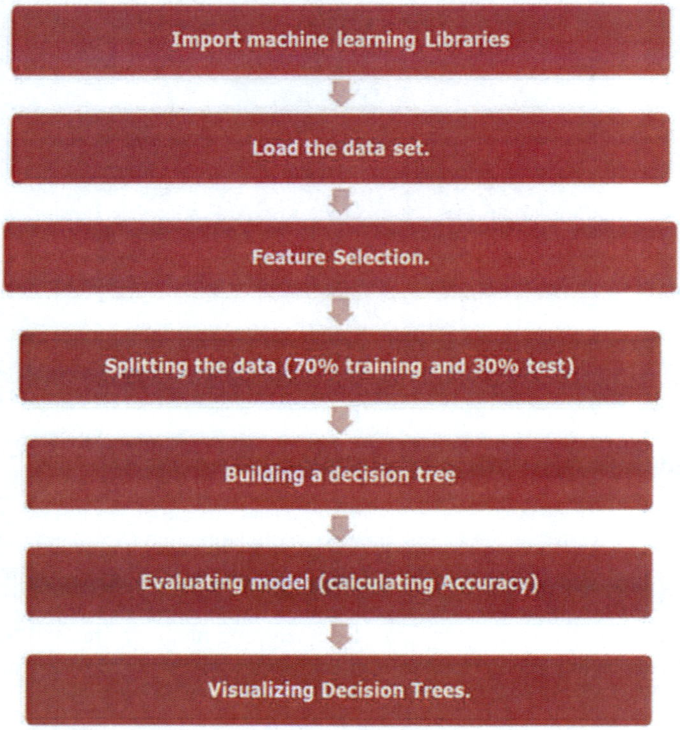

Fig. 11 Steps for an algorithm

algorithms in this paper. Classification is a two-step process: learning step and prediction step. In the learning phase, the model is developed based on given training data. In the prediction step, the model is used to predict the response for shared data. Figure 11 shows the steps of algorithm. We divide our dataset into seven steps to get an output (Fig. 12).

From our whole data, we have considered some columns such as age, learning programs attended, digital devices used, personal benefits, problem faced, and online platforms as features columns (input) and numbers of hours devoted as a label field (output). And found that we are getting 76.47% accuracy with both the algorithms. For splitting, the most popular criteria are 'Gini' for the Gini impurity and 'entropy' for the information gain that can be expressed mathematically as $E(s) = \sum -$ pi log 2 pi. In Fig. 13, class 1 indicates that the number of hours devoted by respondents is one to three hours per day, whereas class 2 shows that the number of hours devoted is more than three hours per day. We can see from Fig. 13 that if age is between 30 and 45, people attended more no. of programs like FDP/workshops and many more. They have good Internet connectivity (not facing any problem), and then the person belongs to the class 1 category. On the other hand, we also noticed from Fig. 13 that if the age is between 30 and 45, people attended more no. of programs like

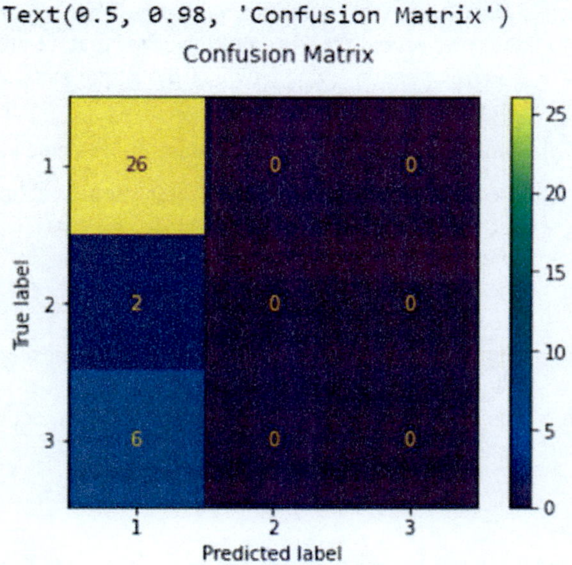

Fig. 12 Confusion matrix of decision tree and SVM

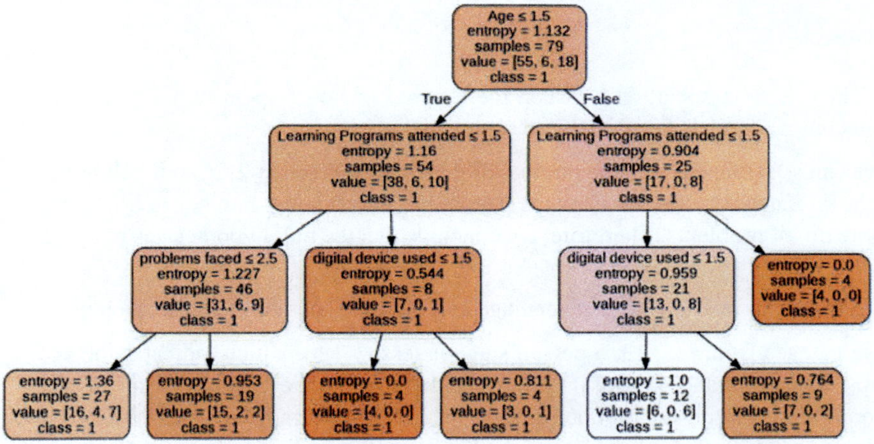

Fig. 13 Output of the decision tree algorithm based on the hypothesis defined in Sect. 3

FDP/workshops and many more. And if they have digital devices available, then the person belongs to the class 1 category.

Confusion matrix is used to know the performance of a machine learning classification. It is represented in a matrix form.

FN: The false negative value for a class will be the sum of values of corresponding rows except for the TP value.

FP: The false positive value for a class will be the sum of values of the corresponding column except for the TP value.

TN: The true negative value for a class will be the sum of values of all columns and rows except the values of that class that we are calculating the values for.

TP: The true positive value is where the actual value and predicted value are the same.

The confusion matrix for the IRIS dataset is as below:

TP: The actual value and predicted value should be the same. So class1, the value of cell 1, i.e., 26 is the TP value. It says that 26 people have invested their time for learning different programs.

FN: The sum of values of corresponding rows except the TP value FN = (cell 2 + cell3) = (0 + 0) = 0.

FP: The sum of values of corresponding column except the TP value. FP = (cell 4 + cell 7) = (2 + 6) = 8.

TN: The sum of values of all columns and row except the values of that class that we are calculating the values for. TN = (cell 5 + cell 6 + cell 8 + cell 9) = 0 + 0 + 0 + 0 = 0.

6 Conclusion of Hypothesis Based on This Decision Tree

Hypothesis

1. There is no association between the number of hours spent in a day and the technical glitches faced by the respondents.

As we can see from Fig. 13, the number of problems is more ($\leq$2.5), so it belongs to the class 1 category, which means the person has devoted less number of hours due to more no. of problems. Therefore, we conclude that the null hypothesis is rejected.

2. There is no relationship between age and the time spent in skills development.

As we can see from Fig. 13, a person's age is less ($\leq$1.5), which indicates the age group between 30 and 45. So, it belongs to the class 1 category, which means the respondents belonging to the young age group have devoted more hours to developing their new skills. Therefore, we conclude that the null hypothesis is rejected.

Based on different features (age, online platform, problems faced, no. of samples considered) of the data set, entropy is calculated, and output (no. hours devoted) is classified as 1 and 2.

7 Findings

1. Most teachers learned how to use digital learning applications by attending different skills development programs during the pandemic that they had not used previously.

2. Most of the teachers became technical experts in different fields.
3. Skills related to engagement of students, interaction with students, etc., by using different tools like gamification sharing videos and taking quizzes in between are developed through these programs.
4. Teachers learned new innovative ways of assessing and evaluating students' performance.
5. Most of the teachers have faced the problem in online teaching for the students' involvement and face-to-face communication.
6. The use of Excel has increased many folds as for the practical subjects, and teachers have started teaching and solving their practical questions in excel.
7. Teachers have also developed their skills of keeping all the essential information in Google Drive.
8. Most of the teachers have chosen the blended teaching method for their teaching–learning process.
9. There is a strong association between the age factor and the number of hours devoted by the faculties on various skill development programs.
10. There is a strong association between the technical glitches and the number of hours devoted by the faculties to various skill development programs.

8 Suggestions

1. Colleges or institutes should give time-to-time training their faculties and students to develop such skills to enhance their teaching–learning process smoothly.
2. Some teachers face problems when it comes to technology. For faculties, detailed training should be provided by the institute itself.
3. The study is restricted to the persons who are teachers by their profession and residing in Mumbai Suburban only.
4. The age group below 30 is also to be considered for this study.
5. The data is fetched from a small sample population, so the results may not be applied to the whole population.
6. Teachers should get resources like laptops/computers from the college to become more efficient in using ICT tools.
7. When teachers come up with unique ideas for the students' engagement and development, then college should appreciate their efforts in the different forms that will motivate faculties to add on more skills.
8. Whatever expenditures are incurred by the teachers for attending these workshops or programs, the colleges should reimburse training amounts to their faculty members.
9. The accuracy of the data can be improved if the size of the data set increases.

9 Conclusion

'Online' is something that lies at the core of everything we do during this pandemic. Educational institutions have also adopted online education as a mode of imparting education to children. This requires a significant number of efforts and skills by the teachers to maintain quality education. This study reveals that the teachers have invested their precious time for their skill development by attending various webinars, workshops, FDPs, orientation/refresher programs, etc. It is noticed that the faculties have developed the number of skills that are required in teaching–learning process, viz communication skills, language skills, listening skills, presentation skills, skills associated with the usage of different digital teaching apps (Zoom, Google Meet), computer skills, skills related to virtual engagement of students, interaction with students, etc. The age group of 30–40 years has been spending two to three hours per day developing such skills by means of webinars in the majority, followed by FDPs. It is also observed that the number of faculties from colleges participating in such skill development programs among women is relatively higher than that of males. After devoting a good number of hours, faculties have developed the skills of using different digital teaching apps (Zoom, Google Meet), presentation skills, virtual engagement of students, and interaction with students the most. And these skills are found to apply in the teaching–learning process to a very large extent. The positive effects of developing such skills are improvement in technical knowledge, greater flexibility in the teaching–learning process, and usage of innovative teaching–learning tools and materials that have become easy. It has also been noticed that even after devoting a great amount of time to developing these skills, the majority of the faculties are not paid the required salary, not given any salary hikes, nor are they promoted. The faculties have been facing difficulties while undergoing such programs as unavailability of digital equipment, low Internet connectivity issues, deterioration of mental health due to overtime, etc. Despite facing many such problems, the faculties are found to be devoting their precious time to developing the required skills.

Apart from these difficulties, most of the faculties have gained more insights into the technology during this pandemic and have learned many innovative teaching techniques. Thus, the majority of them (72.6% respondents) chose to continue with blended teaching methods as their way of teaching in the future as well (post-pandemic).

References

1. Clark-Ibáñez, M., & Scott, L. (2008). Learning to teach online. *Teaching Sociology, 36*(1), 34–41. Retrieved June 14, 2021, from http://www.jstor.org/stable/20058625
2. Ni, A. (2013). Comparing the effectiveness of classroom and online learning: Teaching research methods. *Journal of Public Affairs Education, 19*(2), 199–215. Retrieved June 14, 2021, from http://www.jstor.org/stable/23608947
3. Aisha, N., & Ratra, A. (2020). Influence of home-environment on online teaching-learning during COVID-19 pandemic lockdown among teacher-trainees. *Global Journal of Enterprise*

Information System, 12(4), 33–46. https://ezproxy.svkm.ac.in:2152/10.18311/gjeis/2020

4. Kumar, A., & Pathak, P. (2020). The pros and cons of virtual learning in India: An insight during 'Covid lockdown'. *Adhyayan: A Journal of Management Sciences, 10*(1), 8–13. https://ezproxy.svkm.ac.in:2152/10.21567/adhyayan.v10i1.2
5. Qin, B., Xia, Y., & Li, F. (2009). Dtu: A decision tree for uncertain data. In T. Theeramunkong, B. Kijsirikul, N. Cercone, & T.-B. Ho (Eds.), *PAKDD 2009*. LNCS (Vol. 5476, pp. 4–15). Springer.
6. Uma Pavan Kumar, K., Gandhi, O., Venkata Reddy, M., & Srinivasu, S. V. N. (2021) Usage of KNN, decision tree and random forest algorithms in machine learning and performance analysis with a comparative measure. In *Advances in intelligent systems and computing book series (AISC)* (Vol. 1280). Springer.
7. Dua, D., & Graff, C. (2017). *UCI machine learning repository*. Retrieved February 18, 2019, from http://archive.ics.uci.edu/ml
8. Hehn, T. M., & Hamprecht, F. A. (2018). End-to-end learning of deterministic decision trees. In *German Conference on Pattern Recognition* (pp. 612–627). Springer.
9. Kushwaha, S., Bahl, S., Bagha, A. K., Parmar, K. S., Javaid, M., Haleem, A., & Singh, R. P. (2020). Significant applications of machine learning for covid-19 pandemic. *Journal of Industrial Integration and Management, 5*(4).
10. Lalmuanawma, S., Hussain, J., & Chhakchhuak, L. (2020). Applications of machine learning and artificial intelligence for covid-19 (sars-cov-2) pandemic: A review. *Chaos, Solitons and Fractals*, 110059.
11. Zhu, Q. S., & Cheng, K. (2016). SVM decision-tree multi-classification strategy based on genetic algorithm with cumulative fitness. *Computer Application Research, 33*(291), 64–67.
12. Ayyoubzadeh, S. M., Ayyoubzadeh, S. M., Zahedi, H., et al. (2020). Predicting COVID-19 incidence through analysis of Google trends data in Iran: Data mining and deep learning pilot study. *JMIR Public Health Surveill, 6*(2), e18828.
13. Haruna, A. A., Muhammad, L. J., Yahaya, B. Z., et al. (2019). An improved C4.5 data mining driven algorithm for the diagnosis of coronary artery disease. In *International Conference on Digitization (ICD)* (pp. 48–52). Sharjah, United Arab Emirates.
14. Jebara, T. (2003). *Machine learning: Discriminative and generative*. Springer.
15. Mitchell, T. (1997). *Machine learning*. McGraw Hill. 0-07-042807-7
16. Rustam, F., et al. (2020). COVID-19 future forecasting using supervised machine learning models. *IEEE Access*. https://doi.org/10.1109/access.2020.2997311

Road Lane Line Detection Based on ROI Using Hough Transform Algorithm

Mohammad Haider Syed and Santosh Kumar

Abstract Now-a-days technology has become the means of survival. Automotive Sector is also affected by this technology growth. Driver safety is one of the most important concern for the automobile industry. Lack of attention causes the road accidents and may endanger the driver and co-passenger lives at risk. The stats presented by WHO on road accidents shows that approximately 1.35 million people dies annually as a result of the car accidents. And about 20–50 million peoples suffer from non-fatal injuries, but they may cause lifetime disabilities. These road crashes also impact the economy of the countries. Most of the countries suffers 3% of their GDP due to road accidents. The major challenge is to make the technology available in the commercial sector. So various methods and algorithms are introduced to achieve better performance and robustness. One of the major components of autonomous vehicles are road lane detection. Marking the region of interest (ROI) in which car should be driven. Recent advancement in the technology like image processing and deep learning helps in achieving the aim to detect road lane lines. Autonomous cars are now equipped with cameras, radar and LIDAR for tracking roads and track environment. In this paper, road lane line detection problem has been addressed using Open CV library; also, an approach for finding an efficient way for detecting road lanes precisely and more accurately has been proposed. The road images captured by the camera mounted on the vehicle is processed and region of interest is masked. After masking the ROI, it is converted into a pixel matrix using NumPy library. The Hough Transform is applied on the matrix and lanes are detected in between which vehicle runs.

Keywords Region of interest · Canny edge detection · Grayscale conversion · Hough transform · Computer vision · Lane detection

---

M. H. Syed
College of Computing and Informatics, Saudi Electronic University, Riyadh, Saudi Arabia
e-mail: m.haider@seu.edu.sa

S. Kumar (✉)
Department of Computer Science and Engineering, ABES Engineering College, Ghaziabad, Uttar Pradesh, India
e-mail: santoshg25@gmail.com

P. K. Singh et al. (eds.), *Proceedings of Third International Conference on Computing, Communications, and Cyber-Security*, Lecture Notes in Networks and Systems 421,
https://doi.org/10.1007/978-981-19-1142-2_45

1 Introduction

The increasing number of vehicles on road has gradually increased the traffic problems and also surged the road accidents stats. One of the common road accident causes are driving ruthlessly without being in lane and thus hitting the cars while overtaking. These accidents often occurs because the overtaking driver have a blind spot on the overtaking side as he/she is not aware of vehicle ahead of the preceding vehicle [1]. So, an effort has been made to demonstrate lane detection using OpenCV library. In recent years, with the advancement in the technical field and the availability of various equipment's like digital camera at very low cost and the advancement in the computer vision technology makes it feasible for the commercial use of the proposed system. The feasibility of any road detection system depends on the fact that how good the road quality is or on the road texture so that lanes can easily be identified from the road pavements. So, in both developed and develop countries the researchers are motivated to emphasize their research on finding new ideas and technology like detection with image processing or using the emerging machine learning technology. Since, either detecting road lanes or say boundaries both are critical and difficult task [2, 3]. This includes both estimating the road lanes and the distance of road from the vehicle. Firstly, the research was based on the intensity of colored image as proposed in the paper [4]. As per the above problems, the solution is given using the latest Python OpenCV module which provides the feature and various function for processing the image. Computer vision helps one to detect the surrounding using the camera mounted on car. It is a branch of artificial intelligence that enables software to understand the content of image and video [5]. Back then computer vision methodology was not as much efficient but the recent advancement in artificial intelligence and machine learning has made it an efficient and easy method for detecting road lanes [6]. Many papers [7] based on Hough Transform used to detect road lane by using them with hyperbola-pair model.

Increasing traffic and problems related to it like accidents are becoming more and more serious in most countries. The most common type of accident is due to side impact collision. These accidents mostly occur when vehicle changes the lanes gradually or overtakes the preceding vehicles. In recent years, with the introduction of artificial intelligence and machine learning, and the advanced image capturing tools which are available at very low cost makes the way of the technology in commercial market. The feasibility of automatic road lane detection approaches depends on the factors that the textures of lanes can be identified separately from the background of the pavement surface. In both developed and developing countries the researcher is motivated to do research in this field and explore new ways and methodology for road lane detection. Instead, detection of road lanes or road boundaries is very difficult and rigorous work. As the road detection includes finding the road lanes and the distance between vehicle and road.

This paper presents an approach based on computer vision technology which helps in providing the real-time performance in estimating the road lane lines. The road lanes in the proposed system is detected in multiple steps which includes Grayscale conversion, Edge detection, Marking ROI and at last applying Hough Transform. Hough Transform is a commonly used method for the detection of straight road lane. The reason for using Hough Transform is that it is robust in nature and for detecting the required ROI efficiently even when the image contains lot of noise. The vehicle embedded with the system will move on the marked lines either on straight roads or curvy roads.

2 Related Work

The road lane detection mechanisms have been an interesting field for the researchers. Researchers are finding new and enhanced ways to perform the detection mechanism. The related previous work had been happened in this field. As In [8], for any autonomous vehicle the basic and prime attribute is lane line tracking or lane line detection or lane line warning system. Recent research based on machine vision and image processing categorizes the methods in two different categories, i.e., Image Handling and Acceptable Partition. The paper basically discusses LDWS on three aspects (i) Lane Departure Warning in intelligent Assistant Driving System. (ii) Status of road lanes based on the image processing and computer vision. (iii) Status of road lanes based on Semantic Segmentation Network. In [9], for the last few decades technologies like 3-D Light Detection and Ranging (LIDAR) has been used for road lane detection. The paper proposes new technology which detects road lanes using terrestrial LIDAR data. It includes two phases. First phase includes four steps (i) Converting LIDAR data into intensity range, (ii) Detection of road pixels, (iii) Reconstruction of LIDAR data points, (iv) Component Study. System's second phase consist of the procedures which detects center line and boundary line of the roads by making vertical grids on the exposed surface. The research paper is based on topographic LIDAR data recorded using highly efficient LIDAR laser scanner. The advantage of the proposed research paper are time efficiency and easy implementation. In [10], research proposed road lane detection method based on color intensity segregation. It consists of two steps. In First steps pavements or lanes details in the image frame are utilized to extract region of interest. In Second step there is a filter that uses the color intensity information along with illumination variations, shadows, and scattered backgrounds from extracted region of interest and hence vehicles are detected. This method is based on color intensity differences between vehicles and surrounding. The advantage is that it can output result in various weather conditions and screen resolution. The algorithm proposed is implemented using C++ and NVIDIA GPU (real-time ADS). In [6], several technologies came in the recent years to detect road lanes, but they failed to produce high efficiency accuracy. The research paper proposed machine learning techniques to avoid accidents ensuring the driver safety. The research paper proposes the method which uses convolution neural networks with line detection (CNN-LD) and is tested with and without any data intensifying operation. Under this process metrics evaluation such as accuracy, precision, recall and F-measure are computed, and normalization is done to in order to achieve the best results. It is beneficial for achieving both accuracy and performance when compared to other introduced methods. In [1], article proposing an advanced driver assistance system which provides detailed information of nearby lanes and approaching side vehicles. Lateral vehicles are identified based on length, width, height, and time using lane-based transformations. And recurrent functional NeuroFuzzy networks calculates the distance of vehicles. The region of interest in this paper is marked as green. When the lateral vehicles enter the region of interest it turns to red. In [11], proposed framework based on the computer vision serving road analysis. This is further extended to two sub-processes. First is to recognize road signs. And second is to do lane analysis. It is effective as it has minimal complexity used in real-time. The results are measured on two aspects, (i) Classifying Road Signs and (ii) Lane Analysis and Vehicle Detection. The complexity of the proposed method is $O(n2)$. The MAGMA GPUs if used can optimize the algorithm by reducing time and computational expenses. In [12], method which identify the road lanes by the data from the mobile phone's GPS. This consist of three steps. (i) Identifying nodes and dividing the network to segments. (ii) Dynamic time warping criteria computes dissimilarity matrix which identifies central line. (iii) And at last lanes are identified using the Gaussian mixture method. The method proposed by the authors enables the data fusion from different multiple sources much accurately and estimating state of traffic at a very low cost and much precisely. In [13], the previous research could only be able to detect lanes and vehicles separately. But this paper proposes the approach to fetch information for lane assistant system. In this methodology three cameras are used, two to the right and left side mirrors. Here the vehicles are detected using horizontal edge and Otsu's thresholding while the vertical edge is used to verify the vehicles. Then the track of detected vehicle is estimated using Kalman filter. For lane detection ED lines algorithm are used. The outcomes support advanced lanced lane changing and warning system. In [14], contains 3-D approach for road lanes detection. In this approach to detect lanes line on road B-spline model. The proposed B-spline approach uses Image District Extraction for dividing road image converted into pixels to small Region of Interest (ROI), and then using Hough Transform and Least Squares Method to make lines in the given ROI. Now the road width of image and road are compared by Similarity Transformation Principle and point clouds are generated. This approach gives information for vehicle vision and laser radar, which can also estimate height variation of road surface that can be further used for Active Suspension improving the passenger's comfort. By implementing this model, the accuracy of measuring visual distance also increased. In [2], it focuses on the comparison of different approaches (i) Multilayer Perceptron (ii) Illumination Invariance. And a new approach is also introduced which combines the above-mentioned methods achieving the accuracy and robustness in single method. The basis of comparison of two algorithms are (i) Average time per Image (ii) Minimum Execution Time and (iii) Maximum Execution Time. The robustness desired is achieved by combining these two algorithms while for accuracy IInd algorithm is supported with MLP.

Depending on the problems faced in detecting objects by self-driving vehicles an effort has been made to show road lane detection using Python's OpenCV library.

Computer vision is a mechanism that can help the system embedded with it to make sense of the objects around its surrounding. It is a field in artificial intelligence that help software to detect the information in image and video. Recently advancement in deep learning has make the computer vision a long run technology that one can rely on, enabling the systems to identify the objects present in images and retrieving the needed information. This project shows an approach that is purely based on computer vision mechanism and is able to meet the real-time accomplishment for finding road lanes and tracing the road boundaries with little bit of curvature and low light or shadowy environment. At first the Hough Transform was being used along with hyperbola-pair fit model in order to correctly assess the road lane. As Hough Transform is most common used method for the identification of the object from images. The major advantage of using the Hough Transform is its robust nature and its efficiency to assess the road lanes even when the image consists of the noise. The vehicle embedded with the system is allowed to be driven through straight road or roads with slow curves instead of sharp turns.

3 Proposed Architecture

First, camera captures a test image which is then processed to grayscale. The gray images are masked and canny edge detector, detects the edges then Hough transformation is applied to get the final image for processing. The architecture is shown in Fig. 1.

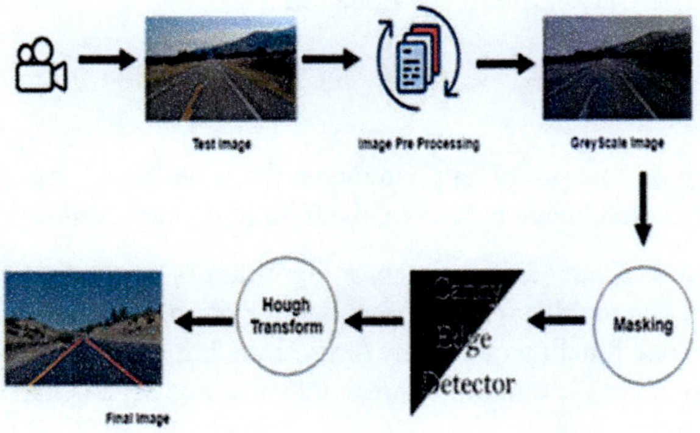

Fig. 1 Architecture of lane detection system

Table 1 Feature set in the dataset

Location	Weather	Light condition	Highway quality

4 Proposed Architecture

4.1 Data Description

For better and accurate results and in calculating the efficiency of the proposed system the data sets are taken at different locations and in various weather condition. The dataset also consists of both high and low light circumstances, i.e., dataset is consider for both day time and night time. The images and videos for the dataset is captured using CANON 1300 D and NIKON 32,100 D on national highway of Ghaziabad, Delhi, Agra, Firozabad, Gorakhpur, Kushinagar, Chandigarh and Mumbai (Table 1).

4.2 Process Flow Diagram

Step-1: Firstly, import all the necessary Python libraries required by the system for implementation.

Step-2: Since dataset contains pictures in RGB format so we needed to convert it into grayscale image. Grayscaling of image can be done in two ways but the grayscaling method we used Weighted Method or Luminosity Method.

The Grayascale Image Equation is

$$((0.3 * R) + (0.59 * G) + (0.11 * B))$$

Step-3: Instead of applying box filter we have applied Gaussian filter on the grayscaled image to remove the high frequency components.

Step-4: Canny Edge detection algorithm is applied on the resulted image. Canny edge is a multi-stage algorithm but in Python's OpenCV module all the functionality is in one function cv2.canny (arg1, arg2, arg3). It consists of three arguments arg1 is image, arg2 is minimum threshold value and arg2 is maximum threshold value.

$$\text{Edge\_ Gradient}\,(G) = |G_x| + \left|G_y\right|$$

Step-5: Now, the main part is to come, i.e., Masking the Region of Interest. ROI refers to the area in which we have to perform the line detection algorithm.

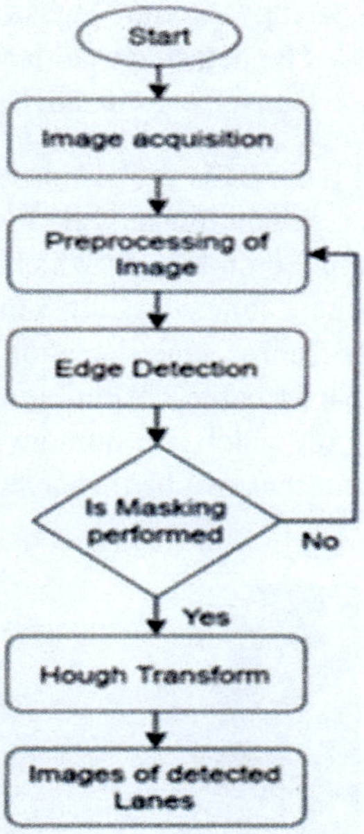

Fig. 2 Flow diagram for detecting road lane lines

Step-6: Once the ROI is marked Hough Transform algorithm is applied for line detection. Once the lines are detected it is displayed on the screen. These steps are shown in Fig. 2.

5 Experimentation and Result

The road detection system is validated on the dataset of 8 different cities and in different weather condition. The procedure followed is as follows.

5.1 Procedure

1. Importing Python Libraries and setting up the environments that are used in the system. These Python libraries are NumPy, OpenCV, MatplotLib, MoviePy.
2. Converting the Video Files into images.
3. The colored images are converted to Grayscale images.
4. Applying Gaussian filter for reducing the noise in the grayscale image.
5. Detecting the edges by applying Canny Edge Detection algorithm.
6. Masking the ROI from the images.
7. Converting the image to co-ordinates using NumPy library. It converts the image into co-ordinates on the basis of the image pixels.

8. Apply Hough Transform on Edge Detected co-ordinates.
9. The detected road lanes will be the output of the Step 8.
10. Converting the single image to video files using MoviePy.

Figure 3a is the testing image on which grayscale operation will be applied and the resultant image will be shown in Fig. 3b. Then Canny Edge detection algorithm is applied on Fig. 3b which will detect all the edges of objects in the image and the resultant image is shown in Fig. 3c. We will mark the ROI on the image obtained after Canny Edge detection algorithm and the resultant image is shown in Fig. 3d. The ROI is the desired area on which the Hough Transform is applied so that the lane on which autonomous vehicle should drive, i.e., shown in Fig. 3e. The visual output that will be displayed on the screen is shown in Fig. 3f.

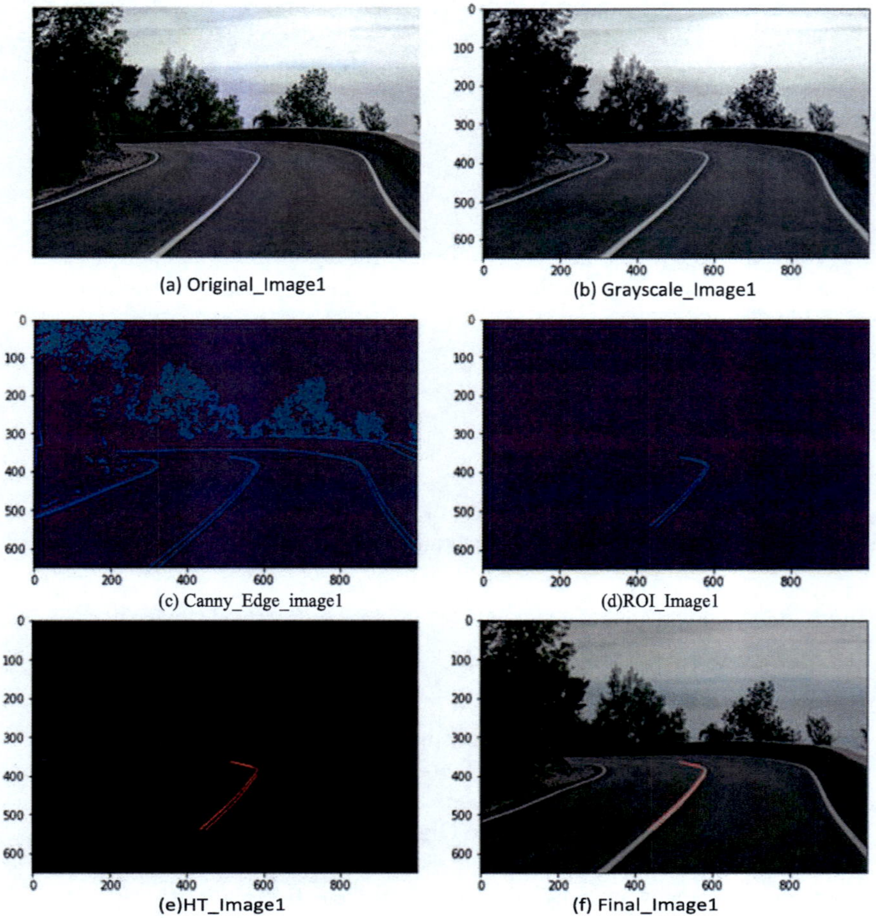

Fig. 3 Road lane detection in hilly areas

The above results are performed on the hilly roads of Mumbai-Lonavla highway which mostly have curvature in the roads and the different steps involved in the proposed system mechanism are shown above. Above dataset is taken during evening time on a normal day.

The above procedure is also performed on the different data set shown in Fig. 4a and the Fig. 4b–f shows the respective operation performed on the dataset that are performed on Fig. 3a.

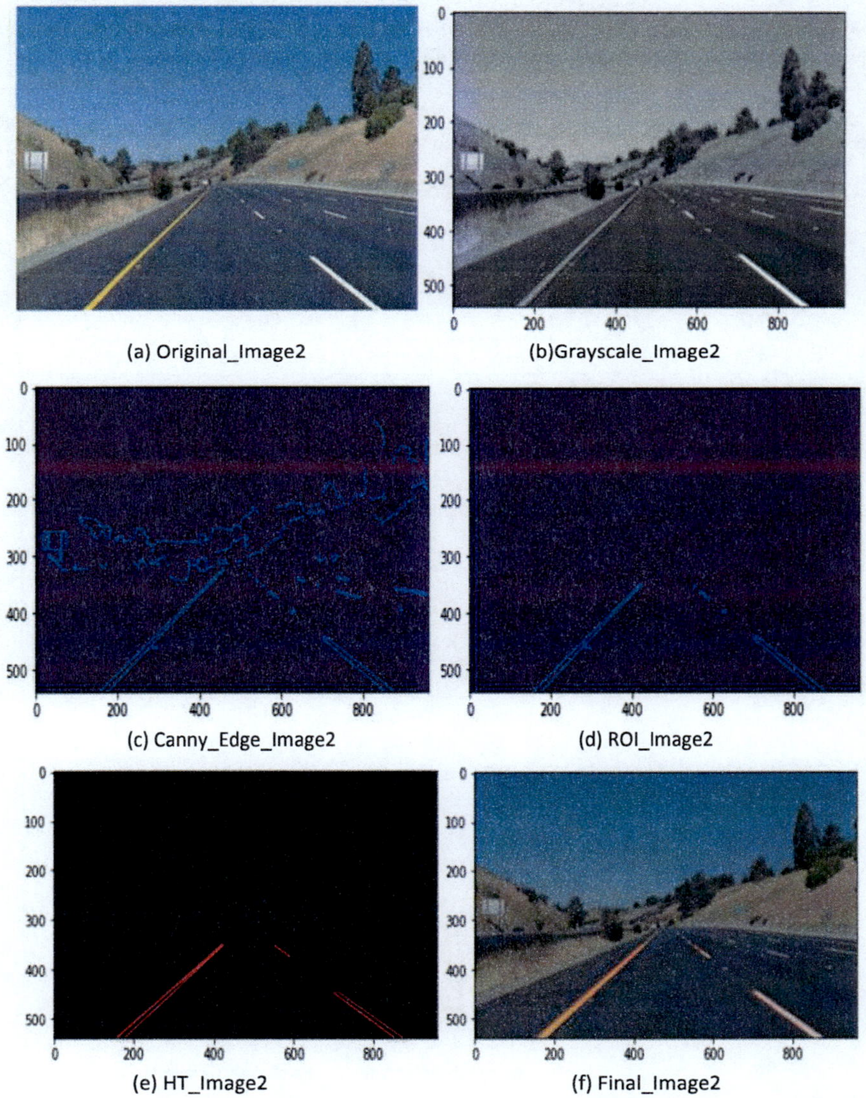

Fig. 4 Road lane detection on national highways

Figure 4 shows the results of the dataset from Yamuna Expressway Agra as the roads are mostly straight and the lines marking on the road pavements surface are clearly visible so it is easier to detect the road lanes in this environment. All the above dataset is taken in day time in as sunny weather. Table 2 contains the miscellaneous datasets recorded during night time on the national highways of Mumbai, Delhi and Ghaziabad. The lane detection rate is calculated for each recorded dataset. The datasets also contain some distorted image to check the robustness of the system if it can keep up with the distorted image with the same efficiency or not (Fig. 5).

Figure 6 gives the accuracy graph of the dataset of Table 2. The accuracy graph is drawn between the dataset number and the efficiency of the system. Zig-Zag manner of the graph shows that system may have quite lagging, but it is durable.

Table 2 Dataset of different cities during night or low light condition

Video no.	No of detected lane	No of correctly detected lane	Lane detection rate (%)
1	260	224	86.15
2	489	407	83.23
3	372	334	89.78
4	710	672	94.65
5	423	340	80.38
6	524	460	87.79
7	320	278	86.88
8	437	418	95.65
9	346	323	93.35
10	208	197	94.71
11	389	342	87.92
12	421	384	91.21
13	148	123	83.11
Total	5047	4502	89.20

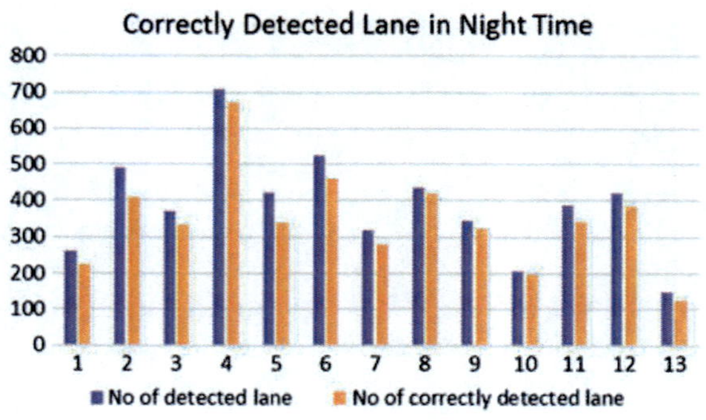

Fig. 5 Correctly detected lanes in night time dataset

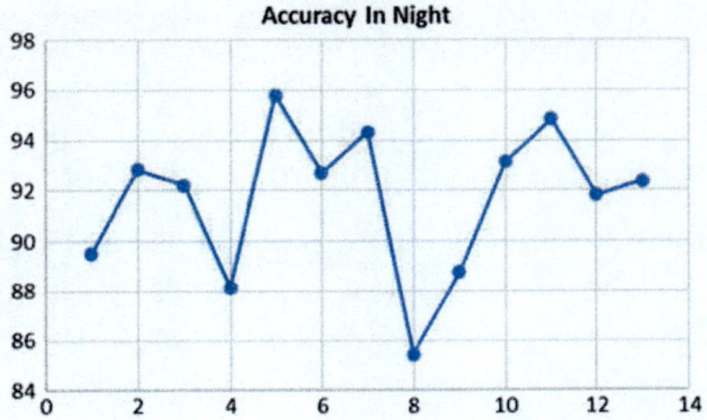

Fig. 6 Accuracy of the system for the dataset taken during night time in Table 2

Table 3 contains the miscellaneous datasets recorded during night time on the national highways of Agra, Kushinagar and Firozabad. Each dataset is used individually and contributed equally to the lane detection efficiency rate calculation. The datasets also contain some images in which there are missing lines at some places and also contains obstruction like pits or object covering the lanes to check the robustness of the system if it can keep up with the distorted image with the same efficiency or not.

Figure 7 shows the graphical representation of the dataset results of Table 3. Here we used the bar chart so that it can be easily differentiate between the detected lines and correctly detected lane. Figure 8 gives the accuracy graph of the dataset of Table 3. As we can see the accuracy graph shows zig-zag form. The roads lane images which are not distorted shows the maximum efficiency represented by trough while the distorted image have less efficiency represented by crest. Table 4 contains

Table 3 Dataset of different cities during day or good light condition

Video no.	No of detected lane	No of correctly detected lane	Lane detected rate (%)
1	258	231	89.53
2	307	285	92.83
3	437	403	92.22
4	287	253	88.15
5	620	594	95.81
6	342	317	92.69
7	478	451	94.35
8	289	247	85.47
9	196	174	88.78
10	394	367	93.15
11	524	497	94.85
12	428	393	91.82
Total	4560	4212	92.37

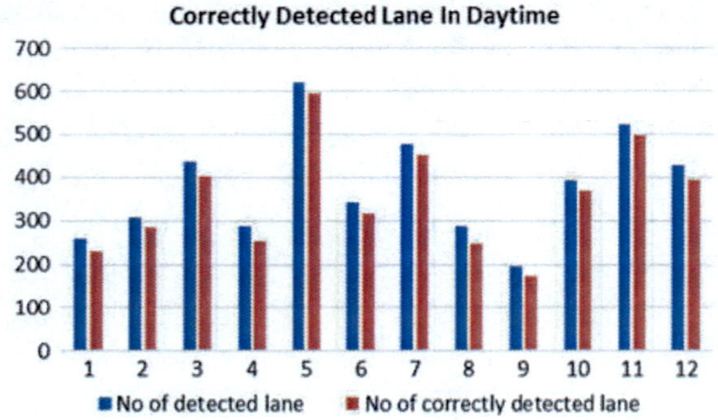

Fig. 7 Bar graph of day time dataset in Table 3

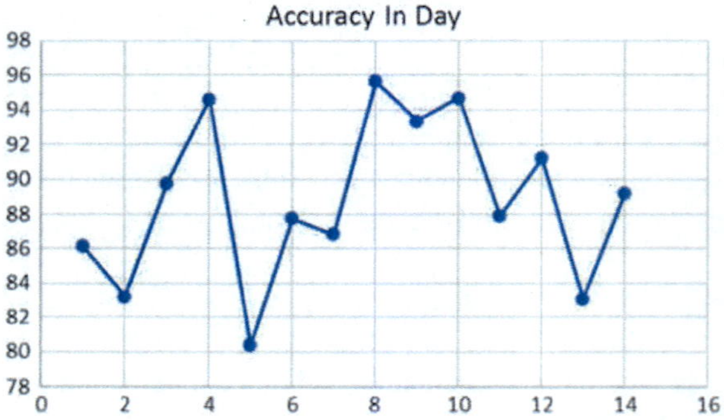

Fig. 8 Accuracy of the system for the dataset taken during night time in Table 3

Table 4 Dataset of different cities in different weather condition

Clip no.	No of detected lanes	No of detection in sunny weather	No of detection in cloudy weather	No of detection in rainy weather
1	478	463	438	448
2	377	362	323	341
3	432	417	374	397
4	279	263	236	258
5	195	179	163	158
Total	1761	1684	1534	1602

the miscellaneous dataset during both day time and night time in different weather condition in cities of Chandigarh and Gorakhpur. In this dataset we did not consider the image which are distorted in one or another way. As the system robustness is assessed on the weather condition parameter. We have considered Sunny, Cloudy and Rainy weather. The lane detection rate is calculated for each dataset.

Figure 9 shows the comparison of the dataset results of Table 4 in different weather conditions. As depicted in the graph we can see that the efficiency of the system in the sunny weather is quite high as compared to the rainy and cloudy weather. And

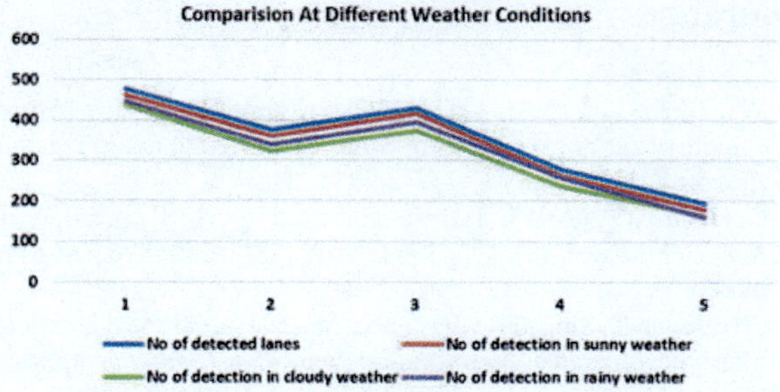

Fig. 9 Comparison of system performance in different weather conditions

the system performance graph in Fig. 8 shows that the system is durable in rainy and cloudy weather. Although its efficiency is less as compared to the sunny weather.

6 Conclusion

In this paper we presented a feasible and affordable mechanism for road lane detection in autonomous vehicle. The system had been checked over various parameters like weather condition or day and night time. The resultant stats show that system performed good during day time and its performance is quite well in night too. When checked on the basis of weather condition system performed good in sunny weather but in cloudy or rainy weather there is some lagging in road lane detection. But as compared to the existing methods for road lane detection which can be either based on LIDAR data or on the basis of Light Intensity the performance of the proposed system is quite well. Existing methods lags when it comes to the distorted image as they cannot correctly assess the lanes. The robustness of Hough Transform has provided us with the advantage that image with little distortion will not affect the system performance. At some point with bigger distortion or missing lanes over a range reduced the system efficiency. So in order to get rid of this problem we propose to apply Image restoration algorithm which uses regression techniques to classify the image contents and then restores the image on the basis of previous and next data. Image restoration algorithm can be applied as soon as the image is captured and then the proposed mechanism is applied on the restored image, thus enhancing the performance and durability of the system.

Future Scope

The proposed work can be extended to bad weather conditions where its hard to predict the lane while driving. The system is capable of achieving an excellent performance in detecting the road lanes and helping the driver with the extracted information to make sure of the safety. The proposed project is for multi-lane roads which suggests the driver to be on the same lane. The number of applications depending on the proposed technology are high, some of them are like in monitoring of traffic in metropolitan cities, ensuring the flow of the traffic and security.

References

1. Wu, C. F., Lin, C. J., Lin, H. Y., & Chung, H. (2013). Adjacent lane detection and lateral vehicle distance measurement using vision-based neuro fuzzy approaches. *Journal of Applied Research and Technology, 11*(2), 251–258.
2. Ochman, M. (2019). Hybrid approach to road detection in front of the vehicle. *IFAC PapersOnLine, 52*(8), 393–396 (2019). ISSN: 24058963. https://doi.org/10.1016/j.ifacol.2019.08.078
3. Srivastava, S., Singal, R., & Lumba, M. (2014). Efficient lane detection algorithm using different filtering techniques. *International Journal of Computer Applications, 88*, 6–11.
4. Sravan, M. S., Natarajan, S., Krishna, E. S., & Kailath, B. J. (2018). Fast and accurate on-road vehicle detection based on colour intensity segregation. *Procedia Computer Science, 133*, 594–603 (2018)
5. Rathore, A. S. (2019). Lane detection for autonomous vehicles using OpenCV library. *International Research Journal of Engineering and Technology, 6*(1), 1326–1332.
6. Satti, S. K., Suganya Devi, K., Dhar, P., & Srinivasan, P. (2020). A machine learning approach for detecting and tracking road boundary lanes. *ICT Express*. https://doi.org/10.1016/j.icte.2020.07.007
7. Khalifa, O. O., Khan, I. M., Assidiq, A. A. M., Abdulla, A. H., Khan, S. (2010). A hyperbola-pair based lane detection system for vehicle guidance. In *Proceedings of the World Congress on Engineering and Computer Science* (Vol. I).
8. Baili, J., Marzougui, M., Sboui, A., Lahouar, S., Hergli, M., Subash Chandra Bose, J., & Besbes, K. (2017). *Lane departure detection using image processing techniques* (pp. 238–241). https://doi.org/10.1109/Anti-Cybercrime.2017.7905298.
9. Husain, A., & Vaishya, R. C. (2018). Road surface and its center line and boundary lines detection using terrestrial Lidar data. *Egyptian Journal of Remote Sensing and Space Science, 21*, 363–374. https://doi.org/10.1016/j.ejrs.2017.12.005
10. Sravana, M. S., Natarajan, S., Krishna, E. S., & Kailath, B. J. (2018). Fast and accurate on-road vehicle detection based on color intensity segregation. *Journal of Procedia Computer Science, 133*, 594–603.
11. Shukla, U., Mishra, A., Jasmine, S. G., Vaidehi, V., & Ganesan, S. (2019). A deep neural network framework for road side analysis and lane detection. *Procedia Computer Science, 165*, 252–258. https://doi.org/10.1016/j.procs.2020.01.081
12. Arman, M., & Tampère, C. (2020). Road centreline and lane reconstruction from pervasive GPS tracking on motorways. ANT/EDI40.
13. Nguyen, V., Kim, H., Jun, S. C., & Boo, K. (2018). A study on real-time detection method of lane and vehicle for lane change assistant system using vision system on highway. *Engineering Science and Technology, an International Journal, 21*(5), 822–833.
14. Xiong, L., Zhenwen, D., Zhang, P., & Fu, Z. (2018). A 3D estimation of structural road surface based on lane-line information. *IFAC-PapersOnLine, 51*, 778–783.

Dimensionality Reduction-Based Discriminatory Classification of Human Activity Recognition Using Machine Learning

Manoj Kumar, Pratiksha Gautam, and Vijay Bhaskar Semwal

Abstract Majority of work in activity recognition using different machine learning and deep learning has shown very challenging results to monitor daily activities. Different datasets available on Web have been used to improve the results, still model fitness need to be verified in terms of different characteristics of matrix and error analysis. Dimensionality reduction (DR) of datasets improves the results of models due to pruning of dataset features. In this paper, we have introduced seven different machine learning models to improve the results. Proposed framework has used principle components analysis (PCA) and linear discriminant analysis (LDA) for dimensionality reduction of UCI-ML dataset. Results show that LDA is better than PCA. Kernel–SVM accuracy has increased from 95.39 to 96.23%. Naïve Bayes has shown 96.78% accuracy with dimensionality reduction. Simple dataset has shown low accuracy while dimensionality reduction has improved the performances of models. We have also introduced different challenges associated with machine learning models, fitness value, and future challenges. At the end of this work, we have done comparative study and error analysis of models.

Keywords Classification · Dimensionality reduction · Machine learning (ML) · LDA · PCA · Accuracy · Human activity recognition (HAR)

Abbreviations

BLSTM Bidirectional LSTM

M. Kumar (✉) · P. Gautam
Amity University Gwalior, Gwalior, Madhya Pradesh, India
e-mail: mannu175@yahoo.com

P. Gautam
e-mail: pgautam@gwa.amity.edu

V. B. Semwal
NIT Bhopal, Bhopal, Madhya Pradesh, India
e-mail: vsemwal@manit.ac.in

P. K. Singh et al. (eds.), *Proceedings of Third International Conference on Computing, Communications, and Cyber-Security*, Lecture Notes in Networks and Systems 421,
https://doi.org/10.1007/978-981-19-1142-2_46

CNN	Convolutional neural network
DT	Decision tree
FA	Factor analysis
KDA	Kernel discriminant analysis
KNN	K mean nearest neighbor
LR	Linear regression
LSTM	Long short-term memory
MDS	Multidimensional scaling
MLP	Multilayer perceptron
RF	Random forest
RNN	Recurrent neural network
SVD	Singular value decomposition
SVM	Support vector machine

1 Introduction [1, 2]

Due to increasing growth of quality sensors and different technologies, daily activities of human have gained interest in the eyes of researchers since few years. Machine learning and deep learning have shown promising results in this domain. This concept has caught attentions in medical fields for monitoring elderly patient's activities in day-to-day life, studying someone daily routines, monitor suspicious activities, mob lynching cases, and new applications based on human centric activities. It will help to identify human gait and later its applicability into bipedal robots. Machine learning and deep learning models have gained popularities to improve results by training and testing phases of datasets.

Dimensionality reduction techniques can be classified into two categories; first is based on only important features, while second is based on creating new variables from old ones (Fig. 1).

In case of small datasets, computation power is less, machine learning performances decreases due to large number of input variables, so it is always advisable to reduce the number of variables. UCI dataset for human activity recognition has many

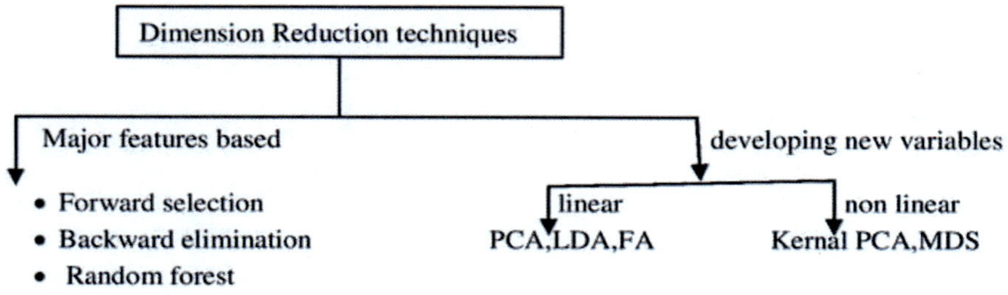

Fig. 1 Classification of dimensionality reduction techniques

attributes which affects performance of models. Dimensionality reduction plays a major role in our work.

In this paper, we have used UCI-ML data from Kaggle repository and trained machine learning models. Seven machine learning models results have been compared, later dimensionality reduction techniques—LDA and PCA have been applied to same dataset and again results of seven machine learning models have been compared to check better resulted model. Dimensionality reduction improves the performance of models by removing unneeded features. There are variety of literature available in activity recognition using deep learning model but very less using machine learning models. We have compared seven different ML-models with and without dimensionality reductions. KNN, logistic regression, SVM, kernel-SVM, Naïve Bayes, decision tree with random forest have been used to predict six human activities. Performances of models have been analyzed by obtaining accuracy, precision, $F1\_score$ of confusion matrix.

Relevance of Work

1. Cyber physical system
2. System simulation, modeling, and visualizations
3. Industrial applications
4. Clinical applications
5. Sport, injury analysis, and recovery management
6. Occlusion construction and gait analysis and detection
7. Transportation solutions through navigation and pedestrian [3]
8. Human machine interaction [4]
9. Physical rehabilitation [4].

Structure of Paper

We have started with introduction section, related work in current years, research gap, and limitations have been discussed in Sect. 2. Section 3 contains proposed framework to improve the results and work activity. Section 4 contains ML-models accuracy outcomes with and without dimensionality reduction. Data visualizations show the possibilities and future expectations in this work. Section 5 has conclusion and future scope of work.

2 Related Work

Authors have [2] compared accuracy of seven different machine learning models, there accuracy and running time have been compared. KNN, neural network, and random forest have shown more than 99% accuracy, while in terms of running time, random forest was better than any others. Dimensionality reduction techniques principle component analysis has tested to improve the result.

Authors [5] have compared ML and DL models classification accuracy on two datasets, named WISDM and Shoaib SA dataset. Overall accuracy of machine learning models shows that KNN is good with dataset 1. Accuracy and loss results of both datasets have been discussed. RNN and CNN deep learning models have been also applied to improve the results. Results show that RNN and CNN are good with dataset 2 by achieving 95.68% and 99.12%, respectively.

In this [6], PCA and frequency-based comparisons using different machine learning approaches have been achieved. Results show that PCA has achieved accuracy of 96.11%, while frequency domain features-based approach has achieved 92.10%.

Authors [7] have discussed how mobile edge computing (MEC) is used to collect data of daily activities of humans. They have proposed architecture for HAR to capture high-volume data; later, five deep learning models on two datasets, named UCI and PAMP2, have been trained and tested. Results show that proposed CNN is good enough over others.

Authors [3] have done a descriptive review of different methodologies, best practices used to identify human activity with possible sensors. Manual and automatic feature selection of datasets with different machine learning and deep learning techniques have been discussed used in different journals, publications. Key findings and future challenges with smart phones and different techniques have been explained with better manner. This paper is good to study for new comer in this domain.

This [8] study shows that getting solution through machine learning is still quite challenging due to inherent limitations of sensors and data proliferation. Authors have used two datasets—wearable sensor data and mobile phone data. Discussion was based on smart home environments covered human activities monitoring.

Hybrid feature selection model [9] has been used in this work. First part do the optimization of feature of dataset, while second part has been used to validate and test dataset through SVM machine learning approach. Proposed approach has shown improvement in results by 6% with 96.81% accuracy using feature selection approach. Future work lies with IoT application in real world with actuator analysis and selection of better controller to improve the accuracy of events.

Authors [10] have proposed neural network architecture segmented convolution gated recurrent neural network (SCGRNN); feature selection using convolution method to improve the accuracy has been done with input data. Reference dataset has been used from MOCAP (simulation based on CMU motion capture database) and supplementary dataset of CC BY license. Experiments results show that their method has better accuracy in terms of fine temporal solution, noise robustness, and generalized solution. Occlusion reconstruction and gait detection could be the future application of this work.

Deep learning has shown better results than traditional methods in activity recognitions. This work [11] shows that hybrid LSTM (H-LSTM) approach has better optimized automatic feature selection of time frequency domain on three UCI datasets. Results achieved from simulators show that H-LSTM has achieved around 99.15% accuracy.

Jaw clenching, loudly speaking, head and eye movements, watching T.V. activities are generally not classified using sensory data. This work [12] focuses on to classify raw EEG signals based on artifacts through proposed framework, named framework for classifying EEG artifacts (FCEA) based on CCN and LSTM-RNN. Suggested concept has shown better improvements in HAR and raw EEG.

UCAml Cup dataset [13] of 24 different classes containing 246 occurrences has been used to identify daily activities in smart home environments scenario. Boosting and bagging techniques for ensemble learning have been used by four base neural networks model. M1, M2, M3, M4 models focused on morning, afternoon, evening activities while M4 focuses hybrid model. Conflicts occur among base models have been discussed and resolved. M4 has shown encouraging results while different challenges during data recording, promising challenges have also discussed in this work.

Increasing number of technologies has raised wide applications of this field [4]. Human machine interaction in smart home environments is getting popular day by day. This work is a comprehensive review work done by authors to understand methods, approaches used in this domain, and getting ready with automatic intelligence system for future challenges. Authors have developed models of different body parts like hand skeletal model, whole body skeletal model, kinematic model of elbow joint, etc. Technical challenges and future prospective of work have been discussed by authors.

Deep learning methods like CNN, RNN, LSTM, and H-LSTM have shown hidden challenges and technical aspects of this domain. This paper [14] contains multi-modality of sensory data and discussion about available datasets publically in different platforms. Later, a discussion on future scope and technical challenges has been illustrated.

Recent years, CNN has shown deep interest by researchers to find results in this domain. Due to different layers and convolution style of it, CNN still needs attention to understand in optimized way to get better results. This work [15] has use light weight filter, named Lego filter and local loss function to train the model. Five different datasets have been used by light CNN, and experiments result shows that proposed CNN architecture has reduced cost and computation power which further emphasis to work on other aspect of CNN in future prospects (Table 1).

Authors Contribution in Research Gap

We have shown recent years work in this domain based on technology and datasets. Review work shows that still there are wide range of technical aspects which needs to be resolve in near future. There is very less work related to performance evaluation of models and error analysis. Mostly work has been done using deep learning, while we lack machine learning models behavior in this domain. This work focuses on the importance of dimensionality reduction and machine learning models performances before jumping into deep learning aspect of era.

During data capturing, activities should be in almost balance proportions, else it will lead to biased results. Camera angle to capture activity in terms of velocity, force, position angle [4] plays a vital role to get numerical values of activity.

Table 1 Latest research literature review on human activity recognition

S. No.	Publication/year/authors	Model used	Dimensionality reduction	Findings	References
1	JITS/2017/M Nabian	LR, SVM, KNN, Naïve Bayes, DT, RF	PCA	RF with 99% accuracy	[2]
2	International journal of machine learning and computing/2018/Sarbagya et al.	RF, DT, KNN, RNN, CNN		DL models are better than ML	[5]
3	IEEE/2018/ASA Sukor et al.	Machine learning classifiers	PCA-based features and time, frequency domain	Higher recognition rate and accuracy in feature and frequencybased PCA	[6]
4	Springer/2019/Shaohua wan et al.	CNN, LSTM, BLSTM, MLP, SVM		Proposed CNN is better	[7]
5	MDPI/2019/W Sausa lima et al.	RNN, CNN, LSTM	KDA	Each has their own kind of advantages	[3]
6	IEEE/2021/BL Alvee et al.	Naïve Bayes, SVM, neural network	PCA, SVD, LDA	SVM is best with 99% accuracy	[16]
7	MDPI/2021/Ankita et al.	CNN, LSTM, hybrid model (CNN + LSTM)		97.89% accuracy by proposed model	[17]

Large datasets with redundant records and attributes produce poor results, more computation power, and time complexity. Features selection approaches to contribute in achieving better results should be picked carefully. Dimensionality reduction has shown a good hope to improve the results in this context (Fig. 2).

Problem Statement

1. To develop a computational model for numerical analysis and evaluation of human activities recognition

Imbalance data
Orientation problem of sensors
Interclass Problem
Data redundancy, noise removal challenges
Feature extraction and segmentation problems
Lack of quality datasets availability

Fig. 2 Different research gaps during human activity recognition

2. To propose a framework to predict various distinct models performances of human activity recognition with and without dimensionality reduction using available datasets.

3 Proposed Framework

- **Conceptual Framework**

(i) **Data Collection and Preparation (UCI-ML-HAR)** [15]:

Dataset has been recorded from 30 people of age 19–48 years keeping variation in age teenage, youth, and elders. Each subject was asked to perform six activities from smartphone containing gyroscope and accelerometer. Three axis data such as X, Y, and Z linear and angular were recorded from phone at the rate of 50 Hz. Signals received from smartphone were further preprocessed to remove noise. This dataset is publically available in Kaggle repository with the name of UCI-ML-HAR. It contains 561 features vector with time and frequency domain variables and 7352 records.

(ii) **Fitting ML—Models Without Dimensionality Reduction**:

We will select seven different ML-models to train and test with this datasets. Best performing model will be picked after many iterations.

(iii) **Fitting ML—Models with Dimensionality Reductions Approaches** [18]:

Sensors record multidimensional data of redundant nature. ML and DL—Models need to be trained carefully to get better results. Missing values, low variance, high correlation, reducing no. of features, and setting few coefficients equal to zero are being practices to get better results.

(iv) **Comparative Study of LDA and PCA Performance**:

We will investigate the ML-models performances applying PCA and LDA on dataset. LDA needs output levels while PCA does not need it. Based on accuracy achieved, better dimensionality reduction approach will be selected. We have selected n_component = none for LDA while n_component = 5 for PCA.

(v) **Final Results Discussion with and Without Dimensionality Reduction**:

Once we will achieve better DR—approach between LDA and PCA, and then it will be compared with UCR-ML-HAR dataset as shown in proposed model-activity framework diagram.

- **Theoretical Concept**
  LDA and PCA are linear approach under dimensionality reduction techniques while nonlinear approaches are cauterized based on global and local properties of datasets.

Principle Component Analysis [19]: PCA is dimensionality reduction techniques that helps us to identify correlations and patterns in data so that it could be transformed into new dataset of lower dimensionality without loss of information.

Algorithm

Step 1: Dataset gathering and organizing in matrix form (M * N).
Step 2: Calculate mean and deviation.
Step 3: Calculate and form covariance matrix.
Step 4: Form Eigen values and Eigen values.
Step 5: Selection of principal component and converting into feature vector form.
Step 6: Deriving new dataset.

Linear Discriminant Analysis [20]: It tries to find best possible separation in samples by their class value between classes and minimum separation within class of samples.

Algorithm

Step 1: Preparation of dataset so called data matrix X.
Step 2: Calculate mean vector for each class.
Step 3: Calculate total mean vector.
Step 4: Calculate within and between class scatters.
Step 5: Compute Eigen decomposition and projection matrix.

Parameter for Feature Selection: [21]

1. For every column of dataset, missing value ratio is being calculated; value above threshold will be drops while for rest of data will be checked for error.

$$\text{ratio} = \frac{\text{No of missing values}}{\text{Total no of observations}} * 100$$

2. Variable having low variance will be drops as that will not have any impact on target data.

$$\sigma^2 = \frac{\sum (x - \bar{x})^2}{n}$$

3. Remove the variable which has low correlation with target variable. Minimize variables with similar correlations.

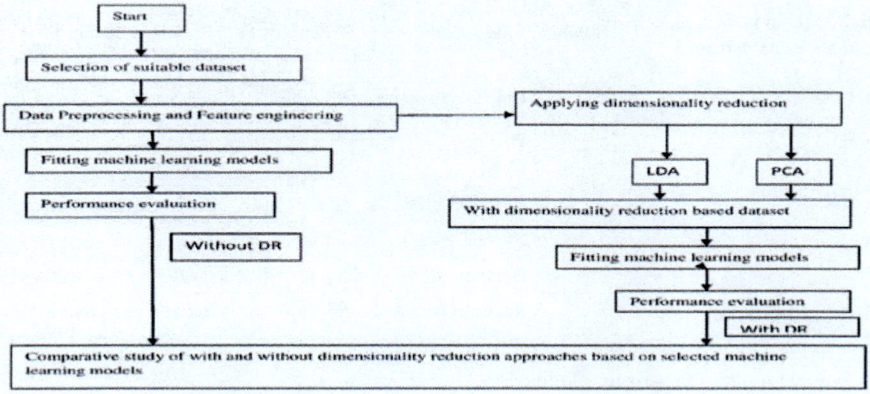

Fig. 3 Proposed conceptual framework

$$r = \frac{\sum (x_i - \bar{x})(y_i - \bar{y})}{\sqrt{\sum (x_i - \bar{x})^2 \sum (y_i - \bar{y})^2}}$$

4. Backward elimination: Eliminate variable one by one during training period of model until it affects model performance.
5. Forward feature selection: Retain those variables which improve performance of model, this can be monitored during training of model one by one (Fig. 3).

Step 1: Without dimensionality reduction approach

In this, dataset will be cleaned and preprocessed, and machine learning models will be applied to selected datasets.

Step 2: With dimensionality reduction

In this case, both linear approaches of dimensionality reduction will be applied to selected dataset. Proposed framework will test the models behaviors in LDA and PCA case both. Initially, LDA and PCA will be compared based on accuracy of models, and then, better techniques will be again compared with step 1 output.

4 Experimental Results

Our work focuses on identifying standing, lying, sitting, walking, walking down stairs, and walking upstairs. Dataset has almost equal no. of activities instances which shows that dataset is in balance form. Figure 5 shows models accuracy on UCI-ML-HAR dataset, and it has been found that Kernel–SVM has highest accuracy than other machine learning models. Tables 2 and 3 show comparison among models without DR and with DR (Figs. 4 and 6).

Table 2 ML-models performance analysis on UCI-ML

S. No.	Model name	Accuracy %	Standard deviation %
1	Linear regression	93.39	4.17
2	KNN	88.09	4.99
3	SVM	93.89	3.62
4	Kernel-SVM	95.39	4.43
5	Naive Bayes	73.56	5.62
6	Decision tree	87.65	4.60
7	Random forest	93.86	2.55

Table 3 Final results comparisons of models

S. No.	Model name	Accuracy %		
Machine learning techniques		Without DR	With DR	
			PCA	LDA
1	Linear regression	93.39	43.81	96.47
2	KNN	88.09	45.47	96.64
3	SVM	93.89	43.57	96.13
4	Kernel-SVM	95.39	44.08	96.23
5	Naive Bayes	73.56	39.57	96.78
6	Decision tree	87.65	37.97	95.72
7	Random forest	93.86	42.59	96.37

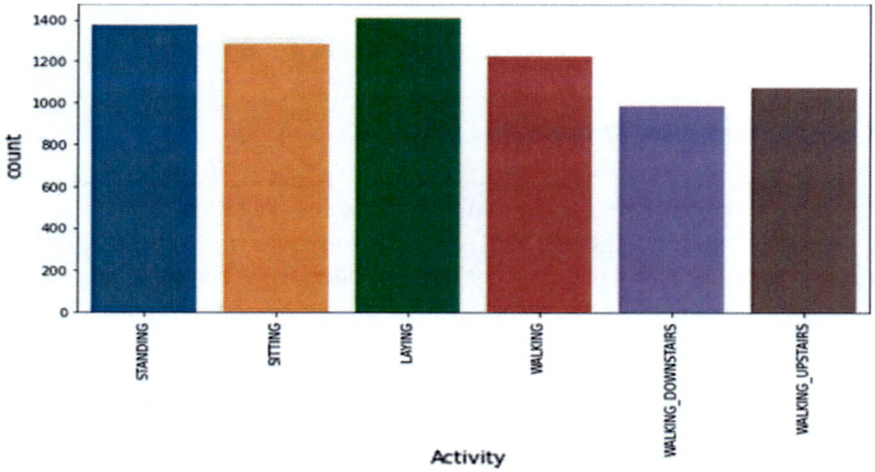

Fig. 4 Activity distribution

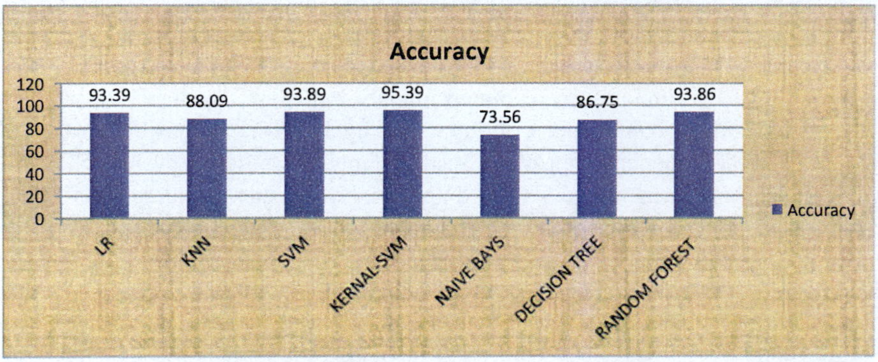

Fig. 5 Comparative study of ML-models without dimensionality reduction (DR)

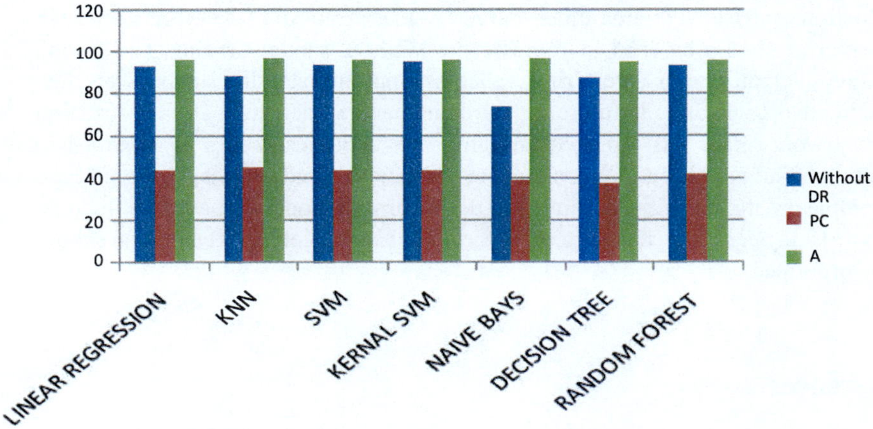

Fig. 6 Accuracy of different models with and without dimension reduction (DR)

In case of without applying dimensionality reduction, Kernel-SVM has shown better results from other. While applying DR concept, LDA has shown better results in each model than PCA. Naïve bayes has shown 96.78% accuracy, while KNN is very near to this performance by securing 96.64% accuracy. LDA is better in both cases, with and without dimensionality reduction approaches. LDA has shown progressive results than other models. LDA has obtained good results in terms of precision, recall, and $F1\_score$.

5 Conclusion and Future Scope

This work shows the importance of dimensionality reduction. Available datasets are often redundant and contain noise during recording. Proposed framework clearly shows how DR techniques can improve the results of models. Initially, seven machine learning models have been trained. Kernel-SVM has shown better results by securing 95.39%. Later, LDA and PCA have been applied to UCI-ML-HAR dataset, and implementations show that LDA has performed better than PCA. Finally, LDA has been compared with seven machine learning models (without DR case) which shows that still LDA is leading with performance.

Future scope of this work can help us to recognize the patient's health habits or wellness, Sport routine of athletes, discovery of activity patterns which variable determines which activity. Predictive model to detect activity should be developed and the corresponding mathematical, computational approach should be designed. Evaluation matrixes like area under curve (AOC), mean absolute error, and mean square error should be used to compare the ML, DL model training, and testing accuracy. Person health monitoring, rehabilitation, and stability analysis are few domain areas to be done in future. Impaired human gait detection and construction are key work under this topic. More number of activities should be recorded to monitor the human daily routines, patient's habits, and elderly assistance. In case of nonlinear data, machine learning and deep learning models behaviors should be analyzed and necessary changes to train the model and to get best fit situation should be incorporated.

References

1. Pan, S. J., Kwok, J. T., & Yang, Q. (2008, July). Transfer learning via dimensionality reduction. In *AAAI* (Vol. 8, pp. 677–682).
2. Nabian, M. (2017). A comparative study on machine learning classification models for activity recognition. *Journal of Information Technology & Software Engineering, 7*(04), 4–8.
3. Sousa Lima, W., Souto, E., El-Khatib, K., Jalali, R., & Gama, J. (2019). Human activity recognition using inertial sensors in a smartphone: An overview. *Sensors, 19*(14), 3213.
4. Meng, Z., Zhang, M., Guo, C., Fan, Q., Zhang, H., Gao, N., & Zhang, Z. (2020). Recent progress in sensing and computing techniques for human activity recognition and motion analysis. *Electronics, 9*(9), 1357.
5. Shakya, S. R., Zhang, C., & Zhou, Z. (2018). Comparative study of machine learning and deep learning architecture for human activity recognition using accelerometer data. *International Journal of Machine Learning and Computing, 8*, 577–582.
6. Sukor, A. A., Zakaria, A., & Rahim, N. A. (2018, March). Activity recognition using accelerometer sensor and machine learning classifiers. In *2018 IEEE 14th International Colloquium on Signal Processing & Its Applications (CSPA)* (pp. 233–238). IEEE.
7. Wan, S., Qi, L., Xu, X., Tong, C., & Gu, Z. (2020). Deep learning models for real-time human activity recognition with smartphones. *Mobile Networks and Applications, 25*(2), 743–755.
8. Subasi, A., Khateeb, K., Brahimi, T., & Sarirete, A. (2020). Human activity recognition using machine learning methods in a smart healthcare environment. In *Innovation in health informatics* (pp. 123–144). Academic Press.

9. Ahmed, N., Rafiq, J. I., & Islam, M. R. (2020). Enhanced human activity recognition based on smartphone sensor data using hybrid feature selection model. *Sensors, 20*(1), 317.
10. Du, H., Jin, T., He, Y., Song, Y., & Dai, Y. (2020). Segmented convolutional gated recurrent neural networks for human activity recognition in ultra-wideband radar. *Neurocomputing, 396*, 451–464.
11. Wang, L., & Liu, R. (2020). Human activity recognition based on wearable sensor using hierarchical deep LSTM networks. *Circuits, Systems, and Signal Processing, 39*(2), 837–856.
12. Salehzadeh, A., Calitz, A. P., & Greyling, J. (2020). Human activity recognition using deep electroencephalography learning. *Biomedical Signal Processing and Control*, *62*, 102094.
13. Irvine, N., Nugent, C., Zhang, S., Wang, H., & Ng, W. W. (2020). Neural network ensembles for sensor-based human activity recognition within smart environments. *Sensors, 20*(1), 216.
14. Chen, K., Zhang, D., Yao, L., Guo, B., Yu, Z., & Liu, Y. (2020). Deep learning for sensor-based human activity recognition: Overview, challenges and opportunities. arXiv preprint arXiv:2001.07416.
15. Tang, Y., Teng, Q., Zhang, L., Min, F., & He, J. (2020). Efficient convolutional neural networks with smaller filters for human activity recognition using wearable sensors. arXiv preprint arXiv:2005.03948.
16. Alvee, B. I., Tisha, S. N., & Chakrabarty, A. (2021, July). Application of machine learning classifiers for predicting human activity. In *2021 IEEE International Conference on Industry 4.0, Artificial Intelligence, and Communications Technology (IAICT)* (pp. 39–44). IEEE.
17. Webber, M., & Rojas, R. F. (2021). Human activity recognition with accelerometer and gyroscope: A data fusion approach. *IEEE Sensors Journal, 21*(15), 16979–16989.
18. https://stackabuse.com/implementing-lda-in-python-with-scikit-learn/
19. Jolliffe, I. T., & Cadima, J. (2016). Principal component analysis: A review and recent developments. *Philosophical Transactions of the Royal Society A: Mathematical, Physical and Engineering Sciences*, *374*(2065), 20150202.
20. https://machinelearningmastery.com/linear-discriminant-analysis-for-dimensionality-reduction-in-python/
21. https://towardsdatascience.com/mathematical-recipe-of-dimensionality-reduction-281ff37957e4

SPECIAL SESSION ON RECENT ADVANCES IN COMPUTATIONAL INTELLIGENCE & TECHNOLOGYS (SS_10_RACIT)

Development of Generic Human Motion Simulation Categorization using Inception based CNN

Ram Kumar Yadav, Subhrendu Guha Neogi, and Vijay Bhaskar Semwal

Abstract Generic human motion is the steps taken by the person for everyday movement. In this paper, the generic framework has been proposed for gait activity recognition using the inception-based convolutional neural network (CNN) model. The gait pattern is a compound of seven sub-phases. However, sequential execution of these seven sub-phases of left and right legs is called gait cycle. Gait features are used to perform the biometric, biomechanics, and human psychology analysis of human beings. Human walking is very challenging to measure due to high variability. It depends on age, gender, walking terrain, walking speed, mental condition, health condition, etc. The analysis of human gait helps identify human activities, diagnose many gait-related diseases like Parkinson's and freezing of gait. Due to variability of gait, biometric is difficult to spool. This paper covers to analyzing the existing available wireless sensor data mining (WISDM) dataset and our gait dataset of human activity (GDOHA) dataset and evaluating the performance. Moreover, we compare performance metrics on datasets applying machine learning and deep learning algorithms. The result achieved 99.03% accuracy using the inception-based CNN model. The proposed computational model helps to early detection of gait abnormality and provides the proper mechanism for recovery from abnormal gait.

Keywords Gait · Biometric · Human activity recognition · Cyber-physical system · Deep learning

R. K. Yadav (✉) · S. G. Neogi
Amity University Gwalior, Gwalior, India
e-mail: yadav20072@gmail.com

S. G. Neogi
e-mail: sgneogi@gwa.amity.edu

V. B. Semwal
MANIT Bhopal, Bhopal, India
e-mail: vsemwal@manit.ac.in

P. K. Singh et al. (eds.), *Proceedings of Third International Conference on Computing, Communications, and Cyber-Security*, Lecture Notes in Networks and Systems 421,
https://doi.org/10.1007/978-981-19-1142-2_47

1 Introduction

Human activity identification (HAI) is a very emerging and effective field of research. It is the method of identifying particular activity, like walking, running, jogging, sitting, standing, slow walk, fast walk, upstairs, downstairs, hill climbing, and hill descending. A recent survey has shown that many older people go too alone without family members' care and can need medical care from their family. This may increase an excellent challenge for the state to develop a good and healthy policy for elders also because it will increase technology-based assistantship. Word will need such applications which could look out of elders and supply a far better help during doing daily activities [1]. This has increased business chances to develop such automated systems, which might be helpful in identifying the particular activity of humans and can inform necessary actions and suggestions taken by particular during activity. Human activity recognition is widely utilized in the computing and engineering field on visualizing and to ascertain more closely; activities must be recorded carefully, and powerful computation power should be used. Human skeleton data are being recorded, and a deep learning model is being applied to detect acts using the image processing concept [2]. HAI [3] is one of the most challenging with the use of machine learning and deep learning approaches. In this procedure, identify and classify particular human activity among multiple activities. Various authors determine and predict the human movement of particular activities like walking, running, and jogging in daily life [4]. Researchers have been defined various human activities in daily life which as walking, catching a train, reading newspapers, eating food, flying kites, etc. The author has been discussed more than hundreds of human activity which performed by a human in daily life. Under this human activity recognition process, almost, all activity recognition and detection have been done with the help of an automatic detection system.

Literature survey was motivated to describe various activities—normal walk, fast walk, running, stair up, stair down [5], hill climbing, hill descending, sitting, eating, typing [6], and standing. These different activities are categorized into spatiotemporal, kinetics, and kinematics information [7]. Human activity classification is one the most powerful area of research application in which authors classified activity and achieved good accuracy and reduced loss factors.

Detection of normal and abnormal gait is another application area of human activity identification in which normality and abnormality of gait are classified using machine learning and deep learning techniques [8].

Assisted and smart home [9] is one more application in which current advances have given inventive methods to improve the living style of older and weak persons. Dynamic and helped living frameworks use activity recognition to monitor and help occupants to ensure their well-being. A brilliant home climate is sensors oriented to improve the security, the degree of independence of life, and the quality of life.

Motoring of health care [9] is one of the most important applications in which the advancement of clinical science and innovation significantly expanded the existing nature of patients. As expressed by Goldstone, 20% future rates will increment drastically in 2050, and roughly, 30% of Americans, Canadians, etc., country persons will be beyond 60 years old. This will prompt greater levels of popularity for the clinical workforce, which might be difficult to be provided soon. Consequently, analysts attempt to upgrade the current medical care checking approaches that would deal with dire clinical circumstances.

Human activity identification is an interesting field in computer science and engineering and electronic communication due to manipulating software tools and inexpensive sensor devices [10]. Various approaches to identify human activity that have been used in the sensory dataset. The authors have covered two types of approaches like model-free and model-based approach. In model-free approach, the static and dynamic information comprehended in the silhouette images is emphasized. Essentially, the model-free approach comprises feature extraction, object detection, silhouette image extraction, and classification [11–13] which analyzes the movement of the human walk and extracted the different features from the human walk. Many researchers work with video taken by the cameras and proposed a new way for much variety of data that distribute non-permanent actions, obstruction, and moving cameras.

In model-based approach, construct the computational model based on previous knowledge to define walking parameters of human dynamics [14]. In the model-based method, human gait is recognized by analyzing the motion of humans and its' various features. Moreover, the low-level measurements of the human image are estimated [15–18] on the basis of the following parameters: angles of limb, joint position, different body parts length, body parts thickness/width, height, and distance vector in the middle of the gravity center of a body component.

2 Related Work

Automatic recognition of gait is often achieved [19] using computer vision techniques to find persons and derive a gait from images. Authors have been used the motion approach for the reorganization of gait. They have defined the context of these approaches; finally, the authors have got experimental results using a motion-based approach with statistical analysis. They have got a 90% rate of recognition based on different conditions.

Boulgouris et al. [20] have explored a complete novel system for gait recognition. Recognition of identity and verification is basically based on comparing linearly time-normalized gait walking cycles. This feature extraction method is additionally proposed for the transformation of human silhouettes into low-dimensional feature vectors consisting of average pixel distances from the middle of the silhouette. By using this methodology, improvements of recognition and verification performance increase 8–20% compared to another well-known method on the "Gait Challenge" dataset.

Tao et al. [21] have developed a new approach for preprocessing named general tensor discriminant analysis (GDTA) in place of linear discriminant analysis (LDA). The authors have compared the proposed GDTA method with the existing preprocessing method and achieved good performance for gait recognition.

In this gait recognition technique, Arai and Andrie [22] have used the wavelet transformation function for feature extraction, and they attempted model-based and model-free preprocessing methods. Using the Chinese Academy of Sciences (CASIA) database, the planned technique is evaluated for correct classification performance and compared with the old human gait recognition technique.

Cheng et al. [23] have investigated how to design an effective sensor system for human activity recognition. Designing this system, author investigates different parameters systematically illustration of result author considered four activities: heart rate, breathing rate, recognition of hand motion, monitoring of swallowing and gait analysis.

A convolution neural network [24] is proposed to perform proficient and powerful HAR utilizing smartphone sensors by abusing the qualities of activity and 1D time series signals, simultaneously giving how to consequently and information adaptively separate vigorous features from data. A more extensive time frame of transient nearby connection can be exploited (1×9–1×14), and a low pooling size (1×2–1×3) is demonstrated to be advantageous. Convents also accomplished a practically ideal grouping on moving activities, particularly fundamentally the same as those previously appeared to be hard to classify. Finally, ConvNets beat other state-of-the-art data preparing strategies in HAR for the benchmark dataset gathered from 30 volunteer subjects, accomplishing a total performance of 94.79% on the test set with crude sensor data and 95.75% on fast Fourier transform of the HAR data.

In the indoor climate, the author [25] proposed a pedestrian activities recognition strategy dependent on a convolution neural network. Another convolution neural network has been intended to gain proficiency with the correct highlights consequently. Investigations show that the proposed technique accomplishes around 98% accuracy in around 2 s in recognizing nine sorts of activities, including still, walk, up elevator, upstairs, downstairs, up escalator, down the escalator, and down the elevator turning. Additionally, the authors have constructed a pedestrian activities dataset containing 6 GB of data of accelerometers, gyroscopes, magnetometers, and barometers gathered with different cell phones.

In this paper, the author [26] assessed two published datasets by comparing different performance parameters; the authors also exhibit our exploratory results. They execute insignificant data preprocessing by fundamental extract features named standard deviation, mean, etc., and experiments performed on datasets to examine its results. We analyze the UCI HAR and WISDM datasets in view of accuracy, precision-recall; furthermore, $F1$-score on nine diverse ML approaches. Both datasets are showing distinct results, but the UCI HAR dataset result is encouraging of them.

Table 1 shows previous ground work which is oriented with our approach.

Table 1 Distinct comparative ground work

Authors	Dataset	Approaches
Cheng et al. [27]	OPP. PAMAP2, UCI HAR	Deep learning model based on RNN
Attal et al. [28]	Own data	Comparing supervised (KNN, SVM, GMM, RF) and unsupervised (k-mean, GMM, HMM) classification technique
Lee et al. [29]	Own data	One-dimensional CNN method
Lawal and Bano [30]	RWHAR	Convolutional neural network (CNN)
Bayat et al. [31]	Own data	ML classification techniques
Bao and Intille [32]	Own data	ML decision tree classifier
Pienaar and Malekian [33]	WISDM	LSTM model
Ordóñez and Roggen [34]	Opportunity, Skoda	Deep convolutional and LSTM-RNN model
Hou [4]	USC-HAD, WISDM	DL (deep learning) and conventional machine learning

Table 2 Performance of existing WISDM and own GDOHA dataset

Performance parameters	WISDM dataset (%)	GDOHA dataset (%)
Accuracy	92.98	99.03
Precision	93.1	95
Recall	96.2	98
$F1$_score	90.13	90

In this part, we have to show more related and available existing research work and also explore various datasets with different performing metrics. It is presented in Table 2.

3 Proposed Framework

Inception-based CNN model is a part of artificial neural network. CNN utilizes activation within the convolution phase for prediction. These activation functions and convolution can fluctuate consistently with the pooling layer, the transformation layer, and completely associated layers of CNN. In completely associated layers, each neuron of a particular layer connects to all opposite neurons in other layers. Figure 1 is designed to classify the generic human motion with the use of optimization technique—inception-based CNN model. The overall framework is comprised of a gait dataset captured by various sensors. The performance of the system is evaluated based on a comparison of GDOHA gait dataset with WISDM. The fitness function is calculated by considering different parameters applied on a different subject. The

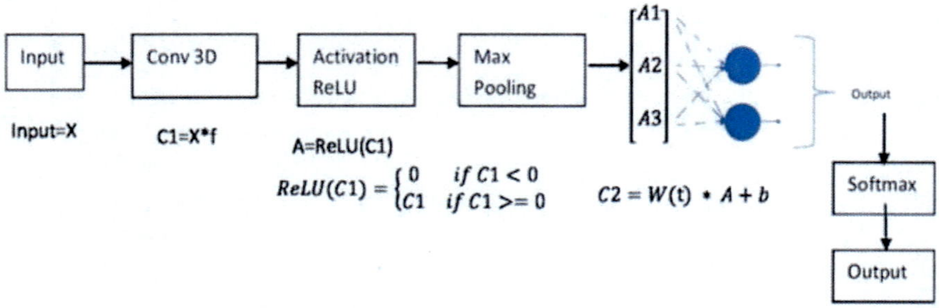

Fig. 1 Inception-based convolution neural network

overall functionality of the framework is explained below. The accuracy of the system is calculated for inception-based CNN model.

3.1 Experimental Setup

Samples were collected through sensors for different age groups (ranges from 20 to 60) by considering 30 subjects that defined complete gait data. The sensor devices and IMU have been used to collect gait data. Seven different activities were executed by individual subjects participants for the collection of gait data. The collection of dataset was further validated by comparison with WISDM [18]. The accuracy of the proposed optimization techniques is based on the fitness function. Fitness function is calculated by subjects and different parameters described above. The classification of the dataset which is based on different activities is further processed by applying different optimization techniques such as—decision tree, random forest, and proposed inception-based CNN model. Figure 2 described the inertial measure-

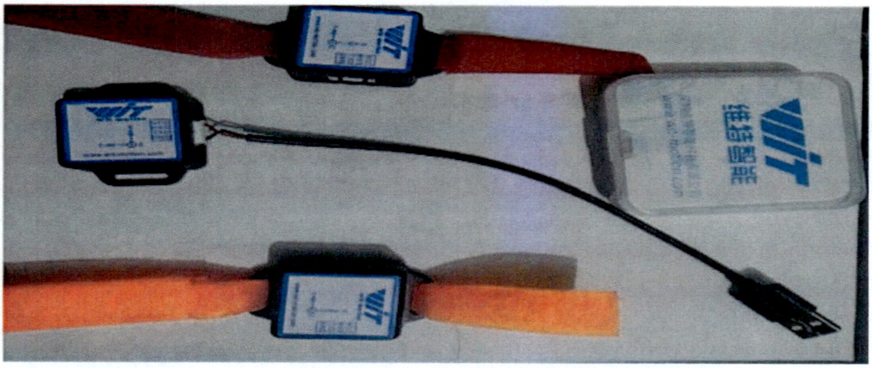

Fig. 2 IMU and sensor devices

Fig. 3 Constrained environment experiment subjects

ment unit (IMU) and sensor which are used to track the activities that generic human motion.

Figure 3 indicates the collection of datasets to various subjects using inertial measurement unit (IMU) and cell phone accelerometer sensors.

3.2 Representation of Dataset

The proposed work has used a multiple-task gait analysis approach with certain time intervals using different tools like wearable devices, kinetic sensors, and mobile phone-based physics tool bar accelerometer. The data were collected from 30 subjects (out of which 20 men and 10 women subjects). We have recorded data of left-hander, right-hander, young, middle-aged, and adult people. Data have been recorded in different conditions like a sunny days, rainy days, and cloudy days. We have considered gait data at different surfaces like a flat surfaces, stairs, and zigzag routes. It consists of seven activities, namely (i) normal walking, (ii) jogging, (iii) walking on toes, (iv) walking on heels, (v) sit-up, (vi) walking upstairs, and (vii) walking downstairs.

In order to validate the created dataset, it is compared with the existing WISDM dataset which contains activities like walk, jog, up stair, down stair, sit, and stand with healthy, young, and adult subject.

3.3 Preprocessing and Architecture

In recorded dataset, preprocessing methods are completed by categorical imputation method is applied on noise presents in excess and less significant value highlighted. Thus, in order to refine and pre-measure the informational index, principal component analysis (PCA) is applied to separate the main highlights. It will support the system by removal of the redundant highlights, further which will be useful for the performance of the proposed framework. As a result, preparation and characterization time will be diminished. In this procedure, the preprocessed dataset is further separated into training and a testing bunch on which diverse classification algorithms have been applied. Figure 4 shows the proposed architecture of inception-based CNN model and model process.

where f is kernel size or filter, $C1$ is convolution, $w(t)$ is weights, and b is bias (constant).

Algorithm

Algorithm 1: Proposed Model Algorithm

Input: GDOHA.csv dataset

```
AccmX, AccmY, AccmZ          #Acceleration data
GyrosX, GyrosY, GyrosZ       #Gyroscope data
kf        #Filter or Kernal size for convolution
nf        #Number of frequency point
st        #Number of time localized points
sc        #step of convolution
wf        #Number of Filters
```

Output: The accuracy prediction of "HAR"

Begin:

1. //Environment development
   Import numpy, pandas, seeborn, scipy, coreml, tensorFlow libraries
2. //Load dataset and do preprocessing process
   (a). Dataset split and Label creation
   (b). Dataset size and activity count visualization
   // Model development

```
a[1], a[2], a[3]   ← AccmX, AccmY, AccmZ ;
g[1], g[2], g[3]   ← GyrosX, GyrosY, GyrosZ ;
for i=1 to wf do
 for t=1 to nf do
      for z=1 to 3 do
```

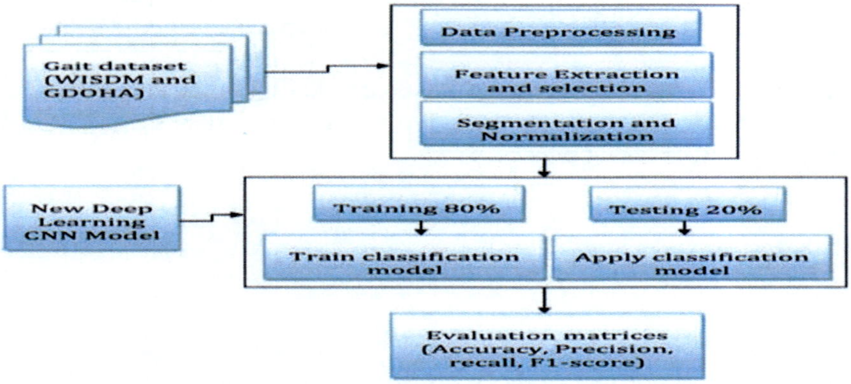

Fig. 4 Proposed model

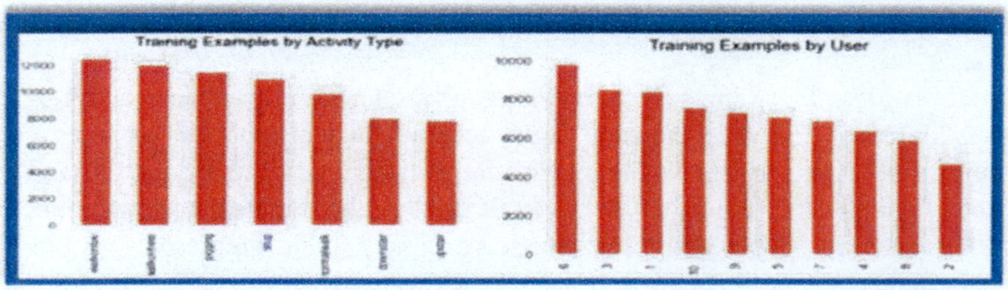

Fig. 5 Training examples by activity and user type

$$o[t]\,[i] += \sum_{j=1}^{st} \cdot \sum_{k=1}^{kf} w[i]\,[j]\,[k] \,*\, g[z]\,[sc \,*\, (t-1)+k]\,[j];$$

$$o[t+nf]\,[i] += \sum_{j=1}^{st} \cdot \sum_{k=1}^{kf} w[i]\,[j]\,[k] \,*\, a[z]\,[sc \,*\, (t-1)+k]\,[j];$$

End for
End for
End for
Fc ← Fully_connected_Neural Network(o);
Result ← softmax(Fc);
3. **Exit**
EndBegin:

3.4 Calculation of Fitness Function

It is the ratio of accurately classified subjects with respect to overall subjects. It is presented by $C(A)$ and given in Eq. 1.

$$C(A) = \frac{S_c}{S_t} \tag{1}$$

where S_c and S_t have represented the accurately classified subjects and the whole number of subjects, respectively.

4 Results and Discussion

We compare the datasets by way of a comparable set of experiments create a training and testing set of data using inception-based CNN method and apply performance metrics named accuracy, precision, recall, and $F1\_score$.

Figure 5 analyzes the training examples by activity and user type of the proposed approach.

All these Figs. 6, 7, 8 images are showing the various training and user activates results. Moreover, Fig. 9 is showing model accuracy and loss of various activity classifications. Figure 10 shows the relationship between performance metrics and datasets. Figure 11 represents the accuracy of used model. Figure 12 shows count of various activities. Table 3 represents models accuracy. Table 4 represents various activity count.

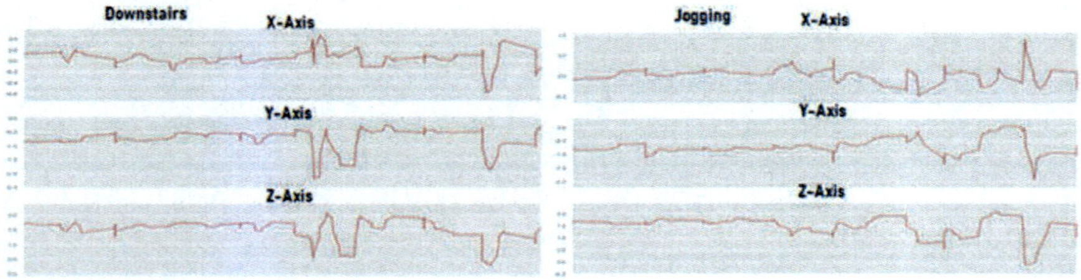

Fig. 6 Downstairs and jogging activity type

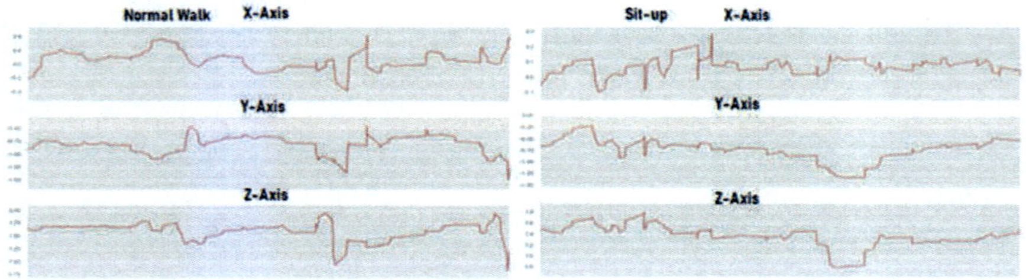

Fig. 7 Normal walk and sit-up activity type

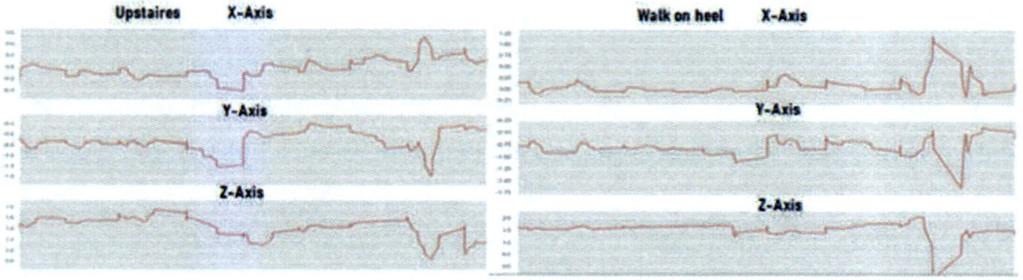

Fig. 8 Upstairs and walk on heel activity type

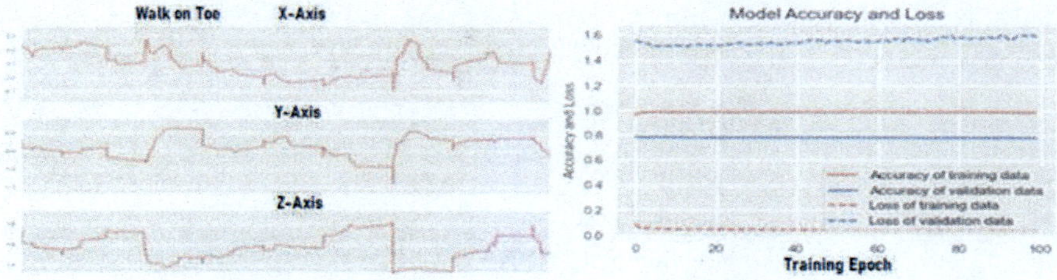

Fig. 9 Walk on toe activity type, model accuracy, and loss

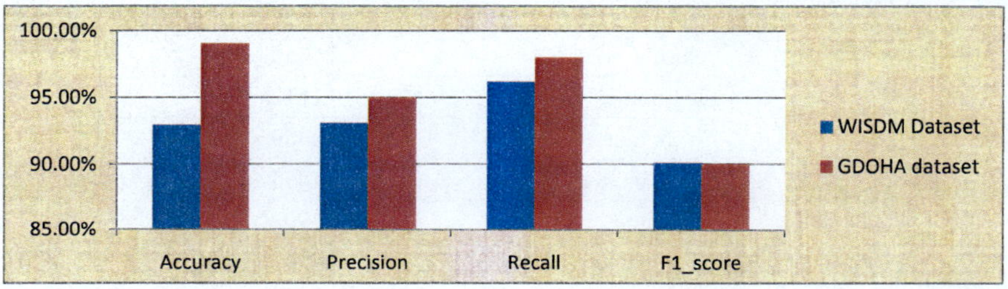

Fig. 10 Performance metrics and datasets

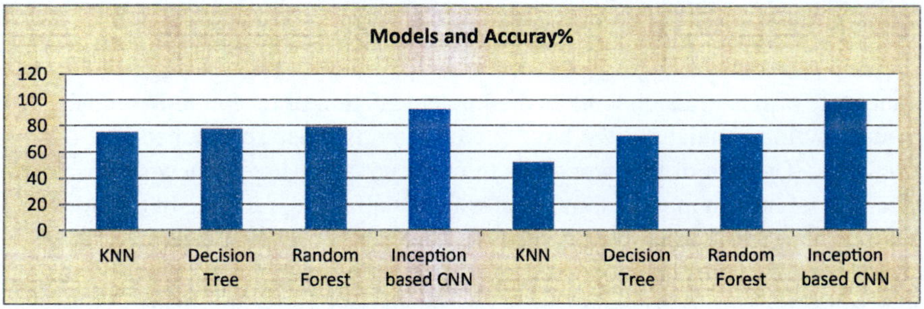

Fig. 11 Accuracy of respective models

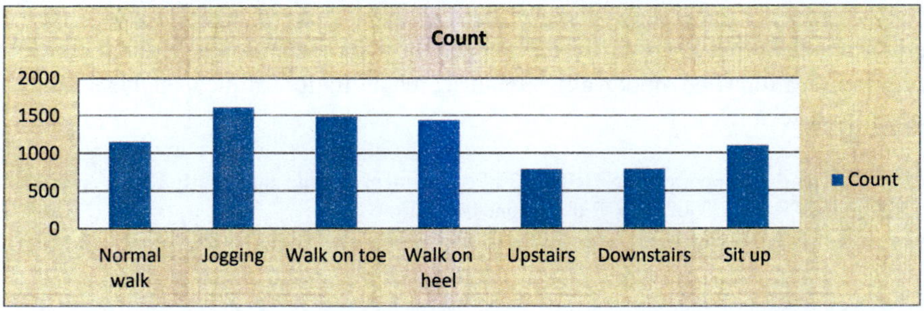

Fig. 12 Activities and respective counts

Table 3 Accuracy of dataset using various models

Model/algorithm	Accuracy%	
	WISDM	GDOHA
KNN	75.0	52.05
Decision tree	77.6	72.09
Random forest	79.3	73.56
Inception-based CNN	92.98	99.03

Table 4 Activity name and respective counts

Activity name	Count
Normal walk	1150
Jogging	1612
Walk on toe	1489
Walk on heel	1435
Upstairs	790
Downstairs	794
Sit up	1100

5 Conclusion and Future Work

The main role of this research work is the implementation of inception-based CNN for human motion simulation. We have explored the two datasets named WISDM and GDOHA. The aim of this work was to construct a generic framework for the classification of human motion simulation which is effective, reliable. In this work, various machine learning models named SVM, KNN, decision tree, and random forest are used for finding the model accuracy. The model's performance is measured using a parameter named accuracy, precision, recall, and $F1\_score$. After analysis, we achieved 99.03% accuracy on the GDOHA dataset using the inception-based CNN model. The GDOHA dataset was collected from the Motion Capture laboratory, which is better than the existing WISDM dataset. This paper has provided the generic framework that can be utilized to expand the gait recognition field and deploy practical applications like robot walk, biometric, rehabilitation, clinical analysis, and the healthcare sector.

Funding This work is supported by SERB, DST of government of India under Early Career Award with DST NO: ECR/2018/000203 ECR dated June 04, 2019.

References

1. Voicu, R. A., Dobre, C., Bajenaru, L., & Ciobanu, R. I. (2019). Human physical activity

recognition using smartphone sensors. *Sensors, 19*(3), 458.
2. Mo, L., Li, F., Zhu, Y., & Huang, A. (2016). Human physical activity recognition based on computer vision with deep learning model. In *2016 IEEE International Instrumentation and Measurement Technology Conference Proceedings* (pp. 1–6). IEEE.
3. Zainudin, M. S., Sulaiman, M. N., Mustapha, N., & Perumal, T. (2015). Activity recognition based on accelerometer sensor using combinational classifiers. In *2015 IEEE Conference on Open Systems (ICOS)* (pp. 68–73). IEEE.
4. Hou, C. (2020). A study on IMU-based human activity recognition using deep learning and traditional machine learning. In *2020 5th International Conference on Computer and Communication Systems (ICCCS)* (pp. 225–234). IEEE.
5. Sunny, J. T., George, S. M., Kizhakkethottam, J. J., Sunny, J. T., George, S. M., & Kizhakkethottam, J. J. (2015). Applications and challenges of human activity recognition using sensors in a smart environment. *IJIRST International Journal of Innovative Science and Research Technology, 2*, 50–57.
6. Lee, H., Ahn, C. R., Choi, N., Kim, T., & Lee, H. (2019). The effects of housing environments on the performance of activity-recognition systems using Wi-Fi channel state information: An exploratory study. *Sensors, 19*(5), 983.
7. di Biase, L., Di Santo, A., Caminiti, M. L., De Liso, A., Shah, S. A., Ricci, L., & Di Lazzaro, V. (2020). Gait analysis in Parkinson's disease: An overview of the most accurate markers for diagnosis and symptoms monitoring. *Sensors, 20*(12), 3529.
8. Luo, J., & Tjahjadi, T. (2020). Multi-set canonical correlation analysis for 3D abnormal gait behaviour recognition based on virtual sample generation. *IEEE Access, 8*, 32485–32501.
9. Ranasinghe, S., Al Machot, F., & Mayr, H. C. (2016). A review on applications of activity recognition systems with regard to performance and evaluation. *International Journal of Distributed Sensor Networks, 12*(8), 1550147716665520.
10. Hussain, Z., Sheng, M., & Zhang, W. E. (2019). Different approaches for human activity recognition: A survey. arXiv preprint arXiv:1906.05074
11. Iqbal, S., Zang, X., Zhu, Y., Saad, H. M. A. A., & Zhao, J. (2015). Nonlinear time-series analysis of different human walking gaits. In *2015 IEEE International Conference on Electro/Information Technology (EIT)* (pp. 025–030). IEEE.
12. Semwal, V. B. (2017). *Data driven computational model for bipedal walking and push recovery* (In Thesis).
13. Iqbal, S., Zang, X., Zhu, Y., Saad, H. M. A. A., & Zhao, J. (2015). Nonlinear time-series analysis of different human walking gaits. In *2015 IEEE International Conference on Electro/Information Technology (EIT)* (pp. 025–030). IEEE.
14. Chadza, T., Kyriakopoulos, K. G., & Lambotharan, S. (2019). Contemporary sequential network attacks prediction using hidden Markov model. In *2019 17th International Conference on Privacy, Security and Trust (PST)* (pp. 1–3). IEEE.
15. Roy, S. K., Krishna, G., Dubey, S. R., & Chaudhuri, B. B. (2019). HybridSN: Exploring 3-D–2-D CNN feature hierarchy for hyperspectral image classification. *IEEE Geoscience and Remote Sensing Letters, 17*(2), 277–281.
16. Hu, C. H., Pei, H., Si, X. S., Du, D. B., Pang, Z. N., & Wang, X. (2019). A prognostic model based on DBN and diffusion process for degrading bearing. *IEEE Transactions on Industrial Electronics, 67*(10), 8767–8777.
17. Vrigkas, M., Nikou, C., & Kakadiaris, I. A. (2015). A review of human activity recognition methods. *Frontiers in Robotics and AI, 2*, 28.
18. Kwapisz, J. R., Weiss, G. M., & Moore, S. A. (2011). Activity recognition using cell phone accelerometers. *ACM SigKDD Explorations Newsletter, 12*(2), 74–82.
19. Nixon, M. S., & Carter, J. N. (2006). Automatic recognition by gait. *Proceedings of the IEEE, 94*(11), 2013–2024.
20. Boulgouris, N. V., Plataniotis, K. N., & Hatzinakos, D. (2006). Gait recognition using linear time normalization. *Pattern Recognition, 39*(5), 969–979.
21. Tao, D., Li, X., Wu, X., & Maybank, S. J. (2007). General tensor discriminant analysis and Gabor features for gait recognition. *IEEE Transactions on Pattern Analysis and Machine Intelligence, 29*(10), 1700–1715.

22. Arai, K., & Andrie, R. (2012). Gait recognition method based on wavelet transformation and its evaluation with Chinese Academy of Sciences (CASIA) gait database as a human gait recognition dataset. In *2012 Ninth International Conference on Information Technology-New Generations* (pp. 656–661). IEEE.
23. Cheng, J., Amft, O., Bahle, G., & Lukowicz, P. (2013). Designing sensitive wearable capacitive sensors for activity recognition. *IEEE Sensors Journal, 13*(10), 3935–3947.
24. Ronao, C. A., & Cho, S. B. (2016). Human activity recognition with smartphone sensors using deep learning neural networks. *Expert Systems with Applications, 59*, 235–244.
25. Zhou, B., Yang, J., & Li, Q. (2019). Smartphone-based activity recognition for indoor localization using a convolutional neural network. *Sensors, 19*(3), 621.
26. Khare, S., Sarkar, S., & Totaro, M. (2020). Comparison of sensor-based datasets for human activity recognition in wearable IoT. In *2020 IEEE 6th World Forum on Internet of Things (WF-IoT)* (pp. 1–6). IEEE.
27. Cheng, X., Chai, D., He, J., Zhang, X., & Duan, S. (2019). In-noHAR: A deep neural network for complex human activity recognition. *IEEE Access, 7*, 9893–9902.
28. Attal, F., Mohammed, S., Dedabrishvili, M., Cham-roukhi, F., Oukhellou, L., & Amirat, Y. (2015). Physical human activity recognition using wearable sensors.
29. Lee, S.-M., Yoon, S. M., & Cho, H. (2017). Human activity recognition from accelerometer data using convolutional neural network. In *2017 IEEE International Conference on Big Data and Smart Computing (BigComp)* (pp. 131–134). IEEE.
30. Lawal, I. A., & Bano, S. (2019). Deep human activity recognition using wearable sensors. In *Proceedings of the 12th ACM International Conference on PErvasive Technologies Related to Assistive Environments—PETRA '19* (pp. 45–48). ACM Press
31. Bayat, A., Pomplun, M., & Tran, D. A. (2014). A study on human activity recognition using accelerometer data from smartphones. *ProcediaComputerScience, 34*, 450–457.
32. Bao, L., & Intille, S. S. (2004). Activity recognition from user-annotated acceleration data. In *International Conference on Pervasive Computing* (pp. 1–17). Springer.
33. Pienaar, S. W., & Malekian, R. (2019). Human activity recognition using LSTM-RNN deep neural network architecture. In *2019 IEEE 2nd Wireless Africa Conference (WAC)* (pp. 1–5). IEEE.
34. Ordóñez, F. J., & Roggen, D. (2016). Deep convolutional and LSTM recurrent neural networks for multimodal wearable activity recognition. *Sensors, 16*(1), 115.

Cryptanalysis on "ESEAP: ECC-Based Secure and Efficient Mutual Authentication Protocol Using Smart Card"

Mohammad Abdussami, Ruhul Amin, and Satyanarayana Vollala

Abstract Very recently, ESEAP mutual authentication protocol was designed to avoid the drawbacks of Wang et al. protocol and highlights that the protocol is protecting all kind of security threats using informal analysis. This work investigates the ESEAP protocol in security point of view and notices that the scheme is not fully protected against stolen verifier attack and does not provide user anonymity. Furthermore, the same protocol has user identity issues, i.e., the server cannot figure out the user identity during the authentication phase. Later we discuss the inconsistencies in the security analysis of ESEAP presented by RESEAP.

Keywords Cryptanalysis · User anonymity · Stolen verifier attack

1 Introduction

Due to the rapid growth of the communication and e-commerce, most of the works today have become online, in which all the works majorly involve accessing the data, as well as transmission of the data. Now here comes the problem of security that is the data may be vulnerable due to attacks; therefore, providing security to the data should be one of the important concerns in this case. Authentication technique is very useful to access the resources safely without the risk of data being modified during transmission, it improves the trust of the clients who are using the network since the data origin will be verified, i.e., whether the data is originating from the authentic users, and it is ensured that the data is not modified during the transmission as well as the entities participating in the conversation are also authenticated. Authentica-

M. Abdussami (✉) · R. Amin · S. Vollala
DSPM IIIT Naya Raipur, Raipur 493661, India
e-mail: mohammad@iiitnr.edu.in

R. Amin
e-mail: ruhul@iiitnr.edu.in

S. Vollala
e-mail: satya@iiitnr.edu.in

P. K. Singh et al. (eds.), *Proceedings of Third International Conference on Computing, Communications, and Cyber-Security*, Lecture Notes in Networks and Systems 421,
https://doi.org/10.1007/978-981-19-1142-2_48

tion is one of the important techniques for the clients participating in the network to access the resources securely in various applications including Telecare Medical Information System, Internet of Things, health care, wireless sensor networks, wireless body sensor networks, smart grids, vehicular ad hoc sensor networks, etc. [1]. It is in 1981 when the authentication protocol based on a password is introduced the first time by Lamport [2]. Later some other password-based schemes [3, 4] were proposed, but they suffered various vulnerabilities. In 1991, Chang and Wu proposed a two-factor authentication scheme for the first time using a smart card where the idea is to improve the security by adding one more authentication parameter that is smart card [5].

1.1 Road Map

In Sect. 2, the related literature review has been discussed followed by the review of the ESEAP protocol in Sect. 3. Section 4 highlights cryptanalysis of the ESEAP protocol. In Sect. 5, the inconsistencies in the cryptanalysis of ESEAP presented by RESEAP are discussed. Section 6 consists of both the conclusion and future work.

2 Related Works

A review on related literature is exposed in this section. An authentication method based on smart card which is designed by Fan et al. [6] is found to be insecure, which is vulnerable to various security attacks and failed in preserving anonymity of the entity. It also fails to establish session key. An authentication scheme given by Juang et al. [7] is stated to be robust in terms of security threats. Later same protocol is analyzed and said to be failed in untraceability and password change property by Sun et al. [8] and Li et al. [9] and is precisely explained that the scheme is not immune to offline and stolen smart card attack. A protocol is put forth by Huang et al. [10] that used smart card and claimed that [7, 8] failed to protect against online and offline attack using smart card. The property of anonymity is to be proved and to be failed in the scheme proposed by Wang and Wang [11] that used two factors for authentication used in wireless sensor networks [12–14]. A two-factor authentication method designed and given by Wang et al. [15] for distributed systems is found to suffer from smart card stolen attack [16] and card revocation attack [17].

The pitfalls of various two-factor authentication schemes [18–21] are analyzed by Wang et al. [22]. The authentication scheme proposed in [18] failed against online/offline guessing attack. The scheme proposed in [19] is found to suffer from attacks through desynchronization, guessing and stolen attack. The protocol proposed in [20] is insecure against smart card loss attack and guessing. Various mistakes are found in formal security proofs. The protocol proposed in [21] is insecure against guessing attack and failed to preserve forward secrecy. A lightweight scheme pro-

Table 1 Survey representing loopholes in the literature

References	Contribution	Loopholes
[6]	Smart card-based authentication scheme	Fails to preserve user anonymity, does not provide session key establishment
[7]	Proposed an anonymous authentication scheme	Lost smart card attack vulnerable to anonymity offline attack
[10]	Smart card-based authentication key agreement scheme	Claimed that [7, 8] failed against offline guessing attack with SC
[15]	Authentication scheme with anonymity property	Claimed that [16] is vulnerable to SC lost attack, desynchronization attack
[23]	Lightweight three-factor scheme for data access in WSN	Vulnerable to replay attack, user anonymity, SC loss attack, forward secrecy
[24]	Lightweight multimedical server authentication for TMIS	Vulnerable to offline password guessing attack, no user anonymity
[25]	Password-based authentication scheme using SC	Fails against to stolen–verifier attack and impersonation attack

posed by Luo et al. [23] that used three factors for authentication which finds its application in wireless sensor networks lacked forward secrecy and found to be vulnerable to sound repairability and reply attack. It is found to be unprotected against server and user linkability attacks.

In [26–28], several authentication schemes based on identity and using two factors for authentication are proved to be weak against offline guessing, anonymity and untraceability. An authentication protocol proposed by Amin et al. [24] is proved to be a failure in preserving user anonymity and mutual authenticity and suffers from several other attacks like offline password guessing attack and no password exposure. Wang et al. [25] designed an authentication protocol based on user secret credentials and smart card that suffer from attacks like guessing the password in offline mode and stolen verifier attack and failed to protect session key. A brief survey representing loopholes in the literature is represented in the form of table, as given in Table 1.

3 ESEAP Protocol

This section presents brief review of Kumari et al.'s authentication scheme. Table 2 indicates the various symbols used in this protocol. ESEA protocol contains the following phases:

Table 2 Symbol descriptions

Symbol	Description
U	The user
SC	The smart card
F_q	The prime finite field
G	Additive ECC group
x	Private key of server
ID_i	The unique identity of ith user/participant
PW_i	Password of ith entity
S	Server
$\mathrm{EC}(F_q)$	Elliptic curve over F_q
K_U	Key generated by user
K_S	Key generated by server
g	Generator point of G
Z_q^*	Additive group of order q
SK_I	Session key of entity I
$\|\|$	Concatenation
$\oplus$	Bitwise XOR operation
$h(.)$	Cryptographic one way secure hash function

3.1 Initialization and Registration Phase

The server S selects an elliptic curve $\mathrm{EC}(F_q)$, and later it selects g as the generator of group G. S randomly picks x as a private key.
U will register with the server in the following way:

Step R1. U chooses ID_i, PW_i, a random value $a \in Z_q^*$ and then calculates $\mathrm{PWU} = h(\mathrm{ID}_i \parallel \mathrm{PW}_i \parallel a)$. U then forwards a registration information $\{\mathrm{PWU}, \mathrm{ID}_i\}$ to S through secure channel.
Step R2. Once the message is received, S selects a random number $b \in Z_q^*$ and sets User_list: $\{\mathrm{ID}_i, b, \mathrm{Honey\_List}\}$. After that S finds/calculates $c = \mathrm{ID}_i \oplus b$, $B_1 = h(\mathrm{ID}_i \parallel x \parallel c \parallel b)$ and $L_1 = B_1 \oplus \mathrm{PWU}$. Finally, S produces a smart card by storing the credentials $\{G, g, h(.), L_1, c\}$ and forwards to U through secured channel.
Step R3. After getting the SC, U will insert P_1, P_2, P_3 into the SC, where $P_1 = a \oplus h(\mathrm{ID}_i \parallel \mathrm{PW}_i)$, $P_2 = h(\mathrm{PWU} \parallel P_1)$ and $P_3 = a \oplus \mathrm{ID}_i \oplus c$. Finally, U will delete c from SC.

3.2 Mutual Authentication Phase

After successful registration, U will send login request to the S, after that mutual authentication process is carried out. The steps for login and mutual authentication are as follows:

Step LA1. U uses SC and puts it into smart card reader and further provides ID_i and PW_i.

Step LA2. SC calculates $a' = P_1 \oplus h(\text{ID}_i \parallel \text{PW}_i)$, $\text{PWU}' = h(\text{ID}_i \parallel \text{PW}_i \parallel a')$, $P_2' = h(\text{PWU}' \parallel P_1)$ and checks whether $P_2' = P_2$ holds or not. After that, SC calculates $B_1^* = \text{PWU}' \oplus L_1$, $C_1 = h(c \parallel \text{ID}_i)$, $C_2 = h(C_1 \parallel a' \parallel \text{ID}_i)$, $K_{U1} = h(P_3 \parallel a' \parallel \text{ID}_i)$, $N = h(T_1 \parallel C_1 \parallel C_2 \parallel B_1^* \parallel \text{ID}_i)$, picks a random number $u \in Z_q^*$, encrypts $E_1 = E_{K_{U1}}(N, u.g, C_1, C_2, P_3, \text{ID}_i, T_1)$ and sends message $M_1 = \{E_1, P_3\}$ to S through unencrypted channel.

Step LA3. On receiving M_1, S calculates $K_{S1} = h(\text{ID}_i \parallel P_3 \oplus b \parallel P_3)$, decrypts $(P_3^*, \text{ID}_i^*, C_1^*, C_2^*, u.g, N^*, T_1) = D_{K_{S1}}(E_1)$ and checks the equality of $P_3^* = P_3$. If the Boolean value is false, S searches the User_list to find $\text{ID}_i^* = \text{ID}_i$, and if the user identity does not match, then it does not allow a new entry request and sets the Honey_List to be Honey_List + 1. If the value of the Honey_List exceeds the mentioned threshold, S will come to know that SC has been cracked and therefore will be suspended until the U registers; else the U will be authenticated by the S. Later S checks the condition $T_2 - T_1 \leq \Delta T$, computes $B_1 = h(\text{ID}_i \parallel x \parallel c \parallel b)$, $N^* = h(\text{ID}_i^* \parallel C_1^* \parallel C_2^* \parallel B_1 \parallel T_1)$ and verifies if $N^* = N$ holds or not. If the condition fails, S sets the Honey_List to Honey_List+1m and if the threshold of Honey_List exceeds the pre-specified value, SC will be blocked until the U re-registers, else the S will compute $B_2 = h(\text{ID}_i^* \parallel x \parallel (P_3 \oplus b) \parallel b \parallel T_1)$, selects $s \in Z_q^*$, computes $K_{S2} = h(C_1^* \parallel C_2^* \parallel B_1 \parallel N^* \parallel T_3)$, session key $\text{SK}_S = h(\text{ID}_i^* \parallel \text{ID}_S \parallel K_{S2} \parallel N^*.g \parallel u.s.g \parallel B_2 \parallel T_3)$, $V = h(T_3 \parallel s.g \parallel u.s.g \parallel u.g)$, $B_2' = B_2 \oplus h(B_1 \parallel \text{ID}_i^* \parallel \text{ID}_S)$, $\text{ID}_{S1} = \text{ID}_S \oplus h(b \parallel N^* \parallel B_1 \parallel B_2)$, encrypts $E_2 = E_{K_{S2}}(\text{ID}_{S1}, s.g, V, B_2')$ and sends message $M_2 = \{E_2, T_3\}$ to U through unencrypted channel.

Step LA4. After getting $M_2 = \{E_2, T_3\}$, U verifies $T_4 - T_3 \leq \Delta T$, computes $K_{U2} = h(C_1 \parallel C_2 \parallel B_1 \parallel N \parallel T_3)$, decrypts $(\text{ID}_{S1}, s.g, V, B_2') = D_{K_{U2}}(E2)$, computes $B_2^* = B_2' \oplus h(B_1^* \parallel \text{ID}_i \parallel \text{ID}_S^*)$ and checks if $V = h(T_3 \parallel s.g \parallel u.s.g \parallel u.g)$ holds or not. If the condition holds, then S will be authenticated by U. Later U computes $\text{ID}_S^* = \text{ID}_{S1} \oplus h((P_3 \oplus a) \parallel N \parallel B_1^* \parallel B_2^*)$ and session key $\text{SK}_U = h(\text{ID}_i \parallel \text{ID}_S^* \parallel K_{U2} \parallel N.g \parallel u.s.g \parallel B_2^* \parallel T_3)$.

4 Cryptanalysis of Kumari et al. Protocol

- Adversary can access the information/messages sent through public broadcasting, and the adversary can inject new messages, retrieve, modify, replay and can drop any messages that are being communicated.
- Adversary cannot get the secret key of the participants.

- Adversary has the knowledge of the identities which are public in nature of all the users and server.
- Adversary can pretend as a genuine user.

Based on the assumptions given in Sect. 4, we concluded that the Kumari et al.'s protocol is vulnerable to some attacks. It has some unavoidable security loopholes. The description of all loopholes is discussed below.

4.1 Stolen Verifier Attack

In step R2 of the registration phase in the same protocol, it can be seen that the server stores User_list: { ID_i, b, Honey_List }. Due to the stolen user_list, the protocol may suffer from several serious issues. The discussion of the issues is given below.

The credentials ID_i, b are known from the stolen User_list, c can be computed using $c = \mathrm{ID}_i \oplus b$. Now the value of a can be computed using $P_3 = a \oplus \mathrm{ID}_i \oplus c$ that is $a = P_3 \oplus \mathrm{ID}_i \oplus c$, where the value of P_3 will be retrieved from the messages that are exchanged publicly. The adversary will get the following credentials $\{G, g, h(.), L_1, P_1, P_2, P_3\}$ from the stolen or lost SC using a power analysis attack [29, 30]. Later, the password can be guessed in the offline mode using $P_1 = a \oplus h(\mathrm{ID}_i \parallel \mathrm{PW}_i)$, since all the credentials except PW_i are known. Therefore, under the assumptions that the User_list, SC are stolen, the adversary can know the secret credentials ID_i, PW_i of the U, and later he/she can change the password or revoke the account of the user or use the SC for the illegal purpose.

4.2 User's Anonymity

Based on the assumption of Sect. 4.1, if the User_list is stolen from the S, then the ID_i, b are known. Now since the ID_is of the users are revealed, the user anonymity property is not preserved.

4.3 Key Computation

Based on the assumption of Sect. 4.1, the key $K_{U1} = h(\mathrm{ID}_i \parallel a' \parallel P_3)$ can be computed since the ID_i, b are known from the stolen User_list and $a = P_3 \oplus b$, where the value of P_3 can be retrieved by the public messages that are exchanged between server and the user.

4.4 ID_i Problem in LA3 of the Protocol

In step LA3 of login authentication phase, on receiving the message M_1, S computes $K_{S1} = h(\mathrm{ID}_i \parallel P_3 \oplus b \parallel P_3)$, but there is no way the server can know the ID_i of the user or that he/she is communicating with that particular U. So, this can be considered one of the major drawbacks of the protocol [31]. Therefore, the S cannot compute the key K_{S1}.

5 Discussions on RESEAP

In this section, we discuss the inconsistencies in the cryptanalysis of Kumari et al. [31] presented by the Safkhani et al. [32].

5.1 Offline Password Guessing Attack

In this attack, the assumptions made by Safkhani et al. [32] are not valid according to the literature. The authors have claimed the possibility of this attack by considering the $\{S_1 = \mathrm{PW}_i\}$ and $\{S_2 = \mathrm{ID}_i\}$, assuming the adversary has extracted the credentials $L_1 = B_1 \oplus \mathrm{PWU}$, $P_1 = a \oplus h(\mathrm{ID}_i \parallel \mathrm{PW}_i)$, $P_2 = h(\mathrm{PWU} \parallel P_1)$ and $P_3 = a \oplus \mathrm{ID}_i \oplus c$ from the SC and will compute the following:

For any $\mathrm{PW}_i^* \parallel \mathrm{ID}_i^* \in \{S_1, S_2\}$

- $a^* = P_1 \oplus h(\mathrm{ID}_i^* \parallel \mathrm{PW}_i^*)$
- $\mathrm{PWU}^* = h(\mathrm{ID}_i^* \parallel \mathrm{PW}_i^* \parallel a)$
- if $P_2 = h(\mathrm{PWU}^* \parallel P_1)$ returns a, PW_i and ID_i

The above attack is claimed based on the assumption that both unknown parameters ID_i and PW_i can be guessed, but according to [33], it is not feasible to guess both the parameters ID_i and PW_i. Hence, the above claim of offline password guessing attack is invalid.

5.2 Smart Card Lost Attack

This attack is claimed based on the assumptions made in Sect. 5.1. Hence, the claim is not valid.

5.3 Desynchronization Attack

This attack is claimed based on the assumptions made in Sect. 5.1. Hence, the claim is not valid.

6 Conclusion

This paper investigates the ESEAP mutual authentication protocol which uses smart card technology concept. We review and analyze that the protocol has several critical issues such as user anonymity, problem for the server in tracing that the message has come from a particular user and stolen verifier attack which leads to password guessing attack also. Hence, the protocol cannot be used in practical implementation. In the future, we will design a new improved protocol which countermeasures the mentioned security drawbacks.

References

1. Amin, R., & Biswas, G. P. (2015). A novel user authentication and key agreement protocol for accessing multi-medical server usable in TMIS. *Journal of Medical Systems, 39*(3), 33–49.
2. Lamport, L. (1981). Password authentication with insecure communication. *Communications of the ACM, 24*(11), 770–772.
3. Shimizu, A. (1991). A dynamic password authentication method using a one-way function. *Systems and Computers in Japan, 22*(7), 32–40.
4. Shieh, S. P., Yang, W. H., & Sun, H. M. (1997). An authentication protocol without trusted third party. *IEEE Communications Letters,1*(3), 87–89.
5. Chang, C. C., & Wu, T. C. (1991). Remote password authentication with smart cards. *IEE Proceedings E (Computers and Digital Techniques),138*(3), 165–168.
6. Fan, C. I., Chan, Y. C., & Zhang, Z. K. (2005). Robust remote authentication scheme with smart cards. *Computers & Security, 24*(8), 619–628.
7. Juang, W. S., Chen, S. T., & Liaw, H. T. (2008). Robust and efficient password-authenticated key agreement using smart cards. *IEEE Transactions on Industrial Electronics, 55*(6), 2551–2556.
8. Sun, D. Z., Huai, J. P., Sun, J. Z., Li, J. X., Zhang, J. W., & Feng, Z. Y. (2009). Improvements of Juang's password-authenticated key agreement scheme using smart cards. *IEEE Transactions on Industrial Electronics, 56*(6), 2284–2291.
9. Li, X., Qiu, W., Zheng, D., Chen, K., & Li, J. (2009). Anonymity enhancement on robust and efficient password-authenticated key agreement using smart cards. *IEEE Transactions on Industrial Electronics, 57*(2), 793–800.
10. Huang, X., Chen, X., Li, J., Xiang, Y., & Xu, L. (2013). Further observations on smart-card-based password-authenticated key agreement in distributed systems. *IEEE Transactions on Parallel and Distributed Systems, 25*(7), 1767–1775.
11. Wang, D., & Wang, P. (2014). On the anonymity of two-factor authentication schemes for wireless sensor networks: Attacks, principle and solutions. *Computer Networks, 73,* 41–57.
12. Fan, R., He, D. J., Pan, X. Z., et al. (2011). An efficient and dos-resistant user authentication scheme for two-tiered wireless sensor networks. *Journal of Zhejiang University (SCIENCE C), 12*(7), 550–560.

13. Xue, K., Ma, C., Hong, P., & Ding, R. (2013). A temporal-credential-based mutual authentication and key agreement scheme for wireless sensor networks. *Journal of Network and Computer Applications, 36*(1), 316–323.
14. Chuang, M. C., Lee, J. F., & Chen, M. C. (2012). Spam: A secure password authentication mechanism for seamless handover in proxy mobile ipv6 networks. *IEEE Systems Journal, 7*(1), 102–113.
15. Wang, D., He, D., Wang, P., & Chu, C.-H. (2014). Anonymous two-factor authentication in distributed systems: Certain goals are beyond attainment. *IEEE Transactions on Dependable and Secure Computing,12*(4), 428–442.
16. Li, C.-T. (2013). A new password authentication and user anonymity scheme based on elliptic curve cryptography and smart card. *IET Information Security, 7*(1), 3–10.
17. Tsai, J. L., Lo, N. W., & Wu, T. C. (2012). Novel anonymous authentication scheme using smart cards. *IEEE Transactions on Industrial Informatics, 9*(4), 2004–2013.
18. Li, X., Niu, J., Khan, M. K., & Liao, J. (2013). An enhanced smart card based remote user password authentication scheme. *Journal of Network and Computer Applications, 36*(5), 1365–1371.
19. Kumari, S., & Khan, M. K. (2014). Cryptanalysis and improvement of a robust smart-card-based remote user password authentication scheme. *International Journal of Communication Systems, 27*(12), 3939–3955.
20. Odelu, V., Das, A. K., & Goswami, A. (2015). An effective and robust secure remote user authenticated key agreement scheme using smart cards in wireless communication systems. *Wireless Personal Communications, 84*(4), 2571–2598.
21. Bin Muhaya, F. T. (2015). Cryptanalysis and security enhancement of Zhu's authentication scheme for telecare medicine information system. *Security and Communication Networks, 8*(2), 149–158.
22. Wang, D., Gu, Q., Cheng, H., & Wang, P. (2016). The request for better measurement: A comparative evaluation of two-factor authentication schemes. In *Proceedings of the 11th ACM on Asia Conference on Computer and Communications Security* (pp. 475–486).
23. Luo, H., Wen, G., & Su, J. (2018). Lightweight three factor scheme for realtime data access in wireless sensor networks. *Wireless Networks,* 1–16.
24. Amin, R., Islam, S. H., Gope, P., Choo, K. K. R., & Tapas, N. (2018). Anonymity preserving and lightweight multimedical server authentication protocol for telecare medical information system. *IEEE Journal of Biomedical and Health Informatics, 23*(4), 1749–1759.
25. Wang, C., Wang, D., Xu, G., & Guo, Y. (2017). A lightweight password-based authentication protocol using smart card. *International Journal of Communication Systems, 30*(16).
26. Ma, C. G., Wang, D., & Zhao, S. D. (2014). Security flaws in two improved remote user authentication schemes using smart cards. *International Journal of Communication Systems, 27*(10), 2215–2227.
27. Madhusudhan, R., & Mittal, R. C. (2012). Dynamic ID-based remote user password authentication schemes using smart cards: A review. *Journal of Network and Computer Applications, 35*(4), 1235–1248.
28. Wang, D., He, D., Wang, P., & Chu, C. H. (2014). Anonymous two-factor authentication in distributed systems: Certain goals are beyond attainment. *IEEE Transactions on Dependable and Secure Computing, 12*(4), 428–442.
29. Kocher, P., Jaffe, J., & Jun, B. (1999). Differential power analysis. In *Annual International Cryptology Conference* (pp. 388–397).
30. Messerges, T. S., Dabbish, E. A., & Sloan, R. H. (2002). Examining smartcard security under the threat of power analysis attacks. *IEEE Transactions on Computers, 51*(5), 541–552.
31. Kumari, A., Jangirala, S., Abbasi, M.Y., Kumar, V., & Alam, M. (2020). Eseap: Ecc based secure and efficient mutual authentication protocol using smart card. *Journal of Information Security and Applications, 51*, 2214–2126.
32. Safkhani, M., Bagheri, N., Kumari, S., Tavakoli, H., Kumar, S., & Chen, J. (2020). RESEAP: An ECC based authentication and key agreement scheme for IoT applications. *IEEE Access.*
33. Amin, R., & Biswas, G. P. (2015). An improved RSA based user authentication and session key agreement protocol usable in TMIS. *Journal of Medical Systems, 39*(8), 79–92.

Modeling, Simulation, and Comparative Analysis of Flyback Inverter Using Different Techniques of PWM Generation

Mangala R. Dhotre, Prashant V. Thakre, and V. M. Deshmukh

Abstract Nowadays, the flyback inverter has increased more interest due to its various advantages. Some of these are its simplicity, low cost, and high efficiency. Applying pulse-width modulation techniques to improve the performance of many types of inverters is under research. In this paper, distinct switching techniques generate the gate pulses for switches of the flyback inverter. The paper deals with the implementation and comparative analysis of three different models of flyback inverters: (1) open-loop flyback inverter with the pulse generator, (2) open-loop flyback inverter with pulse width modulation (PWM), and (3) open-loop flyback inverter with a sinusoidal pulse-width modulation (SPWM) technique. The output voltage and current signals are pure sinusoidal when the SPWM technique generates the gate pulses. The SPWM technique used in model-3 improves the nature of the output AC signal compared to the other two models.

Keywords Flyback inverter · PWM · SPWM · Photovoltaic system

1 Introduction

Nowadays, the demand for solar energy is continuously increasing in India due to its abundant availability. A lot of research work is carrying out on solar technology. Mostly, photovoltaic (PV) systems generate electricity [1–3]. When dealing with these PV systems, different DC–DC converters are used such as boost converter, buck-boost converter, Cuk converter, flyback converter have to be used.

M. R. Dhotre (✉)
SSBT's College of Engineering and Technology, Bambhori, Jalgaon, India
e-mail: mangala_dev@rediffmail.com

P. V. Thakre · V. M. Deshmukh
Kavayitri Bahinabai Chaudhari North Maharashtra University, Jalgaon, Maharashtra, India
e-mail: pvthakre2006@rediffmail.com

V. M. Deshmukh
e-mail: vmdeshmukh947@gmail.com

P. K. Singh et al. (eds.), *Proceedings of Third International Conference on Computing, Communications, and Cyber-Security*, Lecture Notes in Networks and Systems 421,
https://doi.org/10.1007/978-981-19-1142-2_49

Flyback converters are suitable for low output power applications where the circuit produces an output of few watts to 100 W. In the flyback converter, the output voltage remains isolated from the input supply voltage. It needs a smaller number of components compared to the other switch-mode power supply (SMPS) circuits. So, it has a simpler topology. The flyback converter uses unregulated DC voltage directly obtained by rectifying the utility AC voltage followed by a simple capacitor. The circuit produces single or multiple-isolated output voltages and operates over a wide range of input voltage variations. The flyback converter has high energy efficiency than other SMPS circuits, and due to its simple topology and low cost, it became more popular in the low output power range.

Now, the researchers are focusing on implementing the PWM techniques to different types of inverters for performance improvement by considering various factors. The single-phase sinusoidal PWM inverter with bipolar and unipolar switching techniques has been used. By applying the SPWM technique, the inverter gives sinusoidal AC output voltage with higher-order harmonics that is removed using a low-pass filter [4]. The voltage vector switching sequences based on the SVPWM scheme have been implemented for the performance evaluation of a two-level, five-phase voltage source inverter. The switching pattern is configured for quality control of the inverter system. The results show that all of the switching schemes have approximately the same fundamental component magnitude. All the methods are easy to execute and dependable [5]. An advanced PWM technique was applied for a modular multilevel cascaded (MMC), inverter-based grid-integrated solar photovoltaic (PV) system. It offers lower total harmonic distortion (THD) and power losses compared to the other modulation techniques [6]. A new modulation technique has been applied to the three-phase inverter to reduce the THD and inverter power loss. The system shows improved performance in terms of inverter power loss and THD than that of others [7]. Different conventional and advanced PWM techniques such as sinusoidal PWM (SPWM), third harmonic-injected PWM (THPWM), space vector PWM (CSVPWM), trapezoidal PWM (TRPWM), and bus clamping PWM (BCPWM) are applied for inverter switching for a 6-pulse diode rectifier-2 level-3 phase voltage source inverter (VSI) fed IMD to investigate the power quality indices [8]. PWM technique varies the voltage and frequency within the inverter. A comparative study of five different PWM techniques of three-phase inverter for best induction motor drive performance is presented using Simulink simulation. Sinusoidal PWM, sixty-degree PWM, trapezoidal PWM, third harmonic-injected PWM (THIPWM), space vector PWM techniques are presented in the paper. The PWM technique improves the inverter output voltage THD [9].

The study motivates to use of PWM techniques to improve the performance of the flyback inverter. Therefore, different switching techniques have been studied and implemented using Simulink. This paper deals with the implementation and comparison of three models: (1) open-loop flyback inverter with pulse generator (2) open-loop flyback inverter with PWM, and (3) open-loop flyback inverter with SPWM. The PV system produces the DC input of 34.4 V applied to the flyback inverter. The PWM generator generates the switching pulses for triggering the switches which improve the AC output signal. The three models have been implemented in MATLAB, and generated AC output signal is compared with respect to different parameters.

2 Working Principle

This paper consists of three models in which the simple topology of the flyback converter shown in Fig. 1 has been used [10]. The figure explains the operation of the flyback inverter.

The switch plays an important role in how the flyback converter works. When the switch is ON, the current will flow from V_{in} down to the primary ground. This will charge the primary winding and store energy. During this time, the secondary winding has no current flow as the diode is reverse bias. The load demand at this time is supplied by the output capacitance. When the primary switch is cut off, the primary winding will resist sudden change in current and reverses the polarity of the winding. This will result in the forward bias of the output diode. The stored energy in the primary will be transferred to the secondary and to the load via the diode. During this time, the output capacitor will replenish its charge [11, 12].

2.1 Advantages of Flyback Converter

- Output voltage is isolated from the input main supply.
- The overall circuit topology of this converter is considerably simpler than other SMPS circuits.
- Input to the circuit is generally unregulated DC voltage obtained.

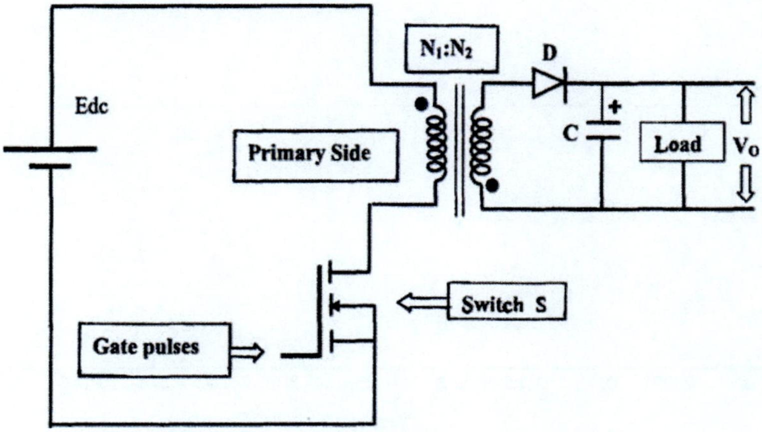

Fig. 1 Flyback inverter

- The circuit can offer single or multiple isolated output voltages and can operate over a wide range of input voltage variations.
- It is cost-effective.

3 Modeling of Open-Loop Flyback Inverter with Pulse Generator

Figure 3 shows the Simulink model of an open-loop flyback inverter [13, 14]. The model consists of the PV system, DC–DC flyback converter, and single-phase bridge inverter. The PV system uses two PV modules in series to produce the DC output voltage of 34.4 V shown in Fig. 2. Both the PV modules use the same PV current, i.e., 4.95 A and the same insolution of 1000 W/m^2. The parameters considered for this PV module are the short circuit current I_{sc}, open-circuit voltage V_{oc}, rated current, i.e., PV current at the maximum power point, and rated voltage under standard test conditions.

The DC–DC flyback converter uses a flyback transformer, MOSFET switch, diodes, inductor, and DC-link capacitor. MOSFET switch controls the functioning of the DC–DC converter. The parameters considered for transformer are voltage at winding 1 (V_{1rms}), the voltage at winding 2 (V_{2rms}), nominal power, and frequency. It uses 34.4 V primary voltage, 230 V secondary voltage, 1000 W nominal power, and 50 Hz frequency. The DC output voltage from the DC–DC converter drives the inverter. The inverter uses a simple bridge topology of the flyback inverter.

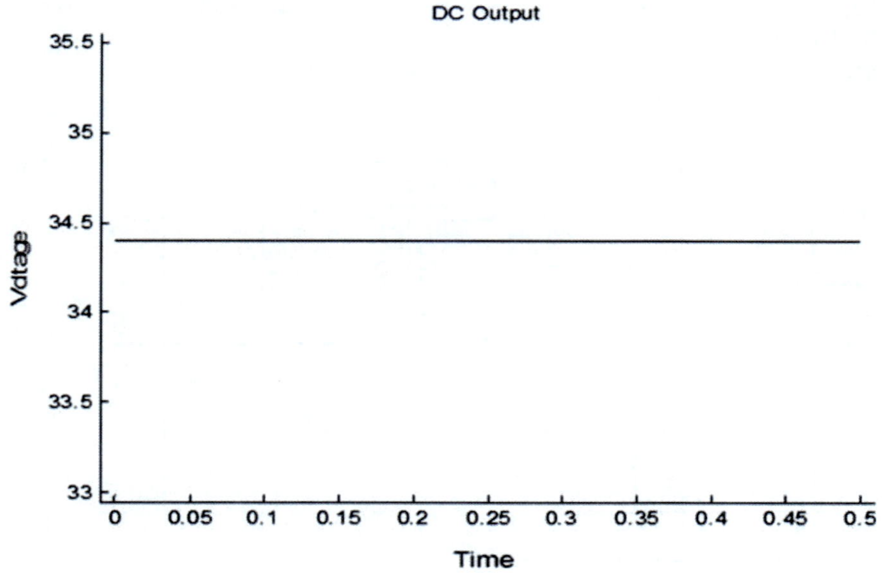

Fig. 2 DC output of PV model

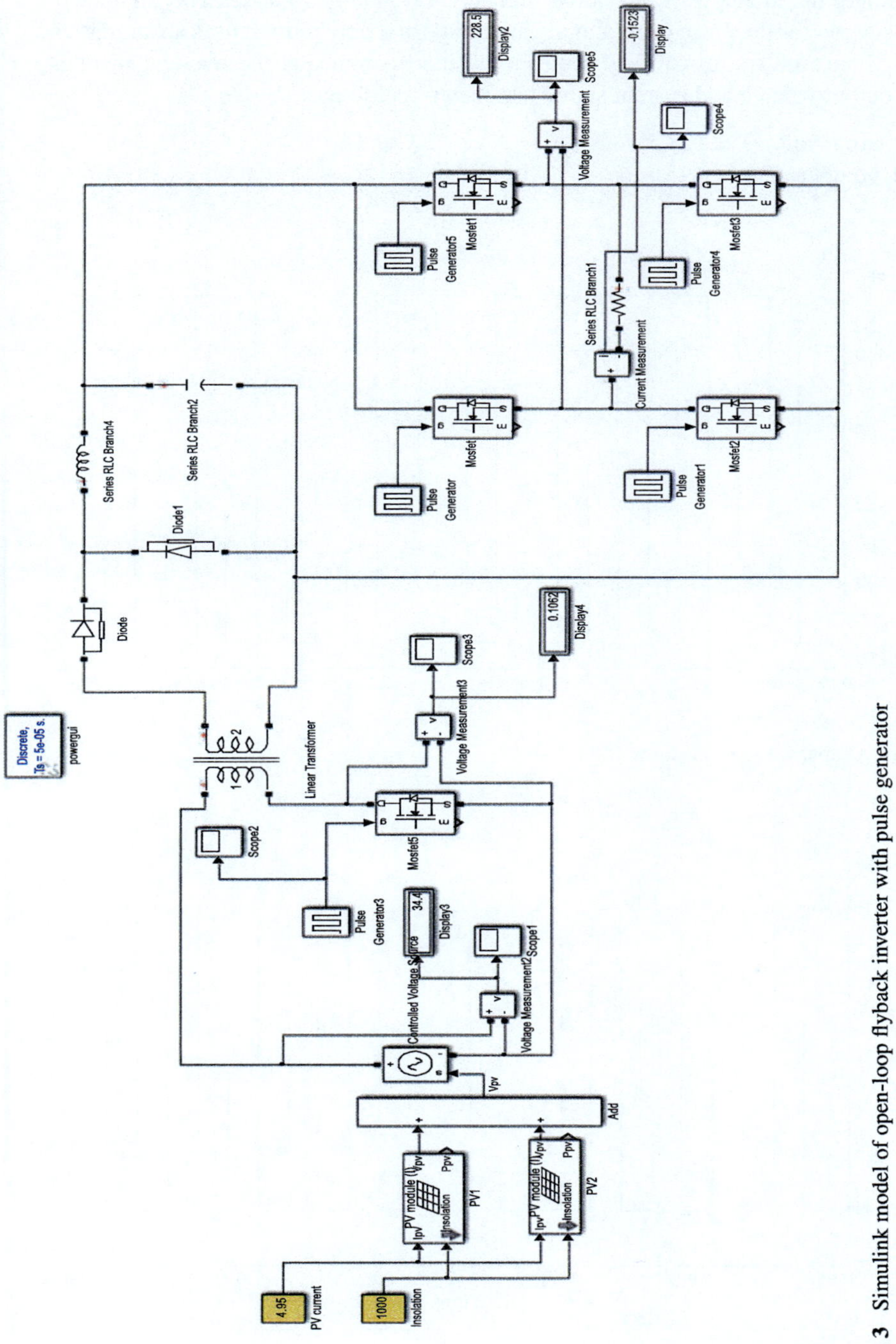

Fig. 3 Simulink model of open-loop flyback inverter with pulse generator

By considering the pulse generator to generate the gate pulses, the flyback inverter generates the square wave AC output signals shown in Figs. 4 and 5. The parameters considered for the pulse generator are amplitude and duty ratio. It uses an amplitude of 1 V and a duty ratio of 50%. By considering these parameters, the inverter generates the output voltage and current signal parameters as follows:

Output voltage = 228.5 V.
Output current = 0.15 A.

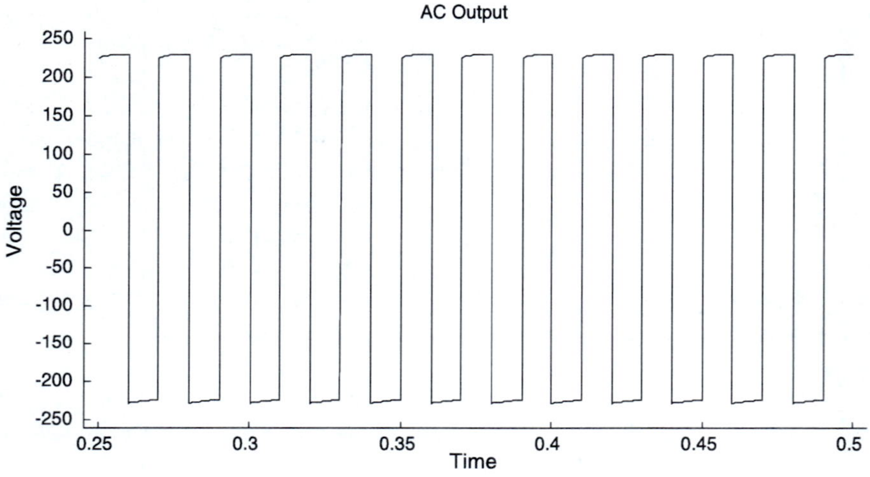

Fig. 4 Output voltage waveform

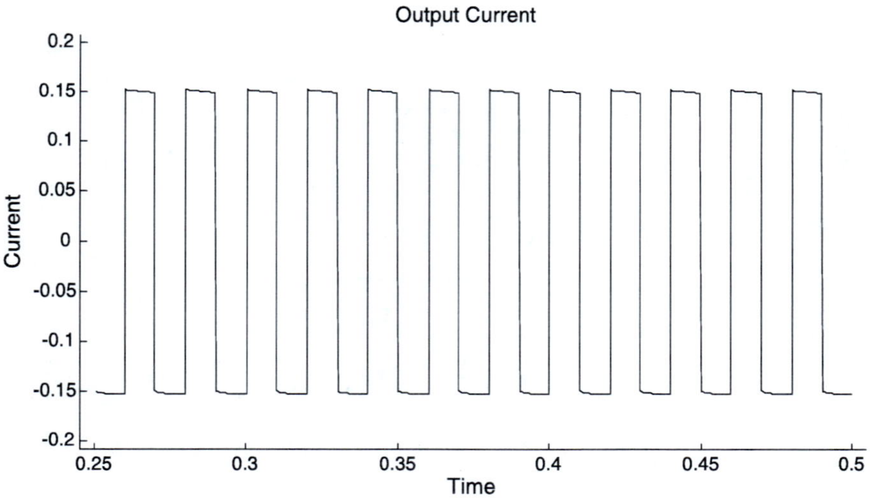

Fig. 5 Output current waveform

Output waveform = Square wave AC signal.

4 Open-Loop Flyback Inverter with Pulse-Width Modulation (PWM) Technique

In this model, the PWM generator produces the gate signal which drives the switches of the flyback inverter [15, 16]. The PWM technique has the following advantages:

1. PWM is a technique that is used to reduce the overall harmonic distortion THD in a load current.
2. Output voltage can be controlled without the addition of any external components.
3. Output can be controlled by adjusting the switching frequency of the transistor.
4. Power loss in switching devices is very low.

Due to these advantages, model-2 uses the PWM technique to generate the gate signal. This model uses a single PWM technique to produce an AC output signal. Figures 6 and 7 show the gate pulses applied to T1, T3, T2, and T4.

Figure 8 shows the Simulink model of an open-loop flyback inverter with PWM. The model uses a PV system, DC–DC flyback converter, simple bridge inverter, and PWM generator. The PWM generator compares the modulating signal and reference signal and generates the PWM signal. The parameters considered for the PWM generator are an amplitude of 1 V, 50% duty ratio, and modulating frequency of 50 Hz.

By considering 34.4 V DC voltage generated from the PV system, 230 V secondary voltage of flyback transformer, 1 V amplitude, and 50% duty ratio of PWM generator, the model generates the AC output voltage of 206.4 V. Figures 9 and 10 show the

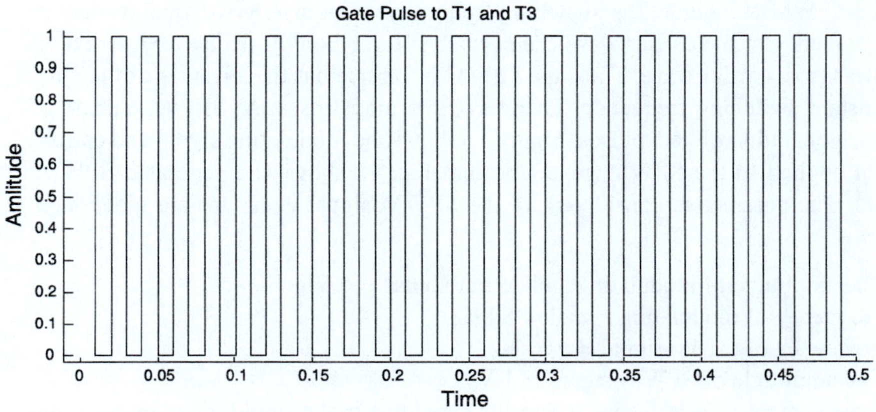

Fig. 6 Gate pulses to T1 and T3

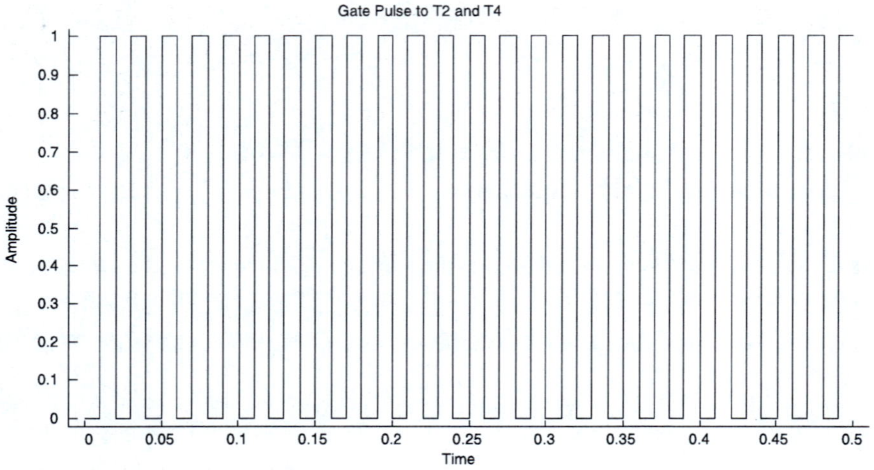

Fig. 7 Gate pulses to T2 and T4

output voltage and current signals. The voltage and current parameters generated output signals are mentioned below:

Output voltage = 206.4 V.
Output current = 2.581 A.
Output waveform = Square wave AC signal.

5 Open-Loop Flyback Inverter with Sinusoidal Pulse-Width Modulation (SPWM) Technique

In the SPWM technique, the switching frequency of an inverter is equal to that of the frequency of the carrier wave. The switch is turned on/off once every period of the triangular carrier wave. Thus, the SPWM technique has the advantage of having a constant switching frequency. Therefore, this model uses the SPWM technique for the generation of gate pulses. Figure 11 shows the modulating signal and carrier signal applied to the SPWM generator, and Fig. 12 shows the generated SPWM signal. The parameters considered for the SPWM signal generator are mentioned below:

The amplitude of modulating and carrier signal = 1 V.
Frequency of modulating signal = 50 Hz.
Frequency of carrier signal = 10 kHz.
The amplitude of SPWM signal = 1 V.
Frequency of SPWM signal = Varying according to the amplitude of modulating signal.

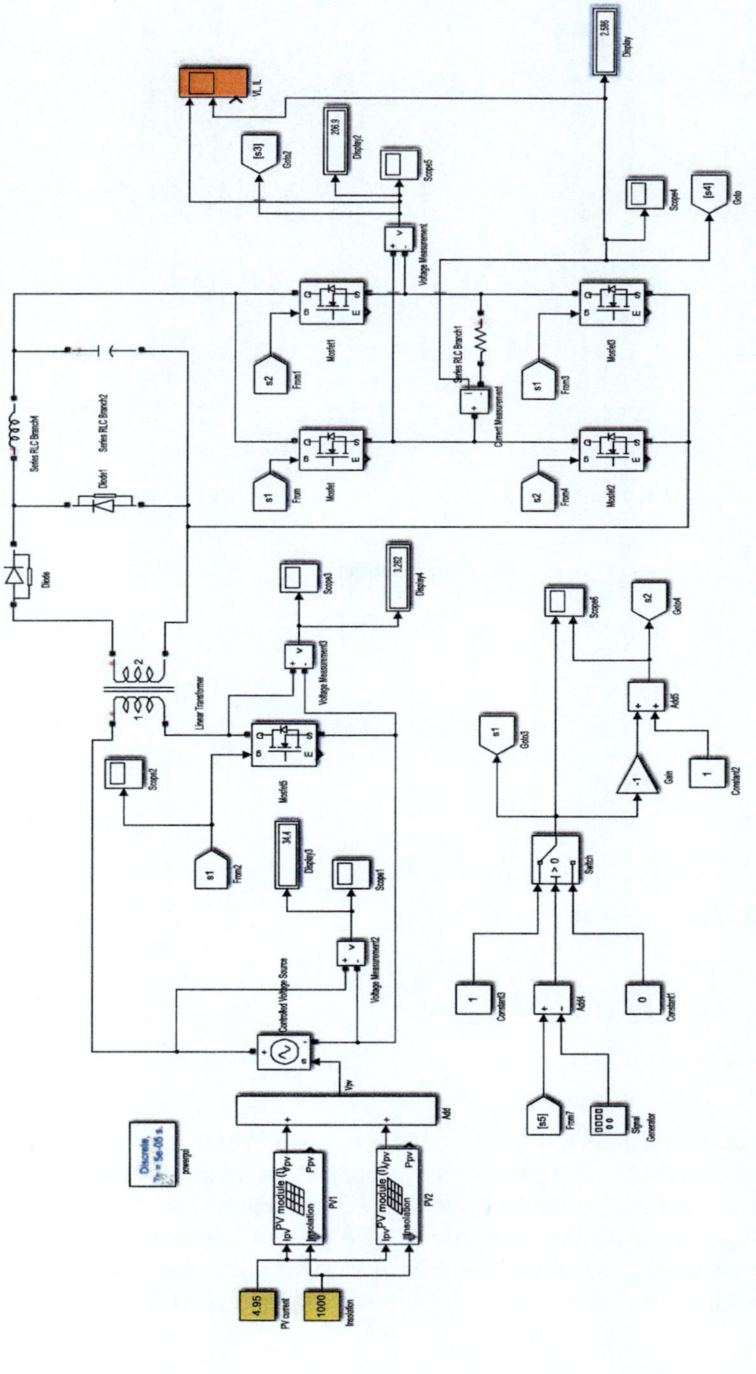

Fig. 8 Simulink model of open-loop flyback inverter with PWM

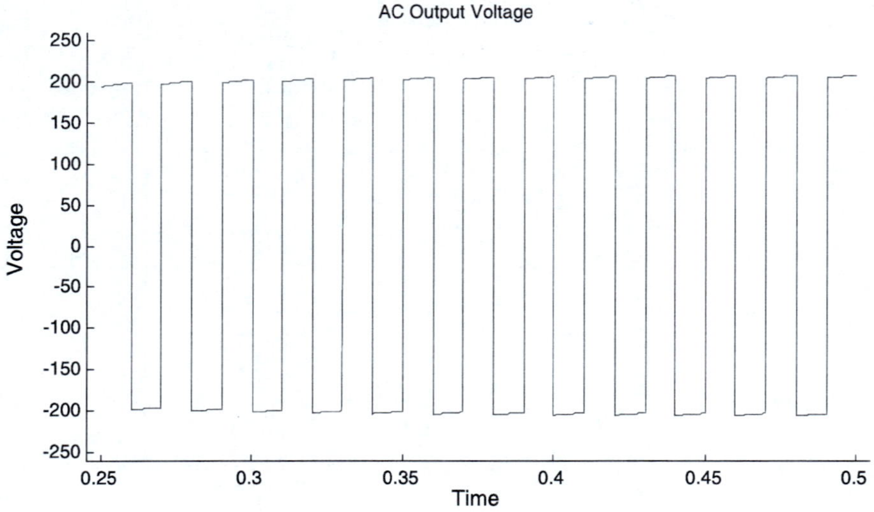

Fig. 9 Output voltage waveform

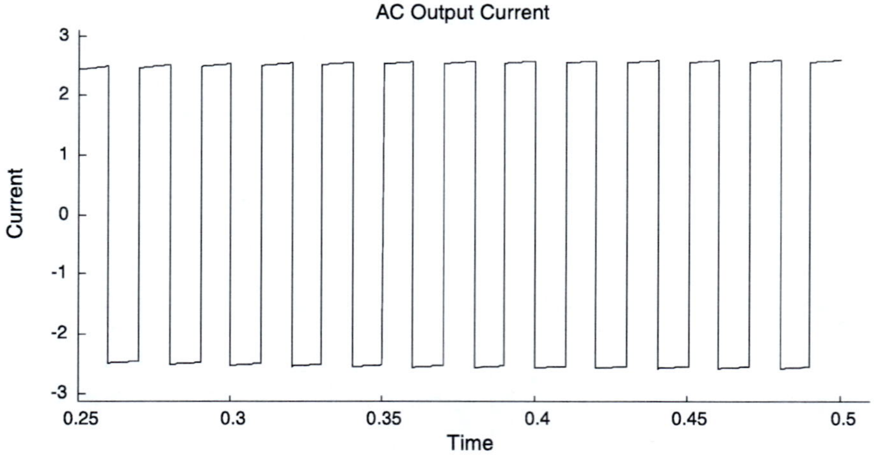

Fig. 10 Output current waveform

Figure 13 shows the Simulink model of an open-loop flyback inverter with an SPWM generator [17]. It uses the same PV system, DC–DC flyback converter, and simple bridge inverter used in model-1 and model-2. The model also uses an additional SPWM generator for the generation of gate pulses. The SPWM generator compares the modulating signal with the reference carrier signal and generator the SPWM pulses whose frequency varies with respect to the amplitude of the modulating signal.

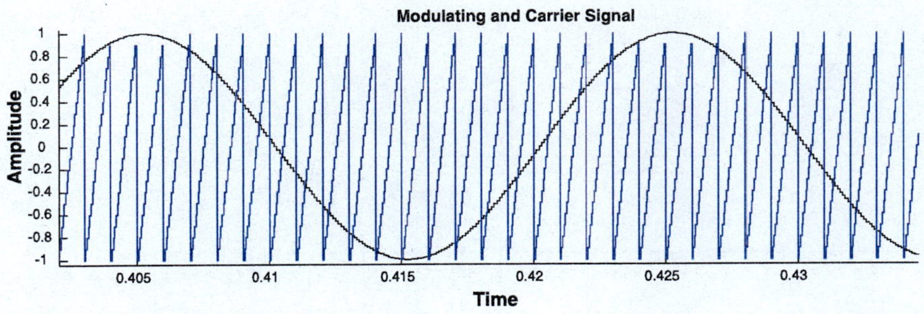

Fig. 11 Modulating and carrier signals

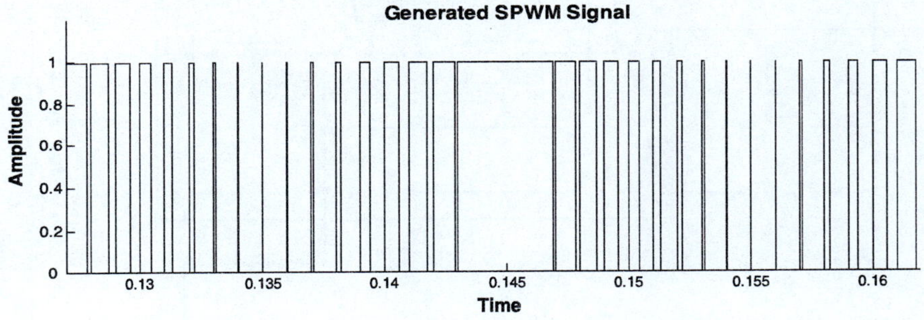

Fig. 12 SPWM signal

By considering the SPWM signal with a carrier frequency of 10 kHz and modulating the frequency of 50 Hz having amplitude 1 V, the flyback inverter generates the sinusoidal ac output signal. The output voltage and current waveforms of the model with SPWM are shown in Figs. 14 and 15. The AC output signal measures the following parameters:

Output voltage = 191.2 V.
Output current = 1.529 A.
Output waveform = Close to sine wave.

6 Results and Discussion

The DC input applied to the flyback inverter is obtained from the PV module which is equal to 34.4 V. Each model is tested with distinct values of insolation. Tables 1, 2 and 3 show the performance of the three models.

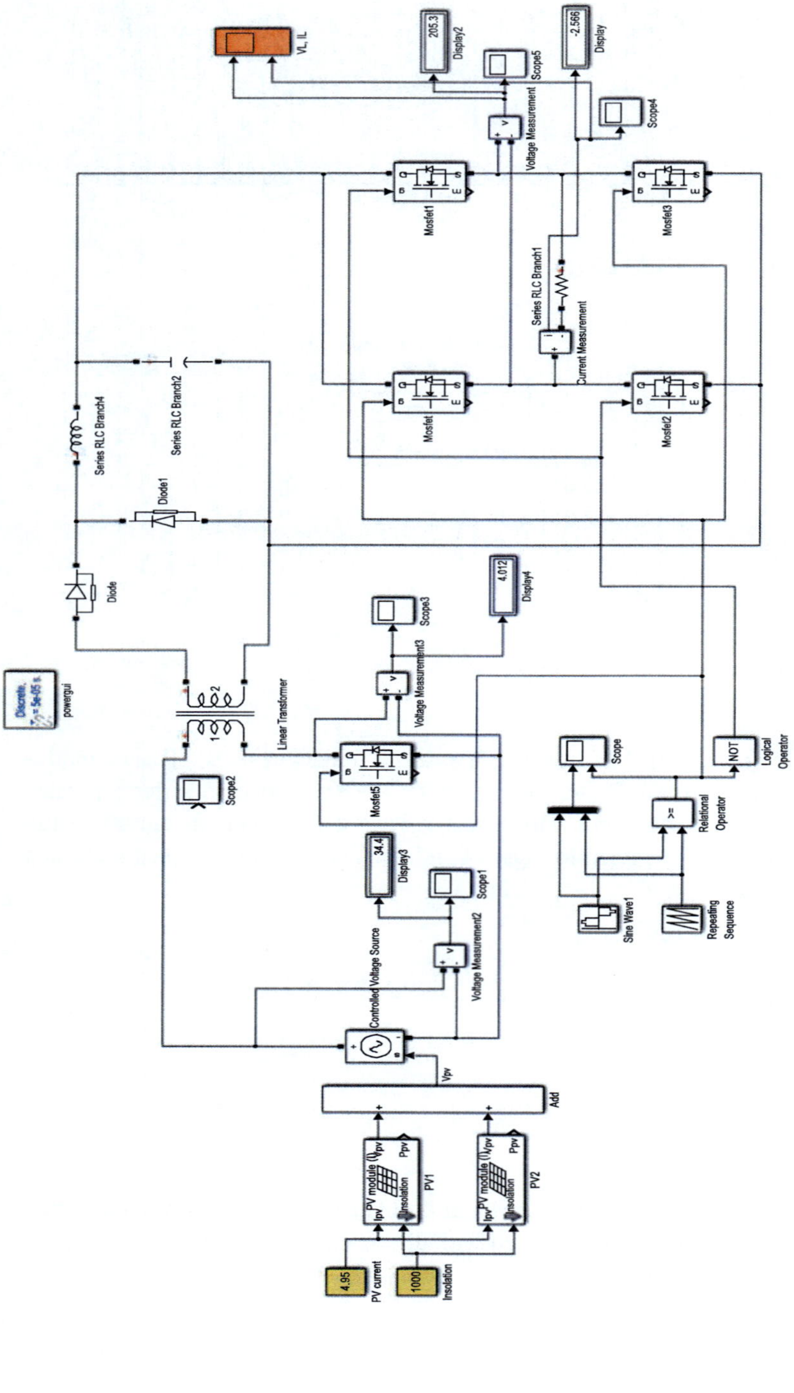

Fig. 13 Simulink model of open-loop flyback inverter with SPWM

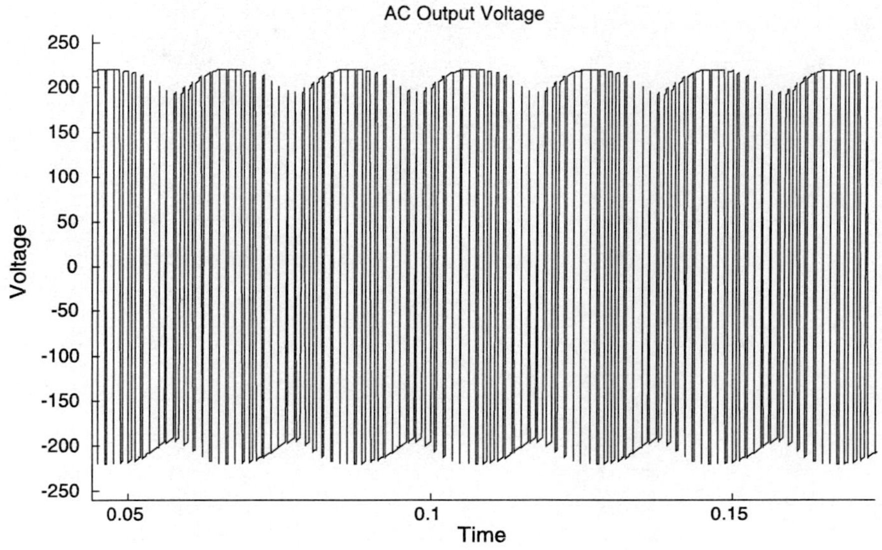

Fig. 14 Output voltage waveform

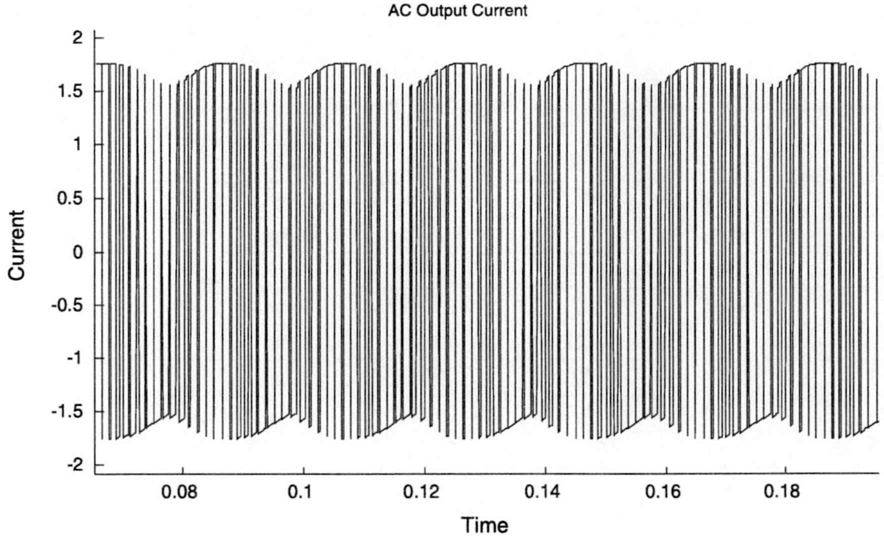

Fig. 15 Output current waveform

Table 1 Parameters for open-loop flyback inverter with pulse generator

Insolation	V_{pv} (V)	V_{out} (V)	I_{out} (A)	Nature of waveform
970	33.03	219.3	0.1462	Square wave AC signal
975	33.34	221.4	0.1476	Same
980	33.6	223.1	0.1488	Same
985	33.83	224.7	0.1498	Same
990	34.04	226.1	0.1507	Same
995	34.23	227.3	0.1515	Same
1000	34.4	228.5	0.1523	Same
1005	34.56	229.5	0.153	Same
1010	34.7	230.5	0.1536	Same
1015	34.84	231.4	0.1542	Same
1020	34.96	232.2	0.1548	Same
1025	35.08	233	0.1553	Same
1030	35.19	233.7	0.1558	Same

Table 2 Parameters for open-loop flyback inverter with pulse-width modulation

Insolation	V_{pv} (V)	V_{out} (V)	I_{out} (A)	Nature of waveform
970	33.03	198.2	2.478	Square wave AC signal
975	33.34	200	2.501	Same
980	33.6	201.6	2.52	Same
985	33.83	203	2.538	Same
990	34.04	204.3	2.554	Same
995	34.23	205.4	2.568	Same
1000	34.4	206.4	2.581	Same
1005	34.56	207.4	2.592	Same
1010	34.7	208.3	2.603	Same
1015	34.84	209.1	2.613	Same
1020	34.96	209.8	2.623	Same
1025	35.08	210.6	2.632	Same
1030	35.19	211.2	2.64	Same

Table 4 shows the comparative analysis of three models. Model-3 with SPWM technique produces the sinusoidal AC output signal with an output voltage of 205.3 V and output current of 2.566 A.

Table 3 Parameters for open-loop flyback inverter with sinusoidal pulse-width modulation

Insolation	V_{pv} (V)	V_{out} (V)	I_{out} (A)	Nature of waveform
970	33.03	183.5	1.468	Close to sine wave
975	33.34	185.2	1.482	Same
980	33.6	186.7	1.494	Same
985	33.83	188	1.504	Same
990	34.04	189.2	1.513	Same
995	34.23	190	1.522	Same
1000	34.4	191.2	1.529	Same
1005	34.56	192	1.536	Same
1010	34.7	192.8	1.543	Same
1015	34.84	193.6	1.549	Same
1020	34.96	194.3	1.554	Same
1025	35.08	195	1.56	Same
1030	35.19	195.6	1.565	Same

Table 4 Comparative analysis of the models

Model	V_{out} (V)	I_{out} (A)	Nature of W.F
Open-loop FBI with pulse generator	225.8	0.15	Square
Open-loop FBI with PWM	206.4	2.581	Square
Open-loop FBI with SPWM	205.3	2.566	Sinusoidal

7 Conclusion

Flyback converters are gaining more popularity in solar photovoltaic applications due to their simplicity, low cost, lightweight, and improved efficiency with proper PWM technique implementation. It has also been observed that the output signal is being controlled with the variation in switching frequency. The proposed model also tested with different values of insolution, and it has been observed that the PWM techniques are efficiently working even with the variation in insolation values which shows the flexibility and viability provided with the model.

References

1. Huang, Y.-P., & Hsu, S.-Y. (2016). A performance evaluation model of a high concentration photovoltaic module with a fractional open circuit voltage-based maximum power point tracking algorithm. *Computers and Electrical Engineering,* 331–342.
2. Putri, R., Wibowo, S., & Rifa'i, M. (2015). Maximum power point tracking for photovoltaic using incremental conductance method. In *2nd International Conference on Sustainable Energy*

Engineering and Application (pp. 22–30).
3. Balamurugan, T., & Manoharan, S. (2012). Fuzzy controller design using soft switching boost converter for MPPT in hybrid system. *International Journal of Soft Computing and Engineering, 2*(5), 87–94.
4. Zala, J., et al. (2019). Sinusoidal pulse width modulation switching technique based single phase inverter. *Journal of Power Electronics & Power Systems, 9*(1).
5. Gupta, S. K., Arif Khan, M., & Singh, O. (2021). Pulse width modulation switching schemes for two-level five-phase voltage source inverter. *European Journal of Electrical Engineering, 23*(2), 137–142.
6. Haq, S., et al. (2021). An advanced PWM technique for MMC inverter based grid-connected photovoltaic systems. *IEEE Transactions on Applied Superconductivity, 31*(8).
7. Biswas, S. P., et al. (2020). A new modulation technique to improve the performance of three-phase inverters. In *Proceedings of 2020 IEEE International Conference on Applied Superconductivity and Electromagnetic Devices*, Tianjin, China, October 16–18, 2020.
8. Biswas, S. P., et al. (2021). Investigation of the impact of different PWM techniques on rectifier-inverter fed induction motor drive. In *Australasian Universities Power Engineering Conference (AUPEC)*, December 2020. IEEE Explorer.
9. Mahbub, M. (2021). Comparative analysis of five different PWM techniques on three-phase voltage source inverter fed induction motor drive. In *2nd International Conference on Robotics, Electrical and Signal Processing Techniques (ICREST'21).*
10. http://nptel.ac.in>IIT Kharagpur>PDF>Lesson 22
11. Shimizu, T., Wada, K., & Nakamura, N. (2002). A flyback—Type single phase utility interactive inverter with low-frequency ripple current reduction on the DC input for an AC photovoltaic module system. In *IEEE 33rd Conference.*
12. Kasa, N., & Iida, T. (2002). Flyback type inverter for small scale photovoltaic power system. In *IECON IEEE 2002 28th Annual Conference of the Industrial Electronics Society* (Vol. 2, pp. 1089–1094).
13. Yang, B., Li, W., Zao, Y., & He, X. (2010). Design & analysis of grid-connected photovoltaic system. *IEEE Transaction on Power Electronics, 25*(4), 992–1000.
14. Khader, S. (2011). Modeling and simulation of various inverter circuits for photovoltaic applications. *International Journal of Electrical and Power Engineering, 5*(2), 74–83.
15. Thakre, P. V., & Rangnekar, S. (2014). Implementation of digital controller TMS320C28027 to MPPT based single phase bidirectional high-frequency link inverter for photovoltaic applications. *International Journal of Renewable Energy Research, 4*(1).
16. Kamalakannan, N., Arulvizhi, A., & Kirthana, C. (2015). Matlab simulink model for flyback inverter with active clamp technique. *International Journal of Innovative Research in Electrical, Electronics, Instrumentation, and Control Engineering, 3*(3).
17. Maheshri, S., & Khampariya, P. (2014). Simulation of single phase SPWM (Unipolar) inverter. *International Journal of Innovative Research in Advanced Engineering (IJIRAE), 1*(9).

Industrial Rod Size Diameter and Size Detection

Swathi Gowroju, N. Santhosh Ramchander, B. Amrita, and S. Harshith

Abstract Thermo-mechanical treatment (TMT) rods are the walloping production of the steel industries, and there are giant machines that will make the task of cutting TMT rods easier for industries. While cutting the rods, photoelectric sensor, manual labor, and complex computing machine are used that need huge maintenance of the machine, and it is a time-consuming process. In the last decade, research on digital image processing and computer vision has seen much progress. In this paper, we propose an adaptable methodology for the industries in measuring the TMT rods much more efficiently, maximizing the efficiency, robust toward cram-full of rods and minimizing the error rate. The captured digital image first undergoes the preprocessing phase, where the first step is image enhancement and then edge detection, which extracts the TMT rods edges then followed by the diameter calculations (pixels per metric). An experiment has been conducted with various challenging conditions to demonstrate the capability of our approach to a good measure of success.

Keywords TMT rod identification · TMT rod diameter · TMT rods detection · TMT rods counting · TMT bar detection

1 Introduction

Steel manufacturing companies produce a walloping amount of the TMT rods, as shown in Fig. 1, which are used heavily in construction sites, and the length and diameter vary according to the structure of the buildings. Industries meet this requirement by using enormous machinery that first identifies the size and cut according to it.

S. Gowroju (✉) · N. Santhosh Ramchander · S. Harshith
Department of CSE, Sreyas Institute of Engineering and Technology, Hyderabad, India
e-mail: swathigowroju@sreyas.ac.in

N. Santhosh Ramchander
e-mail: nsramchander@sreyas.ac.in

B. Amrita
Department of CSE, G. Narayanamma Institute of Technology and Science, Hyderabad, India

P. K. Singh et al. (eds.), *Proceedings of Third International Conference on Computing, Communications, and Cyber-Security*, Lecture Notes in Networks and Systems 421,
https://doi.org/10.1007/978-981-19-1142-2_50

Fig. 1 TMT rods, stack of TMT rods

This process is done by a photoelectric sensor, manual labor, and higher complex counting machines. These methods are tedious, costly, have much human effort, and are highly prone to error. Despite a growing vision-based measuring considering the different positional conditions of the rods, in this paper, we propose a practical and robust methodology that can help in detecting the edges of the rods and automatically measure the diameter of the TMT rods. (This methodology is adopted by considering industrial standards and keeping an ideal situation of the machinery available in the market.) The algorithm is formed in three steps: (1) preprocessing/image enhancement, (2) edge detection, and (3) measuring diameter. In the first step, i.e., preprocessing/image enhancement, a digital image will be given as an input image. We apply Gaussian blurring, which will help us in reducing the noise followed by dilation + erosion. In the second process, we applied the proposed feature detector of 3×3. We will get a future map image that will be an input to the third process, where we will measure the rod diameter using the pixels per metric. Many other concepts are used to determine the object size in an image, but this concept will best suit this process. This proposed methodology works best in identifying the TMT rods edges and finding the diameter of the TMT rods without using any deep learning or machine learning models as an industrial practice that should meet the time complexity factor in the production level grade.

Our methodology has been tested in various conditions such as heat, sawdust, and water. Initially, this project was done using the Raspberry Pi, later adopted with process logic controller (PLC) as the whole machinery of the industries work with the PLC triggering the signals was done with the Ethernet cables connected with the Raspberry Pi 3 and camera attached to the Raspberry Pi. The central ideology of this paper is to provide the methodology and best practices of improving the computer vision capabilities in the area of mechanical industries, and to provide a cost-effective solution to their products, we proposed the best filter (kernel) that gives an outline of the TMT rod which will save the trial and error process of the filters, applying the concepts of pixels per metric for measuring the diameter of TMT rods and counting the rods using the counters reducing the cost of identification from traditional approaches.

2 Literature Survey

Generic approaches that exist in the industries will be using the light addressable potentiometric (LAP) optical sensor, binary template matching [1], lattices detection [2], and automatic detection of counting of circular and rectangular steel bar [3]. There are several methods for identifying the TMT rods, which depend on the point of reference of the TMT rod's image. Adopting the existing methodology in counting the TMT rods using the CNN [3], which will be not suited for this particular problem, while cutting off the TMT rods, the side views will not give the exact size of the rods an uneven cutting will be there if you consider Fig. 2. We took advantage of the top view of the rods. Hummel et al. [4] proposed deblurring Gaussian blur on the images using Hermite polynomials to achieve approximate deblurring. Miyuki and Shun [5] used shape detection algorithm to count the number of pipes in the image., However, this method fails in counting the rods if any other object with the same shape appears in the image. Refs. [6–8] used morphological edge detection on segmented images to generate accuracy over standard edge detection.

As we mentioned, we are using a practical approach of identifying the TMT rod diameter, where the camera is placed inside the cutting machine (an ideal machine available in the market). The digital image captures while cutting the rods, and rods are placed in the conveyor belt, which is about to be cut. The top view of the image captured by the camera inside the machine is shown in Fig. 3. The image will be

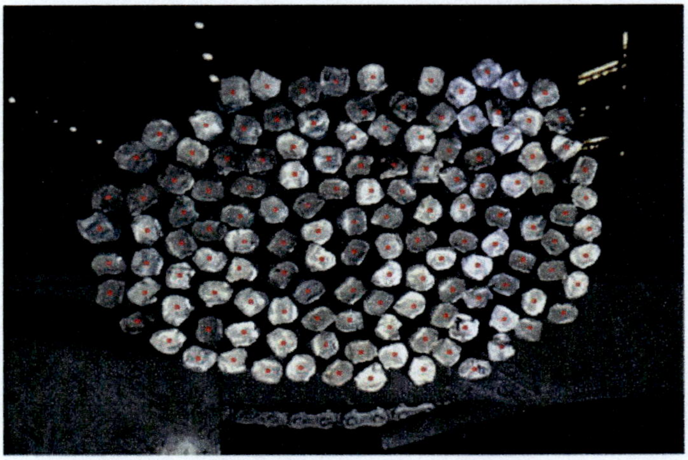

Fig. 2 Centroid of the TMT rod forms the side

Fig. 3 TMT rods placed on conveyor belt view

captured before the cutting process of the TMT rods, and TMT rods are placed on the machine (shown in Fig. 4).

We will be capturing the image and giving a count of rods and size of the rods if the size of the rods is the same. The machine will be given a trigger signal to the machine communicated through the PLC to continue the cutting, and if not, it will stop the cutting. Hence, we design an appropriate diameter estimator to generate a centroid of the rod to cut them into equal size.

Fig. 4 TMT rod cutting machine

3 Proposed Method

In this proposed methodology, the algorithm overview is shown in Fig. 5 is divided into three main steps: (1) preprocessing/image enhancement, (2) edge detection, and (3) measuring diameter, where the below flowchart will be giving an overall idea of how the image is filtered through the process.

Image Preprocessing

In the preprocessing technique, we will be doing the image enhancement techniques such as reducing the noise, correcting the pixels, converting the image from the RGB into the grayscale, calibration of the images, reducing the image size, getting the counters of the subject in the image, and correcting the brightness of the image. Considering the different problems faced by the industries is the sawdust; in our case the heat of the TMT rods, when placed on the machine, making a note of this problem, different image datasets are taken into the consideration while designing this algorithm where the analysis of each image is explained in details in further readings of the paper. The image is captured by the camera, as shown in Fig. 6, and the image is converted into grayscale. After that, a histogram is created, and pixels intensity is observed from the histogram generated in Fig. 7, noticing the $f_{\text{max}} = 6000$ and $f_{\text{min}} = 0$ (as the dark region is observed) calculating the image's dynamic range done by the formulae given below.

$$\text{dynamic range} = 20\log(f_{\text{min}} - f_{\text{max}})$$

The dynamic range of our image is 60 dr, so the dynamic range is low; at this point, making a note of the dynamic range will be used as a parameter consideration for the analysis of the image in further process which now considers Fig. 6, where there is an smooth blend in the curve of the image that indicates that we cannot differentiate the rods and there is intuition for this if you analyze the image where the only pixel that differentiates the rods is the high-intensity pixel of the rod and

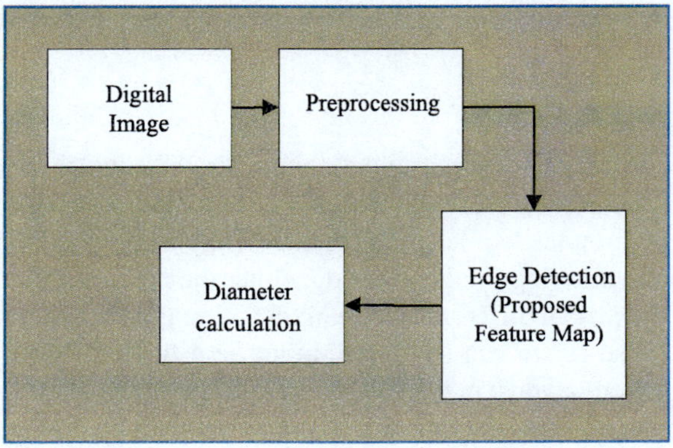

Fig. 5 Flowchart of the overall process

Fig. 6 Input image

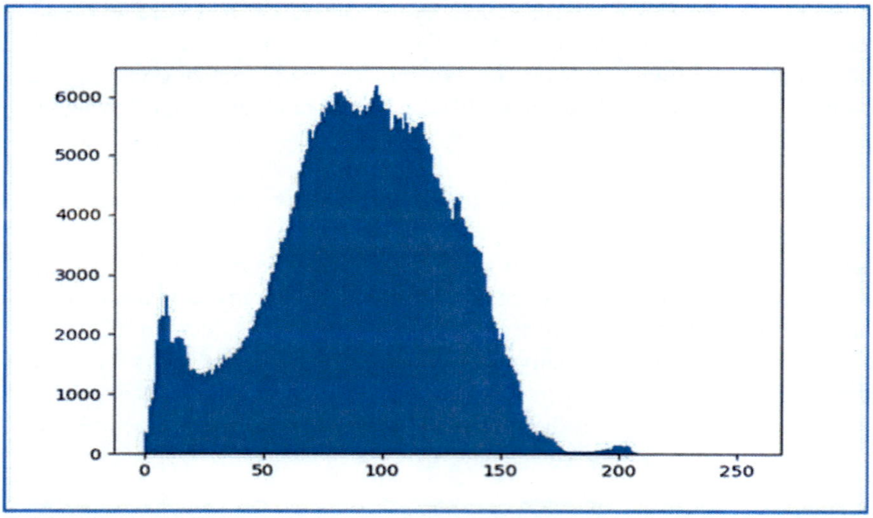

Fig. 7 Histogram of input image

the gaps of low-intensity pixel that make the counter to identify the edges of TMT rods and getting those fluctuation are essential in our case; for that we need to make the high-intensity pixel to max and low-intensity pixel to the lowest. Then process is done by filtering. In this filtering, either the pixel is multiplied, subtracted, or added so that we can differentiate the pixel before going to that process, we need to reduce the noise as we can see there might be a fluctuation in the pixel intensity due to unwanted sawdust in the machine, and we need to smoothen those sawdust and need to reduce those fluctuations.

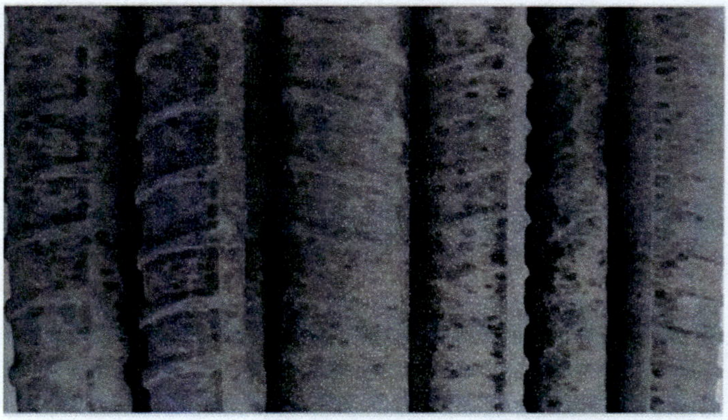

Fig. 8 Morphological closing

As shown in Fig. 7, there is a slight difference in the intensity level compared to Fig. 9 with histogram for every pixel intensity. There are many changes because of the TMT rods threads that do not make the pixel intensity fluctuate. We will further negotiate the pixel's intensity inside the rod, which makes a difference in this process. At the same time, if you go from one rod to another rod, the pixel will be to our intuition as proposed before high-intensity pixel will be at its highest and low pixel be at lowest as the newer image will be somewhat darker as compared to Fig. 6. The dynamic range will be greater than 80 dr in our case so that we need to reduce to match those dynamic ranges. The formula we applied for the morphological closing pulse Gaussian filter and morphological closing is as follows. The resultant image is shown in Fig. 8.

$$G(x) = \frac{1}{2\sqrt{2\Pi\sigma x^2}}.e\frac{-x^2}{2\sigma}$$

Edge Detection

We proposed a kernel-based contour detection method, which gave us accurate rod diameter when performed using the Prewitt y operation. The proposed edge detection is shown in Fig. 10.

The input image is converted into gray and analyzed with histogram features. As the shape of the edges is unclear, we heated the rod and checked with the RGB values. The result of edges has shown significant distortion in the histogram. Hence, we proposed the above algorithm to perform the Prewitt y transform to apply a filter using the proposed kernel. The precise edges were shown in the histogram after applying the filter. On the obtained feature map, the edge detection algorithm is processed to find the definite edges. The complete analysis has been explained in the results section.

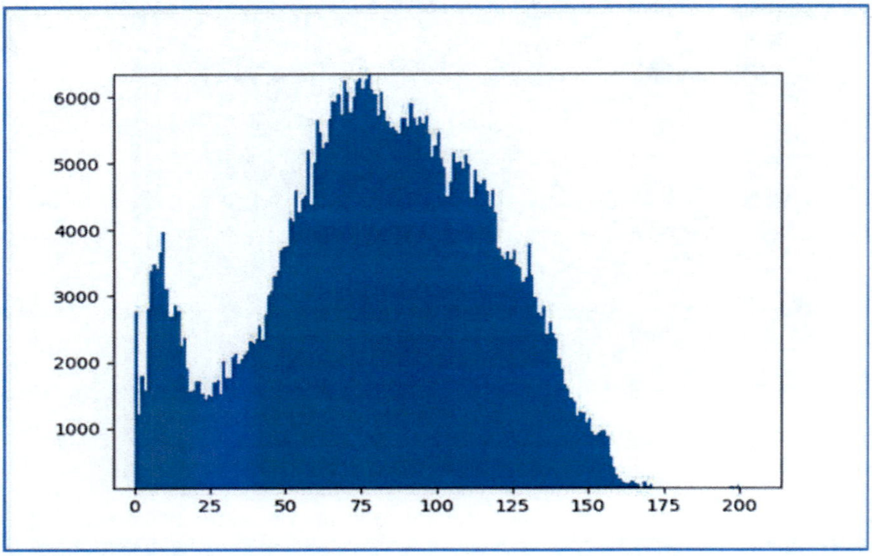

Fig. 9 Histogram of the modified image

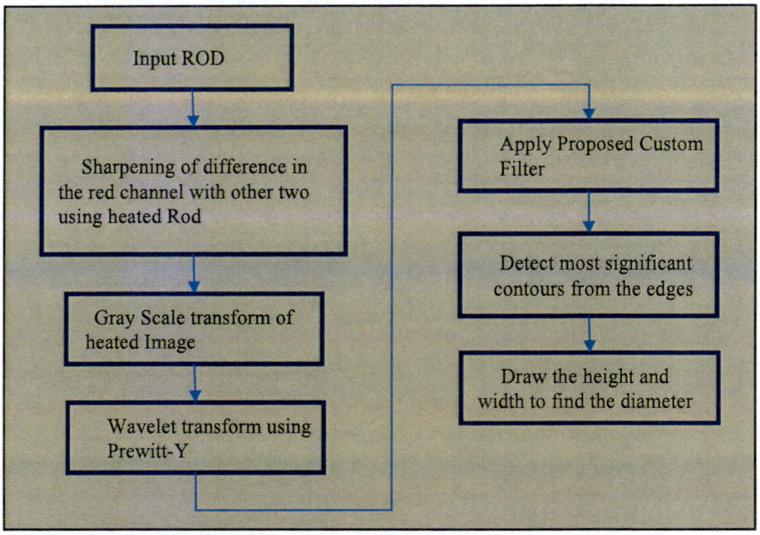

Fig. 10 Proposed system flowchart

4 Proposed Method

At the starting of this methodology, we have mentioned a detailed description of the conversion, where the image is converted from three channel input to one channel.

From the RGB to grayscale image conversion, the followed analysis of the images has been taken. Another parameter that has to be considered for this conversion is that the red hot TMT rods are placed on the conveyor belt while cutting the rod when it is in the hot state. Let us analyze the image when it is in the standard form, as shown in Fig. 6. While seeing the RGB channel, there is the monopoly of the channel in the histogram constructed. However, the red channel might fluctuate as it is hot for the further process, so take this parameter, and see Fig. 11.

TMT rod image is in normal conditions if the RGB channel histogram of the heated TMT rod in Fig. 12 shows the monopoly of the red color in the channels. Figure 12 will not represent the influence of the image for further process in finding the counters. The various factors will not affect the analysis of the image while calculating the TMT rod diameter. The pixel's intensity will remain the same while

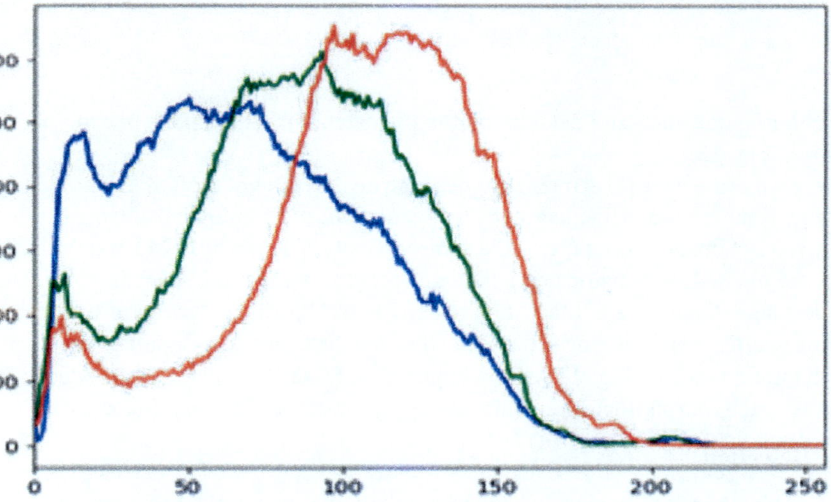

Fig. 11 RGB channel of the input image

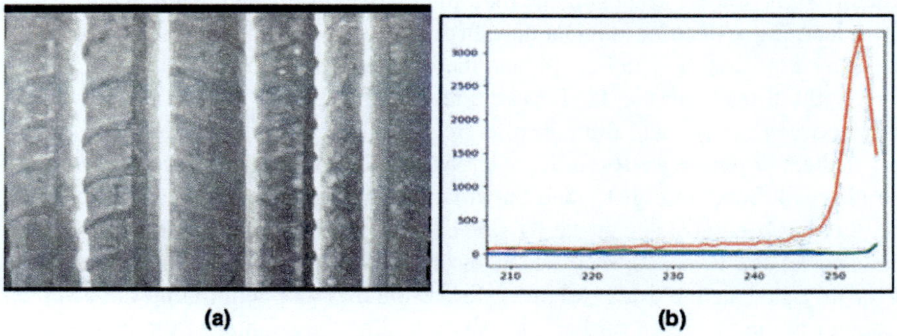

Fig. 12 **a** Heated input rod. **b** RGB analysis of heated rod

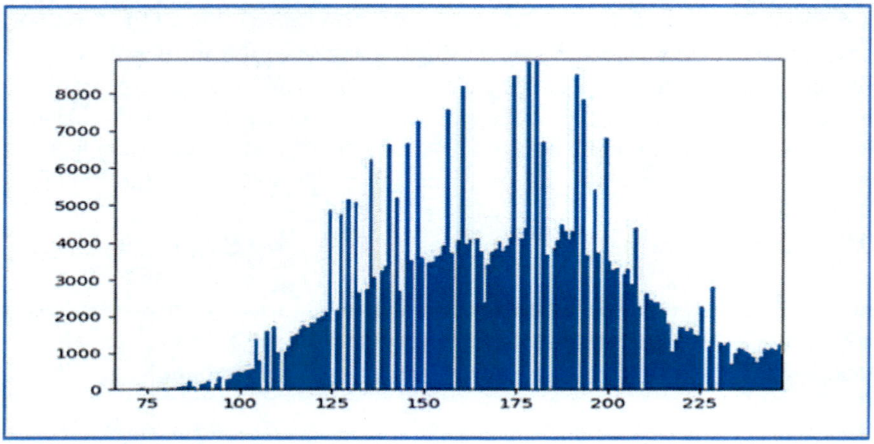

Fig. 13 Histogram of the red hot TMT rod

doing the feature map and calculating the diameter using the pixels per metric, as shown in Fig. 13.

Here we can see the pixel intensity goes high on the edges of the TMT rods. Hence, we need to suppress this fluctuation by using the same morphological closing applied on the red hot image and reporting the same intensity. The red hot TMT rods will not affect the methodology while using the same process in the image enchantment. The output image is sent to the following process known as edge detection of the heated rod image after preprocessing is passed to the edge detection as a digital image input. The dynamic range of Fig. 13 will be higher than 80 dr, and then it also reduced to 60 dr after the morphological closing and applying of the Gaussian filter.

Edge Detection

In this stage, the output image outs the best-suited filter that will find the edges of the TMT rods that will help determine the diameter using the pixel per metric concept. Applying different kinds of filters available and choosing the best out of them will be crucial in this process. Choosing the right filter will help out for not going into the machine learning or deep learning models. In this, we will first apply the well-known filter Canny edge as shown in Fig. 14, and there is no noise in the image.

Due to the threads on the TMT rods, threads that increase the surface area of contact between the bar and the concrete are significant when the rods are in the cram-full state. We have analyzed many already well-known filters, such as Canny edge in Fig. 14, Sobel in Fig. 15, and Prewitt x + y in Fig. 16. A thorough analysis of the filter has been made on a side-by-side comparison. The conclusion was made in choosing of the Prewitt x + y which further digs into that where applying of the Prewitt y gave us the exact rod x-ray dimension that will help to find the TMT rod counter in the image on further analysis in finding the counter of the image as shown in Fig. 17 that will be leading to the low counter, and even after increasing

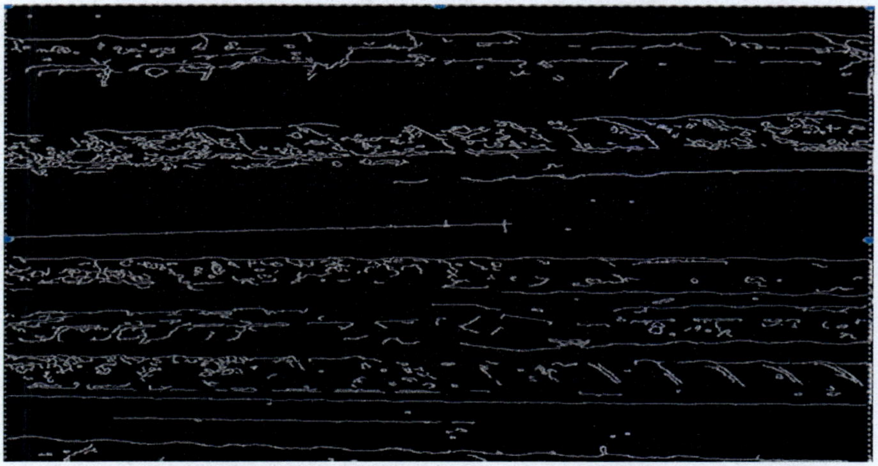

Fig. 14 Canny edge detection outcome

Fig. 15 TMT rods Sobel filter

the counter, there are no expected results. On further thought process and trial and error, we proposed a convolutional matrix or masks that will reduce the contrast and increase the shaping of the images which help to identify the edges of the TMT rods; the proposed kernel as shown in Fig. 18 will help detect outer lines of the TMT rods; the kernel is obtained on slight parameter tuning of the Prewitt y, in Fig. 17, and has helped to find the contours as shown in Fig. 19.

Fig. 16 TMT rods Prewitt x + y

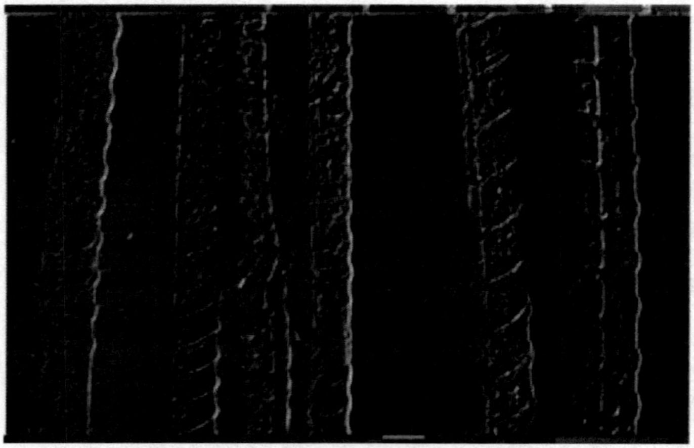

Fig. 17 Applying the Prewitt y

Fig. 18 TMT rods after applying the custom kernel

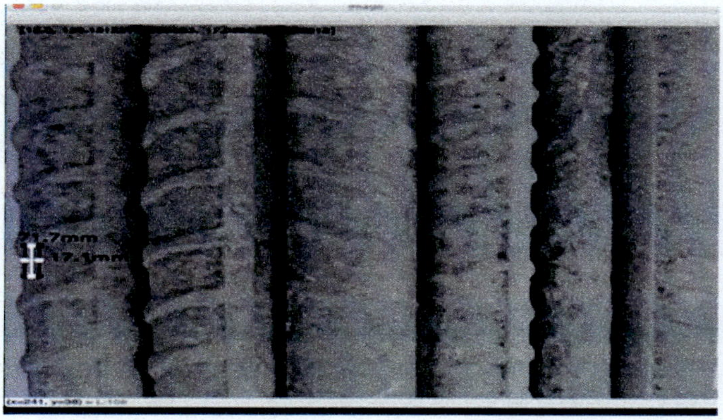

Fig. 19 Low contours

$$\begin{array}{ccc} -.9 & 0 & .9 \\ -.8 & 0 & .7 \\ -1 & 0 & 1 \end{array}$$

Finding the diameter plays a crucial role in our methodology with a camera. We proposed a new methodology in this paper using the pixels per metric concept: If an object length is known in an image, we can calculate other objects' length and height. If an object x of length has XY pixels, then objects in the same frame having XY pixel are calculated using the essential relation [9] by the associative law. However, there will be another problem arriving from this approach: a concave distortion that affects calculating the exact pixels of the object in the image, considering that the camera position is fixed and TMT rods are always placed at the same constant distance the conveyor belt.

To avoid this error, we need to keep the camera 90 to the reference point, maintaining a 90° angle between them. After this, we will be applying the above process to the preprocessed digital image that we have got as an output from the above two steps after applying the above-proposed method. The output of the image is shown in Fig. 9. The contours are not detecting because of the slight noise created by the threads of the TMT rods. The contours output is not expected a reference [10] where they demonstrated how to get the outer counter, and large countries helped to apply the methodology of increasing the contour value in our case to 1000 that helped us to improve in getting larger contour that is the diameter of the TMT rods as shown in Fig. 20.

Further optimization needs to be done while finding the diameter using the pixel per metric ratio to work out the scale of an associated object in an image. First, we must be compelled to perform a "calibration" of the image that removes the concave distortion as earlier mentioned in the 3.2 finding the countries using a reference object. Our reference object ought to have two necessary features.

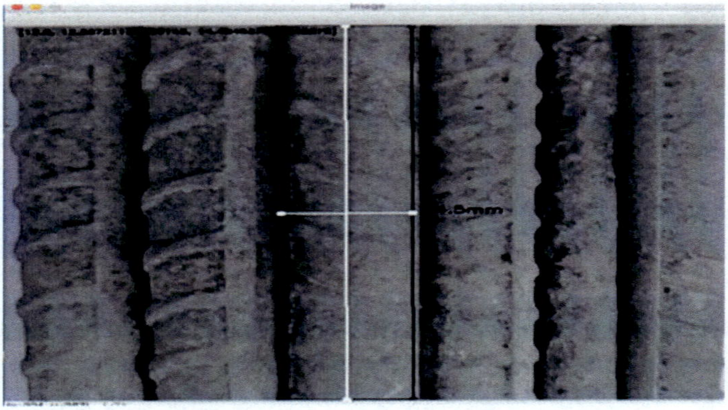

Fig. 20 Final output

Feature 1: We should understand the dimensions of this object (in terms of dimension or height) in a measurable unit (such as millimeters, inches).
Feature 2: We should be ready to notice this reference object in an image, so for these, we had added QR calibrated code within the formula. The formula for finding the pixels per metric ratio is

$$\text{Pixels per metric} = \text{object width/known width}.$$

The evaluation was done by testing images provided by a steel manufacturing company. The TMT rods are placed in the conveyer belt, and the image was taken from the camera inside the machine. The test consists of ten images of the same diameter rod placed under different light and parameters (industrial parameters such as heat, water) for each image in the dataset after the test result (Table 1).

Table 1 Analysis of results an actual TMT rod diameters images

TMT ROD label	Actual diameter (mm)	Predicted diameter (mm)	Error rate
Rod 1	8.00	7.98	−0.02
Rod 2	10.00	9.89	−0.12
Rod 3	10.00	9.67	−0.33
Rod 4	12.00	12.01	+0.01
Rod 5	6	5.67	−0.33
Rod 6	8	7.88	−0.12

5 Conclusion

This paper advises a sensible TMT rod detection methodology that helps each TMT rod reduce the threads. Experiments performed on quite a few challenging situations display the effectiveness and robustness of our method. We trust that our algorithm is of sensible top utilization and may be hired because of the cost-powerful approach to robotically locating and remembering TMT rods in steel industries. In the future, we intend to similarly decorate the robustness of our set of rules toward extraordinarily slanted digital camera angles. Besides, we might enlarge our method to different TMT rods styles, including steel rods and bars.

References

1. Hays, J., Leordeanu, M., Efros, A. A., & Liu, Y. (2006). Discovering texture regularity as a higher-order correspondence problem. In: *European Conference on Computer Vision (ECCV)* (pp. 522–535), Graz, Austria, May 2006.
2. Park, M., Brocklehurst, K., Collins, R. T., & Liu, Y. (2009). Deformed lattice detection in real-world images using mean-shift belief propagation. *IEEE Transactions on Pattern Analysis and Machine Intelligence (PAMI), 31*(10), 1804–1816.
3. Ghazali, M. F., Wong, L.-K., & See, J. (2016). Automatic detection and counting of circular and rectangular steel bars. In *Lecture Notes in Electrical Engineering* (pp. 199–207). http://doi.org/10.1007/978-981-10-1721-6_22
4. Hummel, R. A., Kimia, B., & Zucker, S. W. (1987). Deblurring Gaussian blur. *Computer Vision, Graphics, and Image Processing, 38*(1), 66–80. https://doi.org/10.1016/s0734-189x(87)80153-6
5. Miyuki, O., & Shun, N. (2007). *Algorithm to automatically count the number of steel pipes* (Vol. 41, pp. 25–28). Research reports of Fukui National College of Technology, Natural Science and Engineering.
6. Gowroju, S., & Kumar, S. (2020). Robust deep learning technique: U-net architecture for pupil segmentation. In *2020 11th IEEE Annual Information Technology, Electronics and Mobile Communication Conference (IEMCON)* (pp. 0609–0613). IEEE.
7. Swathi, A., & Kumar, S. (2021). A smart application to detect pupil for the small dataset with low illumination. *Innovations in Systems and Software Engineering*, 1–15.
8. Swathi, A., & Kumar, S. (2021). Review on pupil segmentation using CNN-region of interest. In *Intelligent communication and automation systems* (pp. 157–168). CRC Press.
9. Kakani, V., Kim, H., Lee, J., Ryu, C., & Kumbham, M. Automatic distortion rectification of wide-angle images using outlier refinement for streamlining vision tasks. https://ideas.repec.org/p/arx/papers/1104.0508.html.
10. Zitnick, L., & Parikh, D. *Redmond Proceedings/CVPR, IEEE Computer Society Conference on Computer Vision and Pattern Recognition.* http://doi.org/10.1109/CVPR.2012.6247729

Sentiment Analysis of Twitter Data Using Clustering and Classification

Santanu Modak and Abhoy Chand Mondal

Abstract Data mining helps in collecting and managing data besides performing analysis and prediction analysis. The process that is implemented to discover useful data patterns may have different names. Statisticians, database researchers, and professional organizations were among the first to use term data mining. The fundamental steps for sarcasm detection are dataset collection, feature extraction, and classification. This work puts forward a new model of sarcasm detection formed by fusing K-mean, PCA, and SVM classifiers together. With respect to common evaluation metrics like accuracy, precision, and recall, the architecture designed for this work is especially productive.

Keywords Sarcasm detection · SVM · KNN · K-mean · PCA

1 Introduction

Twitter, a globally famous micro-blogging platform, is shaping and changing the way individuals or establishments receiving information of their interest [1]. By using this platform, users can post status update messages, known as tweets, to let their followers know what they are thinking, doing, or what is going around them. Besides this, users reply to or repost their tweets in order to interact with other users [2]. Since its founding in 2006, Twitter has grown into one of the biggest online social networking platforms globally. In light of the ever-increasing volume of data obtainable from Twitter, the sentiment polarity of mining users articulated in Tweets is one of the most discussed topics nowadays due to its varied applications [3, 4]. For example, based on the polarization analysis of Twitter users' sentiments on political groups and candidates, a number of tools have been devised to make campaigning plans for political elections. Commercial firms also due to the speed and efficiency of Twitter sentiment analysis apply it to track people's sentiments toward their brands

S. Modak (✉) · A. C. Mondal
Department of Computer Science, The University of Burdwan, Burdwan, India
e-mail: modaksantanu@gmail.com

P. K. Singh et al. (eds.), *Proceedings of Third International Conference on Computing, Communications, and Cyber-Security*, Lecture Notes in Networks and Systems 421,
https://doi.org/10.1007/978-981-19-1142-2_51

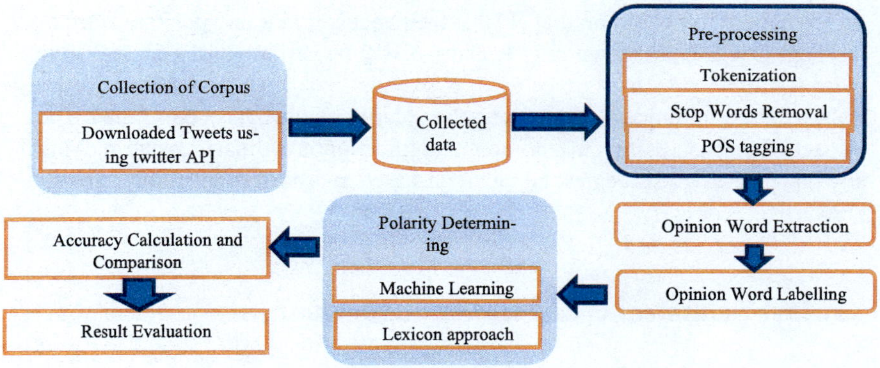

Fig. 1 Twitter sentiment analysis process

and products [5]. Sentiment analysis on Twitter data aims to classify the sentiment polarity of a Tweet as positive, negative, or neutral.

Figure 1 evidences the general process of twitter sentiment analysis [6]. The step involved in twitter sentiment analysis process has been discussed as follows:

1. Corpus Collection: Collecting labeled datasets is one of the crucial challenges in Twitter's sentiment analysis. The Twitter API is used to collect a collection of text posts. Then, these tweet posts are combined together to construct a dataset of three classes [7]: positive emotions, negative emotions, and a group of objective texts (no emotion).
2. Data Cleaning: Data received from Twitter typically consists of multiple HTML objects such as < > &amp, which are embedded in the original data. Therefore, it is essential to remove these objects [8]. The Twitter dataset also contains other information such as retweets, hashtags, usernames, and modified tweets. All of these are overlooked and eliminated from the dataset. These are main steps in data cleaning are tokenization, stop word removal, and POS tagging.
3. Opinion Word Extraction: The Twitter language model has several exclusive features. Some of these features are considered to decrease the feature space [9]. First of all, to initiate process, all unigrams and bigrams in the corpus above a certain limit are extracted. For example, all unigrams and bigrams with frequencies greater than 5 are taken out as candidate features. In general, unigrams and bigrams are selected in word/phrase-level sentiment analysis [10]. It can also be easily stretched using trigrams. Then, for each tweet, the frequency of each candidate features discovered in it is calculated. Accordingly, the following feature vector is created by means of the term frequency for every tweet:

({word1:frequency1, word2:frequency2 ...}, "polarity")

4. Opinion Word Labeling: The aggregated words are then verified on a dictionary of positive words and negative words containing two files, and a polarity is

given to the tweet based on that [11]. If the tweet contains any positive word or hashtag, it will be assigned as +1, and −1 will be assigned for each negative word.

5. Polarity Determining: In the last step, the polarity of a tweet is determined. The impact of preprocessing on sentiment classification is estimated by employing any one of the two strategies, i.e., lexicon-based approach or machine learning [12].

1.1 Twitter Sentiment Classification Algorithms

The researchers in the field of sentiment analysis usually assume that the entire document under analysis involves reliable sentiment polarity by a single holder toward an object. The variety of reviews is a great example of where the assumption is correct. Since tweets are usually shorter, this is also correct for tweet data. It is unnatural for a user to take in complex information in a tweet. Approaches can largely be classified into three classes: lexicon-based, machine learning-based, and rule-based approaches [13, 14].

1. Lexicon-based approach: This approach uses dictionary which already contains tagged dictionaries. The tokenization converts the input text into tokens. After that, all newly obtained tokens are matched for the dictionary in the dictionary [15]. In the case of a positive match, the score is added to the entire set of scores for the input score. Consider the word "dramatic." If this word is a positive match in the dictionary, the total text score is increased. In the second case, the score is reduced, or the word is tagged as negative. Although this approach sounds like a hobbyist, its options are quite decent. The inspiration of vocabulary-based classification algorithms is that the spirit of a document is determined by the key components (words or phrases). Basic plans consist of majority voting, document scoring with thresholding, and simple word count [16].
2. Machine Learning-based approach: Sentiment classification is, by its very nature, a form of binary text classification task. Text classification typically classifies data into many predefined classes [17]. This is an established domain with much matured solutions and applications. Most of the research in both text classification and sentiment analysis come in machine learning-based algorithms. The most common machine learning algorithms for sentiment classification are stated as follows [18]:

   - Maximum Entropy (MaxEnt) Classifier: The motivation behind the maximum entropy (MaxEnt) classifier is that it is required for an individual to select the most similar model that meets a specified limit. Unlike Naive Bayes, MaxEnt does not assume independency for its features [19]. In other words, several features such as bigrams and phrases can be added to MaxAntexclouding the worry of overlapping features. The representation of model takes the following form:

$$P_{ME(c|d,\lambda)=\frac{\exp[\sum_i \lambda_i f_i(c,d)]}{\sum_{c'} \exp[\sum_i \lambda_i f_i(c,d)]}} \quad (1)$$

This formula accepts c as the square, d as the tweet, and a as the weight vector [20]. Weight vectors determine the importance of a feature in classification. A higher weight denotes the features as a powerful indicator for the class. To find the weight vector, numerical optimization of the lambda is applied in order to maximize the conditional probability [21]. Hypothetically, MaxEnt outperforms Naive Bayes as it manages feature overlap superiorly.

- Support Vector Machine: Support vector machines (SVMs) show good performance in sentiment analysis [22]. The SVM is concerned with selecting a hyperplane that extends the difference between the nearest instances of two classes to maximum. It is required to solve the following optimization problem to obtain the best hyperplane.

$$\underset{w,b}{\arg\min} \frac{1}{2} \|w\|^2 \quad (2)$$

$$\text{subject to } y_i(w.x_i + b) \geq 1, \quad i \in [1, n]$$

where x_i is the feature vector and $y_i \in \{+1, -1\}$ is the label of instance i

- Naïve Bayes: Naive Bayes is an easy-to-use model which is well-suited for text categorization. In a multinomial Naive Bayes classifier, class $c*$ is given to tweet d, where [23]:

$$c* = \text{argmax}_c\ P_{NB}(c|d)$$

$$P_{NB}(c|d) = \frac{\left(P(c) \sum_{i=1}^{m} P(f|c)^{n_i(d)}\right)}{P(d)} \quad (3)$$

In this formula, f is a feature, and $n_i(d)$ is the count of feature f_i discovered in tweet d. The total number of features is m. Maximum likelihood estimates are applied to achieve parameters $P(c)$ and $P(f|c)$. Apart from this, and add-1 smoothing is employed for hidden features [24].

3. Rule-based classifiers: In a rule-based classifier, the dataspace is modeled using a rule set [25]. The one on the left represents a position on the feature set, expressed in normal form, while the one on the right is labeled the class. The terms are on the word presence [26]. The term absence is rarely used because it is not informative in sparse data. There are many criteria for making rules, the training phase builds all the rules based on these criteria. The two commonly adopted criteria are support and confidence. The support training dataset contains the absolute number of examples that are relevant to the rule. Confidence is the conditional probability that the right side of the rule is satisfied if the left-hand side is met [27].

2 Literature Review

Md. Rakibul Hasan et al. (2019) constructed a natural language processing-based (NLP) preprocessed data model for filtering the tweets [28]. Thereafter, the sentiment was analyzed by integrating bag-of-words (BoW) and term frequency-inverse document frequency (TF-IDF) models. The latter model assisted in enhancing the accuracy to analyze the sentiments. The simulation outcomes revealed that the constructed approach was effective. This approach provided the accuracy around 85.25% while analyzing the sentiments. An automated system was suggested by Piyush Vyas et al. (2021) for extracting sentiments from tweets and classifying those tweets further using machine learning (ML) methods [29]. This system was a hybrid technique in which a lexicon-based technique was integrated to analyze the sentiment from tweets and labeled with the supervised ML methods in order to classify the tweets. Moreover, the long short-term memory (LSTM) was selected as effective ML method as it yielded the accuracy around 83%. The evaluation results depicted that the suggested hybrid technique was capable of classifying the large tweet volumes in automatic manner. An analysis was conducted by Rosy Indah Permatasari et al. (2018) on the document text related to Indonesian movie review acquired from Twitter [30]. The Naïve Bayes (NB) algorithm was introduced through the ensemble attributes. Various attributes such as twitter specific attributes, textual, part of speech (POS), lexicon-based attributes, and bag of words (BoW) were employed in this ensemble. The experiment outcome revealed that the introduced algorithm provided f-measure up to 0.88 with ensemble attributes. A phenomenon recognized as the sentiment reversal was investigated by Lei Wang et al. (2020) to analyze the sentiment diffusion, and some remarkable properties of sentiment reversals were discovered [31]. An iterative algorithm known as SentiDiff was presented for predicting the sentiment polarities expressed in Twitter messages. The sentiment diffusion patterns were deployed for enhancing the twitter sentiment analysis (SA). A real-world dataset was applied to conduct the experiments. The experimental outcomes demonstrated that the investigated approach offered the PR-AUC improvements up to 8.38% for analyzing the sentiments of twitter. A new attentional bidirectional long short-term memory (LSTM) algorithm was developed by Hanane Elfaik et al. (2021) for determining considerable semantic information and extracting the contextual information in both directions [32]. The impact of the word2vec framework was implemented for generating the word embedding representation and for capturing the semantic information from Arabic tweets. The developed algorithm was authenticated by evaluating it on Arabic sentiment tweets datasets. The experimental outcomes indicated that the developed algorithm performed more efficiently as compared to existing schemes. The word2vec and clustering-based text representation technique was projected by Önder Çoban et al. (2018) in order to analyze the sentiments on twitter [33]. The support vector machine (SVM) algorithm was deployed to classify the tweets. Two diverse datasets having Turkish Twitter feeds were utilized in the experimentation. The experimental outcomes exhibited that the projected technique was effective and performed well with regard to time. This technique assisted in mitigating the feature space but unable to provide attain higher accuracy. An online system was intended by Alaa S. Al Shammari et al. (2018) to analyze and classify the Twitter sentiment in real time [34]. This system assisted the user in entering the query and obtaining a graphical representation of the tweets polarity. The tweets were classified using simple voter and Naïve Bayes (NB) algorithms. The acquired outcomes validated that the intended system provided a superior accuracy with the help of NB classification algorithm. A novel sentiment analysis (SA) technique was presented by Vallikannu Ramanathan et al. (2019) on the basis of common sense knowledge [35]. The Oman tourism ontology was generated in accordance with the ConceptNet. The part of speech (POS) tagger was utilized to recognize the entities from the tweets, and the comparison of these entities was done with concepts in the domain-specific ontology. Moreover, the ensemble sentiment lexicon technique was exploited to classify the sentiment of the extracted entities. The conceptual semantic was utilized as feature along with machine learning (ML) algorithm for boosting the performance of analyzing the sentiments of Oman tourism. A corpus-based technique was established by Hussain AlSalman et al. (2020) in order to analyze the Arabic sentiment of tweets in two categories: negative and positive in twitter social media [36]. This technique was planned on the basis of discriminative multinomial Naïve Bayes (DMNB) technique along with the N-grams tokenizer, stemming, and term frequency-inverse document frequency (TF-IDF) methods. A publicly available dataset of twitter was applied to conduct the experiments so that the established technique was tested on the basis of a set of evaluation parameters while analyzing the sentiments. The experimental outcomes depicted the supremacy of the established technique over the existing techniques which led to enhance the accuracy up to 0.3%. A method was recommended by Mudassir Khan et al. (2020) approach in order to analyze the sentiment for which a Hadoop model and deep learning (DL) classification algorithms were implemented [37]. The data were distributed using Hadoop cluster in order to extract the attributes. Thereafter, the Twitter data were utilized in extracting the considerable attributes. A real-valued review was allocated to each input twitter data, and the input data were classified into two kinds: positive and negative review using deep recurrent neural network (DRNN) algorithm. The recommended technique provided the accuracy around 0.9302, sensitivity up to 0.9404, and higher specificity around 0.9157 for classifying the sentiments. A convolutional neural network-long short-term memory-based (CNN-LSTM) deep learning (DL) technique was suggested by Vishu Tyagi et al. (2020) along with pretrained embedding technique for learning the way of extracting the attributes in automatic manner [38]. Hence, the sentiments were analyzed, and the reviews were classified as positive or negative. The comparison of suggested approach was done with existing techniques. The results depicted that the suggested approach performed well on benchmark dataset. A SegAnalysis model was designed by Mamta Patil et al. (2018) for dealing with diverse issues such as to segment the tweet, detect the event, and analyze the sentiment [39]. The part of speech (POS) tagger was applied on the recent online tweets, which the user had collected, for segmenting the tweet in a batch mode. The events were detected using the Naïve Bayes (NB) classifier and online clustering. These events were useful for enhancing the situational awareness and decision support. The tweets were classified as positive, negative, and neutral on the basis of the sentiment score of a tweet after analyzing the sentiments. The designed model was expanded further to tackle the events related to multiple clusters. The Apache Spark model was introduced by Hossam Elzayady et al. (2018) in which distributed memory abstraction was employed for analyzing the tweets [40]. The machine learning library (MLIB) of this model was emphasized on handling the huge volume of data in efficient manner. The introduced model was more effectual in comparison with other technique as it generated optimal outcomes with the help of Naïve Bayes (NB), logistic regression (LR), and decision tree) (DT). In the end, the introduced model was proved scalable. A mechanism was put forward by Shihab Elbagir Saad et al. (2019) in which the tweets were preprocessed, and a feature extraction technique was utilized to generate an effective attribute [41]. Subsequently, these attributes were scored and balanced under various classes. The sentiment analysis (SA) was classified using diverse algorithms such as SoftMax, support vector regression (SVR), decision tree (DT), and random forest (RF). The presented mechanism was implemented on a Twitter dataset taken from NLTK corpora resources. The experimental outcomes revealed that the DT provided more promising outcomes as compared to other algorithms. A sentiment analysis (SA) technique was formulated by Mariam Khader et al. (2018) in order to analyze the Twitter dataset [42]. The Naive Bayes (NB) algorithm was deployed to classify the text as positive or negative. The deployment of various linguistic and natural language processing (NLP) was done on the dataset. The experimental results exhibited that the formulated technique assisted in enhancing the SA by 5% with the help of NLP and linguistic processing, and its accuracy was calculated 73%.

3 Research Methodology

This research work is conducted on the basis of detecting the sarcasm from the twitter data. Diverse phases utilized to detect the sarcasm from twitter data are defined as:

3.1 Dataset Collection and Preprocessing

In this stage, the dataset is gathered to detect the sarcasm. Tweety API is utilized to extract the dataset. The data are processed using several entities such as credentials of user in the API. The processing of this dataset is done further for eliminating the missing and redundant values from the dataset. This stage focuses on splitting the input data into tokens and processing every token at individual level. During the partition of data, the similes are removed. Finally, this stage assists in removing the stop words and single words from the sentence.

3.2 Feature Extraction and Feature Reduction

The major aim of this stage is to extract the attributes link every attribute with each other. For this, principal component analysis model is implemented to mitigate the attributes. This model is often assisted in changing a group of interrelated factors into a set of linearly unrelated subsets on the basis of a transformation due to which the uncorrelated variables are obtained. It is taken in account as an orthogonal LT which leads to project the initial dataset to another projection system and expects that a projection of 1st coordinate is included in the biggest variance. Moreover, a projection of the 2nd coordinate is comprised in the second biggest variance considering that it is placed vertically to the 1st component. In general, principal component analysis model emphasizes on locating a LT that is defined as $z = W_k^T x$ in which $x \in R^d$, and $r < d$, for improving the variance of the data in the projected space. For a data matrix denoted with $X = \{x_1, x_2, \ldots, x_i\}$, $x_i \in R^d$, $z \in R^r$ and $r < d$, a set of p-dimensional vectors of weights $W = \{w_1, w_2, \ldots, w_p\}$, $w_p \in R^k$ is utilized to characterize the transformation, which is matched with every x_i vector of X to a

$$t_{k(i)} = W_{|(i)}\, T_{x_i} \tag{4}$$

The variance is increased by observing an initial weight $W1$ with the condition defined as:

$$W_i = \mathrm{argmax}_{|w|} = \left\{ \sum_i (x_i \cdot W)^2 \right\} \tag{5}$$

A further expansion of the preceding condition is discussed as:

$$W_i = \mathrm{argmax}_{\|w\|=1}\left\{\|X.W\|^2\right\} = \mathrm{argmax}_{\|w\|=1}\left\{W^{\mathrm{T}} X^{\mathrm{T}} X W\right\} \tag{6}$$

The biggest Eigen value of the matrix is attained to analyze the symmetric grid such as $X^{\mathrm{T}}X$ because W is the related Eigen vector. After attaining the W_1, the projection of first data matrix X is utilized to assume the initial PC onto the W_1 in the space which is generated after the transformation. The acquired components are subtracted after achieving the segments along these lines.

3.3 Clustering

This stage is executed with the implementation of k-mean clustering. In this process, the similar objects are grouped. All these objects are clustered according to their attributes. The instances are grouped generally for the data at which no label is assigned. These clusters have similarity with one another. But, the objects placed in one cluster can be differentiated from the objects of other clusters on the basis of attributes. The intra-clustering is found higher among the objects, and its similarity is found lower among the clusters. Partitioning clustering, hierarchal clustering, etc., are some well-known cluster analysis techniques. But, the KMC technique is an extensively utilized technique by the researchers. This technique is simple and feasible. The implementation of this technique is done on the numerical data, in case, K denotes the center of clusters.

3.4 Classification

This research work deploys several classifiers, namely RF, support vector machine, and k-nearest neighbor in order to detect the sarcasm. The classification and regression rules are learned from the data using support vector machine algorithm which focuses on the notion of statistical learning. This algorithm is utilized to tackle the major issue in indirect manner instead of complicated issue. Two ways are present to implement this algorithm. The initial one employs the mathematical programming, while the kernel functions are exploited in the latter one. The kernel functions are considered for dividing the data: P and N classes. Class P is executed if $y_i = +1$, while class N is executed if $y_i = -1$. A good separating surface called hyperplane is required that has equal distance from each class (Fig. 2).

Random forest is a simple ML algorithm which performs quickly and flexibly. Various tree predictors are grouped in this algorithm, and promising results are generated through this algorithm. It is challenging to improve its performance. RF is useful to handle diverse type of data. Random forest focuses on developing several trees. The optimal outcomes at higher precision are achieved by integrating these trees. ML is majorly utilized to perform classification. This algorithm comprises the hyperparameters having similarity with DT or bagging technique. KNN is a supervised machine learning algorithm which is applicable to perform classification and regression. This model is recognized as lazy algorithm due to absence of any specialized phase for

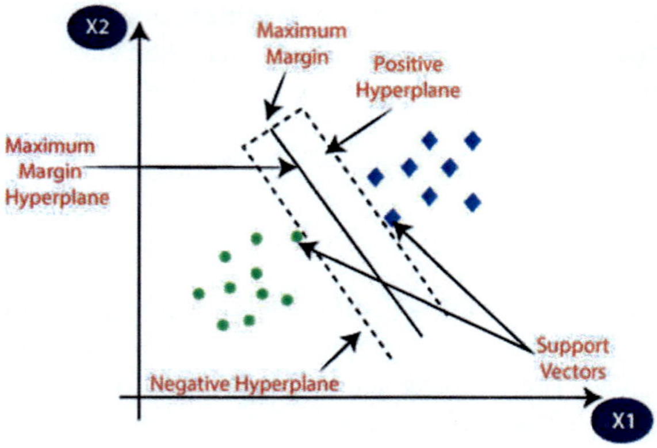

Fig. 2 SVM algorithm

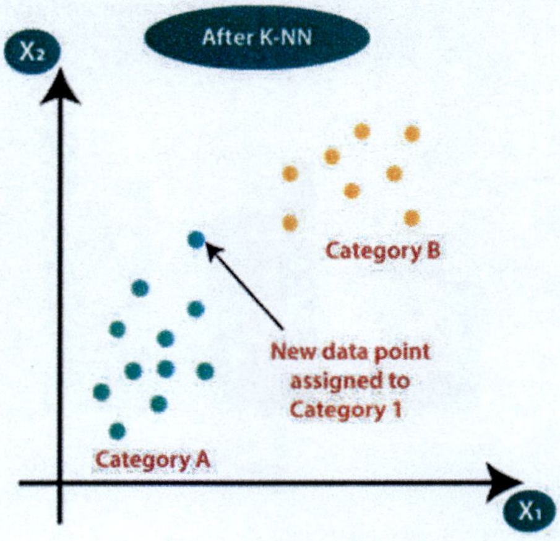

Fig. 3 KNN algorithm

training. The training phase is executed using the entire data. It is also referred as non-parametric learning algorithm as it is incapable of assuming anything in the context of the fundamental data (Fig. 3).

K-nearest neighbor algorithm deploys a feature similarity with the objective of predicting the novel data point values. The novel data point is responsible for assigning a value.

4 Result and Discussion

The sentiments are analyzed using three metrics such as precision, recall, and accuracy. The presented algorithms are analyzed with regard to some metrics which are defined as:

1. Precision: It refers to the degree using which the repeated measurements are utilized to provide same results under static conditions.

$$\text{Precision} = \frac{(\text{True Positive})}{(\text{True Positive} + \text{False Positive})} \tag{7}$$

2. Recall: It is ratio of properly predicted positive observations to overall observations in original class.

$$\text{Recall} = \frac{(\text{True Positive})}{(\text{True Positive} + \text{False Negative})} \tag{8}$$

3. Accuracy: It is the ratio of the accurately labeled subjects to the whole group of subjects.

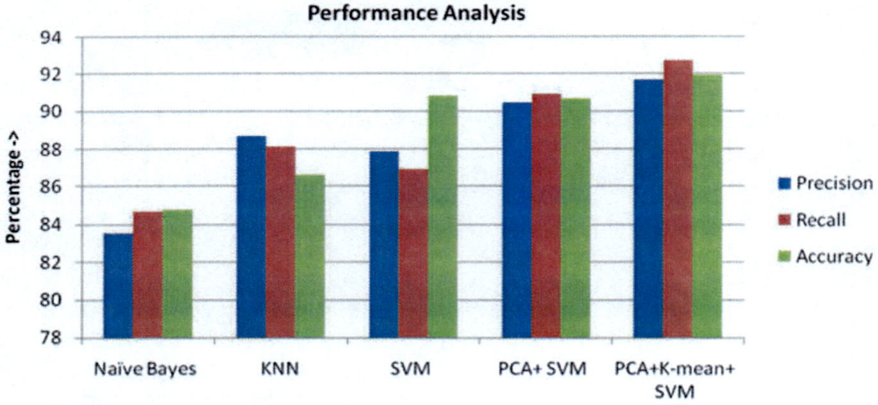

Fig. 4 Performance analysis

Table 1 Comparison table

Parameters	Naïve Bayes	KNN	SVM	PCA + SVM	PCA + K-mean + SVM
Precision (%)	83.56	88.67	87.87	90.45	91.67
Recall (%)	84.66	88.12	86.89	90.89	92.67
Accuracy (%)	84.78	86.56	90.78	90.66	91.90

$$\text{Accuracy} = \frac{\text{No. of points correctly classified}}{\text{Total no. of points}} * 100 \quad (9)$$

Figure 4 represents the performance of all three models (Table 1).

Figure 4 indicates the comparison of several classification algorithms such as SVM, KNN, PCA + SVM, and PCA + K-mean + SVM in order to analyze the performance. This depicts that the integration of principal component analysis, k-mean, and support vector machine performed well in contrast to other algorithms.

5 Conclusion

A significant boost in the use of social media in last decade has increased the popularism of sentiment analysis among a substantial number of individuals with various interests and inspirations. Since users around the world may have different opinions about a variety of topics related to politics, education, travel, culture, commercial goods, or topics of general interest, retrieving knowledge from those data is a matter of great importance and significance. Knowing their sentiments expressed by their messages across different platforms, apart from information related to the sites visited by users, buying priorities, etc., became a crucial aspect for assessing public opinion about a specific topic. Classifying the polarity of a text is one of the most used SA methods these days. Extracting sentiment polarity expressed in Twitter posts is not only substantial but challenging as well. A maximum number of existing methods for analyzing the sentiments from Twitter, only address the textual information of Twitter messages, and may not obtain satisfying performance in the wake of the exceptional features of Twitter messages. Since current studies have depicted that sentiment diffusion patterns are closely related to the sentiment polarity of Twitter messages, existent techniques originally concentrate only on the textual information of Twitter messages but overlook sentiment diffusion information. There is a similarity between the sarcasm detection and OM. The sarcasm can be detected in several stages in which data are collected, features are extracted, and classification is performed. The sentiments are analyzed by comparing diverse classification algorithms such as support vector machine and k-nearest neighbor. The integration of these algorithms performed well in comparison with other techniques. In the results, the combination of PCA, k-means, and support vector machine classifier offers precision of 91.67%, recall of 92.67%, and accuracy of 91.90%.

References

1. Wongkar, M., & Angdresey, A. (2019). Sentiment analysis using Naive Bayes algorithm of the data crawler: Twitter. In *Fourth International Conference on Informatics and Computing (ICIC)* (Vol. 96, No. 2, pp. 555–563).
2. Mandloi, L., & Patel, R. (2020). Twitter sentiments analysis using machine learning methods. In *International Conference for Emerging Technology (INCET)*.
3. Park, C. W., & Seo, D. R. (2018). Sentiment analysis of Twitter corpus related to artificial intelligence assistants. In *5th International Conference on Industrial Engineering and Applications (ICIEA)* (Vol. 6, No. 10, pp. 267–274).
4. Zahoor, S., & Rohilla, R. (2020). Twitter sentiment analysis using machine learning algorithms: A case study. In *International Conference on Advances in Computing, Communication & Materials (ICACCM)*.
5. Bhasin, A., & Das, S. (2021). Twitter sentiment analysis using machine learning and Hadoop: A comparative study. In *2nd International Conference on Secure Cyber Computing and Communications (ICSCCC)*.
6. Prakruthi, V., Sindhu, D., & Kumar, S. A. (2018). Real time sentiment analysis of Twitter posts. In *3rd International Conference on Computational Systems and Information Technology for Sustainable Solutions (CSITSS)*.
7. Pouromid, M., Yekkehkhani, A., Oskoei, M. A., & Aminimehr, A. (2021). ParsBERT post-training for sentiment analysis of tweets concerning stock market. In *26th International Computer Conference, Computer Society of Iran (CSICC)*.
8. Dhawan, S., Singh, K., & Chauhan, P. (2019). Sentiment analysis of Twitter data in online social network. In *5th International Conference on Signal Processing, Computing and Control (ISPCC)*.
9. Karqmibekr, M., & Ghorbani, A. A. (2012). Sentiment analysis of a social issues. In *International Conference on a Social Informatics*, USA.
10. Eirinaki, M., Pisal, S., & Singh, J. (2012). Feature-based opinion mining and ranking. *Journal of Computer and System Sciences, 7*(4), 1175–1183.
11. Haddia, E., Liua, X., & Shib, Y. (2013). The role of text pre-processing in sentiment analysis. *Information Technology and Quantitative Management, 17*(1), 126–133.

12. Ghag, K., & Shah, K. (2013). Comparative analysis of the techniques for sentiment analysis. In *International Conference on Advances in Technology and Engineering (ICATE).*
13. Vuong, Q., & Takasu, A. (2014). Transfer learning for emotional polarity classification. In *International Joint Conferences on Web Intelligence (WI) and Intelligent Agent Technologies (IAT).*
14. Shahana, P. H., & Omman, B. (2015). Evaluation of features on sentimental analysis. In *International Conference on Information and Communication Technologies.*
15. Tripathy, A., Agrawal, A., & Rath, S. K. (2015). Classification of sentimental reviews using machine learning techniques. In *International Conference on Recent Trends in Computing.*
16. Farooq, U., Prasad, T., Nongaillard, A.,Ouzrout, Y., & Qadir, M. (2015). A word sense disambiguation method for feature level sentiment analysis. In *International Conference on Software, Knowledge, Information Management and Applications (SKIMA).*
17. Ahmed, S., & Danti, A. (2016). Effective sentimental analysis and opinion mining of web reviews using rule based classifiers. *Advances in Intelligent Systems and Computing, 41*(8), 1071–1079.
18. Bouazizi, M., & Ohtsuki, T. (2016). A pattern-based approach for sarcasm detection on Twitter. *IEEE Access, 9*(1), 567–574.
19. Tsytsarau, M., & Palpanas, T. (2016). Managing diverse sentiments at large scale. *IEEE Transactions on Knowledge and Data Engineering, 28*(11), 3028–3036.
20. Ankita, & Saleena, N. (2018). An ensemble classification system for Twitter sentiment analysis. In *International Conference on Computational Intelligence and Data Science (ICCIDS). Procedia Computer Science.*
21. Naz, S., Sharan, A., & Malik, N. (2018). Sentiment classification on Twitter data using support vector machine. In *IEEE/WIC/ACM International Conference on Web Intelligence (WI).*
22. Hiremath, S., Manjula, S. H., & Venugopal, K. R. (2021). Unsupervised sentiment classification of Twitter data using emoticons. In *International Conference on Emerging Smart Computing and Informatics (ESCI).*
23. Phand, S. A., & Phand, J. A. (2017). Twitter sentiment classification using Stanford NLP. In *1st International Conference on Intelligent Systems and Information Management (ICISIM).*
24. Rane, A., & Kumar, A. (2018). Sentiment classification system of Twitter data for US airline service analysis. In *IEEE 42nd Annual Computer Software and Applications Conference (COMPSAC).*
25. Ibrahim, M. N., & Yusoff, M. Z. (2015). Twitter sentiment classification using Naive Bayes based on trainer perception. In *IEEE Conference on e-Learning, e-Management and e-Services (IC3e).*
26. Sharma, P., & Moh, T. S. (2016). Prediction of Indian election using sentiment analysis on Hindi Twitter. In *IEEE International Conference on Big Data (Big Data).*
27. Barnaghi, P., Ghaffari, P., & Breslin, J. G. (2016). Opinion mining and sentiment polarity on Twitter and correlation between events and sentiment. In *IEEE Second International Conference on Big Data Computing Service and Applications (BigDataService).*
28. Hasan, M. R., Maliha, M., & Arifuzzaman, M. (2019). Sentiment analysis with NLP on Twitter data. In *International Conference on Computer, Communication, Chemical, Materials and Electronic Engineering (IC4ME2).*
29. Vyas, P., Reisslein, M., Rimal, B. P., Vyas, G., Basyal, G. P., & Muzumdar, P. (2021). Automated classification of societal sentiments on Twitter with machine learning. *IEEE Transactions on Technology and Society, 1*(6), 1348–1356.
30. Permatasari, R. I., Fauzi, M. A., Adikara, P. P., & Sari, E. D. L. (2018). Twitter sentiment analysis of movie reviews using ensemble features based Naïve Bayes. In *International Conference on Sustainable Information Engineering and Technology (SIET).*
31. Wang, L., Niu, J., & Yu, S. (2020). SentiDiff: Combining textual information and sentiment diffusion patterns for Twitter sentiment analysis. *IEEE Transactions on Knowledge and Data Engineering, 13*(78), 110–117.
32. Elfaik, H., & Nfaoui, E. H. (2021). Deep attentional bidirectional LSTM for Arabic sentiment analysis in Twitter. In *1st International Conference on Emerging Smart Technologies and Applications (eSmarTA).*

33. Çoban, O., & Özyer, G. T. (2018). Word2vec and clustering based Twitter sentiment analysis. In *International Conference on Artificial Intelligence and Data Processing (IDAP).*
34. Shammari, A. S. L. (2018). Real-time Twitter sentiment analysis using 3-way classifier. In *21st Saudi Computer Society National Computer Conference (NCC).*
35. Ramanathan, V., & Meyyappan, T. (2019). Twitter text mining for sentiment analysis on people's feedback about Oman tourism. In *4th MEC International Conference on Big Data and Smart City (ICBDSC).*
36. Salman, H. A. (2020). An improved approach for sentiment analysis of Arabic tweets in Twitter social media. In *3rd International Conference on Computer Applications & Information Security (ICCAIS).*
37. Khan, M., & Malviya, A. (2020). Big data approach for sentiment analysis of Twitter data using Hadoop framework and deep learning. In *International Conference on Emerging Trends in Information Technology and Engineering (ic-ETITE).*
38. Tyagi, V., Kumar, A., & Das, S. (2020). Sentiment analysis on Twitter data using deep learning approach. In *2nd International Conference on Advances in Computing, Communication Control and Networking (ICACCCN).*
39. Patil, M., & Chavan, H. K. (2018). Event based sentiment analysis of Twitter data. In *Second International Conference on Computing Methodologies and Communication (ICCMC).*
40. Elzayady, H., Badran, K. M., & Salama, G. I. (2018). Sentiment analysis on Twitter data using Apache Spark framework. In *13th International Conference on Computer Engineering and Systems (ICCES).*
41. Saad, S. E., & Yang, J. (2019). Twitter sentiment analysis based on ordinal regression. *IEEE Access, 23*(7), 5016–5024.
42. Khader, M., Awajan, A., & Al-Naymat, G. (2018). The effects of natural language processing on Big Data analysis: Sentiment analysis case study. In *International Arab Conference on Information Technology (ACIT).*
43. Minab, S. S., Jalali, M., & Moattar, M. H. (2015). A new sentiment classification method based on hybrid classification in Twitter. In *International Congress on Technology, Communication and Knowledge (ICTCK)* (Vol. 7, No. 30, pp. 912–920).

Security and Privacy Issues

Image Distortion Analysis in Stego Images Using LSB

Shubh Gaur, Swati Chaturvedi, Shiavnsh Gupta, Jay Mittal, Rohit Tanwar, and Mrinal Goswami

Abstract One of the most pressing matters at hand of today's communication networks is privacy. Privacy is essential where information to be transmitted to the desired target without being intercepted by third parties or bringing them in a way that they cannot understand. Lossless text encryption technique for gray scale or RGB images using least significant bit (LSB) is the most commonly used method to hide data. LSB technique is preferred as it is difficult to draw a distinction between the cover object and stego object if few LSB bits of the cover object are replaced. However, image distortion is a major issue which needs critical investigation as it can change the original image. This paper proposes an algorithm to embed text in RGB images and also evaluate the distortion in image at various bit positions. Simulation results suggest that distortion in the image is least at 8th-bit and most at 1st-bit position.

Keywords Steganography · LSB · Image distortion · PSNR · SSIM · Encryption · Data hiding · Privacy

1 Introduction

The rapid advancement of the Internet and the digital information revolution resulted in significant cultural shifts. Individuals can disperse large media documents and produce indistinguishable sophisticated duplicates of them using broadband Internet associations that provide almost error-free information transmission [1]. On the other hand, data hiding is essential for network security in today's communication architecture [1]. Sensitive messages and records are exchanged across the Internet in an insecure framework, but everyone has something to hide.

S. Gaur · S. Chaturvedi · S. Gupta · J. Mittal · R. Tanwar (✉) · M. Goswami
School of Computer Science, University of Petroleum and Energy Studies, Dehradun, India
e-mail: rohit.tanwar.cse@gmail.com

M. Goswami
e-mail: mgoswami@ddn.upes.ac.in

P. K. Singh et al. (eds.), *Proceedings of Third International Conference on Computing, Communications, and Cyber-Security*, Lecture Notes in Networks and Systems 421,
https://doi.org/10.1007/978-981-19-1142-2_52

Today, security plays a vital role in data transfer. It is essential to ensure that data are transferred without the involvement of a third party or an attacker, so advanced techniques and technologies should be analyzed to check the accuracy of data transfer and to determine how safe it is to transfer data using that method.

In terms of information, the advancement in the security and various sorts of media has greatly facilitated people's daily lives and work, but it has also exposed an increasing number of security concerns [2]. We can consider an example like when it comes to personal privacy and commercial secrets, as well as military defense security, the penalties of leaking confidential information are immeasurable. Digital data are being communicated more regularly these days due to its ease of access. A broad range of solutions for end-to-end protection is required to address the security dangers in modern communication [3].

Steganography is the art of concealing information and maintaining the privacy of the data. The least significant bit (LSB) substitution is an image steganography technique in which hidden data bits are substituted for the pixels' LSB bits (one, two, three, or four). LSB, or least significant bit, refers to the last position of an 8-bit representation of a single byte, which can be either "0" or "1". Now, if this LSB value changes, the value of that entire byte changes as well, but the difference between the current and the previous values will be minimal. Here, steganography comes into the picture, if changing the LSB bit does not affect the whole byte much, then data hiding is possible here to minimize the distortion in the original file when data are hidden in it.

The progressions made on the advanced pictures because of the substitution of bits is distorted afterward. These distortions lead to visual contrasts or we can say the nature or quality of the image [4]. When data are concealed and the size of the media is really small, the distortion level changes, which is not visible but could be detected with the help of analysis techniques.

Several attempts have been made to improve data security in stenography [1–3, 5–8]. However, none of the work reported till date investigated the impact of image distortion in steganography. The remaining of the paper is structured as follows: some earlier work is mentioned in the next section. The suggested algorithm and its implementation are included in Sect. 3. Section 4 explains the findings and discussion. Finally, in Sect. 5, this paper comes to a close.

2 Literature Review

The study presents various related works on steganography and several network security methodologies for protecting data in cybersecurity. Alexan et al. [1] proposes a message security method with two layers. The message is initially encrypted with AES-128 before being integrated in a three-dimensional image using least significant bit (LSB) substitution. The modified 1D chaotic map was used in [5] to propose a color image LSB steganography technique. They confirmed the accuracy of the suggested color output LSB steganography algorithm through trials. Experimental results proved that the algorithm features a better performance than the previous methods, showing a superb performance against statistical analysis attacks [2]. This study proposes a dual-layered RIH approach based on modified least significant bit (LSB) matching. The goal was to increase the embedding efficiency (EE) by employing a dual-layer-based embedding technique to reduce stego-image distortion and improve its quality. Chatterjee et al. [3] developed a steganographic technique based on optical character recognition (OCR), in which the message is encoded in its feature form within the cover image. They retrieved character-level attributes from photos containing the textual message and embedded these data within the cover image, boosting the steganography's information-hiding goal. Patani and Rathod [9] proposes a 3-bit least significant bits approach for integrating the secrete image within the duvet image.

3 Proposed Implementation

To illustrate the proposed implementation method, we provided a user-friendly algorithm that transforms the secret text into binary data, then encodes the image using LSB over multiple bits, yielding a stego image. After that, we do a histogram analysis to evaluate distortion among images after collecting all of the stego images.

3.1 Algorithm

1. Start
2. Input image
3. Input text
4. Read the cover image and the concealed text message in the cover image.
5. Perform a binary conversion of the text message.
6. Add image into an array
7. Determine the LSB of each pixel in the cover image.
8. Replace nth LSB with message bit of the cover image with each bit of message one by one ($n = 0, 1, 2, 3 \ldots 7$).
9. LSB replacement will stop after replacement of last text bit.
10. Write stego image as output.
11. Analyze stego image using gray scale and RGB image
12. Check SSIM of cover image and stego image
13. Check PSNR value of stego image
14. Stop.

3.2 Flowchart

To describe the implementation, a flowchart was created to graphically explain the same procedure as the algorithm, as well as to display all of the conditional statements that were utilized during the process (Fig. 1).

- **User input**: The user enters the text, which may be of any length, the cover image, which will contain the specified content. Users can use special characters in it as

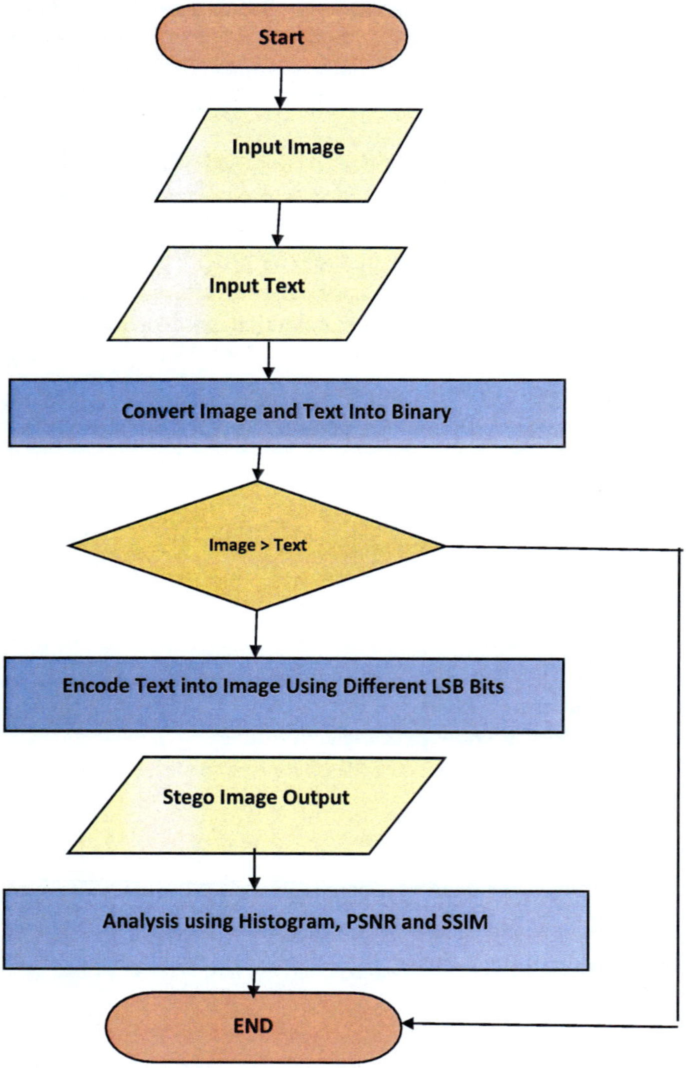

Fig. 1 Proposed methodology

well because the text is considered a single string. Users should provide BMP images; we utilize this format since it does not compress or lose information, allowing for high-quality photographs that are easier to work with.

- **Conditional statement**: The fourth stage comprises a conditional statement that calculates the image's pixel count and the text's character size, then checks if the text size is less than the image size or not to embed, because the text bits size should always be smaller than image bits size. The code below demonstrates the same.

```
//opening Image File
if((fp1 = fopen(argv_2, "r+")) == NULL)
{
printf("Could not open file %s.\n\n", argv_2);
return 1;
}
int size_image = size_of_image(fp1);
printf("Total %d Characters can be stored in %s.\n", size_image,
argv_2);
//opening secret text file
fp2 = fopen(argv_3, "w+");
//Entering secret text to file
printf("Enter the text and Press CTRL + D To Stop : \t");
secret_text(fp2);
int size_txt = secret_text_size(fp2);
printf("\nSize of the entered Message is ==> %d\n", size_txt);
//Comparing Image Size With Text
if(size_image < size_txt)
{
printf("\n***Size of Message is larger then allowed size***\n");
return 1;
}
```

If the size of the message bits is more than the size of the input picture bits, the program will terminate and generate an error. If the text bit size is less than the picture bit size, the program will proceed to the encoding module.

- **Encoding**: The input text (i.e., text bits) is spread out over the image's bits in the fourth stage by changing each bit's rightmost bit (the least significant bit, LSB). In each of the pixel, three bits can be encoded. If the LSB of the pixel value of the cover picture $C(i, j)$ is identical to the message bit of the secret text to be embedded, $C(i, j)$ stays unchanged. Set the LSB of $C(i, j)$ to the input text bit if it is not already set, i.e.;

$$S(i, j) = \begin{cases} C(i, j) - 1, & \text{if LSB}(C(i, j)) = 1 \text{ and MB} = 0; \\ C(i, j) + 1, & \text{if LSB}(C(i, j)) = 0 \text{ and MB} = 1; \\ C(i, j), & \text{if LSB}(C(i, j)) = \text{MB}; \end{cases} \quad (1)$$

"LSB($C(i,j)$)" denotes the LSB of the cover image "$C(i,j)$", and "MB" denotes the next message bit to be inserted. "$S(i, j)$" is the stego image's bit. After the last text bit has been replaced, the replacement process will come to an end.

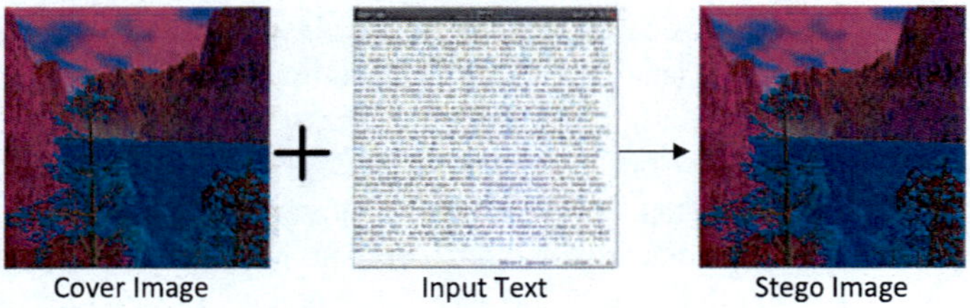

Fig. 2 Embedding process

We can use this technique to encode text in an image, if we replace the last bit of every color's byte with a bit from the message bit. Below given illustration shows how the encoding process works.

Before Encoding

Message A – 01000001

Pixel 1: 11111000 11001001 00000010
Pixel 2: 11111000 11001000 00000010
Pixel 3: 11111000 11001001 00000011

Image Bits Before Message Encryption

After Encoding

Message A – 01000001

Pixel 1: 11111000 11001001 00000010
Pixel 2: 11111000 11001000 00000010
Pixel 3: 11111000 11001001 00000011

Image Bits after Message Encryption

- **Stego output**: In this step, we will receive the stego image (encoded image) as the output of the above process (Fig. 2).

4 Evaluation Parameters

Some standard parameters of image processing have been discussed in this section. The presented work is evaluated using the same parameters.

4.1 *Peak Signal-to-Noise Ratio (PSNR)*

The block of PSNR computes the ratio between two images (i.e., peak signal-to-noise ratio) in decibels. The quality of the PSNR ratio is used to compare the original and compressed images. The higher the PSNR ratio means better the quality of the compress or reconstructed image. The PSNR and mean square error (MSE) are used to compute the quality of compressed images. The PSNR is used to calculate the peak ratio in the image whereas MSE computes a cumulative squared error between the original and compress the image.

To calculate the PSNR, the following equation is used:

$$\text{PSNR} = 10\log_{10}\left(\frac{R^2}{\text{MSE}}\right) \tag{2}$$

In the above equation, R is represent the maximum fluctuations in the cover image.

4.2 *Structural Similarity Index (SSIM)*

The structural similarity index (SSIM) is a fundamental measure that evaluates the image quality degradation due to data compression or loss during transmission. It is a wholly referenced method that necessitates the use of two images: a reference image and a processed image. Unlike PSNR, SSIM is based on the image's visual structure. Furthermore, PSNR is no longer regarded as a reliable method of predicting image quality degradation, but rather as an alternative method of measuring. SSIM is primarily utilized in videography, but it is also a useful statistic in still photography.

5 Result Analysis

In this section, we have analyzed distortion in three different images. Initially, we chose a bitmap image format of 798 kb which is shown in Table 1. We use it to encrypt 3000 characters of secret text. We implement encoding at various bits, ranging from 1st-bit MSB to 8th-bit LSB capturing corresponding PSNR and SSIM value. Although the original image and the encoded image are of the same size, hence distortion in the image is least at 8th-bit encoding and most at 1st-bit encoding.

Again, a bitmap image of 769 kb (shown in Table 2) with an increase text length of 5000 characters has been taken. We use it to encrypt 5000 characters of secret text. We implement same encoding at various bits ranging from 1st-bit MSB to 8th-bit LSB and observed the corresponding PSNR and SSIM values. Although original image and the encoded image are of the same size, distortion in the image is least at 8th-bit encoding and most at 1st-bit encoding.

Table 1 Distortion Analysis for Text Length 3000

	Images	Grey-scale graph	RGB graph	PSNR and SSIM value
Original image		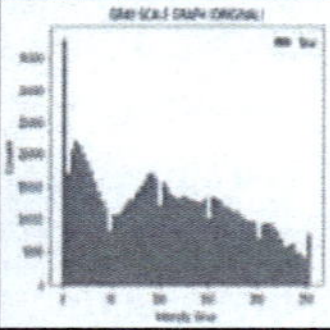	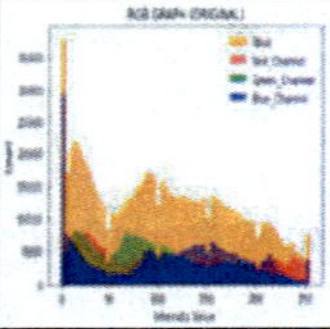	PSNR – **NA** SSIM – **1.0**
Stego image (8th bit)		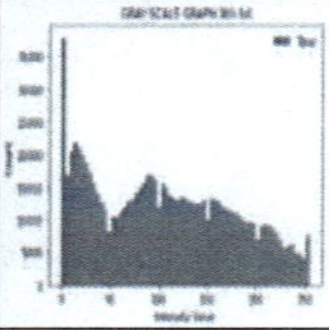	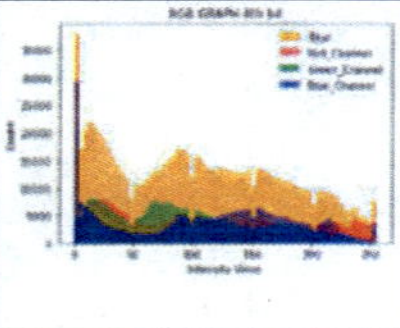	PSNR – **71.34** SSIM – **0.99**
Stego image (7th bit)		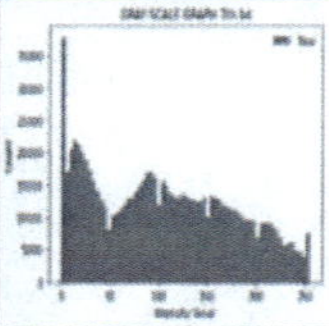	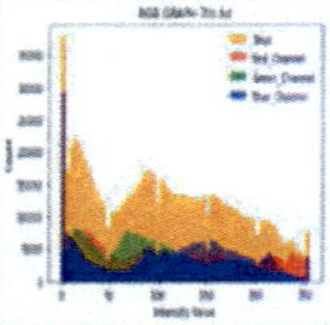	PSNR – **68.83** SSIM – **0.99**
Stego image (6th bit)		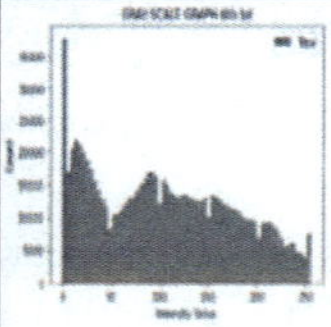	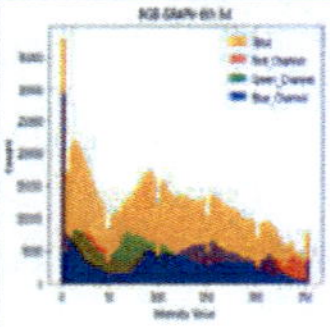	PSNR – **68.63** SSIM – **0.99**
Stego image (5th bit)		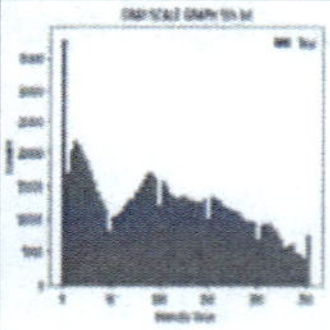	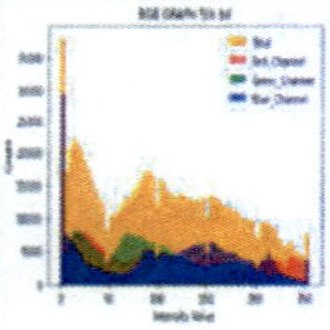	PSNR – **68.86** SSIM – **0.99**
Stego image (4th bit)		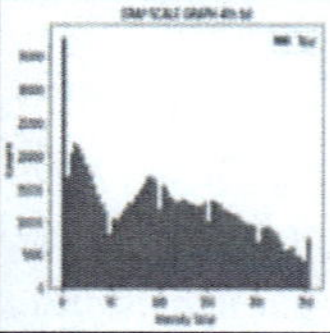	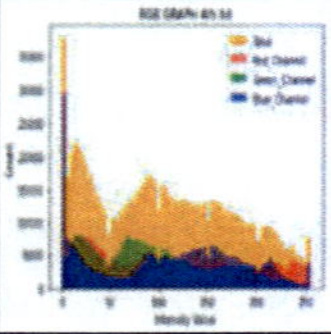	PSNR – **68.97** SSIM – **0.99**
Stego image (3rd bit)		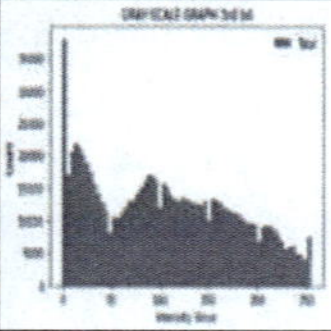	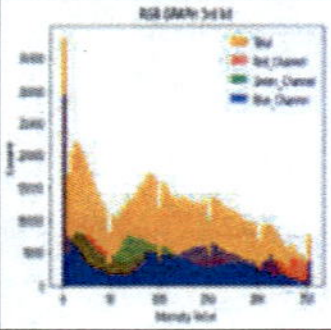	PSNR – **68.84** SSIM – **0.99**

(continued)

Table 1 (continued)

Steg image (2nd bit)		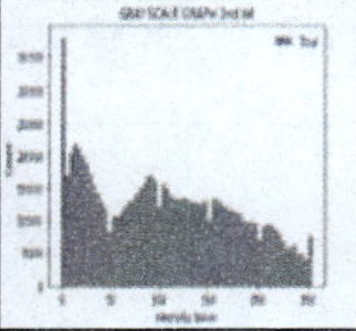	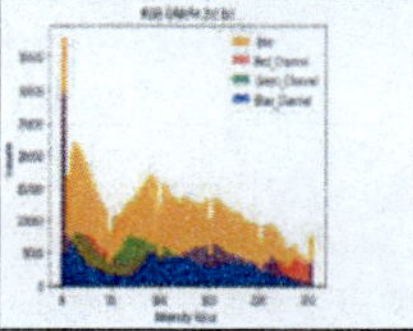	PSNR – 68.87 SSIM – 0.99
Steg image (1st bit)		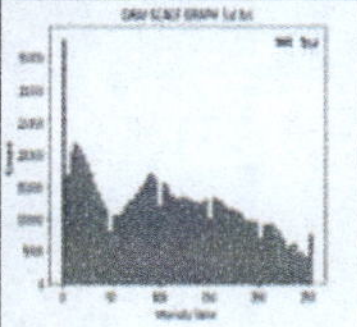	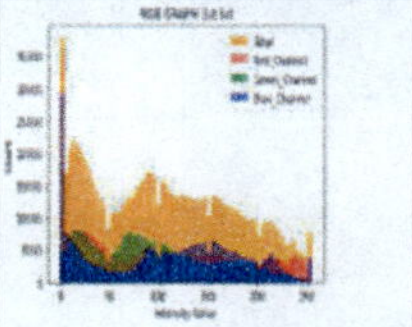	PSNR – 68.87 SSIM – 0.99

Similarly, a bitmap image of 769 kb has been taken as shown in Table 3 to encrypt 7000 characters of secret text. We run the same simulation to encode at various bits ranging from 1st-bit MSB to 8th-bit LSB and found the distortion level in the 8th-bit position is minimal whereas it is maximum at 1st-bit.

6 Conclusion

Although the Internet has numerous benefits, it has also offered a new avenue for hackers and unauthorized users to invade our privacy and intellectual property. Since the emergence of these issues, numerous approaches have been developed. This work employed metrics to examine the distortion levels and discovered that the distortion increases as the number of characters in the data concealment phase increases. As a result, to avoid distortion, the input text should be as brief as possible. The data were encoded using the LSB method, and the changes were detected using the histogram method.

Table 2 Distortion Analysis for Text Length 5000

	Images	Grey-scale graph	RGB graph	PSNR and SSIM values
Original image		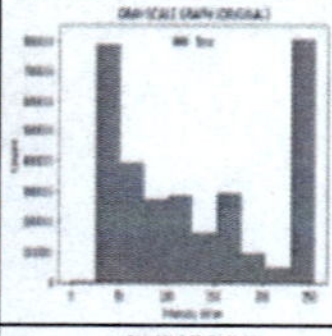	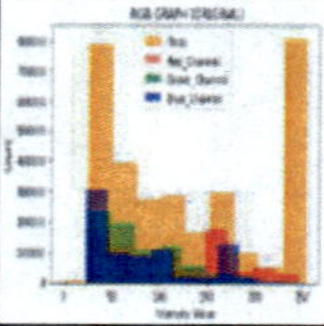	PSNR – **NA** SSIM – **1.0**
Stego image (8th bit)		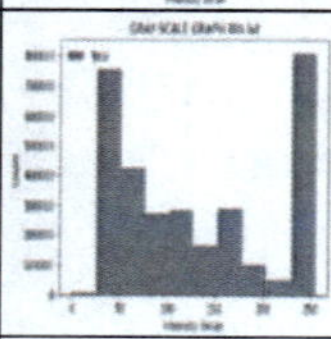	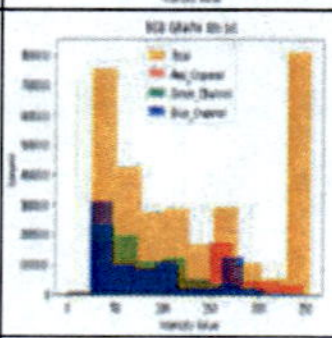	PSNR – **39.71** SSIM – **0.99**
Stego image (7th bit)		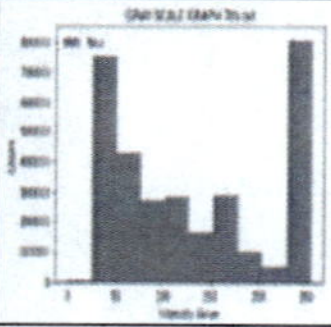	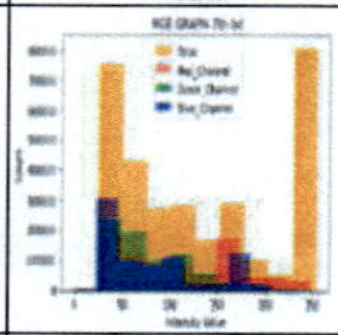	PSNR – **39.96** SSIM – **0.99**
Stego image (6th bit)		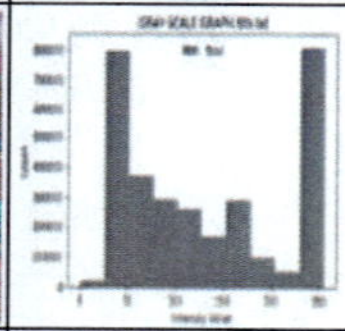	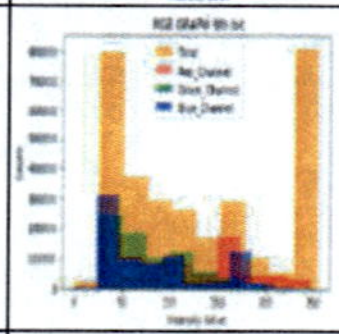	PSNR – **39.68** SSIM – **0.99**
Stego image (5th bit)		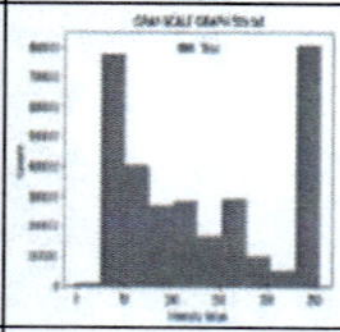	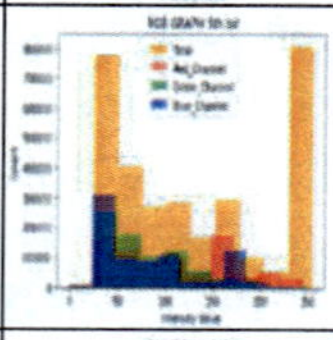	PSNR – **39.39** SSIM – **0.99**
Stego image (4th bit)		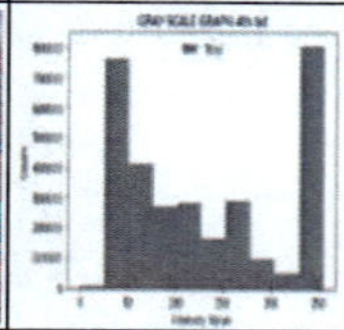	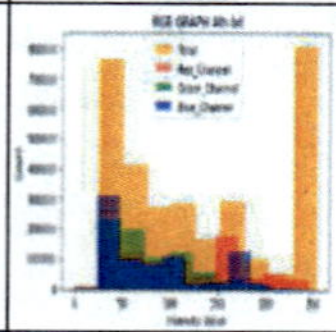	PSNR – **39.38** SSIM – **0.99**
Stego image (3rd bit)		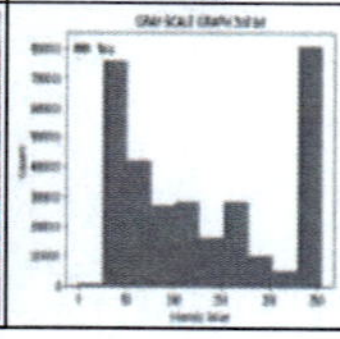	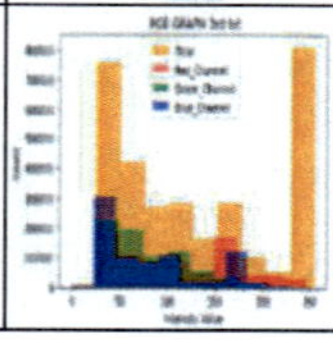	PSNR – **39.36** SSIM – **0.99**

(continued)

Table 2 (continued)

Stego image (2nd bit)		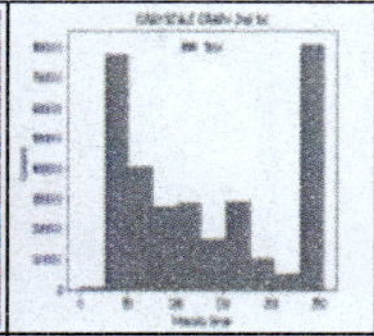	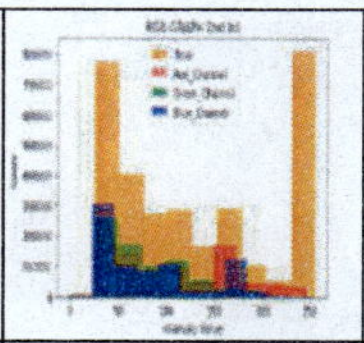	PSNR – **39.37** SSIM – **0.99**
Stego image (1st bit)		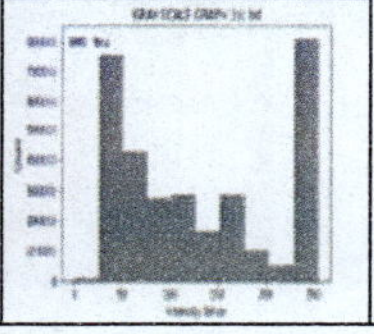	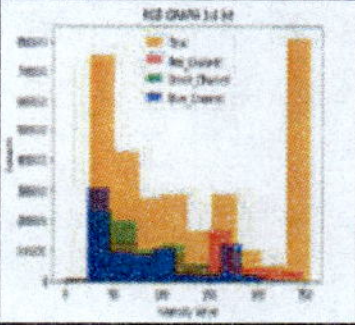	PSNR – **39.44** SSIM – **0.99**

Table 3 Distortion Analysis for Text Length 7000

	Images	Grey-scale graph	RGB graph	PSNR & SSIM value
Original image		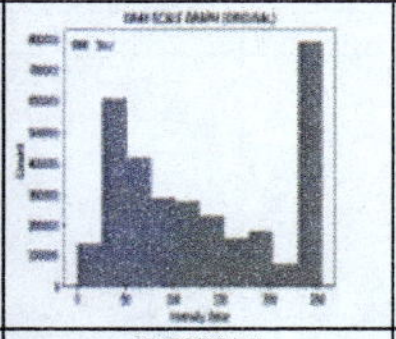	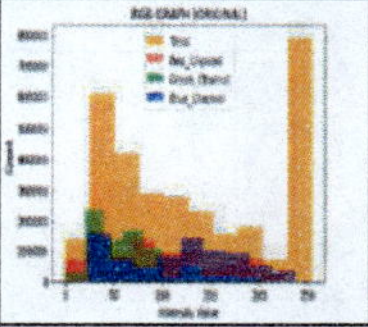	PSNR – **NA** SSIM – **1.0**
Stego image (8th bit)		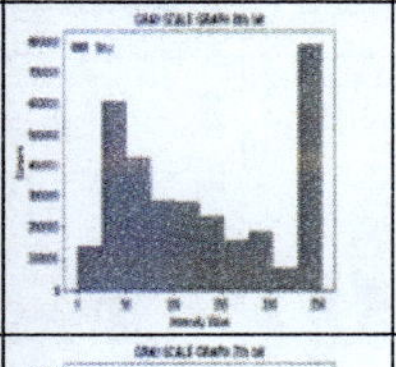	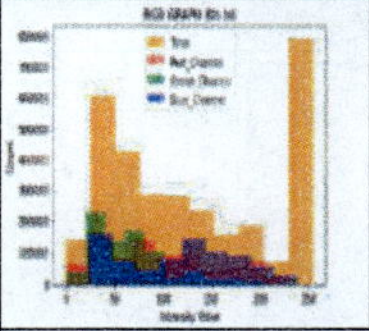	PSNR – **28.10** SSIM – **0.995**
Stego image (7th bit)		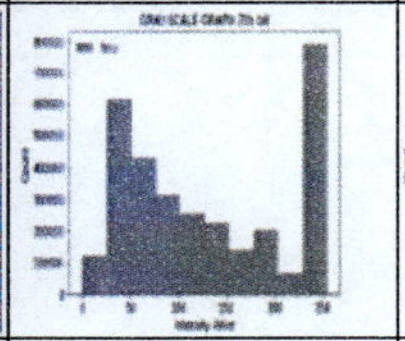	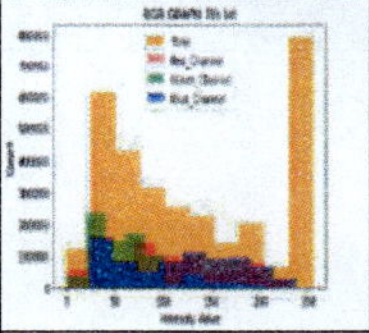	PSNR – **27.88** SSIM – **0.95**
Stego image (6th bit)		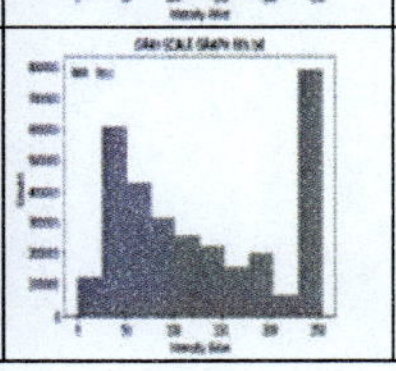	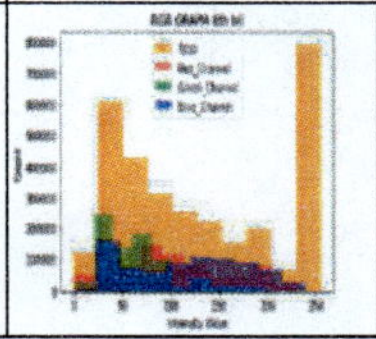	PSNR – **27.56** SSIM – **0.95**

(continued)

Table 3 (continued)

stego image (5th bit)		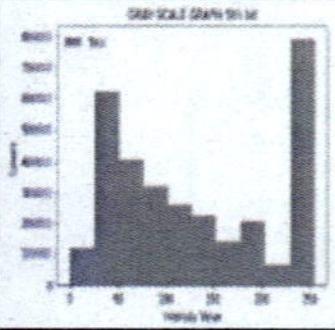	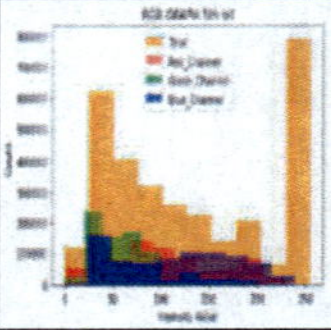	PSNR – **27.60** SSIM – **0.95**
Stego image (4th bit)		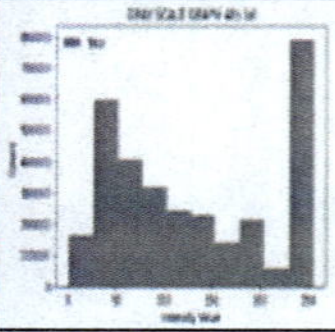	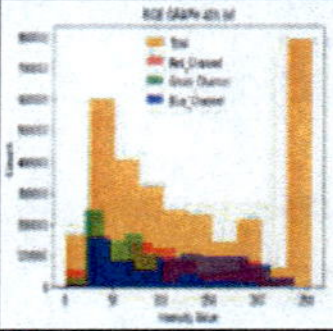	PSNR – **27.68** SSIM – **0.95**
Stego image (3rd bit)		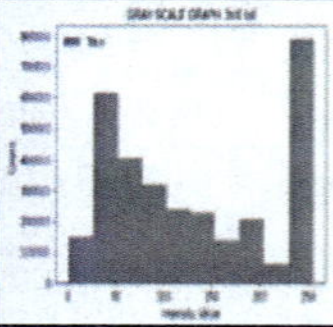	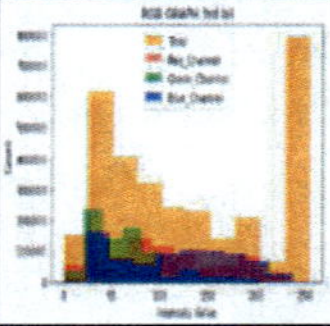	PSNR – **27.64** SSIM – **0.95**
Stego image (2nd bit)		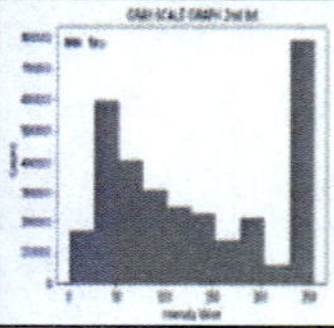	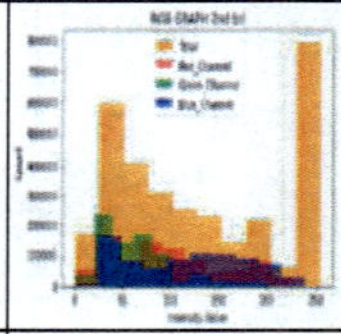	PSNR – **27.61** SSIM – **0.95**
Stego image (1st bit)		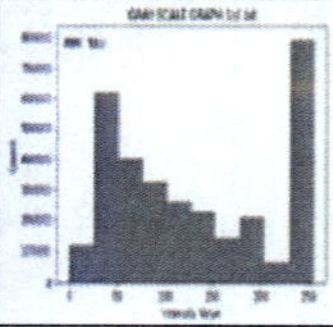	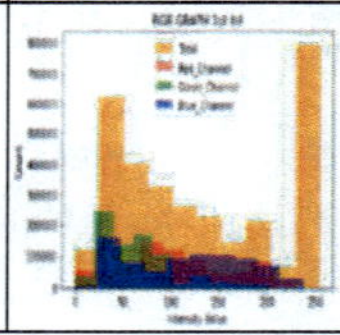	PSNR – **39.93** SSIM – **0.95**

References

1. Alexan, W., El Beheiry, M., & Gamal–Eldin, O. (2020). A comparative study among different mathematical sequences in 3D image steganography. *International Journal of Computing and Digital Systems, 9*(4). ISSN 2210-142X.
2. Sahu, A. K., & Swain, G. (2020). Reversible image steganography using dual-layer LSB matching. *Sensing and Imaging, 21*, 1.
3. Chatterjee, A., Ghosal, S. K., & Sarkar, R. (2020). LSB based steganography with OCR: An intelligent amalgamation. *Multimedia Tools and Applications, 79*, 11747–11765.
4. Kumar, R., & Singh, N. (2021). Concealing the confidential information using LSB steganography techniques in image processing. In D. Gupta, A. Khanna, S. Bhattacharyya, A. E. Hassanien, S. Anand, & A. Jaiswal (Eds.), *International Conference on Innovative Computing and Communications. Advances in Intelligent Systems and Computing* (Vol. 1165). Springer.
5. Pak, C., Kim, J., An, K., et al. (2020). A novel color image LSB steganography using improved 1D chaotic map. *Multimedia Tools and Applications, 79*, 1409–1425.
6. Amarendra, K., Mandhala, V. N., Chetan Gupta, B., Geetha Sudheshna, G., & Venkata Anusha, V. (2019). Image steganography using LSB. *International Journal Of Scientific & Technology Research, 8*(12).
7. Eyssa, A. A., Abdelsamie, F. E., & Abdelnaiem, A. E. (2020). An efficient image steganography approach over wireless communication system. *Wireless Personal Communications, 110*, 321–337. https://doi.org/10.1007/s11277-019-06730-2
8. Desai, P. B., & Bhendwade, P. S. (2016). Image steganography using LSB algorithm. *International Journal of Advanced Research in Electrical, Electronics and Instrumentation Engineering, 5*(8).
9. Patani, K., & Rathod, D. (2021). Advanced 3-bit LSB based on data hiding using steganography. In K. Kotecha, V. Piuri, H. Shah, & R. Patel (Eds.), *Data science and intelligent applications. Lecture Notes on Data Engineering and Communications Technologies* (Vol. 52). Springer.

Towards a Secured IoT Communication: A Blockchain Implementation Through APIs

Rajat Verma, Namrata Dhanda, and Vishal Nagar

Abstract Years ago, intercommunication among systems was not at all possible, but with the development of the application programming interface (APIs-traditional) in the 1960s, this was possible. With the evolution from the 1960s, these APIs have advanced, and the birth of APIs (Modern) took place in the initial 2000s. Many conventional techniques and tools that possess diverse security issues and are operating in a fully functional manner can be solved to a greater extent with the help of APIs if integrated with modern and SMART technologies. One such example is a sub-group of artificial intelligence (AI) i.e., IoT, that possesses diverse issues and challenges and can be solved by the peer-to-peer technology blockchain. Here, blockchain is regarded as the SMART solution. The diverse attacks on IoT spectrum are illustrated in this paper to attain a better scope and security for IoT devices. Moreover, the implementation of blockchain is also illustrated in this paper to depict a convenient approach for solving the issues of IoT with decentralization attribute. The reason is that IoT illustrates a centralized architecture.

Keywords Blockchain · IoT · Decentralization · Enhanced communication

1 Introduction

With the advent of the APIs from the traditional ones to the modern ones (the 1960s–2000s), the world noticed the changed architecture of the IT domain that evolved from being at some places to omnipresent. Communication among applications is possible and efficiently done with the combination of programs along with the set of rules and regulations commonly known as protocols in the presence of the world-renowned

R. Verma (✉) · N. Dhanda
Department of Computer Science and Engineering, Amity University Uttar Pradesh, Lucknow, India
e-mail: rajatverma310795@gmail.com

V. Nagar
Department of Computer Science and Engineering, Pranveer Singh Institute of Technology, Kanpur, Uttar Pradesh, India

P. K. Singh et al. (eds.), *Proceedings of Third International Conference on Computing, Communications, and Cyber-Security*, Lecture Notes in Networks and Systems 421,
https://doi.org/10.1007/978-981-19-1142-2_53

technology, i.e., the Internet [1]. In enhancing the technical sector, the primary factor that has contributed is the APIs. The reason is that with the help of the evolving APIs (1960-the 2000s), impossible inventions also seemed possible in this present era. The success of the APIs began with the essence of creating an illustration (virtual) of an object. That could be hardware which are the tangible parts of the system or the software that are the intangible parts of the system, or it could also be a service that is provided by a recent IT technology, i.e., cloud computing. One of the constituents of the cloud, i.e., IaaS made the processes that are computational, possible across the Internet [2].

The emerging technologies that have been modified in recent times or been developed are primarily a result of the evolution of APIs. Generally, the procedure says that if the client or can be called as a user is performing an operation on an application using his/her cellular phone that has a pre-requisite of accessing the Internet follows a client–server architecture (Traditionally) [3]. The recipient side which can be called the server obtains the data, interprets it line by line into a format, i.e., understandable and forwards it to the user's device. From the application's interface, the user can decode the data and perform the task that was to be done. APIs work in this particular manner [4].

Blockchain, a 23-year-old decentralized invention, is also an advancement of APIs [5]. In the tenure of 60 years (1960–2020), with the development in the spectrum of APIs, a variety of technologies are acting as a solution in solving the conventional challenges. One such scenario is of the IoT systems whose primary challenge is centralization that could be solved by the decentralized characteristic feature of blockchain [6, 7]. Many other issues such as tampering aspect could also be solved by this peer-to-peer technology blockchain. The conventional illustration of Web-programming API communication is shown in Fig. 1 for easy understanding.

Figure 1 shows the communication between the client and server and vice-versa. When the program is implemented in blockchain also, the isolated systems are created. The same architecture is used in blockchain for communication through APIs following the client–server architecture.

The smash-hit tool for implementing APIs is Postman [8]. Diverse types of API calls are available in Postman [9]. In the later sections, the execution of blockchain

Fig. 1 General API communication

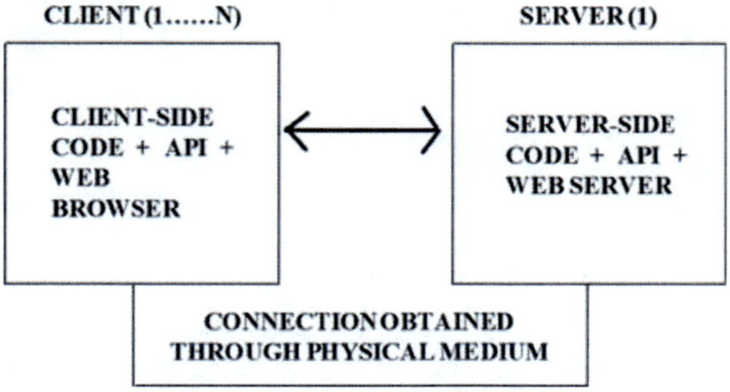

Table 1 Handling IoT issues using blockchain technology (summarized table)

IoT concerns	Handling IoT issues using blockchain
Tampering/breaching	This issue can be handled by blockchain as only append operation is allowed and in between writing is not allowed. Secondly, it follows SHA in addition to ECDSA
Man-in-the-middle	All these issues of IoT are completely solvable by blockchain because of its characteristic features. They are immutability, transparency, and decentralization. It removes the single point of failure
Centralization	
Data attacks on IoT spectrum	
Expression, enforcement, and transparency	

using Python is highlighted, and the review of IoT attacks is done. Additionally, the assumptions and recommendations are also illustrated in the form of Table 1.

2 Material and Methods

2.1 Blockchain Technology

The chief cause for the motivation behind the origination of blockchain technology was the cryptocurrency Bitcoin. Bitcoin was developed in 2008 by a group of people that are popular by a pseudo-name "Satoshi Nakamoto" [10, 11]. Blockchain was integrated as the back end to attain a higher standard of security in the famous currency Bitcoin. A chain is an amalgamated group of blocks that are connected with distinct hash values. Secure hashing algorithm-256 in combination with digital signature algorithm concerning elliptical curves (ECDSA) played a major role in securing this peer-to-peer technology. The structure of a block in a blockchain is depicted in Fig. 2.

Figure 2 illustrates some parts that are highlighted in the middle of the diagram. They are block number (Position of a block), preceding-block hash value, current-block hash value, Nonce (Security), and last but not least the time-stamp [12].

Fig. 2 Block's layout

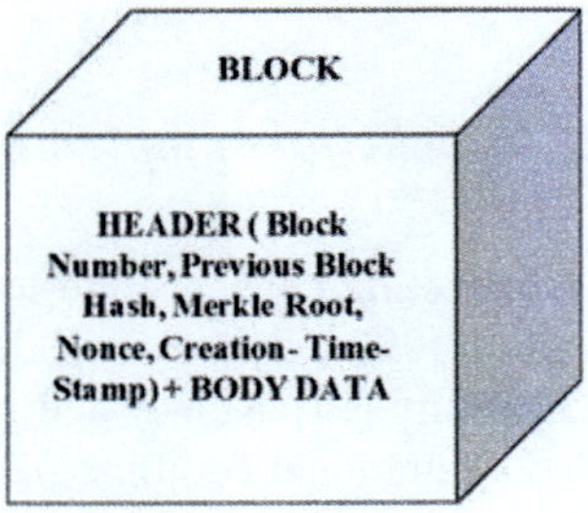

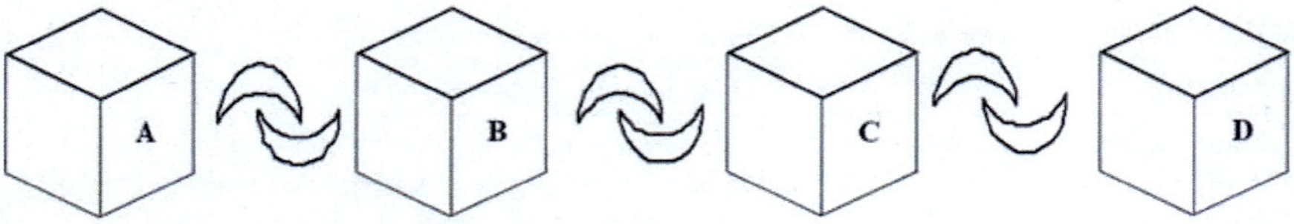

Fig. 3 Illustration of a simple blockchain

The first entity, i.e., the block number, illustrates the sequence of a block (Position) in the blockchain. The time-stamp illustrates the time at which the block was originated or added to the blockchain. There is a necessity of linking the previous block hash value to the current block hash value, as without it the chain will not form. The current block hash value is depicted as the Merkle root, while the previous block hash does not have a specific naming convention. The aspect of security is maintained by the Nonce feature of the block present in the header section [13]. As the blocks are added to the chain, they form a blockchain, and the blocks that are parts of the longest chain are considered as the blocks of the main-chain. If the blocks are not linked with the longest chain, they are termed orphan blocks [14]. The blockchain is known for its three parametric features which are transparency, immutability, and decentralization [15] whose structure is represented in Fig. 3 for easy understanding. A simple illustration of blockchain is highlighted in Fig. 3 that has merely four blocks or entities but can be expanded up to N-blocks if the resources are present and if the memory allows.

Figure 3 illustrates four entities depicted by naming conventions A, B, C, and D where the initial block A is called a genesis block with the missing previous hash value but with the current hash value and other attributes depicted earlier [16]. The following block B is the next node of the sequence whose previous hash value will be equal to the current hash value of A. This is an essential pre-requisite as, without it, the link among diverse nodes of the chain is not possible. An identical phase happens with the other blocks. Here, the other blocks are C and D, but there is a possibility of more nodes as this is only an illustration for understanding the purpose. The nodes are very important in a blockchain environment as they do the process of mining. Classification of blockchain is also important, and the choice of usage will be different from organization to organization. Public, private, hybrid, and consortium are the various types of blockchain.

2.2 Blockchain Implementation

The interesting concepts and techniques are possible with the advent of APIs. In the early 1990s, Python was developed, and at that point, no one thought about its capability [17]. Slowly and gradually, with the add-ins, updates, advancements, the configurational comfortability is also enhanced with the different modules and packages. In my implementation of a local blockchain, Python 3.6 was used as it has the required modules and packages [18]. The specific advantage of implementing blockchain with the Postman API and Python is that the connections of the nodes can easily be made using Postman API platform (in JSON format) using GET and POST call methods. Python allows a very efficient manner of implementing a local blockchain using classes, functions, and arguments. Python also supports flask with which the user can easily build a Web-application. Although Postman API is not providing any security enhancement to the local blockchain by itself because it is just an IDE, it gives a very convenient approach of following things properly, i.e., using GET and POST calls.

In my implementation, eight isolated systems were made on a single machine for development purposes, starting from the loopback address with port number 9000. This initial node with a port number was to run on the Postman platform to connect the rest of the other nodes. The connector system was designated as 127.0.0.1:9000. Similarly, the others ranged from 127.0.0.1:9001 to 127.0.0.1:9007. The first system in the blockchain was 9001, and the last isolated system was 9007. In my implementation, only, seven systems were created but could be extended up to N depending on the resources available. The decentralization concept is highlighted in Fig. 4 for quick and easy understanding.

Figure 4 shows all the isolated systems that were joined using API platform written in JSON format. This corresponds to the working of all isolated systems working together in sync. In the figure, four indexes are there illustrating different things in the API platform as 1 shows the address of the node that is being done using POST method call and is running on 9000 port number to link diverse nodes that are present in the blockchain. The following index, i.e., 2 depicts the nodes or the system that is to be connected. The next index, i.e., 3 shows the output with the measure of connected nodes, i.e., from 127.0.0.1:9001 to 127.0.0.1:9007. The last entity shows the message. It can be customized to attain a clear message to the audience using the system.

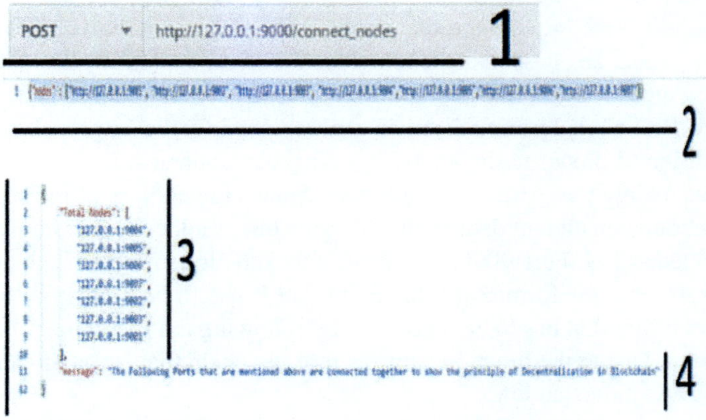

Fig. 4 Illustrating the concept of decentralization in blockchain

Fig. 5 Origination of genesis block: the first block in a blockchain

After the connection is initiated using the connector port (9000 in this case). The following thing to do is to add blocks while using transactions as blockchain is incomplete without transactions. The chain has to be formed on every system ranging from 9001 to 9007. Figure 5 illustrates the origination of the first block on the blockchain. It could be done on a particular port but has to be in sync with the other ports.

Figure 5 shows a few indexes for easy understanding starting from 1 to 4. The first index, i.e., 1 shows the address at which isolated system the block was created. In this particular case, the port number is 9001 but could be any depending on the configuration and need. The following indexes (2, 3) show the function chain in which all the entities are highlighted that should be there for a genesis block to be called. The last index, i.e., 4 depicts the length of the chain at present at a system.

After adding the first block, the next thing to be done is to complete the chain and that is possibly done while forming the main chain that is only the longest chain. The process of addition of blocks is shown in Fig. 6 for easy understanding.

Figure 6 has mainly two parts; the initial part depicts the address of the separate system. The second constituent depicts the 8th node that is added to the 5th isolated system designated as 127.0.0.1:9005. The name of the function in the implementation is mine_block. A chain is formed till this point, but it has to be predicted that the chain is longest or not that has to be made sure by following the principle of longest chain implemented using the function name as replace_chain that is shown in Fig. 7 for quick and easy understanding.

Figure 7 shows the concept of the longest chain in which the initial part focuses on the address of the isolated host/node that is replacing the chain with the main chain. The next part consists of the last node that is added to the chain. Then, the message

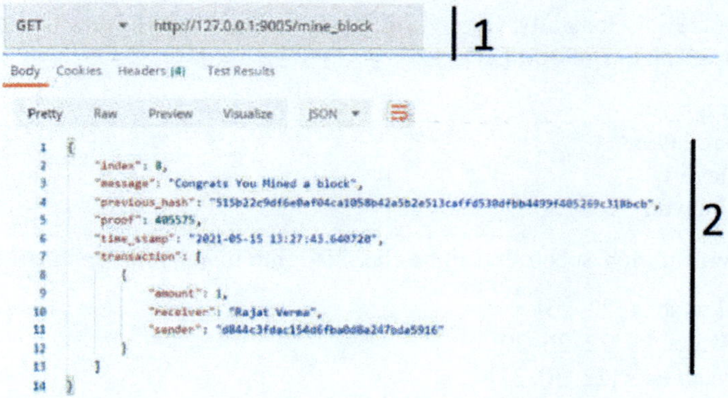

Fig. 6 Blockchain formed with the last block added on the port number 9005

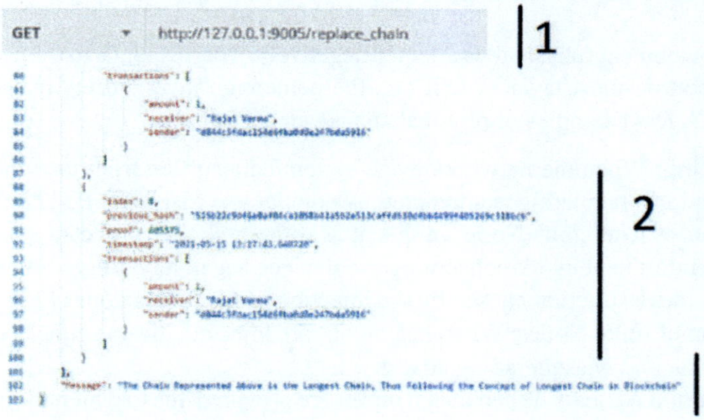

Fig. 7 Main chain or longest chain

appears as "The Chain Represented Above is the Longest Chain, Thus Following the Concept of Longest Chain in Blockchain" depicting the main chain.

It was the implementation of blockchain (Local) using API platform.

2.3 *Internet of Things (IoT) Issues and Concerns*

The IoT has made its name in the past 20 years or two decades and is still growing and that too at a tremendous pace [19]. It was estimated by a researcher that in the next 5 years that is up to 2025; approximately, 75 billion connected devices will be a part of the IoT spectrum [20] that will be exchanging the pieces of information through the unsecured cyberspace while obeying the recommendations of World Wide Web

Consortium (W3C). Generally, the IoT architecture deals with layers that can be categorized as (Top to Bottom) [21]:

- Application layer
- Middleware layer
- Network layer
- Perception layer.

The concerns on IoT spectrum can be classified into major sub-versions:

1. Physical attacks [22–25]
2. Network attacks [22, 26–29]
3. Software attacks [22, 30, 31]
4. Data attacks [22].

2.3.1 Physical Attacks on IoT Spectrum

When the environment (physical) is affected negatively, it corresponds to the physical attack. The bottom-most layer of IoT, i.e., the perception layer suffers from these attacks [22]. A few examples of physical attacks are as follows:

1. **Tampering**: When the network/device/system/data suffers from unauthorized modifications, tampering is attempted. A popular use case is RFID [22, 23].
2. **Injection of Malicious Node**: In this attack, the malicious/infected system is injected into the network concerning the devices negatively. Logic bombs can enhance the destruction caused by the injection of malicious nodes [22, 24].
3. **Injection of Fake Nodes**: When fake nodes are injected, diverse attacks can be invoked such as Masquerading attacks.
4. **Sleep Denial Attacks**: When illegal inputs are provided due to which the battery drains at a rapid rate and the device shuts, sleep denial attacks are invoked [22, 24].
5. **Centralization**: The conventional architecture of IoT follows a centralized architecture thus giving birth to the central point of failure that needs to be eradicated, and at this point, blockchain comes into the picture [25].

2.3.2 Network Attacks on IoT Spectrum

The attacks focus on the single system or a bunch of systems in a network using automated scripts and bots [22, 26]. The second layer of IoT architecture from the bottom, i.e., the network layer is affected using these kinds of attacks. A few of them are depicted below:

1. **Analysis of Traffic**: Passive attacks hold the analysis of traffic under their consideration [27]. In such attacks, the attackers remain silent and only observe the activities [23].

2. **RFID Unauthorized Access and Spoofing**: In this attack, initially, spoofing is done, and then, access (unauthorized) is attempted. In spoofing, the attacker distracts the RFID signal and get access to the data that include the facts and figures. After tricking the RFID, they can inappropriately access the RFID and destruct it in whatever manner they like [22, 28].
3. **Man-in-the-Middle Attack**: In this attack, the attacker ruptures the data transfer among different entities. The cybercriminal pretends to be a real entity but can attack the system or connection actively or passively [22, 23, 29].

2.3.3 Software Attacks on IoT Spectrum

These attacks are done through Malwares [22, 30]. Malware can take any form such as worms, viruses, ransomware, spyware, and Adware. They can have diverse objectives such as infecting the system, stealing sensitive or private data, and launching cyberattacks such as denial of service attacks, or disrupting cloud architectures and infrastructures [31].

2.3.4 Data Attacks on IoT Spectrum

In these attacks, the CIA triad is affected leading to happening of data inconsistency, data breaching, and information rupture, etc. The CIA triad stands for confidentiality-integrity-availability [7, 22].

2.4 Handling of IoT Issues Using Blockchain

Blockchain mechanism can play a key role in eradiating the traditional issues of IoT [22]. The three characteristic features of blockchain that help blockchain to be an extraordinary approach are immutability, transparency, and decentralization [32]. If all these features are amalgamated into the IoT spectrum, then it will have some advantages that are as follows:

1. Trust enhancement [22, 33]
2. Security enhancement [22, 34]
3. Enhanced fairness [22, 34].

Table 1 highlights the issues of IoT and its possible solutions such as centralization that use to happen in IoT can be removed by integrating blockchain into IoT as it follows decentralization. Tampering can be minimized in IoT by integrating both as blockchain follows immutability. Similarly, many issues with their possible solution are highlighted in Table 1. If blockchain is implemented commercially with IoT, it will give a boost to both the network and devices in security aspects. All the solutions will be applicable in real time. A flowchart enhancing the efficiency of IoT using blockchain is shown in Fig. 8.

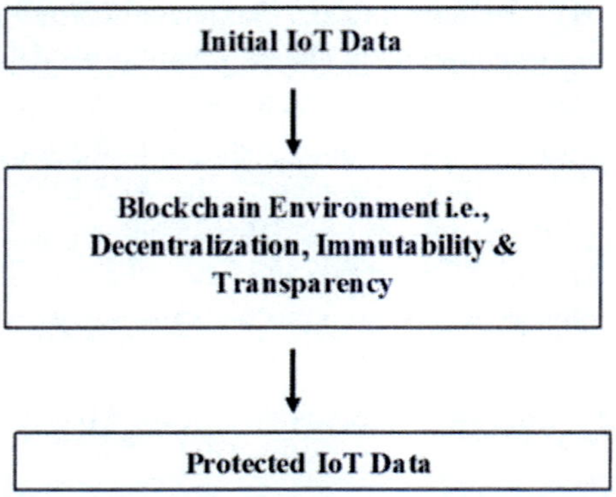

Fig. 8 Simple illustration of securing IoT data

3 Results

With the above text, the prominent statement that can be made is that if blockchain is considered as a commercial solution for IoT issues and challenges, then the efficiency of IoT will be maximized. The same can be observed from the blockchain solutions that has been shown in Table 1 and Sect. 2.4. A review of IoT concerns is shown in Sect. 2.3. Blockchain is implemented in Python and is shown above in Sect. 2.2.

4 Conclusion and Future Scope

IoT has maintained its popularity for the past decades. With the increased popularity, there will be a greater amount of data that will have greater facts and figures. Due to this enhanced data, there will certainly be a larger number of issues and concerns of IoT which is illustrated in this paper. The introduction of blockchain with its implementation through the means of Python and Postman API is highlighted in this paper. The implementation of blockchain along with a review of attacks on IoT is the main contribution of the author. Although, here the assumptions are only given as the properties of blockchain are enough to deal with the concerns and challenges of IoT. The implementation of blockchain can be integrated with the IoT spectrum to improve its efficiency that will be considered as a future scope.

Additionally, it will find some more open concerns of IoT that could be solved by the network technology blockchain.

References

1. Lee, G. M., Crespi, N., Choi, J. K., & Boussard, M. (2013). Internet of things. In *Evolution of telecommunication services* (pp. 257–282). Springer.
2. Ilyas, M. U., Ahmad, M., & Saleem, S. (2020). Internet-of-things-infrastructure-as-a-service: The democratization of access to public internet-of-things infrastructure. *International Journal of Communication Systems, 33*, e4562.
3. Devin, F. (2020). Web-oriented architecture—How to design a RESTFull API. In *TORUS 1—Toward an open resource using services: Cloud computing for environmental data* (pp. 191–206).
4. Gu, X., Zhang, H., Zhang, D., & Kim, S. (2016). Deep API learning. In *Proceedings of the 2016 24th ACM SIGSOFT International Symposium on Foundations of Software Engineering* (pp. 631–642).
5. Palisse, A., Le Bouder, H., Lanet, J. L., Le Guernic, C., & Legay, A. (2016). Ransomware and the legacy crypto API. In *International Conference on Risks and Security of Internet and Systems* (pp. 11–28). Springer.
6. Kumar, N. M., & Mallick, P. K. (2018). Blockchain technology for security issues and challenges in IoT. *Procedia Computer Science, 132*, 1815–1823.
7. Verma, R., Dhanda, N., & Nagar, V. (2020). Addressing the issues & challenges of internet of things using blockchain technology. *International Journal of Advanced Science and Technology, 29*, 10074–10082.
8. Soni, A., & Ranga, V. (2019). API features individualizing of web services: REST and SOAP. *International Journal of Innovative Technology and Exploring Engineering (IJITEE)*, 2278–3075.
9. Neumann, A., Laranjeiro, N., & Bernardino, J. (2018). An analysis of public REST web service APIs. *IEEE Transactions on Services Computing*, 1–14.
10. Nakamoto, S. (2019). Bitcoin: A peer-to-peer electronic cash system. Manubot.
11. Nofer, M., Gomber, P., Hinz, O., & Schiereck, D. (2017). Blockchain. *Business & Information Systems Engineering, 59*(3), 183–187.
12. Saini, H., Bhushan, B., Arora, A., & Kaur, A. (2019). Security vulnerabilities in Information communication technology: Blockchain to the rescue (A survey on blockchain technology). In *2019 2nd International Conference on Intelligent Computing, Instrumentation and Control Technologies (ICICICT)* (Vol. 1., pp. 1680–1684). IEEE.
13. Courtois, N. T., Grajek, M., & Naik, R. (2014). Optimizing SHA256 in bitcoin mining. In *International Conference on Cryptography and Security Systems* (pp. 131–144). Springer.
14. Göbel, J., Keeler, H. P., Krzesinski, A. E., & Taylor, P. G. (2016). Bitcoin blockchain dynamics: The selfish-mine strategy in the presence of propagation delay. *Performance Evaluation, 104*, 23–41.
15. Niranjanamurthy, M., Nithya, B. N., & Jagannatha, S. (2019). Analysis of blockchain technology: Pros, cons and SWOT. *Cluster Computing, 22*(6), 14743–14757.
16. Singh, S., & Singh, N. (2016). Blockchain: Future of financial and cyber security. In *2016 2nd International Conference on Contemporary Computing and Informatics (IC3I)* (pp. 463–467). IEEE.
17. Dalcin, L. D., Paz, R. R., Kler, P. A., & Cosimo, A. (2011). Parallel distributed computing using Python. *Advances in Water Resources, 34*(9), 1124–1139.
18. Alharby, M., & van Moorsel, A. (2019). Blocksim: A simulation framework for blockchain systems. *ACM SIGMETRICS Performance Evaluation Review, 46*(3), 135–138.
19. Gubbi, J., Buyya, R., Marusic, S., & Palaniswami, M. (2013). Internet of Things (IoT): A vision, architectural elements, and future directions. *Future Generation Computer Systems, 29*(7), 1645–1660.
20. Bera, A. (80). Insightful internet of things statistics (infographic). White Paper. Retrieved from https://safeatlast.co/blog/iot-statistics/#gref
21. Li, J., Liu, Y., Xie, J., Li, M., Sun, M., Liu, Z., & Jiang, S. (2019). A remote monitoring and diagnosis method based on four-layer IoT frame perception. *IEEE Access, 7*, 144324–144338.

22. Sengupta, J., Ruj, S., & Bit, S. D. (2020). A comprehensive survey on attacks, security issues and blockchain solutions for IoT and IIoT. *Journal of Network and Computer Applications, 149*, 102481.
23. Andrea, I., Chrysostomou, C., & Hadjichristofi, G. (2015). Internet of things: Security vulnerabilities and challenges. In *2015 IEEE Symposium on Computers and Communication (ISCC)* (pp. 180–187). IEEE.
24. Ahemd, M. M., Shah, M. A., & Wahid, A. (2017). IoT security: A layered approach for attacks & defenses. In *2017 International Conference on Communication Technologies (ComTech)* (pp. 104–110). IEEE.
25. Atlam, H. F., & Wills, G. B. (2019). Technical aspects of blockchain and IoT. In *Advances in computers* (Vol. 115, pp. 1–39). Elsevier.
26. Deogirikar, J., & Vidhate, A. (2017). Security attacks in IoT: A survey. In *2017 International Conference on I-SMAC (IoT in Social, Mobile, Analytics and Cloud) (I-SMAC)* (pp. 32–37). IEEE.
27. Ning, J., Xu, J., Liang, K., Zhang, F., & Chang, E. C. (2018). Passive attacks against searchable encryption. *IEEE Transactions on Information Forensics and Security, 14*(3), 789–802.
28. Khoo, B. (2011). RFID as an enabler of the internet of things: Issues of security and privacy. In *2011 International Conference on Internet of Things and 4th International Conference on Cyber, Physical and Social Computing* (pp. 709–712). IEEE.
29. Navas, R. E., Le Bouder, H., Cuppens, N., Cuppens, F., & Papadopoulos, G. Z. (2018). Do not trust your neighbors! A small IoT platform illustrating a man-in-the-middle attack. In *International Conference on Ad-Hoc Networks and Wireless* (pp. 120–125). Springer.
30. Wang, A., Liang, R., Liu, X., Zhang, Y., Chen, K., & Li, J. (2017). An inside look at IoT malware. In *International Conference on Industrial IoT Technologies and Applications* (pp. 176–186). Springer.
31. Litoussi, M., Kannouf, N., El Makkaoui, K., Ezzati, A., & Fartitchou, M. (2020). IoT security: Challenges and countermeasures. *Procedia Computer Science, 177*, 503–508.
32. Verma, R., Dhanda, N., & Nagar, V. (2022). Security concerns in IoT systems and its blockchain solutions. In J. M. R. S. Tavares, P. Dutta, S. Dutta, & D. Samanta (Eds.), *Cyber intelligence and information retrieval. Lecture Notes in Networks and Systems* (Vol. 291, pp. 485–495). Springer. http://doi.org/10.1007/978-981-16-4284-5_42
33. Hao, W., Zeng, J., Dai, X., Xiao, J., Hua, Q. S., Chen, H., Li, K. C., & Jin, H. (2020). Towards a trust-enhanced blockchain P2P topology for enabling fast and reliable broadcast. *IEEE Transactions on Network and Service Management, 17*(2), 904–917.
34. Pham, H. A., Le, T. K., & Le, T.V. (2019). Enhanced security of IoT data sharing management by smart contracts and blockchain. In *2019 19th International Symposium on Communications and Information Technologies (ISCIT)* (pp. 398–403). IEEE.

Application of Truffle Suite in a Blockchain Environment

Rajat Verma, Namrata Dhanda, and Vishal Nagar

Abstract Advancement is a term that never stops with something so is technology. Blockchain technology is an old term but with updation, it has become a key-buzz term in the technological market. The advancements in blockchain led to the formation of distributed and decentralized applications. This advancement is only possible in the applications phase of blockchain. A simplified ecosystem in which decentralized apps (DApps) can be built is truffle suite. Majorly, three constituents that are completing the truffle suite are truffle, drizzle and ganache. DApps are those that operate among the users and are not monitored by a central authority. In DApps, the ecosystem of a peer-to-peer network works as a complete operating system. This paper focuses on these DApps using truffle suite and its different scenarios. A quick depiction of blockchain with its connection to the truffle suite is also highlighted in this paper. Moreover, this paper also illustrates a popular use case of a DApp with its real-time issues and concerns.

Keywords Blockchain · Security · Cybersecurity · Privacy · Truffle suite

1 Introduction

The blockchain-like-protocol was invented around 39 years ago by a famous computer scientist and cryptographer David Chaum in 1982 when he presented this protocol in his dissertation [1]. In 1991, Stuart Haber and W. Scott Stornetta added the concept of immutable timestamps in documents through the secured series of blocks [2]. A year-later, the concept of Merkle trees was added [2, 3]. The initial conceptualization of blockchain was done in 2008 when it became the backbone of

R. Verma (✉) · N. Dhanda
Department of Computer Science and Engineering, Amity University Uttar Pradesh, Lucknow, India
e-mail: rajatverma310795@gmail.com

V. Nagar
Department of Computer Science and Engineering, Pranveer Singh Institute of Technology, Kanpur, Uttar Pradesh, India

P. K. Singh et al. (eds.), *Proceedings of Third International Conference on Computing, Communications, and Cyber-Security*, Lecture Notes in Networks and Systems 421,
https://doi.org/10.1007/978-981-19-1142-2_54

security behind the famous cryptocurrency Bitcoin [4]. A false-named personality was a representative of it and has been popularly known as Satoshi Nakamoto [4, 5]. The word "blockchain" was a two-letter word in the original white paper but began to combine in the year 2016, and thus became popular as blockchain, a single word [6]. So eventually, the first tenure of blockchain ran from 2008 to 2013 as blockchain 1.0 and was called Bitcoin emergence [7]. The middle phase was focused on Ethereum and was known as Ethereum development (blockchain 2.0) in 2013–2015 [8]. The last phase which is currently happening is the phase of applications that are running from the past 6 years [8, 9]. In the application phase, the things are evolving. Many technologies are being secured by this network peer-to-peer technology blockchain such as the IoT [10]. With the development in the applications phase, this suite (truffle) became a key-buzz expression which is shown below in this paper. With the help of truffle suite, the applications are advancing and making everything possible with its extreme configurational comfortability to the researchers and developers.

2 Material and Methods

2.1 Blockchain Technology

Blockchain works as a ledger that is distributed in nature and provides a decentralized habitat with the additional characteristic features, namely immutability and transparency [11]. Blockchain is a concatenated approach of its constituents known as blocks that are joined using unique cryptographic hash values. Hash values when amalgamated with the elliptic curve digital signature algorithm (ECDSA) provide a greater amount of security and power to blockchain technology in comparison with any conventional technology [12]. The hash values are the results of the hashing algorithms, commonly known as secured hashing algorithms [13]. SHA-256 is a particular type of algorithmic rule that gives the result of 64 hexa-decimal characters whenever input is provided. The input can be given several times, and the output that is obtained will be distinct every time until the identical input is entered. In the case of the same input, the same output will be generated [14, 15]. The block operates in a blockchain environment that is shown in Fig. 1.

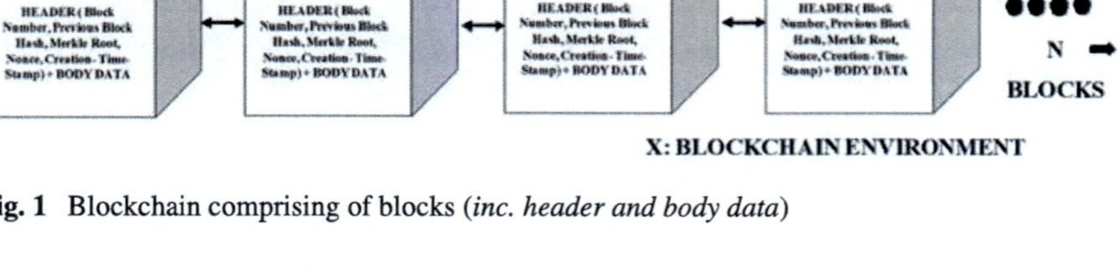

Fig. 1 Blockchain comprising of blocks (*inc. header and body data*)

In Fig. 1, there are four blocks denoted by A, B, C, and D. Although, it can reach N Blocks also. It all depends on how much the availability of resources is present. X denotes the spectrum of the blockchain in which all visible blocks are linked. Block A is known as the genesis block [16] with which all the entries are available as shown in the figure above except the value of the previous block cryptographic hash value. This is so because no block is preceding A. All the other blocks of the chain follow the regular scenario.

The initial attribute of a block is block number. It shows the location of a node/block in the current blockchain. Multiple blocks will be there in a blockchain, so it is necessary to detect the position of an entity in a chain of blocks [17, 18].

The second entity reveals the previous block cryptographic hash value. It means the hash value of the preceding blocks [19].

The third entity deals with the hash value of the current block. The current block's hash value should always be equal to the previous block cryptographic hash value of the next node [19].

The fourth entity depicts the nonce. It is a 32-bit number that acts as a security parameter [20].

The final attribute of the block illustrates the time-stamp of creation. It means the time at which the origin of the block took place. It can vary because of different time zones but will be in sync always [21].

2.2 *Truffle Suite*

With the advancement of blockchain technology (1982–2021), the manner of using blockchain is also diversified. Currently, the era of the application phase is in progress from 2015 [8, 9]. The previous two popular usages of Blockchain are Bitcoin emergence and Ethereum development [7, 8]. Truffle suite is an ecosystem used for the building of DApps [22]. The sub-categories of a truffle suite are shown in Fig. 2.

Figure 2 shows the sub-categories of the truffle suite which are highlighted in the next sub-sections. The methods and tools that act as a pre-requisite in deploying the smart contracts using truffle suite are categorized into two parts:

1. Smart contracts: ganache, truffle, meta mask, solidity.
2. Front end: live server and Web3.js.

For the live server, the following command can be used.

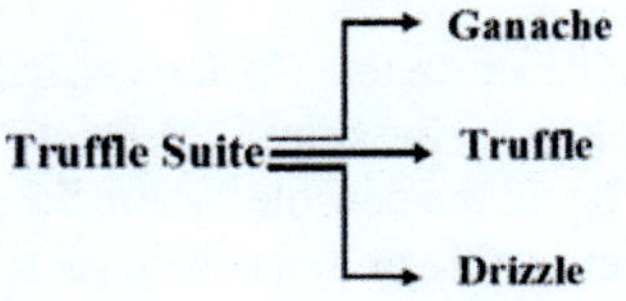

Fig. 2 Truffle suite categories

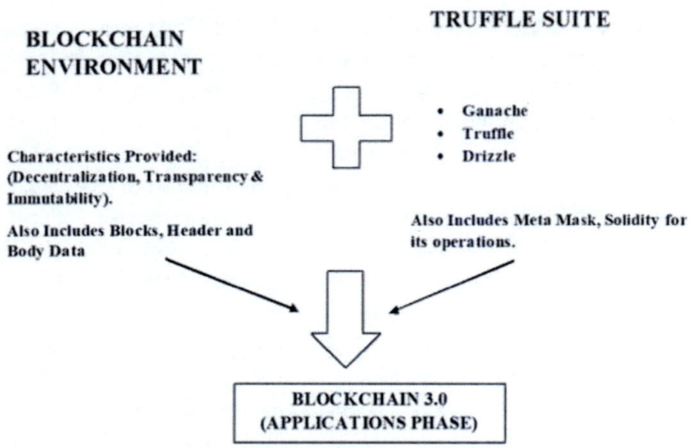

Fig. 3 Origin of third phase of blockchain as blockchain 3.0 (applications phase)

$$\text{npm install -g truffle ganache-cli live-server} \tag{1}$$

After configuring the live server, create the project directory and initiate the truffle using truffle init.

The connection between truffle suite and blockchain is illustrated in Fig. 3 for easy understanding. This connected version of truffle suite and blockchain gave birth to the present phase of blockchain as blockchain 3.0. (applications phase). The previous two versions were blockchain 1.0. and blockchain 2.0.

2.2.1 Ganache

The development in blockchain led to the arrival of applications that do not require a central administrator. This is also a need to eradicate the single point of collapse [23]. Ganache is a personal blockchain that permits the testing of smart contracts [24]. The command of installing and instantiating a ganache network is shown below for easy understanding.

$$\text{npm install -g ganache-cli} \tag{2}$$

$$\text{ganache-cli} \tag{3}$$

The above text (2) depicts the installation of a ganache network using the **Node.js package** using **npm** [25]. It internally utilizes the **ganache-core** for effective installation. In the console, where the ganache network is working must also highlight the other feature of blockchain such as gas price, call gas limit, gas limit, HD wallet, and mnemonics [26].

2.2.2 Truffle

Truffle is an environment that allows the user to build DApps using the Ethereum virtual machine (EVM) [27]. It performs some functions which are depicted below [28]:

1. Framework testing
2. Environment for DApps
3. Asset pipeline
4. Compile contracts
5. Build artifacts.

For installing, compiling, and migrating the contracts, the following commands can be used.

$$\text{\$ npm install -g truffle} \tag{4}$$

$$\text{truffle compile} \tag{5}$$

$$\text{truffle.cmd compile(for Windows OS only)} \tag{6}$$

$$\text{\$ truffle migrate} \tag{7}$$

The features of truffle are shown below [29]:

1. Automated testing of contracts
2. Compatible with Web and console applications
3. Management of package
4. Management of networks.

Truffle can compile contracts written in solidity language [30]. The extension of solidity language is .sol. A variety of logic via smart contracts can be unified with IoT devices to enhance and optimize the security of IoT systems [31].

2.2.3 Drizzle

Drizzle acts as an important pillar of the truffle suite. It is responsible for making the front end of DApps predictable and informative [32]. As CSS is making the HTML document living in a similar manner, drizzle works for DApps. It has a cluster of libraries responsible for the interface. It also supports .Web3, which is a popular package supported in Python 3.9. The synchronization of transactions (base of blockchain), contracts, and data is possible because of drizzle.

2.3 Smart Contracts in Blockchain

In blockchain, smart contracts permit the transactions to happen. It also carries out the agreements by eradicating the central administrator. It also checks the terms and conditions among multiple parties [33]. There are several use cases of smart contracts that are as follows:

1. Insurance [34]
2. Financial industry [35]
3. Mortgage loans [36].

The transactions that are happening are always irreversible, transparent, and traceable.

2.4 Meta Mask

Ethereum needs a graphical user interface for transaction purposes. Here, meta mask gains its importance [37]. Meta mask acts as an add-in for diverse browsers such as Google Chrome, Firefox, and Brave [38]. It was invented by ConsenSys in 2016 [39]. It acts as an alliance between the Web browser and blockchain. It is open-source. It integrates a global API into Websites to read out the data of the users to which blockchains they are connected.

Few features of meta mask are highlighted below [40, 41]:

1. Permits the purchase of built-in coins
2. Storage (local-key)
3. Editable
4. Browser extension is available
5. Simple and secured
6. Generates own passwords and keys
7. Very strong community across the globe
8. Account management
9. Blockchain connection.

Meta mask is a wallet that allows interaction with Ethereum (DApps). It is also available for diverse mobile OS such as iOS and Android [42].

2.5 Application of Truffle Suite in Blockchain

The world is growing at a tremendous pace and so do the applications. With this development, the developers/users can create a variety of applications. It provides full compatibility to develop DApps. This is only possible in the latest phase of blockchain (2015–present) as before this phase, it was not possible. The prior versions only focused on Bitcoin (2008) and Ethereum (2013). These DApps can remove the diverse issues of conventional technologies by integrating blockchain into them. One such technology is a sub-group of artificial intelligence, i.e., Internet of things (IoT) [43]. Since, blockchain has three major characteristics, namely decentralization, transparency, and immutability. The central point of failure can be abolished using blockchain in IoT. The data can be secured using anonymity, i.e., transparency and with the immutability factor, it can eradicate the tampering aspect, fabricated data, data inconsistency, data breach, etc., and can increase the security perspective of IoT [44–46].

2.5.1 A Use Case of a Decentralized App (DApp) in Blockchain

Simulators are a great thing to have when the physical circuits create some sort of problem. A good depiction of a simulator is a GPIO simulator [47]. GPIO simulator is an input–output circuit that can also be visualized as a simulator rather than a physical circuit. Talking about security with blockchain technology. This GPIO simulator can be protected using blockchain. The truffle suite will be a boon in making a GUI-based DApp that will protect the status of pins using blockchain.

Real-Time Challenges in Implementing DApp in Blockchain

In the above scenario of using a GPIO simulator, transactions are done when an operation is implemented, i.e., from on to off or from off to on (status of pins). So, a larger number of demo accounts are needed to be managed in real-time. The scalability of the application also becomes a concern as its needs to be managed in real-time. Also, configurational comfortability needs to be managed in real-time as blockchain consumes very high resources so there must be a lightweight mechanism to run the application on every system.

3 Conclusion and Future Scope

Blockchain, a two-word entity (2008), began to combine in 2016 and became blockchain. This old term with the development and modifications became a buzzword in the ecosystem of technology. From blockchain-like-protocol (1982) to the applications phase (2015), blockchain evolved a lot. With the applications phase, truffle suite that contains ganache, truffle, drizzle came into reality. The truffle suite is illustrated in this paper. Meta mask, a wallet related to Ethereum blockchain, is also highlighted in this paper. A quick depiction of smart contracts is also shown in this paper for an easy grasp. Moreover, the introduction of blockchain with many scenarios is also highlighted in this paper.

Subsequently, will continue to work with a motive to enhance and optimize the security of IoT systems using blockchain.

References

1. Geng, T., Njilla, L., & Huang, C. T. (2021). Smart markers in smart contracts: Enabling multiway branching and merging in blockchain for decentralized runtime verification. In *2021 IEEE Conference on Dependable and Secure Computing (DSC)* (pp. 1–8). IEEE.
2. Elbuz, A., Osmanoglu, M., & Tanriover, O. O. (2019). Designing a secure blockchain-based trading platform for internet of things. *Communications Faculty of Sciences University of Ankara Series A2-A3 Physical Sciences and Engineering, 61*(1), 102–110.
3. Narayanan, A., & Clark, J. (2017). Bitcoin's academic pedigree: The concept of cryptocurrencies is built from forgotten ideas in research literature. *Queue, 15*(4), 20–49.
4. Hawlitschek, F., Notheisen, B., & Teubner, T. (2018). The limits of trust-free systems: A literature review on blockchain technology and trust in the sharing economy. *Electronic Commerce Research and Applications, 29*, 50–63.
5. Nakamoto, S. (2008). Re: Bitcoin P2P e-cash paper. The Cryptography Mailing List.
6. Yang, Y. J., & Hwang, J. C. (2020). Recent development trend of blockchain technologies: A patent analysis. *International Journal of Electronic Commerce Studies, 11*(1), 1–12.
7. Abdulhakeem, S. A., & Hu, Q. (2021). Powered by blockchain technology, DeFi (Decentralized Finance) strives to increase financial inclusion of the unbanked by reshaping the world financial system. *Modern Economy, 12*(01), 1.
8. Manu, M. R., Musthafa, N., Balamurugan, B., & Chauhan, R. (2020). Blockchain components and concept. In *Blockchain technology and applications.*
9. Maesa, D. D. F., & Mori, P. (2020). Blockchain 3.0 applications survey. *Journal of Parallel and Distributed Computing, 138*, 99–114.
10. Verma, R., Dhanda, N., & Nagar, V. (2020). Addressing the issues & challenges of internet of things using blockchain technology. *International Journal of Advanced Science and Technology, 29*, 10074–10082.
11. Pereira, J., Tavalaei, M. M., & Ozalp, H. (2019). Blockchain-based platforms: Decentralized infrastructures and its boundary conditions. *Technological Forecasting and Social Change, 146*, 94–102.
12. Nyame, G., Qin, Z., Agyekum, K. O. B. O., & Sifah, E. B. (2020). An ECDSA approach to access control in knowledge management systems using blockchain. *Information, 11*(2), 111.
13. Luntovskyy, A., & Guetter, D. (2018). Cryptographic technology blockchain and its applications. In *The International Conference on Information and Telecommunication Technologies and Radio Electronics* (pp. 14–33). Springer.
14. Chaves, R., Sousa, L., Sklavos, N., Fournaris, A. P., Kalogeridou, G., Kitsos, P., & Sheikh, F. (2016). Secure hashing: SHA-1, SHA-2, and SHA-3. In *Circuits and systems for security and privacy* (pp. 105–132). Taylor & Francis Group.
15. James, J., Karthika, R., & Nandakumar, R. (2016). Design & characterization of SHA 3–256-bit IP core. *Procedia Technology, 24*, 918–924.
16. Singh, S., & Singh, N. (2016). Blockchain: Future of financial and cyber security. In *2016 2nd International Conference on Contemporary Computing and Informatics (IC3I)* (pp. 463–467). IEEE.
17. Biswas, K., & Muthukkumarasamy, V. (2016). Securing smart cities using blockchain technology. In *2016 IEEE 18th International Conference on High Performance Computing and Communications; IEEE 14th International Conference on Smart City; IEEE 2nd International Conference on Data Science and Systems (HPCC/SmartCity/DSS)* (pp. 1392–1393). IEEE.
18. Dorri, A., Kanhere, S. S., & Jurdak, R. (2017). Towards an optimized blockchain for IoT. In *2017 IEEE/ACM Second International Conference on Internet-of-Things Design and Implementation (IoTDI)* (pp. 173–178). IEEE.

19. Samaniego, M., Jamsrandorj, U., & Deters, R. (2016). Blockchain as a service for IoT. In *2016 IEEE International Conference on Internet of Things (iThings) and IEEE Green Computing and Communications (GreenCom) and IEEE Cyber, Physical and Social Computing (CPSCom) and IEEE Smart Data (SmartData)* (pp. 433–436). IEEE.
20. Hazari, S. S., & Mahmoud, Q. H. (2019). A parallel proof of work to improve transaction speed and scalability in blockchain systems. In *2019 IEEE 9th Annual Computing and Communication Workshop and Conference (CCWC)* (pp. 0916–0921). IEEE.
21. Galiev, A., Prokopyev, N., Ishmukhametov, S., Stolov, E., Latypov, R., & Vlasov, I. (2018). Archain: A novel blockchain based archival system. In *2018 Second World Conference on Smart Trends in Systems, Security and Sustainability (WorldS4)* (pp. 84–89). IEEE.
22. Santosh, S. V. S., Rao, M. K., Sri, P. A., & Hemantha, C. S. (2021). Decentralized application for two-factor authentication with smart contracts. In *Inventive Communication and Computational Technologies* (pp. 477–486). Springer.
23. Helebrandt, P., Bellus, M., Ries, M., Kotuliak, I., & Khilenko, V. (2018). Blockchain adoption for monitoring and management of enterprise networks. In *2018 IEEE 9th Annual Information Technology, Electronics and Mobile Communication Conference (IEMCON)* (pp. 1221–1225). IEEE.
24. Bhosale, K., Akbarabbas, K., Deepak, J., & Sankhe, A. (2019). Blockchain based secure data storage. *International Research Journal of Engineering and Technology (IRJET), 6*(03), 4.
25. Hande, R., Agarwal, T., Monde, R., Sirisha, N. S., & Yadav, R. (2019). Charity chain-donations using blockchain. In *International Conference on Computer Networks and Inventive Communication Technologies* (pp. 606–612). Springer.
26. Lee, W. M. (2019). Testing smart contracts using ganache. In *Beginning ethereum smart contracts programming* (pp. 147–167). Apress.
27. Xu, Q., Song, Z., Goh, R. S. M., & Li, Y. (2018). Building an ethereum and ipfs-based decentralized social network system. In *2018 IEEE 24th International Conference on Parallel and Distributed Systems (ICPADS)* (pp. 1–6). IEEE.
28. Shawn, L. W. M., Murali Mohan, P., Loh Kok Keong, P., & Balachandran, V. (2021). Blockchain-based Proof of Existence (PoE) framework using ethereum smart contracts. In *Proceedings of the Eleventh ACM Conference on Data and Application Security and Privacy* (pp. 301–303).
29. Ramamurthy, B. (2020). *Blockchain in action.* Manning Publications.
30. Patidar, K., & Jain, S. (2019) Decentralized e-voting portal using blockchain. In *2019 10th International Conference on Computing, Communication and Networking Technologies (ICCCNT)* (pp. 1–4). IEEE.
31. Balaji, B. S., Raja, P. V., Nayyar, A., Sanjeevikumar, P., & Pandiyan, S. (2020). Enhancement of security and handling the inconspicuousness in IoT using a simple size extensible blockchain. *Energies, 13*(7), 1795.
32. Anilkumar, V., Joji, J. A., Afzal, A., & Sheik, R. (2019). Blockchain simulation and development platforms: Survey, issues and challenges. In *2019 International Conference on Intelligent Computing and Control Systems (ICCS)* (pp. 935–939). IEEE.
33. Zheng, Z., Xie, S., Dai, H. N., Chen, W., Chen, X., Weng, J., & Imran, M. (2020). An overview on smart contracts: Challenges, advances and platforms. *Future Generation Computer Systems, 105*, 475–491.
34. Gatteschi, V., Lamberti, F., Demartini, C., Pranteda, C., & Santamaría, V. (2018). Blockchain and smart contracts for insurance: Is the technology mature enough? *Future Internet, 10*(2), 20.
35. Cong, L. W., & He, Z. (2019). Blockchain disruption and smart contracts. *The Review of Financial Studies, 32*(5), 1754–1797.
36. Hamilton, M. (2020). Blockchain distributed ledger technology: An introduction and focus on smart contracts. *Journal of Corporate Accounting & Finance, 31*(2), 7–12.
37. Jain, A., Tripathi, A. K., Chandra, N., & Chinnasamy, P. (2021). Smart contract enabled online examination system based in blockchain network. In *2021 International Conference on Computer Communication and Informatics (ICCCI)* (pp. 1–7). IEEE.

38. Bhanushali, D., Koul, A., Sharma, S., & Shaikh, B. (2020). Blockchain to prevent fraudulent activities: Buying and selling property using blockchain. In *2020 International Conference on Inventive Computation Technologies (ICICT)* (pp. 705–709). IEEE.
39. Roopa, C., Suganthe, R. C., & Shanthi, N. (2020). Blockchain based certificate verification using ethereum and smart contract. *Journal of Critical Reviews, 7*(9), 330–336.
40. Mehta, R., Kapoor, N., Sourav, S., & Shorey, R. (2019). Decentralised image sharing and copyright protection using blockchain and perceptual hashes. In *2019 11th International Conference on Communication Systems & Networks (COMSNETS)* (pp. 1–6). IEEE.
41. Laurence, T. (2019). *Blockchain for dummies*. Wiley.
42. Ruby Benita, K., Ganesh Kumar, S., Murugamantham, B., & Murugan, A. (2020). Authentic drug usage and tracking with blockchain using mobile apps. *iJIM, 14*(17), 21.
43. Razzaq, M. A., Gill, S. H., Qureshi, M. A., & Ullah, S. (2017). Security issues in the Internet of Things (IoT): A comprehensive study. *International Journal of Advanced Computer Science and Applications, 8*(6), 383.
44. Dedeoglu, V., Jurdak, R., Dorri, A., Lunardi, R. C., Michelin, R. A., Zorzo, A. F., & Kanhere, S. S. (2020). Blockchain technologies for IoT. In *Advanced applications of blockchain technology* (pp. 55–89). Springer.
45. Verma, R., Dhanda, N., & Nagar, V. (2022). Security concerns in IoT systems and its blockchain solutions. In *Cyber intelligence and information retrieval* (pp. 485–495). Springer.
46. Lazaroiu, C., & Roscia, M. (2017). Smart district through IoT and blockchain. In *2017 IEEE 6th International Conference on Renewable Energy Research and Applications (ICRERA)* (pp. 454–461). IEEE.
47. Bachrudin, Z., Widodo, C. E., & Adi, K. (2017). Simulator input-output sistem kontrol menggunakan Raspberry Pi. *Youngster Physics Journal, 6*(3), 272–279.

Assessment of Compliance of GDPR in IT Industry and Fintech

Pankaj Pathak, Parashu Ram Pal, Rajesh Kumar Maurya, Rishabh, Mayur Rahul, and Vikash Yadav

Abstract The general data protection regulation (GDPR) is the tough privacy and security policy toward protection of data. The policy was drafted by the European Union but imposed obligations for any organization which collects data related to the people in the European Union. This policy came into effect from 2018. If any organization violets, the law will levy huge fines. Consumer driven companies in the areas like IT services and Fintech likely to affected by the GDPPR and have to comply. The research paper seeks to explore the implication of GDPR on these two industries. The challenges faced by the two industries in planning, implementation, and complying to GDPR, the overlaps and contradictions with the existing industry frameworks which will last post GDP6R implementation, the pre- and post-GDPR scenario analysis, and lastly the trial process for the data breach of the two industries. Based on the comprehensive study and research on the aforementioned areas, this research paper then delves into building a hypothesis through qualitative and quantitative data gathered which provides a solution for the two industries to prepare, plan, implement, and comply with GDPR across industry level with respect to user data management centers and Fintechs. We have used empirical methodology and

P. Pathak
Symbiosis Institute of Digital and Telecom Management Symbiosis International (Deemed University), Pune, India
e-mail: pankajpathak@sidtm.edu.in

P. R. Pal
SAGE University, Bhopal, Madhya Pradesh, India

R. K. Maurya
ABES Engineering College, Ghaziabad, Uttar Pradesh, India

Rishabh
Galgotias College of Engineering and Technology, Greater Noida, India

M. Rahul
Department of Computer Application, CSJM University, Kanpur, India

V. Yadav (✉)
Department of Technical Education, Kanpur, Uttar Pradesh, India
e-mail: vikas.yadav.cs@gmail.com

P. K. Singh et al. (eds.), *Proceedings of Third International Conference on Computing, Communications, and Cyber-Security*, Lecture Notes in Networks and Systems 421,
https://doi.org/10.1007/978-981-19-1142-2_55

collected responses through the questionnaire. Through the research study, we have found that how important is data encryption, not only because it is mandated in GDPR but also since any sort of data revelation to a criminal party can cause a lot of damage.

Keywords GDPR · Data protection · Fintech · IT industry

1 Introduction

1.1 GDPR

The implementation of general data protection regulation or GDPR is transforming the digital landscape globally [1]. GDPR applies to any citizen irrespective of nationality or location if the company has its establishment in the EU and where personal data are processed "in the context of the activities" of such an establishment and to any EU citizen regardless the location of the company. Hence, a large number of citizens and organizations across the globe fall under the purview of GDPR as we live in a more interconnected and interdependent world. The application of GDPR spans across all the industries and organizations, from public to private sector, that involve processing of data of the EU citizens (or personal data subjects) [2]. This research paper focuses on application of GDPR on the user management data centers in information technology and Fintech industries.

1.2 Aims and Objectives of the Research Study

- To intensively study the GDPR frameworks
- To capture the impact of GDPR on Fintech and unified data management (UDM) in IT industry
- List out the key areas and pain points of impact of GDPR on Fintech and unified data management (UDM) in IT
- Understand the organizational, operational, legal, and technical perspectives of GDPR w.r.t Fintech and UDM in IT.

2 Literature Review

The literature review starts with a clarification of how the earlier law, the data protection directive from 1995, works and the thought behind it. We at this point, in short, clarify the thought behind elaboration of GDPR. We additionally clarify the most critical parts of the GDPR in connection to information forms and what sort of impacts can diverse perspectives of IT and Fintech are expected to have. This is done to give a superior comprehension of different directions within data handling organizations and its conceivable effects. In conclusion, we clarify the hypothesis behind consistency in data frameworks.

The present enactment that secures the rights concerning individual information of residents inside the EU is the data protection directive from 1995. This mandate has then confirmed an endeavor to blend the way information can be put away and handled inside the EU, however, has been translated contrastingly in all European states as per Lynskey [3] and Wong [4]. In this way, Lynskey contends that it is indistinct whether the data protection directive is accurately embraced and executed by the national courts. In spite of this, she composes that the order introduced a few purposes and definitions which stay in the GDPR. Key ideas like information controller, information processor, and individual information are characterized and have changed little in the imminent directions. Data collection and processing [5] are growing at a rapid rate in present business models. The data protection regulations introduce [6] specific ways to create and empower the digital ecosystem as well as assessing the effects of digital privacy and consumer welfare.

Regardless of whether the data protection directive is generally a traditional approach, Lynskey [3] focuses on the way that the mandate is out of date in a few angles. Since the law was composed when just a small amount of the natives of the EU utilized the Web routinely, the data protection directive tends to be deficient with regards to securing the EU resident's rights. While a few sections of the GDPR and the data protection directive are firmly related, issues like the reuse of individual information and uniform appropriation by European states are not dealt with in the present enactment.

A regulation needs to be engulfed with range of penalties to ensure a proper conduct is followed within the organization. These regulations are bounded by laws and jurisdictions, where organizations are penalized for misconduct in practicing the approved legislation. An organization that does not comply with the GDPR is liable to a fine of up to 4% of annual global turnover or €20 million, whichever is the highest according to the articles mentioned in the regulation. The data protection directive failed to state proper consequences for cases of data violation. However, GDPR ensures stricter methods for avoiding data violation and involves the EU in providing legal tools to enforce the new regulation with clearer and homogenous penalties [7].

Since 25th May 2018, many legal cases were filed against many top multi-national companies are fined under and outside GDPR framework because of data-breaches. Many large IT organizations are facing trial under GDPR in most of the EU countries as of Jan 2019. In United Kingdom, total 6281 cases were filed between 25th may 2018 and July 3rd 2018.

Also, many e-commerce sites faced GDPR lawsuits due to credit card information leakage-related issues. One of the biggest global tech giants has been fined close to $57 USD in France and also fined over 44 million Euro because of advertising using personal data [8].

3 Research Methodology

"Assessment of compliance of GDPR for user data management in IT industry and Fintech" is based on clarification of critical parts of GDPR in connection to form of information and impact on them in diverse perspective. In conclusion, it is an approach to clarify the hypothesis behind consistency in data framework [9].

Methodologies followed for this research are

Primary Research

1. End-to-end GDPR framework analysis to understand its articles and clauses in comparison with data protection directive.
2. Build a questionnaire with approval from faculty and the corporates and share it with corporates. The recorded data will help to understand the way organizations have implemented GDPR, challenges, and limitations faced while implementing GDPR and the organizations will handle the data breach and its impact on it.
3. Build a survey with approval from guiding faculty and corporate and share it with corporates and record the observations. By applying statistical techniques, the observation output will then be analyzed.
4. Interviews with DPOs, legal experts, CXOs, and GDPR experts which will help to analyze the implementation, challenges in implementation, and its impact in IT and Fintech user data management.

Secondary Research

1. Connected with the corporates through live webinar.
2. This helped to understand the current status of GDPR in different IT and Fintech organizations and how they are handling the painful and unnoticed areas of implementing GDPR.
3. Attend GDPR workshop in the institute.
4. Research done by reading different Website blogs, connecting with GDPR expertise through social media to discuss with them the about various aspects of implementing, pain areas, and based on the methodologies [10].

Research Opportunities and Challenges

When the pre-GDPR data protection regulations were first developed, the concept of user data in a digital format was in its infancy. It was a major challenge to seek the organizations and people who really worked under GDPR constraints. But as technology evolved and reliable, high-speed Internet access became more widely available, the range of options, and different services continued to grow. The implementation of general data protection regulation or GDPR will transform the digital landscape globally. GDPR applies to any citizen irrespective of nationality or location if the company is an EU company and to any EU citizen if the company is a non-EU company. Hence, a large number of citizens and organizations across the globe fall under the purview of GDPR as we live in a more interconnected and interdependent world [11, 12].

Fintech

Firms providing financial services are adopting biometric and digital authentication systems, which acquire fingerprints and eye scans of data subjects to identify their customers. In addition to obtaining the data subject's explicit consent [7] while obtaining his biometric data, Fintech companies must also have separate controls in place to protect them, to achieve necessary compliance of GDPR. These controls must guarantee that the data controllers take the technical, specialized, and authoritative measures necessary to prevent this special data from being uncovered, as a consequence of their systems being poorly managed previously.

All financial firms are built around IT systems. With the digital era under great influence [13], financial service providers are being more and more reliant on IT systems. This leads to data flowing across multiple IT systems which create a complexity due to increasing trend of outsourcing development and support functions. This in turn leads to data being exposed to a broader supply chain without guaranteeing necessary vigilance measures. The data flow becomes more vulnerable when it flows through cross border organizations, thereby causing a threat to end-to-end data management under GDPR.

4 Result Analysis

Though the questionnaires, we have ask several questions and recorded response from the companies who are in process to implement GDP.

In the study from Table 1 responses, we found that most of the Fintech companies have completed GDPR implementation process, i.e., approx. 81%. But few of the companies could not implement it which is mentioned as dropout, they have started with implementation but could not succeed (Fig. 1; Table 2).

Table 1 GDPR implementation completion

Viewed	Started	Completed	Completion rate	Drop outs (after starting)	Average time to complete the survey
36	31	25	80.65%	6	3 min

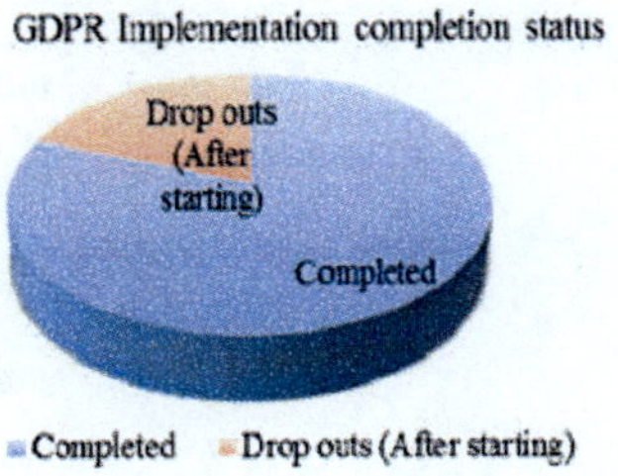

Fig. 1 GDPR implementation completion

Table 2 Consent to process or store personal data

S. No.	Answer	Count	Percent
1	We do not ask for consent	3	12.50%
2	There is a clause in the contract (s) that is signed	7	29.17%
3	We have a separate consent form or a policy on the Website	14	58.33%
4	Other	0	0.00%
	Total	24	100%
Mean: 2.458	Confidence interval @ 95%: [2.170–2.747]	Standard deviation: 0.721	Standard error: 0.147

1. **How do you ask for consent to process and/or store personal or sensitive data?**

From the above question's response, it is clear that the most of the Fintech and IT companies who have implemented GDPR take the consent from their clients to store their personal or sensitive data. Some companies included the clause for storing the sensitive data in the contract itself. But it is found that most of the clients want to include it as a separate document. Some companies who have not implemented GDPR do not ask any consent to store sensitive data (Fig. 2; Table 3).

2. **Do you store personal or employee data outside of the European economic area?**

From above question's response, it is clear that companies are very clever. They have not stored the sensitive information of clients in the European Union area where GDPR is in act. But they store the sensitive information outside the European area where GDPR is not implemented (Fig. 3; Table 4).

3. **Do you store or transfer personal data to companies outside of your direct control?**

Fig. 2 Consent to process or store personal data

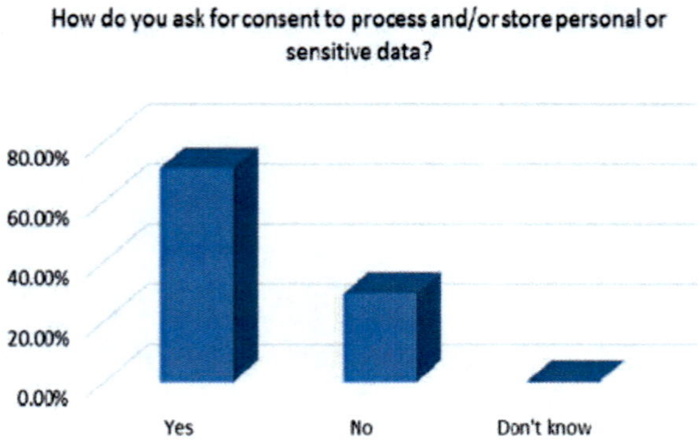

Table 3 Store data outside European economic area

S. No.	Answer	Count	Percent
1	Yes	15	62.50%
2	No	9	37.50%
3	Do not know	0	0.00%
	Total	24	100%
Mean: 1.375	Confidence interval @ 95%: [1.177–1.573]	Standard deviation: 0.495	Standard error: 0.101

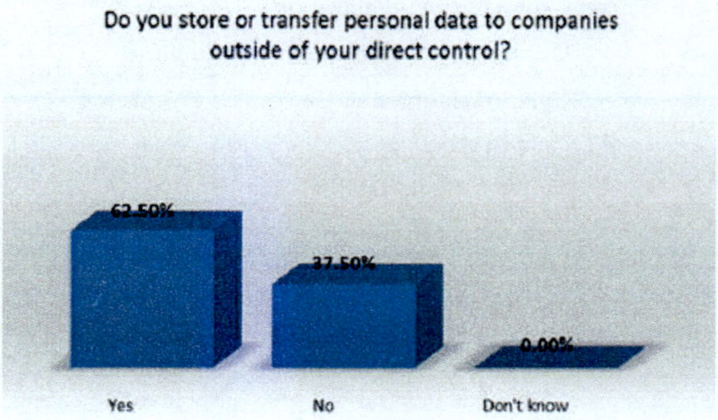

Fig. 3 Store data outside European economic area

Table 4 Store or transfer data outside companies

S. No.	Answer	Count	Percent
1	Yes	17	70.83%
2	No	7	29.17%
3	Do not know	0	0.00%
	Total	24	100%
Mean: 1.292	Confidence interval @ 95%: [1.106–1.477]	Standard deviation: 0.464	Standard error: 0.095

From the above question's response, it is very critical situation that companies not only storing sensitive information of clients but they are transferring personal information to the companies which are not in their direct control. It is very dangerous. Some of the companies following the regulation and not sending the sensitive information outside their direct control (Fig. 4; Table 5).

4. **Which of the below rights of GDPR has the highest priority in your organization?**

Above question is very interesting when asked about highest priority about the rights of GDPR from the organization then most of the companies agreed on right to data

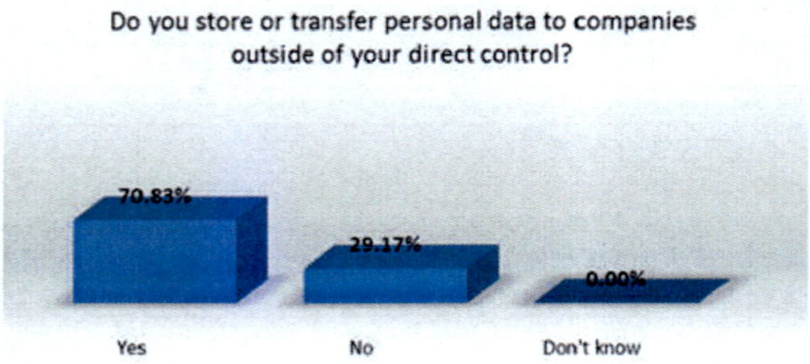

Fig. 4 Store or transfer data outside companies

Table 5 Priority of GDPR in organization

S. No.	Answer	Count	Percent
1	Right to be forgotten	7	29.17%
2	Right to access	4	16.67%
3	Right to data portability	8	33.33%
4	Right to rectification	2	8.33%
5	Do not know about the rights or management policies	3	12.50%
	Total	24	100%
Mean: 2.583	Confidence interval @ 95%: [2.044–3.123]	Standard deviation: 1.349	Standard error: 0.275

portability, some of give their consent to right to forgotten the data. GDPR has given the "right to be forgotten" to its citizens which gives them power to erase their data and stop its processing immediately which is a great step toward empowering their citizens. Some also wants access to the data and some companies do not know about the data management policies (Fig. 5).

Fig. 5 Priority of GDPR in organization

5 Discussion and Analysis

The analysis of the survey was undertaken with the help of QuestionPro. A total of 36 respondents were surveyed of which 31 respondents took the survey and 25 respondents managed to complete the survey with the completion rate of 80.65% and average time of 3 min per survey. The respondents included corporate professionals, researchers, consultants, and legal experts. All the respondents are either professionals in Fintech and IT or have extensive knowledge of the two industries. The responses of the respondents were recorded on nominal and interval scales. The survey includes questions which comprise of Article 12 to Article 23 of GDPR which includes the rights of the data subjects under GDPR. In order to develop the basic understanding of the regulations and to check whether the systems and processes are in place, the variables include a simple yes/no/maybe. The survey concludes with the most challenging aspect of GDPR for Fintech and IT firms.

In [14], software of GDPR compliance is discussed. In our research study, we have focused on Fintech and IT industry, and it will be very helpful in designing the perfect GDPR compliant software.

Automated individual decision-making, including profiling: the respondents were asked if they inform the data subject regarding the data being collected and having a consent procedure at every stage. Demographic variables were considered when asked if the data were stored outside the European economic area and outside their direct control. Companies were also asked if they have a mechanism to report data breach. This was asked to check the technical compliance with the GDPR. The respondents were then asked in the end as to which of the data subject rights would they give the highest priority in the organization.

6 Conclusion and Limitation

It has been a really challenging endeavor to study the impacts of GDPR on industries like the Fintech and the IT, since these industries deal with data day in and day out.

Some important points of concluding the research

- GDPR has given accountability to data processors and controllers and has brought out the role of data protection officers who will ensure that no data breach takes place.
- GDPR has earned trust among the citizens of EU as it has spoken of sensitive topics like consent and privacy. With globalization raising its gigantic head and with the advent of social media, issues like consent and privacy are not considered that important and are not given much thought. Also, GDPR has given the "right to be forgotten" to its citizens which gives them power to erase their data and stop its processing immediately which is a great step toward empowering their citizens.

- One of the most compelling things for its conformance would be the heavy penalties that would be levied to firms who do not comply by it. Such strict rules would always promote tighter laws which would never allow personal data to be taken less seriously.
- Through the research, we have found that how important is data encryption, not only because it is mandated in GDPR but also since any sort of data revelation to a criminal party can cause a lot of damage.

Despite of all hypes, some organizations found it really difficult to comply with GDPR and faced lots of challenges in being cent percent compliant. However, GDPR has been the lighting guide to a lot of firms who deal with personal data. Not only are more and more nations finding ways to conform to it but are also working on a large-scale end-to-end implementation.

Despite the challenges addressed in the study which are common for all the industry, the study focused on the Fintech and IT industry. Other industry may have some specific challenges that are not covered, and it is the future scope for the research.

References

1. Albrecht, J. P. (2016). How the GDPR will change the world. *European Data Protection Law Review, 2*, 287.
2. Goddard, M. (2017). The EU general data protection regulation (GDPR): European regulation that has a global impact. *International Journal of Market Research, 59*(6), 703–705.
3. Lynskey, O. (2015). *The foundations of EU data protection law*. Oxford University Press.
4. Wong, R. (2012). The data protection directive 95/46/EC: Idealisms and realisms. *International Review of Law, Computers & Technology, 26*(2–3), 229–244.
5. Laurer, M., & Seidl, T. (2021). Regulating the European data-driven economy: A case study on the general data protection regulation. *Policy & Internet, 13*(2), 257–277.
6. Kira, B., Sinha, V., & Srinivasan, S. (2021). Regulating digital ecosystems: Bridging the gap between competition policy and data protection. *Industrial and Corporate Change, 30*(5), 1337–1360.
7. Utz, C., Degeling, M., Fahl, S., Schaub, F., & Holz, T. (2019). (Un)informed consent: Studying GDPR consent notices in the field. In *Proceedings of the 2019 ACM SIGSAC Conference on Computer and Communications Security* (pp. 973–990).
8. Sanchez-Rola, I., Dell'Amico, M., Kotzias, P., Balzarotti, D., Bilge, L., Vervier, P. A., & Santos, I. (2019, July). Can i opt out yet? GDPR and the global illusion of cookie control. In *Proceedings of the 2019 ACM Asia Conference on Computer and Communications Security* (pp. 340–351).
9. Tankard, C. (2016). What the GDPR means for businesses. *Network Security, 2016*(6), 5–8.
10. IT Governance Privacy Team. (2020). *EU general data protection regulation (GDPR)—An implementation and compliance guide*. IT Governance Ltd.
11. Li, H., Yu, L., & He, W. (2019). The impact of GDPR on global technology development.
12. Linden, T., Khandelwal, R., Harkous, H., & Fawaz, K. (2020). The privacy policy landscape after the GDPR. *Proceedings on Privacy Enhancing Technologies, 2020*(1), 47–64.
13. Nouwens, M., Liccardi, I., Veale, M., Karger, D., & Kagal, L. (2020, April). Dark patterns after the GDPR: Scraping consent pop-ups and demonstrating their influence. In *Proceedings of the 2020 CHI Conference on Human Factors in Computing Systems* (pp. 1–13).

14. Ryan, P., Crane, M., & Brennan, R. (2021). GDPR compliance tools: Best practice from RegTech. In J. Filipe, M. Śmiałek, A. Brodsky, & S. Hammoudi (Eds.), *Enterprise Information Systems. ICEIS 2020. Lecture Notes in Business Information Processing* (Vol. 417). Springer. http://doi.org/10.1007/978-3-030-75418-1_41

Digitally Signed Document Chain (DSDC) Blockchain

Udai Bhan Trivedi and Santosh Sharma

Abstract This paper suggested the framework of digitally signed document chain (DSDC) blockchain which focuses on digitalization and decentralization of educational certificates storage, authentication, authorization, confidentiality, ownership, and privacy. Blockchain generation has lately emerged as a capacity suggests for authenticating the record verification and a sizeable device to fight record fraud and misuse. This research paper diagnosed the safety subject matters required for file verification with inside the blockchain. This paper also suggested the framework for modifications and deletions of data under circumstances by competent authority only which is against the principal of immutability of blockchain by utilizing chameleon hashing.

Keywords Blockchain · Chameleon hashing · Document chain

1 Introduction

Blockchain is a decentralized peer-to-peer ledger that is used to record cryptographically signed transactions in a series of blocks. Each block in the chain holds the preceding block's hash value, resulting in a chain of blocks. A genesis block is the first block of any block. All other participants in the peer-to-peer distributed ledger are called nodes. Each node in the network will keep a copy of the ledger, propose and verify transactions based on the type of blockchain, and participate in the consensus algorithm [1].

Blockchain network users submit candidate transactions to the blockchain network through software. These transactions are routed by the software to one

U. B. Trivedi (✉)
Pranveer Singh Institute of Technology, Kanpur, India
e-mail: udaibhantrivedi@gmail.com

S. Sharma
PSIT College of Higher Education, Kanpur, India

P. K. Singh et al. (eds.), *Proceedings of Third International Conference on Computing, Communications, and Cyber-Security*, Lecture Notes in Networks and Systems 421,
https://doi.org/10.1007/978-981-19-1142-2_56

or more nodes in the blockchain network. The selected node can be a complete non-publishing node or a publishing node. The transmitted transaction is subsequently propagated to additional network nodes, but this does not place the transaction on the blockchain. For various blockchain applications, as soon as the pending transaction is shipped to the node, it has to wait with inside the queue till the publishing node provides it to the blockchain [2, 3].

When the publishing node publishes the block, the transaction is added to the blockchain. A block consists of a block header and data. The block header contains the block's metadata (data about data). A list of confirmed and genuine transactions that have been transmitted to the blockchain network is contained in block data. This ensures that the transaction format is proper and that the digital asset supplier has cryptographically signed the transaction to confirm its legitimacy and authenticity. This validates if the transaction's digital asset supplier has access to the private key with which the accessible digital asset may sign. Other full nodes will verify the validity and authenticity of all transactions in the published block, and if the block contains invalid transactions, the block will not be accepted. Many blockchain implementations use the following data fields [4, 5]:

- Block header
  1. Block number
  2. Previous block hash value
  3. Hash of present block (can be calculated with the help of Merkle tree)
  4. Timestamp.
  5. Block size
  6. Nonce value.
- Block data
  - consist of all transaction which occur recently and ledger events within block
  - Other data may be present.

1.1 Chaining Blocks

The blockchain is built up by set of blocks. Each block includes the hash digest of the previous block's header. If a previously published block was changed, a new hash would be produced. As a result, all subsequent blocks will have different hashes since they integrate the hash of the preceding block. This facilitates the identification and rejection of manipulated data [1] (Fig. 1).

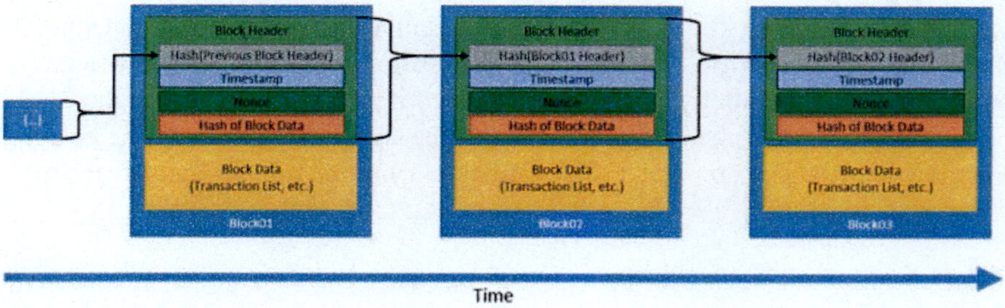

Fig. 1 Generic chain of blocks [6]

1.2 Hash Functions

The hash function is a compression function that receives a message represented as an arbitrary length string of bits and converts it to a fixed-length string known as a digest. The hash function is designed as a one-way function, which means that reversal is practically infeasible, and the only way to find the original message is to search through all possible inputs by brute force. The cryptographic secure hash function is a hash function $h()$ that meets the three requirements listed below [7]:

1. Antigenic image: If, given the digest y, discovering x such that $y = h(x)$ is computationally impossible, then hash function $h()$ is resistant to the original picture.
2. The second resistance prior to the image. Given a summary y and a message x such that $y = h(x)$, finding another message $\hat{x} \neq x$ such that $y = h(\hat{x})$ is computationally impossible.
3. Anti-collision: If two messages x and $\hat{x}$, $\hat{x} \neq x$, are found so that $h(\hat{x}) = h(x)$ is computationally infeasible, then this is collision resistant hash function $h()$.

The use of cryptographically secure hash functions in the blockchain ranges from the creation of a unique identifier to the protection and connection of data blocks [7]. A hash pointer connects the data block, and the hash pointer leads to the data storage location, which is the preceding block in the chain. The hash pointer can be used to verify if the block has been tampered with, thus ensuring the integrity (immutability) of the data [8, 9].

1.3 Digital Signature

The digital signature technique is divided into three sections [8]. The first is the key creation algorithm, which is used to creating a pair of keys. The signer uses his private key (which is only known to the user) to sign a message, and the signature may subsequently be confirmed using the public key. The second component is the signature algorithm. The digital signature algorithm uses the digest of the message

$h(x)$, the signer's private key sk, and a random number to generate the signature s. Once a party receives the signature, the verification algorithm (the third core component) verifies its validity. The verification algorithm uses message x, signature s, and sender's public key pk to determine whether the signature is genuine. The goals of digital signature algorithms are identity verification, non-repudiation, and integrity [6].

1.4 Asymmetric-Key Cryptography

Asymmetric-key cryptography (also known as public-key cryptography) is a type of cryptography that employs a pair of keys: a public key and a private key [7]. Although the two keys are theoretically linked, obtaining the private key from the public key should be impossible. The public key can be made public without jeopardizing the algorithm's security.

The non-public key, on the other hand, should be kept private. The two keys are interchangeable because you can either (i) encrypt the plaintext using the private key and decode the cipher text with the public key, or (ii) encrypt using the public key and decode with the private key. In case (i), the algorithm is used to ensure the message's integrity and to establish the message's authentication. On the contrary, in case (ii), the algorithm is used to guarantee the confidentiality of the message [6].

2 Background and Literature Review

Blockchain is a peer-to-peer ledger that stores cryptographically signed transactions in blocks. Each block in the chain holds the hash of the preceding block. Immutable property of the blockchain is provided by hash function. More precisely, the blockchain can be described as immutable chain of blocks. The immutability of the blockchain is undoubtedly one of its most powerful features. However, the inability to change or delete data can be an unwanted feature in certain situations, posing another challenge to using blockchain when personal data are under threat. Article 16 and article 17 of general data protection regulation (GDPR) introduce the right of correction and deletion of related parties, assuming that data can always be modified and deleted. Therefore, there may be situations where the data are compulsorily deleted (or modified). Failure to do so will result in a system breach [10, 11].

To facilitate the possibility of compliance, we propose and design the architecture of a digitally signed document chain (DSDC) blockchain that allows for modifications and deletions of data under circumstances by competent authority. To upgrade the standard hash capabilities used in blockchains, we employ chameleon-hash capabilities with ephemeral trapdoors. Each freshly generated hash contains a unique ephemeral trapdoor, allowing for targeted and fine-grained collision computation [7,

12]. Few research papers have been reviewed, and summary sheet has been prepared to get insight of blockchain security and chameleon hash for implementation in DSDC (Table 1).

3 Purpose of Digitally Signed Document Chain (DSDC)

The transmission of personally identifiable information and proprietary information has always been the main concern of enterprises and professionals. Using blockchain technology, digitally signed document chain (DSDC) is able to create a historical ledger of document accessibility. DSDC allows users to audit the history of user access for documents by date and time. By creating multiple layers of accessibility and encryption history, businesses and professionals can operate under full responsibility for document access.

DSDC provides a solution to minimize security risks when sharing documents using blockchain.

- Save the encrypted document.
- Transfer access to documents safely and securely among the required users.
- Access audit to the current and historical user.
- Check that it does not review the document.
- History of loading, download, change documents.

3.1 Document Storage and Security

Any university/autonomous organization/an entity which is entitled to give degree/certificate/diploma/mark sheet, etc., to the students needs to provide degree in encrypted format (signed by private key of concern body whose public key is available to decrypt the certificate), and we call this encrypted degree/certificate/diploma/mark sheet etc., as native document (Fig. 2).

1. Using DSDC, application user can encrypt the native document with a single key-document's key.
2. The encrypted native document is sent to a node and stored in the blockchain.
3. Simultaneously, the newly created document's key encrypted with the user's private key, successfully securely sharing the document's key.

3.2 Document Sharing and Authentication

User 1 wants to transfer the document to user 2 (Fig. 3).

1. The application encrypts the documents key using the user 1 private key.

Table 1 Literature review of related papers

Author	Research paper	URL	Year of publication	Publisher/conference
S. Nakamoto	Bitcoin: a peer-to-peer electronic cash system	https://bitcoin.org/bitcoin.pdf	2009	Cryptography mailing list at https://metzdowd.com
Central idea: His research paper suggests utilizing a peer-to-peer network to overcome the double-spending problem. The study proposed using community timestamp transactions to create a never-ending hash-based completely proof-of-work chain, resulting in a file that cannot be changed without redoing POW. Author also demonstrates that the longest chain may be employed to establish not only that the series of events were seen, but also that it originates from the broadest group of CPU capabilities. Messages will do their best to transfer, and nodes can leave and rejoin the network at any time, taking the longest chain evidence as proof of what happened when they left [2]				
Sri Aravinda Krishnan Thyagarajan, Dominic Deuber, Bernardo Magri	"Redactable blockchain in the permission less setting"	https://ieeexplore.ieee.org/stamp/stamp.jsp?tp=&arnumber=8835372	2019	Security and privacy (SP) IEEE symposium
Central idea: The author presents the first efficient editable blockchain for permission less setups that integrates seamlessly into Bitcoin and does not rely on hefty encryption technologies or trust assumptions. Its protocol employs consensus-based voting and is governed by rules that establish review criteria and constraints; if the review receives enough votes, the operation is carried out on the chain. The protocol also includes public verifiability and accountability for the publication chain as more of an extra advantage [13]				
Rui xue, Rui Zhang, and Ling Liu	Security and privacy on blockchain	scholar.google.com/scholar_url?url=https://dl.acm.org/doi/pdf/10.1145/3316481&hl=en&sa=T&oi=ucasa&ct=ufr&ei=IwgQYa-5D8yQ6rQP9pqIqAo&scisig=AAGBfm3sTrc7KcROpilyEyQ-9PHPgpTH3Q	2019	ACM computing surveys 52.3 (July 2019)

(continued)

Table 1 (continued)

Author	Research paper	URL	Year of publication	Publisher/conference
Central idea: The author proposed the notion of blockchain and its use in the context of online transactions analogous to Bitcoin. The article also outlines the fundamental requirements and core security features provided by building blocks as a monetary system comparable to Bitcoin, before introducing the extra security and privacy attributes required by many blockchain applications. Finally, the author examines the security and privacy technologies used for the blockchain-based applications to accomplish these security features, such as representative hash chain storage, consensus algorithms, anonymous signatures, hybrid protocols, and non-interactive zero-knowledge proofs [8]				
David Derler, Kai Samelin, Daniel Slamanig, and Christoph Striecks	Fine-grained and controlled rewriting in blockchains: chameleon-hashing gone attribute-based	https://www.ndss-symposium.org/ndss-paper/fine-grained-and-controlled-rewriting-in-blockchains-chameleon-hashing-gone-attribute-based/	2019	26th annual network and distributed system security symposium, NDSS 2019 at: San Diego, California, USA
Central idea: The author of this article determined that replacing the standard hash function with a chameleon hash will not solve the problem of controlled rewriting of the blockchain because this method is too coarse-grained. The author reviewed this idea and introduced a new concept of party computing chameleon-hashes (PCH). PCH generalizes the idea of chameleon hash through giving part of computing a hash the capacity to companion a get entry to coverage with the generated hash. Anyone with enough privileges to fulfill the coverage can also additionally come across. They then extended this approach to transaction-level rewriting within the blockchain to enable detailed and controlled modification of blockchain objects [14]				
Diogo Duarte	An introduction to blockchain technology from a legal perspective and its tensions with the GDPR	https://papers.ssrn.com/sol3/papers.cfm?abstract_id=3545331	2019	CIJIC—cyber law journal (2019)
Central idea: The purpose of the author's research is to provide the first method to help lawyers, researchers, and students to gain understanding of blockchain technology, how it works, and its impact on data protection requirements, especially in mapping stakeholders in terms of responsibilities and rights. This research mainly focuses on the decentralization and immutability of blockchain technology, as well as the complexity and uncertainty caused by its centralized operation of the GDPR. They also proposed some solutions that can be implemented in blockchain-based application design to achieve certain goals of the GDPR [10]				

(continued)

Table 1 (continued)

Author	Research paper	URL	Year of publication	Publisher/conference
Jan Camenisch, David Derler, Stephan Krenn, Henrich C. Pöhls, Kai Samelin, and Daniel Slamanig	Chameleon-hashes with ephemeral trapdoors—and applications to invisible sanitizable signatures	https://doi.org/10.1007/978-3-662-54388-7%5C_6	2017	Lecture notes in computer science. Springer, 2017
Central idea: In this research article, the author introduces the concept of a chameleon-hash function with ephemeral trapdoors. This form of hash function includes an extra, that is, temporary, trapdoor that is chosen by the portion that calculates the hash value. As a result, unless a temporary trapdoor used to calculate the hash value is likewise given with the ephemeral trapdoor used to compute the hash value, the main trapdoors cannot locate the second preview picture of the hash value				
Adi Shamir	"How to share a secret"	http://web.mit.edu/6.857/OldStuff/Fall03/ref/Shamir-HowToShareASecret.pdf	1979	Communications of the ACM
Central idea: The author presents a technique for dividing data D into n blocks so that D may be easily reconstructed from any k blocks, but even knowing everything about k − 1 block would never give anything about D. This technology allows the construction of robust key management. Even if half of the part is unfortunately destroyed, the security breach also exposes all parts except the remaining part, but still can operate safely and reliably [15]				

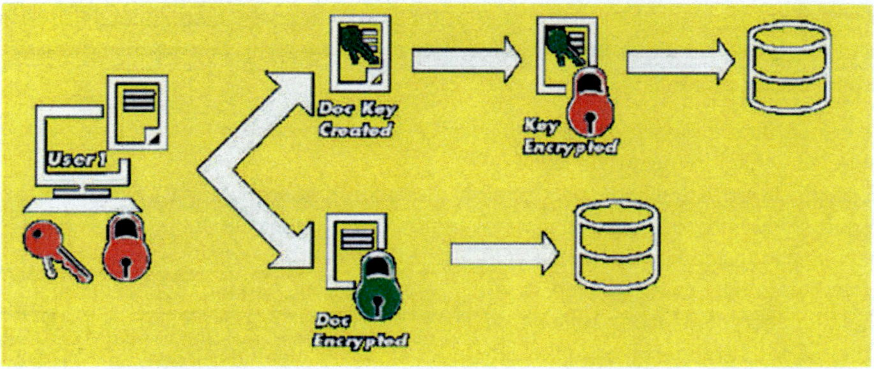

Fig. 2 Proposed model for documents storage and security

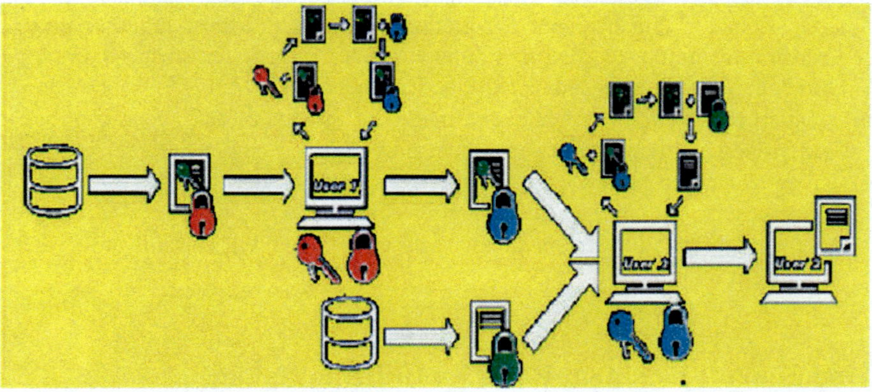

Fig. 3 Proposed model for document sharing and authentication

2. The application then encrypts the document key with user 2 public key, sharing it with user 2.
3. User 2 can now decrypt the document key using their private key.
4. User 2 can use the decrypted documents key to decrypt the document itself as user 1 public key is available to anyone on the blockchain network.
5. Once user 2 got the documents key, user 2 can easily decrypt the message and find the native message.
6. This native message if decrypted public key of university/autonomous organization/an entity which is entitled to give degree/certificate/diploma/mark sheet, etc., this authenticate that the degree/certificate/diploma/mark sheet, etc., is issued by concern organization.

DSDC can be implemented at on premise of any organization, off-site data center, or on cloud platform, but at national level project, it must be govern by government agency. Transferring blockchain statistics to the DSDC is a very theoretical discussion, at this point. No blockchain has been built in this idea yet. Few points need to take under considerations.

- Are there unique assets had to archaically save blockchain records, including node/network administration skills?
- Since the information in the blocks might also additionally encompass quite a few report sorts gathered from a couple of transactions, might DSDC be capable of get entry to the codecs contained with inside the blocks?
- How will DSDC manage, keep, or offer get right of entry to the blockchain that includes encrypted inaccessible parts? The blocks cannot be eliminated due to the fact this will invalidate the blockchain, however, they cannot be accessed because of the manner, the blockchain policies are set.
- Personal documents are a transactional data, i.e., data which can be change over a period. More records can be added to the block as

  (a) Qualification enhancement ex. graduate can become post graduate and doctorate; employees also continuously pursue various certificates in order to enhance their knowledge, skill, and attitude.
  (b) Qualification improvements ex. marks of graduate final year can be improved by giving improvements examination.
  (c) New education policy which converts the educational qualification from certificates, diploma, and degree after completion of one, two, and three years of graduations, respectively. Any individuals will certainly want to update it on proposed DSDC.

4 Updation of Document Block by Utilizing Enhance Chameleon-Hash Function with Trapdoor Implementation

A chameleon hash is a cryptographic pseudo-random function that consists of a mystery trapdoor. Without the trapdoor, it is far hard to discover collisions, however, with the understanding of the trapdoor, hash collisions may be generated efficiently [13, 14].

An enhanced chameleon-hash function consists of 4-tuple known as <KeyGen; HashGen; HashVer; HashCol> defined as follows.

(hk, tk) ← KeyGen(1k): Key generation technique that uses the security parameter k to produce the public key hk and the secret trapdoor key tk.

(h, t) ← HashGen(hk, m): Hash generation technique that takes hk, m, and a random r as inputs and returns a pair (h, t) consisting of the hash value h and a check string t.

d = HashVer(hk, m, (h, t)): Hash verification algorithm that takes a message m and a candidate (h, t) as input and outputs a bit d equal to 1 if HashGen(hk, m) = h.

$t' \leftarrow$ Hash collision (tk, (h, m, t), m′): Finding method that accepts tk as input, a valid tuple (h, m, t), and a new message m′ as output and provides a new check string t′ such that HashVer(hk, m, (h, t)) = HashVer(hk, m′, (h, t′)) = 1.

We present our implementation of a enhance chameleon-hash function with ephemeral trapdoor-based RSA public-key encryption [7, 15, 16] (Table 2).

Table 2 Chameleon-hash function with trapdoor implementation using RSA

Key generation phase	**Algorithm**

The ParGen function in Algo 1 generates the public exponent RSA e as a random prime number with a bit length of λ. In our implementation, λ is set to 2048 bits. This function will run when users join the network and assign them a hash key. This step is performed by a certification authority (CA) that is responsible for generating identities for users when enrolling on the network	Algo1: Generation of RSA public exponent
	func ParGen (1^{λ})
	{
	e: = randomPrime(λ)
	Return e
	}

The KeyGen function in Algo2 generates RSA modulo n and private exponent d. The CA will calculate the KeyGen when the user logs into the network. The existence of the second ephemeral trapdoor is provided by the order service that executes this function during the hashing process	Algo 2: Generation of RSA modulo and private exponent
	func KeyGen (λ, e)
	{
	repeat
	{
	p1: = randomPrime(λ/2)
	p2: = randomPrime(λ/2)
	n: = p1 * p2
	φ(n): = (p1 − 1) * (p2 − 1)
	} until e > n AND is Coprime(φ(n), e),
	d = inverseModulo(e, φ(n))
	Return n, d
	}

Hash generation phase	**Algorithm**

Algorithm 3 shows the hash function executed by the order service to use the user's public key to hash the transaction, i.e., the RSA public exponent. The randomNumber() function is designed to provide a random number with a maximum of λ digits. The *H* * *n*() function is a random SHA256 hash function, with an output of 256 bytes. This function calculates a random number r and uses the public exponent e, randomness r, and hash function *H* * *n*() to process the message. To insert the second ephemeral trapdoor, it is sufficient to execute the same function a second time using the second RSA modulo generated through the execution of KeyGen	Algo 3: Hashing Process
	func Hash (λ, e, n, tx)
	{
	r: = randomNumber(λ)
	h1: = H_n * (tx)
	hash: = h1 * r^e mod n
	return hash, r
	}

Hash verification and collision	**Algorithm**

(continued)

Table 2 (continued)

The algorithm listed as func Aadapt (Algo. 4) is used to find a collision, i.e., to calculate the new value of R (new_R), the new transaction data newTx allows to find collision for the original hash value. The private exponent d (owned by the authority) allows restoring the initial exponentiation as in any standard RSA	Algo 4: Hash verification and Collision
	func Adapt (old_Hash, n,·tx, newTX, d)
	{
	if newTX == tx return r
	h1: = H_n* (newTx)
	new_R: = (old_Hash * inverseModulo(h1, n))^d mod n
	return new_R
	}

5 Conclusion

Digitally signed document chain (DSDC) blockchain ensures authentication, confidentiality, ownership, and privacy of the data with the help of public-key cryptography. DSDC takes advantages of all blockchain features like decentralization, transparencies between all parties, and immutability. DSDC also allows modifications and deletions of data under particular circumstances by competent authority, one can perceive as some compromise on immutability but this is possible only by competent authority for storing all personal data on same block without disturbing the hash chain. DSDC employs chameleon-hash functions with ephemeral trapdoor as a substitute for the standard hash functions used in the blockchain. The ephemeral trapdoor is distinct for every newly created hash, which lets in for focused and fine-grained collision computation. With the help of chameleon hash between the blocks, the competent authority (who knows the ephemeral trapdoor) can change the data of blocks without having change in hash value which facilitates the update and deletion of personal information.

References

1. UK Government Office for Science. (2016). Distributed ledger technology: Beyond blockchain.
2. Nakamoto, S. (2009). Bitcoin: A peer-to-peer electronic cash system. Cryptography mailing list at https://metzdowd.com
3. Zheng, Z., Xie, S., Dai, H.-N., Chen, X., & Wang, H. (2017). An overview of blockchain technology: Architecture, consensus, and future trends. http://doi.org/10.1109/BigDataCongress.2017.85
4. Merkle, R. C. (1980). Protocols for public key cryptosystems. In *Proceedings of 1980 Symposium on Security and Privacy* (pp. 122–133). IEEE Computer Society.
5. Chelsea, P., Victoria, L., & Chris, R. (2021). Multidisciplinary blockchain research and design: A case study in moving from theory to pedagogy to practice' part of the *Lecture Notes in Computer Science* book series (LNCS, Vol. 12645).
6. European Union Agency for Network and Information Security. (2016). Distributed ledger technology & cybersecurity. http://doi.org/10.2824/80997
7. Camenisch, J., Derler, D., Krenn, S., Pöhls, H. C., Samelin, K., & Slamanig, D. (2017). Chameleon-hashes with ephemeral trapdoors and applications to invisible sanitizable signatures. In S. Fehr (Ed.), *Public-Key Cryptography—PKC 2017—20th IACR International*

Conference on Practice and Theory in Public-Key Cryptography (Part II, Vol. 10175, pp. 152–182), Amsterdam, The Netherlands, March 28–31, 2017. *Lecture Notes in Computer Science*. Springer. http://doi.org/10.1007/978-3-662-54388-7%5C_6

8. Zhang, R., Xue, R., & Liu, L. (2019). Security and privacy on blockchain. In *ACM computing surveys 52.3*. ISSN: 0360-0300. http://doi.org/10.1145/3316481.103104
9. Jianyu, N., Ziyu, W., Fangyu, G., & Chen, F. (2020). *Incentive analysis of Bitcoin-NG*, revisited. Elsevier. http://doi.org/10.1016/j.peva.2020.102144
10. Duarte, D. (2019). An Introduction to Blockchain Technology From a Legal Perspective and Its Tensions With the GDPR (September 30, 2019). *Cyberlaw Journal of the Cyberlaw Research Centre of the University of Lisbon School of Law -CIJIC*. Available at SSRN: https://ssrn.com/abstract=3545331 or http://dx.doi.org/10.2139/ssrn.3545331
11. Florian, M., Henningsen, S., Beaucamp, S., & Scheuermann, B. (2019). Erasing data from Blockchain nodes. In *2019 IEEE European Symposium on Security and Privacy Workshops (EuroS PW)* (pp. 367–376). http://doi.org/10.1109/EuroSPW.2019.00047
12. Lee, N., Yang, J., Onik, M. M. H., & Kim, C. (2019). Modifiable public blockchains using truncated hashing and sidechains. *IEEE Access, 7*, 173571–173582. ISSN: 2169-3536. http://doi.org/10.1109/ACCESS.2019.2956628
13. Deuber, D., Magri, B., & Thyagarajan, S. A. K. (2019). Redactable blockchain in the permissionless setting. In *2019 IEEE Symposium on Security and Privacy (SP)* (pp. 124–138). IEEE.
14. Derler, D., Samelin, K., Slamanig, D., & Striecks, C. (2019). Fine-grained and controlled rewriting in Blockchains: chameleon-hashing gone attribute-based. In *26th Annual Network and Distributed System Security Symposium, NDSS 2019*, San Diego, California, USA, February 24–27, 2019. The Internet Society.
15. Shamir, A. (1979). How to share a secret. *Communications of the ACM*, *22*(11), 612–613. http://doi.org/10.1145/359168.359176
16. Cai, X., Ren, Y., Zhang, X. (2020) "Privacy-Protected Deletable Blockchain". *IEEE Access, 8*, pp. 6060–6070. http://doi.org/10.1109/ACCESS.2019.2962816

Algorithms of AI in Deciding Optimum Mix Design of Concrete: Review

Rajat Verma, Uzair Khan, Binod Kumar Singh, and Rizwan A. Khan

Abstract The preparation of mix design of concrete requires a knowledge of design mix proportioning. Various properties like slump value, compressive strength are considered while preparing mix design. This traditional mix proportion method is a time-consuming and costly process. It is also done manually which may lead to different errors. To overcome this, use of artificial intelligence has been brought into this field to predict the design mix of concrete in limited time, low cost and minimum error due to use of computational algorithms as compared to traditional methods. In this paper, the studies using different algorithms of artificial intelligence are reviewed. Estimation of properties like compressive strength, slump value of concrete is done. Further, this paper also presents comparative analysis between different algorithms of AI. This research paper will be of great help to concrete technologist to explore future possibilities of AI techniques in concrete industry.

Keywords Artificial intelligence (AI) · Concrete mix · Algorithms · Design mix

1 Introduction

Concrete is a commonly used building material obtained by the mixing of cement, water, fine aggregates and coarse aggregates. Admixtures are generally added to change the properties of concrete. Concrete possesses property like flowability; i.e. it can take any desired shape when it is wet, and on getting hardened, development of strength takes place.

R. Verma · U. Khan (✉)
Department of Civil Engineering, ABES Engineering College, Ghaziabad, Uttar Pradesh, India
e-mail: uzairkhan2890@gmail.com

B. K. Singh
Structural Engineering, School of Planning and Architecture, Delhi, India

R. A. Khan
Department of Civil Engineering, Z.H College of Engineering and Technology, Aligarh Muslim University, Aligarh, India

P. K. Singh et al. (eds.), *Proceedings of Third International Conference on Computing, Communications, and Cyber-Security*, Lecture Notes in Networks and Systems 421,
https://doi.org/10.1007/978-981-19-1142-2_57

One of the important characteristic properties of concrete is its compressive strength. It is obtained by doing standard curing of the period of 28 days and followed by testing the specimen of the concrete mix under compressive testing machine. The testing of design mix of concrete is done as per the specifications given by Indian Standard (IS) codes. The strength of the concrete mix is determined as per methods laid down by IS 516 (1959). The theoretical method of finding compressive strength involves calculation of design mix which is categorized as statistical analysis [1]. The theoretical calculation of the design mix of concrete is done according to the guidelines which are given by IS 10262 (2009) code.

Mix design involves knowing the composition of ingredients that are used in concrete. After preparing design mix, the major task is to select the most optimum concrete mix from the experiment. To have the optimum mix design concrete, several trials are performed. This often leads to the wastage of various ingredients of building materials. It also includes human labour, cost and time.

Also, the production of concrete is a complex process and involves various steps and analysis of parameters like strength, workability. So, sometime problem arises due to inaccurate linear and nonlinear relationship used in analysis.

To overcome such issues, various algorithms of artificial intelligence (AI) are used to predict the behaviour of different parameters used in the concrete mix. This technique has proved to be a great tool in deciding the optimum design mix of the concrete in a shorter period of time [2].

AI is a tool that uses machines to perform analysis of complex functions with the help of human intelligence. Artificial intelligence was first used in Dartmouth College in 1956 at a conference. It was discovered by John McCarthy. The interaction of computer science, cybernetics, linguistics and information technology has led to the development of artificial intelligence [3]. This tool can do calculations much faster than humans. The inherent benefits of this technique have attracted the attention of various researchers [4].

Artificial intelligence works on various algorithms to solve various complex problems. These include genetic algorithm, support vector machine, artificial neural networks, fuzzy logic, bee colony algorithm, firefly algorithm, cuckoo search algorithm. The algorithms of AI ensure that every step in it must be processed in an orderly manner to diminish the chance of any error [5, 6].

Recent advancements in computing algorithms have enabled engineers to use these methods over existing traditional ones. These computing algorithms result in minimum approximations and make little assumptions. This tool will be utilized for an evaluation on the design mix of concrete and proved to be successful tool in the coming era [7].

For selection of mix design of concrete, these computing tools enable to have more accurate and scientific approach. The paper presents a critical review on various computing algorithms available for finding the most optimum mix design of concrete.

2 Novelty of Work/Research Significance

The novelty/research significance of the present study is as given below:

1. The use of artificial intelligence has been brought up in concrete industry to deal with on-site applications like design mix study, prediction of compressive strength, slump.
2. To bring out analysis between the various algorithms of AI in the design mix study of concrete.
3. The conventional method of selecting optimum mix design requires more efforts and takes larger time to proceed. To overcome these problems, various algorithms of AI are studied to find out the benefits of using the technique when compared with conventional method. This helps in dealing the real-time challenges at construction site.

3 Literature Review

Artificial intelligence reflects human thinking and presents it in different computational algorithms like ANNs, fuzzy tool, back propagation, neural network, adaptive network fuzzy technique. Due to its adaptability, higher efficiency, time-saving approach and accuracy, it has various applications. It is used in research, medicals, engineering, sports, etc. as shown in Fig. 1. AI is also brought in civil engineering with the purpose to do an assigned tasks in a limited time frame with higher accuracy rate. Algorithms adopted by AI nowadays are used in selecting the optimum mix design of concrete. These algorithms ensure that minimum error is obtained when design mix is prepared for experiment [8–10].

Artificial intelligence has been brought up in different areas of civil engineering. These include concrete industry, hydraulics, structural health monitoring, prediction

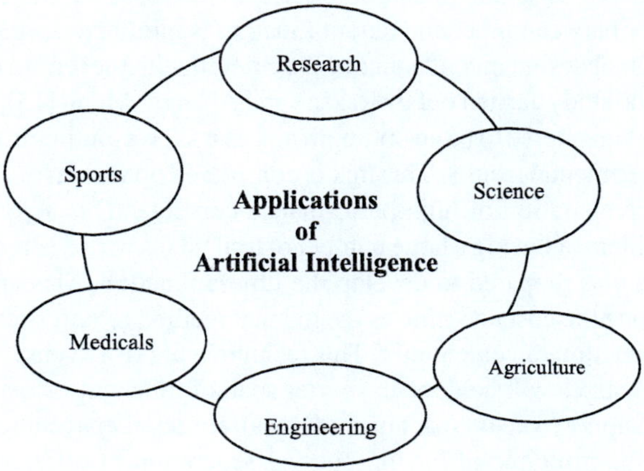

Fig. 1 Multi-disciplinary fields of artificial intelligence

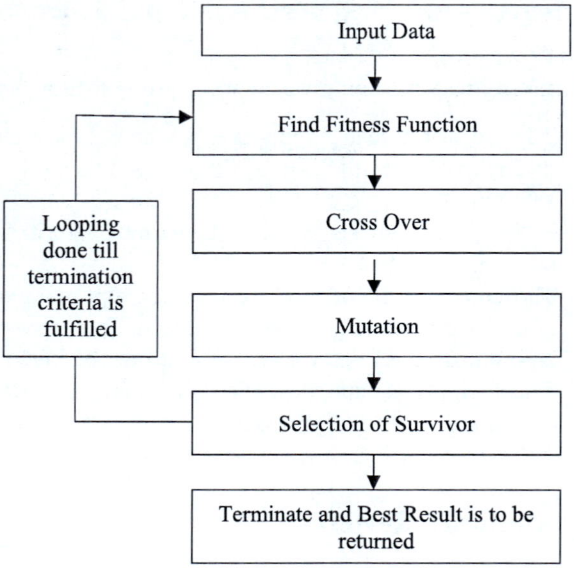

Fig. 2 Flow chart of genetic algorithm

of compressive strength, geotechnical engineering, geopolymer, concrete, earthquake engineering [7, 11].

Generally, for solving the complex problems of concrete research, two popular approaches of AI are used, namely genetic algorithm and support vector machines [12, 13]. Genetic algorithm is an evolutionary algorithm of AI which is based on the theory of Darwin. Genetic algorithm mainly involves four steps of operation—selection, crossover, mutation and sampling. The flow chart of model of genetic algorithm is shown in Fig. 2 [14].

To have an optimum design mix of concrete, genetic algorithm is adopted. For this, input and output functions are defined. Compressive strength of concrete mix is taken as input, while slump values obtained were taken as output functions. Fitness function is increased as the operation of selection, crossover and mutation proceeds. As the error between input and output function is minimized, fitness would be satisfied and optimal design mix of concrete is achieved with the termination of entire programme. In the study carried out by Rezaee and Ahangar Mohd [15], mix having error of 0.069 was considered to be an optimum mix as it shows minimum error when compared with experimental results. This mix is considered optimum for designing concrete mix for the preparation of high-performance concrete. This algorithm of AI decreases the problem of having a large number of trial mixes with desired properties. Experimental data was prepared to develop the fitness function. This approach was developed to reduce the cost and time as the manual method requires more time and is costlier for production of cement mix. This technique of AI will ensure that preparation of fresh trial mixes will be done in shorter span of time and is cost-effective.

Support vector machines (SVMs) are another technique of AI which is based on the principle of having optimal separation of different classes. This technique is generally based on selection of optimum concrete mix design among various mix designs by selecting the one which generates a minimum error in results when compared to calculation of mix design by manual method. Two user-defined parameters are required for making decisions. These include radial basis function kernel and polynomial kernel. SVMs tool is proved to be a powerful computational tool in selecting optimum concrete mix design [16].

Fuzzy logic is another technique of AI that is used in the field of risk management. In calculation of design mix, chance of having optimum design mix of concrete is uncertain. The use of this tool ensures that design mix obtained must fulfil desire criteria and with more accuracy [17].

Artificial neural networks (ANNs) are one of the algorithms of AI used in the analysis of studying the nonlinear and complex material to have an optimum design mix of concrete. ANN model is defined by node characteristics, topology and learning or training rules. It is divided into two models—supervised and unsupervised models of data [18, 19].

ANN is useful in understanding the problems related to the properties of materials used in concrete mix. These networks are much useful in the analysis of compressive strength. ANN works on the same principle as that of processing of the brain. The neuron is the processing unit of entire network and consists of input data, hidden layer and an output. The method predicts the optimum mix design using accuracy of strength validation and mean square error. Feedforward network is developed to predict the optimum mix design and its compressive strength [20, 21]. ANN generally divides collected data into training and does final validations after processing. Following this, normalization of data collection takes place. Network architecture is drawn, training parameters are discussed, and at last, final evaluation takes place. This is clearly shown in Fig. 3.

In the study by Gupta [2], artificial neural networks were used to predict the compressive strength at the curing age 7, 14 and 28 days. The multilayer perceptron is implemented for studying the mix properties. The input parameters taken were cement, coarse aggregates, fine aggregates, admixtures and water. For the hidden layers, hyperbolic tangent and log-sigmoid transfer functions have been kept. Linear transfer function was assigned for output layer. Training datasets were prepared with the aim to have an optimum mix design that shows minimum error. After predicting compression values, experimental testing was done to check whether ANN predicted the right value or not. It was found that the value of error was lying in the range from −3.5 to 3.7 for the curing period. The analysis showed that there is a little marginal difference between the predicted and actual values.

Figure 4 shows regression plot between the predicted value of slump and the observed value of slump. The predicted value of slump is that slump value which is predicted by the researcher before conducting an experiment. The observed value of slump is that value of slump which is recorded by researcher after an experiment is done. An increasing nature of curve is obtained; i.e. positive trend is observed between the predicted and observed values of slump. This means the value predicted by ANN and value obtained after experiment show minimum error. The figure also

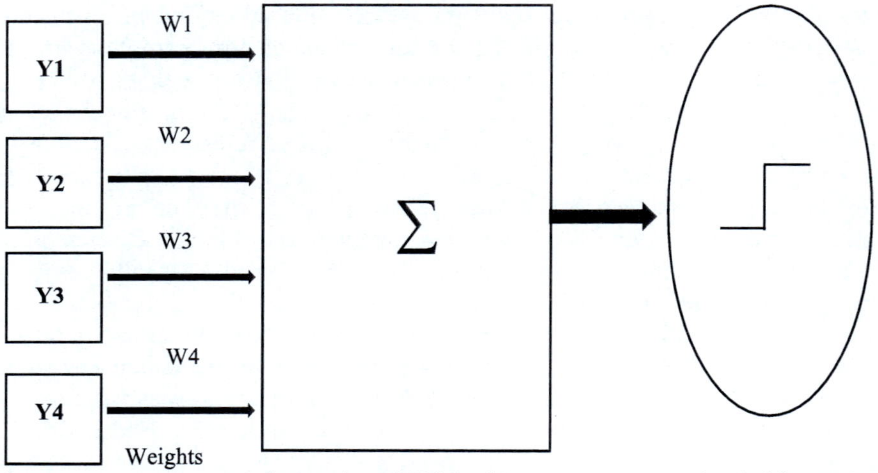

Fig. 3 Model of artificial neural network

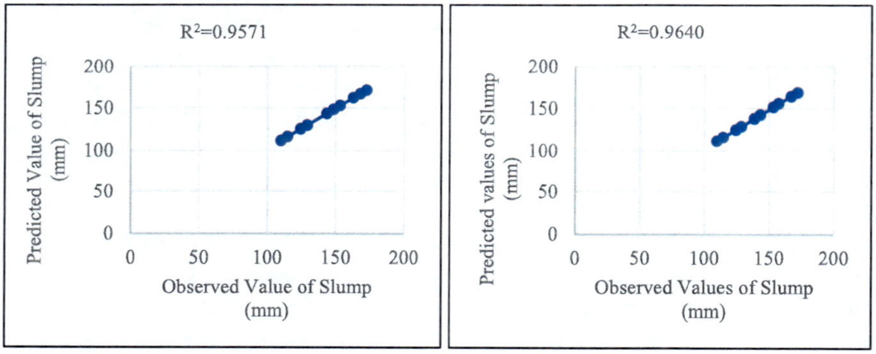

Fig. 4 Regression plot between predicted value of slump and observed values of slump of AI model [22]

clearly shows that training parameters evolved in algorithms of AI have fewer error statistics when training and validation are done.

The prediction accuracy of slump is much higher as compared to the conventional method of determining slump value of concrete mix. The error while testing the mix design with conventional method comes out to be just 1.0841% in the AI model. A linear variation is obtained between the predicted and observed values of slump. The result obtained shows that maximum accuracy is obtained in predicting optimum mix design.

ANN is also useful in predicting the complex behaviour of the concrete when quick lime is there. It is proved to be a decision-making tool for having concrete mix by easily predicting slump value of concrete mix design. It avoids multiple trials as it directly gives the slump value as per the conditions applied on the engineering site. The technique is useful as it focuses on adaptive learning, self-organization and fast computing. This could also be found that ANN may be used as a modelling tool for the civil engineers in the area of design mix. These tools are beneficial in capturing nonlinear interactions between different parameters. ANN is a more explicit model and is a user-friendly technique of AI when compared with basic genetic algorithms [22, 23].

Another algorithm of AI, adaptive network-based fuzzy interface system (ANFIS), is proved to be more effective tool in predicting the optimum mix design of concrete. ANFIS model is based on providing learning techniques for extraction of information from the input and output dataset.

Further setting of antecedent takes place. It is mainly a multilayered feed forward network. There is the number of nodes connected through links. Node acts as a processing unit that uses node function to give output. Normally, ANFIS model consists of five layers. The model of ANFIS algorithm of AI is clearly shown in Fig. 5.

For estimation of optimum concrete mix design, a detailed study is carried out by Tesfamariam and Najjaran [24]. First of all, preparation of training data is done. Parameters include water, cement, fine aggregate and coarse aggregates. Input parameters are divided into two groups—absolute variables and relative variables. Next step deals with the identification of the structure and parameters of various design mixes prepared. This is done with the help of sensitivity analysis. Model validation is done in last to carry out analysis on obtained desired output mix for input datasets.

Water-cement ratio came as the most dominant parameter towards the variation in concrete strength. It was also found that mix having coarse aggregate nearly 60% and 40% of fine aggregates for design mix gives highest compressive strength among all mixes present. For mixes having less than 30% coarse aggregate content of total concrete mix, a decreasing trend was observed in compressive strength. A similar trend was seen when ANFIS model was carried out for the same dataset.

Fig. 5 ANFIS model

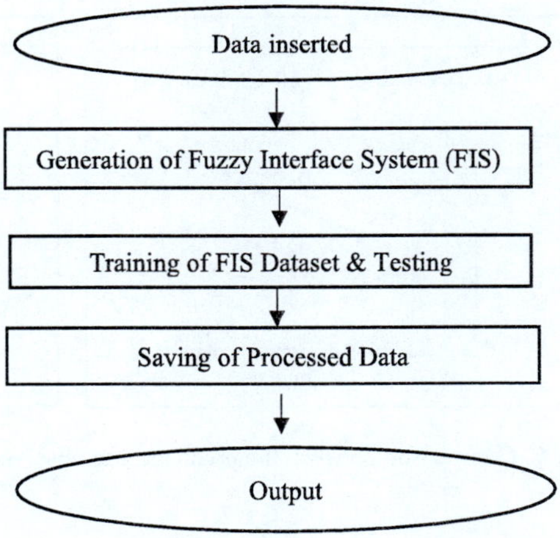

Figure 6 shows that ANFIS model analysis is carried out to predict the compressive strength of concrete. Regression plot is drawn between the predicted and observed value of compressive strength. The plot shows an increasing trend in the predicted and observed values of compressive strength. This means that for both absolute and relative models, ANFIS is proved to be right decision-making tool for the selection of optimum mix design of concrete. The plot also helps in identification of various dominant parameters that help to bring stringent monitoring and quality control of concrete mix design. It is also clear that there is little variation in predicted and observed values of compressive strength of concrete. It is concluded that concrete mix proportioning is a nonlinear process and depends on different parameters like water-cement ratio, fine aggregates, coarse aggregates.

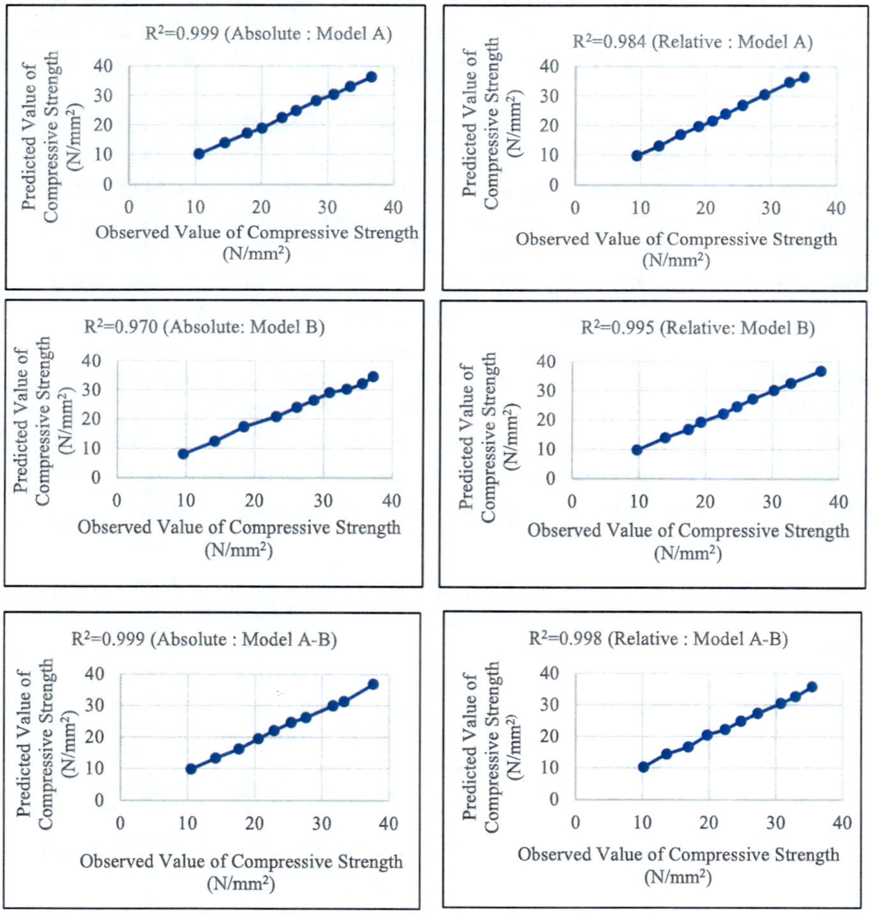

Fig. 6 Comparison between the predicted and observed value of compressive strength [24]

ANFIS algorithm of AI presents more appealing and understandable model as compared to ANN because it used intuitive experience of designer and also numerical information included in the datasets. This model is more useful than ANN as it allows post-modelling adjustment of datasets; i.e. after output model is generated, it can regenerate results by removing errors in case if there is any error left. It can also be used at different stages of production of concrete in concrete industry [25]. After the experiment using computational tool ANFIS model, it was found that it has become beneficial as it incorporates external factors like quality of construction, environmental factors. This tool reduces the risk of having faulty concrete mix during construction, ensures durability and provides safety to construction industry by avoiding risk of failure. These factors were not considered while designing ANN models.

The use of another algorithm of artificial intelligence, i.e. back propagation neural network (BP) is proved to be potential tool for predicting the design mix of concrete mix. It is widely used algorithm in studying supervised learning. It generally involves the use of steep gradient descend analysis to give details on approximations. The performance of the concrete mix is based on the nature of materials used in the composition. BP involves three layers, input, hidden and output layers. After providing data to be inserted, layers are formed to perform analysis to study the properties of concrete mix. Neuron numbers are assigned to predict performance, and graph is plotted to study the effect of strength of various concrete mixes. Cascading called doing connection between various layers is done to predict accuracy between two nodes of graph. After this, calculation of BP neural networks with parameters like root means square error (RMSE) and coefficient of determination (R2) is obtained. The relationship between the two can help select the optimum concrete mix. It is found that back propagation is mainly affected by external factors. So, it is suggested to determine its performance by properly analysing the different concert mix available for study [26–28].

In the study carried out by Heidari [29] for analysing the back propagation algorithm, two ratios of sand to cement were considered, i.e. 1 and 1.5. Total of 11 mix designs was prepared for each phase. The variation of uses of ordinary sand and waste sand ranges from 0 to 100%. In phase one, sample having 20% waste sand gave the highest strength, and for second phase, 30% waste sand has the highest compressive strength. Now to obtain optimum mix design through back propagation algorithm, five kinds of input, two hidden layers and three output-based values of compressive strength are taken for data fitting. To minimize errors, data is imported into network and network training is processed. Regression analysis is now performed. It was found that two-layer back propagation gave same output as that was obtained through direct compression test. For phase one, the highest compression strength was obtained for 20% replacement of waste sand, and for second phase, it is 30% replacement of sand waste. Thus, BP algorithm gives more accurate results as compared to ANN due to its ability to detect minute errors through RMSE method. Its processing speed of execution of algorithm is also faster than previous ANN models.

Gene expression programming with free coefficient (GEP-FC) is an algorithm of artificial intelligence which is useful in detecting the optimum concrete mix. This algorithm is generally the hybrid form of genetic algorithm and genetic programming. This algorithm is categorized as population-based heuristics and is based on the principle of evolution. In this algorithm, population of individuals is considered, selection is made according to fitness function, and introduction of genetic variation takes place. It generally uses a linear and symbolic string of fixed length [30].

Generally, in the study of GEP-FC done by Funke et al. [31], when parameters are inserted and data is provided, decoding of individuals takes place. Mathematical structure is adopted for the operation of mutation, transposition, inversion and crossover. GEP-FC procedure is used to generate mathematical procedure for computing formulation of concrete mix strength. Different mixes are prepared, and algorithm is processed to calculate properties of concrete mix. FRC composition is varied in only one direction. Water-binder ratio varied from 0.28 to 0.42. Amorphous aluminosilicate was used as a pozzolanic binder, and polycarboxylate ether was a high-performance super-plasticiser. Further, ten mixes were prepared for analysing the mechanical strength. Compressive strength and bending tensile strength were done. It was observed that mix having water-binding ratio of 0.36 was found to have the highest compressive strength and tensile strength. Now to determine the optimum concrete mix through GEP-FC algorithm of AI, 30 independent runs of algorithms were performed. The graph is plotted for water-cement ratio with compressive strength, flow spread and bending tensile strength from the GEP-FC model. Figure 7 shows the plot between the compressive strength and water-cement ratio. It is observed that there is an increase up to a water-binder ratio of 0.36, and after that, it decreases. The conclusions were drawn that the parametric GEP-FC model gave the same result when compared to conventional method. It is also observed in conventional method that on increasing water-binder ratio, compressive strength increases up to some extent, after which it begins to decrease as studied by Elnemr [32]. Initially, there is increase in compressive strength due to adequate hydration process. But later, there is decrease in water content due to inadequate hydration process which finally lowers the compressive strength of concrete mix.

Similarly for the same study by Funke et al. [31], Fig. 8 is plotted between the flow spread and water-cement ratio. The graph obtained from GEP-FC model shows that flow spread is maximum for water-cement ratio of 0.36. The study by Chen et al. [33] showed that higher water-cement ratio would give rise to the larger values of the flow spread. For low water content, there is little flow spread due to the inadequate hydration process and improper mixing resulting in low slump values.

Fig. 7 Compressive strength versus water-binder ratio [31]

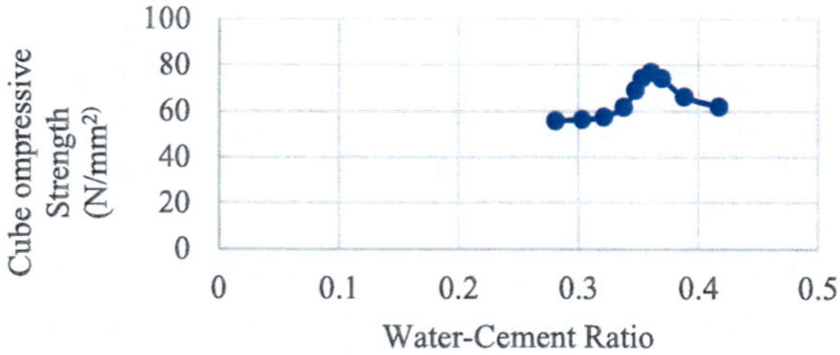

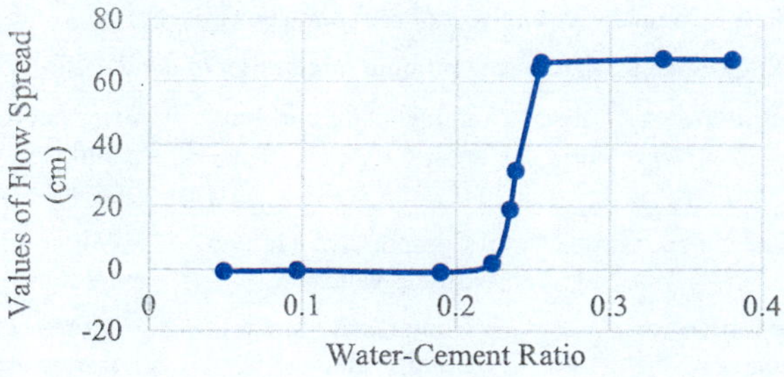

Fig. 8 Flow spread versus water-cement ratio [31]

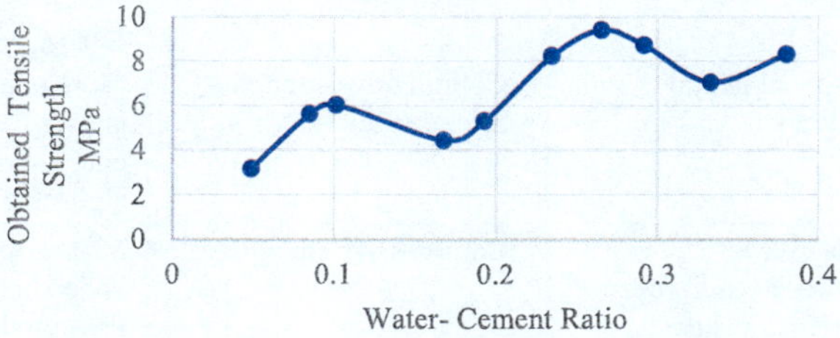

Fig. 9 Bending tensile strength versus water-cement ratio [31]

Figure 9 shows the plot between the tensile strength and water-cement ratio as plotted in the study Funke et al. [31]. The graph obtained from this model shows that there is much variation in bending tensile strength with little variation in water-cement ratio. The study by Elnemr [32] shows that increase in water-cement ratio decreases the split tensile strength.

The graphs obtained from GEP-FC model show that the algorithm is advantageous as it saves time and gives accurate results with minimum error because it has the ability to do automatic analysis for complex model, thus resulting in explicit expression of mathematics. Good correlations are obtained with measured data when GEP-FC model is processed, and thus, it is more advantageous than other models. It reduces experimental and analytical efforts and, thus, gives optimum concrete mix design for good quality construction [31].

The algorithms of artificial intelligence discussed above give a clear view that AI is useful in detecting optimum concrete mix design by reducing human efforts and with minimum error. Artificial intelligence due to its adaptability helps to decide the optimum mix design of concrete available for the construction [34, 35]. The comparative analysis between the various algorithms of AI is discussed in Table 1.

4 Conclusions

With the coming of computational tools, i.e. artificial intelligence optimum concrete mix design can be selected. Various algorithms of artificial intelligence are helpful in selecting optimized concrete mix which will bring sustainability in construction

Table 1 Comparative analysis of different algorithms of AI

Problem statement: to have optimum mix design of concrete			
Algorithms of AI used	Parameters of concrete	Brief findings (Comparison with traditional method)	References
Genetic algorithm (GA)	• Compressive strength • Slump value	• Minimum errors obtained • Less trials	[12, 15]
Support vector machines	• Compressive strength • Slump value	• More accurate and powerful tool than GA	[16]
Fuzzy logic	• Compressive strength • Slump value	• Desire mix is easily obtained • Used in risk management	[17]
Artificial neural networks (ANNs)	• Compressive strength • Slump value	• More explicit model and is user-friendly technique compared to other algorithms	[2, 22, 23]
Adaptive network-based fuzzy interface system (ANFIS)	• Compressive strength	• More appealing and understandable model • Post-modelling of dataset is allowed	[25]
Back propagation neural network (BP)	• Compressive strength	• Supervised learning • Minute errors are detected	[29, 28]
Gene expression programming with free coefficient (GEP-FC)	• Compressive strength • Bending tensile strength • Flow spread • Water-cement ratio	• Automatic analysis of complex model • Reduces experimental and analytical efforts	[31]

in the coming years. AI is proved to be economical and can compute results within a limited time frame as its efficiency rate to do a task is much higher as compared to the conventional method of selecting concrete mix. This tool reduced human labour. The above study shows that algorithms of artificial intelligence can work with minimum errors and are subjected to reliability to work upon.

From the present review, following conclusions can be drawn:

1. It is found that ANN comes out to be the explicit model and is a user-friendly technique of AI when compared with basic genetic algorithms in selection of optimum design mix of concrete.
2. The ANFIS algorithm of AI to select optimum mix design of concrete presents to be more appealing and understandable algorithm as compared to ANN because it uses intuitive experience of designer, and also, numerical information is included in the datasets.
3. BP algorithm gives more accurate results as compared to ANN due to its ability to detect minute errors through RMSE method. Its processing speed of execution is also faster.

4. It is observed that GEP-FC model of AI reduces experimental and analytical efforts and thus gives optimum concrete mix design for good quality construction. It is considered to be the finest algorithm of AI in selecting optimum mix design of concrete as it establishes good correlation with experimental and algorithm processed data.
5. Algorithms of AI also help in risk management by detecting the faults if present in the available concrete mix.

References

1. Yeh, I.-C. (1998). Modeling of strength of high-performance concrete using artificial neural networks. *Cement and Concrete Research, 28*(12), 1797–1808.
2. Gupta, S. (2013). Concrete mix design using artificial neural network. *Journal on Today's Ideas—Tomorrow's Technologies, 1*(1), 29–43.
3. Russell, S. J., & Norvig, P. (2003). *Artificial intelligence: A modern approach.* Prentice Hall.
4. Kalogirou, S. A. (2001). Artificial neural networks in renewable energy systems applications: A review. *Renewable and Sustainable Energy Reviews, 5*(4), 373–401.
5. Chang, K., & Cheng, C. (2020). Learning to simulate and design for structural engineering. In *Proceedings of the 37th International Conference on Machine Learning* (pp. 1–11), Vienna, Austria.
6. Reich, Y. (1996). Machine learning techniques for civil engineering problems. *Computer-Aided Civil and Infrastructure Engineering, 12*(4), 1–27.
7. Khan, U., Verma, R., Khan, R. A., Kumar, A. S., & Varshney, H. (2020). Application of machine learning and artificial intelligence in civil engineering: Review. In *4th International Conference (Online) on Recent Trends in Communication & Electronics (ICCE-2020)* (pp. 1–7).
8. Mitra, S. (2017). *Applications of machine learning and computer vision for smart infrastructure management in civil engineering* (Master's theses and Capstones), p. 1138.
9. Muliauwan, H. N., Prayogo, D., Gaby, G., & Harsono, K. (2020). Prediction of concrete compressive strength using artificial intelligence methods. *Journal of Physics: Conference Series, 1625*, 1–10.
10. Varshney, H., Khan, R. A., Khan, U., & Verma, R. (2020). Approaches of artificial intelligence and machine learning in smart cities: A critical review. In *1st International Conference on Computational Research and Data Analytics, IOP Conference Series: Material Science and Engineering*, jointly organized by Suleyman Demirel University and Isparta Applied Sciences University, Turkey and College of Engineering Roorkee (pp. 1–12).
11. Dao, D. V., Ly, H.-B., Trinh, S. H., Le, T.-T., & Pham, B. T. (2019). Artificial intelligence approaches for prediction of compressive strength of geopolymer concrete. *Materials, 12*(983), 1–17.
12. Gupta, S. M. (2007). Support vector machines based modelling of concrete strength. *Engineering and Technology, 36*, 305–311.
13. Lim, C. H., Yoon, Y.-S., & Kim, J.-H. (2004). Genetic algorithm in mix proportioning of high-performance concrete. *Cement and Concrete Research, 34*(3), 409–420.
14. Yadav, P. K., & Prajapati, N. L. (2012). An overview of genetic algorithm and modeling. *International Journal of Scientific and Research Publications, 2*(9), 1–4.
15. Rezaee, A., & Ahangar Mohd, R. H. (2012). Mix proportioning of high-performance concrete by applying the GA and PSO. *International Journal of Smart Electrical Engineering, 1*, 1–8.
16. Gupta, S. M. (2001). *Experimental studies on the behaviour of high strength concrete* (Ph.D. thesis), Kurukshetra University, Kurukshetra, India.

17. Vengadeshwari, R. S., & Reddy, H. N. J. (2013). Optimum concrete mix design using heuristic techniques. *International Journal of Scientific & Engineering Research, 4*(8).
18. Dias, W. P. S., & Pooliyadda, S. P. (2001). Neural networks for predicting properties of concretes with admixtures. *Construction and Building Materials, 15*(7), 371–379.
19. Suryadi, A., & Triwulan, A. (2011). Artificial neural networks for evaluating the compressive strength of self compacting concrete. *Journal of Basic and Applied Scientific Research, 1*(3), 236–241.
20. Faruqi, M. A., Agarwala, R., Sai, J., & Francisco, A. (2015). Application of artificial intelligence to predict compressive strength of concrete from mix design parameters: A structural engineering application. *Journal of Civil Engineering Research, 5*(6), 158–161.
21. Kukreja, H., Bharath, N., Siddesh, C. S., & Shiruru, K. (2016). An introduction to artificial neural network. *International Journal of Advance Research and Innovative Ideas in Education, 1*(5), 27–30.
22. Chandwani, V., Agrawal, V., & Nagar, R. (2014). Modeling slump of ready-mix concrete using genetically evolved artificial neural networks. *Advances in Artificial Neural Systems*, 1–9.
23. Yeh, I.-C. (2006). Exploring concrete slump model using artificial neural networks. *Journal of Computing in Civil Engineering, 20*(3), 217–221.
24. Tesfamariam, S., & Najjaran, H. (2007). Adaptive network-fuzzy inferencing to estimate concrete strength using mix design. *Journal of Materials in Civil Engineering, 19*(7), 1–46.
25. Kim, J. I., Kim, D. K., Feng, M. Q., & Yazdani, F. (2004). Application of neural networks for estimation of concrete strength. *Journal of Materials in Civil Engineering, 16*(30), 257–264.
26. Duan, Z. H., Kou, S. C., & Poon, C. S. (2013). Prediction of compressive strength of recycled aggregate concrete using artificial neural networks. *Construction & Building Materials, 40*, 1200–1206.
27. Shin, Y. S., & Kim, G. H. (2013). Predicting compressive strength of recycled aggregate concrete by multiple regression analysis. *Applied Mechanics & Materials, 253*, 546–549.
28. Tan, K. (2018). Predicting compressive strength of recycled concrete for construction 3D printing based on statistical analysis of various neural networks. *Journal of Building Construction and Planning Research, 6*, 71–89.
29. Heidari, A., Hashempour, M., & Tavakoli, D. (2017). Using of backpropagation neural network in estimating of compressive strength of waste concrete. *Journal of Soft Computing in Civil Engineering, 1*(1), 54–64.
30. Ferreira, C. (2001). Gene expression programming: A new adaptive algorithm for solving problems. *Complex Systems, 13*(2), 87–129.
31. Funke, H. L., Ulke-Winter, L., Gelbrich, S., & Kroll, L. (2018). A numerical description of a fibre reinforced concrete using a genetic algorithm. *International Journal of Civil and Environmental Engineering, 12*(3), 341–346.
32. Elnemr, A. (2019). Role of water/binder ratio on strength development of cement mortar. *American Journal of Engineering Research, 8*(1), 172–183.
33. Chen, J. J., Ng, P. L., Li, L. G., & Kwan, A. K. H. (2016). Production of high-performance concrete by addition of fly ash microsphere and condensed silica fume. In *Modern Building Materials, Structures and Techniques, MBMST 2016, Procedia Engineering* (Vol. 172, pp. 165–171).
34. Duan, J., Asteris, P. G., Nguyen, H., Bui, X.-N., & Moayedi, H. (2020). A novel artificial intelligence technique to predict compressive strength of recycled aggregate concrete using ICA-XGBoost mode. *Engineers with Computers*, 1–18.
35. Huang, Y., Li, J., & Fu, J. (2019). Review on application of artificial intelligence in civil engineering. *Computer Modeling in Engineering & Sciences CMES, 121*(3), 845–875.

A Review of Integration of Data Warehousing and WWW in the Last Decade

Priyanka Bhutani, Anju Saha, and Anjana Gosain

Abstract The data warehouse (DW) is a powerful technology to store and analyse huge volumes of historical data supporting business intelligence. The World Wide Web, or simply the Web, has revolutionized the way to author, share, search and access information. In the past few decades, a significant amount of research has been done in both the DW and Web domains. Interestingly, the integration of data warehousing and the World Wide Web has led to a variety of new opportunities as well as challenges for the researchers and the industry. The main motivation to conduct this systematic review of the relevant research works integrating DW and the Web in the last decade is to provide the groundwork for the research advancement in this field. A total of 27 relevant research works were identified for the research. An in-depth analysis was performed to find the problems addressed, the most relevant research categories, the tools or techniques applied and the application domains of these research works. Encouragingly, our results yielded seven categories and four sub-categories of research employing the integration of DW and Web. On the other hand, we found some open research issues, and the future research works should focus on generalized solutions for handling semantic heterogeneity, change propagation and quality analysis of identified Web sources for the DW.

Keywords Data warehouse · WWW · Web · Web warehouse · XML DW

1 Introduction

In the last few decades, the data warehouse (DW) technology has clearly risen to become the backbone of the decision-making process in most of the medium to large-scale organizations [1]. This powerful and useful technology, used to store and analyse large volumes of historical data supporting the business intelligence, has been a topic of interest for the research community as well as the industry

P. Bhutani (✉) · A. Saha · A. Gosain
University School of Information and Communication Technology, Guru Gobind Singh Indraprastha University, Sector-16C, Dwarka, Delhi 110078, India
e-mail: priyanka.b@ipu.ac.in

P. K. Singh et al. (eds.), *Proceedings of Third International Conference on Computing, Communications, and Cyber-Security*, Lecture Notes in Networks and Systems 421,
https://doi.org/10.1007/978-981-19-1142-2_58

[2]. W. H. Inmon, considered to be the "Father of Data Warehouse", classically defines it as a subject-oriented, integrated, time-variant and non-volatile collection of data to support the management's decision-making process [3]. This integrated and processed data store in one place is most commonly used for business analysis later [4]. Another pioneer of the field of DW, R. Kimball, defines DW as a copy of the transaction data that is specifically structured for querying and analysis [5]. Hence, the huge repository of data in the DW is optimized to provide integrated viewing and analysis, much different from the traditional transactional databases with online analytical processing (OLAP) [2].

In around the same timeline as the rise of DW, a remarkable change in the access, search and authoring of information has been brought about by the revolution of the World Wide Web (WWW). The medium of WWW is undoubtedly most prevalent, extensively used as well as very diverse [6]. As on 18 September 2021, the estimated size of the World Wide Web is at least 3.93 billion pages, according to statistics of the Website worldwidewebsize.com [7]. Being an open and independent platform, the heterogenous information provided and accessed via the WWW can be from organizations, government agencies or individuals. Also, being very dynamic, WWW is constantly in a state of flux with Webpages being added, removed or modified daily [8]. With the advent of semantic annotations in the Web like metadata, ontologies and RDF, the retrieval and analysis of information are becoming more effective [6].

The WWW opened an extensive array of attractive opportunities as well as challenges for the data warehousing domain. As the Web developed, e-commerce over the Web became the major push towards making the business process more broad, complex and fast [8]. With this development, not only the amount of data generated by business processes became bigger than ever, but even the types of data became diverse [4]. The Web usage logs showing clickstream data of customers, e.g. started being used to personalize the e-commerce portal or company's Website for the customer. The typically used enterprise's internal structured transactional data became insufficient for the strategic decision-making in the DW [8]. Hence, the suitable external data from the Web started being integrated in the DW (Fig. 1). The Web also became an enabling platform to provide access to the DW for the enterprise's suppliers, customers or any other beneficiaries [2]. The WWW also brought about interesting changes in the modelling, querying or OLAP in the DW [2].

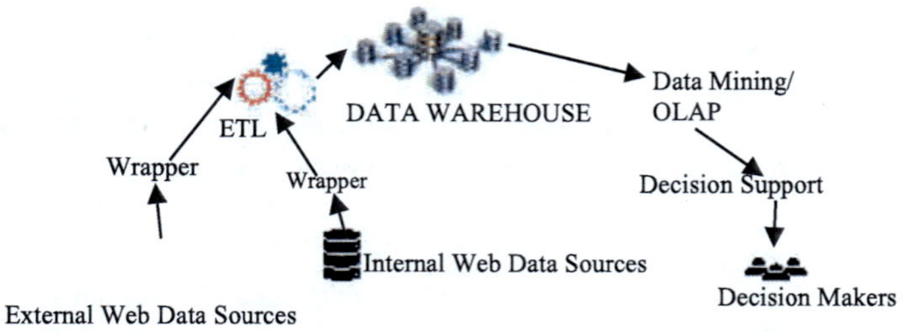

Fig. 1 Integration of external data from the Web in the DW (Adapted from [9])

This research study provides the review of the research works that are using the integration of data warehousing and WWW for various needs in the last decade. The rest of the paper is organized as follows: Section 2 presents a brief overview of the systematic literature review process followed in this work. Section 3 provides the classified summary of the 27 research works integrating DW and WWW from the last decade that were searched and finalized in the SLR. Section 4 summarizes the answers to the research questions. Section 5 presents the conclusion.

2 Systematic Literature Review (SLR) Process

The systematic literature review (SLR) process in this work has been carried out in accordance with the guidelines of the work of Kitchenham [10]. These guidelines suggest a systematic methodology to be employed in the process of SLR. The applications of the steps of this methodology are briefly presented below.

2.1 Aim of Research

We formulated the aim of the research in the form of research questions in order to direct the search in extraction of useful data from the studies while formulating answers to these questions. The aim of our current work is to answer the following research questions:

Research Question-1: What are the most relevant research categories in the integration of WWW and the data warehouse?
Research Question-2: What are the major application domains in which this synergy of WWW and the data warehouse was employed?
Research Qustion-3: What are open research issues related to the current trends of the integration of these technologies?

2.2 Search Process

Sources of information. The research works for our study were searched from the major online research databases (Google Scholar [11], IEEE Xplore [12], Springer LNCS [13], Science Direct [14] and ACM digital library [15]).

Search criteria. The last decade, i.e. from 2010 till date, was the timeline set by us for search of research works. We set the search string for extracting the relevant works as ([(Data Warehouse) <or> (DW) <or> (Web Warehouse) <or> (warehousing)] <and> [(WWW) <or> (Web) <or> (XML)]).

Criteria for work selection. The exclusion criteria set for the works retrieved from the above-stated search criteria were to NOT include content from magazines, newsletters and educational courses. We also excluded the papers which were irrelevant papers based on the perusal of their titles and abstracts. Finally, after reading of full texts, 27 studies were found relevant and providing a fairly comprehensive view of the research in the last decade in the field of integration of data warehousing and WWW.

Data extraction strategy. The next step was the thorough exploration of papers to find the answers to the research questions. The research categories and application domain of the work was summarized for each work. Additionally, any specific technique/model/tool/technology applied in the work was also identified.

3 State of the Art

Table 1 presents the classified summary of the 27 research works integrating DW and WWW from the last decade that were searched and finalized for our work as described in the previous section.

4 Literature Review Results

4.1 Research Question-1: Most Relevant Research Categories

The distribution of the 27 identified research works (Fig. 2a) selected for our research, as summarized in the last section, shows that almost half, i.e. 13 of these were published in journals, 13 of them were from conference proceedings and one was a doctoral thesis. The year-wise publication (Fig. 2b) encouragingly shows a steady flow of research has been going on in the research area of integration of DW and Web in the last decade, although much remains to be explored in the synergy of these two domains.

The research in the integration of WWW and data warehouse can be categorized broadly in the following categories (Fig. 3a):

Website ClickStream DW. DW has been used as a repository for Web usage data and analysis. The mining of Web usage logs or clickstream data provides valuable information for Website improvement and personalization. To analyse this data, automatic and semi-automatic approaches have been proposed to map this data, store it in clicks stream DW and improve the performance of its analysis [17, 18].

Building Web/XML data-based DW. Many authors have proposed building DW with Web/XML data and dealt with various aspects related to it (Fig. 3b). Some works deal with providing the architecture of such a DW [16, 19], while others

Table 1 Classified summary of research works integrating DW and WWW in last decade

Author(s)	Year	Problems addressed/work done	Other techniques/models applied or technologies/tools used	Research categories	Any specific application domain of work
Liu and Luo [16]	2010	e-governance solution using DW under a European project	PostgreSQL; RDF	Building Web/XML data-based DW: overall architecture	e-governance
Sudhamathy [17]	2010	Automated approach for mining Web logs for Website personalization using DW	WEKA tool; W3C extended log files	Website clickstream DW	Website improvement
Chen et al. [18]	2010	Mining Web logs and improving performance of clickstream DW	Quotient space granular computing	Website clickstream DW	Website improvement
Moya et al. [19]	2011	Sentiment analysis Web data integration in DW	Clustering	Building Web/XML data-based DW: Architecture; Modelling; Extraction of Web data	Business intelligence
Nguyen et al. [20]	2011	Building XML warehouse using mailing list Web data	Altova XML-spy tool	Building Web/XML data-based DW: modelling; extraction of Web data	Sociological analysis
Marotta et al. [21]	2012	Data and service quality-based Web warehouse platform	Service-oriented architecture	Quality of Web/XML DW: quality-aware architecture	
Lv et al. [22]	2012	Cloud DW design, ETL and OLAP	Hadoop ecosystem	DW hosted on the cloud: ETL, OLAP	Business intelligence
Ali et al. [23]	2013	Semiautomatic Web XML data transformation to star schema for DW using schema matching	XSLT and XPATH	Building Web/XML data-based DW: transformation	

(continued)

Table 1 (continued)

Author(s)	Year	Problems addressed/work done	Other techniques/models applied or technologies/tools used	Research categories	Any specific application domain of work
Domingues et al. [24]	2014	DW to support Website automation and monitoring	WUMPrep tool; Edmate system	Building Web/XML data-based DW: modelling, ETL of Web data; XML-based DW query processing	Website improvement
Mehmood et al. [25]	2014	Security in Web warehouse architecture	Cryptography	Security in Web-based warehouse: architecture	
Samuel [26]	2014	Mediation ETL approach for feeding Web data in DW	Web services	Building Web/XML data-based DW: Web ETL	Business intelligence
Kavitha and Vydehi [27]	2014	XML DW querying and pattern matching techniques	Pattern matching algorithms like TreeMatch	XML-based DW query processing	
Delgado and Marotta [28]	2015	Web DW construction business process modelling	BPMS Activiti tool	Building Web/XML data-based DW: modelling	
Jiang et al. [29]	2015	Reconstruction of XML DW for OLAP analysis based on knowledge graph	Semantic Web tools Protégé and KGELP	Web/XML data OLAP analysis	
Mehmood et al. [30]	2015	Secure Web service-oriented architecture of the WW	Internet of Things (IoT)	Security in Web-based warehouse: architecture	
Om Sharan Sinha [31]	2016	External Web source selection for DW	MCDM technique TOPSIS	Quality of Web/XML DW: quality-based Web source selection	
Nikam et al. [32]	2016	Web DW conceptual modelling	Representational state transfer (REST)	Building Web/XML data-based DW: modelling	Education

(continued)

Table 1 (continued)

Author(s)	Year	Problems addressed/work done	Other techniques/models applied or technologies/tools used	Research categories	Any specific application domain of work
Ravat and Song [33]	2016	Unified OLAP analysis of DW and Web data	Linked open data (LOD)	Web/XML data OLAP analysis	Business intelligence
Gupta et al. [34]	2017	Web patent data extraction algorithm for DW	Java, MySQL and JSoup; MATLAB's biograph	Building Web/XML data-based DW: extraction	
Alrefae and Cao [35]	2017	Active XML-based dynamic Web data extraction for DW	Service-oriented architecture	Building Web/XML data-based DW: extraction	Business intelligence; public health
Gupta et al. [36]	2018	Web ETL transforming Web patent data for Hadoop cluster based DW	Hadoop	DW hosted on the Cloud: Web ETL	
Strand and Syberfeldt [37]	2019	External Web data in BI solution	Web services	Building Web/XML: identification, acquisition and integration	
Walha et al. [38]	2019	Modelling of Web social data ETL process for DW	UML; Talend Open Studio (TOS)	Building Web/XML: modelling, Web ETL	
Agapito et al. [39]	2020	Italy's COVID-19, pollution and climate Web data automatic ETL and OLAP in DW	Python	Building Web/XML: ETL, Web/XML data OLAP analysis	
Sellami et al. [40]	2020	Web social media data-based NoSQL DW modelling	NoSQL database Neo4j; Java; TOS (Talend Open Studio)	Building Web/XML: modelling, NoSQL DW	
Bhutani et al. [41]	2021	Quality-based Web source evaluation and selection for DW	MATLAB; MCDM; Website testing tools like Google Lighthouse	Quality of Web/XML DW: quality-based Web source selection	

(continued)

Table 1 (continued)

Author(s)	Year	Problems addressed/work done	Other techniques/models applied or technologies/tools used	Research categories	Any specific application domain of work
Bhutani et al. [9]	2021	Quality model for Web source selection for DW	Empirical validation technique	Quality of Web/XML DW: quality model validation	

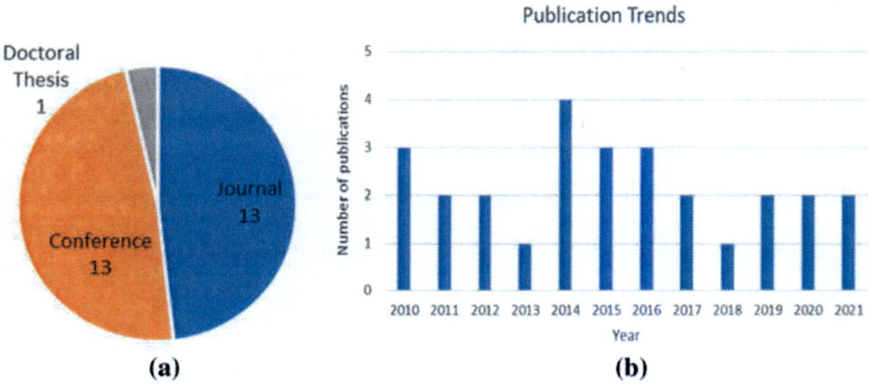

Fig. 2 **a** Distribution of research papers, **b** publication trends in the research area of integration of DW and WWW

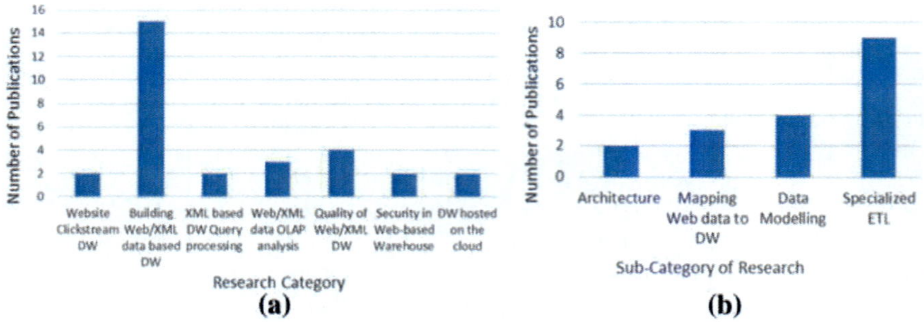

Fig. 3 **a** Distribution of research papers by identified research categories, **b** distribution of research papers in the sub-categories of "Building Web/XML data-based DW"

provide solutions for mapping Web data to the DW [34, 37, 40]. The modelling of schemas [19, 20, 28, 32] and specialized extraction, transformation and loading (ETL) [19, 20, 23, 24, 26, 35, 36, 38, 39] of Web/XML-based data warehouses have also been dealt with in significant number of research works.

XML-based DW query processing. The works which have dealt with the querying and optimization of query processing of the XML-based DW are in this category [24, 27].

Web/XML data OLAP analysis. The specialized OLAP analysis of the DW with XML data has been dealt with, in the research works of Jiang et al. [29], Ravat and Song [33] and Agapito et al. [39].

Quality of Web/XML DW. Marotta et al. [21] have proposed an architecture of the Web warehouse such that each stage of the data warehousing of Web data is a quality-aware process. For enhanced quality of Web-based DW, researchers have proposed approaches for the quality-based Web source selection from the identified possible candidate Web data sources [31]. This process has been carried out by the assessment of the Web data sources on various quality parameters formed into quality models like the Web QMDW model [41] which has also been empirically validated [9].

Security in Web-based warehouse. There have been few research works in the direction of providing better security in the Web-based data warehouse [25, 30].

DW hosted on the cloud. The hosting of the data warehouse on cloud platforms like the Hadoop ecosystem provides distributed storage capabilities and parallel computing abilities for vast amount of data [22]. It gave rise to an array of research problems, like design, ETL, data mining and OLAP for cloud-based DW, which have been dealt with in published research literature [22, 36].

4.2 Research Question-2: Major Application Areas

The synergy of DW and WWW technologies has been leveraged in numerous application areas. In the last decade, the major areas identified are as follows (as depicted in Table 1): e-governance, Website improvement, business intelligence, public health, public waste management, sociological analysis, and education.

4.3 Research Question-3: Open Research Issues Related to Current Trends

The synergy of data warehousing and WWW has been researched and applied for many application areas. However, the full potential of this synergy has not been realized till date which leads to open lines of research. One big challenge is that the current focus of the approaches of the integration of the DWs and WWW is mostly oriented towards specific problem scenarios. The approaches, whether semi-automatic or automatic, cannot be applied directly to the upcoming large-scale integration requirements of the Web and data warehousing. Hence, the big challenge is the need for generalized solutions, techniques and approaches, which can handle the various levels of semantic heterogeneity as well to suit various application domains.

Another issue of "change propagation" which comes up to the forefront is due to the highly dynamic nature and lack of authority/control over the sources over the WWW. The traditional data warehousing needs to be geared up with solutions to change the schema as well as the materialized views of the DW in response to the fast-changing Web sources.

Also, since the Web is a huge repertory of information from variety of sources, the seemingly "relevant" data for the DW could be actually of bad quality. The integration of such data in the DW could lead to the compromised quality, consistency and integrity of data warehouse as well as the resultant decision-making process. But the quality analysis of identified candidate Web sources before their integration in DW, although an extremely critical fundamental step, is yet not that widely researched issue.

5 Conclusion

In this work, we have presented the systematic literature review of the research literature employing the integration of data warehouse and the World Wide Web in the last decade. A set of 27 works were finalized and studied for identifying the problems addressed, the most relevant research categories, the tools or techniques applied and the application domains of these research works. The results indicated that the synergy of the Web and DW has led researchers to provide solutions to a variety of issues in the varied application areas like business intelligence, education and public health. Seven categories of research – Website ClickStream DW, Building Web/XML data-based DW, XML-based DW query processing, Web/XML data OLAP analysis, Quality of Web/XML DW, Security in Web-based warehouse, and DW hosted on the cloud, were identified. The majority of the research works were found in the category of "building of the Web/XML data-based DW" and its sub-category of "specialized Web ETL". The review highlighted the existing open research issues that were identified during the course of research—the need for generalized approaches effectively handling semantic heterogeneity, refined solutions for change propagation from dynamic Web sources to DW and in-depth, viable quality analysis of identified candidate Web sources before their integration in DW.

References

1. Chandra, P., & Gupta, M. K. (2018). Comprehensive survey on data warehousing research. *International Journal of Information Technology, 10*, 217–224.
2. Perez, J. M., Berlanga, R., Aramburu, M. J., & Pedersen, T. B. (2008). Integrating data warehouses with web data: A survey. *IEEE Transactions on Knowledge and Data Engineering, 20*,

940–955.
3. Inmon, W. H. (2005). *Building the data warehouse*. Wiley.
4. Brajkovic, H., Jaksic, D., & Poscic, P. (2020). Data warehouse and data quality—An overview. In *Central European Conference on Information and Intelligent Systems 2020* (pp. 17–24).
5. Kimball, R., & Ross, M. (2002). *The data warehouse toolkit*. Wiley.
6. Bhutani, P., & Saha, A. (2019). Towards an evolved information food chain of world wide web and taxonomy of semantic web mining. In S. Bhattacharyya, A. E. Hassanien, D. Gupta, A. Khanna, & I. Pan (Eds.), *International Conference on Innovative Computing and Communications* (pp. 443–451). Springer.
7. WorldWideWebSize.com | The size of the World Wide Web (The Internet). https://www.worldwidewebsize.com/, last accessed 2021/09/20.
8. Zhu, Y., & Buchmann, A. (2002). Evaluating and selecting web sources as external information resources of a data warehouse. In *Proceedings of the Third International Conference on Web Information Systems Engineering, 2002. WISE 2002* (pp. 149–160). IEEE.
9. Bhutani, P., Saha, A., & Gosain, A. (2021). Empirical validation of WebQMDW model for quality-based external web data source incorporation in a data warehouse. *IJACSA, 12*.
10. Kitchenham, B., Pearl Brereton, O., Budgen, D., Turner, M., Bailey, J., & Linkman, S. (2009). Systematic literature reviews in software engineering—A systematic literature review. *Information and Software Technology, 51*, 7–15.
11. Google Scholar. https://scholar.google.co.in/, last accessed 2021/09/20.
12. IEEE Xplore. https://ieeexplore.ieee.org/Xplore/home.jsp, last accessed 2021/09/20.
13. Springer—International Publisher Science, Technology, Medicine. https://www.springer.com/gp/, last accessed 2021/09/20.
14. ScienceDirect.com | Science, health and medical journals, full text articles and books. https://www.sciencedirect.com/, last accessed 2021/09/20.
15. ACM Digital Library. https://dl.acm.org/, last accessed 2021/09/20.
16. Liu, X., & Luo, X. (2010). A data warehouse solution for e-Government. *International Journal of Research and Reviews in Applied Sciences, 4*, 101–105.
17. Sudhamathy, G. (2010). Mining web logs: An automated approach. In *Proceedings of the 1st Amrita ACM-W Celebration on Women in Computing in India* (pp. 1–4).
18. Chen, X., Wu, Y., & Cheng, H. (2010). Quotient space granular computing for the Click-stream data warehouse in web servers. In *2010 International Conference on Computer and Communication Technologies in Agriculture Engineering* (pp. 93–96). IEEE, Chengdu, China.
19. Moya, L. G., Kudama, S., Cabo, M. J. A., & Llavori, R. B. (2011). Integrating web feed opinions into a corporate data warehouse. In *Proceedings of the 2nd International Workshop on Business intelligence and the WEB—BEWEB '11* (pp. 20–27), Uppsala, Sweden. ACM Press.
20. Nguyen, B., Vion, A., Dudouet, F.-X., Colazzo, D., Manolescu, I., & Senellart, P. (2011). XML content warehousing: Improving sociological studies of mailing lists and web data. *Bulletin of Sociological Methodology/Bulletin de Méthodologie Sociologique, 112*, 5–31.
21. Marotta, A., González, L., & Ruggia, R. (2012). A quality aware service-oriented web warehouse platform. In *Proceedings of the 2012 Joint EDBT/ICDT Workshops on EDBT-ICDT '12* (p. 29), Berlin, Germany. ACM Press.
22. Lv, H. L., Van, A. M., Cheng, V. L., & Wang, F. V. (2012). Design of cloud data warehouse and its application in smart grid. In *International Conference on Automatic Control and Artificial Intelligence (ACAI 2012)* (pp. 849–852), Xiamen, China. Institution of Engineering and Technology.
23. Ali, A. A., Abdelrahman, T. A., & Mohamed, W. M. (2013). Using schema matching in data transformation for warehousing web data. *International Journal of Information Technologies and Knowledge, 7*, 230–240.
24. Domingues, M. A., Soares, C., Jorge, A. M., & Rezende, S. O. (2014). A data warehouse to support web site automation. *Journal of the Brazilian Computer Society, 20*, 11.
25. Mehmood, R., Shaikh, M. U., Ma, L., & Bie, R. (2014). Enhanced web warehouse model: A secure approach. In *2014 International Conference on Identification, Information and Knowledge in the Internet of Things* (pp. 88–91), Beijing, China. IEEE.

26. Samuel, J. (2014). *Feeding a data warehouse with data coming from web services. A mediation approach for the DaWeS prototype* (Doctoral thesis), Université Blaise Pascal-Clermont-Ferrand II.
27. Kavitha, P., & Vydehi, M. S. (2014). Query processing of XML data warehouse using XML pattern matching techniques. *International Journal of Engineering Research, 3.*
28. Delgado, A., & Marotta, A. (2015). Automating the process of building flexible web warehouses with BPM systems. In *2015 Latin American Computing Conference (CLEI)* (pp. 1–11), Arequipa, Peru. IEEE.
29. Jiang, Y., Shao, Z., Guo, Y., Zhang, H., & Sun, L. (2015). Building XML data warehouse with data reconstruction by knowledge graph. In *2015 IEEE Fifth International Conference on Big Data and Cloud Computing* (pp. 314–320), Dalian, China. IEEE.
30. Mehmood, R., Shaikh, M. U., Bie, R., Dawood, H., & Dawood, H. (2015). IoT-enabled web warehouse architecture: A secure approach. *Personal and Ubiquitous Computing, 19*, 1157–1167.
31. Om Sharan Sinha, H. (2016). An improvised Topsis approach to select web source as external data source for web warehousing. *Indian Journal of Science and Technology, 9.*
32. Nikam, R. V., Shirwaikar, S., & Kharat, V. S. (2016). Conceptual model for a data warehouse on the web. In *2016 IEEE Bombay Section Symposium (IBSS)* (pp. 1–6), Baramati, India. IEEE.
33. Ravat, F., & Song, J. (2016). Enabling OLAP analyses on the web of data. In *2016 Eleventh International Conference on Digital Information Management (ICDIM)* (pp. 215–224), Porto, Portugal. IEEE.
34. Gupta, G., Kumar, N., & Chhabra, I. (2017). Data acquisition based web scrapping algorithm for extraction of data sets from patent portal. In *International Conference on Communication, Computing and Networking (ICCCN-2017)*, Chandigarh, India. NITTTR.
35. Alrefae, A., & Cao, J. (2017). Intensional XML-enabled web-based real-time decision support system. In *2017 International Conference on Computing Networking and Informatics (ICCNI)* (pp. 1–10), Lagos. IEEE.
36. Gupta, G., Kumar, N., & Chhabra, I. (2018). Optimised transformation algorithm for hadoop data loading in web ETL framework. *ICST Transactions on Scalable Information Systems*, 160600.
37. Strand, M., & Syberfeldt, A. (2019). Incorporating external data into a BI solution at a public waste management organization. *International Journal of Business Intelligence Research, 10*, 36–56.
38. Walha, A., Ghozzi, F., & Gargouri, F. (2019). From user generated content to social data warehouse: Processes, operations and data modelling. *IJWET, 14*, 203.
39. Agapito, G., Zucco, C., & Cannataro, M. (2020). COVID-WAREHOUSE: A data warehouse of Italian covid-19, pollution, and climate data. *IJERPH, 17*, 5596.
40. Sellami, A., Nabli, A., & Gargouri, F. (2020). Graph NoSQL data warehouse creation. In *Proceedings of the 22nd International Conference on Information Integration and Web-based Applications & Services* (pp. 34–38), Chiang Mai, Thailand. ACM.
41. Bhutani, P., Saha, A., & Gosain, A. (2020). WSEM_{QT} : A novel approach for quality-based evaluation of web data sources for a data warehouse. *IET Software, 14*, 806–815.

WeScribe: An Intelligent Meeting Transcriber and Analyzer Application

Mohammad Aftab Alam Khan, Maryam AlAyat, Jumana AlGhamdi, Shahad Mohammed AlOtaibi, Maha AlZahrani, Malak AlQahtani, Atta-ur-Rahman, Mona Altassan, and Farmanullah Jan

Abstract In all existing organizations regardless of their type or size, meetings are conducted on the regular basis to invite discussions for organizational decision making. While many organizations, even large ones, still hire employees to perform these tasks, there is no doubt that the results are exposed to human error. Documenting meetings' minutes is essential for its success and keeping track of the work progress and decisions flow, approvals, while keeping it complete, consistent, and coherent. This project idea was proposed by Aramco to develop a suitable solution for a hectic problem. The process of documenting and taking minutes can be tedious, so we aim

---

M. A. A. Khan
Department of Computer Engineering, College of Computer Science and Information Technology, Imam Abdulrahman Bin Faisal University, P.O. Box 1982, Dammam 31441, Saudi Arabia
e-mail: mkhan@iau.edu.sa

M. AlAyat · J. AlGhamdi (✉) · S. M. AlOtaibi · M. AlZahrani · M. AlQahtani · Atta-ur-Rahman · M. Altassan · F. Jan
Department of Computer Science, College of Computer Science and Information Technology, Imam Abdulrahman Bin Faisal University, P.O. Box 1982, Dammam 31441, Saudi Arabia
e-mail: 2180006622@iau.edu.sa

M. AlAyat
e-mail: 2170008191@iau.edu.sa

S. M. AlOtaibi
e-mail: 2170003084@iau.edu.sa

M. AlZahrani
e-mail: 2170000238@iau.edu.sa

M. AlQahtani
e-mail: 2160001752@iau.edu.sa

Atta-ur-Rahman
e-mail: aaurrahman@iau.edu.sa

M. Altassan
e-mail: maltassan@iau.edu.sa

F. Jan
e-mail: fzmjan@iau.edu.sa

P. K. Singh et al. (eds.), *Proceedings of Third International Conference on Computing, Communications, and Cyber-Security*, Lecture Notes in Networks and Systems 421,
https://doi.org/10.1007/978-981-19-1142-2_59

to automate audio meeting transcription with the use of technologies that convert speech to text while recognizing the speaker and then process and analyze the most valuable information tagged based on persons in the meeting. This goal can be accomplished through the development of an app that uses speech recognition for conversion, voice recognition for identification of speakers, and natural language processing (NLP) for analysis and then combines them all in a transcription form with considerable accuracy. Further, the proposed system identifies potential events, deadlines, and follow-ups and adds them to the speaker's calendar upon approval. In the future, we aspire to expand it with some features such as increasing the number of meeting members, creating special sections for each department in the company which adopt WeScribe, and feed our NLP model with more data to develop its performance and increase its accuracy.

Keywords NLP · Meeting transcriber · Named entity recognition · Information extraction · Text to speech

1 Introduction

These days, no company, governmental or private organization, or even a small working group is working without having meetings in which important topics are discussed and decisions are taken. Taking the minutes of meetings is one of the success factors of the gatherings. These meeting minutes are a historical record that contains the important points, decisions, and agreements that have been discussed in a meeting. These minutes can help both meeting's participants and those members who could not attend the meeting to revise the main important ideas and assignments of tasks, as well as benefiting the decision-makers to view the previous meetings, and then conclusions are determined [1]. Companies assign the task of manual documentation to an employee or even external hiring for the sake of documentation, thus consuming effort, money, and time in a task that can be automated technologically. Therefore, we decided to develop a transcriber system using machine learning (ML) and natural language processing (NLP) techniques to convert the audio record of a meeting to written text that facilitates and makes the process of taking minutes and notes of the meetings in a more precise and comprehensive manner. The text can be viewed in a transcript form noting the speaker of the segment using speech recognition technologies. This automation can be helpful in many ways. Like reducing the effort of a secretary or member of the meeting to take the minutes, possibly that person might not be from the area of the discussion, and it may not be feasible to add such a person due to nature confidentiality of the meeting, in addition to covering the user experience of the proposed solution by adding the feature of sharing meetings with the attendees or customized list of members in the organization, while providing the essential summary notes, such as emphasizing on important deadlines, events and dates specified during the meeting, organizations and countries mentioned, and overall summary text. Moreover, in this work a software prototype has been developed, implementing the initial main features, and later it can be taken up as a functional enterprise level software and even it can be extended to the hardware level. Rest of the paper is organized as follows. Section 2 is dedicated to the literature that comprehends approaches close to the proposed one. Section 3 is dedicated to proposed system description. Section 4 highlights the core functionalities and limitations of the proposed system, while Sects. 5 and 6 conclude the paper after stating the result outcomes.

2 Literature Review

The workflow of the meeting transcription system is mostly similar beginning with uploading an audio recorded from a meeting and then converts the audio to text after that extracting the information from the text. Later, we will separate the text as per the participant's name. The implementation of the idea is different according to the huge number of technologies that exist. From discovering existing works, we can have a better vision of the idea and learn about the various techniques they used in implementing the system.

2.1 Voice Recognition

All meeting attendees must provide their voiceprint for speaker identification to the system before starting the meeting for better transcribing [2]. Typically, voice authentication is performed as a two-stage process over a fixed telephone network (e.g., a public switched telephone network (PSTN), telecommunications networks like orthogonal frequency division multiplexing (OFDM)-based systems [3, 4] and even in satellite communication [5, 6]). The enrollment stage includes saving a recorded sample for a person's voice to extract their unique voice characteristics and encrypt them. The authentication stage includes receiving a voice sample of a person to be authenticated over the phone and matching the voice characteristics of the obtained sample with those of the recorded voice sample, using a voice authentication engine [7]. Such a voice authentication system needs significant infrastructure and high cost in terms of the security service provider.

2.2 Information Extraction

Information extraction is among the hottest areas of research in data mining especially in text mining [8–12]. Information extraction involves several subtasks, and they are as follows: named entity recognition (NER), named entity linking (NEL), coreference resolution (CR), temporal information extraction, relation extraction (RE), knowledge base construction, and reasoning [13]. There are many available tools for information extraction like OpenNLP (Apache OpenNLP-Java machine learning toolkit for NLP), machine learning for language toolkit (Mallet), natural language toolkit (suite of Python libraries for NLP), and DBpedia Spotlight (open-source tool for named entity recognition and named entity linking) [13]. There are several applications for automatic information extraction for various domains like extracting the information from published research articles and books for sake of better indexing and visibility of the articles to the scholars and readers in the same area in a different language [14–17]. The information extraction is also potentially utilized for sake of better classification of the published document [18–21]. Moreover, it is potentially used for establishing digital libraries to build knowledge-based systems [22–24]. In this project, we will apply the NER subtask. NER are systems required to identify the named entities that appear in the text, particularly to find person (PER), organization (ORG), location (LOC), and geopolitical entities (GPE). In many ways, Web 3.0 technologies are supportive for establishing the security functions that are significantly approved from Web 2.0 [25, 26].

3 Proposed System Methodology

N-tier layered infrastructure considerably improves security and application flexibility. It can operate the additions and update process without changing the entire application and that leads to a sustainable system. As for security, the independence of layers support identifies security threats in a single layer which ease solving them. The architecture used for this system has three layers with separate functionalities.

3.1 Presentation Layer

It interacts with the user to display data by taking inputs from users via a graphical user interface (GUI). This layer is all widgets and tier states. We are going to use business logic component (BLoC) which are separated into three core components as follows (Fig. 1):

- States are to deliver values (variables) to the widgets.
- Events are equal to methods inside. These activate logic inside the BLoC and can carry some raw data (like a string from a TextField).
- BLoC is not a part of the presentation layer, but it implements logic depend on the incoming events.

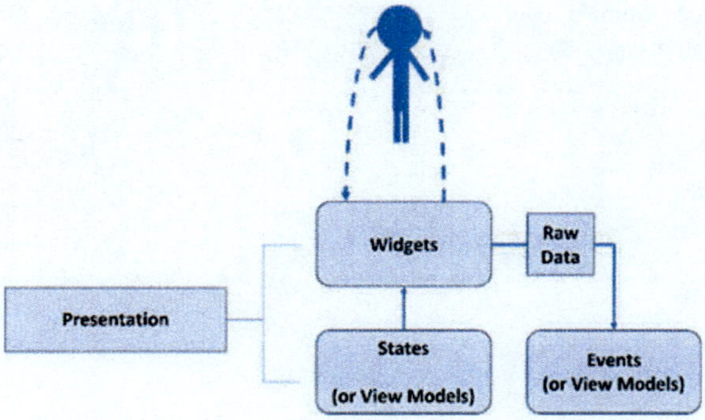

Fig. 1 Presentation layer diagram

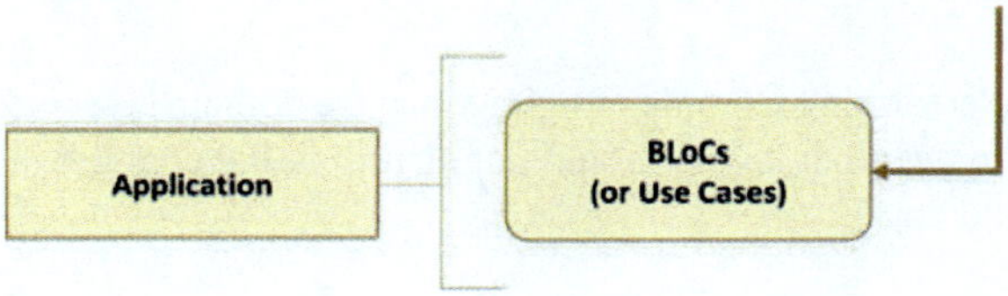

Fig. 2 Application layer diagram

3.2 Application Layer

The responsibility of the application layer is to decide what to do next with the data. It does not execute any complex logic, but instead, it just assures the validation of user input (taken from the domain layer) and manages contributions to infrastructure data (Fig. 2).

Events are sent from the presentation layer.

3.3 Domain Layer

It uses to perform and manage user requests by behaving as a link between the interfaces and the database. Then, the database will be used to retrieve the required data by the user and save the updated data. The domain layer is an independent layer that is the center of the app. Similar approaches are used for the video data as well [27] (Fig. 3).

This is where the app's logic is, including:

- Validating Data: It will encapsulate a string value and assure the string is not empty and validate the characters' length limitation. This kind of validation is usually the UI task; however, we are going to take advantage of the Dart type system by applying the extra validation.
- Transforming Data: between user and database, and vice versa.
- Grouping or uniquely identifying data that fit in together.

Fig. 3 Domain layer diagram

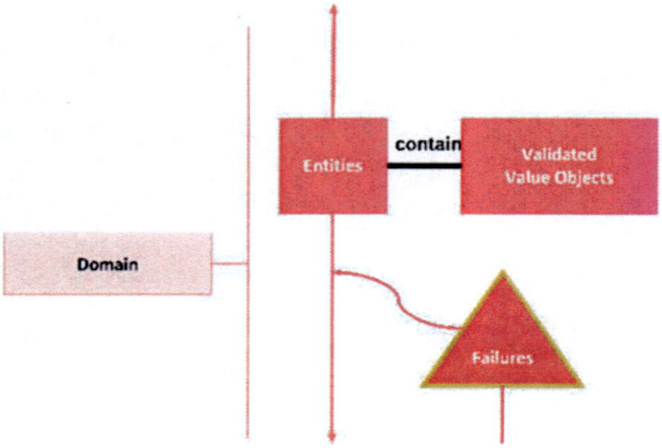

- Performing Complex Logic: Since the domain layer is the heart of the app, changes in the other layers should not disturb it, but changes in the domain affect all other layers.

3.4 Data Layer (Infrastructure)

The user sends queries to the database NoSQL, and then the database sends the outcomes back. Just as in the presentation, this layer is at the boundary of our app, but instead of dealing with the user input\output, it deals with APIs, libraries, and databases (Fig. 4).

The infrastructure layer is consisting of two parts: low-level data sources and high-level repositories. Also, this layer maintains data transfer objects (DTOs) whose only purpose is to convert data from the domain layer. Figure 5 shows design architecture.

Data sources are used in the lowest level. Remote data sources match JSON response strings from the server into DTOs and operate server requests with DTOs converted to JSON. Likewise, local data sources retrieve data from a local database. Repositories have the important task of being the border between the domain and application layers. It is the repository's job to do the caching logic and coordinate data from the remote data source to the local source.

Subsystem Architecture. In this section, we will provide a detailed data flow diagram (DFD) and major processes about the system. It shows the functions of the system and how to flow and store data in the system. Eventually, this section describes the association between the users with the system.

Meeting Transcription Subsystem. After the recording or uploading process, the application can transcribe the audio file by sending it to the converting model via a server. The application splits the meeting audio into segments, then converts each segment to a text, and recognizes who said it. The speaker's name, speaking time, and the segment audio will be assigned to each speaker's transcribed text (Fig. 6).

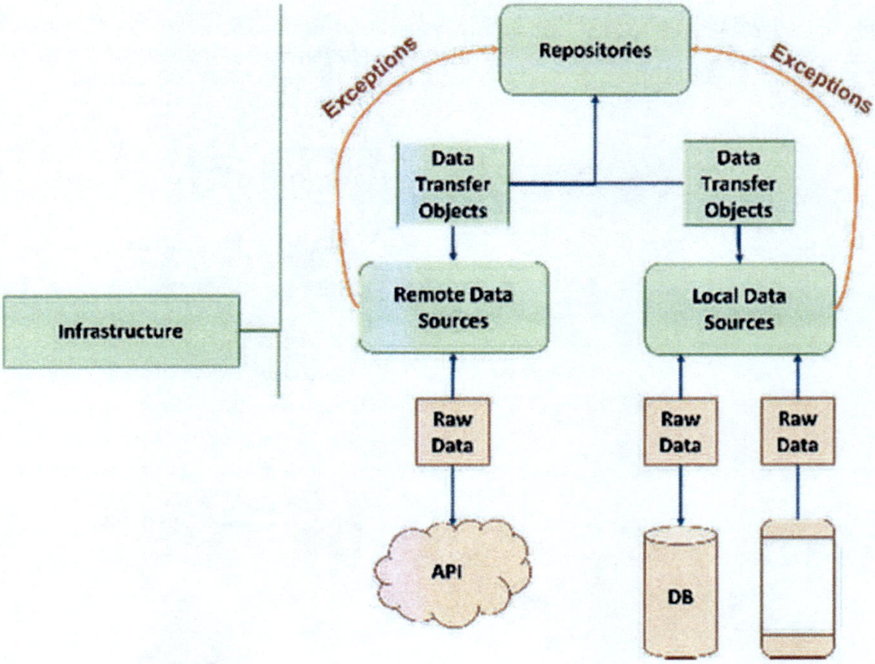

Fig. 4 Data layer diagram

3.5 General View of the System

DFD below explains how the system operates and how the users exchange information within the system. Context DFD also shows the application's limitations and its relations with its environment (Fig. 7).

4 Application Functionalities and Limitations

The following list is comprised of the potential application functionalities:

- Create Account: A new user can create an account by providing the required information.
- Login: Registered users can log in to their account by username and password.
- Update Profile: Registered users can change their personal information.
- Add a Meeting: After providing the required meeting information, registered users can upload or record a meeting.
- Transcribe a Meeting: A meeting is transcribed by converting the audio to text and structured into a transcript format.

Fig. 5 N-tier infrastructure

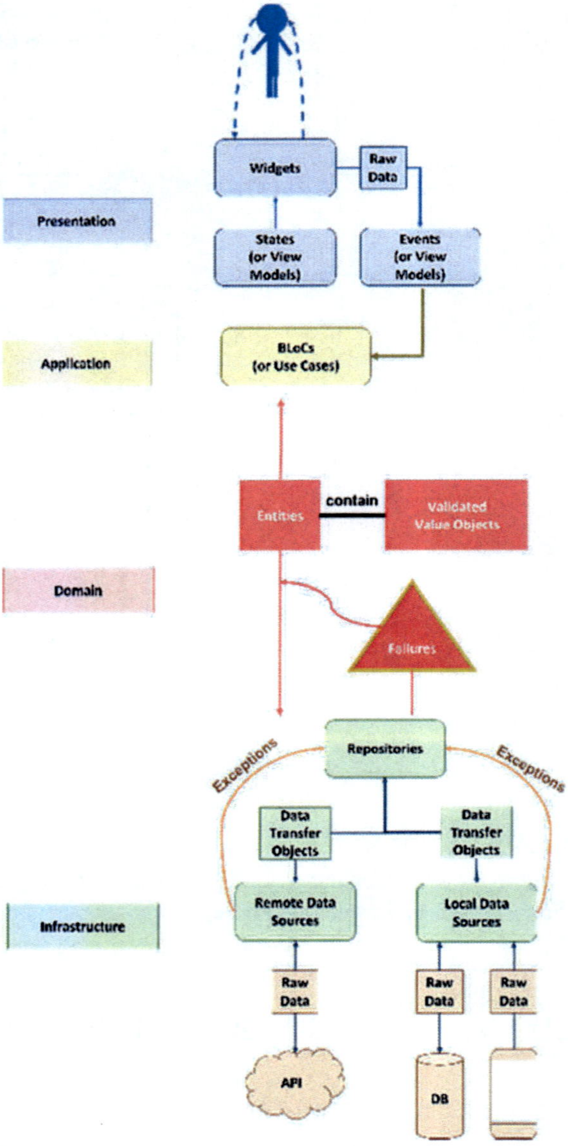

- Analyzing a Meeting: The important information is extracted from the meeting transcription.
- View Meetings and View a Meeting: Registered users can view all their meetings and view their transcript and summary analysis.
- Edit Meeting Transcript: Registered users can edit the meeting transcript.
- Search Meetings: Registered users can search for a meeting by the title.
- Delete a Meeting: The user can delete a meeting.

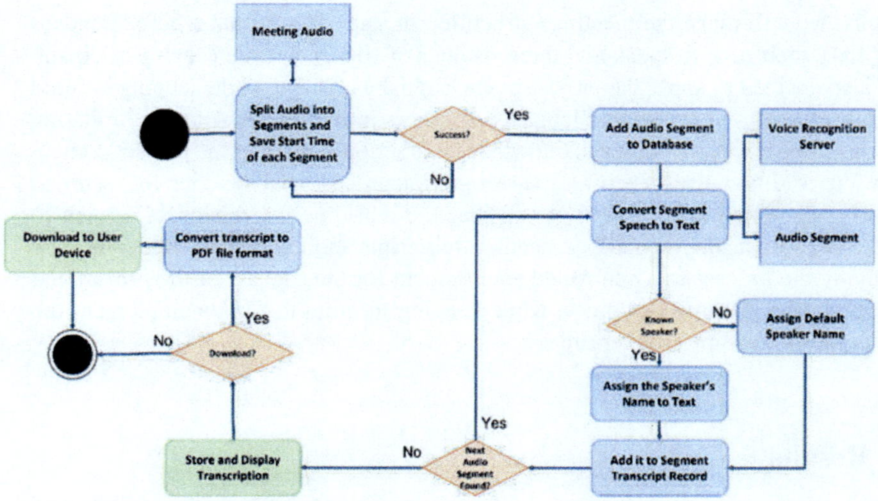

Fig. 6 Detailed diagram of the subsystem

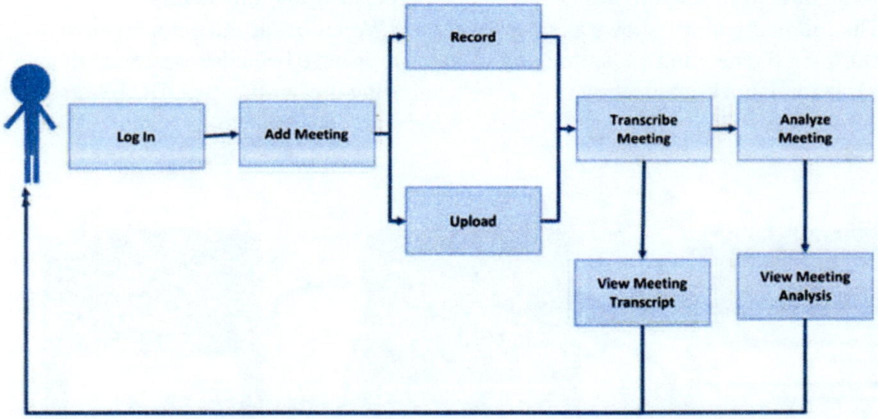

Fig. 7 Flow of the process

- Approval of minutes: It can be used for minutes' approval by the attendees.

Meeting transcripts differ significantly from normally written texts that some differences will have a negative impact on the process of keyword extraction such as lacking the structure of sentences or the whole text, having different speaking styles participants, and containing incomplete thoughts and sentences. In addition to the possibility of a high error rate in conversion from speech to text that will lead to complex results in understanding the data [28], the challenges arising from the conversion of speech to text may delay our project, so we made few decisions to simplify the project to reach the outcome in the given parameters. Thus, for better results, we will make the meetings structured in a specific format where attendees will have their time to speak and three seconds of silence between each participant. We also decided to apply the project in the English language while keeping in mind to further deploy it for Arabic language speakers in the future. Another limitation for this project when it comes to knowledge extraction is making the project domain focused. Additionally, speech-to-text recognition process shall be done for recorded audio only. This project does not consider the real-time conversion of speech to text. We are focusing on making meeting transcripts and their outline efficiently by applying the process to a controlled environment for the success of the project and achieving the learning outcomes, while keeping in mind to further implement the prototypes may improve over time.

5 Results

WeScribe application aims to serve meeting-based organizations, it also harnesses its features for the benefit of the customer and presents it as a product, and the integration between these different models had been done efficiently and effectively.

The following figure shows a few interfaces of WeScribe, as the main features of transcribing the meeting and extracting its key words have been demonstrated in an example for a mock-up meeting run by the team using pre-written text. IBM Watson speech-to-text API has been used, with an average accuracy up to 80% scalable (Fig. 8).

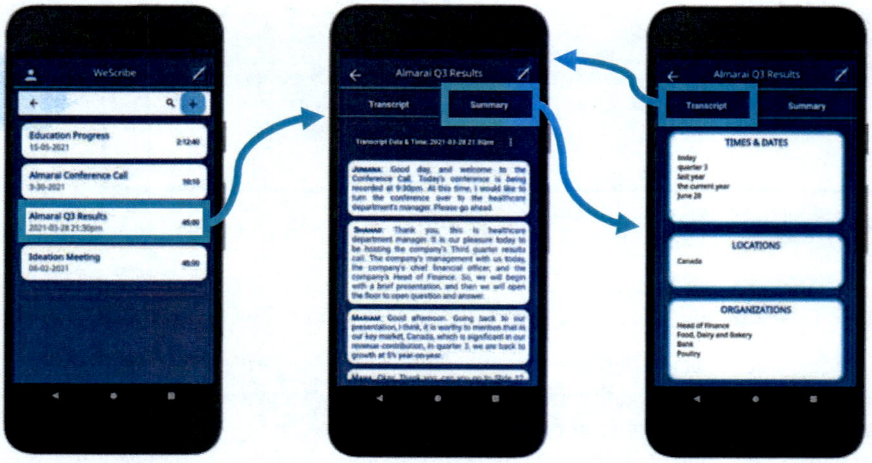

Fig. 8 Click on meeting to view, by switching between tabs, you can view transcript and summary

6 Conclusion

This paper focuses on the presentation of the proposed idea of a working meeting documentation solution called WeScribe that converts audio records of meetings to transcripts and provides its summary. The application's purpose is to facilitate notes and minutes taking of the regular meetings of any category for companies and individuals. For sake of implementation, the combination of several technologies has been investigated. That includes processing audio to text with voice recognition of speakers in the meeting and natural language processing to summarize important notes and highlight the important entities. Those entities are events, dates, locations, and organizations mentioned, in addition to counting the most used/frequent words overall. The application documents meetings in a sufficient pattern accessed by all meeting members providing significant results.

References

1. Gutmann, J. (2006). Taking minutes of meetings.
2. Rahman, A., Qureshi, I. M., Malik, A. N., & Naseem, M. T. (2014). Dynamic resource allocation for OFDM systems using DE and fuzzy rule base system. *Journal of Intelligent & Fuzzy Systems (JIFS), 26*(4), 2035–2046. http://doi.org/10.3233/IFS-130880
3. Rahman, A., Qureshi, I. M., Malik, A. N., & Naseem, M. T. (2014). A real time adaptive resource allocation scheme for OFDM systems using GRBF-neural networks and fuzzy rule base system. *International Arab Journal of Information Technology (IAJIT), 11*(6), 593–601.
4. Rahman, A. (2020). GRBF-NN based ambient aware realtime adaptive communication in DVB-S2. *Journal of Ambient Intelligence and Humanized Computing, 2020*(12), 1–11.
5. Rahman, A., Dash, S., & Luhach, A. K. (2021). Dynamic MODCOD and power allocation in DVB-S2: a hybrid intelligent approach. *Telecommunication Systems, 76,* 49–61. http://doi.org/10.1007/s11235-020-00700-x
6. Summerfield, W., & Moss, J. (2019). US10304464B2—Voice recognition system and methods.
7. Ahmad, M., Farooq, U., Rahman, A., Alqatari, A., Dash, S., & Luhach, A. K. (2019). Investigating TYPE constraint for frequent pattern mining. *Journal of Discrete Mathematical Sciences and Cryptography, 22*(4), 605–626.
8. Rahman, A., Sultan, K., Aldhafferi, N., & Alqahtani, A. (2018). Educational data mining for enhanced teaching and learning. *Journal of Theoretical and Applied Information Technology, 96*(14), 4417–4427.
9. Aldhafferi, N., Alqahtani, A., Rahman, A., & Azam, M. (2018). Constraint based rule mining in patient claim data. *Journal of Computational and Theoretical Nanoscience, 15*(2), 1064–1071.
10. Rahman, A., & Dash, S. (2017). Big data analysis for teacher recommendation using data mining techniques. *International Journal Control Theory and Applications, 10*(18), 95–105.
11. Rahman, A., & Dash, S. (2017). Data mining for students' trends analysis using Apriori algorithm. *International Journal Control Theory and Applications, 10*(18), 107–115.
12. Singh, S. (2018). Natural language processing for information extraction. https://arxiv.org/abs/1807.02383%0A. Accessed October 19, 2020.
13. Zaman, G., Mahdin, H., Hussain, K., Rahman, A., Abawajy, J., & Mostafa, S. A. (2021). An ontological framework for information extraction from diverse scientific sources. *IEEE Access, 9*, 42111–42124. https://doi.org/10.1109/ACCESS.2021.3063181
14. Alamoudi, A., Alomari, A., Alwarthan, S., & Rahman, A. (2021). A rule-based information extraction approach for extracting metadata from PDF books. *ICIC Express Letters: Part B: Applications, 12*(2), 121–132.

15. Zaman, G., Mahdin, H., Hussain, K., & Rahman, A. (2020). Information extraction from semi and unstructured data sources: A systematic literature review. *ICIC Express Letters, 14*(6), 593–603.
16. Musleh, D., Ahmed, R., Rahman, A., & Alhaidari, F. (2019). A novel approach to arabic keyphrase extraction. *ICIC Express Letters, 10*(10), 875–884.
17. Rahman, A., Dash, S., Luhach, A. K., Chilamkurti, N., Baek, S., & Nam, Y. (2019). A neuro-fuzzy approach for user behavior classification and prediction. *Journal of Cloud Computing, 8*(17).
18. Rahman, A. (2016). Knowledge representation: A semantic network approach. In *Handbook of research on computational intelligence applications in bioinformatics* (1st ed., Chap. 4). IGI Global.
19. Rahman, A., Alrashed, S. A., & Abraham, A. (2017). User behavior classification and prediction using FRBS and linear regression. *Journal of Information Assurance and Security, 12*(3), 86–93.
20. Rahman, A., & Alhaidari, F. A. (2018). Querying RDF data. *Journal of Theoretical and Applied Information Technology, 26*(22), 7599–7614.
21. Rahman, A., & Alhaidari, F. A. (2019). The digital library and the archiving system for educational institutes. *Pakistan Journal of Information Management and Libraries (PJIM&L), 20*(1), 94–117.
22. Zaman, G., Mahdin, H., Hussain, K., Rahman, A., Ibrahim, N., & Safar, N. Z. M. (2020). Digital library of online PDF sources: An ETL approach. *IJCSNS International Journal of Computer Science and Network Security, 20*(11), 172–181.
23. Faisal, H. M., Ahmad, M., Asghar, S., & Atta-ur-Rahman. (2017). Intelligent Quranic story builder. *International Journal of Hybrid Intelligent Systems*, 1–8.
24. Saqib, N. A., Salam, A. A., Rahman, A., & Dash, S. (2021). Reviewing risks and vulnerabilities in web 2.0 for matching security considerations in web 3.0. *Journal of Discrete Mathematical Sciences and Cryptography, 24*(3), 1–17.
25. Rehman, S. U., Mahmud, M., Rahman, A., Haq, I. U., & Safdar, M. (2021). Information security in business: A bibliometric analysis of the 100 top cited articles. *Library Philosophy and Practice (e-journal), 5354.*
26. Dilawari, A., Khan, M. U. G., Al-Otaibi, Y. D., Rehman, Z., Rahman, A., & Nam, Y. (2021). Natural language description of videos for smart surveillance. *Applied Sciences, 11*(9), 3730. https://doi.org/10.3390/app11093730
27. Alhaidari, F. A., Rahman, A., Alghamdi, A., & Dash, S. (2019). Motion detection in digital video recording format with static background. In *SEAHF 2019*. http://doi.org/10.1007/978-3-030-22964-1_2
28. Liu, Y., Liu, F., & Liu, F. (2011). A supervised framework for keyword extraction from meeting transcripts. *IEEE Transactions on Audio, Speech, and Language Processing*. http://doi.org/10.1109/TASL.2010.2052119
29. Yoshioka, T., et al. (2019). Meeting transcription using asynchronous distant microphones. In *Proceedings of Annual Conference of the International Speech Communication Association, INTERSPEECH* (Vol. 2019, pp. 2968–2972). http://doi.org/10.21437/Interspeech.2019-3088

Customer Churn Prediction in Banking Industry Using Power Bi

Awe M. Oluwatoyin, Sanjay Misra, John Wejin, Abhavya Gautam, Ranjan Kumar Behera, and Ravin Ahuja

Abstract The development of technology in our modern day has led to the generation of huge data. This is evident by the 2.5 quintillions of data generated by persons connected to the Internet per day in 2020. With the expectation of 5.3 billion Internet users by 2023, complex and efficient tools, models, or approaches that will explore, analyze, and produce meaningful hidden information from huge data are needed. In recent years, machine learning techniques such as logistic regression, decision trees, and clustering are beginning to gain relevance, especially in churn prediction. Customer churn prediction is the process of determining the proportion of clients who avoid or might stop using or subscribing to a product or service offered by an organization or company. Though various prediction models have been proposed, most research attention has been given to measuring the efficiency of prediction models, rather than identifying its application for sustainable economic development. In this paper, we investigate the determining factor for customer attrition in the banking sector using Power BI. Dataset from United Bank of Africa (UBA), Nigeria was preprocessed with four key customer variables were used. The decision

A. M. Oluwatoyin · J. Wejin
Center of ICT/ICE, CUCRID, Covenant University, Ota, Nigeria
e-mail: jonathan.oluranti@covenantuniversity.edu.ng

J. Wejin
e-mail: john.wejinpgs@stu.cu.edu.ng

S. Misra (✉)
Department of Computer Science and Communication, Østfold University College, Halden, Norway
e-mail: sanjay.misra@hiof.no

A. Gautam
Guild Insurance Group, Brandon, Canada

R. K. Behera
Birla Institute of Technology, Mesra, India
e-mail: ranjan.behera@bitmesra.ac.in

R. Ahuja
Delhi Skill and Entrepreneurship University, New Delhi, India

P. K. Singh et al. (eds.), *Proceedings of Third International Conference on Computing, Communications, and Cyber-Security*, Lecture Notes in Networks and Systems 421,
https://doi.org/10.1007/978-981-19-1142-2_60

tree algorithm available in the Power Bi software was employed for training and testing. The results show that customer account balance is a key determining variable for churning. Furthermore, the results show that churning occurs less in male than female clients. This work will provide banks with useful knowledge on building effective customer retention strategies. Building an effective and accurate customer churn prediction model is an important research problem for both academics and practitioners.

Keywords Customer churn · Machine learning · Decision trees

1 Introduction

Technology has improved the way humans live in various facets of life endeavors. This is evident by the 2.5 quintillions of data generated per person in the year 2020 [1]. Furthermore, it is estimated that by 2023, there will be 5.3 billion Internet users, and over, 70% of the world's human population will have mobile connectivity [2]. With huge data generation on the increasingly complex and efficient tools, models, or approaches that will explore, analyze, and produce meaningful hidden information, for productivity and business efficiency. One of the approaches in exploring the hidden information in data is machine learning-based data mining techniques. Data mining is a technique that extracts meaningful facts from data. This approach has been applied in various fields of human endeavor [3] which include the medical sector [4], retail services [5], and financial analysis and forecasting [6]. In recent years, data mining using machine learning techniques such as logistic regression, decision trees, and clustering is beginning to gain relevance, especially in churn prediction [7].

Customer churn or customer attrition [8] can be defined as the proportion of consumers who avoid using or subscribing to a product or service offered by an organization or company. Managing customer attrition is very vital, especially for the banking industry where huge data are analyzed, with the intention of removing information for efficient and profitable activities [9].

In this paper, we investigate the determining factor for customer attrition in the banking sector using Power BI. Our study, using the machine learning technique, aims to provide the banking industries with knowledge on how to build an effective strategy for customer retention.

The rest of the paper is organized as follows: We review related works in Sect. 2. The materials and methodology used in this study is described in Sect. 3. Findings from our study are discussed in Sect. 4, while the conclusion and future research directions are presented in Sect. 5.

2 Related Works

The authors in [10] proposed a clustering algorithm based on a Hadoop map-reduced structure. Their study demonstrates that leveraging on Hadoop structure enhances the accuracy of clustering algorithms in churn prediction. Hossam in [11] put forward a smart neural network model for customer attrition prediction based on particle swarm optimization (PSO) and feed-forward neural network. This model proposed handles the imbalanced class dissemination of the information using a progressed oversampling strategy. Compared to other modern classifiers, their model achieves an improvement in the inclusion rate of churn.

The study in [12] used 3333 clients' data to investigate the churn rate using support vector machines (SVMs). The proposed model performs best with a gain measure metrics of 88.56%.

Qureshi in [13] compared the accuracy of are decision trees, regression analysis, artificial neural networks (ANNs), and KNN in churn prediction using a dataset containing 106,000 client's traffic data. The findings show that decision trees are usually the most reliable algorithm classifier for the identification of possible churners in the dataset used.

In the Olle and Cai research [14], a consumer churn analysis was performed with a hybrid model strategy using a 6-month dataset with 2000 subscriber records and 23 features. The prediction of the (hybrid) cross-breed model outperforms that of a lone method. There are several other works related to customer churn [15] available in the literature.

From the reviewed studies, it can be observed that most researchers tend to explore the accuracy or performance of a particular algorithm or model rather than determining factors that influences customers to attrite from using a particular product or service. This paper seeks to bridge this gap.

3 Methodology

The methodology used in this research is shown in Fig. 1. The dataset used in this paper was collected from customer database of Union Bank of Africa (UBA) in Nigeria. It consists of 1,048,576 clients' with more than eleven features. To increase accuracy, features irrelevant to our study were removed. After data cleaning, four features: age, gender, account balance, and status of account were used. Microsoft PowerBI™ was used for the data preparation.

First, a connection to the data source was made. The power query editor was used for data profiling and transformation. The high number of missing values in the datasets was removed. The dataset columns were then mapped to elements of a common. The dataset was split into a training and test datasets with a ratio of 70:30. Since our study uses a yes or no approach to determine whether a feature is

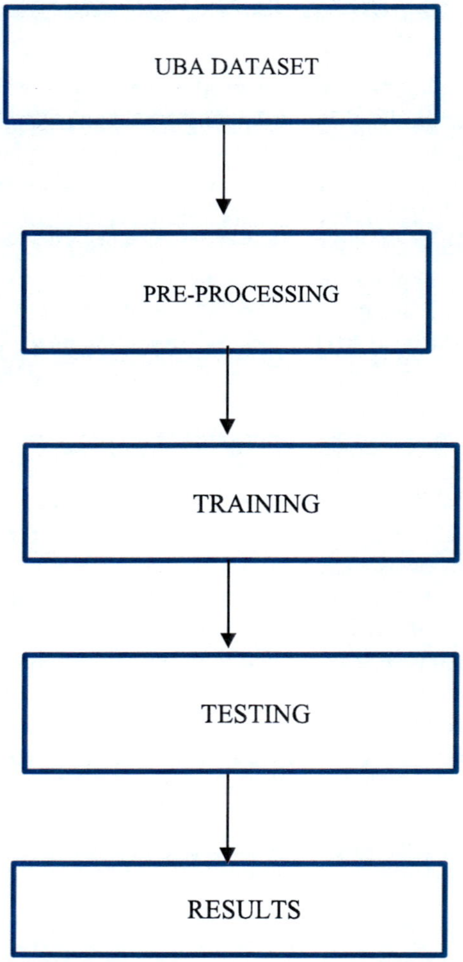

Fig. 1 Churn prediction flow process

churn or not, logistic regression was used for validation using Python. Decision trees algorithm was run on each training.

4 Results and Discussion

This study uses a machine learning algorithm called the decision tree in PowerBI™ to predict customer churn rate in the banking industry. The result in Fig. 1 gives a general outlook on the churn status of the customers. It indicates that 20.21% of customer transactions is churn, while 79.79% of customer transactions was churn.

The target variable was set to churn status against customers' balance. As shown in Fig. 2, customers with a balance of less than 2 million in their account were churned.

Fig. 2 Overview of customer churn

Churn Status	Gender	Balance	Transcations
No	Female	₦242,307,781.770	2219927
No	Male	₦336,963,016.480	2970780
Yes	Female	₦101,412,732.230	736800
Yes	Male	₦84,175,362.400	577781
Total		**₦764,858,892.880**	**6505288**

This indicates that the account balance of a customer is a determining factor in the churning rate of a customer in the banking industry.

Figure 3 shows that the customers within the age bracket less than 43 years have a lesser churn rate. The age feature gains slight significance only when the customer is within the 43–46 yrs and has a lesser amount in his account. When the target features are balance, age and gender or balance age, gender, and tenure as indicated in Fig. 4, gender feature has more effect on churning rate than the balance and age. It can also be seen that the tenure feature doesn't have any effect on the churning rate. Furthermore, when the target variables include balance, age, gender, tenure, and transaction, the balance feature becomes the most determinant factor of the customer churning rate. It can be concluded from the results that the customer balance is a key factor that determines the churning rate in the banking industry (Figs. 5 and 6).

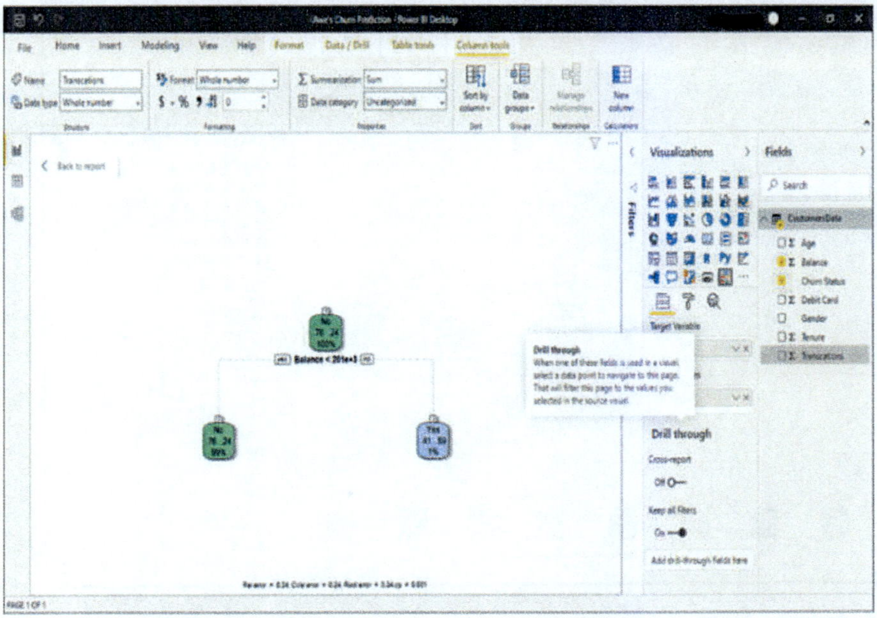

Fig. 3 Churn status versus balance

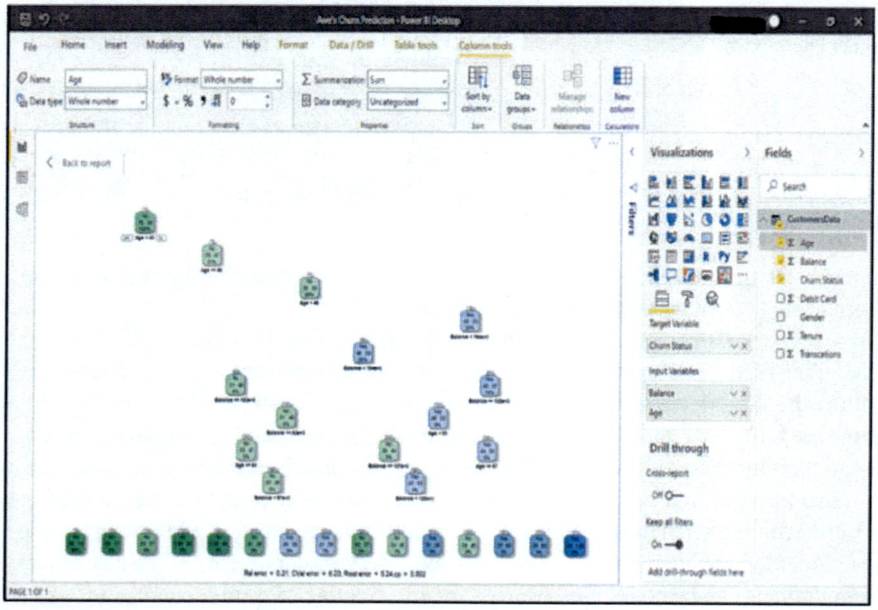

Fig. 4 Churn status versus age and balance

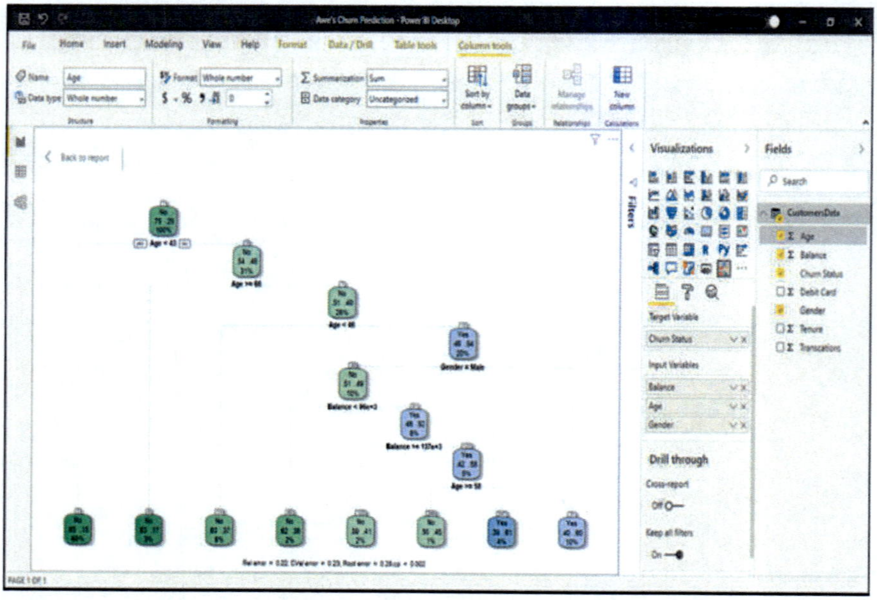

Fig. 5 Churn status versus age and balance and gender

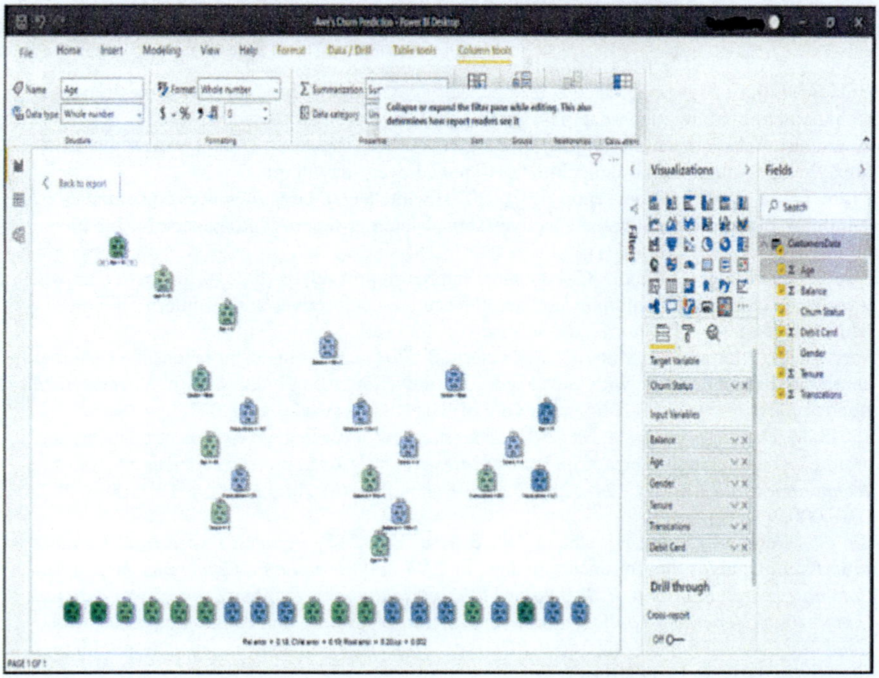

Fig. 6 Churn status against age, balance, gender, tenure, transaction, and debit

5 Conclusion and Future Work

We live in an era where data are the fuel that drives business innovation and capital base for businesses. This makes churn predictive models and approaches a vital tool for efficient and profitable market campaign strategies. This paper investigates the most determining factor that influences customers in the banking industry to churn. Churning dataset from the UBA dataset was preprocessed, and important customer features were used. The decision tree algorithm available in the Power Bi software was employed for training and testing. The results show that customer account balance is a key determining factor for churning. Furthermore, the results show that churning occurs less in male than females. This work will provide banks with useful knowledge on building effective customer retention strategies. Building an effective and accurate customer churn prediction model is an important research problem for both academics and practitioners.

References

1. Bulao, J. (2020). How much data is created every day in 2020? *techjury*. Available at: https://techjury.net/blog/how-much-data-is-created-every-day/#gref. Accessed 3/21/2021.
2. How much data is created every day in 2021? [You'll be shocked!] (*techjury.net*). Available: https://techjury.net/blog/how-much-data-is-created-every-day/#gref
3. Cisco. Cisco annual internet report (2018–2023) white paper. https://www.cisco.com/c/en/us/solutions/collateral/executive-perspectives/annual-internet-report/white-paper-c11-741490.html. Accessed 3/21/2021.
4. Cisco Annual Internet Report. Cisco annual internet report (2018–2023) white paper. https://www.cisco.com/c/en/us/solutions/collateral/executive-perspectives/annual-internet-report/white-paper-c11-741490.html. Last accessed on 15.11.21.
5. Verbeke, W., Martens, D., Mues, C., & Baesens, B. (2011). Building comprehensible customer churn prediction models with advanced rule induction techniques. *Expert Systems with Applications, 38*(3), 2354–2364. https://doi.org/10.1016/j.eswa.2010.08.023
6. Tafish, M. H., & El-Halees, A. M. (2018). Breast cancer severity degree predication using data mining techniques in the Gaza strip. In *2018 International Conference on Promising Electronic Technologies (ICPET)* (pp. 124–128), Deir El-Balah, Palestine. http://doi.org/10.1109/ICPET.2018.00029
7. Lin, T., Wen, B., Chang, H., Chang, W., & Hsu, S. (2017). A rapid incremental frequent pattern mining algorithm for uncertain data. In *2017 5th International Conference on Applied Computing and Information Technology/4th International Conference on Computational Science/Intelligence and Applied Informatics/2nd International Conference on Big Data, Cloud Computing, Data Science (ACIT-CSII-BCD)* (pp. 284–288), Hamamatsu, Japan. http://doi.org/10.1109/ACIT-CSII-BCD.2017.61
8. Liu, J., & Huang, Z. (2018). Study on the application of data mining method in enterprise financial management. In *2018 IEEE International Conference on Advanced Manufacturing (ICAM)* (pp. 332–334), Yunlin, Taiwan. http://doi.org/10.1109/AMCON.2018.8614814
9. Olaniyi, A. S., Olaolu, A. M., Jimada-Ojuolape, B., & Kayode, S. Y. (2020). Customer churn prediction in banking industry using K-means and support vector machine algorithms. *International Journal of Multidisciplinary Sciences and Advanced Technology, 1*(1), 48–54.
10. KDnugget. Customer churn prediction using machine learning. KDnugget, 2019 white paper. Last accessed on 15.11.21. https://www.kdnuggets.com/2019/05/churn-prediction-machine-learning.html
11. Hossam, F. (2018). A hybrid swarm intelligent neural network model for customer churn prediction and identifying the influencing factors. *Information Journal, 9*(288), 2–18. https://doi.org/10.3390/info9110288
12. Bi, W., Cai, M., Liu, M., & Li, G. (2016). A big data clustering algorithm for mitigating the risk of customer churn. *IEEE Transactions on Industrial Informatics, 12*(3), 1270–1281.
13. Qureshi, S. R. A. Q. A. K. A. (2013). Churn prediction model using machine learning. In *International Conference on Digital Information Management* (Vol. 8, pp. 131–136).
14. Olle, G., & Cai, S. (2014). A hybrid churn prediction. *International Journal of e-Education, e-Business, e-Management and e-Learning, 4*(1), 52–62.
15. Odusami, M., Abayomi-Alli, O., Misra, S., Abayomi-Alli, A., & Sharma, M. M. (2020, December). A hybrid machine learning model for predicting customer churn in the telecommunication industry. In *International Conference on Innovations in Bio-Inspired Computing and Applications* (pp. 458–468). Springer.

Issues in Credit Card Transactional Data Stream: A Rational Review

Rinku, Sushil Kumar Narang, and Neha Kishore

Abstract Online transactions are trending worldwide now and in future developments. A big amount of transactional data is generated like networking, stock market, telecommunications, and weather forecasting. This data can be classified for the knowledge extraction and learning. Credit card nowadays is very easy methods for physical and online transactions. Transactions using a credit/debit card having some advantages and flaws. Some of the problems with the credit card transactions are also highlighted here, further in the paper focused on the extensive studies of the various learning methods used by various authors on the imbalanced data stream of the credit card transactions.

Keywords Credit card transaction · Online transaction · Imbalanced data streaming · Learning methods

1 Introduction

With the development of digitalization, deals through the Web got one of fundamental business strategies for the organizations, relations, and government offices to expand their effectiveness in global transactions. In case of online business, the use credit card is increasing day by day due to the security and effectiveness. As we say about the online transactions of the money from one account to the other account, there we additionally need to think about the fraudulent transactions. As credit card transaction is the most well-known technique for payments in the ongoing years.

Rinku (✉)
Chitkara University School of Computer Applications, Chitkara University, Solan, Himachal Pradesh, India
e-mail: rinku.cse@chitkarauniversity.edu.in

S. K. Narang · N. Kishore
Chitkara University School of Engineering and Technology, Chitkara University, Solan, Himachal Pradesh, India

P. K. Singh et al. (eds.), *Proceedings of Third International Conference on Computing, Communications, and Cyber-Security*, Lecture Notes in Networks and Systems 421,
https://doi.org/10.1007/978-981-19-1142-2_61

2 History of Credit Cards

Prior to the introduction of credit cards, the main installment strategies were money, check, credit extension, or a credit account. In the mid-1900s, oil organizations and other corporate monsters presented card spending by means of their own exclusive cards [1]. The exclusive cards were classified "credit cards," and they extraordinarily decreased administrative bookkeeping blunders and expanded client spending [2]. The goal behind giving these cards was to give accommodation to their clients as well as to energize client dependability. Around the 1940s, money related go between entered the credit card industry. Their administrations included dealing with the bookkeeping and bill gathering for business and client exchanges. These go between would then work with different vendors around town which expanded clients' possibilities for shopping. Toward the day's end, organizations would turn in their business slips to the bank, including the "credit" transactions, and afterward, the bank would gather the cash from the client toward the month's end. In 1946, the main bank card was presented. As indicated by MasterCard (2017), a New York broker by the name of John Biggins chose to give a card that when utilized by qualified clients at taking part organizations, the bank would repay the vendor and would then gather the cash from the client toward the month's end. Qualified clients were clients who were at Biggins' bank. In the event that a client did not save money with Biggins, at that point, they could not utilize this administration. The Diners Club, the principal eating/travel credit card, was created during the 1950s by Frank McNamara. While going out to eat with his significant other, he understood he left his wallet in another suit. His significant other took care of the bill yet McNamara was so humiliated by this that he went to the proprietor of the eatery and got some information about tolerating a multi-reason credit card as installment. The proprietor was available to the thought and McNamara called different speculators to back and build up the Diners Club card in its first year, the Diners Club had more than 20,000 individuals and in its subsequent year, its part base developed to more than 42,000. Obviously, charging merchandise and enterprises turned out to be progressively well known, and now, it is the favored strategy for installment by most purchasers around the globe.

3 Credit Card Processing

There are four primary players associated with credit card handling. The first is the giving bank. The giving bank (in future alluded to as the backer) is answerable for stretching out the credit extension to the client. Normally, backers are banks themselves, and they decide a client's credit extension, loan cost, and last endorsement for products and enterprises bought with the card. The subsequent player is the processor. The processor gives a system to the client's Visa data to be transmitted from the trader to the guarantor. The processor can be thought of as the go between during the 1940s. The four significant processors in the United States are Visa, MasterCard, American

Express, and Discover. The third player is the guarantor's misrepresentation security group. This administration is customarily performed by an outsider. Without a doubt, the calculations are intricate and are unquestionably safely secured. The main individuals aware of such safety efforts are officials. Infrequently, if at any time, are any delegates at the giving bank mindful of which standard is utilized while deciding if a transaction is fake. The fourth player is the procuring bank. The procuring bank (in future alluded to as the acquirer) is the dealer's bank, and they are liable for sending the client's installment data to the processor. When the exchange gets endorsement from the processor, the client is permitted to leave with the products or have the administrations rendered. Credit card handling experiences six phases. The first is the point at which a client buys products or administrations and pays with a credit card. The second is the point at which the dealer runs the credit card. The client's data are transmitted by means of the vendor's terminal to the acquirer's system. The acquirer advances this data to the processor. The processor's system at that point speaks with the guarantor's misrepresentation recognition organization and trusts that the exchange will breeze through the fraud assessment. When the exchange is considered non-fake, it is sent on to the backer for conclusive endorsement. At the point when this endorsement is sent, the client leaves with the merchandise or gets the administrations gave by the shipper. The backer at that point sends a bill to the client.

4 Types of Credit Card Transactions

Credit card transactions fall under two classifications. The first is card present. All transactions made by swiping or embedding are a card fall under this sort. Because of the ongoing chip security highlight, on the off chance that a vendor does not swipe a card with the chip include, at that point, the trader is exclusively dependable if that exchange is deceitful. On the off chance that they utilized the chip security include, at that point, the guarantor is capable. The subsequent kind is known as card not present. All Web-based business transactions fall under this class, and they are the most vulnerable to credit card fraud. In the event that a deceitful transaction is card not present, at that point, the shipper retains the expenses.

5 Types of Credit Card Fraud

Fraud using the credit card can be classified into two forms. In the two kinds, fraudsters caught touchy credit card data having a place with cardholders. What separates them is the means by which cardholders' data was blocked. The principal type is the most self-evident a normal individual some way or some way or another got cardholders' data. They are a solitary element and are not running a venture or some likeness thereof. These individuals will in general be servers in cafés, programmers, or trash jumpers. The other sort of fraud is acquiesced by the traders themselves. Shippers will take the cardholder's data and adventure it for their potential benefit.

5.1 *Card Not Present*

If a card is not truly present when a customer makes a purchase, the shipper must rely upon the cardholder, or someone showing to be in this way, presenting card information in an indirect way, whether or not by means of mail, telephone, or over the Internet.

5.2 *Skimming*

ATM skimming is a procedure when hoodlums place a gadget on the essence of an ATM, which resembles a piece of the machine as shown in Fig. 1.

It is practically unthinkable for individuals to distinguish the distinction except if they are sharp security onlookers, or the skimmer is of low quality. The lawbreakers regularly shroud a little pinhole camera in a handout holder close to the ATM. It is generally done to remove the focused on casualty's PIN. The camera is avoided the view in this way, when an obscure casualty utilizes their card to make an exchange, the card subtleties, including the pin code number, are caught. The gas siphon tricks are similarly defenseless against this sort of fraud. An electronic methodology for getting a setback's up close and personal information used by character hoodlums. The skimmer is a little contraption that yields a Visa and stores the information contained in the alluring strip.

Fig. 1 ATM card skimming. *Source* https://vajiramias.com/current-affairs/atm-card-skimming/5cde79c71d5def11fddf8e1b/

5.3 Phishing

Phishing email, endpoints, and calls are expected to take currency. Cybercriminals can do this by offering poisonous software program on your PC or removing singular information from your PC. Phishing is such a social structuring attack normally used to take customer data, including login affirmations and Master card numbers (Fig. 2).

Some other frauds related to credit cards are

- Cards stealing
- Cards forging
- Cards fraud by mail
- Credit card footprints
- Card—ID theft
- Spotless fraud
- Friendly fraud.

6 Fraud Detection System

Fraud is as old as possible take a limitless assortment of different structures. Also, the advancement of new innovations gives extra manners by which hoodlums may submit fraud. Fraud detection is, given a bunch of credit card transactions, the way toward distinguishing if another approved transaction has a place with the class of deceitful or certifiable transactions.

7 Based on Supervised Learning

Few techniques which researchers [3–6] have used in credit card fraud detection are

Fig. 2 Phishing. *Source* https://www.duocircle.com/content/protection-from-phishing

- Discriminant analysis
- Decision trees and random trees
- Radial basis function networks
- Meta-classifier
- Bays minimum risk classifier
- Random forest
- Bayesian network classifier
- Artificial neural network
- Deep learning
- Decision tree-based classifier
- Hybrid supervised approach.

8 Based on Unsupervised Learning

Unsupervised learning is valuable in contemplates that need to distinguish changes in conduct or bizarre exchanges. Real named deceitful and ordinary exchanges are not accessible. An underlying arrangement of exchanges considered as would be expected is utilized to begin the classification procedure. Few unsupervised learning discussed by various researchers [7–9]:

- Peer group analysis
- Break point analysis
- Self-organizing map
- Improved competitive learning network
- Adversarial learning.

9 Based on Nature Inspired

- Genetic algorithm
- Artificial immune system.

10 Literature Review

Credit card fraud is perhaps the most severe peril to occupational establishments today. Irrespective, to fight the deception feasibly, it is basic to initially understand the frameworks of executing a cheat. Credit card fraudsters use an incalculable regular strategy to submit blackmail. In essential terms, credit card fraud is described as "exactly when an individual uses another individuals' cards information for singular reasons while the owner of the card and the card patron do not think about how the card is being used" [10].

Frauds using credit cards are given in the supplementary manners:

- A show of illegal precariousness (mislead with desire) by use of unapproved interpretation or conceivably singular information.
- Unlawful or unapproved usage of records for singular get.
- Deception of record evidence to get items and moreover benefits (Table 1).

Table 1 Literature review

Author	Summary
Liu et al. [11]	In this papers, authors discussed in concept of learning with the concept drift from the data stream. Learning with concept drift and the multiclass imbalance is the major focus of the authors
Voican [12]	In this paper, author tries to highlight the need to evaluate the credit card frauds using transactional dataset. The author uses the random forest and support vector machine learning methods for the detection of the credit card frauds
Wu et al. [13]	In this paper, authors introduced a new method for tracking and monitoring the drift into the data stream. They also discussed a mechanism of linked list query and method reliability for the large datasets
Mînăstireanu and Meșniță [14]	In this paper, authors discussed about the various techniques to handle the unbalanced data. The authors precisely implement the classification techniques to find the sensitivity, specificity, precision, recall, and accuracy in the unbalanced dataset in credit card fraud dataset
Lucas et al. [5]	In this research work, authors proposed an HMM-based element designing technique that permits them to join successive information in the exchanges as HMM-based highlights. These HMM-based highlights empower a non-consecutive classifier (random forest) to utilize successive data for the grouping
Gianini et al. [15]	Authors proposed an approach based on estimating the individual rule contribution to the complete pool performance through the Shapley value (SV)
Aldasoro et al. [16]	In this paper, authors discussed about few conflicts of interest in the banking domain. The authors highlighted the conflicts in the financial transactions required to be address
Rtayli et al. [17]	The authors proposed an credit card risk identification (CCRI) model for the affectability of fraud detection. The primary focal points of SVM dependent on RFC are: initially, the model has a decent precision rate to 95%. Furthermore, it diminishes the quantity of bogus positive exchanges by improving the affectability rate to 87% in a huge and lopsided dataset where the pace of fraud is low (<0.17%), which is exceptionally advantageous for the organizations to limit the high credit of examination movement. At last, it has a high rate (91%) in term of grouping execution

(continued)

Table 1 (continued)

Author	Summary
Husejinović [18]	Generally discussed as indicated by the PRC area by and large best performing calculation is packing with a C4.5 choice tree as a base student with a pace of 1.000 for class 0 and 0.825 for class 1. Most elevated review paces of 0.978 for class 0 and 0.829 for class 1 are recorder in the exhibition of the Naive Bayes model. Most elevated accuracy paces of 1.000 for class 0 and 0.927 for class 1 are recorder in the exhibition of the C4.5 choice tree model. In the event that it remain that the dataset is very imbalanced PRC paces of 1.000 for class 0 and 0.825 for the class are very encouraging
Shah et al. [19]	In this paper, a mechanized framework that uses MPV alongside the random forest learning calculation for recognizing fraud proposed. The pre-preparing stage is basic and is all around characterized utilizing clamor taking care of and resizing activity. The got dataset is taken care of into the prepared system for include confusion matrix extraction utilizing the random forest learning calculation and arrangement is performed utilizing MPV. The half and half methodology followed give better outcomes. The fundamental goal of the proposed writing is to make enhanced recognition utilizing random forest for better precision. Higher exactness is accomplished by the utilization of said writing. Later on, the proposed methodology can be inspected against the constant datasets for better assessment of the exactness
Yousefi et al. [20]	In this review, authors saw that administered learning strategies have been utilized more much of the time than solo techniques. Logical regression, SVM, decision tree, and artificial neural network are the generally used techniques for the fraud findings
Jain and Singh [21]	In this paper, according to the authors, credit card fraud can be categorize into two way—application level cheating and transaction level cheating. Author implemented the machine learning technologies for application level fraud
Sangers et al. [22]	Existing algorithms for detection of fraud would exceptionally benefit as of a coordinated effort between business organizations. In any case, the trading of important data is frequently constrained, or not even conceivable, because of protection limitations or confidentiality. This paper showed that protected multiparty calculations can help hold this test
Kim et al. [10]	In the business of the credit, it was a set-up standard to build up the system of detection of fraud as a troupe of various models. The authors led a top to the bottom relative investigation between profound learning and cross breed troupe with different commonsense assessment measurements to figure out which models achieve better than the other on enormous true exchange information

(continued)

Table 1 (continued)

Author	Summary
Carcillo et al. [23]	This paper proposes the execution of a hybrid methodology that utilizes solo anomaly scores to broaden the list of capabilities of a fraud detection. The granularity being referred to ranges from the card level to the global level, thinking about the middle of the road levels of card total through grouping
Shah and Kumar [24]	In this paper, authors discussed about the methods used to identify fraud inside the online transactions. This perspective is basic since the cutting-edge period is moving toward cashless transactions. This viewpoint in spite of the fact that it is improving so does the danger of fraud by malicious clients. This paper gives the subtleties of methods used to distinguish such frauds alongside a bit of flexibility and problem of each. The authors inferred that missing information taking care of alongside constrained uses of detection system causes higher mistake rate and low-characterization exactness
Fiore et al. [9]	In this work, a system is introduced to manage the issue of class lopsidedness in the utilization of administered classification to the recognition of credit card misrepresentation. Given a preparation set, an enlarged set is created, containing more instances of the minority class as for the first
Wang et al. [25]	In this paper, the author talks about a system for a shipper to forestall untrustworthy credit card transactions. They first look at two systems that are utilized by and by no counteraction and utilizing the machine learning identification model for all transactions
Akila et al. [26]	Fraud discovery in credit card transactions is an issue requiring arrangements adjusted to the business objectives of associations, specifically regarding cost. In any case, not very many examination commitments approach the issue from this point of view. The proposed RIBIB design is a troupe-based model joining three commitments in the area of stowing groups. The RIBIB model proposes a compelled sack creation approach that has been specifically intended for taking care of imbalanced information, a hazard incited likelihood-based base student model for diminishing the expected cost and a cost-touchy weighted democratic combiner for second-level cost decrease
Ryman-Tubb et al. [27]	The authors discussed two clarifications for the dissatisfaction of the learning techniques (1) That there is little business motivating force to improve them while fraud stages are decided as an expense of business and are viewed as regularizing (2) Scholarly work around there is troublesome furthermore, minimized regarding subsidizing

(continued)

Table 1 (continued)

Author	Summary
Nami and Shajari [28]	Fraud using payment card is an enormous issue for the financial area. Henceforth, a successful fraud location framework for card installments is required by any bank or money-related organization to lessen the harms brought about by fake exercises. In this paper, the author tried to fraud identification with the type of use of payment card
Mohammed et al. [29]	Ongoing credit card fraud location is a difficult issue due to profoundly imbalanced enormous information. This research paper depends on assessments that thought about a few mainstream machine learning techniques and researched their appropriateness as an "adaptable calculation" when working with exceptionally imbalanced huge or "big" datasets
Khare et al.	In this paper, author implemented the multiple learning mechanism and compare the exactness in the detection of the fraud into the credit card transactional dataset
Carcillo et al. [31]	The paper introduced SCARFF, a unique adaptable framework for the fraud detection using spark
Vikrant Agaskar et al. [32]	This paper has proposed another way to deal with transactions observing and credit card fraud location utilizing unaided learning. It empowers the computerized production of exchange checking rules in a learning procedure and makes conceivable their ceaseless improvement in a domain of powerfully changing data in a mechanized framework
Abdallah et al. [33]	Fraud cases have expanded lately, especially in significant and touchy specialized zones. Thus, there is a critical need to battle fraud. This study article has investigated the best in class fraud recognition frameworks in five regions of frauds. The most normally utilized fraud recognition method is artificial neural networks, decision trees, support vector machines rule-acceptance procedures, strategic relapse, and meta-heuristics, for example, hereditary calculations

Financial organizations such as banks, insurance, and loan companies are found at very high risk of frauds as per the previous studies. The above studies show a vast demand and need to work in financial fraud detections specially in credit card fraud detection because it deals directly to the individuals and the organizations.

11 Critical Areas for Attentions

Building a powerful, constant, and adaptable calculation-based mechanized FDS is exposed to a few difficulties and difficulties counted as follows.

11.1 Concept Drift

FDS focusing on odd conduct experience the ill effects of the way that in reality, the profile of ordinary and deceitful conduct changes with time [30]. For computational procedures, this prompts a non-stationary impact in displaying the connection among ward and target factors [3].

11.2 Imbalanced Classes

Transactions of the credit card data show the highly imbalanced data. There are a huge amount of transactions takes places in the unit of time and containing very few entries need to be checked for fraud [10]. Generally, the percentage of fraudster transactions is found less than one percent, and the traditional methods of learning may not perform as desired [3].

11.3 Deficiency of Real-Time FDS

Multiple FDSs are available those are working on the dataset learning. But frauds can be in the real-time transactions [15]. This detection is compelling in a restricted way to recognize and square false transactions continuously [34].

11.4 Fraud Detection Cost over Expenses

Many-related examinations helpfully overlook the overheads in actualizing FDS [28]. Cost is anyway significant thought while assessing the viability of any arrangement [35].

11.5 Deficiency of Adaptability

Conduct examination-based fraud identification strategies define typical conduct from past real exchanges of a client. Numerous a period client conducts may advance because of outside elements like family conditions, an expansion or reduction in pay, and incessant voyaging [7]. Existing administered and unaided methodologies utilized in misrepresentation location frameworks are not versatile to evolving datasets. In this manner, the efficiency of distinguishing new examples of ordinary and false practices becomes difficult [36].

11.6 Deficiency of Availability of Know-How

Existing fraud detection systems are not made open because of the dread of them being lesser compelling [37]. In this manner, everybody needs to re-concoct the haggle information cannot be utilized [38].

11.7 Unavailability of Open Datasets for Testing

Financial industries do not release their noticeable datasets for open examination. Numerous computational strategies depend on gaining from datasets [39]. Indeed, even a couple datasets that are openly accessible are really a handled type of genuine datasets to conceal genuine factors and their relations [27].

11.8 Deficiency of Aggregation Possibility to Leverage Cross User Data

Ideally utilizing exchange information across card-giving organizations and sorts of cardholders are unrealistic because of an absence of trust among card-giving organizations [40].

11.9 Unavailability of Standard FDS Framework

There are many methods and algorithms present to detect the fraud due to credit card usage. Some authors proposed the frameworks according to their implementation of FDS [10]. There is no fixed framework available according to that the fraud detection system can be designed and implemented [5].

11.10 Problem of Conflict

Credit card transactions are generated with high rate and having conflicts in the transactions [41]. Due to the conflict, many times transactions are declined by either merchant or the server processing the transactions [42].

12 Result and Conclusion

This paper introduced the extensive review of the imbalanced data streaming while credit card transactions are going on. Here, we have highlighted the various issues found in the transaction and the processing of the credit card data. Some of the major issues are discussed by the various authors are highlighted into the above section. This paper enlightens the various transactional issues and their detection methods in case of the credit card transactions in the imbalanced data streaming.

References

1. Gerson, E. S., & Woolsey, B. (2016). *The history of credit cards*. CreditCards.com. http://www.creditcards.com/credit-cardnews/credit-cards-history-1264.php. Accessed January 12, 2017
2. Olaechea, D. (2014). NerdWallet. https://www.nerdwallet.com/blog/credit-cards/issued-first-credit-card/. Accessed January 18, 2019
3. Dal Pozzolo, A., Boracchi, G., Caelen, O., et al. (2018). Credit card fraud detection: A realistic modeling and a novel learning strategy. *IEEE Transactions on Neural Networks and Learning Systems, 29*, 3784–3797. https://doi.org/10.1109/TNNLS.2017.2736643
4. Patil, S., Nemade, V., & Soni, P. K. (2018). Predictive modelling for credit card fraud detection using data analytics. *Procedia Computer Science, 132*, 385–395. https://doi.org/10.1016/j.procs.2018.05.199
5. Lucas, Y., Portier, P. E., Laporte, L., He-Guelton, L., Caelen, O., Granitzer, M., & Calabretto, S. (2020). Towards automated feature engineering for credit card fraud detection using multi-perspective HMMs. *Future Generation Computer Systems, 102*, 393–340.https://doi.org/10.1016/j.future.2019.08.029
6. Mandal, P., Mahata, A., Biswas, B., et al. (2016). A complete literature review on financial fraud detection applying data mining techniques. *International Journal of Trust Management in Computing and Communications, 3*, 336. https://doi.org/10.1504/ijtmcc.2016.10005490
7. Mittal, S., & Tyagi, S. (2019). Computational techniques for real-time credit card fraud detection. *Handbook of computer networks and cyber security—Principles and paradigms* (pp. 653–681). https://doi.org/10.1007/978-3-030-22277-2_26
8. Suman, & Kumar, D. (2016). Performance analysis of various credit card fraud detection approaches: A review. *International Journal of Advance Research in Science and Engineering*, 120–126.
9. Fiore, U., De Santis, A., Perla, F., Zanetti, P., & Palmieri, F. (2019). Using generative adversarial networks for improving classification effectiveness in credit card fraud detection. *Information Sciences, 479*, 448–455. https://doi.org/10.1016/j.ins.2017.12.030
10. Kim, E., Lee, J., Shin, H., et al. (2019). Champion-challenger analysis for credit card fraud detection: Hybrid ensemble and deep learning. *Expert Systems with Applications, 128*, 214–224. https://doi.org/10.1016/j.eswa.2019.03.042
11. Liu, W., Zhang, H., Ding, Z., et al. (2021). A comprehensive active learning method for multi-class imbalanced data streams with concept drift. *Knowledge-Based System, 215*, 106778. https://doi.org/10.1016/j.knosys.2021.106778
12. Voican, O. (2021). Credit card fraud detection using deep learning techniques. *Informatica Economica, 25*, 70–85. https://doi.org/10.24818/issn14531305/25.1.2021.06
13. Wu, Y. m., Chen, L. s., Li, S. b., Chen, J. d. (2021). An adaptive algorithm for dealing with data stream evolution and singularity. *Information Sciences (New York), 545*, 312–330. https://doi.org/10.1016/j.ins.2020.07.010

14. Mînăstireanu, E.-A., & Meșniță, G. (2020). Methods of handling unbalanced datasets in credit card fraud detection. *Brain. Broad Research in Artificial Intelligence and Neuroscience, 11*, 131–143. https://doi.org/10.18662/brain/11.1/19
15. Gianini, G., Ghemmogne Fossi, L., Mio, C., et al. (2020). Managing a pool of rules for credit card fraud detection by a Game Theory based approach. *Future Generation Computer Systems, 102*, 549–561. https://doi.org/10.1016/j.future.2019.08.028
16. Aldasoro, I., Gambacorta, L., & Giudici, P. (2020). The drivers of cyber risk.
17. Rtayli, N., & Enneya, N. (2020). Selection features and support vector machine for credit card risk identification. *Procedia Manufacturing, 46*, 941–948. https://doi.org/10.1016/j.promfg.2020.05.012
18. Husejinović, A. (2020). Credit card fraud detection using naive Bayesian and C4.5 decision tree classifiers. *Periodicals of Engineering and Natural Sciences, 8*, 1–5. https://doi.org/10.21533/pen.v
19. Shah, Y. A., Kumar, S., & Scholar, P. G. (2020). Detecting frauds from credit card transaction using improved
20. Yousefi, N., Alaghband, M., & Garibay, I. (2019). *A comprehensive survey on machine learning techniques and user authentication approaches for credit card fraud detection* (pp. 1–27).
21. Jain, A. S., & Singh, A. (2019). Adaptive credit card fraud detection techniques based on feature selection method. *Recent Advances in Computer Science and Communications.*
22. Sangers, A., van Heesch, M., Attema, T., Veugen, T., Wiggerman, M., Veldsink, J., Bloemen, O., & Worm, D. (2019). Secure multiparty pagerank algorithm for collaborative fraud detection. In *International Conference on Financial Cryptography and Data Security.*
23. Carcillo, F., Le Borgne, Y. A., Caelen, O., et al. (2019). Combining unsupervised and supervised learning in credit card fraud detection. *Information Sciences (New York).* https://doi.org/10.1016/j.ins.2019.05.042
24. Shah, Y. A., & Kumar, E. S. (2019). Online transaction fraud detection mechanisms: a comparative analysis. *Journal of the Gujarat Research Society.*
25. Wang, D., Chen, B., & Chen, J. (2019). Credit card fraud detection strategies with consumer incentives. *Omega (United Kingdom), 88*, 179–195. https://doi.org/10.1016/j.omega.2018.07.001
26. Akila, S., & Srinivasulu Reddy, U. (2018). Cost-sensitive Risk Induced Bayesian Inference Bagging (RIBIB) for credit card fraud detection. *Journal of Computer Science, 27*, 247–254. https://doi.org/10.1016/j.jocs.2018.06.009
27. Ryman-Tubb, N. F., Krause, P., & Garn, W. (2018). How Artificial Intelligence and machine learning research impacts payment card fraud detection: A survey and industry benchmark. *Engineering Applications of Artificial Intelligence, 76*, 130–157. https://doi.org/10.1016/j.engappai.2018.07.008
28. Nami, S., & Shajari, M. (2018). Cost-sensitive payment card fraud detection based on dynamic random forest and k-nearest neighbors. *Expert Systems with Applications, 110*, 381–392. https://doi.org/10.1016/j.eswa.2018.06.011
29. Mohammed, R. A., Wong, K. W., Shiratuddin, M. F., & Wang, X. (2018). Scalable machine learning techniques for highly imbalanced credit card fraud detection: A comparative study. In *Pacific Rim International Conference on Artificial Intelligence.*
30. Dornadula, V. N., & Geetha, S. (2019). Credit card fraud detection using machine learning algorithms. *Procedia Computer Science, 165*, 631–641. https://doi.org/10.1016/j.procs.2020.01.057
31. Carcillo, F., Dal Pozzolo, A., Le Borgne, Y. A., et al. (2018). SCARFF: A scalable framework for streaming credit card fraud detection with spark. *Information Fusion, 41*, 182–194. https://doi.org/10.1016/j.inffus.2017.09.005
32. Vikrant Agaskar, P., Babariya, M., Chandran, S., & Giri, N. (2017). Unsupervised learning for credit card fraud detection. *International Research Journal of Engineering and Technology, 4*, 2395–2456.
33. Abdallah, A., Maarof, M. A., & Zainal, A. (2016). Fraud detection system: A survey. *Journal of Network and Computer Applications, 68*, 90–113. https://doi.org/10.1016/j.jnca.2016.04.007

34. Subbulakshmi, T., Mathew, G., & Shalinie, M. (2010). Real time classification and clustering of ids alerts using machine learning algorithms. *International Journal of Artificial Intelligence, 1*, 1–9.
35. Dreibholz, T., Mazumdar, S., Zahid, F., et al. (2019). Mobile edge as part of the multi-cloud ecosystem: A performance study. In *Proceedings—27th Euromicro International Conference on Parallel, Distributed and Network-Based Processing (PDP)* (pp. 59–66). https://doi.org/10.1109/EMPDP.2019.8671599
36. Tariq, N. (2018). Impact of cyberattacks on financial institutions. *Journal of Internet Banking and Commerce, 23*, 1–11.
37. Richardson, J., Technologies, B., Jones, M., & Knowledge, G. (2020). *Fraud & Security*. https://doi.org/10.1016/S1361-3723(20)30045-2
38. Suresh, G., & Raj, R. J. (2018). A study on credit card fraud detection using data mining techniques. *International Journal of Data Mining Techniques and Applications, 7*, 21–24. https://doi.org/10.20894/ijdmta.102.007.001.004
39. Misra, S., Thakur, S., Ghosh, M., & Saha, S. K. (2020). An autoencoder based model for detecting fraudulent credit card transaction. *Procedia Computer Science, 167*, 254–262. https://doi.org/10.1016/j.procs.2020.03.219
40. Jurgovsky, J., Granitzer, M., Ziegler, K., et al. (2018). Sequence classification for credit-card fraud detection. *Expert Systems with Applications, 100*, 234–245. https://doi.org/10.1016/j.eswa.2018.01.037
41. Song, C., Wang, T., & Hu, M. Y. (2019). Referral reward programs with scarcity messages on bank credit card adoption. *International Journal of Bank Marketing, 37*, 531–544. https://doi.org/10.1108/IJBM-12-2017-0260
42. Jung, H. (2020). The impact of ambient fine particulate matter on consumer expenditures. *Sustainability, 12*, 1855. https://doi.org/10.3390/su12051855

Artificial Intelligence-Based Smart Packet Filter

Mohit Dayal, Ameya Chawla, Manju Khari, and Aparna N. Mahajan

Abstract Packet filtering is a fundamental feature for the firewall as the security of the whole system depends on it, but on the other hand, it should also be fast and efficient and should be able to process queries as fast as possible due to the large number of queries sent to the server together. There is a need for an artificial intelligence-based smart packet filter that can act on the packet in a fraction of seconds of receiving the packet. This research proposes an ensemble model (smart learner) made up of a random forest of max depth 11. The average accuracy obtained by the model is 99.76% using the stratified K-gold cross-validation technique. Earlier papers published on packet filtering are dividing the dataset into test and train where their test and train are fixed which can cause overfitting on the dataset and in real life testing can reduce the accuracy of the model, that is why this research focusses on using K-fold cross-validation technique and to support this technique highest accuracy was achieved with ensemble model of random forest. This research can be implemented in the firewall to make them faster.

Keywords Packet filter · Smart learner · Random forest · K-fold cross-validation · t-SNE

M. Dayal
Dr Akhilesh Das Gupta Institute of Technology and Management, New Delhi, India

A. Chawla
Guru Tegh Bahadur Institute of Technology, Guru Gobind Singh Indraprastha University, New Delhi, India

M. Khari
Jawaharlal Nehru University, New Delhi, India

A. N. Mahajan (✉)
Maharaja Agrasen University, Solan, Himachal Pradesh, India
e-mail: directormait2014@gmail.com

P. K. Singh et al. (eds.), *Proceedings of Third International Conference on Computing, Communications, and Cyber-Security*, Lecture Notes in Networks and Systems 421,
https://doi.org/10.1007/978-981-19-1142-2_62

1 Introduction

Packet filtering is a technique used in firewalls as it helps in monitoring incoming packets and outgoing packets to check whether any packet incoming or outgoing can lead to data leakage from the server. Action is taken after monitoring the packet, and there are mainly 4 actions taken for the packet [1]. Packet filtering works generally on the network layer and is classified on basis of IP address, but it can also work on higher layers of the OSI model where filtering is done based on the port numbers. This research proposes a model in which packet filtering is done based on the ports of the sender and receiver with information about the data size, data packets, elapsed time, packets sent, and packets received during the communication.

There is a need of a smart packet filter which works on application-level gateway as the hacker tries to gain access using these application ports which can be lethal to the sensitive information of the user/organization. Objective of this research paper is to implement a smart packet filter using machine learning algorithms which works on application-level gateway which filters packets accurately.

Every firewall has rules set to check whether the packet received can satisfy necessary conditions to go further beyond the firewall. In a traditional firewall, every data packet is checked for all the rules which can decrease efficiency if there are many data packets flowing in a fraction of seconds so to improve this performance an artificial intelligence-based smart learner can be implemented where it will be trained to filter data packets efficiently (Table 1).

There are 5 types of firewalls which protect the system from different kind of threats.

1.1 Packet-Filtering Firewall

Packet-filtering firewalls are software-based firewalls where they filter the packet based on many parameters like packet type, packet count, IP address, etc. These firewalls filter the packet; if they find the packet harmful, then the packet is dropped, else passed on to the internal mechanism of the system.

Table 1 Actions taken for packets by the firewall

Action	Description
Allow	Allows the packet to travel
Drop	Silently drops the traffic
Deny	Blocks the traffic
Reset-both	Reset TCP and send to both

1.2 Circuit-Level Gateway

Circuit-level gateways are software-based firewalls which monitor the protocols used between the client and server. These firewalls never analyze the data packets used in the communication; instead, they check whether the sessions initiated are genuine or not; based on this decision, it takes decision to terminate the session with the remote system.

1.3 Stateful Inspection Firewall

Stateful inspection firewalls are software-based firewalls which monitor both the session established and packet for enhanced security in comparison with the previous two types of firewalls. This firewall trades of security for speed as it takes a load on performance when the firewall must analyze both the session and the packet.

1.4 Application-Level Gateway

Application-level gateways are software-based firewalls which monitor the requested service and other characteristics. This firewall focusses on port number which tells which service requested and other characteristics of packet. The dataset used in this research is also an example of application-level gateway. These firewalls are considered more secure but increase usage of performance of the system.

1.5 Next-Generation Firewall

Next-generation firewalls are software-based firewalls which combine all the previous traditional firewall features with advanced device filtering, application filtering, and deep packet filtering. These firewalls are considered most secure firewalls which use high performance.

1.6 Deployment

This research proposes AI-based smart packet filtering which is faster than the traditional packet-filtering method where it checks each rule for each and every packet. This smart packet filter will be added in application layer as it will filter packet based on the application port numbers and data specifications and it will be helpful to be added after the IP-based filtering of packets to add another secondary firewall to protect the local area network as there can be mistake by IP filtering, but this hidden firewall will be behind it as another level of check on base of application port number access of the data packet.

2 Dataset

UCI machine learning repository dataset of firewall log files was used for this research. The dataset has 11 parameter values, for example, 65,532; based on these parameters, the firewall takes action on the data packet like allow, deny, drop, and reset-both. Stratified K Fold cross-validation was used for validation, and the dataset was split into a 9:1 train-test split ratio with an equal proportion of classes in the split is chosen at helps to make more samples to be trained and increases the accuracy of the model [2]. For stratified K-fold cross-validation, the training set was divided into 10 subsets where each had approximately 6553 samples to test and others to train.

3 Data Visualization

The firewall log file contained 11 features, and plotting them together would not be possible, so the dataset was dimensionally reduced for visualization purposes, t-SNE is used for dimensionality reduction, and plot was generated after dimensionally reducing the data from 11 to 2 components. It is very difficult to plot 11 features together as it will be 11-dimensional graph plus adding individually creating graphs of one parameter versus another would also result in ${}^{11}C_2$ number of graphs which is 55 graphs which cannot be shown properly that is why t-SNE plot was chosen to visualize how separate these classes are and give an idea which model will be best fit for this problem (see Fig. 1) [3].

The plot in Fig. 1 indicates the separation between each action class, and a nonlinear function can easily fit on this dataset to solve the classification problem.

4 Features of Dataset

Packet filtering consists of 11 parameters based on which the model predicts the action taken for a given data packet; the following are those parameters source port, destination port, NAT source port, NAT destination port, bytes, bytes sent, bytes received, packets, elapsed time, packets sent, and packets received.

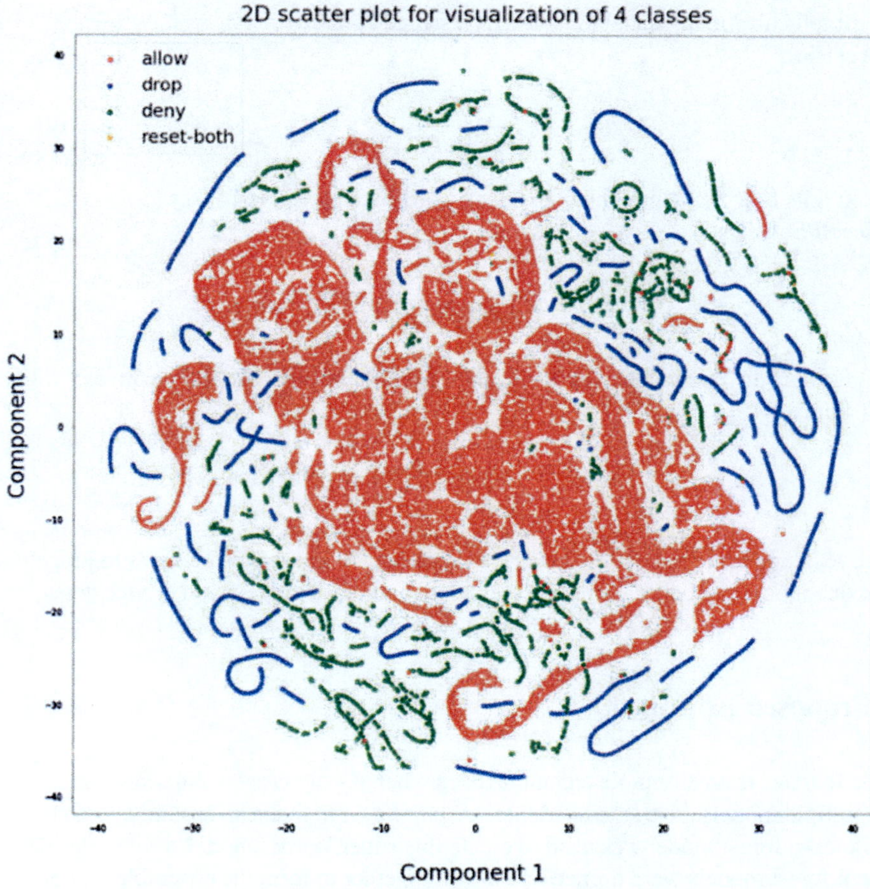

Fig. 1 t-SNE plot of two components after dimensional reduction

5 Validation Technique

Stratified K-fold cross-validation technique is used where first the dataset is randomly shuffled so that there should be no majority of a particular class in the testing dataset. After shuffling, the dataset is divided into K groups of equal size and the model is then trained K times where each time model is training for data excluding those K samples and after training the model it is tested against those K samples, and accuracy, precision, recall, and F1 score are calculated [10].

- **Accuracy**

$$\frac{\mathrm{TN} + \mathrm{TP}}{\mathrm{TN} + \mathrm{TP} + \mathrm{FP} + \mathrm{FN}} \tag{1}$$

It tells how many data points were classified correctly.

- **Precision**

$$\frac{\mathrm{TP}}{\mathrm{TP}+\mathrm{FP}} \tag{2}$$

It tells how many predicted positives were actual positives.

- **Recall/Sensitivity**

$$\frac{\mathrm{TP}}{\mathrm{TP}+\mathrm{FN}} \tag{3}$$

It tells how many positives were predicted correctly out of total positives.

- **$\mathbf{F_1}$ Score**

$$\frac{2 * R * P}{R + P} \tag{4}$$

where R stands for recall and P stands for precision. F_1 is used in the cases to judge when one model has better score in either recall and lacks in precision or vice versa.

6 Proposed Approach

Super learner is an ensemble technique used when K-fold cross-validation is used, and multiple models are trained and then are combined to make an ensemble model. The random forest model was used to create this super learner model where total K random forest models were trained and joined together to form the ensemble model. The highest average accuracy was obtained with a max depth of 11. Gini is used as criterion while splitting as it is faster than entropy, and our objective was to build a fast and efficient packet filter [9] (Table 2).

Algorithm defined to create the ensemble model.

Algorithm 1 Algorithm to construct ensemble model

Table 2 Hyperparameter values for random forest model

Hyperparameter	Value
N estimators	100
Max depth	11
Criterion	Gini
Minimum sample split	10

```
Input: Dataset
Output: Ensemble model and accuracies
K = 10 // Number of Folds
K1, K2 . . . K10 = Split(Dataset) //Divided Dataset into 10 parts
K_1[10] = {K1, K2 . . . K10} // Storing all dividing indexes in K_1
i=0 //Iterator
while (i< k )
        Modeli =RandomForest(Hyperparameters)
        Modeli .train(Dataset-Ki)
        Accuracyi=Modeli.test(Ki)
        i++
end
EnsembleModel= Maxvote(Model1, Model2, . . . Model10)
Accuracies={Accuracy1, Accuracy2 . . . Accuracy10}
```

7 Results

The average accuracy obtained by the super learner model based on random forest is 99.76%. The confusion matrix of the matrix explains on average how many samples were classified correctly for each class while testing (see Fig. 2).

Figure 2 plots graph between accuracy and *K* value where *K* value determines the subset of dataset as dataset was divided into 10 equal parts. Highest accuracy 99.81% was obtained on the second and third subset, while lowest accuracy 99.71% was obtained on fifth subset. The average accuracy of the model is 99.76%.

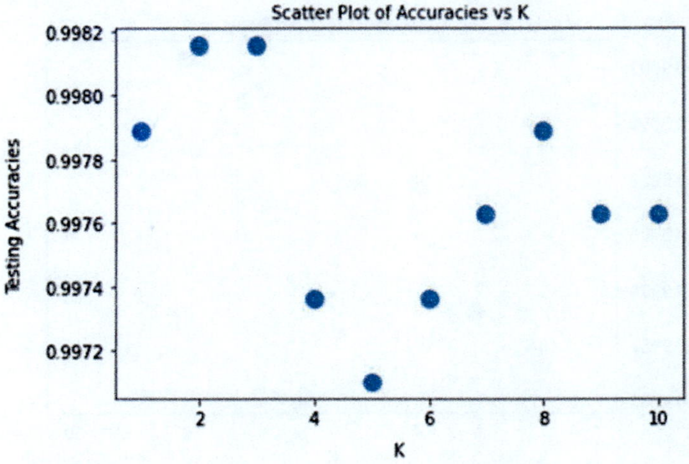

Fig. 2 Scatter plot of testing accuracy versus *K*

Figure 3 plots graph between recall score and K value where K value determines the subset of dataset as dataset was divided into 10 equal parts. Highest recall 99.92% was obtained on the sixth subset, while lowest recall score 99.8% was obtained on fifth subset. The average recall score of the model is 99.85%.

Figure 4 plots graph between precision score and K value where K value determines the subset of dataset as dataset was divided into 10 equal parts. Highest precision 99.92% was obtained on the sixth subset, while lowest precision score 99.8% was obtained on fifth subset. The average precision score of the model is 99.86%.

Figure 5 plots graph between F1 score and K value where K value determines the subset of dataset as dataset was divided into 10 equal parts. Highest F1 score 99.92%

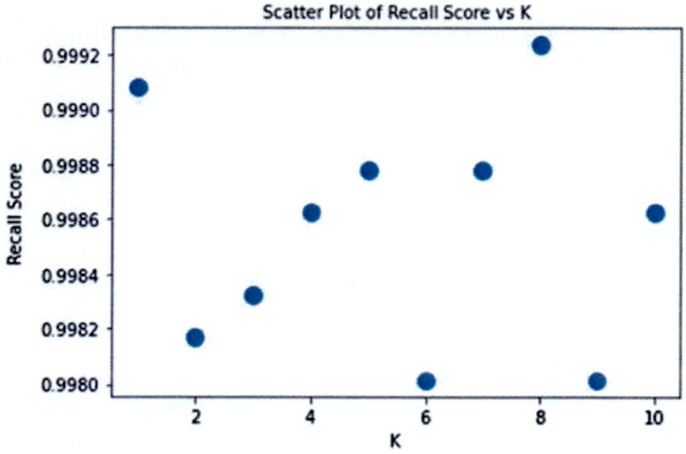

Fig. 3 Scatter plot of recall score versus K

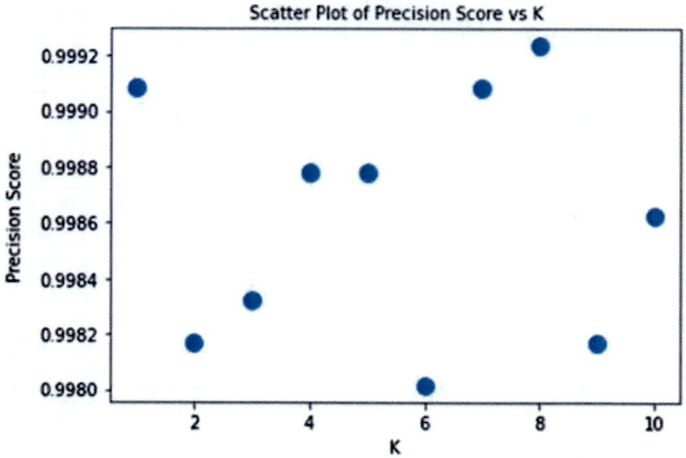

Fig. 4 Scatter plot of precision score versus K

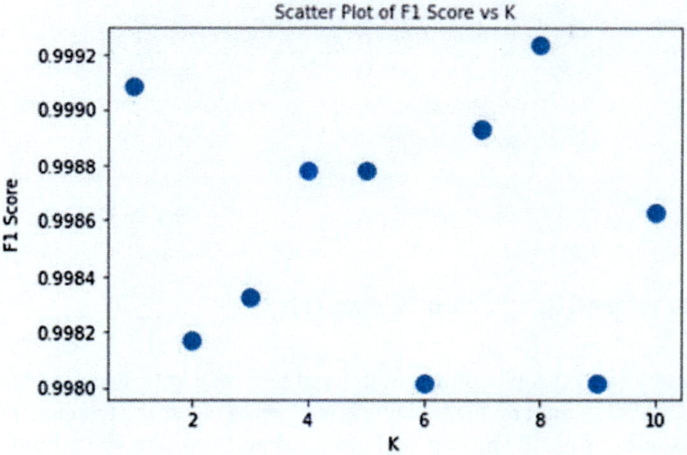

Fig. 5 Scatter plot of F1 score versus *K*

was obtained on the sixth subset, while lowest F1 score 99.8% was obtained on fifth subset. The average F1 score of the model is 99.85%. All three precision, recall, and F1 scores have equal average, and their graph is almost similar; their values are just different in seventh and ninth fold.

Figure 6 shows only 7 times there was wrong classification and rest 6540 times the model classified correctly which accounts for the higher accuracy of the model.

Fig. 6 Confusion matrix

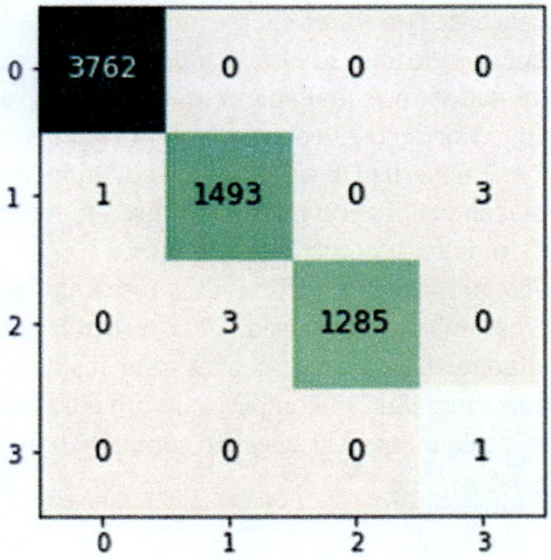

Table 3 Comparison with existing research

Existing paper scores	My research scores
Recall—98.5% [6]	Recall—99.8–99.92%
Precision—67.5% [6]	Precision—99.8–99.92%
F1—76.4% [4]	F1—99.8–99.92%
Accuracy—99.8% [2]	Accuracy—99.71–99.81%
F1—99.9% [2]	F1—99.8–99.92%

8 Comparison with Previous Research

Existing research on this topic all use traditional test train split which reduces the chance of the model to train on the remaining test which can make the model overfit if the train/test ratio is small and can lead too good performance while training but reduced performance when testing in real life and to solve that problem previously made in research this K-Fold Cross-Validation technique with Random Forest is suggested (Table 3).

In comparison with exiting research, the proposed model has increased the accuracy of classification of the data packets.

9 Conclusion

Firewalls need to have smart packet filters on application-level gateway to filter out packets based on which source port they arrived and which destination port they want to access. Due to need of fast and efficient packet filter, this research was conducted where dataset was first analyzed using t-SNE plot, which helped to understand how separated or segregated are these 4 classes and after testing the best approach was selected. Stratified K-fold cross-validation technique was used as to validate the results, and confusion matrix was used to measure accuracy for each subset created by K-fold cross-validation technique.

The proposed model in this research was successful, and the objective of this research was achieved. This research can be used to implement an artificial intelligence-based smart packet filter inside the firewall software which can filter packets efficiently in comparison with a traditional system where each rule defined in the packet filter was checked for every individual packet.

References

1. Ahmad, Z., Khan, A. S., Shiang, C. W., Abdullah, J., & Ahmad, F. (2020). Network intrusion detection system: A systematic study of machine learning and deep learning approaches. *Transactions on Emerging Telecommunications Technologies, 32*(1). https://doi.org/10.1002/ett.4150
2. Al-Haija, Q. A., & Ishtaiwi, A. (2021). Machine learning based model to identify firewall decisions to improve cyber-defense. *International Journal on Advanced Science, Engineering and Information Technology, 11*(4), 1688. https://doi.org/10.18517/ijaseit.11.4.14608
3. As-Suhbani, H. E., & Khamitkar, S. (2019). Classification of firewall logs using supervised machine learning algorithms. *International Journal of Computer Sciences and Engineering, 7*(8), 301–304. https://doi.org/10.26438/ijcse/v7i8.301304
4. Bibalbenifa, J., Krishnann, S., Long, H., Kumar, R., & Taniar, D. (2021). Performance analysis of machine learning and pattern matching techniques for deep packet inspection in firewalls.https://doi.org/10.21203/rs.3.rs-260788/v1
5. Buhari, M., Habaebi, M., & Ali, B. (2005). Artificial neural system for packet filtering. *Journal of Computer Science, 1*(2), 259–269. https://doi.org/10.3844/jcssp.2005.259.269
6. Ertam, F., & Kaya, M. (2018). Classification of firewall log files with multiclass support vector machine. In *2018 6th International Symposium on Digital Forensic and Security (ISDFS)*. https://doi.org/10.1109/isdfs.2018.8355382
7. Huang, Y., Nazir, S., Ma, X., Kong, S., & Liu, Y. (2021). Acquiring data traffic for sustainable IoT and smart devices using machine learning algorithm. *Security and Communication Networks, 2021*, 1–11. https://doi.org/10.1155/2021/1852466
8. Sharma, N., & Arora, B. (2020). Review of machine learning techniques for network traffic classification. *SSRN Electronic Journal*. https://doi.org/10.2139/ssrn.3747605
9. Ehwerhemuepha, L., Danioko, S., Verma, S., Marano, R., Feaster, W., Taraman, S., & Chang, A. (2021). A super learner ensemble of 14 statistical learning models for predicting COVID-19 severity among patients with cardiovascular conditions. *Intelligence-Based Medicine, 5*, 100030.https://doi.org/10.1016/j.ibmed.2021.100030
10. Zhou, Z. (2021). Model selection and evaluation. *Machine Learning*, 25–55.https://doi.org/10.1007/978-981-15-1967-3_2

Preserving Privacy in Internet of Things (IoT)-Based Devices

Dheeraj Sharma and Amit Kumar Tyagi

Abstract According to the global risk reports, data breaches and cyberattacks are in the top 5 deliberate risks. We all are aware of the rapid advancement and deployment of the IoT. Because these technologies are so tightly linked to individuals, privacy and security are important problems in today's world. Attackers who try to target IoT must constantly expose communication relations to capture transferred data and identify subtle data since they always rely on formerly gathered information to launch their attacks. Sleep is one of the crucial activities to our health. Depression, difficulty in concentrating, and irritability are a few important concerns that are caused by sleeping disorders. Using a sleep tracker may help a person understand their sleeping behavior and detect many important concerns. There are several dangers connected with information gathering since these IoT-based gadgets, including tracking device stowage, data transmission across a system, and information storage in the cloud. The information gathered by IoT instruments can expose the users' everyday activities, location, and other delicate statistics. Hackers usually try to attack these and when gadgets or the stowage is hewed, they may get confidential information and facts about their personal belonging and that data can be future used for phishing or advertisements. As a result, the privacy of data gathered by IoT-based devices must be protected.

Keywords Privacy-preserving · IoT · Man-in-the-middle attack · ARP poisoning · SSL strip · DNS spoofing · IoT-based sleep trackers · IoT healthcare

D. Sharma
School of Electronics Engineering, Vellore Institute of Technology, Chennai, India
e-mail: dheeraj.sharma2019@vitstudent.ac.in

A. K. Tyagi (✉)
School of Computer Science and Engineering, Vellore Institute of Technology, Chennai, Tamil Nadu, India

Centre for Advanced Data Science, Vellore Institute of Technology, Chennai, Tamil Nadu, India

P. K. Singh et al. (eds.), *Proceedings of Third International Conference on Computing, Communications, and Cyber-Security*, Lecture Notes in Networks and Systems 421,
https://doi.org/10.1007/978-981-19-1142-2_63

1 Introduction

We have seen the Internet of things (IoT) explode in popularity in the previous decade. It is popular in the technological world because it is easy and has a pre-programmed nature, and IoT provides essential services to users by sensing the proper environment. Several sensors and associated gadgets to build and deliver an IoT interface that provides advantages. Despite this, it has numerous advantages in terms of wide-ranging applications. For example, in the automated administration of home applications, as well as in the transportation control and healthcare systems, several privacy and security concerns were discovered. Each hub at the Internet of things produces part of the information. It comprises touchy data around the client's conduct. Henceforth, it is vital toward secure such data after the escape. Another critical problem is the capacity of IoT data. To protect the information, it requires genuine confirmation and a control device [1].

IoT-based frameworks for healthcare are items that range from accessible to versatile embedded systems that collect persistent bio signals, movement, or pertinent data. Among many frameworks that measure heart rate and electrocardiogram (ECG). None of the aforementioned do well when it comes to obtaining a large amount of data. As a result, we now have frameworks that can integrate bio signals, movement, and pertinent data like sleep monitors. Here, the author mentioned, "it is supported by the fact that the IoT is not constrained to mechanical and digital machines, but also covers other objects, animals, and indeed individuals that are given unique identifiers which have the capacity to transfer data over an interface." [2]. These days, it is frequently utilized the "Internet of everything," which is a concept with the broadest meaning. Any human movement is not necessary during this encounter. It is valuable when smart gadgets, such as televisions or smartwatches, are associated with the Web and get or deliver the information requested by the customer. However, information that has lately arrived at their most recent aims may transit through all four phases of the TCP/IP system, exposing them to all known hazards. "With the broad use of cloud storage, it can be said that there showed up to the fifth layer, the cloud layer. The possibility of this layer being attacked is high, beginning with the brute force attacks at password-based attacks." [2]. Moreover, it is conceivable toward altering information at the assembly level utilizing man-in-the-middle attacks. One ought to not leave locate of the "sniffer attacks, denial of service (DoS)" attacks. Every IT expert has heard of the man-in-the-middle attacks, and this sort of attack is habitually depicted in completely different articles. Alongside the organizing innovation development, with cloud computing, the Internet of things (IoT) and bring your own technology (BYOT) [3].

Organization of the work: Sect. 2 discusses the literature survey on privacy and security concerns related to IoT-based devices, man-in-the-middle attacks in IoT, and DNS spoofing, address resolution protocol, SSL strip. Further, Sect. 3 describes the existed solutions and various approaches of privacy-preserving of IoT-based devices and proposed solutions on how to provoke the man-in-the-middle attacks by means of a few precautions and techniques. Then, Sect. 4 discusses the Ettercap tool, virtualBox, procedure to initiate the attacks, ARP poisoning simulation steps. In the last, Sect. 5 discusses the conclusion and future work in brief.

2 Literature Survey

In this smart era, we require innovative solutions to issue related to health. How IoT can help older citizens to help them to track their sleeping (deep sleep) time. In this section, we will provide a few existing solutions toward the IoT role as a sleep tracker.

2.1 IoT Sleep Trackers

The health service urgently needs reform in order to reduce costs, improve quality, and expand access. A large percentage of healing facility costs is spent on therapy determination. Restorative check schedules can be moved from a healing center (hospital-cantered) to the domestic (home-cantered) of the calm using technology. The Internet of things, or IoT, is an underutilized concept that has vast applications in a variety of fields, including healthcare. The full implementation of this paradigm in healthcare may be a source of mutual optimism, as it allows therapeutic centers to function more efficiently and patients to receive far better care. There are distinct advantages to using this innovation, which appear to improve the quality and productivity of medications and as a result, advance in understanding wellness [4]. Regularly, analyzing an individual's rest needs an instant rest assessment or "polysomnography (PSG)" which permits the checking for a few physical capacities other than rest cycles. Even though the PSG is identified as "the gold standard for sleep" observing, giving actual and precise data approximately rest, it is awkward, costly, and time-consuming. Sleep tracking devices help us down-to-earth applications in picking up quality rest, progressing survives by measurement of heartrates and developments. They may have extreme safety and protection vulnerabilities. The paper [5, 6] describes "since the sleep tracker users can end up an attack to malware by downloading the unreliable third-party apps and in this way gives consent to the potential attacker to get to access of the devices remotely."

Hence, Fig. 1 discusses the privacy-preserving data collection model for IoT in the healthcare sector.

2.2 Privacy and Security Concerns

There are concerns associated with collecting data from IoT trackers, such as gadget memory, transmission via the Web, and data storage in the cloud for analysis. A

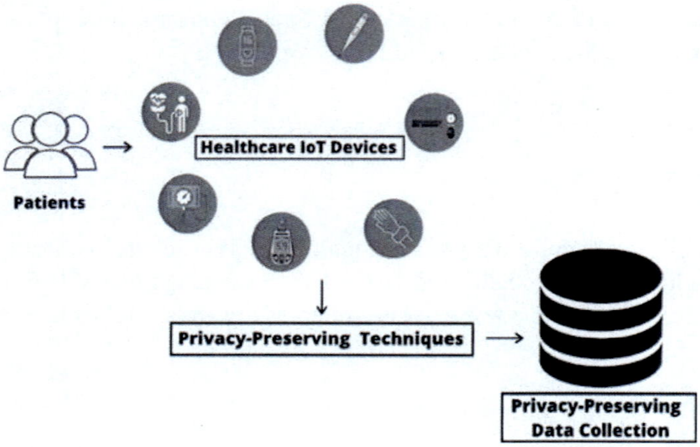

Fig. 1 Privacy-preserving data collection model for IoT in the healthcare sector [23]

similar danger exists with the IoT sleep monitor architecture, in which these devices are used to gather data while users are resting, which is then transferred to the cloud through a remote connection. The majority of IoT devices connect over an open interface [7]. While the information would be transmitted to the Web, the adversary can be caught by using "botnets, denial of service (DoS), and man-in-the-middle (MITM) attacks" on the communication channel. There are also concerns about data breaches, because an adversary may remotely access information stored in the cloud by infecting it with malware attacks. A hacker can obtain the user's private information and credentials after the equipment has been compromised. A security flaw of this kind could have a significant negative impact on a participant's credibility.

2.3 *Man-in-the-Middle Attacks*

From ancient times, long back there existed the man-in-the-middle attack, even before the computers were discovered. If we look in the nuts and bolts of the man-in-the-middle attacks, it could be utilized a case like a man say delivery person opens the parcels and get to know what is inside before handing those to the addressees. To offer secure services, most Internet apps attempt to leverage encrypted relations provided by SSL/TLS protocols at the submission level [8]. SSL/TLS can establish a mutual faith association, however, due to the organization's complexity, SSL/TLS is most commonly utilized when one member accepts the linking. Figure 2 explains a scenario of the ARP poisoning process in a man-in-the-middle attack.

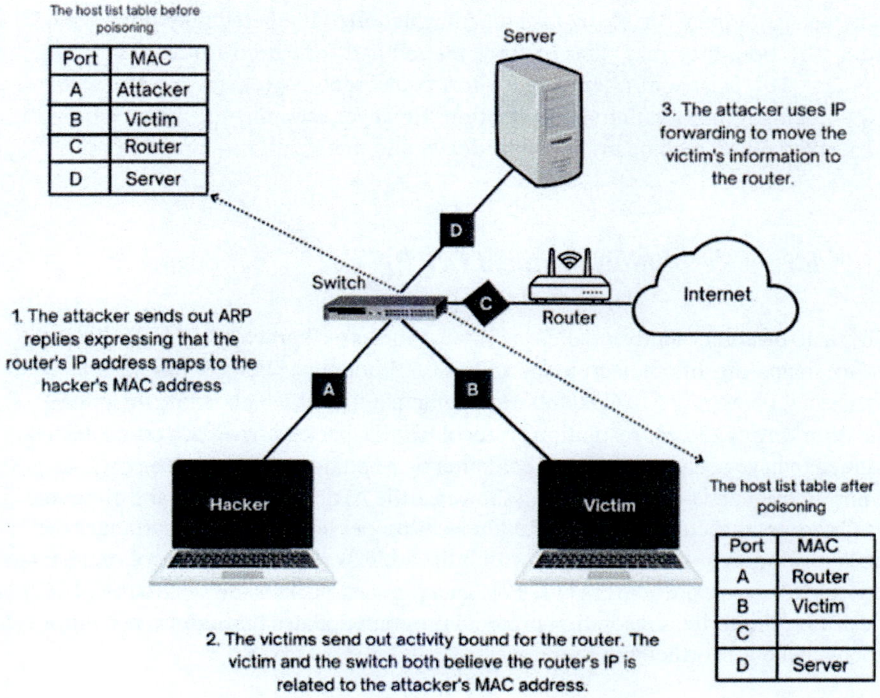

Fig. 2 An illustration of ARP poisoning process in a man-in-the-middle attack

2.4 *MIT-IoT*

It is visible that the expanding sector of the Internet of things, man-in-the-middle attacks may be ended up a better and greater encounter. There exist a kind of man-in-the-middle in IoT which mainly focuses on mobile phones and was produced by the destitute approval of credentials. Nevertheless, there exists an instance, additional thoroughly to domestic gadgets, maybe Internet of things-based refrigeration system which shows unwanted information about the user. The SSL certificate was not got approved for this case [9]. It is predicted that currently billions of IoT gadgets are susceptible, and the count is increasing day by day. Foremost among them are continuously twisted on and exist in the insecure interface. In the event that these interfaces permit superfast networks, a cooperated gadget could be a basis of expansive distributed denial of service (DDoS) attack activity. The paper [2] discusses "the embedded Internet of things (IoT) botnets are our present and future. They were used for DDoS attacks, spam-sending, MITM credentials hijacking, Internet chaos making, and other malicious activities. One of the massive attacks involving compromised IoT devices against dynamic domain title service providers." The man-in-the-middle attack can also be attempted in all the other devices and appliances which are connected to the IoT network or interface. According to a basic assumption, a malicious participant can easily fake the information of temperatures from a supervised application, they may alter the information and forward it to the monitoring equipment [10]. Heat controllers might allow equipment to overheat and as a result, stop producing if they get untrue information. However, stopping the manufacturing can harm the running company both physically and financially.

2.5 Address Resolution Protocol (ARP)

A physical machine address called a MAC address is known within the nearby interface maps the Internet protocol address (IP address). For comparing the IP addresses and mapping MAC addresses, a table is utilized which is known as ARP cache. Working of address resolution protocol RE: If a packet arrives at a portal that is presumed to have been received by a computer on a certain local area network (LAN) on a portal, the gateway may inquire to showcase the ARP program look and discover a MAC address that consists of the IP address. A proper length of packet arranges the address resolution protocol program within the address resolution protocol cache in case it may discover the address [11, 12]. Then, related packets are sent to the MAC address after the address resolution protocol program updates the address resolution protocol cache for forthcoming reference.

SSL Strip

Moxie Marlinspike, a security analyst formed SSL strip. Moxie uncovered the implementation within the "Black Hat Briefings in 2009 held at Washington D.C." He moreover found numerous susceptibilities in SSL usage. Using a man-in-the-middle attack through an SSL strip, the device ensures that the browser should not caution the client of almost an untrue certificate or a lapsed record. So, the client will not see any of the warning indications, and the lock icon from HTTPS will be missing from the top address bar. A large part of the user who uses the Internet on a regular basis do not enter HTTP or HTTP in the address bar; instead, naturally, HTTP diverts to access secure Websites. By serving as an intermediate, the SSL strip redirects all activity automatically [13]. For instance, if a client opens Gmail and as we all know, Gmail normally redirects to an SSL enabled Webpage login, if the SSL strip is enabled, it will detect that redirect, strip the SSL-enabled Webpage login, and instep provides the client with a non-SSL-enabled adaptation of the same location Gmail. With a packet sniffer like Ettercap, the attacker will be able to observe any movement that is communicated across the user's unprotected HTTP association. "SSL strip does not demonstrate a flaw in SSL encryption, but rather takes advantage of consumers who are unable to find reliable SSL encryption while transferring sensitive data over the Internet."

DNS Spoofing

Domain name server (DNS) spoofing (also identified as DNS cache poisoning) is classified as an attack in which changed the user's DNS records are utilized to divert

Web-based activities into a false site which takes after the proposed endpoint. After that the users are asked to enter the credentials into (what they admit to being) their account, therefore allowing attackers to take their access to the sensitive data plus the credentials which they just entered. Besides, the malevolent site has frequently been utilized to mount viruses into the user's personal computer, providing long-time access to the personal computer and the information to the attacker [14].

DNS spoofing is challenging to preserve since it is typically passive. The victim may never realize he has been confronted if the attacker is intensely focused on him. However, there are a couple of ways that may be executed to be secure:

- To secure internal machines
- There is no reliance on DNS for secure systems
- Using of IDS
- Using domain name system security extensions (DNSSEC).

Moreover, a few ways were discussed in [2] "A few things can be done to defend against session hijacking

- doing online banking from home
- being cognizant and keeping an eye out for things that seem unusual
- securing own internal machines; such attacks are mostly executed from inside the network.

SSL hijacking is virtually undetectable from the server-side, but a few things can be done from the client's side

- insurance of secure connections using HTTPS
- by doing online banking from home
- securing own internal machines".

3 Existed and Proposed Solutions for Privacy-Preserving in the Internet of Things

There are a few methods for privacy-preserving in the Internet of things, which can be summarized in this section. By using different anonymization and cryptographic strategies, the privacy break may be protected in the IoT framework. The ways for protecting IoT gadgets are demonstrated toward protecting the confidentiality of the Internet of things created information. Protected information can be used for academic and research determinations. "The different privacy-preserving techniques in the IoT framework are depicted in this part. Anonymization methods trusted third-party computation, and homomorphic encryption is the techniques under discussion." [15, 16].

1. **Techniques for Anonymization**

   A data-processing method that eliminates or alters personally identifying data is known as anonymization. The author discusses "the sensitive information over IoT nodes is secured through the aggregation of the information. However, there arise other problems such as computational delay, invalid results, software bugs, etc. This scheme collects the data from untrusted nodes and then performs aggregation. The trustworthiness of the data can be verified publicly using the proposed tuple algorithms. Nevertheless, some data owners have dropped out of this scheme." [16].

2. **Homomorphic Encryption on IoT**

   "Homomorphic encryption (HE) may be a promising way that can empower computation to be performed over encrypted information without recovering the plaintext." The headways in the homomorphic encryption have made it somewhat possible. It was discussed in the paper "the fully homomorphic encryption (FHE) can give privacy protection by completely supporting homomorphic operations over encrypted information" [17].

3. **Blockchain-based Privacy Preservation**

   Blockchain's anonymous nature innovation can be utilized to defend the protection of operators. Blockchain-based privacy is most extremely troublesome to gather the operation accounts subsequently by the hackers till the system is secured since analyzing those attacks.

4. **Trusted Third Party on IoT**

   By sustaining the spatial k-anonymity property for gathering users' preview inquiries, this technique protects the operator's location privacy. It stands up to induction attacks on the location-based service provider and in this way, secures the location privacy of individual users. It employs the best average closest neighbor method to maintain the separation among users while concealing the participant's true location data [18, 19].

Figure 3 depicts various models for preserving privacy in IoT or IoT-based applications.

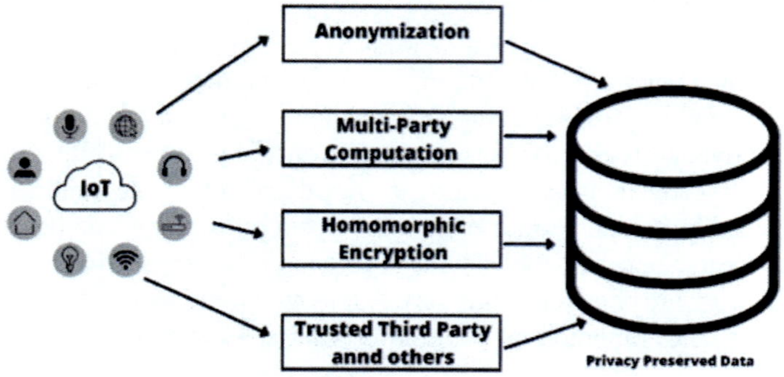

Fig. 3 Privacy-preserving in IoT model [16]

How to confront the Man-in-the-Middle Attack

There are many studies that found that there is an extraordinary variation of probable man-in-the-middle attacks. The wide-ranging elimination may be a very complex job, nevertheless, an aware user can suggestively reduce the underlying risks. Due to the incredible possibility of structures that each device retains, there are dissimilar sorts of man-in-the-middle-attacks, but too diverse ways of attacks that could be practical. Though not any enchantment baton could ensure a system from all types of attacks, among all the approaches one of the most appropriate ways is to work on the protection of the existing stage of producing the network, and after that, the system patches should be updated on a regular basis. It was noticed that most IoT devices are exclusive, made for precise resolves, and thus the security in particular and to some amount, could be less demanding. In any case, the quantity and differences of gadgets and endeavoring to economical arrangements be moreover an annoying issue. These gadgets are regularly discharged with safety susceptibilities and maybe they could incline to man-in-the-middle-attack [20]. The robust encryption of SSL certificates between the user, and the server is ought to utilize. If it is found that the network communication is not shifting, fairly, it is achievable to form a post of inactive address resolution protocol sections and use automated scripts that are deployed by the user. This may ensure that applications do not depend on the on-address resolution protocol requests or replies rather it should depend upon the local address resolution protocol cache.

Within a situation of convenient implementation, it is conceivable to avoid man-in-the-middle-attacks after the gadgets are not ever associated straightforwardly with Wi-Fi switches for unreliable systems. In such networks, an extra shield ought to be utilized at whatever point possible, for illustration, HTTPS all over or force TLS. To guarantee a trustworthy recognizable proof of gadgets, the applicants in statement inside the IoT, IoT obligation to provide the better request of "public-key cryptography" (PKC). The key factor while using PKC is approving if the open key is reliable and has a place in a certain individual or was supplanted by a hacker. A digital record is needed from a trusted "certification authority" (CA) as a confirmation. At the time when the communication begins securely, the dangers related to the man-in-the-middle attack are altogether decreased. In general, the best possible way to avoid a man-in-the-middle attack is to use a strong encryption method between the client and the server [21]. Here, a digital certificate is approved and that is first established by the client's request and after that only the connection could be completed. It was noticed that IoT producers ought to consider uniqueness and verification while creating gadgets and market them out. We got to know that these types of man-in-the-middle attacks are all around transferring false data and affectation as a gadget to another kind of gadgets and individual, one wants something to demonstrate that gadgets and individuals truly are whom they mention they are when they interconnect with one another.

Firmware Updates: Attackers are continuously in the look of locating weak points via which to attack people, which can include mobile applications and smart appliances. Obsolete smartphone applications, for example, are by far the most exposed to security and privacy concerns. Production houses and service providers should offer regular updates on mobile apps and ensure the availability of devices in the latest firmware. End-users should maintain their gadget apps and maintain IoT devices' firmware updated to avoid any data breaches, as security organizations are unaware of security risks.

Installation of the Software: Here, the author discusses how important is a software installation, "after ensuring the mobile app and IoT devices are updated. End-users should also refrain from downloading any untrusted third-party computer program, applications, or click on any adware link by doing so, and they are welcoming the malware into their mobile devices." [6]. End-users, for instance, get any well-being-related advertising by ticking on a linkage or installing malicious software, and IoT device users provide the attacker access, allowing the hacker to monitor their privacy remotely. After gaining access to the smart device user device, an attacker can steal the private data that most IoT device users are unaware of. To avoid malware infection on their mobile devices, IoT device users should validate their source or validity before installation of any software or clicking such associations.

4 Simulation Results, Conclusion, and Future Work

Ettercap

Ettercap is one of the most popular programs for attacks like MITM. A permitted and open-source tool, which works on different uniplexed information computing system (UNICS)-like working frameworks counting, Mac OS X, Linux, and Solaris. "Ettercap is a sniffer, interceptor, and logger" for shared local area networks that may be used for a variety of tasks. Because it can conduct appeal injection, packet sifting, and terminate any connection apart from MITM attacks, it has evolved into a versatile organized control device. Ettercap can secure and analyze every contact between the different victims once it has placed itself in the middle of an exchanged connection, and then, see whether the attacker can take advantage of the scenario (refer to Fig. 4).

VirtualBox setup: VirtualBox is a massive open-source tool for virtualization, which is developed by Oracle Corporation, that supports the creation and management of virtual machines [22]. Here, the man-in-the-middle attack was executed on the VirtualBox. We make a "virtual environment" with a couple of virtual servers:

- The hacker's computer, which runs the Kali Linux working framework.
- The target PC, which is running the Windows operating system.

The virtual machine that is both the victim and the attacker's setup settings must be adjusted to "bridged network." Instead of communicating through the primary machine, the virtualized device's IP address will be used. This same Ettercap program

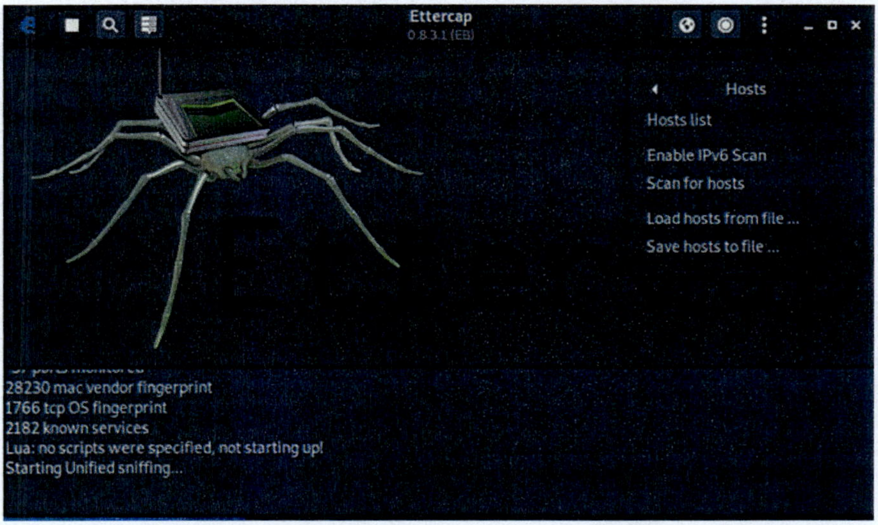

Fig. 4 The Ettercap tool, when unified sniffing started

may now check for the victim's IP address. The attack target was a "Windows 10 operating system," while the attacker had been a "Kali Linux operating system" [22].

- In Kali Linux, the Ettercap graphical unit may be accessed for apps. Within the Ettercap graphical interface, the sniffing begins.
- Select "ETH0" as the network interface in the unified sniff section.
- Ettercap might begin the bound together sniffing presently; at that point, we got to filter to select the targets properly. In this manner, we move to the host's list and choose filters aimed at the hosts. When the filter has been chosen, the computer program remains to progress to begin perusing within the arrange which can be appeared by way of "scanning the whole netmask for 255 hosts."
- The marked device that is "Windows 10" is chosen as "Target 1," and "Target 2" as the gateway. By using the "ifconfig" command, we got the IP as "192.168.1.44."
- At that point, begin the MITM attack with "ARP poisoning" by choosing "ARP poisoning."
- The MITM attack has commenced, and the Ettercap program will now display all of the client authorizations typed in the target's browser.

Hence, Fig. 5 shows the hosts list in the Ettercap, especially when the sniffing started and the credentials are captured by the attacker. Further, readers/researchers can refer to articles [24–28] to know more about attacks, and their nature and effect on various applications (with solutions).

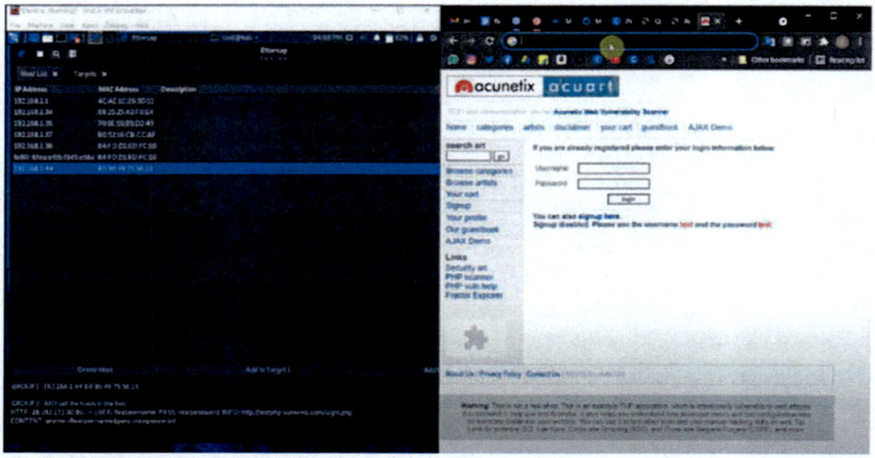

Fig. 5 Hosts list in the Ettercap when the sniffing started and the credentials are captured by the attacker

5 Conclusion and Future Work

There are numerous factors that affect the security and privacy of IoT devices, to begin with, the makers of these devices and the perception of the device privacy and security, moving on to the users and awareness of the attacks and vulnerabilities. All these factors are important but the most conspicuous is the security of IoT devices, which depends on numerous factors, starting from the creators of the device and their conception of the device security, to the end-user and their awareness of possible vulnerabilities and needs for device software patching whenever an update is rolled out. Device identification and encryption of communication between the device and its user are the most visible among all aspects. To differentiate each device in such conversations, certificates are required. Because there are and will be billions of devices, it is to be assumed that Internet of things-based gadgets may be unreliable on a regular basis, at the very least from the time a device is connected to the network until susceptibility is found. In many cases, the attackers have an edge in terms of information and expertise at the time of transfer. Because of its unique characteristics, man-in-the-middle attacks remain a viable technique for carrying out attacks and obtaining illicit advantages. Despite the fact that they are done in different ways, they are all based on the same concept.

The man-in-the-middle attacks are regularly coupled with or developed into other threats. The study has demonstrated the promise of the Internet of things, as well as the hazards that would arise from inadequate protection, and that, at the very least, for Internet of things users, it is a good idea to avoid using public Wi-Fi in circumstances when delicate and/or private information is being exchanged. When entering sensitive information, the user should always look for the HTTPS permit. The hacker may easily remove it with the use of an SSL strip. As a result, whenever a user enters critical info over the Internet, he or she should enquire to see if SSL is present. False SSL certificates should be avoided by users. We have shown how quickly a man-in-the-middle attack can be carried out using inexpensive and open-source devices like Ettercap and SSL strip. If the user gets alarmed when looking at the Web, these attacks can be predicted. The customer has to double-check for encoding or an SSL certificate, especially when sending delicate statistics over the Web. If a person is not vigilant when using the Internet, information might be negotiated.

References

1. Sadek, I., Demarasse, A., & Mokhtari, M. (2020). Internet of things for sleep tracking: Wearables vs. non wearables. *Health and Technology, 10*(1), 333–340. https://doi.org/10.1007/S12553-019-00318-3
2. Čekerevac, Z., Dvorak, Z., Prigoda, L., & Čekerevac, P. (2017). Internet of Things and the man-in-the-middle attacks—Security and economic risks. *MEST Journal, 5*(2), 15–25. https://doi.org/10.12709/MEST.05.05.02.03
3. Internet of Things Global Standards Initiative. https://www.itu.int/en/ITU-T/gsi/iot/Pages/default.aspx. Accessed August 18, 2021
4. Surantha, N., Putra Kusuma, G., & Isa, S. M. Internet of Things for sleep quality monitoring system: A survey.
5. Jia, J., et al. (2018). Intelligent and privacy-preserving medication adherence system. *Smart Health, 9–10*, 250–264. https://doi.org/10.1016/J.SMHL.2018.07.012
6. Sadek, I., Rehman, S. U., Codjo, J., Abdulrazak, B. (2019). Privacy and security of IoT based healthcare systems: concerns, solutions, and recommendations. *Lecture notes in computer science (including subseries Lecture notes in artificial intelligence and Lecture notes in bioinformatics)* (Vol. 11862, pp. 3–17). LNCS. https://doi.org/10.1007/978-3-030-32785-9_1
7. Seliem, M., Elgazzar, K., & Khalil, K. (2018). Towards privacy preserving IoT environments: A survey. *Wireless Communications and Mobile Computing, 2018*. https://doi.org/10.1155/2018/1032761
8. Agyemang, J. O., Kponyo, J. J., & Acquah, I. (2019). Lightweight man-in-the-middle (MITM) detection and defense algorithm for WiFi-enabled Internet of Things (IoT) gateways. *Information Security and Computer Fraud, 7*(1), 1–6. https://doi.org/10.12691/ISCF-7-1-1
9. Lounis, K., & Zulkernine, M. (2020). Attacks and defenses in short-range wireless technologies for IoT. *IEEE Access, 8*, 88892–88932. https://doi.org/10.1109/ACCESS.2020.2993553
10. Celiktas, B., Serkan Tok, M., & Unlu, N. (2018). Man in the middle (MiTM) attack detection tool design. *International Journal of Engineering Sciences & Research Technology*. https://doi.org/10.5281/zenodo.1336698
11. Liu, Y., Dong, K., Dong, L., & Li, B. (2008). Research of the ARP spoofing principle and a defensive algorithm.
12. Hammouda, S. (2009). An enhanced secure ARP protocol and LAN switch for preventing ARP based attacks. In *IWCMC '09*.
13. Nikiforakis, N., Younan, Y., & Joosen, W. HProxy: Client-side detection of SSL stripping attacks.
14. Sharma, B. (2014). Review paper on prevention of DNS spoofing. *International Journal of Engineering and Management Research, 3*.
15. Prakash Jayaraman, P., Yang, X., Yavari, A., Georgakopoulos, D., & Yi, X. (2017). Privacy preserving Internet of Things: From privacy techniques to a blueprint architecture and efficient implementation. *Future Generation Computer Systems, 76*, 540–549. https://doi.org/10.1016/j.future.2017.03.001

16. Privacy preserving internet of things: Survey of techniques and applications. *International Journal of Engineering and Advanced Technology*, 3230. https://doi.org/10.35940/ijeat.F8830.088619
17. Ren, W., et al. (2021). Privacy-preserving using homomorphic encryption in mobile IoT systems. *Computer Communications, 165*, 105–111. https://doi.org/10.1016/J.COMCOM.2020.10.022
18. Almohaimeed, A., Gampa, S., & Singh, G. (2019). Privacy-preserving IoT devices. In *2019 IEEE Long Island Systems, Applications and Technology Conference, LISAT 2019*, May 2019. https://doi.org/10.1109/LISAT.2019.8817349
19. Nicolazzo, S., Nocera, A., Ursino, D., & Virgili, L. (2020). A privacy-preserving approach to prevent feature disclosure in an IoT scenario. *Future Generation Computer Systems, 105*, 502–519. https://doi.org/10.1016/J.FUTURE.2019.12.017
20. Mallik, A., Ahsan, A., Shahadat, M. M. Z., & Tsou, J.-C. (2019). Man-in-the-middle-attack: Understanding in simple words. *International Journal of Data and Network Science, 3*, 77–92. https://doi.org/10.5267/j.ijdns.2019.1.001
21. Sowah, R. A., Ofori-Amanfo, K. B., Mills, G. A., & Koumadi, K. M. (2019). Detection and prevention of man-in-the-middle spoofing attacks in MANETs using predictive techniques in Artificial Neural Networks (ANN). *Journal of Computer Networks and Communications, 2019*. https://doi.org/10.1155/2019/4683982
22. Pingle, B., Mairaj, A., Javaid, A. Y. (2018). Real-world man-in-the-middle (MITM) attack implementation using open-source tools for instructional use. In *IEEE International Conference on Electro Information Technology*, May 2018 (pp. 192–197). https://doi.org/10.1109/EIT.2018.8500082
23. Onesimu, J. A., Karthikeyan, J., & Sei, Y. (2021). An efficient clustering-based anonymization scheme for privacy-preserving data collection in IoT based healthcare services. *Peer-to-Peer Networking and Applications, 14*, 1629–1649. https://doi.org/10.1007/s12083-021-01077-7
24. Tyagi, A. K., Aswathy, S. U., Aghila, G., Sreenath, N. (2021, October). AARIN: affordable, accurate, reliable and innovative mechanism to protect a medical cyber-physical system using blockchain technology. *IJIN, 2*, 175–183.
25. Nair, M. M., Tyagi, A. K., & Goyal, R. (2019). Medical cyber physical systems and its issues. *Procedia Computer Science, 165*, 647–655. ISSN 1877-0509. https://doi.org/10.1016/j.procs.2020.01.059
26. Tyagi, A. K., & Aghila, G. (2011, November). A wide scale survey on Botnet. *International Journal of Computer Applications (ISSN: 0975-8887), 34*(9), 9–22.
27. Tyagi, A. K. (2016, March). Cyber physical systems (CPSs)—Opportunities and challenges for improving cyber security. *International Journal of Computer Applications, 137*(14), 19–27.
28. Rekha, G., Malik, S., Tyagi, A. K., & Nair, M. M. (2020). Intrusion detection in cyber security: Role of machine learning and data mining in cyber security. *Advances in Science, Technology and Engineering Systems Journal, 5*(3), 72–81.

A Sentiment Analysis-Based Recommender Framework for Massive Open Online Courses Toward Education 4.0

Akhil Bhatia, Anansha Asthana, Pronaya Bhattacharya, Sudeep Tanwar, Arunendra Singh, and Gulshan Sharma

Abstract The emergence and confluence of progressive technologies like artificial intelligence, Internet of things, and automation in Industry 4.0 have also driven parallel domains like the education sector. Today's digital education aligns with the progressive dynamics of Industry 4.0, and with the increasing mix of information and communication technology (ICT), we have entered the era of Education 4.0. The ICT tools gather a lot of data content, which is generated through data generation in the form of text, audio, images, and video in online social networks (OSNs), blogs, posts, and many others. Usage of ICT has facilitated the conduction of open courses to masses of people connected through heterogeneous networked applications. Such courses termed as massive open online course (MOOC) platforms have grown significantly and have reaped high profits. However, users browsing for suitable courses in MOOC platforms are faced with challenges of selecting and filtering courses, based on current demands, effectiveness, and pre-requisite knowledge. Scientifically, it is observed that due to incorrect course selection, users are many times not satisfied with the MOOC course, which results in high dropouts. In the past, researchers have addressed the issue through recommender systems for users, but recommendation systems require effective filtering mechanisms for proper results. Thus, to address the research gap, in this paper, we propose an approach that is based on skills information from users' LinkedIn profiles combined with ratings and review data of courses. For

A. Bhatia · A. Asthana
Indian Institute of Technology, Jodhpur, Rajasthan 342037, India

P. Bhattacharya (✉) · S. Tanwar
Institute of Technology, Nirma University, Ahmedabad, Gujarat 382481, India
e-mail: pronoya.bhattacharya@nirmauni.ac.in

S. Tanwar
e-mail: sudeep.tanwar@nirmauni.ac.in

A. Singh
Pranveer Singh Institute of Technology, Kanpur, Uttar Pradesh 208005, India

G. Sharma
Durban University of Technology, Durban 4001, South Africa
e-mail: gulshans1@dut.ac.za

P. K. Singh et al. (eds.), *Proceedings of Third International Conference on Computing, Communications, and Cyber-Security*, Lecture Notes in Networks and Systems 421,
https://doi.org/10.1007/978-981-19-1142-2_64

experimental validation, we consider a Udemy MOOC user public dataset and apply natural language processing (NLP) to contextually organize user reviews, skill-set keywords from LinkedIn and refine search keywords. The proposed results indicate the efficacy of the framework toward correct MOOC recommendations for active learners and users.

Keywords Crowd mining · Long short-term memory · Massive open online courses · Review mining · Recommender systems

1 Introduction

A recommendation system considers certain inputs, analyzes like or disliked items by a set of similar users, which is then considered as input to refine the search from a large set of data. For example, E-commerce giant Amazon uses RS for recommending products and movies on its prime video platform [1]. In Education 4.0, we have shifted toward massive online open courses (MOOC), and MOOC platforms provide recommendations considering the user's search and history.

Users get help using recommender systems when they start looking for content out of humongous content from different sources. Recommender systems show users the right direction, although its performance may vary upon datasets. The recommender system considers the user's preference to endorse something. It must be done by considering what users say, what users prefer, and what the users do [1].

Education institutions all over the world face challenges with an increasing number of students from diverse backgrounds. Considering this situation, many institutions have considered MOOCs as an important part of educational strategy. Realizing the potential of MOOCs, many universities have started to offer MOOCs which are available to students and the public as well. E-learning is moving from conventional from online PowerPoint presentations or animations to a more flexible and distributed learning environment that supports self-paced and collaborative learning [2].

There is a rapid expansion in the development of MOOCs with an increasing number of platforms and tools, and there is an influential growth in intense research on MOOCs due to its openness and massiveness [3]. Learners must search and select certain courses on all the MOOC platforms which may cause the learner to acquire knowledge of multiple unlike courses, or miss out on basic courses, or choose courses that are coincidental [4]. Ample platforms are providing various learning items and it generates complexity since there are several providers, topics, tuition language, etc.

Taking professional career as a focus area, one of the popular professional online social networks (OSNs) is LinkedIn [5]. As per LinkedIn statistics, there are around 575+ million users with more than 260 million active users, and it is the best social network for lead generation. Active users on online social networks (OSNs) are growing exponentially, which brings out the challenging issue of collecting data from

OSNs [6]. There are set of application programming interface (API) which enables the connection between many applications. In LinkedIn, APIs allow applications to access public profiles with certain limitations. Data collection is carried out in two ways, the first is by using LinkedIn APIs [7], and the second is Udemy Affiliate APIs [8].

MOOCs can be of different categories, and some of the MOOCs can be of intersected types which makes recommendations complex. For example, the SQL Bootcamp course can be part of the development category and subcategory as databases. It can be part of the business category subcategory being data and analytics. While performing sentiment analysis, the reviews given to the user are essential because there are few instance reviews where the rating provided by the user was average, but the comments were positive. Also, fake user reviews remain a matter of concern while performing sentiment analysis on the data.

In Fig. 1, we present the impact that MOOC courses have on a global scale on Education 4.0. Figure 1a displays the MOOC courses that have increased progressively over the years, and the data in the figures excludes the data taken from China [9], and Fig. 1b shows the various renowned MOOC platforms that are operational in the country [10]. This shows that 2013 brought a transformation in the learning process. Even in a critical situation like complete lockdown, e-learning would be a choice for everyone to brush up on skills and keep the learning curve rising.

The remaining sections of the paper are organized into four parts. Section 2 of the paper explains the work which is related to MOOCs recommendations using a different methodology. Section 3 of the paper explains the proposed architecture of the recommendation system for MOOCs. Finally, Sect. 4 shows the algorithms designed for the proposed architecture and the methods which have been used to implement the system. The architecture has been developed using ensemble models for review mining of courses.

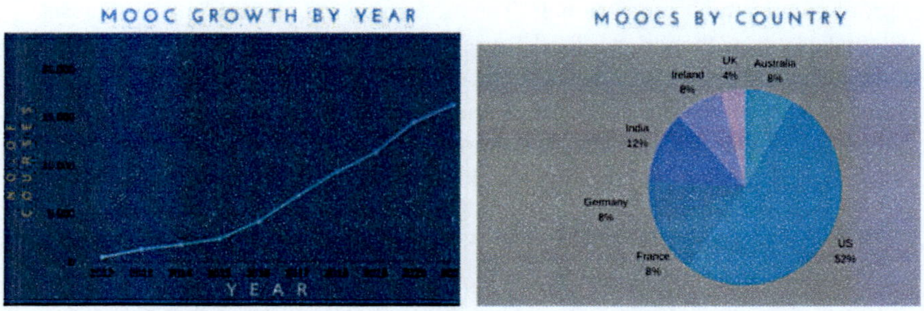

Fig. 1 Impact of MOOC courses in Education 4.0

2 Related Work

There has been a significant contribution in the field of MOOCs and recommendation systems. In ScienceDirect [11], if you search term MOOCs, then there are around 324 review and research articles in the year 2021. This shows MOOCs research is a challenging area that needs to be explored more considering its massiveness. There are many advantages of MOOCs:

i. Anywhere and anytime accessibility.
ii. Learning material can be created and upgraded easily (open for review as well as per feedback).
iii. Language flexibility ends language barriers worldwide for many students.

Sara et al. [12] proposed a semantic-based MOOC recommender framework. They proposed a three-layer architecture with where first being the user layer, the second layer is a semantic layer to structure collected data, and the third layer is the intelligence layer to provide predictions based on weights of semantics and return intelligent feedback to the user. Kulkarni et al. [13] presented a survey on various recommendation systems in the e-learning domain. This work recommends the use of advanced techniques to improve the recommendation system's accuracy and its relevance.

Users participating in MOOCs as learners or as reviewers can also contribute to bringing improvements and revisions in material and syllabi. Koukis et al. [2] proposed a framework for teachers' professional development by using principles of case study learning, peer-supported learning, or collaborative learning. This framework is aimed at improving peer interaction and support and collaborative creation of educational content. Pang et al. [14] proposed an adaptive learning model by analyzing learners' activity and learning duration to predict recommendations for similar learners. Dai et al. [3] propose a model to recommend MOOCs on active user's profiles and similar professionals. The model is based on the Coursera dataset and also makes use of the job market that needs to display results.

Avdoshin et al. [15] proposed MOOCs recommender system using job classification in the market. In the proposed work, author recommends users to enter skills for which they are looking for courses and has integrated with some renowned MOOC portals. Zhao et al. [16] proposed a model which depends on users' historical enrolled courses. The model constructs a knowledge graph based on learning feedback and maps it to a keywords graph to recommend courses to users.

3 Proposed Architecture

Figure 2 represents the proposed architecture for MOOCs recommender system. This framework combines the approach of mining user review data and similarity between active users and the job markets. The details are presented as follows.

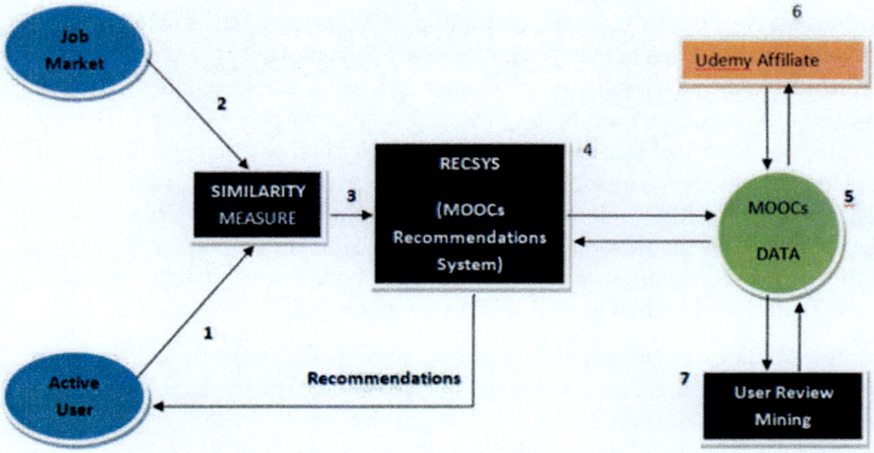

Fig. 2 Proposed architecture for MOOCs recommender system

Module 1: Module1 represents the skill set of an active user extracted from a LinkedIn profile. The information of LinkedIn members whose profile is public can be retrieved using profile API. The response of this API contains skills fields, test score fields, and certification fields. Rating of skills can be performed based on certification or test scores related to any skill. Holding the assumption that skill with test score or certification is strongest among other skills. The skills can be rated as shown in Table 1.

Module 2: Module 2 represents the skill set required for a certain job in the market. The data would be extracted by making use of two APIs. Using recommended jobs API [17] which will return recommended jobs for a user and using that job Id, further skills and experiences can be extracted using job lookup API [18]. To improve more on jobs matching and its data, job search API [19] considering the limitations of recommended jobs API. Job search API allows searching jobs based on various attributes like skills and experience, location, etc.

Module 3: Module 3 is the conclusion of a similarity measure. Leftover keywords after calculating similarity measures and skills marked as other skills in the active user profile will be submitted to recommendation system module (RECSYS). Similarity measures are considered to check if a user matches with that job to a certain extent.

Table 1 Skills categorization

Skill type	Condition	Rating
Top skills	Having certification in that skill	1.0
Industry knowledge	Holding decent test scores in that skill	0.8
Tools and technology	Mentioned in profile	0.6

Also considering the case, if the market needs new skills, those can be recommended to active users. Since job demands are constantly changing over time, users can be benefitted from the adaptability of this system [2]. For example, company A needs a software developer with skills like Java, Python, Go, Apache Tomcat, Django. LinkedIn Member Z holds Java with certification, Python, and Django, the matrix would look like as shown in Table 2.

The scoring of skills of job requirement would be on the order of skills required in the job post. This matrix would help us calculate similarity measures for job and user. It would also help us out with keywords. The above matrix would forward Go, Apache Tomcat, and Python as skill set for further analysis.

Module 4: Module 4 refines search for the set of keywords in the database on the basis of user ratings, ordering, and cost of the course. The SQL query would be applied on database to bring out data matching keywords in courses and also ordering them on the basis of different attributes. Figure 3 shows two sets; i.e., SS' and SS would be resultant MOOCs which would be considered for further analysis.

Module 5: Module 5 is the database part of the application which is refreshed on an hourly basis using Module 6. It contains required details in RDBMS tables.

Table 2 Similarity matrix calculation

M	Java	Python	Go	Apache Tomcat	Django
Z	1	0.9	0.8	0.7	0.6
A	1.0	0.6	0	0	0.6

Fig. 3 Venn diagram for skill-set keywords

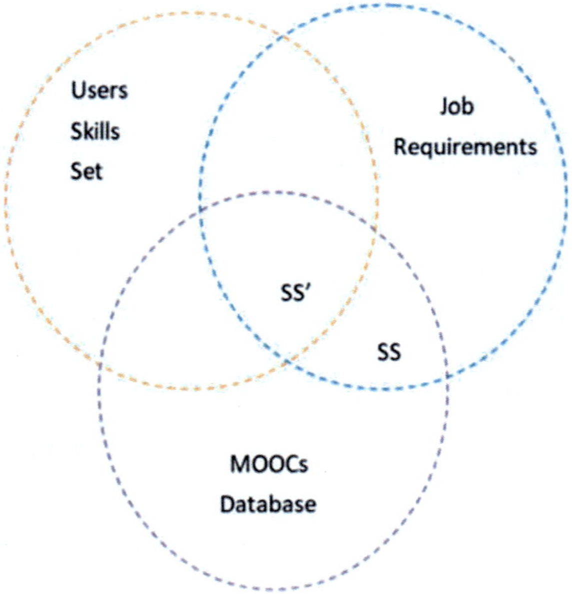

Table 3 Udemy set of APIs

API name	Description
GET coursereviews-list	Returns the reviews of courses by user using numeric course identifier
GET courses-detail	Returns course details like title, instructor, etc., using course identifier
GET courses-list	Returns the list of courses on Udemy
GET publiccurribulum-list	Returns the curriculum item against course ID

Module 6: Module marked 6 is the set of APIs provided by Udemy to access its MOOC data. There are 4 APIs which take authorization as a header and require parameters to return a response as shown in Table 3.

Module 7: The response from MOOCs aggregator is sent as a request to module 8. This module considers a set of reviews from users and returns an aggregated response whether the review for this course is positive or negative using natural language processing (NLP). MOOCs aggregate top 5 courses for a keyword on the basis of conditions explained in Module 4, and then, review data are submitted to Module 7 which returns positive or negative for a course.

Top 3 course for each keyword sorted on the basis of all conditions in modules is then returned to RECSYS which recommends the course to the user.

4 Algorithm Design

Algorithm 1 MOOCs Recommendation System

1. userData← Get Active User Skill Set
2. jobData←Get Market Job Data
3. simiMeas←SimilarityMeasure(userData, jobData)
4. recResp←RecSys(simiMeas)
5. return recResp

Algorithm 1 is step-by-step implementation of complete system. The following is the explanation for variables used in pseudo code:

userData: Collecting skill set using profile API from LinkedIn APIs[7]. The user details can be extracted using its name or person ID of a user (person ID is the ID in profile URL).

jobData: We are making use of combination of job search API according to user's skill and then job lookup API for getting details of that job using Job ID.

simiMeas: Both user's skill data and multiple job skill data would be passed to calculate similarity measure and extracts recommended skills.

recResp: The response with final set of courses to the user.

Algorithm 2 SimilarityMeasure (userData, JobData)

Similarity measure has been introduced in architecture because we have multiple jobs for which user can match according to skills. But only few of them fits them according to skills. Here, we have made use of Pearson correlation to perform so.

1. similMeas = PearCorr(userData, jobData)

Pearson Correlation: It is a measure of the linear relationship between two variables. For feature vector X and Y, it is computed as

$$s(X, Y) = \frac{\sum_i (X_i - \overline{X})(Y_i - \overline{Y})}{\sqrt{\sum_i (X_i - \overline{X})^2}\sqrt{\sum_i (Y_i - Y)^2}} \quad (1)$$

Algorithm 3 RecSys (similMeas, userData, jobData)

1. if(similMeas > thresholdValue)
2. A similarity matrix is calculated userData, jobData
3. simiKey⟵ Get Keywords on the basis of matrix
4. courseData⟵SearchCourse(simiKey)
5. for courseReviewData⟵ 1 to length(courseData)
6. revResp⟵ReviewMining(courseReviewData)
7. aggrResp⟵collectResponse(revResp)
8. return aggrResp

A threshold value would be considered to be compared to the result of this similarity measure (**similMeas**). **userData** and **jobData** would be used to calculate similarity matrix between job and user skills which will give keywords to search for courses.

The resultant course would bring it out then reviews of each course on and collect it in **aggrResp** object.

Algorithm 4 ReviewMining(courseReviewData)

Here, we perform sentiment analysis of review given by users on courses using ensemble of CNN and bi-LSTM models. The dataset that has been considered to train model is SST2 dataset [20], and it will be tested on filtered response of reviews given by users. Figure 4 shows the block diagram of model.

5 Results

The system was made in Java programming language with the use of the Spring Boot Framework and Python Keras Library. There was limitation to data extracted from

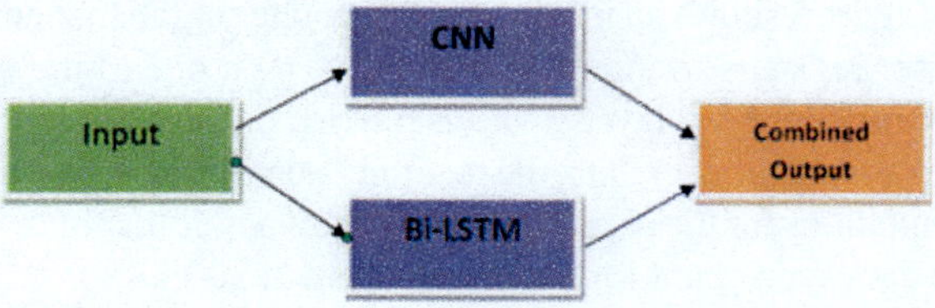

Fig. 4 Model for review mining

user's profile due to authorization limitations from LinkedIn APIs. We were still able to collect recommendations for around 10 unique users.

The procedure to produce results was as follows:

1. The system was fetched with user name of multiple users simultaneously. This was done to check course number of courses in database for improving recommendations.
2. For example: We came across a profile which mentioned PowerPoint as skill and other skills were PHP, Java, and related frameworks. This profile was matched with various IT companies job but recommendation to power point was not bought to him since we were not considering the dataset of this category (office productivity in Udemy) as per our work.
3. Some of the renowned skills which we found to be common among these profiles were to be teamwork, leadership, or dance. There was dataset for MOOCs of such skills, and they can be considered in similarity measure but in recommendations can be low prioritized. We also came across profile with no skills added, and such profiles were ignored as our framework relies on this data.
4. There were no profiles with certification so the rating format was considered accordingly. In that case, the rating was done from high to low staring from 0.7 to 0.3. Data were fetched aggregator after assigning weights and sorting out the set of keywords required.

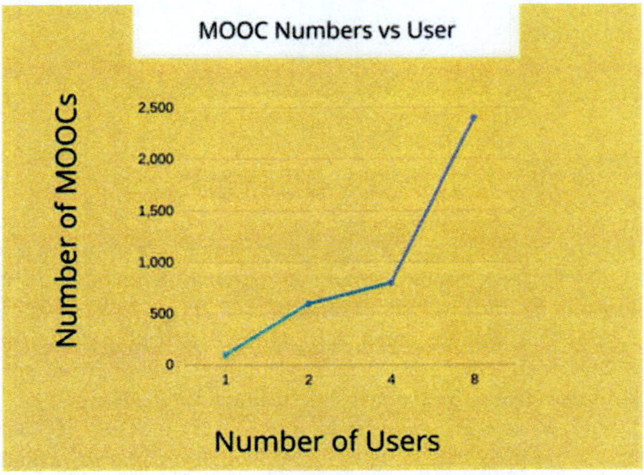

Fig. 5 No. of user versus no. of MOOCs

Figure 5 shows an increase with a greater number of enrolled users. Initially, when a user registers to the MOOC content, then due to the new account usage, limited categories are displayed to the user based on the skill sets of the user. Once the system learns from the user data, the recommendation process builds specific MOOC recommendations for the user. There might be a possibility of fewer recommendations since the data sync count for each category was 300.

The paper aims to represent recommendations, so we have not presented sentiment analysis results and its accuracy measures.

6 Conclusion and Future Work

We have presented a system which analyzes sentiments of reviews given users on courses of Udemy. Such a system can save time in finding relevant courses. The collection of courses counted to 2500, but there were similar courses in different categories. The recommendation of such courses needs more skill collection from profile to result with ideal course as per category. Such results were outcome of multiple skill keywords in single course name. For example: A user holds skills in the finance sector and also holds basic knowledge of Python, then they would be recommended with a Python course for finance and financial analysis. As per the data, there were around courses in 40 languages in development category. Among these courses in English are maximum followed by Portuguese 1942 and then Spanish with count 1534 and so on. This system can be extended by integrating it to multiple MOOCs platforms and improving the recommendations by comparing similar sets of courses across platforms. Also making use of more information from LinkedIn data, the system can make better predictions. An improvement to the approach of similarity measures can also be introduced if we make use of advertisements of jobs for a user. This can help us gain which job role users aspire to shift from their current job role. Extracting such things can present personalized recommendations of courses to users with a better set of options (in terms of cost and reliability) to choose among.

References

1. Thukral, R., & Ramesh, D. (2018). Ensemble similarity based collaborative filtering feedback: A recommender system scenario. In *2018 International Conference on Advances in Computing, Communications and Informatics (ICACCI)* (pp. 2398–2402). IEEE.
2. Koukis, N., & Jimoyiannis, A. (2019). Investigating participants' collaborative patterns in a MOOC for teacher professional development. In *European Conference on e-Learning* (pp. 303–312). Academic Conferences International Limited.
3. Dai, K., Vilas, A. F., & Redondo, R. P. D. (2017). A new MOOCs' recommendation framework based on LinkedIn data. In *2017 Innovations in Smart Learning* (pp. 19–22). Springer.

4. Dang, F., Tang, J., & Li, S. (2019). MOOC-KG: A MOOC knowledge graph for cross-platform online learning resources. In *2019 IEEE 9th International Conference on Electronics Information and Emergency Communication (ICEIEC)* (pp. 1–8). IEEE.
5. Dai, K., Nespereira, C. G., Vilas, A. F., & Redondo, R. P. D. (2015). Scraping and clustering techniques for the characterization of LinkedIn profiles. arXiv preprint arXiv:1505.00989.
6. Mind-Blowing LinkedIn Statistics and Facts. (2020). Kinsta Managed WordPress Hosting, 10 April 2020. kinsta.com/blog/linkedin-statistics/. Accessed May 04, 2020.
7. Tonyxu-Io. (n.d.). LinkedIn API documentation—LinkedIn. https://docs.microsoft.com/en-us/linkedin/. Accessed May 04, 2020.
8. Online courses—Anytime, anywhere. https://www.udemy.com/developers/affiliate/. Accessed May 04, 2020.
9. By the numbers: Moocs in 2020—class central. The Report by Class Central. https://www.classcentral.com/report/mooc-stats-2020. Accessed October 18, 2021
10. List of MOOC providers. Wikipedia. https://en.wikipedia.org/wiki/List_of_MOOC_providers. Accessed October 18, 2021
11. ScienceDirect.com | Science, health and medical journals, full text articles and books. https://www.sciencedirect.com. Accessed October 18, 2021
12. Assami, S., Daoudi, N., & Ajhoun, R. (2020). A semantic recommendation system for learning personalization in massive open online courses. *International Journal of Recent Contributions from Engineering, Science & IT (iJES), 8*(1), 71–80.
13. Kulkarni, P. V., Rai, S., & Kale, R. (2020). Recommender system in eLearning: A survey. In *Proceeding of International Conference on Computational Science and Applications* (pp. 119–126). Springer.
14. Pang, Y., Liu, W., Jin, Y., Peng, H., Xia, T., & Wu, Y. (2018). Adaptive recommendation for MOOC with collaborative filtering and time series. *Computer Applications in Engineering Education, 26*(6), 2071–2083.
15. Avdoshin, S., Pesotskaya, E., & Kuruppuge, D. (2021). A modern approach to MOOC recommender systems using ICT job classification. In *Future of Information and Communication Conference* (pp. 837–847). Springer.
16. Zhao, Y., Ma, W., Jiang, Y., & Zhan, J. (2021). A MOOCs recommender system based on user's knowledge background. In *International Conference on Knowledge Science, Engineering and Management* (pp. 140–153). Springer.
17. Recommended Jobs: LinkedIn Developer Network. LinkedIn Developers. https://developer.linkedin.com/docs/guide/v2/jobs/recommended-jobs. Accessed May 04, 2020
18. Job Lookup API and Fields: LinkedIn Developer Network. LinkedIn Developers. https://developer.linkedin.com/docs/v1/jobs/job-lookup-api-and-fields. Accessed May 04, 2020
19. Job Search API: LinkedIn Developer Network. LinkedIn Developers. https://developer.linkedin.com/docs/v1/jobs/job-search-api. Accessed May 04, 2020
20. Minaee, S., Azimi, E., & Abdolrashidi, A. (2019). Deep-Sentiment: Sentiment analysis using ensemble of CNN and Bi-LSTM models. arXiv preprint arXiv:1904.04206.

Lung Cancer Detection Using Textural Feature Extraction and Hybrid Classification Model

Jasbir Kaur and Meenu Gupta

Abstract Medical image processing (MIP) offers powerful and promising key developments in modernistic three-dimensional (3D) medical imaging based on science and medicine due to the creation of hi-tech images. Image processing is used to detect lung cancer. Detecting a cancer nodule consists of three levels. CT scans are generally adopted to identify the incidence of cancer affected nodules. To improve the interpretation of information in an image to a human audience, the step of image enhancement is enforced. The next step of segmentation involves segmenting the required area into many sub-areas. The output of this step is used as input for the next step of feature extraction. Cancer, at this stage, is detected on the basis of the abstracted features. This work implements GLCM with a hybrid classifier model to localize and classify the cancer affected area from the CT scan. The hybrid classifier framework constructed by integrating KNN, SVM, and decision tree classifiers is an efficient cancer detection framework work. This work takes three parameters (i.e., accuracy, precision, and recall) under consideration to evaluate the designed hybrid classifier model.

Keywords CT scan · KNN · SVM · Decision tree · GLCM · Segmentation

1 Introduction

1.1 Lung Cancer

The expansion of anomalous and unrestrained cells in body is called cancer. It can damage surrounding tissue extending away from its source. Some cancer forms are benign, while some are malignant. Malignant tumors are cancerous and can spread to nearby tissue. The malignant may lead to death, and it can multiply from every cell form in the body. Lung cancer is one of the main causes of most of the deaths due to

---

J. Kaur · M. Gupta (✉)
Chandigarh University, Mohali, Punjab, India
e-mail: Meenu.e9406@cumail.com

P. K. Singh et al. (eds.), *Proceedings of Third International Conference on Computing, Communications, and Cyber-Security*, Lecture Notes in Networks and Systems 421,
https://doi.org/10.1007/978-981-19-1142-2_65

cancer across the world. On time, treatment of lung cancer post-detection can improve the chances of patient's recovery. Lung cancer has two types, i.e., small cell lung cancer (SCLC) and non-small cell lung cancer (NSCLC). Chain smokers are more prone to small cell lung cancer (SCLC) [1]. This cancer occurs less commonly in comparison with non-small cell lung cancer (NSCLC). NSCL encompasses all forms of lung cancer. NSCLCs include squalors cell carcinoma, a den carcinoma, and large cell carcinoma. Several factors such as smoking, exposure to second-hand smoke, family history of cancer, and former radiation therapy can increase the risk of lung cancer. The result of NLST test reflected that three yearly screening levels of high-risk subjects by low-dose computed tomography (CT) significantly decrease mortality. CT is a 3D imaging method generally employed for the detection and diagnosis of lung cancer. 3D images are rebuilt from thousands of 2D X-ray transmission prognoses. Figure 1 shows the CT scan of a patient who is diagnosed with lung cancer. Advanced 3D reconstructions were developed for better image quality and diagnostic accuracy [2, 3]. Unlike chest X-ray examination, X-ray CT is used to detect the lung cancer more clearly with the characteristic shade of pathological variation seven in its early stage. The two prevalent computational frameworks designed to support radiologists include computer-aided detection (CADe) system and computer-aided diagnosis (CADx) system. CADe models use medical images to detect lesions while CADx models aim at measuring the lesion type, for instance, ascertaining the cancer's malignancy with its stage.

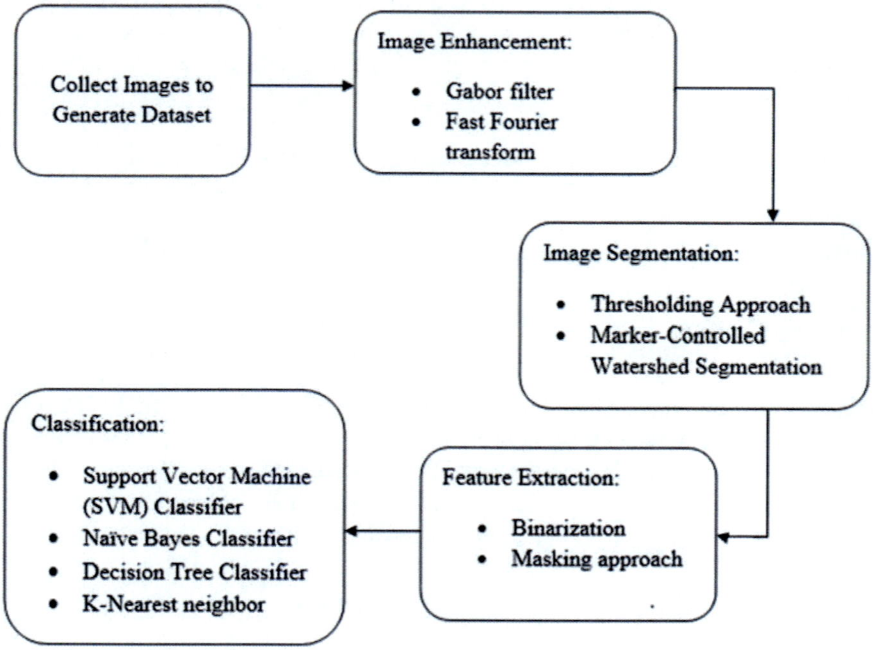

Fig. 1 Lung cancer detection framework

1.1.1 Role of Image Processing in Medical Domain

Medical images contain a lot of information about bodily compositions that play a vital role for accurate and timely diagnosis, opting the appropriate treatment method, and examining treatment result. Thus, processing and interpretation clinical scans accurately, like many other data in the medical domain, prompt to discover the association between these data and developments in medical services. In general, image processing is the process to be applied on a digital image to manipulate the gray level information presents in its pixels [4]. Besides, owing the human intelligence and visual limits in precise image processing, computer-based image processing with detailed analysis and recognition of the image may assist the early-stage detection of any abnormal changes in body cells. The digital image processing system has six stages: image acquisition, pre-processing, feature extraction, associative storage, knowledge base, and recognition Acquiring or capturing a digital image is the foremost stage in image processing. A digitized image is an image $f(x, y)$, wherein both the spatial coordinates and the intensity are in digital form. The components of a digital range are known as pixels or image constituents. This step involves sensors to click images. The sensor may take the form of a camera or a scanner. The next level called pre-processing copes with intensity vision along the image restoration and rebuilding. Image restoration is concerned with approximating an original image from a corrupted one. Restoration methods make up for system degrading that the image can pass through. Current neural network (NN) frameworks have established image restoration [5]. The feature extraction aims at reducing the data by quantifying some of the features that differentiate the input patterns. The features are extracted by selecting a subset of the noted input vector. It also converts the input observing vector. To obtain this, an input image representing very correlated data is used. Reducing dimensions while retaining maximum information makes use of the observation vector to be mapped to a feature space area, the data in the transformed field are ranked based on the scale of importance of the content and the superiority of the extracted pattern. Associative memories address content [6]. It is the potential to pass from one internal illustration to another. It is also to infer a compound illustration from a part on the promise of associative memory. Its underlying task is to save associative patterns pairs upon performance of the matching stimulus pattern. The recognition phase is related to classification. In this phase, a label is assigned to an object on the basis of the descriptors' information [7].

1.2 Lung Cancer Detection Process

Overall, since the early 1970s, computer-supported image analysis has been used to predict and diagnose disease, mainly cancer. Likewise, digital image processing aids in timely and accurate diagnosis of cancerous tissue and reduces redundant surgeries. According to different studies, computerized image processing improves approximate 20% rate of cancer detection. This amount is enough to reduce the death rate. This approach has the potential to assist in cancer detection and precise therapy for less experienced clinicians and learners. Image processing may help increase lung cancer survival rates through early identification [8]. It is possible to use digital image processing for tissue differentiation, detection of lung lesions and nodules, tumor classification, and tumor growth count. Some of the steps in image processing for lung cancer detection include image capture, image enhancement, image segmentation, and feature extraction. The application of image enhancement can wipe out noise, degradation, and interfering of images. The next step, image segmentation, plays a significant part in recognizing the objects' details in critical portions.

Figure 1 represents a schematic diagram of lung cancer detection based on CT images.

The phases involved in the detection of lung cancer are defined as

a. **Image Capture**: CT scans are widely used to detect lung cancer. Computer-processed combinations of different X-ray images taken from different angles are used in CT scans to generate cross-sectional images of specific areas of the scanned object to allow the user to see inside the object without cropping. CT scans have many upsides unlike classic two-dimensional clinical radiography, for example, resolution with greater contrast, less noise [9].
b. **Image Enhancement**: The image pre-processing stage is initiated with the image enhancing procedure. This stage aims to enhance the interpretation or perception of the information available in the image for a human audience or provides enhanced input for other computerized image processing techniques. To accomplish this, 3 techniques have been implemented which are defined as follows:

   - Gabor Filter: An excellent local and multi-scale decomposition are composed when rendering an image based on the Gabor function, which are at the same time localized in the space and frequency domain [10] with respect to the logon. This is a linear filter. A harmonic function is used to describe its impulse response and is multiplied by a Gaussian function. The Fourier transform of the impulse response of this filter is due to the FT of the harmonic function as well as the convolution theorem of the Gaussian function.
   - Auto-enhancement: This technique is dependent on subjective observation and statistical functions such as mean and variance calculations.
   - Fast Fourier Transform: This method is applied on the FT of a provided image. The frequency domain is a space that represents the scale of variation in intensity levels of image "I" at a specific distance relative to F by using each image value at the location of the image, represented by F. The image filtering process is accomplished using the FFT [11].

c. **Image Segmentation**: This is a very important process for most of the subsequent works of analyzing an image [12]. In segmentation process, the image is divided into its constituent regions or objects. Slice by slice includes various useful applications for the medical professional when segmenting medical images into two-dimensional slices. These applications are used to visualize objects of interest, detect abnormalities, quantify tissue, and classify. This approach emphasizes simplifying and transforming the depiction of the image into something meaningful and easy to analyze. Images that implement image segmentation have objects and borders placed in them. Furthermore, this process is done to label each pixel in an image so that similarly labeled pixels share some of their visual characteristics. The product of image segmentation is a set of segments covering the entire image in a collective manner or a set of contours can be extracted from the image. The similarity is found among all the pixels in a given area related to some feature or computed property, i.e., color, intensity, or texture. Nearby regions differ considerably in the context of the same characteristic. The following are the two popular approaches of image segmentation [13]:

- Otsu's Thresholding Algorithm: This algorithm is implemented to automatically accomplish image thresholding based on clustering or to convert a gray level image into a binary image. It is assumed in Otsu's thresholding algorithm that there are two types of pixels contained in an image before the bimodal histogram [14]. The optimal threshold is determined for dividing the two types with the aim of minimizing their inter-class variation, even though their combined prevalence was minimal. When the distance between the background and the target is large, the difference between the two parts is maximized.
- Marker-Controlled Watershed Segmentation Approach: This approach is performed to extract seeds that indicate that the object or background is available at specific locations in the image. After that, the marker locations are set as regional minima in the topological surface, and the implementation of the watershed algorithm is completed. The challenging image processing operation is to isolate touching objects in an image that is specifically WT to deal with such an issue. There are two types of this approach [15]. The external type is associated with the background, and the internal is associated with objects of interest.

d. **Features Extraction**: This stage is very essential in which different desired parts or shapes of a given image are detected, and different algorithms and methods are applied. Binarization and masking are two techniques that can be used to predict the likelihood of lung cancer, based on facts that relate to information on lung anatomy and lung CT imaging.

- Binarization Approach: This approach is inspired from the idea that in normal lung images, black pixels outnumber white pixels. The average is obtained by counting the black pixels for normal or abnormal images. This average is treated as a range. Image is normal when the number of black images was found to exceed the threshold. However, a smaller number of thresholds is specified as the abnormal image.

- Masking Approach: This is linked to the truth that the percentage of cancer presence increases when the appearance of masses is similar to the white areas associated with the area of interest. The normal case is described by the appearance of blue, and the presence of cancer is indicated by the appearance of RGB masses. The TAR and FAR is this technique are counted as 85.7% and 14.3%, respectively [16].

  Whether the condition is normal or abnormal, it is decided on the basis of binarization and masking approach. These two perspectives are integrated, and decisions are made on behalf of the beliefs of these perspectives. The image is considered abnormal if there are a higher number of black pixels than white pixels but is it considered abnormal if the number of black pixels is less.

e. **Classification**: In this step, an array of classifiers is implemented to perform classification by means of driven features. The following are the most common algorithms for classifying lung cancer [17].

Support Vector Machine (SVM): The support vector machine (SVM) was devised by Vapnik and his fellows. This approach is employed to classify the linear and nonlinear data. SVM classifiers use nonlinear mapping to transform the original training data into greater dimensions. In this novel dimension, it seeks the linear best partition of the hyperplane which was called the "decision limit." A "decision boundary" can differentiate one class of tuples from another. The hyperplane can always divide the data into both classes when the data are mapped in suitably high dimensions with high nonlinearity. To find hyperplane, SVM uses support vectors and margins. This classifier generally performs well on classification problems. The SVM attempts to achieve the optimum compromise between the complexity of the model and the ability to learn based on partial sample information [18].

Naive Bayes (NB): The Naive Bayes is a probabilistic machine learning model. This model is more suitable for classification. The motivation behind this classifier is Bayes theorem. Bayes' theorem takes the form:

$$P(A \setminus B) = \frac{P(B \setminus A)P(A)}{P(B)}$$

The probability of occurrence of A, given that B has occurred, is found using this theorem. In this theorem, the proof is represented by B and hypothesis A. This classifier is based on the idea that predictors used are independent in nature. It specifies that the presence of one specific attribute does not affect the other characteristic. Thus, this algorithm is termed as "Naïve" [19].

Artificial Neural Network (ANN): Artificial neural network (ANN) is a kind of information processing system. These networks are rested upon the mathematical generality of the human neuron. The most non-complex configurations of an ANN include only one hidden layer. But this configuration closely resembles that of a brain neuron network, because this structure includes vast interconnections between neurons in successive layers. The figure shows a typical composition of ANN. In

Fig. 4, all the circular nodes of the neural network generate their own artificial neuron. Here, arrows indicate the possible connection medium between neurons. These arrows model the paths through which information passes. In this way, these arrows simulate the neuronal functions of various biological organisms, such as the human brain. Full connectivity and persistent communication between all neurons describe the ANN model, and this is done to enable each neuron to receive input and perform basic operations on the input data. The output activation of neurons is transferred through links to other neurons. This connection or link allocates load to a signal to modify its strength. It is important to note that information only flows in one direction. This specifies that the connection does not get any feedback of the information. The knowledge of the ANN is stored in a differently distributed manner post training to establish connections between neurons, much like synopsis in a biological brain. The information saved like this can be dealt with from a particular point of view. Furthermore, an input, similar to a trained one, which can produce similar results, is interrogated from this point of view. Thus, ANNs are capable of learning and generalizing. The weight optimization base on the cost function achieves the ability to maximize the overall forecast efficacy.

2 Literature Review

2.1 Lung Cancer Detection Using Deep Learning

Fang et al. [20] suggested a new deep learning (DL)-based system in order to detect the lung cancer which provided accuracy and stability. The transfer learning (TL) model was utilized to develop the convolutional neural network (CNN) model similar to a GoogLeNet. Unlike the previous methods, multi-view attributes of three-dimensional (3D) computed tomography (CT) scans were comprised using median intensity projection (MIP). The LIDC-IDRI dataset containing lung nodule images was applied to quantify the suggested system, and 100-fold data augmentation was carried out for ensuring the training efficacy. The suggested system had provided the accuracy around 81%, sensitivity of 84%, and specificity of 78% in comparison with other programs. An artificial neural network backpropagation (ANN-BP)-based gray level co-occurrence matrix (GLCM) was introduced by Anifah et al. [21] for detecting the lung cancer. The lung data that had 50 CT images were gathered from the cancer imaging archive database. The images were divided as normal or cancer affected. The image was pre-processed and segmented, attributes were extracted, and a backpropagation neural network (BPNN) was implemented to detect the lung cancer. The results depicted that the introduced system yielded the accuracy up to 80% while detecting the lung cancer. A fully convolutional network (FCN)-based technique was developed by Chunran et al. [22] for detecting and segmenting the lung cancer. First of all, the developed technique employed the lung CT images to segment the lung. Afterward, the threshold technique and other image processing methods were utilized to detect the lung nodules in the lung region. In the end, the level set technique and threshold method had applied for segmenting the detected lung nodules and their spiculation on the basis of coordinate system transformation. The experimental result exhibited that the developed technique was efficient in detecting and segmenting the lung nodules with 100% accuracy. A three-dimensional convolution neural network (3D-CNN) was formulated by Jin et al. [23] for detecting the lung cancer using segmented CT lung volumes as training and testing samples. The novel system was assisted in extracting and projecting 3D attributes to the following hidden layers so that the temporal relation was preserved among the neighboring CT slices. The model was useful to classify the patient as cancer infected or healthy. ReLU nonlinearity and sigmoid function were used as activation techniques and classifiers. The formulated system offered the accuracy around 87.5% using the biomedical images themselves as the input dataset. The formulated system attained error rate around 12.5% and enhanced the classic AlexNet architecture by 2.8%.

2.1 Comparison Table

Author	Year	Technique used	Dataset	Results
Tiantian Fang et al.	2018	Convolutional neural network (CNN)	LIDC-IDRI	The suggested system had provided the accuracy around 81%, sensitivity of 84% and specificity of 78%
Lilik Anifah et al.	2017	Artificial neural network backpropagation-based GLCM	Cancer imaging archive database	The introduced system yielded the accuracy up to 80% while detecting the lung cancer
Yang Chunran et al.	2018	Fully convolutional network (FCN)	LIDC database	The developed technique was efficient in detecting and segmenting the lung nodules with 100% accuracy
Taolin Jin et al.	2017	3D CNN network architecture	Kaggle	The formulated system offered the accuracy around 87.5% using the biomedical images themselves as the input dataset

2.2 *Lung Cancer Detection Using Machine Learning*

Krishna et al. [24] established multi-layered perceptron backpropagation neural network (MLP-BPNN) on the basis of scale invariant feature transform (SIFT) to extract and attributes. The bag of words (BoW) method was put forward for detecting the lung cancer. Around 300 lung images were gathered from the Rajiv Gandhi Cancer Institute and Research Center, Delhi as a dataset in which 100 images were employed in the testing phase and rest of the images were utilized in the training stage. The established algorithm provided the accuracy rate around 89% in comparison with other methods for detecting the lung cancer. A new neural network (NN)-based algorithm recognized as entropy degradation method (EDM) was designed by Wu and Zhao [25] with the objective of detecting small cell lung cancer (SCLC) from computed tomography (CT) images. This approach was useful in detecting the lung cancer at premature phase. The CT scans of lung were obtained from the National Cancer Institute (NCI) for data utilized to train and test the approach. The selection of 12 lung CT scans was done from the library in which half images were related to healthy lungs, and rest were patients of SCLC. The results indicated that the accuracy attained from the designed approach was calculated 77.8%. An efficient algorithm of detecting the lung cancer was intended by Alam et al. [26] in which multi-class support vector machine (SVM) scheme was deployed. The cancer was detected using multi-stage classification. The image was enhanced and segmented in the classification process. Threshold and marker-controlled watershed-based technique were adopted to segment the image. Support vector machine (SVM) binary classification model was put forward for classifying the lung image. The intended scheme was proved more efficient with regard to superior accuracy while detecting the lung cancer. A three-dimensional (3D) deep multi-task convolutional neural network (CNN) was put forward by Khosravan and Bagci [27] for detecting the lung cancer. This technique was capable of tackling the drawbacks of labeled data to accomplish the 3D segmentation task. LUNA16 dataset was utilized to test the presented technique. The results revealed that the presented technique provided the dice similarity coefficient (DSC) around 91% as accuracy and a score around 92%. Moreover, this technique was more effective in contrast to two baselines.

2.2 Comparison Table

Author	Year	Technique used	Dataset	Results
Azmira Krishna et al.	2018	MLP-BPNN based on SIFT	MATLAB 2018	The established algorithm provided the accuracy rate around 89%
Qing Wu et al.	2017	Entropy degradation method (EDM)	MATLAB	The accuracy attained from the designed approach was calculated 77.8%

(continued)

(continued)

Author	Year	Technique used	Dataset	Results
Janee Alam et al.	2018	Multi-class support vector machine (SVM)	LIDC database	The intended scheme was proved more efficient with regard to superior accuracy
Naji Khosravan et al.	2018	3D deep multi-task CNN	LUNA16 dataset	The presented technique provided the dice similarity coefficient (DSC) around 91% as accuracy and a score around 92%

2.3 *Lung Cancer Detection Using Image Processing*

Vas and Dessai [28] recommended a mechanism which focused on constructing an automated system in order to detect the lung cancer with the help of mathematical morphological operations so that lung region of interest (ROI) was segmented. These operations provided the Haralick attributes. Application of median filter was proved effective in removing the impulse noise in the images. The morphological operations also generated optimal outcomes in the process of segmentation. The artificial neural network (ANN) was employed to classify the images with higher accuracy. The results acquired on hospital database indicated that the recommended mechanism yielded the accuracy of 92% to detect the lung cancer. A CT scan-based system was devised by Firdaus et al. [29] to detect the lung cancer. The lung cancer was classified as benign and malignant using the devised system. For this purpose, the CT scan images were deployed. These images were processed to assist the clinical sector in diagnosing the lung cancer effectively. The results validated that the devised system attained 83.33% accuracy on the system decision to determine the lung image as cancerous or non-cancerous. An innovative fast thresholding-based lung segmentation was suggested by Huidrom et al. [30] in which threshold value was not investigated. The accuracy obtained from this method was found near to the accuracy of the previous technique. However, this technique worked more quickly in comparison with the existing technique. Thus, the results exhibited that the suggested technique had provided superior accuracy while segmenting the images for detecting the lung cancer. An automated method which deployed computed tomography (CT) images was constructed by Hoque et al. [31] with the purpose of detecting the lung cancer at its premature phase. This method emphasized on providing greater accuracy. A novel model was also put forward to diagnose the lung cancer with the help of a variety of features extracted from CT images in which image was enhanced and segmented, and attributes were extracted using support vector machine (SVM). In the end, the results of experiments demonstrated that the constructed performed well with regard to accuracy.

2.3 Comparison Table

Author	Year	Technique used	Dataset	Results
Moffy Vas et al.	2017	Mathematical morphological operations	Hospital database	The recommended mechanism yielded the accuracy of 92% to detect the lung cancer
Qurina Firdaus et al.	2020	CT scan-based image-based lung cancer detection system	Open CV	The devised system attained 83.33% accuracy on the system decision
Ratishchandra Huidrom et al.	2017	A new fast thresholding-based lung segmentation	Lung image database consortium (LIDC)	The suggested technique had provided superior accuracy while segmenting the images for detecting the lung cancer
Ariful Hoque et al.	2020	An automated approach	MATLAB	The results of experiments demonstrated that the constructed performed well with regard to accuracy

3 Research Methodology

In this section of the paper, the research methodology is discussed which is used for the lung cancer detection. This section is divided into dataset description and proposed methodology which are described below.

3.1 Dataset Description

The dataset is designed to allow for different methods to be tested for examining the trends in CT image data associated with using contrast and patient age. The basic idea is to identify image textures, statistical patterns, and features correlating strongly with these traits and possibly build simple tools for automatically classifying these images when they have been misclassified (or finding outliers which could be suspicious cases, bad measurements, or poorly calibrated machines). The data are a tiny subset of images from the cancer imaging archive. They consist of the middle slice of all CT images taken where valid age, modality, and contrast tags could be found. This results in 475 series from 69 different patients.

3.2 Proposed Model

This research work deals with the detection of lung cancer from the CT scan image with the help of image processing methods. The suggested mechanism has 4 stages to localize and characterize the lung cancer. The stages to detect the lung cancer are defined as (Fig. 2):

1. **Pre-processing**: It is the initial stage that takes CT scan image as an input. The method of image de-noising is implemented for eliminating the noise from the input image.
2. **Segmentation**: This stage makes the deployment of region-based segmentation for segmenting the same and distinct regions from the CT scan image. A digital image is segmented into diverse sections while segmenting the image such that sets of pixels are known as super-pixels. This stage emphasizes on changing the image exhibition easily. The position of objects, limits, and borders is recognized

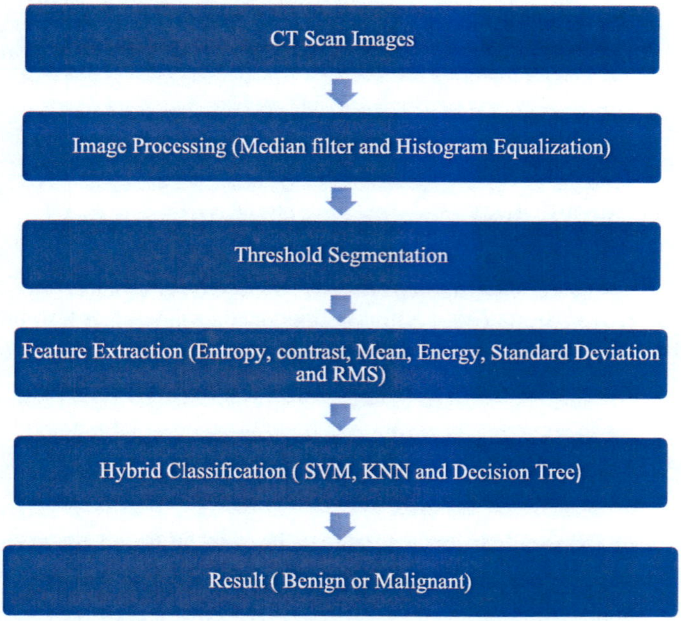

Fig. 2 Proposed model

in the images using this approach. This process is executed to assign every pixel in an image and share the definite attributes through the pixels having the identical label. The image can be segmented using a number of methods.

3. **Feature Extraction**: This stage employs a gray level co-occurrence matrix (GLCM) in order to extract the attributes from the CT scan image. The attributes acted efficiently in the image processing. The attributes are acquired using various methods, namely binarization, thresholding, normalization, masking approach, etc., on the sampled image. Afterward, several methods of feature extraction are utilized to obtain the attributes so that the descriptions are classified and detected. A number of methods are implemented to extract the attributes. The textual features are extracted from an image using GLCM. There are six attributes extracted to detect the lung cancer.
4. **Classification**: It is the final stage that deploys a hybrid classification approach with the objective of categorizing and localizing the cancer region. In this approach, the support vector machine (SVM), K-nearest neighbor (KNN), and decision tree (DT) are integrated. The SVM classifier is planned on the basis of the optimization theory. This algorithm assists in increasing the margin; thus, it is called a binary classification model. The best hyper plane is exploited to split all the data points of an individual class. The classification performed using SVM recognizes this hyper plane. The greatest margin amid two classes are considered to define the significant hyper plane. The interior data points are not present in case of maximum width amid the slabs parallel to the hyperplane called margin. The support vector machine is adopted to split the maximum margin in hyperplane. The training sets are described using n-dimensional arithmetic qualities. Each training sample exposes a point in the n-dimensional region. The integration of more effective component of the training samples is done into an n-dimensional sample space in addition to these lines. The K-nearest neighbor classification algorithm is recognized for the sample space for the k training samples which are present nearby to the unidentified model when training sample is unknown. The concept of Euclidean distance is utilized to define the closeness. These classification algorithms are considered as non-parametric supervised learning methods and assisted in classifying and explaining the images. This technique focuses on constructing a framework in order to predict a target variable accurately on the basis of various key variables. This technique allows every core nodule to communicate with other key variables. The value of the intended variable is depicted by every side. This technique exploits a key trait to assign label to each interior nodule. Every probable characteristic value is utilized to assign a label to the rounded sections whose generation is done from a nodule called a trait. The bagging process is executed to put together the output of KNN, SVM, and DT. The bagging process chooses an output of an efficient classification method to detect the lung cancer.

4 Experiment Result Analysis

A dataset containing the medical record of four patients is collected from the hospital. These subjects are detected with a specific type of brain tumor. All these subjects have malignant tumor, and they are not undergone surgery yet. Data sourced from Midnapore Diagnostics Pvt. Ltd., R G Kar Medical College and Hospital, Kolkata (West Bengal). Midnapore Diagnostics Pvt. Ltd. is a joint project with the Department of Health and Family Welfare, Government of India. Of West Bengal. Overall, 20 MR scans screening brain tumors have been discovered the entire database. Finally, we use the implementation under MATLAB software. To implement new approach, the tools used are robot vision and neural networks (NNs). To examine performance level, the parameter considered is accuracy.

Figure 3 depicts the processing of the interface. To do so, the different operations in hybrid classification are performed to detect the lung cancer. The methods implemented to diagnose lung cancer from the CT scans are histogram equalizer, segmentation, filtering, diluting and filtering (Table 1).

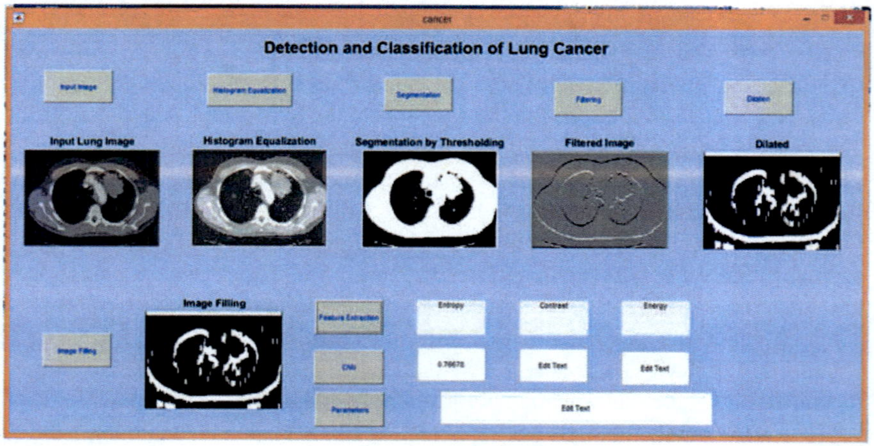

Fig. 3 Marking of tumor

Table 1 Performance analysis

Parameter	Baseline approach (%)	New approach (%)
Accuracy	89	95
Precision	85.4	89
Recall	86	88

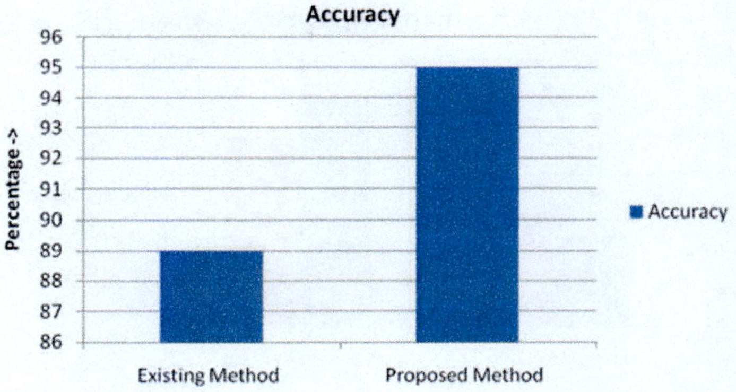

Fig. 4 Accuracy analysis

4.1 Accuracy Analysis

The accuracy of the proposed model which is the hybrid classification model and it is the combination of decision tree, KNN, and SVM for the lung cancer detection is compared with existing SVM classification model.

Figure 4 depicts comparison between the new and the baseline approaches with respect to the accuracy. The hybrid classifier architecture achieves better accuracy in compassion with the baseline approach. The hybrid classifier is an automated framework that uses hog and surf features to train the system in order to detect the malignant area.

4.2 Precision and Recall Analysis

The precision and recall value of the proposed model are compared with the existing model for the lung cancer detection. The hybrid classification is the combination of decision tree, SVM, and KNN classifier.

Figure 5 depicts comparison between the new and the baseline approaches with respect to the precision and recall. The hybrid classifier architecture achieves better precision and recall rates in compassion with the baseline approach. The parameters of precision and recall define the reliability of the newly constructed framework. The reliability of the new framework refers to how accurate it is for detecting lung cancer.

5 Conclusion

A significant growth in the cancer affected people has been noticed in the last few years. Way of living is the major cause of cancer in the present scenario. Lung cancer

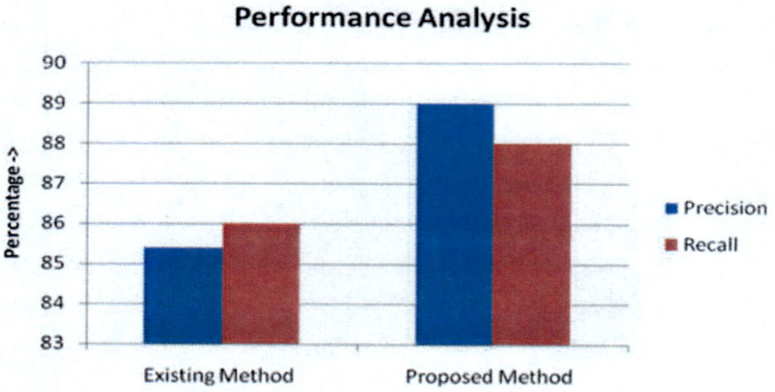

Fig. 5 Precision and recall analysis

has become the second most common cancer-causing death in both ladies and gents. From preliminary baseline reports on cancer patients, it has been observed that lung cancer causes a higher number of deaths than other cancers. The factors that cause lung cancer include smoking (or second-hand smoke), or less often exposure to other ecological dynamics, is therefore preventable. This research identifies and classifies region of interest by enforcing three steps of segmentation, feature extraction, and hybrid classification. This work takes three parameters (i.e., accuracy, precision, and recall) under consideration to evaluate the designed hybrid classifier model. According to the analytic outcomes, the new architecture outperforms those proposed in the state of the art.

References

1. Roy, K., Chaudhury, S. S., Burman, M., Ganguly, A., Dutta, C., Banik, S., & Banik, R. (2019). A comparative study of lung cancer detection using supervised neural network. In *International Conference on Opto-Electronics and Applied Optics (Optronix).*
2. Jony, M. H., Johora, F. T., Khatun, P., & Rana, H. K. (2019). Detection of lung cancer from CT scan images using GLCM and SVM. In *1st International Conference on Advances in Science, Engineering and Robotics Technology (ICASERT).*
3. Günaydin, Ö., Günay, M., & Şengel, Ö. (2019). Comparison of lung cancer detection algorithms. In *Scientific Meeting on Electrical-Electronics & Biomedical Engineering and Computer Science (EBBT).*
4. Devarapalli, R. M., Kalluri, H. K., & Dondeti, V. (2019). Lung cancer detection of CT lung images. *International Journal of Recent Technology and Engineering (IJRTE).*
5. Shukla, A., Parab, C., Patil, P., & Sangam, S. (2018). Lung cancer detection using image processing techniques. *International Research Journal of Engineering and Technology (IRJET).*
6. Mithuna, B.N., Ravikumar, P., & Arpitha, C. N. (2018). A quantitative approach for determining lung cancer using CT scan images. In *Second International Conference on Electronics, Communication and Aerospace Technology (ICECA)*
7. Lobo, P., & Guruprasad, S. (2018). Classification and segmentation techniques for detection of lung cancer from CT images. In *International Conference on Inventive Research in Computing Applications (ICIRCA).*

8. Makaju, S., Prasad, P. W. C., Alsadoon, A., Singh, A. K., Elchouemi, A. (2018). Lung cancer detection using CT scan images. *Procedia Computer Science.*
9. Kaucha, D. P., Prasad, P. W. C., Alsadoon, A., Elchouemi, A., & Sreedharan, S. (2017). Early detection of lung cancer using SVM classifier in biomedical image processing. In *IEEE International Conference on Power, Control, Signals and Instrumentation Engineering (ICPCSI)*
10. Anifah, L., Harimurti, R., Permatasari, Z., Rusimamto, P. W., & Muhamad, A. R. (2017). Cancer lungs detection on CT scan image using artificial neural network backpropagation based gray level co-occurrence matrices feature. In *International Conference on Advanced Computer Science and Information Systems (ICACSIS).*
11. Fule, S. (2017). Lung cancer detection using image processing techniques. *International Research Journal of Engineering and Technology (IRJET).*
12. Abdillah, B., Bustamam, A., & Sarwinda, D. (2016). Image processing-based detection of lung cancer on CT scan images. In *The Asian Mathematical Conference.*
13. Dhaware, B. U., & Pise, A. C. (2016). Lung cancer detection using Bayesian classifier and FCM segmentation. In *International Conference on Automatic Control and Dynamic Optimization Techniques (ICACDOT).*
14. Avinash, S., Manjunath, K., & Senthil Kumar, S. (2016). An improved image processing analysis for the detection of lung cancer using Gabor filters and watershed segmentation technique. In *International Conference on Inventive Computation Technologies (ICICT).*
15. Usman, M., Shoaib, M., & Rahal, M. (2013). Lung cancer detection using digital image processing. In *PIERS Proceedings*, Stockholm, Sweden.
16. Al-Tarawneh, M. S. (2012). Lung cancer detection using image processing techniques. *Leonardo Electronic Journal of Practices and Technologies.*
17. Chaudhary, A., & Singh, S. S. (2012). Multiresolution analysis technique for lung cancer detection in computed tomographic images. *International Journal of Research in Engineering & Applied Sciences, IJREAS.*
18. Al-Tarawneh, F. S. (2012). Lung cancer detection using image processing techniques. *Leonardo Electronic Journal of Practices and Technologies.*
19. Bandyopadhyay, S. K. (2012). Edge detection from CT images of lung. *International Journal of Engineering Science & Advanced Technology.*
20. Fang, T. (2018). A novel computer-aided lung cancer detection method based on transfer learning from GoogLeNet and median intensity projections. In *IEEE International Conference on Computer and Communication Engineering Technology (CCET).*
21. Anifah, L., Harimurti, R., Permatasari, Z., Rusimamto, P. W., & Muhamad, A. R. (2017). Cancer lungs detection on CT scan image using artificial neural network backpropagation based gray level coocurrence matrices feature. In *International Conference on Advanced Computer Science and Information Systems (ICACSIS).*
22. Chunran, Y., Yuanvuan, W., & Yi, G. (2018). Automatic detection and segmentation of lung nodule on CT images. In *11th International Congress on Image and Signal Processing, BioMedical Engineering and Informatics (CISP-BMEI).*
23. Jin, T., Cui, H., Zeng, S., & Wang, X. (2017). Learning deep spatial lung features by 3D convolutional neural network for early cancer detection. In *International Conference on Digital Image Computing: Techniques and Applications (DICTA).*
24. Krishna, A., Srinivasa Rao, P.C., & Basha, C. Z. (2020). Efficient computerized lung cancer detection using bag of words. In *7th International Conference on Smart Structures and Systems (ICSSS).*
25. Wu, Q., & Zhao, W. (2017). Small-cell lung cancer detection using a supervised machine learning algorithm. In *International Symposium on Computer Science and Intelligent Controls (ISCSIC).*
26. Alam, J., Alam, S., & Hossan, A. (2018). Multi-stage lung cancer detection and prediction using multi-class SVM classifier. In *International Conference on Computer, Communication, Chemical, Material and Electronic Engineering (IC4ME2).*

27. Khosravan, N., & Bagci, U. (2018). Semi-supervised multi-task learning for lung cancer diagnosis. In *40th Annual International Conference of the IEEE Engineering in Medicine and Biology Society (EMBC)*.
28. Vas, M., & Dessai, A. (2017). In *International Conference on Computing, Communication, Control and Automation (ICCUBEA)*
29. Firdaus, Q., Sigit, R., Harsono, T., & Anwar, A. (2020). Lung cancer detection based on CT-scan images with detection features using gray level co-occurrence matrix (GLCM) and support vector machine (SVM) methods. In *International Electronics Symposium (IES)*
30. Huidrom, R., Chanu, Y. J., Singh, K. M. (2017). A fast automated lung segmentation method for the diagnosis of lung cancer. In *IEEE Region 10 Conference*.
31. Hoque, A., Ashek Farabi, A. K. M., Ahmed, F., & Islam, M. Z. (2020). Automated detection of lung cancer using CT scan images. In *IEEE Region 10 Symposium (TENSYMP)*.

Overview of Security Approaches Using Metamorphic Cryptography

Lokesh Negi and Lalit Negi

Abstract Initially, researchers employed only one information security technique either cryptography or steganography to secure the communication. But later, researchers stress on the amalgamation of both cryptography and steganography, and this amalgamation is popularly known as metamorphic cryptography. Steganography can be classified on the basis of cover medium. This paper surveys the different metamorphic cryptography approaches which uses image as cover media for securing the data. This paper also covers general concepts of cryptography, steganography, classification of metamorphic cryptography, and evaluation parameters like PSNR, MSE.

Keywords Metamorphic cryptography · Steganography · Image steganography · Symmetric key metamorphic cryptography · Asymmetric key metamorphic cryptography · PSNR · MSE

1 Introduction

In today's world, millions of people exchange their confidential data through different mediums. As we all know, these mediums may be secure or not. Because of this uncertainty in the security of the medium, people prefer to send confidential data in the encoded form. People mainly use two different data hiding techniques to achieve the same. In the past, cryptography and steganography are two techniques which have been used alone by the researchers to protect their confidential data. Now, researchers emphasize on the use of metamorphic cryptography for securing the data. Steganography can be of different types, and the classification is based on the cover medium like image steganography which uses an image as a cover medium.

L. Negi (✉)
CSE Department, Netaji Subhas University of Technology, Delhi, India
e-mail: lokeshnegi97@gmail.com

L. Negi
IT Department, Netaji Subhas University of Technology, Delhi, India

P. K. Singh et al. (eds.), *Proceedings of Third International Conference on Computing, Communications, and Cyber-Security*, Lecture Notes in Networks and Systems 421,
https://doi.org/10.1007/978-981-19-1142-2_66

2 Metamorphic Cryptography

Cryptography deals with the techniques which are used for secure communication and helps in preventing the eavesdroppers from understanding the message that is shared between source and destination. Cryptography provides the confidentiality of data by transforming the plaintext message into unreadable format called ciphertext. This transformation is carried out in sender site using key and appropriate algorithm, and to thwart the effect of transformation in receiver site, receiver applies the same key (or associated key to sender key in case of asymmetric encryption) to the algorithm and gets the original plaintext that the sender wants to send in place of ciphertext. This transformation is popularly known as encryption. Ciphertext can only be deciphered by that party which has the possession of secret key, and knowledge of algorithm is used in its encryption process.

Steganography means the covered writing. It is defined as the activity of concealing the secret data in a cover or various kind of container media so that its presence can be traced only by intended party not by anyone else. Text, image, audio, and video are used as cover media for holding the secret input message. Steganography techniques embed the secret data into the cover media, and then the resultant embedded media is shared to the other communication party instead of secret data.

Metamorphic cryptography is the combination of cryptography and steganography. Philjon et al. [1] in 2011, proposed the first metamorphic cryptography approach for fool-proof information security. Generally, metamorphic cryptography (see Fig. 1) first encrypts the data using the cryptography technique, and then, the appropriate steganography technique is chosen to embed the confidential information. Metamorphic cryptography is more secure than using the cryptography or steganography alone.

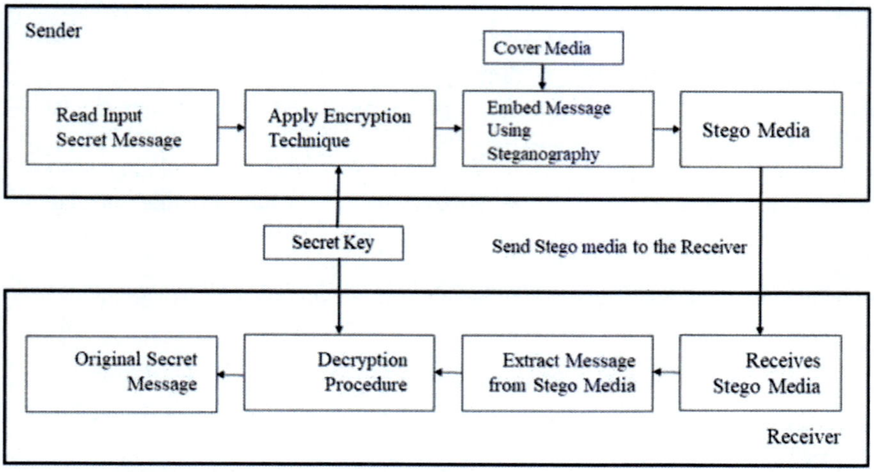

Fig. 1 General metamorphic cryptography system

2.1 Terminologies

- Message: Confidential data that it is meant to be shared.
- Cover or Container Media: Cover media is the media which is used to conceal or hide the message. Cover media may be image, audio, or video.
- Stego-Media: The media file that is generated after embedding of message into the cover media is called as stego-media. Sender sends stego-media to receiver for sharing the message.
- Stego-Key: Key which is used to encrypt or decrypt the message.
- Embedding Algorithm: Embedding algorithm is a defined procedure through which the message is to be hidden inside chosen cover media.
- Extraction Algorithm: Operation of extraction algorithm is just apposite to embedding algorithm, unlike embedding, extraction unhides the data from stego-media.

2.2 Evaluation Parameters

PSNR and MSE are the parameters which are often used in measuring the quality of two images. PSNR is used as measure which evaluates the degradation in stego-image with respect to cover image or simply say PSNR tells the similarity between images. PSNR can be calculated using the

$$\mathrm{PSNR} = 10\log_{10}\left[\frac{I^2}{\mathrm{MSE}}\right] \tag{1}$$

where I defines the maximum intensity value of image and MSE is the mean square error between the stego-object and container image. Higher PSNR gives an indication of higher image quality [2]. Unit of PSNR is decibels and is inversely proportional to the mean square error. MSE can be mathematically computed using the below equation

$$\mathrm{MSE} = \frac{1}{[P \times Q]^2}\sum_{i=1}^{P}\sum_{j=1}^{Q}\left(X_{ij} - Y_{ij}\right)^2 \tag{2}$$

In MSE calculation, X_{ij} and Y_{ij} represent the intensity of ijth pixel of container image and image generated after embedding, respectively. P is total count of rows present in cover image, and Q is total count of columns in image.

2.3 Classification

Metamorphic cryptography can be classified into two types: one is symmetric metamorphic cryptography and other one is asymmetric cryptography. In symmetric key

Table 1 Symmetric key metamorphic cryptography versus asymmetric key metamorphic cryptography

	SKMC	ASKMC
Keys	Single key	Two keys
Security principles	Provides confidentiality	Confidentiality, authenticity, non-repudiation
Computational efficiency	Very fast	Slow
Size of ciphertext	Smaller than plaintext	Same or larger than plaintext
Hardware	Requires inexpensive hardware	Expensive hardware
Application	Bulk encryption	Key distribution or small amount of data

metamorphic cryptography (SKMC), only single key is used to cipher or decipher the text. Receiver uses the same key for decryption which sender used for encryption process. This single key can be known as public key or symmetric key.

In asymmetric metamorphic key cryptography (ASKMC), two keys are used by the communicating principles. One key is used by sender for encrypting the plaintext, which is known as public key, and the receiver uses the another key known as private key for decrypting the ciphertext. Table 1 lists some differences between the SKMC and ASKMC.

3 Literature Survey

Joshi and Yadav in [3] used a new steganographic method which uses LSB and circular shift. In the LSB-S steganographic algorithm, two operations, one is circular left shift and second is XOR operation which is performed on the last 4-bits of the pixel. The proposed system first encrypts the data using the transpositional vernam cipher after this LSB-S conceals the data in the chosen carrier image. The extraction algorithm is performed to retrieve the covered text and then retrieved text is decrypted with the Vernam cipher as the Vernam cipher is symmetric cipher, so the cipher used for decryption is same. The new LSB-S is advantageous as it gives 100% hiding capacity with the benefits of perceptual degradation.

Alotaibi et al. [4] proposed a new metamorphic approach to ensure confidentiality, integrity, and authenticity. The proposed approach has been used in the security authentication systems in mobile devices. The proposed system takes the password and a key and an image from the users. The SHA-256 algorithm computes the hash of the password which later encrypted by AES algorithm with the username as a key. The user selects the image to embed the data in it using the popular steganographic algorithm LSB. The output showed that all the three attributes of information security, i.e., confidentiality, integrity, and authentication have been achieved.

The affine transform and the DWT have been used by Sharma et al. in [5] for the encryption of the secret message. The proposed system first divides the message into groups, and then, each group is encrypted using the affine transform with a key. For encrypting each group, there is a different key used. After encrypting, the ciphered data is now transformed into the frequency domain using the DWT. Finally, the obtained data and the RGB component of the cover image are coupled by storing these at LSB of the image using the LSB steganography. To retrieve the secret message, first retrieve the information from the image and apply the inverse DCT on the data, and then at last, apply the affine transport to get the original message. The result showed the output of cover-stego-image looks similar with the input cover image. The proposed system is more robust than the existing key as the robust of a system depends on the size of key.

Al-Qwider and Salameh [6] presented the crypto stego system which uses the modified Jamal encryption algorithm for encrypting the secret message. The MJEA divides the messages into 64-bit blocks, and then, each block is encrypted separately and performs eight rounds along with the initial and final permutation. The initial permutation outputs the 64-bit blocks into eight groups and each group is of 8 bits. In each round, the XORed operation is performed in all these 8 bits and the final permutes combine these eight groups into the 64-bit block to produce the ciphertext. The least 3-3-2 bits of the RGB components is selected to hold the ciphertext of the cover image. Embedding of other information like the encrypted steg key, and the message length is done at the last rows of the cover image. The proposed system is also resistant to the attacks that use the histogram analysis. The weakness of algorithm: it uses 24-bit color image so it has less embedding capacity than system using the 32-bit image.

Shanthakumari and Malliga [7] presented a dual-layer security method for the data. ECC scheme is used in the first layer for encryption. Elliptic curve cryptography (ECC) takes two coefficients and a prime to generate the curve points and after generating the curve points, the alphabets, numerals, and special characters are mapped with the elliptic curve points and then takes the secret data and maps the data character by character with the elliptic curve points. The second layer uses the concept of steganography to conceal the message into the image. Least significant bit inversion (LSBI) algorithm embeds the elliptic curve points in the carrier image. In the LSBI, each 4-pixels of cover image holds 6-bits of information. The proposed system has a time and space complexity of $o(n)$. Also, the proposed strategies have been checked through the two popular steganalysis (histogram analysis and chi-square) attack.

A new metamorphic cryptography proposed by Ahmed and Ahmed in [8] uses the double-XOR operations and LSB as a steganographic algorithm. The proposed algorithm converts the message into the decimal values based on ASCII standard, and then, binary value is calculated for the decimal value. Take the MSB values of image pixels having the size same as the message, and these values work as the key for encryption. XORing of secret data with the generated key is used to compute the value for encrypted data, and then, it is embedded inside the image using LSB. The extraction algorithm inputs the message size to extract the hidden information. MSB of binary values of stego-image pixels are XOREd with the hidden information to get the original message.

In the presented paper [9], Candra et al. used three different algorithms: OTP, LSB, and Canny edge detection. The proposed algorithm uses the edge of the cover image to hide the text as a small change in the pixel value in this area cannot be easily detected by humans. The algorithm first converts the message into the binary form using ASCII codes. With the help of Canny algorithm, detect the edge in the cover image and store it into a variable (v). Generate a key for the OTP encryption with the help of a random matrix. And finally, encrypt binary values of the secret message with the help of key and OTP cipher to get the encrypted message and locate the image edge area using the variable (v). Finally, the embed the encrypted message into the 8th bit to the image edge area.

Marwa et al. modified the AES algorithm for the encryption in [10]. The AES has been modified using the MPK encoding. The MPK represents each character by using the 2 digits instead of 8 bits. The modified AES first represents the secret message using the mobile phone encoding and encrypts it using the AES and then produces the output also in the form of MPK. So, the steganographic algorithm takes the MPK encoded encrypted data as input and stores into the cover image using the modified substitute last digit in pixel using the mobile phone keypad and pixel value difference schemes. In the steganographic algorithm, the cover image is divided into blocks of non-consecutive fashion the consecutive pixels, and then, there is a difference for each block, and this difference is checked in the range table whether it belongs to lower level or not. If it belongs to the lower level, then embed the MPK encoded encrypted data into the image using the PVD-MPK method, otherwise embeds into the image using the MSLDIP-MPK method.

Islam et al. [11] introduced and proposed a new method for selecting the pixel in which the message is embedded. The proposed pixel selection algorithm is the pixel filtering method. The pixel filtering method maintains the counter for each RGB components. The counter for each RGB component is incremented if the MSB for corresponding component is 1. At last, the component which has the high value is selected. After selecting the pixel, the symmetric block AES encrypts message. The embedding procedure divides password into blocks of 3 bits and calculates its corresponding decimal value (P). The XORed operation is performed between the P and the message bits, and then, according to the result, set the value of the component to 0 or 1, and perform the above procedure again until all the messages are embedded. The algorithm has high PSNR as well as provides the high security.

Shanthakumari and Malliga in [12] developed a metamorphic cryptography in which IDEA algorithm has been used for the encryption part and group LSB for the steganography. The IDEA algorithm takes 16-bit plaintext and performs four cycles of transformation along with the bitwise XOR and two modular operations, i.e., addition modulo and multiplication modulo with the help of a 32-bit key. IDEA encrypted data is hidden using the LSBG image steganography in the image. The LSBG converts the encrypted message into the stream of bits using the ASCII code and partitions the cover image into the four 8-bit planes for the embedding purpose. The proposed algorithm has been evaluated through three assessments: one is for quality which has been evaluated through the parameters including PSNR, MSE, SSIM, AD and others, and the second one is for the capacity or payload, i.e., the maximum capacity of message that can be hidden into the cover media without distorting the media and the third is for resistance to attacks. Steganalysis attacks, visual analysis are done in the third assessment along with the chi-square. The result of the proposed algorithm successfully completed all the evaluation assessment.

The system developed by Murad et al. in [13] produces two output images instead of a single image. The proposed system uses the two cryptographic algorithms: Blowfish, visual cryptography. The proposed system is very different from the existing system as it converts the text into two images, and then, two stego-images are generated by hiding the 1 generated image into the 1 cover image. The system divides the algorithm into three phases: E1, E2, and E3. In the E1 phase, the Blowfish algorithm generates the cipher, and then, this cipher or encrypted text is fed as input to the E2. The E2 converts the encrypted message into two images using the (2, 2) visual cryptography. Finally, the last phase E3 or the steganographic phase takes these two generated images and two cover images and embeds one generated image inside the 1 cover image using the LSB algorithm and similar for another generated image, and this phase outputs two stego-images. The system is resistant to the visual attacks.

A secure communication is utmost need nowadays. To fulfill this need, Alexan and Hemeida [14] designed a system by incorporating steganography with cryptography which come up with good results for evaluation parameter. The proposed system used strong and fast encryption algorithm Blowfish for ciphering the secret message with help of 256-bit key. Other reasons for using Blowfish algo are its simple implementation and high throughput. System firstly encrypts the secret message using Blowfish before its concealment in image. In second layer, encrypted data is being hidden inside the carrier image by using the LSB substitution with arithmetic sequence and infinite sequence, namely Rudin Shipro sequence. Value of pixel location where data is to be hidden is calculated by using arithmetic sequence with common difference 5 and Rudin Shipro sequence computes which color channels is to be used embedding the data. Embedding algorithm of the system finally hides the encrypted secret data in the appropriate pixels and color channel of image. The merit of system lies in the fact that it provides high embedding capacity while preserving good quality for stego-image.

In paper [15], Shivani et al. presented a metamorphic cryptography in which the process of encryption is done on the input secret text, and then, the steganographic technique is performed. The encryption process encrypts the data by performing the XOR between the three inputs. Before encryption, the three-digit input random key is converted into a single-digit random key by using the folding method in which the digit of the sum is calculated recursively until a single digit does not come, and then, this single-digit random key and ASCII values of the input message and the length of message are XORed. The concept of raster scan is used along with the LSB algorithm in the embedding procedure. The authors tested the system using the quantitative analysis through the parameters like PSNR and MSE and qualitative analysis to show the visualization of both the images (input image or output image): cover image and stego-image is same.

The Hasan et al. used the Bats algorithm and Sobel filter along with the Blowfish and LSB in the [16]. In the key generation process, the author applies the Blowfish algorithm on cover image to generate key. The generated key and the input plaintext is XORed with each other to find the ciphertext. After this, the author finds the pixels where the data is to be stored using sobel algorithm and metaheuristic bats algorithm. Through the sobel algorithm, the edges are detected while the bats algorithm finds random pixel which combinedly constitute the pixel locations. The pixels' locations found by the bats algorithm are then used to store the encrypted message using the LSB. The author calculates the PSNR and MSE for the obtained output stego-image and claimed the proposed method achieves the higher performances in the data transmission.

Sharma et al. [17] proposed an integrated model of cryptography scheme and wavelet-based steganography. Here, the researchers used substitution technique for encryption as cryptographic scheme. For encryption, the plaintext is divided into blocks of 15 bytes size. The encryption is n round process and, in each round, new value for plaintext is computed by subtracting old plaintext from 32 ($P = P - 32$). A new matrix is computed by applying mode 95 to the product of key matrices and plaintext matrices, and this new matrix is added with 32. Ciphertext is generated by translating this numerical value obtained from matrix into alphabet. Stego-image is generated by applying addition operation on the cover and secret image (which is formed from the ciphered message using the text-to-image generator) after applying DWT. This algorithm produces stego-image having 40 db PSNR and less average difference is observed too by using the proposed method.

A robust method is given by Muhammad in [18] which based on the concept of transposition, bits shuffling, bitxoring, and secret key to design an advanced steganographic system. The proposed method provides three levels of security. First, all three channels (RGB) of input carrier image are transposed. Secret key and data is encrypted using multiple encryption algorithm at second level of security and at last level of security gray-level modification method is applied for mapping of secret data to blue channel of carrier image. Two modules of the given method, namely encryption module and mapping module have been used for mapping the secret data to one of channels of carrier image. The encryption module performs the encryption on the secret key and secret data. Bits of secret key is XORed with logical 1 and after that shuffling operation is applied on these encrypted bits of key. Now, conditional-based XOR is performed on secret data bits with logical 1. Second module, i.e., mapping module divides image into RGB channels, and image transform is applied on all three channels. Now, all pixel's value of blue channel is converted to odd or even by adding one to it. Secret data is mapped to the pixel value using the secret key bits & pixel values. At last, stego-image is created by taking the combination of all three transposed plane.

In [19], cryptography, steganography, and compression algorithms are combinedly used in the proposed system which provides complete security to the public cloud model. In the proposed system, first, the confidential data is encrypted using hybrid RSA & AES cryptographic algorithms, and then, the LempelZiv-Welch (LZW) compression algorithm is applied to the encrypted data. Then, the compressed encrypted is concealed into the image using the LSB algorithm. To achieve the integrity, the hash value of the image is calculated before uploading to the image.

Tabassum and Mahmood in [20] proposed a multiple layer encryption method which uses both cryptography and steganography for data protection. The system is segregated into two units: encoding unit and decoding unit. The encoding unit of the system takes secret message and cover media as inputs and generates the cipher image. Encoding unit employs double-layer encryption using the Blowfish and AES algorithm and uses LSB algorithm to encode encrypted data into image, and lastly, the image is encrypted using the genetic algorithm. The decoding unit is just the reverse of the encoding unit.

A efficient hybrid method is proposed for data protection in [21], which uses the RSA algorithm, DWT compression, Huffman coding, and LSB encoding. Firstly, the secret information is encrypted using asymmetric key algorithm RSA, and then, in the next step, encrypted information is compressed using the Huffman coding. Using LSB encoding, compressed information is embedded into selected DWT decomposed subbands of cover media. This hybrid method provides higher storage capacity with good stego-image quality.

4 Result and Discussion

From Table 2, it is concluded that mostly researchers use the symmetric encryption in metamorphic cryptographic systems for the cryptographic part rather than asymmetric encryption because of its speed of execution and their suitability for encrypting large chunks of data. In the papers [3–7, 10–14, 16–18], researchers use the various symmetric key encryption algorithm like Vernam, IDEA, AES, Affine, and Blowfish for ciphering the data in their proposed approaches. All the approaches surveyed in this paper supported the RGB images for hiding the data which is good thing as today mostly RGB images are used. Nowadays, the metamorphic cryptography can be used for application like authentication. As in [4], Muneera et al. use the metamorphic cryptography along with hash function for authenticating the systems. Nowadays, cloud environment uses the concept of metamorphic cryptography to sort out the data security challenges in the cloud storage like Shanthakumari and Malliga in [12], and Abbas et al. [19] use the metamorphic cryptography for cloud environment. Researchers prefer the edges of the cover images for storing the data as it increases imperceptibility of the stego-image like Candra in [9] uses the canny edge detection algorithm to select the edges of the image for storing the data. Metaheuristic approaches and randomization are used for selecting the random pixels which results in high robustness. For example, in [16], author used the metaheuristic Bats algorithm to find the random pixels for storing the encrypted message. Table 3 presents the value of PSNR and MSE for various approaches.

Table 2 Techniques used in various metamorphic cryptography approaches

Paper	Cryptography technique	Steganography technique	Compression
[3]	Vernam cipher	LSB with shifting (LSB-S)	No
[4]	AES, hash function	LSB	No
[5]	Affine transform	DCT, LSB	Yes
[6]	MJEA	Enhanced LSB	Yes
[7]	Elliptic curve cryptography	LSB inversion	No
[8]	Double XOR operations	LSB	No
[9]	OTP operations	LSB	No
[10]	AES_MPK	Enhancing PVD image steganography	No
[11]	AES	LSB using user defined password	No
[12]	IDEA	LSBG	No
[13]	Blowfish	Visual cryptography, LSB	No
[14]	Blowfish	LSB bit-cycling with mathematical sequences	No
[15]	Variable block size data encryption	Modified LSB technique	No
[16]	XOR encryption, blowfish	Bats algorithm, LSB	No
[17]	Substitution technique	Wavelet based steganographic encoding	No
[18]	Transposition	Bitxoring, bitshuffling	No
[19]	RSA, AES	LSB	Yes
[20]	Blowfish, AES, GA operators	LSB algorithm	No
[21]	RSA	LSB encoding	Yes

Table 3 Evaluation parameters

	Range	Paper
MSE	0–0.4487	1, 6, 7, 9, 10, 12, 13, 16, 20
	0.7565–0.9969	3, 5, 10, 15
	1.9194–2.717	4, 14, 21
PSNR	10.790–48.13	2, 4, 5, 6, 8, 15, 20, 21
	51.16–58.0468	1, 3, 10, 12
	61.94–72.54	7, 11, 13, 19
	78.2502–93.6632	9, 14

5 Conclusion

This paper gives the general overview of metamorphic cryptography for data security. Here, we discussed about the classification and evaluation parameters for approaches. This paper also presented a comprehensive survey of a data security approaches which uses metamorphic cryptography. Furthermore, we have evaluated the approaches using the standard evaluation parameters like PSNR, MSE. We have listed the cryptography and steganography techniques name used in various metamorphic cryptography in the tabular form. Our survey helps the researchers in making new approaches or modifying the existing approaches.

References

1. Philjon, J. T. L., & Venkateshvara Rao, N. (2011). Metamorphic cryptography—A paradox between cryptography and steganography using dynamic encryption. In *2011 International Conference on Recent Trends in Information Technology (ICRTIT)*, Chennai, India.
2. Hambouz, A., Shaheen, Y., Manna, A., Al-Fayoumi, M., & Tedmori, S. (2019). Achieving data integrity and confidentiality using image steganography and hashing techniques. In *2019 2nd International Conference on new Trends in Computing Sciences (ICTCS)*, Amman, Jordan.
3. Joshi, K., & Yadav, R. (2015). A new LSB-S image steganography method blend with cryptography for secret communication. In *2015 Third International Conference on Image Information Processing (ICIIP)*, Waknaghat, India.
4. Alotaibi, M., Al-hendi, D., Roithy, B. A., Ghamdi, M. A., & Gutub, A. (2019). Secure mobile computing authentication utilizing hash, cryptography and steganography combination. *Journal of Information Security and Cybercrimes Research, 2*(1).
5. Sharma, H., Mishra, D. C., Sharma, R. K., & Kumar, N. (2018). Crypto-stego system for securing text and image data. *International Journal of Image and Graphics, 18*(4).
6. Al-Qwider, W. H., & Salameh, J. N. B. (2017). Novel technique for securing data communication systems by using cryptography and steganography. *Jordanian Journal of Computers and Information Technology, 3*(2), 110–130.
7. Shanthakumari, R., & Malliga, S. (2020). Dual layer security of data using LSB inversion image steganography with elliptic curve cryptography encryption algorithm. *Multimedia Tools and Applications, 79,* 3975–3991.
8. Ahmed, A., & Ahmed, A. (2020). A secure image steganography using LSB and double XOR operations. *International Journal of Computer Science and Network Security, 20*(5), 139–144.
9. Irawan, C., Setiadi, D. R. I. M., Sari, A., & Rachmawanto, E. (2017). Hiding and securing message on edge areas of image using LSB steganography and OTP encryption. In *2017 1st International Conference on Informatics and Computational Sciences (ICICoS).*
10. Saleh, M. E., Aly, A. A., & Omara, F. A. (2016). Data security using cryptography and steganography techniques. *International Journal of Advanced Computer Science and Applications (IJACSA), 7*(6).
11. Islam, M. R., Tanni, T. R., Parvin, S., Sultana, M. J., & Siddiqa, A. (2016). A modified LSB image steganography method using filtering algorithm and stream of password. *Information Security Journal: A Global Perspective, 7*(6).
12. Shanthakumari, R., & Malliga, S. (2019). Dual-layer security of image steganography based on IDEA and LSBG algorithm in the cloud environment. *Sādhanā, 44*(5).
13. Murad, S., Gody, A., & Barakat, T. (2018). Enhanced security of symmetric encryption using combination of steganography with visual cryptography. *International Journal of Engineering Trends and Technology, 65*(3).

14. Alexan, W., & Hemeida, F. (2019). Security through blowfish and LSB bit-cycling with mathematical sequences. In *2019 Signal Processing: Algorithms, Architectures, Arrangements, and Applications (SPA)*.
15. Chauhan, S., Jyotsna, Kumar, J., & Doegar, A. (2017). Multiple layer text security using variable block size cryptography and image steganography. In *2017 3rd International Conference on Computational Intelligence & Communication Technology (CICT)*.
16. Hasan, N., Ahmed, R., Abed, H., & Alkhazraji, A. (2018). Multilevel hiding text security using hybrid technique steganography and cryptography. *International Journal of Engineering and Technology (UAE), 7*(4), 3674–3677.
17. Sharma, V. K., Mathur, P., & Srivastava, D. K. (2018). Highly secure DWT steganography scheme for encrypted data hiding. In *Information and Communication Technology for Intelligent Systems*. Smart Innovation, Systems and Technologies.
18. Muhammad, K., Ahmad, J., Sajjad, M., & Zubair, M. (2015). Secure image steganography using cryptography and image transposition. *NED University Journal of Research*, 81–91.
19. Abbas, M. S., Mahdi, S. S., & Hussien, S. A. (2020). Security improvement of cloud data using hybrid cryptography and steganography. In *2020 International Conference on Computer Science and Software Engineering (CSASE)* (pp. 123–127).
20. Tabassum, T., & Mahmood, M. A. (2020). A multi-layer data encryption and decryption mechanism employing cryptography and steganography. *2020 Emerging Technology in Computing, Communication and Electronics (ETCCE)* (pp. 1–6). https://doi.org/10.1109/ETCCE51779.2020.9350908
21. Wahab, O. F. A., Khalaf, A. A. M., Hussein, A. I., & Hamed, H. F. A. (2021). Hiding data using efficient combination of RSA cryptography, and compression steganography techniques. *IEEE Access, 9*, 31805–31815.

A Bibliometric Analysis to Unveil the Impact of Digital Object Identifiers (DOI) on Bibliometric Indicators

Parul Khurana, Geetha Ganesan, Gulshan Kumar, and Kiran Sharma

Abstract Digital object identifier (DOI) is often used as an important identifier of scientific contributions. It raises the readers' awareness toward genuine and authentic work of authors, organizations, and journals. The aim of this study is to identify the scientific contributions with and without DOI information associated with them in multidisciplinary indexing databases such as Web of Science (WoS). This study also sheds light on the contribution of self-citations in calculating the author, organization, and journal informetrics. The result shows that at author level, 82.2% of publications and 81.6% of citations are with DOIs, at organization level, 76.3% of publications and 73.5% of citations are with DOIs, and at journal level, 83.9% of publications and 62.1% of citations are with DOIs. Author level has 7.7% of self-citations, organization level has 13.7% of self-citations, and journal level has 10.3% of self-citations. Decreases in publications and citations have resulted in an average downfall of h-index 2.9 in author data, 15 in organization data and 12.3 in journal data. Finally, stakeholders are encouraged to review the publications, and citations data with DOIs, and note on the self-citations before considering for final informetrics of authors, organizations and journals.

Keywords DOI · Publications · Citations · Self-citations · h-index

---

P. Khurana (✉)
School of Computer Applications, Lovely Professional University, Phagwara, Punjab 144411, India
e-mail: parul.khurana@lpu.co.in

G. Ganesan
Advanced Computing Research Society, Porur, Chennai 600116, India
e-mail: gitaskumar@yahoo.com

G. Kumar
School of Computer Science and Engineering, Lovely Professional University, Phagwara, Punjab 144411, India
e-mail: gulshan.16865@lpu.co.in

K. Sharma
School of Engineering and Technology, BML Munjal University, Gurugram, Haryana 122413, India
e-mail: kiran.sharma@bmu.edu.in

P. K. Singh et al. (eds.), *Proceedings of Third International Conference on Computing, Communications, and Cyber-Security*, Lecture Notes in Networks and Systems 421,
https://doi.org/10.1007/978-981-19-1142-2_67

1 Introduction

Digital object identifier (DOI) presented in URLs (uniform resource locator) form is conceived as a generic standard for the identification of various types of objects or metadata such as documents, images, and audios in the Internet environment. It is designed as a robust linking option for sharing an actionable identification with the interested users or community [1]. The major benefits include permanent identification and uniqueness of objects, interoperability, persistence, and network accessibility in various applications. Since 2000, DOI has been seen as presenting the metadata in its physical and electronic forms in digital environments. Once assigned, the DOI remains unchanged through its lifetime, whereas metadata may observe changes with the ongoing time span [2]. As such, the length of the DOI name is unrestricted. DOI is a character string composed of two components presented as prefix and suffix, separated by forward slash '/'. The prefix portion denotes a unique naming authority (usually represents organization), and suffix portion denotes any user-entered string (usually represents actual identity). After combining both components, the identifier component becomes an actionable link just like any URL [3, 4].

DOI has gained importance in the scientific publication industry as a crucial emergent. The ongoing momentum of DOIs has grown as a global consortium with DataCite in 2009 by issuing DOIs to scientific publications and research datasets [5]. Various indexing databases like Scopus, Web of Science (WoS), Google Scholar, etc., use DOIs in their scientometrics for achieving the accuracy of scientific data. DOIs are often used by them for referencing and sharing publication information with the scientific community. Availability of DOIs among different indexing databases generates their potential stability of scientific data as well [6].

Many stakeholders like academic institutions, research organizations, government bodies, promotion committees, ranking, and accreditation agencies have keen interest to measure the research contributions of an individual or a group [7]. Maybe for the individual hiring, promotion, tenure, releasing of grants, or for the search of literature etc. Such stakeholders always rely on these indexing databases for the retrieval of genuine informetrics like publications, citations, and h-index of an author, organization, journal [8]. Retrieved informetrics from indexing databases may have various disguised accelerations like consideration of publications and citations with no DOI information associated with them and, secondly, consideration of self-citations for the undue gain of citations and rise in h-index [9–12].

1.1 Research Objectives

The motivation behind this study is to identify the availability of informetrics such as publications and citations with and without DOIs in the different indexing databases such as WoS so that the accurate potential of author, organization, and journal may be presented to its stakeholders in terms of valid publications, citations, and self-citations [13–16]. Therefore, our research questions focus on:

- To identify the difference in number of publications, citations, and h-index with and without DOIs.
- To compute the citations and h-index after eliminating self-citations.

The contribution of our study sheds the light on the impact of DOI-based informetrics at three levels, i.e., author, organization, and journal. Study encourages the fact that DOI-based informetrics are genuine and comparable across different indexing databases. Such comparisons present the unified informetrics to its stakeholders in a reliable way for various purposes.

2 Data Description

2.1 Data Selection

In order to fetch the details of documents with and without DOI information, we accessed the Web of Science (WoS) database with Python-based queries [17]. The choice of the database was arbitrary and was based on the availability of the data. The analysis was performed for three entities: (1) Authors, (2) Organizations, and (3) Journals. For authors, we visited various university websites to fetch author details like their name, orcid id, discipline etc., but we did not find the suitable details as per the requirement of the study. Then, after accessing various university websites, we found that 6316 staff profiles were available on the website of Monash University in Australia. For organizations in India, we visited the website of University Grants Commission (UGC), Ministry of Human Resource Development (MHRD), and National Institutional Ranking Framework (NIRF). We found that NIRF as a MHRD initiative ranks Indian organizations on various parameters. Hence, we selected NIRF ranking 2020 as a reliable source for the requirement of organization information for the study. For journal information, we accessed the master journal list of WoS. There is a common list available as a search result on the website of Clarivate, where we can access the various parameters of the journal information.

2.2 Data Filtration

In order to perform the analysis at author level, we manually checked the various author profiles on the website of Monash University [18]. Further, we selected 400 author profiles carrying required information of author name, orcid id, discipline, or subject categories. These 400 author profiles were queried on WoS database and publication, citation, and h-index information were retrieved. 400 author profiles were divided into five disciplines such as life sciences, engineering, sciences, social sciences, and humanities. For the analysis at organization level, we retrieved the list of top 100 organizations listed in NIRF-2020 under the category of overall [19]. Thus, organization rank, name, type, id, city, and state information was collected. 100 organizations were divided into four types such as Universities, IITs, IIEST, IISC & IISER, and NITs. WoS database was queried and their publication, citation and h-index information was downloaded. For the analysis at journal level, we divided the journal list into five disciplines such as engineering, social sciences, life sciences, sciences, and humanities. Journal name, ISSN, and discipline were retrieved [20], and then, WoS database was queried to download their publication, citation, and h-index information.

3 Results

3.1 At Author Level

Figure 1 shows the comparative analysis between (a) documents (with and without DOI), (b) citations (with and without DOI), (c) self-citations and citations (with DOI), (d) h-index (with and without DOI and self-citations). The analysis is performed on 400 authors; there are 90.5% authors where document count is changed. Study shows that after considering documents with DOIs, document count has decreased to 26,101 from 31,732, equivalent to 82.3% of total documents. Total citation count for 400 authors is 1,024,808, and it is decreased to 835,962 after considering citations with DOI only which is 81.6% of total citations. Citation count change is observed for all authors. On an average, 19% citations have been decreased per author. Out of 400, 13 authors have been identified with more than 50% decrease in citations due to consideration of DOI-based citations only. Initial analysis of self-citations reveal that there are 13 authors with 0 self-citations which is 3.3% of total authors, 263 authors with less than 10% self-citations which is 65.8% of total authors, 101 authors with less than 20% self-citations equivalent to 25.3% of total authors, and 23 authors with more than or equal to 20% self-citations equivalent to 5.8% of total authors with maximum self-citations of 38.1%. Minimum h-index is 1 and maximum h-index is 95 if we consider self-citations, DOI and non-DOI-based documents, but if we consider only DOI-based documents and exclude self-citations also, then there is no change in the minimum h-index with 13 point change in maximum h-index, which comes to 82. 70 authors do not observe any change in the h-index if we follow DOIs and exclude self-citations, for such authors minimum h-index is 1, and maximum is 23. 314 authors (78.5% of authors) observe the decrease of 1–9 points with minimum h-index as 1, maximum as 64, and average h-index as 17.4. 16 authors out of 400 (4% of authors) have observed the change of 10–16 points, with minimum h-index as 4, maximum as 82, and an average h-index of 36.9.

The result of additional analysis of 400 authors based on five disciplines such as *Life Sciences*, *Engineering*, *Sciences*, *Social Sciences*, and *Humanities* is presented

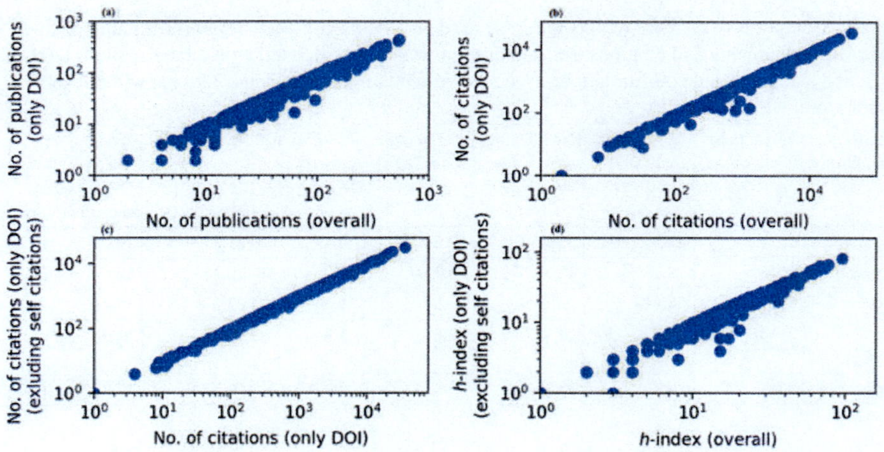

Fig. 1 Comparative analysis of 400 authors, **a** based on DOI information associated with documents. **b** Based on DOI information associated with citations. **c** Based on self-citation information associated with citations (with DOI). **d** Based on h-index information associated with citations (with DOI and excluding self-citations)

in Table 1. Among all, *Sciences* is at top with 88.1% documents with DOI and *Engineering* at bottom with 78.8% documents with DOI. *Humanities* has received 88.6% citations with valid DOIs, followed by *Life Sciences*. *Engineering* has received highest self-citations equivalent to 11.0%, followed by *Sciences*. *Social Sciences* has received minimum self-citations as 5.3% among all. Average h-index of *Sciences* is 23.2 including self-citations and considering all documents, i.e., with DOI and without DOI, followed by *Life Sciences* with 21.8. Average h-index after excluding self-citations and considering only DOI-based documents is 19.4 for *Sciences*, followed by 19.0 in *Life Sciences*. Overall, *Engineering* has recorded average downfall of 4.4 in h-index, followed by 3.8 in *Sciences*, 2.8 in *Life Sciences*, 1.9 in *Humanities*, and 1.5 in *Social Sciences*.

3.2 At Organization Level

Figure 2 shows the comparative analysis between (a) documents (with and without DOI), (b) citations (with and without DOI), (c) self-citations and citations (with DOI), and (d) h-index (with and without DOI and self-citations). The analysis is performed on 100 organizations, and difference in documents is observed in all 100 organizations, with minimum difference as of 45 documents, maximum difference of 10,944 documents, and average difference of 1893 documents per organization. 11 organizations have observed the difference of 5000 or more documents with DOI and without DOI, and 4 organizations have observed a decrease of less than 100 documents per organization. The average number of documents are 7971.6 as com-

Table 1 Comparative analysis of 400 author's based on five disciplines for DOI information associated with documents and citations, self-citation information associated with citations (with DOI only), and average h-index information associated with citations (with self-citations, without self-citations, and with DOI only)

Author disciplines	No. of pubs	(%) of pubs (only DOIs)	No. of cites	(%) of cites (only DOIs)	(%) of self cites (only DOIs)	Avg. h-index	Avg. h -index (only DOIs, exc. self cites)
Life sciences	18,257	82.2	631,244	82.8	7.1	21.8	19.0
Engineering	5658	78.8	13,8631	73.3	11.0	20.9	16.6
Sciences	3187	88.1	121,752	81.5	9.4	23.2	19.4
Social sciences	3113	80.5	94,195	82.6	5.3	13.2	11.6
Humanities	1517	86.9	38,986	88.6	8.2	19.8	17.9

pared to 6079.4 documents with DOI only. The minimum number of documents received by an organization is 569 and maximum number of documents received by an organization is 52,779 as compared to 478 as minimum number of documents with DOI, and maximum number of documents as 41,997 with DOI. Total citations received by 100 organizations with DOI are 6,866,250 as compared to 9,337,059 which is 73.5% of total citations. Minimum citations received by an organization are 849, maximum citations received by an organization are 876,753, and average citations are 93,370.6. While considering only DOI-based citations, minimum citations are 636, maximum citations are 656,860, and average citations are 68,662.5. 13.7% citations are self-citations received by 100 organizations with an average of 9372.9 per organization. There are 16 organizations (16%) which have received less than 1000 self-citations, 39 (39%)organizations which have received less than 5000 self-citations, and 45 (45%)organizations which have received more than 10,000 self-citations. Minimum self-citations are 100, and maximum self-citations are 85,490. Average h-index including self-citations and all documents, i.e., with DOI and without DOI is 81.5 with minimum h-index as 12 and maximum h-index as 246. If we consider only documents with DOI and exclude self-citations as well, then minimum h-index will come as 8, maximum h-index will come as 203 and average h-index will come as 66.5. This shows that there is a decrease in a h-index by 4 points as minimum, 43 points as maximum, and 15 points as an average. Study also reveals that there are 30 organizations (30%) where h-index difference point is less than 10, 69 organizations (69%) where h-index difference point is less than 50 and 1 organization (1%) where h-index difference point is greater than 50.

The result of additional analysis of 100 organizations based on four types such as *Universities*, *IITs*, *IIEST, IISC & IISER*, and *NITs* is presented in Table 2. Among all, *IIEST, IISC & IISER* are at top with 81.7% of documents with DOI, followed by *IITs* as 79.9% and *NITs* at the bottom with 72.7%. 76.6% of citations received by *NITs* are with valid DOIs, followed by 74.9% by *IITs*, and 72.6% by *Universities* at the bottom. 13.9% of citations received by *IITs* are self-citations, followed by 13.7%

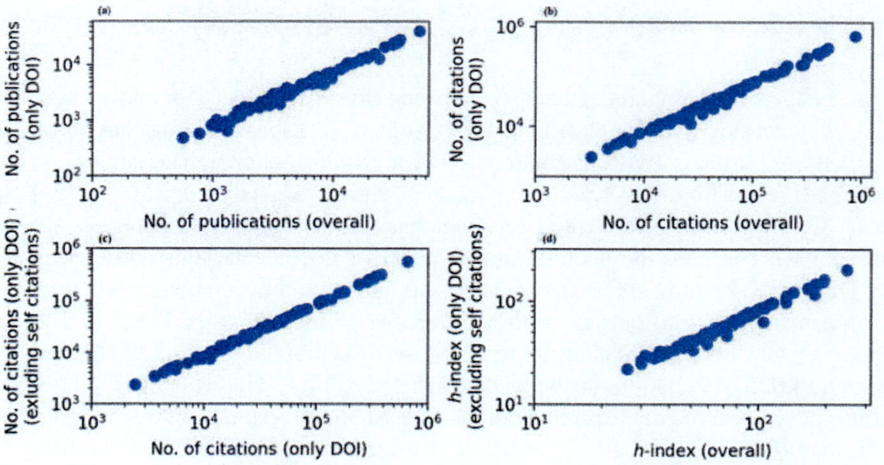

Fig. 2 Comparative analysis of 100 organizations, **a** Based on DOI information associated with documents. **b** Based on DOI information associated with citations. **c** Based on self-citation information associated with citations (with DOI). **d** Based on h-index information associated with citations (with DOI and excluding self-citations)

Table 2 Comparative analysis of 100 organizations based on four categories for DOI information associated with documents and citations, self-citation information associated with citations (with DOI only), and average h-index information associated with citations (with self-citations, without self-citations, and with DOI only)

Organization types	No. of pubs	(%) of pubs (only DOIs)	No. of cites	(%) of cites (only DOIs)	(%) of self cites (only DOIs)	Avg. h-index	Avg. h-index (only DOIs, exc. self cites)
Universities	451,489	73.8	4,917,831	72.6	13.7	74.5	62.0
IITs	236,547	79.9	2,993,534	74.9	13.9	108.9	88.3
IIEST, IISC & IISER	70,063	81.7	1,107,018	73.4	13.2	91.6	74.7
NITs	39,059	72.7	318,676	76.0	12.0	64.4	54.4

self-citations by *Universities* and 12.0% received by *NITs* at the bottom. Average h-index considering self-citations and all documents (with and without DOIs) of *IITs* is 108.9, *IIEST, IISC & IISER* is 91.6, followed by *NITs* as 64.4 at the bottom. If we consider only DOI-based documents and excludes self-citations, then average h-index of *IITs* will be decreased by 20.6, *IIEST, IISC & IISER* by 16.9, and *NITs* as 10.0 at the bottom.

3.3 At Journal Level

Figure 3 shows the comparative analysis between (a) documents (with and without DOI), (b) citations (with and without DOI), (c) self-citations and citations (with DOI), and (d) h-index (with and without DOI and self-citations). The analysis is performed on 1000 journals. 1,415,093 documents have been analyzed, and we found that 1,187,692 documents are with DOIs, which is 83.9% of total documents. 77.6% journals have observed the decrease in the count of documents after considering only DOI-based documents, such as 45.2% journals with a difference of less than 100 documents, 32.4% journals with a difference of more than or equal to 100 documents, and so on. Total citations received by 1000 journals are 22,570,461, out of which 14,005,489 citations are with DOIs, which is 62.1% of total citations. 99.9% journals have observed decrease in citations such as 36.7% journals with a decrease of less than 1000 citations, 29.1% journals with a decrease of less than 5000 citations, 13% journals with a decrease of less than 10,000 citations, and 21.1% journals with a decrease of more than or equal to 10,000 citations. For self-citations, there are 95.2% journals which have received self-citations such as 54.2% journals with less than 500 self-citations, 14.4% with less than 1000 self-citations, 23.8% with less than 5000 self-citations, and 7.6% journals with more than 5000 self-citations as well. Average h-index including self-citations and all documents, i.e., with DOI and without DOI is 43.9, with minimum h-index as 2 and maximum h-index as 344. If we consider only DOI-based documents and exclude self-citations, then average h-index would be 31.6, with minimum h-index as 1 and maximum h-index as 236. Similarly, 52.6% of journals have observed the decrease of less than 10 points in their h-index, 28.5% of journals have observed the decrease of less than 20 points in their h-index, and 18.9% of journals have observed the difference of greater than 20 points in h-index.

The result of additional analysis of 1000 journals based on five disciplines such as *Engineering*, *Social Sciences*, *Life Sciences*, *Sciences*, and *Humanities* is presented in Table 3. Initial study reveals that *Engineering*, *Sciences*, *Humanities*, and *Social Sciences* disciplines have more than 80% of documents with DOIs, whereas *Life Sciences* has only 60.6% documents with DOIs. *Engineering* discipline has highest citations among all, but *Sciences* discipline has highest citations with DOIs, i.e., 73.4%. On the other hand, *Engineering* discipline has lowest citations with DOIs, i.e., only 60.5%. *Life Sciences* is the discipline where 60.6% documents are with DOIs and 67.0% citations with DOIs which is the closest difference as compared to other disciplines. Sciences is at the top with 12.6% self-citations, followed by *Engineering* and *Social Sciences*. Comparatively, *Engineering* has received highest citations, but *Sciences* has received highest self-citations. Average h-index including self-citations and all documents (with and without DOIs) is highest for *Life Sciences* and lowest for *Social Sciences*. *Engineering* observes highest decrease in average h-index, i.e., 13.2 and *Social Sciences* observes a lowest decrease in average h-index, i.e., 7.6. Average decrease in h-index is of 10 points among all disciplines.

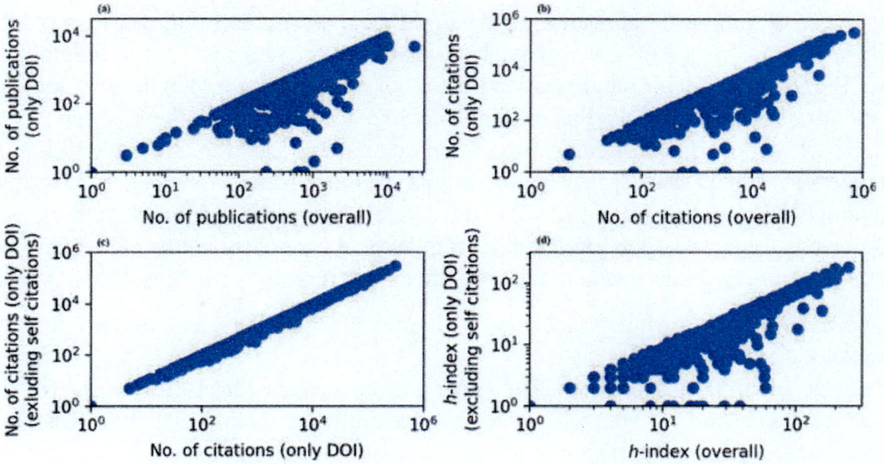

Fig. 3 Comparative analysis of 1000 journals, **a** based on DOI information associated with documents. **b** Based on DOI information associated with citations. **c** Based on self-citation information associated with citations (with DOI). **d** Based on h-index information associated with citations (with DOI and excluding self-citations)

Table 3 Comparative analysis of 1000 journals based on five disciplines for DOI information associated with documents and citations, self-citation information associated with citations (with DOI only), and average h-index information associated with citations (with self-citations, without self-citations, and with DOI only)

Journal disciplines	No. of pubs	(%) of pubs (only DOIs)	No. of cites	(%) of cites (only DOIs)	(%) of self cites (only DOIs)	Avg. h-index	Avg. h-index (only DOIs, exc. self cites)
Engineering	1,179,771	85.7	19,176,940	60.5	10.5	45.5	32.3
Social sciences	86,719	80.4	1,268,836	70.7	9.5	29.6	21.9
Life sciences	68,142	60.6	858,220	67.0	7.1	49.9	39.1
Sciences	53,458	81.0	783,108	73.4	12.6	48.8	38.6
Humanities	27,003	84.3	483,357	72.0	9.0	46.3	37.5

4 Discussion and Conclusion

This study reports the existence of differences in number of publications, and citations with and without DOIs. Author level shows that there are 17.8% of publications and 18.4% of citations without DOIs, organization level shows that there are 23.7% of publications and 26.5% of citations without DOIs, and journal level shows that there are 16.1% of publications and 37.9% of citations without DOIs. In terms of self-citations, author level has 7.7% of self-citations, organization level has 13.7% of self-citations, and journal level has 10.3% of self-citations. Thus, surprisingly, if we do not consider publications, citations without DOIs, and we exclude self-citations also from the author, organization, and journal data, we observe the decrease of average h-index by 2.9 in author data, 15 in organization data and 12.3 in journal data [21]. Hence, indexing databases should improve and update the DOI information of scientific research work so that stakeholders like academic institutions, research organizations, government bodies, promotion committees, ranking, and accreditation agencies should have a clear representation of the research contribution of an individual or group or journal [22, 23]. The above study is limited to 400 authors, 100 organizations, and 1000 journal data from WoS. As we observe the differences in publications, citations, and h-index, one can extend this study further as follows:

- One should also experiment this study with other indexing databases like Scopus and Google Scholar.
- The size and time span of the study sample may be increased for further analysis.
- The study may be done year wise to analyze the growth or downfall of DOIs from one year to another with regard to various categories.
- Study may carry the data from more disciplines for insight views.

Conflict of Interest The authors declare that they have no conflict of interest.

References

1. Chandrakar, R.: Digital object identifier system: An overview. *The Electronic Library* (2006)
2. Gorraiz, J., Melero-Fuentes, D., Gumpenberger, C., & Valderrama-Zurián, J. C. (2016). Availability of digital object identifiers (DOIs) in Web of Science and Scopus. *Journal of Informetrics, 10*(1), 98–109.
3. Mugnaini, R., Fraumann, G., Tuesta, E. F., & Packer, A. L. (2021). Openness trends in Brazilian citation data: Factors related to the use of DOIs. *Scientometrics, 126*(3), 2523–2556.
4. Homenda, N. (2021). Persistent urls and citations offered for digital objects by digital libraries. *Information Technology and Libraries, 40*(2).
5. Carreiro, E. (2010). Electronic books: How digital devices and supplementary new technologies are changing the face of the publishing industry. *Publishing Research Quarterly, 26*(4), 219–235.
6. Mooney, S. (2001). Digital object identifiers for ebooks: What are we identifying? *Publishing Research Quarterly, 17*(1), 29–36.
7. Agarwal, A., Durairajanayagam, D., Tatagari, S., Esteves, S. C., Harlev, A., Henkel, R., Roychoudhury, S., Homa, S., Puchalt, N. G., Ramasamy, R., et al. (2016). Bibliometrics: Tracking research impact by selecting the appropriate metrics. *Asian Journal of Andrology, 18*(2), 296.
8. Roldan-Valadez, E., Salazar-Ruiz, S. Y., Ibarra-Contreras, R., & Rios, C. (2019). Current concepts on bibliometrics: A brief review about impact factor, Eigenfactor score, CiteScore, SCImago Journal Rank, source-normalised impact per paper, h-index, and alternative metrics. *Irish Journal of Medical Science (1971-), 188*(3), 939–951.
9. Fowler, J., & Aksnes, D. (2007). Does self-citation pay? *Scientometrics, 72*(3), 427–437.
10. Norris, M., & Oppenheim, C. (2010). The h-index: A broad review of a new bibliometric indicator. *Journal of Documentation*.
11. Szomszor, M., Pendlebury, D. A., & Adams, J. (2020). How much is too much? The difference between research influence and self-citation excess. *Scientometrics, 123*(2), 1119–1147.
12. Craigle, V. (2021, forthcoming). Adopting DOI in legal citation: A roadmap for the legal academy. Legal Citation: A roadmap for the Legal Academy (March 8, 2021). Legal Reference Services Quarterly (2021, forthcoming), University of Utah College of Law Research Paper.

13. Meho, L. I., & Yang, K. (2006). A new era in citation and bibliometric analyses: Web of Science, Scopus, and Google Scholar. arXiv preprint cs/0612132.
14. Bartneck, C., & Kokkelmans, S. (2011). Detecting h-index manipulation through self-citation analysis. *Scientometrics, 87*(1), 85–98.
15. Franceschini, F., Maisano, D., & Mastrogiacomo, L. (2015). Errors in DOI indexing by bibliometric databases. *Scientometrics, 102*(3), 2181–2186.
16. Khurana, P., Ganesan, G., Kumar, G., & Sharma, K. (2021). A weighted unified informetrics based on Scopus and WoS. arXiv preprint arXiv:2106.01232
17. Bacis, E. (2019). enricobacis/wos. https://github.com/enricobacis/wos
18. University, M. (2019). Find profiles. https://research.monash.edu/en/persons/
19. Ministry of Education, G.o.I. (2020). India rankings 2020: Overall. https://www.nirfindia.org/2020/OverallRanking.html
20. Web of Science Group. (2020). Master journal list. https://mjl.clarivate.com/search-results
21. Simoes, N., & Crespo, N. (2020). Self-citations and scientific evaluation: Leadership, influence, and performance. *Journal of Informetrics, 14*(1), 100990.
22. Bakkalbasi, N., Bauer, K., Glover, J., & Wang, L. (2006). Three options for citation tracking: Google Scholar, Scopus and Web of Science. *Biomedical Digital Libraries, 3*(1), 1–8.
23. Adriaanse, L. S., & Rensleigh, C. (2013). Web of Science, Scopus and Google Scholar: A content comprehensiveness comparison. *The Electronic Library*.

Cyber Attack Modeling Recent Approaches: A Review

Neha and Anubha Maurya

Abstract The advancement in cyber technology has enhanced user convenience tremendously hence accelerated its uses. But at the same time, cyber frauds, threats, and attacks have increased with same pace. So, to protect our cyber system and devices from them, cyber attack modeling is quite essential and challenging task. It provides us the chance to detect and protect our system by applying suitable security measures to them. There are many attack modeling techniques available today. This paper provides an elaborate discussion on the two very popular graphical attack modeling techniques, that is *Attack graph and attack tree*-based approaches. A comparative analysis of various works done in these techniques is presented here.

Keywords Attack · Vulnerability · Threat · Attack modeling · Cyber security

1 Introduction

The use of electronic devices (computer, laptop, mobile phones, tablets etc.) has increased tremendously in the current era. This trend has been accelerated by the high computational capacity and lucrative network applications provided by these devices. This enables a user of these modern devices with a feeling of having information, communication, and digitization power at their finger tip. But with the increased power of these devices, it has also become vulnerable attack points for different threats, leading to cyber attacks. According to [1] FBI's Internet Crime Complaint Center (IC3), a total of 791,790 Internet crimes have been reported in 2020. This shows a rise of 300,000 more complaints in comparison to 2019. This all has reported looses exceeding $42 billion. In many cases, these crimes also affect user's social and

Neha (✉) · A. Maurya
Department of Computer Science and Engineering, National Institute of Technology Patna, Patna, Bihar, India
e-mail: neha.cse14@nitp.ac.in

A. Maurya
e-mail: anubhamaurya@nitp.ac.in

P. K. Singh et al. (eds.), *Proceedings of Third International Conference on Computing, Communications, and Cyber-Security*, Lecture Notes in Networks and Systems 421,
https://doi.org/10.1007/978-981-19-1142-2_68

personal safety. So, cyber security has become a challenging task today. Advanced security measures and threat analysis models are necessary to detect these threats and apply protective means to secure the system. Research in area of cyber attack modeling has gained immense emphasis these days [6–11]. In this paper, the author has presented a review of cyber attack modeling. This paper will provide a beginner in this research area a complete understanding of this research domain. It presents a comparative analysis of various attack modeling techniques available and their limitations. The organization of the paper is as follows: Sect. 2 presents a cyber attack modeling framework; it categorizes various steps in cyber attack modeling. Section 3 describes Attack graph in detail. Section 4 introduces Attack tree. Section 5 concludes the paper.

2 Cyber Attack Modeling Framework

According to "Internet Security Glossary, Version 2" [2] attack can be declared as "An intentional act by which an entity attempts to evade security services and violate the security policy of a system".

For proper detection and apply suitable means to avoid the attack and minimize its effect, attack modeling is helpful. A complete attack modeling process can be configured as shown in Fig. 1.

2.1 Network Information Collection

Before performing attack modeling of any network system, it is important to collect various network topology information, network configuration, network connectivity, and list out the possible vulnerabilities. For listing out various vulnerabilities, the security alerts generated by the intrusion detection system, firewalls or any other alert generation system can be correlated. Many alert correlation systems have been proposed [13–16]. In [13–15], proposed alert correlation system depends upon prior knowledge and consequences of alerts. But these are not able to detect new attacks. Whereas [16], proposed system that used time series and statistical analysis for alert correlation. This consists of four steps.

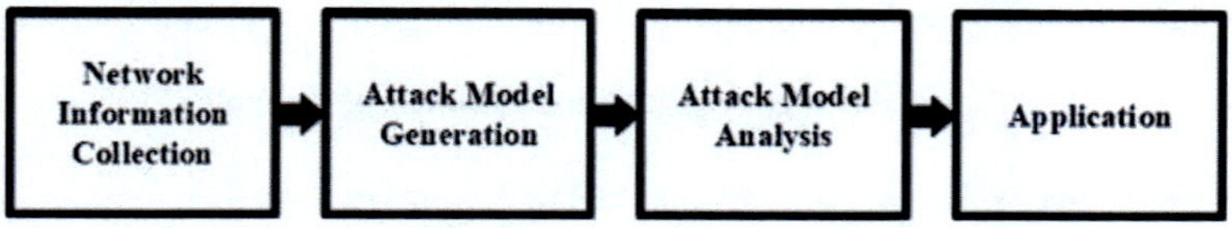

Fig. 1 Framework for attack modeling

- First, the alert correlation and clustering is performed to convert low-level alerts into aggregated alerts.
- The next step is the prioritization of alerts depending upon various factors, like relation with network, host, and goals of attack.
- Third step was alert time sires variable creation.
- And finally, attack scenario construction is performed.

Various alerts generated system requires manual intervention to correlate them for a complex multistage attack scenario to be recognized. Chang et al. [13] in his work proposed an automated attack representation modeling process that identify the relations between system alerts and develop an attack scenario recognition system.

2.2 *Attack Model Generation*

Understanding cyber attack clearly demands better techniques and methods that aid the perception and assessment of cyber attack. Attack modeling techniques are used to analyze, understand even complex cyber attacks by visualizing well-designed diagrams and graphical system representation. The various attack modeling technique can be categorized as:

- Usecase method
- Temporal methods
- Graph-based methods.

This paper has been targeted to graphical-based attack modeling techniques only. The various graph-based modeling techniques we are discussing here are, *Attack graph and Attack tree*-based approaches.

Attack Graph Attack graph model was first proposed by Swiler et al. [3] in 1997. It has a strong ability to graphically represent a network attack behavior elaborately. To accomplish an attack, first, it is initialized by compromising a single node, and gradually infiltrate to other nodes, which finally reaches to target node to do the desired harm to the system. An attack graph model is designed to describe a network topology having nodes, paths, and affect of network attack in a directed acyclic graph. The *node* in an attack graph represents the state (host, vulnerability, or network device). Whereas, the edges represent transition from one state to other.

So, basically, an attack graph represents whether an attacker can reach the final goal state by penetrating the security holes of system from initial state. An example of attack graph is shown in Fig. 2. In this figure, S represents the starting node, G represents the goal node, and V_i represents the intermediate nodes. The nodes are connected to each other through various vertices.

Attack Tree Inspired by techniques of fault tree [4], *Attack Tree* was proposed in 1999 by Schneier [5]. It is a powerful technique for analyzing and modeling the vulnerability of an information system. It consists of three parts: root, leaves, and

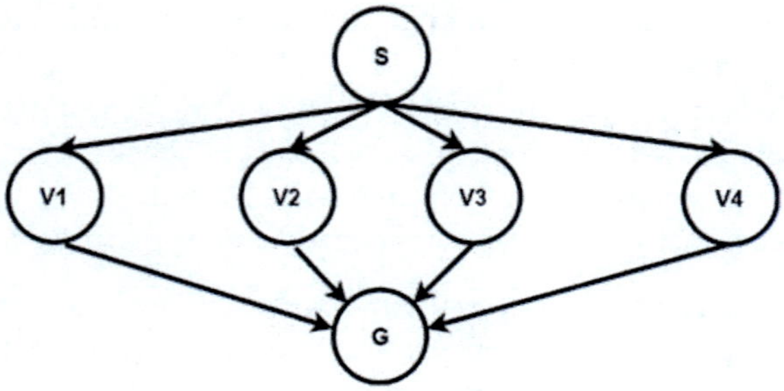

Fig. 2 Attack graph

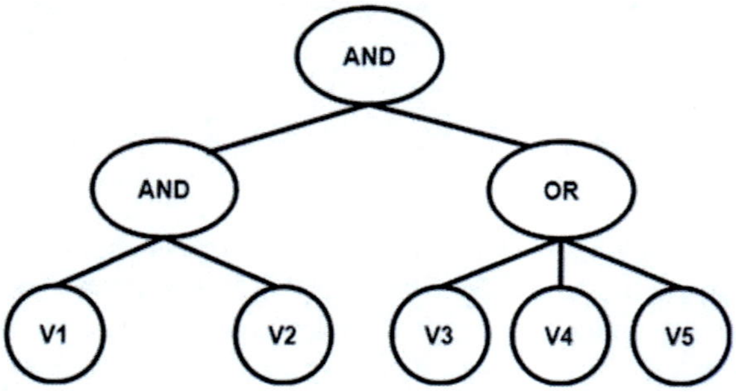

Fig. 3 Attack tree

sub-trees. The root node represents the ultimate goal of an attacker or malicious user. Whereas, the leave nodes represent primary or simple individual actions. The sub-tree (non-leaf node) represents intermediate sequence of combined attacks of non-leaf nodes that steps toward ultimate goal. The sub-tree components can be associated as "AND" (Conjunctive) or "OR" (Disjunctive) node in the attack tree. In case of conjunctive node, all the intermediate node will be in action to achieve the goal. Whereas, in disjunctive, any one of the attack is sufficient to achieve the goal.

So an attack tree visualizes an attack as a hierarchy of sub-goals, leading to ultimate goal. An example of attack tree is shown in Fig. 3. The node at level 0 is the root or goal node, nodes at level 1 are intermediate nodes and nodes V1, V2, V3, V4, V5 are the leave nodes.

2.3 *Attack Model Formalization*

While representing the various nodes, vulnerability, network structure, and possible attacks graphically aid to the security analysis to simple user easily by providing clearer visualization of present system. But to present a graphical model for a complex system, formal interpretation of these graphical modeling approaches is quit essential. The formal representation helps in solving many problems, like to identify whether two models represent the same security scenario or not, finding more informative model among given two for a system, finding attack vector path in the given model.

Many researches in this direction has been performed. Formalization of attack tree is presented in [12] (propositional semantics). Multi-set semantics is presented in [9].

2.4 Application

The applications of security modeling approaches can be summarized as follows:

- *Security assessment metrics*: The various graph-based attack modeling techniques can be used to evaluate the security metrics for the given network or system. An assessment of security level achieved can be used to determine weather a given system is under attack or not. The various nodes and edges must be provided the probability value of occurrence and expected damage value on their occurrence. This value will be aggregated to measure the expected risk zone in the network.
- *Security hardening*: The network security hardening can be achieved with the help of graph-based security models. These models map out the network structure detecting more vulnerable node with there attack and defense cost. Then, the suitable hardening technique can be applied.

3 Attack Graph

The attack graph modeling can be performed broadly in following phases:

- Reachability Test
- Attack graph generation
- Attack graph building.

3.1 Rechability Test

It is done before attack graph generation phase. It is used to check the rechability of target host from the attacker's current host position. It means it determines whether the attacker and attacked host can be access each other or not. A rechability matrix is constructed among hosts. The value into this matrix can be Boolean value or network protocol between the corresponding hosts. The matrix is then optimized to speed up the traversal during attack graph building phase. The main direction of rechability information provided are *rechability scope* and *rechability content*. The scope deals the scope of the host network among which the rechability is determined. So the scope of rechability can be determined for whole network, or it can be determined for sub-domains. Whereas, the *rechability content* defines the network security element that are used to determine the rechability. So, the content can be either the firewall filtering rules, routers access control rules, or the signatures in intrusion detection sensors.

3.2 *Attack Graph Generation*

This phase starts with formulating *attack model* and then *attack graph model*. An *attack model* is a collection of correlated *attack templates*. *Attack templates* generally provides the specification of condition for performing an attack by attackers as well as the preferences gained but the attack after successful execution of the attack. Based upon the way of formulation of attack template, attack model can be of following three types.

- Attack template created manually by experts
- The attack template related based on past data of system log, intrusion alerts of target network.

After attack model generation the attack graph model is constructed that basically shows the attack instances representation of target host and connected nodes.

3.3 *Attack Graph Building Phase*

In this phase, the core algorithm used for attack graph construction is applied and pruning of some attack paths are performed. It helps in selecting only the most critical attack paths and decreasing the graph generation time. The various works in attack graph-based modeling are summarized in Table 1.

4 Attack Tree

The attack tree presents cyber attack bottom-up modeling approach. It means decomposing a complex problem to smaller sub-problems that can be modeled easily. It is a model to hierarchy represent an attack scenario. The various challenges in attack tree base modeling are,

- Formal representation
- Model generation
- Quantitative security analysis.

4.1 *Formal Representation*

In an attack tree, the root represents the main goal of the attacker. To achieve this goal, if sum sub-goal is required to be achieved, then the root nodes of this sub-goal are called the refinements. Whereas the leaf nodes represent attack which cannot be further refined. The refinement can be basically of two types, OR and AND. If

Table 1 Classification of works in attack graph-based modeling

Author	Rechability	Attack graph model	Graph core building	Application
Ammann et al. [28]	• Whole network reachability • Filtering and access control rule modeling, trust relationship	• Manually defined attack templates • State, vulnerability, host-based attack graph	Goal-oriented attack path pruning	Network hardening
Ritchey and Ammann [29]	• Whole network reachability • Trust relationship, filtering, and access control rule modeling	• Manually defined attack templates • State- and vulnerability-based attack graph	–	–
Jajodial and Noel [30]	• Atomic domain reachability • Filtering and access control rule modeling, trust relationship, application relationship	• Text processing-based, state-based, vulnerability-based attack graph	Graph-based method	• Network security metric computation • Network hardening
Noel et al. [31]	• Atomic domain reachability • Filtering and access control rule, IDS, trust relationship, application relationship-based modeling	• State-based attack graph, vulnerability-based attack graph	Graph-based method	• Network security metric computation • Network hardening
Wang et al. [32–34]	• Atomic domain reachability • Filtering and access control rule modeling, trust relationship	• Text processing-based attack tree	Graph-based method	Network hardening
Berkers et al. [35]	• Whole network reachability • Filtering and access control rules, trust relationship-based modeling	• Manual defined attack tree • State-based attack graph	• Graph-based method • Probability-based path pruning	Network security metric computation
Azqa Nadeem et al. [41]	• Whole network reachability	• SAGE Automated generation tool	• Intrusion-alert driven attack graph extractor	Ranking attacks

the node is of AND refinement, then to achieve the main goal, the sub-goal of all its children need to be achieved. Whereas to achieve OR node refinement goal, it is required to achieve goal of any sub-node children. One more refinement is SAND. For achieving the goal of SAND node, the goal of all its sub-children need to be achieved in a particular order. Various works have been proposed in the direction of formal representation of attack tree. The work proposed by Jhawer et al. [7] present a serial–parallel interpretation of attack tree. It is a concise extended version of multi-set semantics. Later, a more expressive semantics using *linear logic* have been proposed for SAND attack tree by Horne et al. [6]. In this work, attack tree interpreted using logical semantics. These two works are contrary to each other. The work proposed by Jhawar et al., AND, and SAND refinements are not related to each other, whereas in Horne et al. work, the one refinement specializes the other depending on the application used. Latter Audinot et al. [20] proposed a work, which checks the validity of OR/AND/SAND attack tree of a given system.

4.2 Model Generation

Attack tree generation is generally performed either manually or by using some automated tools. The manual generation is performed by the experts in this field. But there are a few issues regarding manual generation.

- The model construction is quite subjective as it is dependent on expert's knowledge. In this case attack model for same system by different experts may vary from each other in size and structure.
- The manual construction is a very tedious task and chances of occurrence of error is there which can result in erroneous attack tree.

Considering these issues, many automated approaches for attack tree generation have been proposed. Vigo and Nielson [21, 22] first proposed a process algebra-based automated attack tree generation technique. In this work, first, the value passing quantity calculus is applied on a target location to generate an AND/OR attack tree. The value passing quantity calculus is a type of process algebra where the system is considered as a set of process, which can run sequentially or parallelly. Latter Pinchinate et al. [23, 24] proposed a semiautomatic ATaYRA (Attack tree synthesis for risk analysis) tool-based attack tree generation. In this work, the behavioral modeling is done by using domain-specific language (DSL) using ASTaYRA tool and parsing and merging technique are used to factorize or generate human understandable tree from the tool generated attack path. Various other approaches were proposed in this direction are summarized in Table 2.

Table 2 Classification of works in attack tree-based modeling

Author	Formal modeling	Generation approaches	Security analysis
Jawahel et al. [7]	Mathematical foundation of attack tree with SAND	Automated SP tool	–
Horne et al. [6]	Causal attack tree	Linear logic interpretation	–
Vigo and Nielson [21, 22]	Value passing quality calculus	Process algebra	Cost structure-based optimization
Pinchinate et al. [23, 24]	SAND tree modeled in DSL	ASTyRA semiautomatic tool used for generation	–
Gadyatskaya [25]	Uses labeled transition system	Factorization	Cost optimization
Aslanyane and Nielson [26, 27]	Socio technical system modeled	Recursive policy invalidation	
Ivanova et al. [36]	AD tree	–	Multi-parameter optimization Pareto efficient strategy
Kordy and Widel [37]	AD tree	–	Optimization using integer linear programming (ILP)
Singh and Ujjwala [39]	Named data network	Manual generation	Cost, probability, time, and technical difficulty
Meyur [40]	Bayesian Network	–	Probability, time for attack efficiency, and impact on power system

4.3 Security Analysis

When a system is modeled as an attack tree, the level of security provided by them is analyzed with the help of some quantitative properties of the modeled system. The quantitative properties can be *minimal cost, probability of accessing root node, time needed for reaching root goal*. The general approach to evaluate these quantities are through simple bottom-up algorithm. This approach is quite fast but have two major drawbacks. Firstly, it can only quantify one property at a time, and secondly, it assumes all node present in the tree to be independent. So, many works on formal framework have been proposed for improvement of this classical approach. The work by Aslanyan and Nielson [36] proposed a multi-parameter optimization for AD tree. This uses *Pareto efficiency strategy* for this work. This approach has high complexity, and it does not capture the multiple time execution of single action. Latter Kordy and Widel [37] proposed a scheme of selecting optimal set of countermeasure using *integer linear programming*. After that, Kordy and Widel [38] proposed a quantitative analysis for repeated actions.

5 Conclusion

Cyber attack modeling is very important and challenging field. It provides the user means to detect the various possible attacks and opportunity to apply suitable measures to mitigate its effects. This paper provides a discussion on a few graphical attack modeling techniques. A common framework for attack modeling is presented here. A review of works done in various directions of two very known graph-based modeling technique: attack graph and attack tree are presented here. These modeling techniques have a few limitations which can be overcome by net-based modeling technique.

References

1. 2020 Internet Crime Report.
2. Shirey, R. (2007). Internet security glossary, version 2.
3. Swiler, L. P., Phillips, C., & Gaylor, T. (1998). *A graph-based network-vulnerability analysis system* (No. SAND-97-3010/1). Sandia National Labs.
4. Haasl, D. F., Roberts, N. H., Vesely, W. E., & Goldberg, F. F. (1981). *Fault tree handbook* (No. NUREG-0492). Nuclear Regulatory Commission.
5. Schneier, B. (1999). Attack trees. *Dr. Dobb's Journal, 24*(12), 21–29.
6. Horne, R., Mauw, S., & Tiu, A. (2017). Semantics for specialising attack trees based on linear logic. *Fundamenta Informaticae, 153*(1–2), 57–86.
7. Jhawar, R., Kordy, B., Mauw, S., Radomirović, S., & Trujillo-Rasua, R. (2015). Attack trees with sequential conjunction. In *IFIP International Information Security and Privacy Conference* (pp. 339–353). Springer.
8. Kordy, B., Mauw, S., Radomirović, S., & Schweitzer, P. (2014). Attack-defense trees. *Journal of Logic and Computation, 24*(1), 55–87.
9. Mauw, S., & Oostdijk, M. (2005). Foundations of attack trees. In *International Conference on Information Security and Cryptology* (pp. 186–198). Springer.
10. Hermanns, H., Krämer, J., Kršál, J., & Stoelinga, M. (2016). The value of attack-defence diagrams. In *International Conference on Principles of Security and Trust* (pp. 163–185). Springer.
11. Roy, A., Kim, D. S., & Trivedi, K. S. (2012). Attack countermeasure trees (ACT): Towards unifying the constructs of attack and defense trees. *Security and Communication Networks, 5*(8), 929–943.
12. Jürgenson, A., & Willemson, J. (2015). Computing exact outcomes of multi-parameter attack trees. In *OTM Confederated International Conferences On the Move to Meaningful Internet Systems* (pp. 1036–1051). Springer.
13. Cheung, S., Lindqvist, U., & Fong, M. W. (2003). Modeling multistep cyber attacks for scenario recognition. In *Proceedings DARPA Information Survivability Conference and Exposition* (Vol. 1, pp. 284–292). IEEE.
14. Valdes, A., & Skinner, K. (2001). Probabilistic alert correlation. In *International Workshop on Recent Advances in Intrusion Detection* (pp. 54–68). Springer.
15. Porras, P. A., Fong, M. W., & Valdes, A. (2002). A mission-impact-based approach to INFOSEC alarm correlation. In *International Workshop on Recent Advances in Intrusion Detection* (pp. 95–114). Springer.
16. Qin, X., & Lee, W. (2003). Statistical causality analysis of INFOSEC alert data. In *International Workshop on Recent Advances in Intrusion Detection* (pp. 73–93). Springer.

17. Petri, C. A. (1962). *Communication with automata* [Ph.D. thesis]. Technische Universitat Darmstadt.
18. Zakrzewska, A. N., & Ferragut, E. M. (2011). Modeling cyber conflicts using an extended Petri Net formalism. In *2011 IEEE Symposium on Computational Intelligence in Cyber Security (CICS)* (pp. 60–67). IEEE.
19. Petty, M. D., Whitaker, T. S., Bland, J. A., Cantrell, W. A., & Mayfield, K. P. (2019). Modeling cyberattacks with extended petri nets: Research program overview and status report. In *Proceedings of the International Conference on Modeling, Simulation and Visualization Methods (MSV)* (pp. 27–33). The Steering Committee of The World Congress in Computer Science, Computer Engineering and Applied Computing (WorldComp).
20. Audinot, M., Pinchinat, S., & Kordy, B. (2017). Is my attack tree correct? In *European Symposium on Research in Computer Security* (pp. 83–102). Springer.
21. Vigo, R., Nielson, F., & Nielson, H. R. (2014). Automated generation of attack trees. In *2014 IEEE 27th Computer Security Foundations Symposium* (pp. 337–350). IEEE.
22. Vigo, R., Nielson, F., & Nielson, H. R. (2016). Discovering, quantifying, and displaying attacks. arXiv preprint arXiv:1607.07720.
23. Pinchinat, S., Acher, M., & Vojtisek, D. (2014). Towards synthesis of attack trees for supporting computer-aided risk analysis. In *International Conference on Software Engineering and Formal Methods* (pp. 363–375). Springer.
24. Pinchinat, S., Acher, M., & Vojtisek, D. (2015). ATSyRa: An integrated environment for synthesizing attack trees. In *International Workshop on Graphical Models for Security* (pp. 97–101). Springer.
25. Gadyatskaya, O., Jhawar, R., Mauw, S., Trujillo-Rasua, R., & Willemse, T. A. C. (2017). Refinement-aware generation of attack trees. In *International Workshop on Security and Trust Management* (pp. 164–179). Springer.
26. Ivanova, M. G., Probst, C. W., Hansen, R. R., & Kammüller, F. (2015). Attack tree generation by policy invalidation. In *IFIP International Conference on Information Security Theory and Practice* (pp. 249–259). Springer.
27. Ivanova, M. G., Probst, C. W., Hansen, R. R., & Kammüller, F. (2015). Transforming graphical system models to graphical attack models. In *International Workshop on Graphical Models for Security* (pp. 82–96). Springer.
28. Ammann, P., Wijesekera, D., & Kaushik, S. (2002). Scalable, graph-based network vulnerability analysis. In *Proceedings of the 9th ACM Conference on Computer and Communications Security* (pp. 217–224).
29. Ritchey, R. W., & Ammann, P. (2000). Using model checking to analyze network vulnerabilities. In *Proceeding 2000 IEEE Symposium on Security and Privacy, S&P 2000* (pp. 156–165). IEEE.
30. Jajodia, S., & Noel, S. (2010). Topological vulnerability analysis. In *Cyber situational awareness* (pp. 139–154). Springer.
31. Noel, S., Elder, M., Jajodia, S., Kalapa, P., O'Hare, S., & Prole, K. (2009). Advances in topological vulnerability analysis. In *2009 Cybersecurity Applications & Technology Conference for Homeland Security* (pp. 124–129). IEEE.
32. Wang, L., Albanese, M., & Jajodia, S. (2014). Attack graph and network hardening. In *Network hardening* (pp. 15–22). Springer.
33. Wang, L., Albanese, M., & Jajodia, S. (2014). Minimum-cost network hardening. In *Network hardening* (pp. 23–38). Springer.
34. Wang, L., Albanese, M., & Jajodia, S. (2014). Linear-time network hardening. In *Network hardening* (pp. 39–58). Springer.
35. Beckers, K., Krautsevich, L., & Yautsiukhin, A. (2014) Analysis of social engineering threats with attack graphs. In *Data privacy management, autonomous spontaneous security, and security assurance* (pp. 216–232). Springer.
36. Aslanyan, Z., & Nielson, F. (2016). Pareto efficient solutions of attack-defence trees. In *International Conference on Principles of Security and Trust* (pp. 95–114). Springer.
37. Kordy, B., & Wideł, W. (2017). How well can I secure my system? In *International Conference on Integrated Formal Methods* (pp. 332–347). Springer.

38. Kordy, B., & Wideł, W. (2018). On quantitative analysis of attack-defense trees with repeated labels. In *International Conference on Principles of Security and Trust* (pp. 325–346). Springer.
39. Singh, V. P., & Ujjwal, R. L. (2019). Privacy attack modeling and risk assessment method for name data networking. In *Advances in Computer Communication and Computational Sciences* (pp. 109–119). Springer.
40. Meyur, R. (2020). A Bayesian attack tree based approach to assess cyber-physical security of power system. In *2020 IEEE Texas Power and Energy Conference (TPEC)* (pp. 1–6). IEEE.
41. Nadeem, A., Verwer, S., Moskal, S., & Yang, S. J. (2021). Sage: Intrusion alert-driven attack graph extractor. arXiv preprint arXiv:2107.02783.

A Secure DBA Management System: A Comprehensive Study

Khushboo Jain, Umesh Jangid, Princy Kansara, Smita Agrawal, and Parita Oza

Abstract The current world is transforming into a digitalized world. Here, every information is taken care of as online data. Data are growing unquestionably rapidly and irksome in managing them moreover overall taking care of securely. Nonetheless, information is put away in gigantic sums, so consequently, security is similarly significant. For this situation, it is the test for the question author to produce an inquiry that aids in shielding information from unapproved access and malignant assault on data sets. Database management system (DBMS) rules accompany a simple illustration of a question generator; by time, database management system (DBMS) is overwhelmed by relational database management system (RDBMS) for their productive work. After some time, the standard method of inquiry writing in social information bases is confronting many new difficulties because a high measure of information comes from different places additionally with blunders, infections, and in numerous others in vindictive structure. The database administrator (DBA) management framework is made to watch out for the security of information/data. This paper presents various security aspects of the database administrator (DB) management system. The article also gives an analysis of various security threats of the last two decades.

Keywords Database · Security · Threats · Database administrator (DBA)

K. Jain · U. Jangid · P. Kansara · S. Agrawal (✉) · P. Oza
CSE Department, Institute of Technology, Nirma University, Ahmedabad, India
e-mail: smita.agrawal@nirmauni.ac.in

K. Jain
e-mail: 19mca006@nirmauni.ac.in

U. Jangid
e-mail: 19mca007@nirmauni.ac.in

P. Kansara
e-mail: 19mca009@nirmauni.ac.in

P. Oza
e-mail: parita.prajapati@nirmauni.ac.in

P. K. Singh et al. (eds.), *Proceedings of Third International Conference on Computing, Communications, and Cyber-Security*, Lecture Notes in Networks and Systems 421,
https://doi.org/10.1007/978-981-19-1142-2_69

1 Introduction

Security in the database management system assumes a significant part in everyday data/information, which is put away in the association's data set framework. Each datum, either close to home or ordinary information, contains delicate and classified data. On the off chance that unapproved or pernicious activities happen, the association faces inconvenience to recuperate them. Therefore, the data set is a highly pivotal piece of any association. Through the data set, anybody can get to an association's delicate and classified data of their representatives, partners, trustees, customers, projects, and so forth.

Moreover, the present world is currently moving toward the online stockpiling of data like on the cloud utilizing Kubernetes, dockers, and other numerous stages. For this situation, data sets must get and have the appropriate assurance from malevolent digital assaults and infections. Simply envision if the data set is not secure; how can individuals store their information in a data set for their utilization?

Information security is the principal worry of each association. Database administrator (DBA) is answerable for the turn of events and support of data set security. Here, the database administrator is responsible for dealing with every one of the parts of the information base like climate, execution, recuperation, and central security. "Security stands up from the requirement for conservancy from the unapproved access and debasement of sensory information. The vital objective of safety is to safeguard the framework from interruption" [1, 2].

Those requirements permit the framework to control the ploy that clients are allowed to perform.

There are two kinds of client advantages.

1. Framework advantages
2. Article advantages.

Database management system (DBMS) performs information recuperation activities of data sets and homomorphism control capacity. These day's associations need data sets to store each sort of information required because the moderate expense, quick, and superior data set are well known among the associations. For instance, clients can trust PC reports that are put away in the data set rather than physically examining exchanges. Rather than entering stock data physically, scanners can be utilized to save data in the information base [3]. The data set can give productivity in the cutting-edge working environment. One more inquiry for any association "is data obtained utilizing a data set?" Security these days is one of the essential and resisting undertakings that individuals are confronting around the world. Numerous data set security experts do not completely comprehend hazard and security identified with different data sets. Data sets are uphill to get a complete insight into the information base properties because there are exceptional advances in execution and mechanics for data sets. Data set security is a wide range of information security that ensures data sets against interior assault or outside assault, against mollification of data set accessibility and classification. Security includes various sorts of controls like regulatory and specialized controls [2, 4].

1.1 *Objective of Secure DBA System*

The objective of a secure database administrator (DBA) system is to comprehend the parts of information base security. How information base security frameworks work and how we can further develop them by giving limitations. The principal center is here; how secure is our data set? Which security controls would we be able to use to make it safer and ensured? Step-by-step instructions to screen security, database management system (DBMS) arrangements, framework solidifying. These security controls help to figure out how to evade security conventions. In any database administrator (DBA) board framework, security assumes the primary part to ensure that it is secure and any confirmed individual can utilize it from any place. Security consistently starts things out for any individual, dealer, vendor, government area, banking, association, media, or anywhere where men put their information or delicate data. Database administrator (DBA) security has three essential things: authentication, information administration, and access controls. These three things cover the complete protection in the database administrator (DBA) board framework idea. The following three points are the common goal for any secure database administrator (DBA) system.

- Consider data set security issues with regards to thoughts and general security standards.
- Examine issues identified with both data set administration framework correspondence with various applications and data set stockpiling.
- Get a light on security issues in an overall data set framework climate.

The rest of the paper is organized as follows: Sect. 2 describes the security mechanism for the system. Then, in Sect. 3, we discuss database security threats. Finally, we end with the conclusion in Sect. 4.

2 The Security Mechanism

Variation methods of information security procedures are encryption. Encryption is utilized to ensure delicate data, or information is communicated over the Web or through satellite. Some calculation encodes this information or data for a reason [5–7]. If the unauthenticated client needs to get to the information, they need to unscramble this because it is in a scrambled structure; verified individuals can decode information without much of a stretch as and when they sense, they have calculations to decode data [2].

The idea of encryption depiction is simple. It gives conspicuous insinuation on the information in which you apply encryption. This message is decoded by just the people who know the way to unscramble.

Fig. 1 Symmetric encryption

2.1 *Symmetric Encryption*

In symmetric encryption, there is a common key for encryption and decryption of data. So it is the most extreme shot at getting that key to unapproved individuals. Figure 1 shows the pictorial representation of how symmetric encryption works with shared keys.

2.2 *Asymmetric Encryption [5–7]*

In asymmetric encryption, there are two keys: public and private. One for the sender, which is shared, and 1 for a collector, which is private, so here, it will make the base shot at getting the way to unapproved individuals. Since the person who is the recipient can just peruse the implication of the sender, no other can see it to the extent, it is a private key. Figure 2 shows the pictorial representation of how asymmetric encryption works with the public and private keys.

Fig. 2 Asymmetric encryption

According to the cryptography concern, symmetric encryption is more significant than topsy-turvy encryption due to having a solitary key and getting quick execution; however, then again, deviated encryption is important simultaneously because it gives them the security of unscrambling.

3 Database Security Threats

In data sets, clients have various advantages. Information base security and honesty dangers are frequently shocking [4]. Security dangers can have different wellsprings of start like internal, external, and accomplice [8]. There are many sorts of information base security dangers that can influence any kind of task [9–11]. Table 1 presents rank-wise threats in the last two decades.

Legitimate Privilege Abuse

Clients may likewise mishandle genuine information base advantages for unapproved purposes. Consider a theoretical maverick medical care laborer with benefits to see individual patient records utilizing a custom Web application. The design of the Web application regularly restricts clients from reviewing a singular patient's medical services history—different documents cannot be seen all the while, and electronic duplicates are not permitted. Notwithstanding, the maverick laborer might dodge these constraints by associating with the information base utilizing an elective customer like MS Excel. Using MS Excel and his real login qualifications, the specialist might recover and save every persistent record [12, 13].

Database Communications Protocol Vulnerabilities

A developing number of safety weaknesses are being recognized in the information base correspondence conventions of all data set sellers. Four out of seven security fixes in the two latest IBM DB2 fix packs address convention vulnerabilities1. Additionally, 11 out of 23 data set weaknesses fixed in the latest Oracle quarterly fix identify with conventions. False movement focusing on these weaknesses can go from unapproved information admittance to information debasement to administration refusal. The SQL Slammer2 worm, for instance, exploited a defect in the Microsoft SQL server convention to drive refusal of administration. No record of these misrepresentation vectors will exist in the local review trail to exacerbate the situation since local information-based review systems do not cover convention activities [13, 14].

Platform Vulnerabilities

Weaknesses in hidden working frameworks (Windows 2000, UNIX, and so on) and extra administrations introduced on an information base server might prompt unapproved access, information debasement, or refusal of administration. The Blaster worm, for instance, exploited a Windows 2000 weakness to make a refusal of administration conditions [13].

Table 1 Rank-wise threats in last two decades

Ranking	2010 top threats	2013 top threats	2019 top threats	2020 top threats
1	Excessive privilege abuse	Excessive and unused privileges	Cloud database configuration errors	Injection
2	Legitimate privilege abuse	Privilege abuse	Structure query language (SQL) injection	Broken authentication
3	Privilege elevation	Structure query language (SQL) injection	Weak authentication	Sensitive data exposure
4	The exploitation of vulnerable, misconfigured databases	Malware NEW	Privilege abuse	Extensible markup language (XML) external entities (XXE)
5	Structure query language (SQL) injection	Weak audit trail	Excessive privilege	Broken access control
6	Weak audit trail	Storage media exposure	Inadequate logging and weak auditing	Security misconfiguration
7	Denial of service	The exploitation of vulnerabilities and misconfigured databases	Denial of service	Cross-site scripting (XSS)
8	Database communication protocol vulnerabilities	Unmanaged sensitive data	Exploiting unpatched services	Insure deserialization
9	Unauthorized copies of sensitive data	Denial of service	Insecure system architecture	Using components with vulnerabilities
10	Backup data exposure	Limited security expertise and education new	Inadequate backup	Insufficient logging and monitoring

Excessive Privilege Abuse

At the point when clients (or applications) are conceded information base access advantages that surpass the necessities of their work, these advantages might be mishandled for pernicious purposes. For instance, a college overseer whose work requires just the capacity to change understudy contact data might exploit unnecessary data set update advantages to change grades [13, 15].

Privilege Elevation

Assailants might exploit data set stage programming weaknesses to change access advantages from a conventional client to those of a director. Faults might be found in input-away techniques, inherent capacities, convention executions, and even SQL proclamations. For instance, a product designer at a monetary establishment may exploit a weak ability to acquire the data set regulatory advantage. With authoritative advantage, the maverick designer might wind down review systems, make false records, move reserves, and so on [13, 14].

SQL Injection

In a SQL infusion assault, a culprit commonly embeds (or "infuses") unapproved data set explanations into a weak SQL information channel. For example, normally designated information channels to incorporate put away techniques and Web application input boundaries. These infused explanations are then passed to the data set, where they are executed. Thus, utilizing SQL infusion, aggressors might acquire unhindered admittance to a whole data set.

Denial of Service

Denial of service (DoS) is an overall assault classification wherein expected clients are denied admittance to organize applications or information. Denial of service (DoS) conditions might be made by means of numerous strategies—a significant number of which are identified with recently referenced weaknesses. For instance, DoS might be accomplished by exploiting an information base stage weakness to crash a server. Other regular DoS strategies incorporate information debasement, network flooding, and server asset over-burden (memory, CPU, and so forth). Asset over-burden is exceptionally regular in information base conditions. The inspirations driving DoS are also different. DoS assaults are often connected to coercion tricks in which a far-off aggressor will over and over crash servers until the casualty stores assets to a global ledger. Then again, DoS might be followed by worm contamination. Whatever the source, DoS addresses a genuine danger for some associations [13, 15].

Backup Data Exposure

Reinforcement information base stockpiling media is regularly totally unprotected from assault. Thus, a few high-profile security breaks have involved the robbery of information base reinforcement tapes and hard circles [13].

Weak Authentication

Weak authentication plans permit aggressors to expect the character of actual data set clients by taking or in any case acquiring login qualifications. An aggressor might utilize quite a few systems to get qualifications [13].

Weak Audit Trail

Mechanized recording of all touchy and uncommon data set exchanges ought to be essential for establishing fundamental any data set organization. Weak data set review strategy addresses a genuine authoritative danger on many levels [13].

Figure 3 depicts Azure SQL database [16]. In Azure structure query language (SQL), this layer will clarify the profundity of the structure query language (SQL)

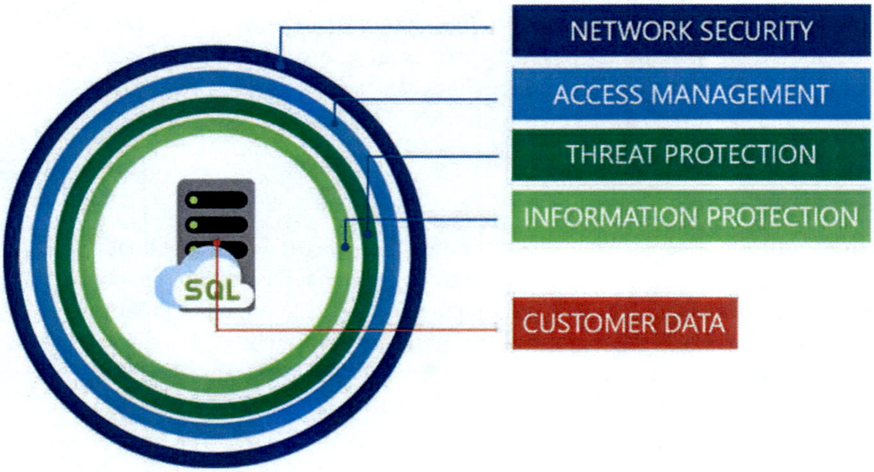

Fig. 3 Azure SQL database

information base security levels. There are five layers: network security, access management, threat protection, information protection, and customer data.

In network security, a firewall ensures the association's information over the Wutilizing an Internet protocol (IP) address and sky blue virtual insurance. Few virtual organization firewall rules expand virtual organizations over purplish-blue and empower sky blue structure query language (SQL) data set personality in subnets where traffic starts. To arrive at the traffic purplish-blue, utilize the structure query language (SQL) administration labels to permit outbound traffic in network security gatherings [17, 18].

Column-level security, validation, and approval are the central issues of access to the executives. In column-level security, clients will deal with the straight information organization. Clients need to get to the information base to table in succession and play out the activities like embed, select, update, erase from the verified way. Validation, the client initially goes to the data set security and takes a look at the basics or the information base and plays out the activities identified with the confirmed individual who is the administrator. Here, just confirmed individuals can get to the data set because they have the authorization to get to the information base [18]. In approval, "a worker administrator login with username and secret word should be determined when the worker is being made. A worker administrator can verify any data set on that work as the information base proprietor. After this, extra structure query language (SQL) logins and the clients can be made by the administrator, which can empower clients to associate utilizing the username and secret key."

4 Limitations

Stolen Database Backups

There are two sorts of dangers to your data sets: outside and inner. There are situations when organizations battle with internal threats much more than with outer. Entrepreneurs can never be 100% certain of their representatives' dedication, regardless of PC security programming they use and how dependable they appear to be. Anyone who approaches touchy information can take it and offer it to outsider associations for benefit. In any case, there is a way of dispensing with the danger: scramble information base documents, carry out severe security norms, apply fines in the event of infringement, utilize network protection programming, and ceaselessly increment your groups' mindfulness utilizing corporate gatherings and individual counseling.

Test Website Security to Avoid SQL Injections

This is a significant road obstruction en route to data set insurance. Infusions assault the applications, and information base overseers are compelled to tidy up the wreck of pernicious codes and factors that are embedded into the strings. Web application security testing and firewall execution are the ideal choices to ensure Web-confronting information bases. Anyway, this is a significant issue for online organizations; it is not one of the significant versatile security challenges, which is an extraordinary benefit for the proprietors who just have a portable adaptation of their application.

Poor Encryption and Data Breaches Come Together

You should seriously think about the information base as a back-end part of your setup and center more around the disposal of Internet-borne dangers. Unfortunately, it does not actually work that way. There are network interfaces inside the information bases that programmers can handily follow if your product security is poor. To stay away from such circumstances, use TLS or SSL encoded correspondence stages.

5 Future Enhancements

There are many exciting research topics in database administrator security, for example, the application of computer immune systems to the distributed database system. An immunological model of distributed detection has been designed and developed for computer security [19].

Ex. Consider a database that is distributed across many locations in a network. These locations can consist of computers connected in LAN or individual computers. The communication between systems at different locations may pass through multiple intermediate local systems. The natural system protects from some of the attacks. So in the database, we have to identify the self user (authorized) and nonself (not authorized user). We have to research which can identify the non-authorized users of the database in any form like trojan horses or corrupted data. This is a fascinating and helpful research topic in the database security area.

6 Conclusion

Security in the database administrator (DBA) management system is vital, particularly for an association's data set. The data set framework assumes a significant part, so security concerns should be there. As we all realize, the present world is digital assaults, slacking of data that is not helpful for any country. Additionally, in light of assaults, different crimes have occurred so far that we need to make our data set as solid as possible, handle malevolent assaults, and offer security to our valuable information. We presented a comprehensive study of various security aspects of the database administrator (DBA) management system, secure mechanisms, and types of threats that can influence clients' sensitive information.

References

1. Ali, A., & Afzal, M. M. (2017). Database security: Threats and solutions. *International Journal of Engineering Inventions, 6*(2), 25–27. e-ISSN: 2278-7461, p-ISSN: 2319-6491.
2. Thuraisngham, B. M., Ford, W. R. B., Collins, M. S., & O'Keeffe, J. P. (1994). System for multilevel secure database management using a knowledge base with release-based and other security constraints for query, response and update modification. U.S. Patent 5,355,474, issued October 11, 1994.
3. Jorge, D. C. Basic principles of database security.
4. Thuraisingham, M. B. (1987). Security checking in relational database management systems augmented with inference engines. *Computers & Security, 6*(6), 479–492.
5. Shah, Y., Joshi, S., Oza, P., & Agrawal, S. (2019). An insight of information security: A skeleton. *International Journal of Recent Technology and Engineering (IJRTE), 8*(3). ISSN: 2277-3878
6. Patel, N., Oza, P., & Agrawal, S. (2019). Homomorphic cryptography and its applications in various domains. In *International Conference on Innovative Computing and Communications*. Springer, Singapore.
7. Antonopoulos, P., et al. (2020). Azure SQL database always encrypted. In *Proceedings of the 2020 ACM SIGMOD International Conference on Management of Data*.
8. Oza, P., & Malvi, P. (2016). Encryption algorithm using Rubik's cube principle for the secure transmission of multimedia files. In *Third International Conference on Multidisciplinary Research and Practice IJRSI* (Vol. 4).
9. Win, A. M., & Myint, K. L. (2019). Database security model using access control mechanism in student data management. *International Journal of Trend in Scientific Research and Development (IJTSRD), 3*(3), 529–531. e-ISSN: 2456-6470
10. Verizon. (2019). Data breach investigations report. https://enterprise.verizon.com/resources/reports/2019-data-breach-investigations-report.pdf
11. Jain, S., & Chawla, D. (2020). A relative study on different database security threats and their security techniques. *International Journal of Innovative Science and Research Technology, 5*(5), 794–799.
12. Igure, V. M., & Williams, R. D. (2008). Taxonomies of attacks and vulnerabilities in computer systems. *IEEE Communications Surveys & Tutorials, 10*(1), 6–19.

13. Malik, M., & Patel, T. (2016). Database security attacks and control methods. *International Journal of Information, 6*(1/2), 175–183.
14. Basharat, I., Azam, F., & Muzaffar, A. W. (2012). Database security and encryption: A survey study. *International Journal of Computer Applications, 47*(12).
15. Rohini, K., Kasturi, K., & Vignesh, R. (2020). Method for simulating SQL injection and DOS attack. In *Intelligent computing and innovation on data science* (pp. 793–801). Springer.
16. Aravindharamanan, S., Ramasubbareddy, S., & Govinda, K. (2019). Legitimate privilege abuse and data security in database. In *Innovations in computer science and engineering* (pp. 175–181). Springer.
17. Vieira, M., & Madeira, H. (2005). Towards a security benchmark for database management systems. In *2005 International Conference on Dependable Systems and Networks (DSN'05)* (pp. 592–601). IEEE.
18. Bertino, E., & Haas, L. M. (1988). Views and security in distributed database management systems. In *International Conference on Extending Database Technology* (pp. 155–169). Springer.
19. Ke, M. *Computer database security and Oracle security implementation.* The University of Montana Graduate student theses.
20. Dwyer, P. A., Jelatis, G. D., & Thuraisingham, B. M. (1987). Multi-level security in database management systems. *Computers & Security, 6*(3), 245–251.

Education 4.0: Hesitant Fuzzy SWARA Assessment Approach for Intelligent Selection of Research Opportunities

Pooja Khanna, Pragya, Ritika Gauba, and Sachin Kumar

Abstract Reforms and revolutions are periodic in nature, a cyclic process they are and with innovations in technology and analytics, the idea of learning education has transformed into experiencing the education. With every Industrial Revolution, education pattern, its content and delivery got new dimension. Education in India has been categorically segregated as primary, junior, middle, senior, undergraduation, post-graduation, and research degrees. Aspirants are usually aware and well informed till they reach research degrees level and beyond. With macro- and micro-options within specializations, scholarships, center for excellence, availability of experts, eligibility criteria, research trends, technological advancements, procedural steps during the course, live projects associated, financial aspects, and education policies, the whole process becomes a huge unsolved mystery and aspirants might miss better opportunities which otherwise he might have gone for. The work carried is an effort to identify a fuzzy-assisted pattern for the aspirant to make an informed decision about the research option he or she can avail for enhanced growth. Study proposes a data analytics model for the informed intelligent selection toward optimum research carrier. The selection is assisted by number of weighted parameters which an aspirant should take into consideration while selecting a particular domain. SWARA technique was used for estimating weights of criteria based on experts' preferences further to establish feasibility of the technique proposed, an empirical study of sustainable organization selection taken under hesitant fuzzy (HF) environment. The data used for the research work was obtained from Zenith Ph.D. Training & Consultancy (ZPTC), Jaipur, proposed technique was tested on data from 300 universities. The organization

P. Khanna · Pragya · S. Kumar (✉)
Amity University, Lucknow Campus, India
e-mail: skumar3@lko.amity.edu

P. Khanna
e-mail: pkhanna@lko.amity.edu

MVPG College, Lucknow, India

R. Gauba
Zenith Ph.D. Training and Consultancy, Jaipur, India
e-mail: contact@zenithphd.in

P. K. Singh et al. (eds.), *Proceedings of Third International Conference on Computing, Communications, and Cyber-Security*, Lecture Notes in Networks and Systems 421,
https://doi.org/10.1007/978-981-19-1142-2_70

focusses and specializes in doctoral fellowships, examinations, processes, and regulations in India and has huge database for the same. Priority order G^* was computed, and degree of utility was estimated (λ_i) for the proposed analysis. The degree of utility came as 99.8% for K_3, 93.1% for K_2, and 86.4% for K_4. With estimated values, it was established that aspirants preferred organizations in following order $K_3 > K_2 > K_4 > K_1 > K_5 > K_6$.

Keywords SWARA · Dimension · AHF-D · Decision expert · Hesitant fuzzy number

1 Introduction

With Education 2.0, technological innovations have majorly moderated the education process that leads to infiltration of more user-generated Internet information, laying the foundation of Education 3.0. Learners now readily available with virtual platforms to gain knowledge and can easily connect with trainers and pear group for exchange of information. Education was no longer limited to classroom study in a closed group, instead took a more networked approach, now students have direct access to variety of different information sources. This personalized way of learning and independence is more appreciated and adopted by the learners. Though education domain is evolving with exponential technological changes and a new phase; with Fourth Industrial Revolution, education domain witnessed new processes and methodologies, i.e., Education 4.0. Smart technology, artificial intelligence, and robotics have major impact on every part of lives including normal working of university system. University system needs to incorporate technological changes in traditional education system helping students to get prepare for a world where these cyber-physical systems are prevalent across all industries. Cyber-physical systems are steadily becoming more integrated into various industries, inevitably affecting the skills requirements for employees. With evolution of Education 4.0, it is essential for higher education institutions, to understand the gap between learning system and industrial need. However, with growth and availability of possible diverse domains, lack of awareness about potential opportunities that can be explored remains one of the major drawbacks that student experience. Though with pandemic, COVID-19 not only heavily burdened health sector but also led to economy crises owing to lockdowns, supply, and demand issues, but there were sectors which did find new ways to tackle the situation. Government took many initiatives especially with education sector to establish a constant growth. There were several initiatives taken by the government to handle the uncertain situation [1, 2].

Government of India has initiated many projects for supporting use of ICT as a means of mass education. CLASS (Computer Literacy and Social on Schools) project is one such step launched in 1984 to make computer literacy a compulsory project for higher middle and higher middle classes. In the seventh five-year plan and eight five-year plans thousands of schools started computer literacy, respectively [3].

Satellite like EDUSAT has been launched devoted to education sector on the 20th of September 2004 by the Indian Space Research Organization (ISRO) [4]. SWAYAM, an integrated platform and online portal, is initiated by Ministry of Human Resources Development (MHRD) and All India Council for Technical Education (AICTE) with the help of Microsoft [5, 6]. Indian government has analyzed the need of e-Learning platforms and introduced initiatives likes; Consortium for Educational Communication (CEC), National Program on Technology Enhanced Learning (NPTEL), Indira Gandhi National Open University (IGNOU), Online Education Broadcast and Virtual Classrooms, Sakshat, Institute of Lifelong Learning (ILLL), and School of Open Learning (SOL) E-learning Gateways. In current scenarios when most of the countries all over the world struggling to start with normal life, the Indian government launches DIKSHA (Digital Infrastructure for Knowledge Sharing) in September 2017, proved to be one of the tools the government leveraged in the COVID-19 era [1].

Government took many e-learning initiatives post COVID-19 as depicted in Fig. 1, and these include MHRD initiatives like dedicated channels to broadcast high-quality educational programs, course contents for National Institute of Open Schooling, Radio broadcasting for learning-based activities. Direct-to-Home (DTH) channel, Digitally Accessible Information System (DAISY), YouTube, and NIOS website are few more initiatives in this direction. Apart from all these projects,

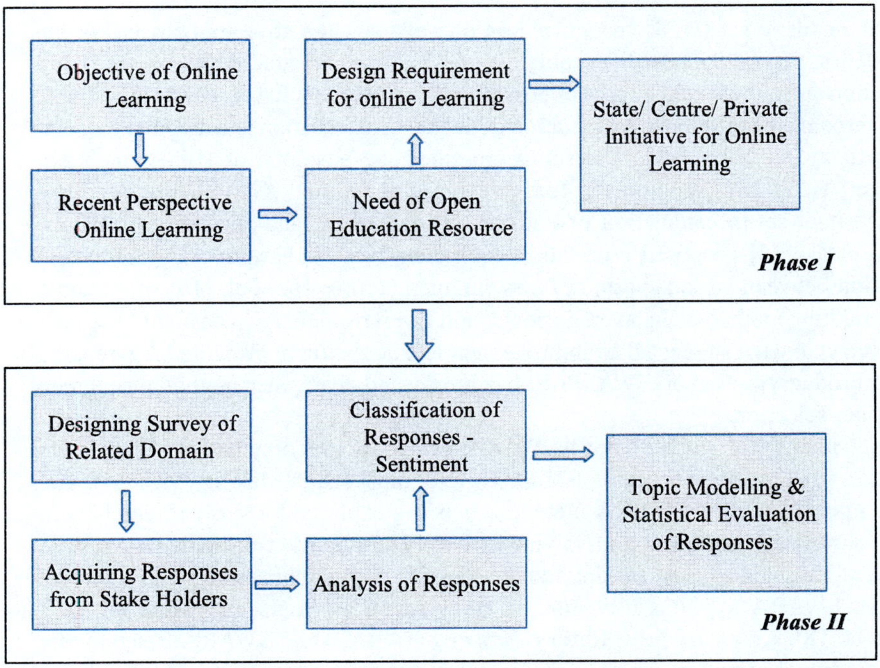

Fig. 1 E-learning initiatives in response to COVID-19 crisis

e-Pathshala web portal Electronic for textbooks, National Repository of open educational resources (NROER) for e-content of The National Council of Educational Research and Training (NCERT).

The work carried is an effort to identify a Fuzzy-assisted pattern for the aspirant to make an informed decision about the research option he or she can avail for enhanced growth. Study proposes a data analytics model for the informed intelligent selection toward optimum research carrier. The selection is assisted by number of weighted parameters which an aspirant should take into consideration while selecting a particular domain. SWARA technique was employed for estimating weights of the criteria based on experts' preferences further to establish practicability of the proposed methodology, an empirical case study of sustainable organization selection taken under hesitant fuzzy (HF) environment. Algorithm is also subjected to sensitivity analysis to validate stability of the presented methodology. The data used for the research work was obtained from Zenith Ph.D. Training & Consultancy (ZPTC), Jaipur. The proposed technique was tested on data from 300 universities. The organization focuses and specializes in doctoral fellowships, examinations, processes, and regulations in India and has huge database for the same [9, 10].

2 Motivation

Hesitant fuzzy set (HFS) recognized as one of the alternative tools to tackle the vagueness occurred in real-life problems [11, 12]. It is characterized by a membership function and represented by a set of possible values. Hesitant fuzzy set was introduced to overcome the problem of calculation of degree of association of a member element into fuzzy set Torra [11]. The study presented the envelope of HFS, which can convert HFSs into intuitionistic fuzzy sets (IFSs). Xu and Xia [13] proposed use of Hesitant set-in calculation of entropy, cross-entropy, and similarity measures. Liao et al. [14] proposed correlation coefficient-based measures for establishing relation between various hesitant fuzzy linguistic terms. He et al. [15] suggested a new ranking method using average power and Bonferroni mean in hesitant fuzzy set. Mishra et al. [16] suggested an improvement in calculation of Weighted Aggregated Sum Product Assessment (WASPAS) by entropy and divergence method ideal green supplier selection.

Kersuliene et al. studied a novel SWARA method for the computation of subjective criteria weights in the process of Multi-Criteria in Decision Making (MCDM). The ease in calculation in SWARA method it is being explored by many researchers in various domain-based problem-solving issues [17]. Dehnavi et al. studied a landslide susceptible region based on data collected from Geographic Information System (GIS). SWARA methods used for the study identify criteria and value for each criteria [18]. A new Additive Ratio Assessment (ARAS) and SWARA-based model for personnel selection problem within FS context are suggested in the study by Karabasevic et al. [19]. Literature indicates successful use of SWARA method in solving various fuzzy set-based approaches. Nakhaei et al. assessed the vulnerability of buildings against explosion a hybrid approach of simple Multi-Attribute Ranking Technique (SMART) and SWARA [20]. A study discussed the evaluation of MCDM problems using Operational Competitiveness Rating Analysis (OCRA) and SWARA methods Isik et al. [21]. For selection of personnel in the tourism business, SWARA- and WASPAS-based approach was suggested by study Urosevic et al. [22].

3 Material and Methods

Hesitant fuzzy SWARA assessment approach has been employed for intelligent selection of research opportunities, to address the stated problem. The estimation process proposed is presented in the following steps:

Step 1: Identify options based on favorable criteria.

A cluster of DEs (C1, C2, C3, …, CN) estimates the group of k alternatives B = (B1, B2, B3, …, B_k) and m criteria H = (H1, H2, H3, …, H_m), respectively. Assume that $X = \left(X_{ij}^{(k)}\right)_{k\times m}$, $i = 1, 2, 3, \ldots, k$ and $j = 1, 2, 3, \ldots, m$, is the HF decision matrix presented by kth DE where $X_{ij}^{(k)}$ is the assessment of option G_i for the criteria F_j in terms of HFN.

Step 2: Evaluate the crisp value of weights. The estimation is performed employing the expression of kth DE's weight:

$$\lambda_l = \frac{(1 - e(\hbar_l))}{\sum_{l=1}^{k} 1 - e(\hbar_l)}, \quad l = 1(1)k$$

Clearly $\lambda_l \geq 0$ and $\sum_{l=1}^{\ddagger} \lambda_l = 1$.

Step 3: To rationalize and aggregate the values in matrix, the HFWA operator is employed and then $O = \left(\xi_{ij}\right)_{m\times n}$ be the required AHF-D.

Matrix in which

$$\xi_{ij} = U_{i1\in h_1, i2\in h_1, \ldots, in\in h_n}\left\{1 - \prod_{l=1}^{\ddagger}\left(1 - \hbar_{ij}^{k}\right)^{\lambda_l}\right\}$$

Step 4: SWARA technique is employed to estimate the criteria weights. Steps involved using SWARA scheme are:

Step 4-a: Score values $S(\xi_{ij})$ are estimated using following expression

$$S(\hbar) = \frac{1}{g_\hbar}\sum_{i\in\hbar} i$$

The equation generates the number of objects in $\hbar$.

Step 4-b: Criteria are sorted according to DE's opinions. Most significant are ranked first to least significant.

Step 4-c: Evaluate the degree of comparative significance (S_j) for each criteria.

Step 4-d: Coefficient of comparison (K_j) is estimated employing following relation

$$k_j = \begin{cases} 1, & j = 1 \\ s_j + 1, & j > 1 \end{cases}$$

Step 4-e: Weight is recalculated weight (p_j) employing following relation

$$p_j = \begin{cases} 1, & j = 1 \\ \frac{p_{j-1}}{k_j} s_j + 1, & j > 1 \end{cases}$$

Step 4-f: Estimate the weight of each criteria employing following relationship:

$$w_j = \frac{p_j}{\sum_{j=1}^{n} p_j}$$

Step 5: Criteria values are added to evaluate overall profit and loss parameters.

Assume that $\lambda_1 = \{1, 2, 3, \ldots, l\}$ is the beneficial criteria type, the highest index value for each option is given as

$$\sigma_i = \oplus_{j=1}^{\ddagger} w_j \xi_{ij}, \quad i = 1, 2, 3, \ldots, m$$

Assume that $\lambda_2 = \{l + 1, l + 2, l + 3, \ldots, n\}$ profit type of criteria. The highest index value for each option is given as

$$\upsilon_i = \oplus_{j=1+\ddagger}^{n} w_j \xi_{ij}, \quad i = 1, 2, 3, \ldots, n$$

where '*l*' depicts the total of criterion types that are beneficial and '*n*' depicts the total of criteria in totality.

Step 6: Relative weight (θ_i) is estimated using following relation for each option:

$$\theta_i = \Upsilon s(\sigma_i) + (1 - \Upsilon) \frac{\sum_{i=1}^{m} s(\upsilon_i)}{s(\upsilon_i) \sum_{i=1}^{m} \frac{1}{s(\upsilon_i)}}, \quad i = 1, 2, 3, \ldots, M$$

Here, $S(\sigma_i)$ and $S(\upsilon_i)$ represent the score degrees of σ_i and υ_i, respectively.

Step 7: Priority order for the option is estimated employing following expression:

$$G^* = \max_i \theta_{i,}, \quad i = 1, 2, 3, \ldots, m$$

Criteria with highest relative weight are ranked first and therefore an ideal choice.

Step 8: Estimate the degree of utility, the evaluation is performed employing following relation:

Table 1 *Categories* of organization offering doctoral programs

S. No.	Organization category	Denotation
1	State government university	K_1
2	Central government university	K_2
3	Institute of national importance	K_3
4	Private university	K_4
5	Deemed university	K_5
6	Open university	K_6

$$\lambda_i = \frac{\theta_i}{\theta_{\max}} \times 100\%, \quad i = 1(1)m$$

Step IX: End.

A case study was conducted on dataset of 300 students obtained from Zenith Ph.D. Training & Consultancy' (ZPTC), Jaipur. The organization focuses and specializes in PhD fellowships, examinations, processes, and regulations in India and has huge database for the same [18–22].

The students were subjected to diversified preferences and options for selecting a particular organization and stream for doctoral degree. Dataset consisting of aspirants were asked to fill questionnaire consisting of dimensions and criteria as given in Table 1; parameters were weighted according to the rating of dimension as filled by the candidate in the questionnaire.

In accordance with the previous operation reviews, the organizations were divided into six main categories (K_1, K_2, K_3, K_4, K_5, and K_6 depicted in Table 1); options were considered for the analysis after preconsideration of dependent parameters. Organizations were adjudged with dimensions and criteria with features as depicted in Table 2.

The variability just mentioned was incorporated in marking scheme to standardized variance. Methodology adopted had four dimensions as

1. Economic: Aspirants are usually ambiguous with finance involved with doctoral programs; there are programs which are fully funded or partially funded, and there are variations in fees structure with government, state, or private universities. The macro-parameters were divided and were categorized as

   i. Partially or fully funded (E1)
   ii. Government, state, or private universities (E2)
   iii. Fee's structure (E3)

   Finance also takes into scholarship schemes available with the program; these can be categorized as

   i. Government/university/category/industry schemes (E4)
   ii. Number of seats available (E5)

   Expenses also extend to conveyance; the aspirant has to commutate daily

   i. Ease of access/conveyance availability (E6)

Table 2 Criteria details for sustainable organization selection (SOS)

Dimension	Criteria	Meaning	Type
Economic	Finance (C1)	Fully funded, partially funded, variations—state/central/private, fees	Cost
	Scholarship (C2)	Seat available, government schemes, university schemes, category schemes, industry schemes	Cost
	Distance (C3)	Ease of access, conveyance availability, same city, across city, different state	Cost
Academic	University ranking (C4)	University grading, accreditation, faculty resource	Benefit
	Eligibility criteria (C5)	Seats available, entry-level minimum marks, essential degree requirement	Benefit
	Submission prerequisites (C6)	Duration, publication requirement, type of publication required conference/journal—UGC/SCOPUS/SCI, course work requirement	Benefit
Resource	Center of excellence (C7)	Support schemes, infrastructure available, time boundation	Benefit
	Research lab (C8)	Working hours boundation, simulation software available, plagiarism software, workstation configuration	Benefit
	Journal access (C9)	Journals subscribed, ease of access, degree of access, diversity of access	Cost
Brand	Reputation (C10)	How old is organization, central/state/private, UGC grading, NIRF ranking, international accreditations, alumni support, institutions of national importance	Benefit

ii. Same city/across city/different state (E7)

2. Academic: Aspirants always aim for the best possible options while opting for doctoral program, excellence in work not only depends upon academics of research scholar but also upon organization standing as center of academic excellence, broadly following parameters can include; first one identified was university ranking:

i. University grading by NAAC/UGC (A1)
ii. National/International accreditation university has undergone (A2)
iii. Rich faculty resource (A3).

Eligibility criteria for doctoral programs establish one of the quality checks on intake

i. Seats availability for domain (A4)
ii. Minimum entry level marks (A5)
iii. Essential degree requirement for domain (A6).

Every university program has certain prerequisites for doctoral program

   i. Minimum duration requirement for doctoral program (A7)
   ii. Number of publications required (A8)
   iii. Type of publication required conference/journal—UGC/SCOPUS/SCI/Others (A9)
   iv. Course work requirement (A10)

3. Resource: Resource acts as one of the potential contributors in literature access, an enriching resource automatically raises the bar of the work proposed, likely components include center of excellence.

   i. Support schemes running under center of excellence (R1)
   ii. Infrastructure support for research work, i.e., Research Labs/Internet/Workstations (R2)
   iii. Access and duration of availability of resources (R3)

   Availability of research labs

   i. Working hours available for the infrastructure (R4)
   ii. Availability of simulation software's/Sample testing Labs (R5)
   iii. Free/paid availability of Plagiarism software for similarity checks (R6)
   iv. Number/configuration/compatibility of workstations (R7)

   Ease of literature access in terms of journal papers and conference papers

   i. Category of journals subscribed (R8)
   ii. Ease and degree of access subscribed (R9)
   iii. Diversity of literature subscribed (R10)

4. Brand value of organization plays a major role in acceptance of work conducted, following components form an essential part in building reputation.

   i. How old is the organization (B1)
   ii. Organization comes under state/center/private funding (B2)
   iii. UGC grading/NIRF ranking of organization (B3)
   iv. International accreditation organization is holding (B4)
   v. Alumni support organization is having (B5)
   vi. Institutions of national importance with special privileges (B6).

Dimensions defined had components comprising of criteria for different domains, which are further categorized as E, A, R, and B series; these were framed as questionnaire and floated to 300 aspirants, and next concern was weight allocation to the series; weight allocation was accomplished via feedback from faculties and students pursuing doctoral degree.

Tables 3 and 4 represent the linguistic values (LVs) and related HFNs for establishing the rating of the relative importance of criteria.

The conclusion derived from three decision experts has been evaluated from step 3, and following aggregated hesitant fuzzy decision matrix (AHF-D) matrix is generated as depicted in Table 5.

SWARA algorithm considers the recommendations of domain experts, as one of the contributors of weight criteria for assessment. The DE allocates preferences to

Table 3 Linguistic values for decision expert risk preference

Linguistic values	Hesitant fuzzy number	Decision expert risk preference		
		Confirmed	Doubtful	Declined
Very high	[0.8, 1.0]	0.80	0.90	1.00
High	[0.70, 0.9]	0.70	0.80	0.90
Medium	[0.60, 0.70]	0.60	0.65	0.70
Low	[0.40, 0.55]	0.40	0.50	0.60
Very low	[0.20, 0.40]	0.20	0.30	0.40

Table 4 Linguistic values for decision expert risk preference

Linguistic values	Hesitant fuzzy number	Decision expert risk preference		
		Confirmed	Doubtful	Declined
Extremely preferable	[0.9, 1.0]	0.9	0.95	1.00
Strong preferable	[0.75, 0.9]	0.75	0.825	0.9
Preferable	[0.6, 0.75]	0.6	0.675	0.75
Moderate	[0.45, 0.6]	0.45	0.525	0.6
Undesirable	[0.35, 0.45]	0.35	0.4	0.45
Strong undesirable	[0.2, 0.35]	0.2	0.275	0.35
Extremely undesirable	[0.0, 0.15]	0.00	0.075	0.15

Table 5 Aggregated hesitant fuzzy decision matrix for SOS problem

	K_1	K_2	K_3	K_4	K_5	K_6
C_1	0.726	0.822	0.873	0.744	0.612	0.567
C_2	0.634	0.792	0.856	0.678	0.512	0.498
C_3	0.727	0.755	0.754	0.720	0.673	0.604
C_4	0.827	0.849	0.912	0.811	0.747	0.713
C_5	0.648	0.705	0.788	0.631	0.649	0.592
C_6	0.664	0.698	0.756	0.672	0.585	0.543
C_7	0.682	0.754	0.802	0.683	0.638	0.538
C_8	0.773	0.822	0.843	0.734	0.727	0.672
C_9	0.684	0.722	0.748	0.682	0.603	0.575
C_{10}	0.744	0.800	0.801	0.740	0.674	0.610

every criterion based on their information; the criteria with the highest estimate of significance degree are ranked first, the sequence is arranged in descending order, with steps iv and v final estimate of weights were estimated as:

$$w_j = \{0.1001, 0.1061, 0.1114, 0.1302, 0.1201, 0.1411, 0.1253, 0.1127,$$

0.1127, 0.1423}

The weights for the criteria were employed for assessing SOS ranking.

4 Result and Discussion

Table 6 depicts the results evaluated by SWARA technique with step-by-step weight assessment ratio method for SOS.

Using equations in steps v, vi, vii, and viii were evaluated the estimated values of σ_i, $S(\sigma_i)$, $S(\upsilon_i)$ υ_i, θ_i, and γ_i depicted in Table 7.

From Table 7, the preference ordering of the organization options is K_3 is the best SOS choice. From the SWARA analysis, it was concluded that K_3 appears to be the most influential dimension; therefore, Institute of National Importance stands tall satisfying most of the criteria with highest significance ratio for most of the index parameters; parameters estimated included Finance (C1), Scholarship

Table 6 Weights estimated by SWARA technique

Criteria	Crisp values	Comparative significance of criteria value (s_j)	Coefficient (k_j)	Recalculated weight (p_j)	Criteria weight (w_j)
C_1	0.712	–	1.000	1.000	0.1542
C_5	0.703	0.051	1.051	0.951	0.1493
C_6	0.678	0.023	1.023	0.923	0.1378
C_2	0.655	0.011	1.011	0.811	0.1345
C_{10}	0.618	0.037	1.037	0.794	0.1293
C_3	0.523	0.471	1.471	0.746	0.1257
C_4	0.517	0.001	1.001	0.713	0.1221
C_9	0.485	0.028	1.028	0.696	0.1175
C_8	0.412	0.017	1.017	0.643	0.1058
C_7	0.395	0.053	1.053	0.627	0.1002

Table 7 Estimated values of σ_i, $S(\sigma_i)$, $S(\upsilon_i)$ υ_i, θ_i, and λ_i

SOS option	$S(\sigma_i)$	$S(\upsilon_i)$	θ_i	λ_i (%)	Ranking
K_3	0.612	0.191	0.341	99.8	1
K_2	0.523	0.234	0.327	93.1	2
K_4	0.501	0.172	0.315	86.4	3
K_1	0.472	0.201	0.273	84.1	4
K_5	0.447	0.126	0.256	77.4	5
K_6	0.317	0.178	0.211	74.9	6

(C2), Distance (C3), University Ranking (C4), Eligibility Criteria (C5), Submission Prerequisites (C6), Center of Excellence (C7), Research Lab (C8), Journal Access (C9), and Reputation (C10), from the aggregated hesitant fuzzy decision matrix for SOS problem, comparative significance of criteria value (s_j), and weights were estimated. Here, $S(\sigma_i)$ and $S(\upsilon_i)$ denote the score degrees of σ_i and υ_i, respectively. Lastly, priority order G^* was computed, and degree of utility was estimated (λ_i). The degree of utility came as 99.8% for K_3, 93.1% for K_2, and 86.4% for K_4. With estimated values, it was established that aspirants preferred organizations in following order $K_3 > K_2 > K_4 > K_1 > K_5 > K_6$.

5 Conclusion

The work carried is an effort to identify a fuzzy-assisted pattern for the aspirant to make an informed decision about the research option he or she can avail for enhanced growth. Study proposes a data analytics model for the informed intelligent selection toward optimum research carrier. The selection is assisted by number of weighted parameters which an aspirant should take into consideration while selecting a particular domain. SWARA technique was used for estimating criteria weights based on experts' preferences further to establish practicability of the proposed methodology, an empirical case study of sustainable organization selection taken under hesitant fuzzy (HF) environment. From the SWARA analysis, it was concluded that K_3 appears to be the most influential dimension; therefore, Institute of National Importance stands tall satisfying most of the criteria with highest significance ratio for most of the index parameters; parameters estimated included Finance (C1), Scholarship (C2), Distance (C3), University Ranking (C4), Eligibility Criteria (C5), Submission Prerequisites (C6), Center of Excellence (C7), Research Lab (C8), Journal Access (C9), and Reputation (C10), from the aggregated hesitant fuzzy decision matrix for SOS problem, comparative significance of criteria value (s_j), and weights were estimated, Here, $S(\sigma_i)$ and $S(\upsilon_i)$ denote the score degrees of σ_i and υ_i, respectively. Lastly, priority order G^* was computed, and degree of utility was estimated (λ_i). The degree of utility came as 99.8% for K_3, 93.1% for K_2, and 86.4% for K_4. With estimated values, it was established that aspirants preferred organizations in following order $K_3 > K_2 > K_4 > K_1 > K_5 > K_6$.

References

1. Radha, R., Mahalakshmi, K., Kumar, V. S., & Saravanakumar, A. R. (2020). E-learning during lockdown of Covid-19 pandemic: A global perspective. *International Journal of Control and Automation, 13*(4), 1088–1099.
2. Pal, D., & Vanijja, V. (2020). Perceived usability evaluation of Microsoft teams as an online learning platform during COVID-19 using system usability scale and technology acceptance model in India. *Children and Youth Services Review, 119*, 105535.

3. Chandwani, A., Lihitkar, S., & Anilkumar, S. (2010). E-learning initiatives in India. In *Modern Practices in Library and Information Services*, Nagpur (India), December 30, 2010.
4. Khanchandani, V., Kumar, M., & Kumar, R. (2015). E-learning initiatives in India and libraries. In *International Conference on Grey to Green* (pp. 517–527).
5. Nayek, J. (2018). A survey report on awareness among LIS professionals/students about SWAYAM: A government of India initiative on E-learning. *Knowledge Librarian. An International Peer Reviewed Bilingual E-Journal of Library and Information Science, 5*(01), 39–45.
6. Kaveri, A., Gupta, D., Gunasekar, S., & Pratap, M. (2016). Convergence or divergence: MOOCs and legacy of higher education outcomes. In *2016 IEEE 4th International Conference on MOOCs, Innovation and Technology in Education (MITE)* (pp. 20–24).
7. Singh, M., Adebayo, S. O., Saini, M., et al. (2021). Indian government E-learning initiatives in response to COVID-19 crisis: A case study on online learning in Indian higher education system. *Education and Information Technologies*. https://doi.org/10.1007/s10639-021-10585-1
8. Albalawi, A., & Badawi, M. (2008). Teachers' perception of E-learning at the University of Tabuk. In *E-Learn: World Conference on E-Learning in Corporate, Government, Healthcare, and Higher Education* (pp. 2434–2448). Association for the Advancement of Computing in Education (AACE).
9. Al-Rahmi, W. M., Yahaya, N., Aldraiweesh, A. A., Alamri, M. M., Aljarboa, N. A., Alturki, U., & Aljeraiwi, A. A. (2019). Integrating technology acceptance model with innovation diffusion theory: An empirical investigation on students' intention to use E-learning systems. *IEEE Access, 7*, 26797–26809.
10. Bezovski, Z., & Poorani, S. (2016). The evolution of E-learning and new trends. *Information and Knowledge Management, 6*(3), 50–57.
11. Torra, V. (2010). Hesitant fuzzy sets. *International Journal of Intelligent Systems, 25*, 529–539.
12. Torra, V., & Narukawa, Y. (2009). On hesitant fuzzy sets and decision. In *Proceedings of the 18th IEEE International Conference on Fuzzy Systems*, Jeju Island, Korea, August 20–24, 2009.
13. Xu, Z. S., & Xia, M. (2012). Hesitant fuzzy entropy and cross-entropy and their use in multiattribute decision-making. *International Journal of Intelligent Systems, 27*, 799–822.
14. Liao, H., Xu, Z. S., & Zeng, X. J. (2015). Novel correlation coefficients between hesitant fuzzy sets and their application in decision making. *Knowledge-Based System, 82*, 115–127.
15. He, Y., He, Z., Wang, G., & Chen, H. (2015). Hesitant fuzzy power Bonferroni means and their application to multiple attribute decision making. *IEEE Transactions on Fuzzy Systems, 23*, 1655–1668.
16. Mishra, A. R., Rani, P., Pardasani, K. R., & Mardani, A. (2019). A novel hesitant fuzzy WASPAS method for assessment of green supplier problem based on exponential information measures. *Journal of Cleaner Production*.
17. Kersuliene, V., Zavadskas, E. K., & Turskis, Z. (2010). Selection of rational dispute resolution method by applying new stepwise weight assessment ratio analysis (SWARA). *Journal of Business Economics and Management, 11*, 243–258.
18. Dehnavi, A., Aghdam, I. N., Pradhan, B., & Morshed Varzandeh, M. H. (2015). A new hybrid model using stepwise weight assessment ratio analysis (SWARA) technique and adaptive neuro-fuzzy inference system (ANFIS) for regional landslide hazard assessment in Iran. *CATENA, 135*, 122–148.
19. Karabasevic, D., Zavadskas, E. K., Turskis, Z., & Stanujkic, D. (2016). The framework for the selection of personnel based on the SWARA and ARAS methods under uncertainties. *Informatica, 27*, 49–65.
20. Nakhaei, J., Bitarafan, M., LaleArefi, S., & Kaplinski, O. (2016). Model for rapid assessment of vulnerability of office buildings to blast using SWARA and SMART methods (a case study of swiss re tower). *Journal of Civil Engineering and Management, 22*, 831–843.
21. Isık, A. T., & Adalı, E. A. (2016). A new integrated decision-making approach based on SWARA and OCRA methods for the hotel selection problem. *International Journal of Advanced Operations Management, 8*, 140–151.

22. Urosevic, S., Karabasevic, D., Stanujkic, D., & Maksimovic, M. (2017). An approach to personnel selection in the tourism industry based on the SWARA and the WASPAS methods. *Economic Computation and Economic Cybernetics Studies and Research, 51*, 75–88.

Author Index

P. K. Singh et al. (eds.), *Proceedings of Third International Conference on Computing, Communications, and Cyber-Security*, Lecture Notes in Networks and Systems 421,
https://doi.org/10.1007/978-981-19-1142-2